Springer Series in Biomaterials Science and Engineering

Volume 16

Series Editor

Min Wang, Department of Mechanical Engineering, The University of Hong Kong, Pokfulam Road, Hong Kong

The Springer Series in Biomaterials Science and Engineering addresses the manufacture, structure and properties, and applications of materials that are in contact with biological systems, temporarily or permanently. It deals with many aspects of modern biomaterials, from basic science to clinical applications, as well as host responses. It covers the whole spectrum of biomaterials–polymers, metals, glasses and ceramics, and composites/hybrids–and includes both biological materials (collagen, polysaccharides, biological apatites, etc.) and synthetic materials. The materials can be in different forms: single crystals, polycrystalline materials, particles, fibers/wires, coatings, non-porous materials, porous scaffolds, etc. New and developing areas of biomaterials, such as nano-biomaterials and diagnostic and therapeutic nanodevices, are also focuses in this series. Advanced analytical techniques that are applicable in R&D and theoretical methods and analyses for biomaterials are also important topics. Frontiers in nanomedicine, regenerative medicine and other rapidly advancing areas calling for great explorations are highly relevant. The Springer Series in Biomaterials Science and Engineering aims to provide critical reviews of important subjects in the field, publish new discoveries and significant progresses that have been made in both biomaterials development and the advancement of principles, theories and designs, and report cutting-edge research and relevant technologies. The individual volumes in the series are thematic. The goal of each volume is to give readers a comprehensive overview of an area where new knowledge has been gained and insights made. Significant topics in the area are dealt with in good depth and future directions are predicted on the basis of current developments. As a collection, the series provides authoritative works to a wide audience in academia, the research community, and industry.

This book series is indexed by the EI Compendex and Scopus databases.

If you are interested in publishing your book in the series, please contact Dr. Mengchu Huang (Email: mengchu.huang@springer.com).

More information about this series at http://www.springer.com/series/10955

Tuhin Subhra Santra · Loganathan Mohan
Editors

Nanomaterials and Their Biomedical Applications

Springer

Editors
Tuhin Subhra Santra
Department of Engineering Design
Indian Institute of Technology Madras
Chennai, Tamil Nadu, India

Loganathan Mohan
Department of Mechanical Engineering
Toyohashi University of Technology
Toyohashi, Aichi, Japan

ISSN 2195-0644 ISSN 2195-0652 (electronic)
Springer Series in Biomaterials Science and Engineering
ISBN 978-981-33-6251-2 ISBN 978-981-33-6252-9 (eBook)
https://doi.org/10.1007/978-981-33-6252-9

This Springer imprint is published by the registered company Springer Nature Singapore Pte Ltd.
The registered company address is: 152 Beach Road, #21-01/04 Gateway East, Singapore 189721, Singapore

Preface

Nanomaterials are defined based on their size, which is one-billionth of a meter (10^{-9} m) in any one of its dimensions. In other words, "the materials with any external dimension or having an internal structure or surface structure in the nanoscale range." In such extremely small size, materials exhibit unique and spectacular performance due to an increment of the surface to volume ratio. Materials, usually in the size range of 1–100 nm, show different physicochemical properties from their bulk. In the past two decades, several engineered nanomaterials have been established for different application purposes. After the rapid development of micro and nanotechnologies, fabrication of nanoscale structures and devices with exceptional sights of the supramolecular association has grown interest into a systematic way, and this has enhanced their perspective for an innovative connection in a huge variety of research areas, such as biology, chemistry, physics, engineering, computer science, etc. Nanoscale materials are progressively creating a foremost influence on human health, and they are used increasingly further for therapeutic and diagnostic applications. Nanotechnology plays a vital role in the current technological developments and plays an important role in diagnosing diseases, drug delivery, designing drugs, etc. This book intends to bring science and advanced nanotechnology together and their applications in cell biology and biomedical engineering. From a nanoscale and nanomaterial perspective, it highlighted different nanostructured material design, synthesis, processing, characterization, and potential applications. This book also covers different nanoscale and nanostructured materials for biomedical applications such as therapeutics, diagnostic, prosthesis, implant, drug discovery, and drug delivery, etc. An overview of nanomaterials' progress and prospects and its biomedical applications are discussed with a series of reasoning and practical considerations.

This book contains 17 chapters, covering an extensive range of the critical aspects of nanomaterials for biomedical applications. Each chapter is contributed by professionals in their fields, and these chapters deliver technical information based on their valuable knowledge and skills. Possible glitches and challenges, as well as potential keys, are also deliberated with importance on prospects. Chapter "Nanomaterials: An Introduction" narrates brief advancement of nanotechnology to date. The different health-related problems that arise due to the application of nanotechnology in medicine, food, agriculture, etc., are reported. Environmental nano pollution and its effect on society, social–economic disruption due to the rapid use of nanotechnology, safety and security of nanotechnological developments, and its future direction is also discussed. Chapter "Metallic Nanoparticles for Biomedical Applications," discusses the top-down and bottom-up approach and current trends in the synthesis of *metallic nanoparticles* for biomedical purposes. Further, it describes how the parameters can be tuned to get metallic nanoparticles with the desired shape, size, and crystallinity. Chapter "Size and Shape-Selective Metal Oxide Nanomaterials: Preparation, Characterization and Prospective Biomedical Applications," describes the different preparation methods for *metal-oxide nanomaterials* (MONMs), characterization techniques, and MONMs usage in different biomedical applications. The mechanism of interaction between nanomaterials and internal structures of microorganisms and institutes working on the nanomaterials standardizations is deliberated. Chapter "Nanofibers and Nanosurfaces" discusses the different synthesis routes associated with the development of nanofibers for cartilage regeneration. The fabrication and effect of nanosurfaces on metallic implants for enhanced chondrocyte conductivity are also highlighted. Chapter "Nanoceramics: Synthesis, Characterizations and Applications," narrates in detail the nanoceramics, their preparation methods, available characterization techniques, their unique properties, and their widespread biomedical applications arising due to their excellent properties. Chapter "Biomedical Applications of Carbon-Based Nanomaterials," discusses briefly carbon-based nanomaterials like Nanodiamonds (NDs), Carbon nanotubes (CNTs), Buckminsterfullerene (C_{60}), Carbon quantum dots (CQDs), Carbon nanohorns (CNs), and its biomedical applications. Chapter, "Solution Combustion Synthesis of Calcium Phosphate-Based Bioceramic Powders for Biomedical Applications," gives an overview of the solution combustion synthesis of pure and doped hydroxyapatite powders and their characterization. This chapter also discusses the solution combustion synthesis of plasma sprayable hydroxyapatite powder, fabrication of coating, and characterization of the developed plasma-sprayed coating. Chapter "Nanomaterials in Medicine," outlines the various types of nanomaterials and their applications in the field of nanomedicine. The clinical applications of the nanomaterials in sepsis therapy, chemotherapy, and applications of nanomaterials in heart, kidney, lungs, brain diseases, etc., are discussed. Chapter "Hydrogels: Biomaterials for Sustained and Localized Drug Delivery," illustrates the synthesis, functionalization, tailoring mechanism of hydrogel matrix, followed by *in-vitro, ex-vivo,* and *in-vivo* characterization and drug loading/delivery efficiency. The different classifications of the hydrogel, along with its crosslinking chemistry, hydrogel nanocomposites biomedical perspective, and hydrogel applications ranging from lab scale to industrial level, are also discussed. Chapter "Nanomaterials: Versatile Drug Carriers for Nanomedicine" highlights the recent advancements and applications of nanocarriers for drug delivery in medicine, especially wound healing therapeutics and also different approaches to enhance drug cargo capacity, to improve cell entry efficiency, to avoid host immune systems and to achieve specific tissue targeting. Chapter "Nanomaterials for Medical Implants," discusses the general consideration of using nanomaterials in implantable devices, dental implants/prosthodontics, spinal orthopedic implants, and hip and knee replacements, cardiovascular implants, others—phakic intraocular lens and cosmetic implants. Chapter "Fabrication of Nanostructured Scaffolds for Tissue Engineering Applications" addresses different categories of biomaterials used to fabricate nanostructured scaffolds for tissue regeneration applications. The desired properties required for various tissue engineering scaffolds and its fabrication methods, merits and demerits, current development, and future directions of these methodologies are discussed. In addition to that, the special emphasis given on three dimensional (3D printing) technologies to manufacture tissue engineering scaffolds using various nanomaterials are also discussed in this chapter. Chapter "Nanomaterials for Medical Imaging and In Vivo Sensing," is devoted to nanomaterials for medical imaging techniques. The extensive information about the current developments in imaging systems, its advantages & disadvantages, the science behind individual imaging systems and the basic instrumentation, how the nano-based contrast agents are helpful in various biomedical applications are systematically discussed. Chapter "Nanomaterials: Surface Functionalization, Modification, and Applications," presents the various surface modifications and functionalization of nanomaterials, including metallic nanoparticles, carbon nanomaterials, nano-ceramics, and self-assembled materials for biomedical applications. Chapter "Laser-Induced Micro/Nano Functional Surfaces on Metals for Biomedical Applications" emphasizes the recent research works on laser processing such as selective laser melting (SLM), laser surface melting (LSM), laser surface patterning of various metallic implant materials and surface characteristics, and biomedical applications of processed surfaces are summarized. Chapter "Surface Nanostructuring of Metallic Materials for Implant Applications," provides detailed coverage of surface mechanical attrition treatment (SMAT) of stainless steels, Ti alloys, Ni-Ti alloy, CoCrMo alloy, and how the nanostructured surface enables an improvement in the characteristic properties that are suitable for biomedical applications. Chapter "Tailoring the Surface Functionalities of Titania Nanotubes for Biomedical Applications," deliberates the key electrochemical factors that control the nanotube geometry and demonstrate various surface functionalization approaches for tailoring the surface properties of TiO_2 nanotubes to develop new functional biomaterials for biomedical applications.

We hope this book can be pleasant reading material and, at the same time, a handy resource for students, scientists in academia, and professionals in industries working on various traits of nanomaterials.

Chennai, India
Toyohashi, Japan

Tuhin Subhra Santra
Loganathan Mohan

Contents

Nanomaterials: An Introduction

Tarun Kumar Barik, Gopal Chandra Maity, Pallavi Gupta, L. Mohan, and Tuhin Subhra Santra

Abstract Nanotechnology offers a significant advantage in science, engineering, medicine, medical surgery, foods, packing, clothes, robotics, and computing from the beginning of the twenty-first century. As the potential scientific discovery always contains some good and bad effects on human civilization and the environment, nanotechnology is not an exception. The major drawbacks include economic disruption along with imposing threats to security, privacy, health, and environment. The introduction of the chapter discusses the historical background of nanotechnology. Later it also discusses the advancement of nanotechnology to date with its benefits. Major drawbacks of nanotechnology arise in human health due to the enormous involvement in medicine, food, agriculture, etc. This chapter also deals with environmental nano pollution and its effect on society, highlighting the social-economic disruption due to the rapid use of nanotechnology. Nano pollution affects not only human beings but also other living beings like microorganisms, animals and plants, which are briefly reviewed. This chapter also demonstrates the safety and security of nanotechnological developments, current policy and regulation status, challenges, and future trends. Finally, it is concluded, while nanotechnology offers more efficient power sources, faster and modern computers and technologies, life-saving medical treatments, but due to some negative impacts, it bounds us to think twice before any further advanced technological applications.

T. K. Barik (✉)
Department of Physics, Achhruram Memorial College, Jhalda, Purulia, West Bengal, India
e-mail: tarun.barik2003@gmail.com

G. C. Maity
Department of Chemistry, Abhedananda Mahavidyalaya, Sainthia, Birbhum, India

P. Gupta · L. Mohan · T. S. Santra
Department of Engineering Design, Indian Institute of Technology Madras, Chennai, India

L. Mohan
Department of Mechanical Engineering, Toyohashi University of Technology, Toyohashi, Japan

T. S. Santra and L. Mohan (eds.), *Nanomaterials and Their Biomedical Applications*, Springer Series in Biomaterials Science and Engineering 16,
https://doi.org/10.1007/978-981-33-6252-9_1

Keywords Advantages and disadvantages of nanotechnology · Nanotoxicity · Nanopollution · Economic disruption · Security · Privacy · Health · Environment

1 Introduction

Nanotechnology is an emerging field of science and technology with numerous applications in biomedical and manufacturing engineering [1–3]. In the last two decades, nanotechnology integrates with mechanical and electronic engineering to develop Micro/Nano-electromechanical systems (MEMS/NEMS) devices, which have diverse applications in different fields of science and engineering. These devices are potentially applicable for various sensing, actuating, and biomedical analysis purposes [4–13]. Recently, quantum dots have increased much attention in biological fields due to their unique size, tunable light absorption, and emission properties [14]. Further, biocompatible nanomaterials have many applications in biomedical purposes such as orthopedic, cardiovascular, contact lenses, catheter, prosthetic replacement, etc., [15–21]. Among noble metals, Ag and Au nanoparticles synthesis via marine algae are used as a broad-spectrum antimicrobial agent towards a variety of pathogens in the biomedical field [22]. Nowadays, nanomaterials are produced by industries for commercial applications with enormous benefits. While there lies a vast potential of nanomaterials for fulfilling individual requirements, it also represents potential risks to human health [23].

The green synthesis of nanoparticles attracts many researchers and industries. Many microorganisms are utilized for the synthesis of nanoparticles. Biosynthesis of nanoparticles has been reported using photoautotrophic microorganisms such as cyanobacteria, eukaryotic algae, and fungi. The biogenic fabrication of nanoparticles via microalgae is a non-toxic, and eco-friendly, green chemistry method with a large variety of compositions and physicochemical properties. Biosynthesis of nanoparticles by plant extracts is currently under exploitation. Plant extracts are a better source of nanomaterials compared to the various biological processes often considered eco-friendly substitutes of chemical and physical methods [1, 17]. Seaweeds contain different organic and inorganic substances that can benefit human health [24]. The green seaweed is used widely in agriculture, pharmaceutical, biomedical, and nutraceutical industries for its presence of a high amount of vitamins and minerals [25]. Among several genera of microalgae, *Spirulina platensis* is blue-green algae of the cyanobacteria family grown in temperate water in the whole world. A blue-green alga has served as food with high protein content and nutritional value from ancient times [26]. The algae produce novel and potentially useful bioactive compounds [27, 28]. The bioactive materials have gained significant attention in recent years and have been used considerably in developing new pharmaceutical products, food products, renewable bio-energy, and biomedical applications [29–31]. However, a new global health problem has been arisen as in discriminant antibiotic use and the remarkable ability of bacteria to acquire resistance to lower these drugs' effectiveness via genetic mutation or gene acquisition. Therefore, new classes of antibiotics with novel structures are needed to combat this trend. Food preservation is now dealing with the severe concern of microorganisms mediated spoilage and fall in quality and nutrition worldwide [32]. Hence, increasing the continuous demand for pathogen control measures to combat resistant microorganisms against multiple antimicrobial agents. However, nanoparticles own large surface area to volume ratio, unique quantum size, magnetic properties, heat conductivity in addition to some catalytic and antimicrobial properties [33]. In this regard, nanomaterials, including metal nanoparticles, carbon nanotubes, quantum dots, and other active nanomaterials can be used to develop biosensors against a broad spectrum of microorganisms for the formulation of a new generation of antimicrobial agents.

2 Historical Background of Nanotechnology

The first experiment of nanotechnology was shown in 1857 when Michael Faraday introduced 'gold colloid' samples to the Royal Society. He added phosphorous to a solution of gold chloride and, after a short while, noted that the blue color of the solution changed to a ruby red dispersion, without knowing the actual cause of color changing. Indeed, the resulting suspension of nanosized gold particles in solution appeared transparent at some frequencies, but others could look colored (ruby, green, violet, or blue). Since then, many experiments and theoretical studies have been carried out to explain similar systems' unique properties, which in today's terminology are called low-dimensional systems. Nearly after 100 years, in 1959, Richard Feynman inspired the field of nanotechnology in his lecture at the American Physical Society (APS) meeting, Caltech, saying the meaningful words "There's Plenty of Room at the Bottom." From the late 1980s, we find there is a growth of activity on these low dimensional materials. In general, low dimensional systems are categorized as follows: (a) two dimensional (2D) systems, in which the electrons are confined in a plane (e.g., Layered structures, quantum wells and superlattices); (b) one dimensional (1D) systems, in which electrons are free to move only in one dimension (e.g., linear chain-like structures, semiconductor quantum wires), and (c) zero-dimensional (0D) systems, where electrons are confined in all three dimensions (e.g., quantum dots, clusters, and nanosized colloidal particles) [34–41].

The dimension of these materials in the direction of confinement lies in the nanometer scale, given the name nanomaterials. In this length scale, classical physics fails to explain the behavior of these materials. Instead, one needs quantum mechanical concepts. Interestingly, due to quantum effects, the physical properties of nanomaterials change drastically from their corresponding bulk behavior. This unique feature of nanomaterials has been exploited by modern technology in various applications. The link between human life and nanotechnology is as old as Ayurveda, a 5000-year-old Indian medicine system.

Moreover, twenty-first century modern science marks the beginning of nanoscience, while it existed from ancient times of Vedas, much before even the term "nano" was coined [42, 43]. As per strict nanometer terminology, any objects with dimensions in the nm range can be termed as a nanoparticle or a "nano" object, as TiO_2 dust in the study mentioned above [44]. Nanotechnology not only combines engineering, physics, and chemistry but also integrates with biology [45]. A physicist generally tries to identify and quantify nanomaterials' fundamental interactions with different surrounding systems such as the thermodynamics, the interface of the nanoparticles with the liquid, and the role of mechanical properties (e.g., stiffness, elasticity, adhesion), etc.

Past three decades, extensive work has been performed to develop new drugs from natural products, because of the resistance of microorganisms to the existing drugs [46]. Researchers from the Indian Institute of Technology Bombay, India, have discovered that the age-old complementary medicines of Homeopathic pills and Ayurvedic Bhasmas are having metal nanoparticles such as gold, silver, copper, platinum, tin, and iron [46, 47]. Metallic nanoparticles (mainly silver and gold) have unique optical, electrical, and biological properties, that have attracted significant attention due to their potential use in many applications, such as catalysis, ultra-sensitive chemical and biological sensors, bio-imaging, targeted drug delivery and nanodevice fabrication [13, 48–57]. Recently, various industries like electronics, aerospace, cosmetics, textile, and even food use nanoparticles. Consequently, the chance of human exposure to nanoparticles rises, heading towards the time when nanoparticles are eventually present in blood circulation and interacting with immune blood cells.

Nanoparticles can be synthesized via various chemical and physical routes such as chemical reduction, [58–60] photochemical reduction, [61–65] electrochemical reduction, [66, 67] heat evaporation, [68, 69], etc. In all the above-mentioned methods, the reagents can be from different properties, i.e. inorganic such as sodium or potassium borohydrate, hydrazine, and salts of tartrate, or organic ones like sodium citrate, ascorbic acid, or amino acids, capable of getting oxidized. Various options are also available to work as a stabilizing agent. Several studies have reported shape and size dependency of silver nanoparticles formation on capping agents such as dendrimer, [70] chitosan, [71] ionic liquid, [72], and poly (vinylpyrrolidone) PVP [73]. These capping agents control the nanoparticle growth via reaction confinement within the matrix or preferential adsorption on specific crystal facets. Since these approaches are costly, hazardous, toxic, and non-environment friendly, hence, evaluation of the risk of these nanoparticles to human health becomes critical. Multiple studies have shown the increase in the number of leukocytes, mainly neutrophils, in the lungs and bronchoalveolar lavages during airway exposure of nanoparticles *in-vivo* models of inflammation. The neutrophil counts act as biomarkers for inflammation. Therefore, the selection of a synthesis route that minimizes the toxicity and increases nanoparticle stability leads to enhanced biomedical applications of silver and gold nanoparticles. The development of better experimental procedures for the synthesis of nanoparticles employing a variety of chemical compositions and controlled polydispersity offers considerable advancement [74]. Methods of nanoparticle production through different physical and chemical routes, as stated above, have their demerits as they produce enormous environmental contaminations and hazardous byproducts. Thus, there is a need for "green chemistry" that ensures clean, non-toxic, and environment-friendly nanoparticles production [75].

In recent years, environment-friendly approaches have been developed to fabricate stable nanoparticles with well-defined morphology and configured constricted sizes [76]. Additionally, owing to the high demand for precious metals (like silver and gold) and metal oxides in electronics, catalysis, medical, and other industrial applications, its recovery from primary and secondary sources is of considerable significance and interest. Biological recovery of these precious metals by preparing their nanoparticle is a green alternative to the conventional physical and chemical methods [77, 78]. Bio-inspired synthesis of nanoparticles is an advanced, cost-effective, environment-friendly approach over chemical and physical processes, without any inclusion of high pressure, energy, temperature, and toxic chemicals [79]. For example, the plant leaf extract is used for the biosynthesis of silver and gold nanoparticles for pharmaceutical and biomedical applications, without employing any toxic chemicals in the synthesis protocols [80]. An environmentally acceptable solvent system, eco-friendly reducing and capping agents are considered to be an essential element for an ultimately "green" synthesis [81]. The green synthesis techniques are generally utilizing relatively non-toxic chemicals to synthesize nanomaterials. The fabrication process also includes the use of non-toxic solvents such as water, biological extracts, biological systems, etc. In this technique, generally, microwave maintains a constant temperature of solvent systems. The conventional extraction technique using hexane, ethanol, and water was used to collect bioactive molecules [82]. However, they are immensely problematic due to instability as well as environmental and health hazards [83]. To overwhelm this problem, researchers developed a new approach, i.e., supercritical fluid (SCF) extraction technology for avoiding toxic organic solvents in green technology. SCF possesses physical properties intermediate between CO_2 gas and a liquid at a temperature and pressure above its critical point. Since supercritical CO_2 is non-polar, non-toxicity, non-flammability, and low critical temperature.

3 Benefits of Nanotechnology

Recently, research and development in nanotechnology have seen exponential growth due to advantages in different fields, i.e., drug delivery, cell imaging, material improvement, and medical devices for diagnosis and treatment. More powerful computers are being designed using nanomaterials in a smaller size, faster in speed, and consuming very less power, having long-life batteries. Circuits consisting of carbon nanotubes can maintain the computer system advancement. Carbon nanotubes are also commercially used in sports equipment, to increase their strength while maintaining a low weight. Nanoparticles or nanofibers in fabrics improve the water-resistance, stain resistance, and flame resistance, without putting on extra weight, stiffness, or thickness of the material. Nanoparticles are used in medical products for dermal, oral or inhalation applications, and pharmaceuticals. These are also used in various consumer products, including cosmetics, food, and food packaging. The nanomaterials having potential uses in cosmetics include nanosilver, nanogold, nanoemulsions, nanocapsules, nanocrystals, dendrimers, fullerenes, liposomes, hydrogels, and solid lipid nanoparticles. Smaller the size, corresponding to the higher surface area of nanomaterials offer greater strength, stability, chemical, physical, and biological activity. Nanomaterials present in the human environment can be primarily classified into four categories: carbon-based nanomaterials, metal-based nanomaterials, dendrimers, and composites. The carbon-based nanomaterials (fullerenes and nanotubes) are employed in thin films, coatings, and electronics.

The metal-based nanomaterials (i.e., nanosilver, nanogold, and metal oxides (i.e., titanium dioxide (TiO_2)) are useful for food, cosmetics, and drug-related products. The dendrimers are nano-polymers, an ideal candidate for drug delivery. Composites such as nanoclays are formed with a combination of nanoparticles with other nanosize or larger particles. Many beverage bottles are made up of plastics with nanoclays. The nanoclay reinforcement increases permeation resistance to oxygen, carbon dioxide, moisture, and thus retaining carbonation, pressure with increased shelf life by several months. Nanoclays are also being used in packaging materials.

Different classes of nanomaterials are composed of nanoparticles with different shape, size, and chemistry and biology. Nanotechnology helps to improve vehicle fuel efficiency and corrosion resistance by building vehicle parts from Diamond-Like-Nanocomposite (DLN) materials that are lighter, stronger, and more chemically resistant than metal [84–88]. The DLN film exhibits biocompatibility in nature, which has potential applications as a coating material for biomedical purposes [89, 90].

A few nanometers wide water filters can remove nanosized particles, including virtually all viruses and bacteria, which can revolutionize the water filtration method. These cost-effective, portable water-treatment systems are ideal for the improvement of drinking water quality in developing countries. Nowadays, most sunscreens also contain nanoparticles for effective absorption of light, including the more dangerous ultraviolet range and passing the other wavelengths, which is healthy for the skin. Recently, nanosensors can be programmed to detect a particular chemical at low levels, such as a single-molecule detection, out of billions of molecules. This capability is ideal for security systems and surveillance at labs, industrial sites, and airports. In medical science, the detection of single biomolecules has tremendous DNA/RNA sequencing and disease analysis applications. The nanobiosensors can be used to precisely identify particular cells or substances in the body for different diagnostics purposes. Current research is focused on preparing the smaller, highly sensitive, and cost-efficient biosensors. The new biosensors are updated to even detect odors specific diseases for medical diagnosis, pollutant detection, and gas leaks for environmental protection. Figure 1 shows the technological tsunami that occurs due to nanotechnology in energy storage, defense & security, metallurgy & materials, electronics, optical engineering & communication, biomedical & drug delivery, agriculture & food, cosmetics & paints, biotechnology, textile, etc. [91]. According to Zion market research analysis in 2017 [92], there is a rapid increase of global nanomaterials market volume (in kilo tons) and revenue (in USD Billion), which is estimated from 2014 to 2022, is shown in Fig. 2a. Other statistical surveys

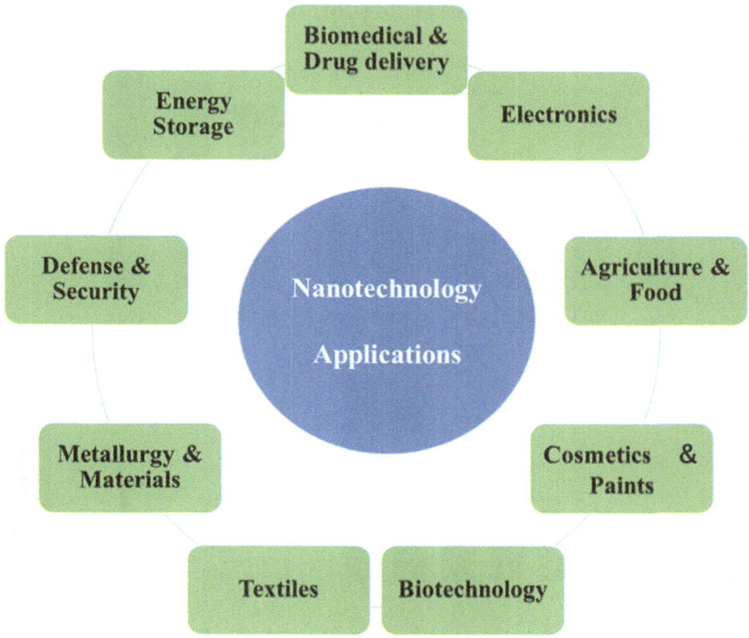

Fig. 1 Technological tsunami due to nanotechnology. Redrawn from [91]

from two different agencies (see Fig. 2b and c (BCC research)) also confirmed the rapid increase of the global nanotechnology market of nanomaterials, nanotools, and nanodevices, etc. [93, 94].

4 Nanotechnology in Health

4.1 Potential Routes for Nanomaterials to Enter into the Human Body

Nanomaterials can enter into the human body in various ways. Potential routes nanomaterials enter the human body are ingestion, inhalation, and skin absorption [95–97]. Many nanomaterials are employed in drug transport or cell imaging via intravenous entry to the human body. In the body, nanomaterials are translocated throughout the body by blood circulation. For the purpose, the nanoparticles must fulfill the requirement of permeability across the barrier of the blood vessel wall. Absorption through the skin serves as an alternate route of entry for nanoparticles inside a human body. The skin is the largest organ of the human body, provides a large surface area for interactions with the external environment. TiO_2 nanoparticles can take either

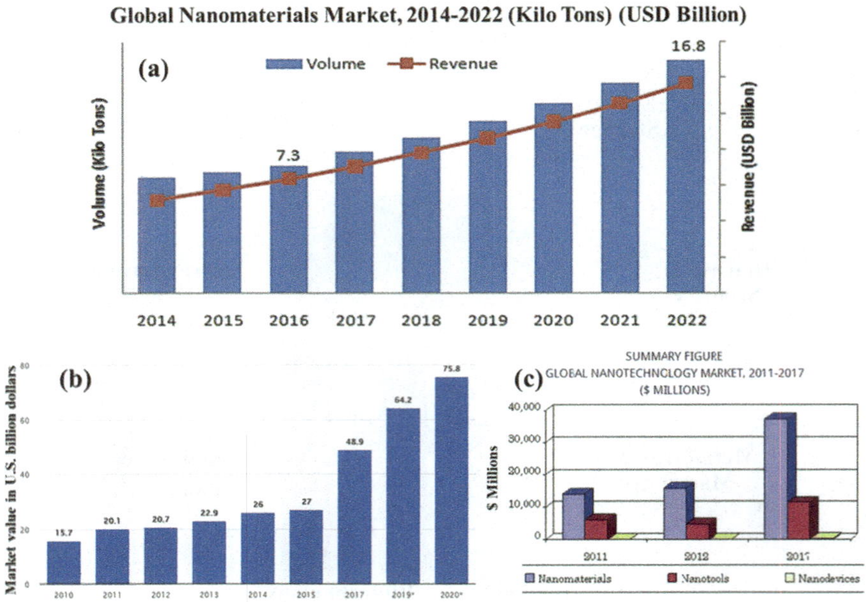

Fig. 2 **a** Rapid increase in global nanomaterials market volume (Kilo Tons) and revenue (USD Billion) for the period of 2014–2022 [92]. *Source* Zion market research analysis, 2017. **b** The global market value of nanotechnology from 2010 to 2020 (in Billion USD) [93]. **c** Global nanotechnology review for nanomaterials, nanotools, and nanodevices market from 2011 to 2017 (in Million USD). *Source* BCC Research [94]

route for entry, i.e., the lungs or gastrointestinal tract. Nanomaterials can enter the body through the skin for various reasons, such as the use of medicine, cosmetics, ointments, and use of clothes containing nanomaterials, occupational contact in the industry, etc. Soaps, shampoos, toothpaste, hair gels, creams, and some cosmetics containing the nanosilver, which can enter into the body through the skin.

Cream or solution containing Silver nanoparticles is used to treat wounds, burns, etc. to prevent infections and damaged skin. The penetrating ability depends on the size of the nanoparticles. The smaller the nanoparticle, has the more exceptional penetrating ability. The inhaled particulate matter gets accumulate in the human respiratory tract, while one significant portion of those inhaled particles gets deposited in the lungs. Nanoparticles also can travel across the placenta in pregnant women to the fetus along with other organs, i.e., brain, liver, and spleen. The effects of inhaled nanoparticles in the body may include lung inflammation and heart disease problems [95]. The pulmonary injury and inflammation resulting from the inhalation of nano-sized urban particulate matter appear due to the oxidative stress imposed by these particles in the cells [98–101]. The first reported nanoparticle is nanosilver, which can damage DNA molecules. Silver nanoparticles have the most harmful effects on the most sensitive biological groups [98, 102–105]. This nanoparticle can enter into the blood through the skin. Silver binds with the thiol group of some proteins. If silver complexes with thiol groups are located near-skin region, it gets readily available to get reduced either by visible or UV light into metallic nanosilver particles. Therefore, the immobilization of silver nanoparticles takes place in the skin. Further, the effect of nano copper-induced renal proximal tubule necrosis in kidneys has been reported by Liao and Liu [106].

4.2 Nanomaterials for Therapy and Diagnostics

Nanoparticles in pharmaceutical products facilitate improved absorption within the human body and easy delivery, often in association with medical devices. For example, magnetite, a metal oxide, has high potential applications in nanomedicine. Nanoparticles can assist the targeted delivery of chemotherapy drugs to specific cells, i.e., cancer cells. Superparamagnetic iron oxide nanoparticles (SPIONs) and ultrasmall superparamagnetic iron oxide (USPIO) have also proved its significance for targeted drug delivery [107]. Nanoparticles improve the solubility of poorly water-soluble drugs, increase drug half-life, modify pharmacokinetics, improve bioavailability, diminish drug metabolism, assist controlled and targeted, and combined drug delivery [98, 108–111]. According to the International Agency for Research on Cancer (IARC) data, estimates of nearly 13.1 million deaths due to cancer by 2030. It is evident that the low survival rate occurs not because of the scarcity of potent, natural, or synthetic antitumor agents but owing to inadequate drug delivery systems. Hence develops the requirement of technology advancement to establish carriers and delivery systems capable of targeted and efficient delivery of the chemotherapeutic agents without unwanted systemic side effects [112]. The solid lipid nanoparticles and nanoemulsions are the most employed lipid-based drug delivery particles. However, nanosilver based commercial products are capturing the market. The newly developed nanomaterials for theranostics are being employed alone or in association with "classical" drugs, e.g., cytostatic drugs, or antibiotics. Theranostics is a combined term for nanomaterials with diagnostic and therapeutic properties [111].

5 Drawbacks of Nanotechnology

Nanomaterials are being employed in different industries and everyday life. Therefore, the interplay of nanomaterials and social surroundings is worth scientific exploration. Nanomaterials with several benefits can be toxic. Various studies also confer the effects, as mentioned above, indicating the potential toxicological effects on the human environment [98]. Different toxic and hazardous effects of nanotechnology are briefly discussed below.

5.1 Toxicity of Nanomaterials

Greater human exposure of nanomaterials presents in the environment; more significant is the harmful effect on human health. The assessment of the cytotoxicity of nanomaterials assists in the proper elucidation of the biological activity. Gerloff et al. reported the cytotoxicity of various nanoparticles, such as zinc oxide (ZnO), SiO_2, and TiO_2, on human Caco-2 cells [113]. Shen et al. [114] showed the human immune cells are prone to toxicity due to ZnO nanoparticles [115]. The ZnO nanoparticles damage mitochondrial and cell membranes in rat kidney, ultimately leading to nephrotoxicity [115]. Generally, the nanomaterial toxicity mechanism comprises reactive oxygen species formation and genotoxicity. However, as described earlier, the toxicity of ZnO nanoparticles mainly affects immune cells. Various nanomaterials with their diverse sizes alter mitochondrial function. For example, ZnO nanoparticles generate Zn^{2+} ions, which disrupts charge balance in the electron transport chain in the mitochondria and therefore triggers reactive oxygen species generation. Nanosilver particle has a genotoxic effect. A 20-nm nanosilver has a genotoxic effect on human liver HepG2 and colon Caco2 cells. It has also increased mitochondrial injury and the loss of double-stranded DNA helix in both cell types [116]. Inhalation of TiO_2 nanoparticles resulted in pulmonary overload in rats and mice with inflammation [117, 118]. The cytotoxic and genotoxic effects of TiO_2 nanoparticles on the human lung were reported by Jugan et al. [119]. TiO_2 nanoparticles are genotoxic, and it can induce pathological damage of the liver, kidney, spleen, and brain. Du et al. reported cardiovascular toxicity of silica nanoparticles in rats [120]. The surface coating of quantum dots causes toxicity to the skin cells, including cytotoxicity and immunotoxicity [121]. Nanosilver is used in wound dressings, affects both keratinocytes and fibroblasts. Fibroblasts show higher sensitivity towards nanosilver than by keratinocytes. Again, iron oxide nanoparticles rapidly get endocytosis on cultured human fibroblasts and interrupt the function. Citrate/gold nanoparticles have shown toxicity on human dermal fibroblasts [122]. Carbon nanotubes have high toxicity and produce harmful effects on humans. The nanoparticles can penetrate the lungs, then reached the blood and acted as a barrier for the circulation of blood into the brain. They can also enter inside other organs like bone marrow, lymph nodes, spleen, or heart. Sometimes, nanoparticles can incite inflammation, oxidant and antioxidant activities, oxidative stress, and change in mitochondrial distribution. These effects depend on the type of nanoparticles and their concentrations [101]. Copper nanoparticles (diameters 40 nm and 60 nm) harm brain cells at low concentrations. It activated the proliferation of the endothelial cells in brain capillaries. Ag nanoparticles (25, 40, or 80 nm) influenced the blood-brain barrier, causing a pro-inflammatory reaction, which might induce a brain inflammation with neurotoxic effects [123]. Smaller Ag nanoparticles (25 nm and 40 nm diameter) can induce cytotoxic effect at a higher rate than larger nanoparticles. Nanoparticles also have harmful effects on the brain cell of the mouse and rat. The high concentration of nanoparticles can affect brain blood fluxes, with consequent cerebral edema. Pathogenic effects of Ag-nanoparticles (25, 40, and 80 nm diameter), Cu-nanoparticles (40 and 60 nm), and Au-nanoparticles

(3 and 5 nm) on the blood-brain barrier of the pig have been reported [124]. Silver nanoparticles (45 nm) influenced the acetylcholine activity via nitric oxide generation; it induces hyperactivity of rat tracheal smooth muscle [125]. It is also reported that Ag- nanoparticles (25 nm) produced oxidative stress after the injection into the mouse. The nanoparticles were aggregated in the kidneys, lungs, spleen red pulp, and the nasal airway, with no observable morphological changes apart from the nasal cavity [126].

Very few cells do not undergo morphological changes after withstanding the air-liquid interface culture for an extended duration. Au-nanoparticles (5 nm and 15 nm diameter) penetrated the mouse fibroblasts, where they remained stocked. Only the presence of 5 nm Ag-nanoparticles disrupted cytoskeleton resulting in narrowing and contraction of cells. Many engineered nanomaterials, such as TiO_2, magnetite iron, CeO_2, carbon black, SWCNTs, and MWCNTs, also might cause different levels of inflammatory reactions, including enhanced pro-inflammatory cytokines expression, target inflammation-related genes, and micro-granulomas formation [127, 128]. The intra-tracheal administration of MWCNTs with variable length and iron content in hypertensive rats Led to the lung inflammation with increased blood pressure and lesions in abdominal arteries along with accumulation in multiple organs i.e., liver, kidneys, and spleen post seven days and 30 days exposure [129]. Maneewatttanapinyo et al. studied acute toxicity of colloidal silver nanoparticles administered in laboratory mice and observed no mortality any acute toxicity symptoms after a limited dose of 5.000 mg/kg post 14 days of oral administration. No differences could be observed among groups after hematological and biochemical assessment and the histopathological study. The instillation of silver nanoparticles at the concentration of 5.000 ppm developed a transient eye irritation for 24 h. The application of these nanomaterials on the skin did not produce any micro or macroscopic toxicity [130]. The schematic mechanism of silver nanoparticle's toxicity in the human body is shown in Fig. 3 [131]. The liver and spleen are maximum exposed organs to nanomaterials owing to the prevalence of phagocytic cells in the reticuloendothelial system. Also, the organs with high blood flow, such as kidneys and lungs, can be affected.

5.2 *Health Hazards in Human*

Despite having many benefits and using nanomaterials, it may cause health hazards to humans due to a tiny size. The broad absorption surface of the lung, the thinner air–blood barrier, and comparatively less inactivation of enzymes leads to faster entry for particles into the systemic blood circulation at higher drug concentrations. Additionally, intended uptake, exposure of airborne particles from the environment, and nanoparticles released during the manufacturing process may also cause health hazards for humans. Usually, nanomaterials' biological effects are based on their size, composition, shape, and even on their electronic, magnetic, optical, and mechanical

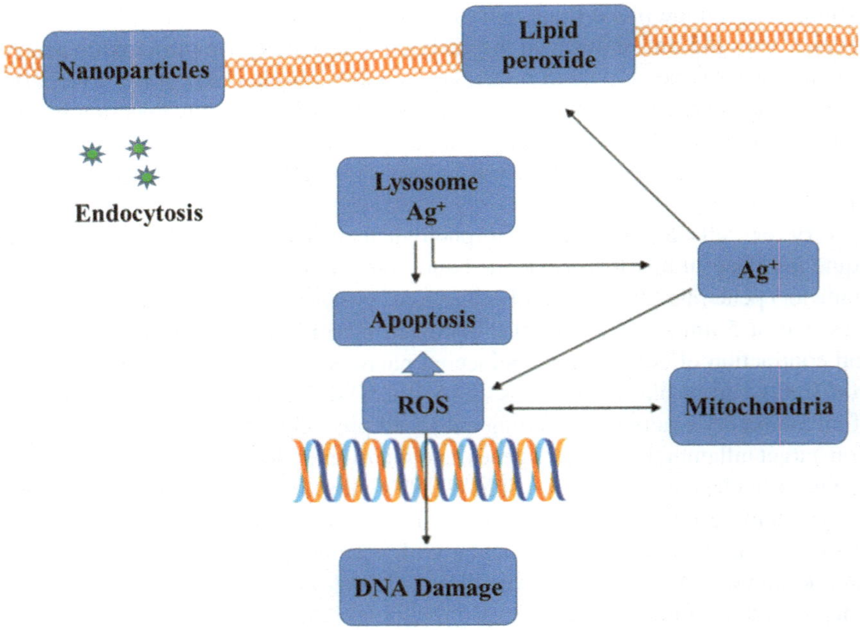

Fig. 3 Mechanism of silver nanoparticles toxicity (Abbreviations: NPs—nanoparticles; ROS—reactive oxygen species; Ag+—silver ions) Redrawn from [131]

properties. Presently, the influence of nanotechnology on human health and the environment is still controlled. Most of the studies assessed the outcomes of unintentional and accidental exposure (inhalation, medical procedures, or accidental ingestion) and focused only on local effects [98, 99]. Though, along with introducing nanomaterial-based biomedical methods, it is mandatory to analyze their toxicity at a systemic level. Centuries before, Paracelsus said, "everything is a poison, and nothing is a poison, it is only a matter of a dose." For nanomaterials, it is applicable in both the aspects of dose and particle size [100]. There is a massive demand for nanomaterials in various applications, ranging from diagnostic technology, bio-imaging, to gene/drug delivery [132–145]. Therefore, intended or unintended human exposure to nanomaterials is unavoidable and has higher prospects of exposure. Thus, a branch of science is developing, named "nanotoxicology", the study of the toxicity of nanomaterials. Nanotoxicology assesses the role and safety of nanomaterials on human health. Several anthropogenic sources, like power plants, internal combustion engines, and other thermo-degradation reactions also generate nanoparticles and develop the need to assess them [101].

5.2.1 Hazards in Nanomedicine

The nanomaterials represent a variety of biomedical applications. However, there is some potential risks factor related to the toxic issue. For example, oxidative stress, cytotoxicity, genotoxicity, and inflammation have been reported on in vitro and in vivo models for testing nanoparticles. The difference in the size of nanomaterial and bulk comes with the differences in properties and toxicity. Nanomaterials are tremendously beneficial yet can be toxic. Ag, ZnO, or CuO nanoparticles are frequently used as bactericides [102]. Nevertheless, waste disposal in the environment can also negatively affect non-target organisms.

5.2.2 Hazards in Medical Instrumentation

Nanomaterials are involved in medical interventions like prevention, diagnosis, and treatment of diseases. More functional and accurate medical diagnostic equipment are being designed for easy and safe operation. The lab-on-a-chip technology facilitates real-time point-of-care testing, enhancing the standards of medical care. Nanomaterial based thin films on implant surfaces improve the wear and resist infection. However, until now, these medical nanodevices are not 100% hazard free due to manufacturing processes, not following guidelines of nanotoxicity, and operating without the assessment of long term effects of nanotoxicity.

5.2.3 Hazards in Food Product

Nanotechnology is used to produce advanced food products and smart packaging technology [146–148]. In this way, the possibility of direct exposure to nanomaterials with human beings is enhanced, and different types of long-term or short-term toxicity may occur [149–151]. Nanoparticles and diamond-like nanocomposite (DLN) thin films are used in food packaging to reduce UV exposure and prolonged shelf life. Due to very few articles being reported in this area, further research is needed to fully explore the potential use of these nanoparticles for food products and medical treatments.

6 Environmental Nanopollution and Its Effect in Society

Environment conservation is a challenging task. Its vastness and complexity make this even more difficult. As nanomaterials' production is growing, multiple issues concerning nanotechnology arise as environmental pollution and industrial exposure. Nanoparticles serve as pollutants in diesel exhaust or welding fumes, presenting new toxicological mechanisms [152, 153]. It also makes us face pollution in macro, micro, and nanoscale. New branches of electronics are also creating new sources of occupational exposure hazard. The circumstances produce new challenges for both classical toxicology and nanotoxicology. Though nanotechnology improves the living standard, a simultaneous increase in water and air pollution has also occurred. As the origin of this pollution lies in nanomaterials hence termed "Nanopollution." Nanopollution is exceptionally lethal to both underwater flora and fauna and organisms living on soil. The pollutants can enter the human body in multiple ways. Cellular mechanisms can get affected by nanomaterial toxicity, which mainly comprises reactive oxygen species generation and genotoxicity [153–155]. The nanoparticle's exposure on humans can occur accidentally by environmental particles (e.g., air pollution) and intentionally because of a variety of consumer products, cosmetics, and medical products containing nanoparticles. The release of nanoparticles during the manufacturing process may result in exposure to workers via dermal, oral, and inhalation routes. Exposure to air pollutants, such as ultrafine particles, is known to cause inflammatory airway diseases and cardiovascular problems in humans [156]. Pope et al. [157] stated that even low levels of ambient nanoparticle exposure have a significant effect on mortality. To decrease nano pollution, scientists and researchers used nanotechnology to develop nanofilters, eliminating almost all airborne particles [158].

7 Social-economic Disruption Due to Rapid Use of Nanotechnology

As the speed of nanotechnology development is growing, as a consequence, the job opportunities are decreasing, arising the problem of unemployment in fields like industrial sector, manufacturing, and traditional farming [159, 160]. Nanotechnology-based devices and machines have replaced humans to furnish the job more rapidly and efficiently, which has pointed out the importance of human resources in practical work. Increasing growth and instant performance of nanotechnology have compromised the worth of commodities like diamond and oil. As an alternative technology, i.e., Nanotechnology has a detrimental effect on demand as substitutes have more efficiency and do not need fossil fuels. Diamonds are losing the worth due to greater availability from nanotechnology-based fabrication methods. Currently, manufacturing companies are equipped to produce the bulk of these products at a molecular scale, followed by disintegration to create new components.

At present, nanotechnology involves high investment technologies, raising the cost daily. The high price is the result of intricate molecular structure and processing charges of the product. The whole process makes it difficult for manufacturers to produce dynamic products using nanotechnology randomly. Currently, it is an unaffordable business owing to the massive pricing of nanotechnology-based machines. Hence, nanotechnology can also bring financial risks as manufacturers have to invest a large sum of money for setting up nanotech plants. The manufacturers have to face a considerable loss if, by any chance, the manufactured products fail to satisfy the customers. Alternate options such as the recovery of the original product or maintenance of the nanomaterials are also a costly and tedious affair.

Further, nanotechnology does not leave any byproducts or residues, generally based on small industries, therefore creating a considerable risk of extinction for small scale industries. As an outcome, the quantity of sub-products of coal and petroleum is deteriorating. Another massive threat, like the Covid-19 pandemic situation, may be born with the arrival of nanotechnology. It can make the easy accessibility of biochemical weapons or nano-bio engineered biological weapons. Nanotechnology is making these weapons more powerful and destructive. Unauthorized criminal bodies or corrupt politicians can steal the formulations and may reach these dangerous weapons easily, and they can quickly destroy our civilization [161].

8 Effect of Nanotechnology on Microorganisms, Animals, and Plants

Some nanomaterials are hazardous to human beings and are also harmful to the existence of different microorganisms, animals, and plants. Human-made nano pollution is very unsafe for living microorganisms, animals, and plants under the water or on the earth. As a result, many of microorganism's families have entirely disappeared from the world. Due to the rapid application of nanotechnology in the agriculture sector without proper nanotoxicological analysis, many plants are directly exposed to nanotoxicity, and animals are indirectly exposed. Thus, in the last two decades, a vast number of valuable plants and animals are entirely disappeared from our world.

9 Safety and Security of Nanotechnological Developments

Nanotechnology is an extensively expanding field. Researchers, scientists, and engineers are getting high success in producing nanoscale materials and taking advantage of enhanced properties, such as higher strength, lighter weight, increased electrical conductivity, and chemical reactivity compared to their larger-scale equivalents [162, 163].

Human health concerns are also growing due to nanomaterials. The attempts of technological manipulations raise the vocational risk to the workers in case of accidental exposures. The ethical issues regarding the poisoning of mass material are processed at a nanoscale, causing adverse effects on the health and industry. Mass poisoning occurs in the case of toxic micro particles coatings on the products. These microparticles penetrate inside the brain, while in contact with humans. Academic and industry experts suggest that there exists ambiguity regarding the toxic effects of releasing nanoparticles into the environment. It is also noteworthy that there is a lack of knowledge of nanoparticles interactions with humans and the environment. Similar to most of the emerging technologies, nanotechnology, and nanochemistry industries have both benefits and challenges. To obtain maximum benefits, the problems must be overcome, managed, and endured. In combination with other inorganic or organic counterparts, mesoporous silicates have been extensively explored for targeted drug delivery and cancer treatment. Even though the long-term toxicity of the nanoparticles is subjected to controversies and doubts, the use of gold and silver nanoparticles have provided more advantages in comparison to other actual alternatives (cytostatics).

Consequently, there is a growing interest in developing in vitro assays for nanotoxicology study [164]. It is strongly encouraged to use primary human cells as a source for in vitro study with nanoparticles since different origins of cancerous cell lines complicate data interpretation for human risk evaluation. Till now, the environmental effects and the toxicity of nanomaterials to organisms are in the infancy state. The evaluation methods need to be cost-effective rapid, and quantity efficient.

10 Current Policy and Regulation Status

The social implications of nanotechnology comprise many fundamental aspects like ethics, privacy, environment, and security. Occasionally, the negative impacts on the environment are too averse to handle that the people simply give up. However, nanoscience researchers are still optimistic about seeing the light of hope on the other side of the tunnel. Environmental clean-up is possible via the design and manipulation of the atomic and molecular scale of materials. It would develop cleaner energy production, energy efficiency, water treatment, and environmental remediation. Nanoscale fluid dynamics decipher the flow of nanoparticles in the environment as a result of interactions with biological and ecological systems. Researchers are keen to understand the transportation of nanomaterials in association with environmental contaminants through groundwater systems. For food authenticity, safety, and traceability, every food company should need to use smart labels at more robust and innovative functional lightweight packaging. Each developed and developing countries have a separate policy and regulation for the use of nanotechnological products and applications. Explicit initiatives on nanotechnology must be needed to promise that the opportunity provided by nanotechnology is not misused, and research does not become fragmented. The uncertainty, complexity, and diversity of nanotechnology mean that any initiative should not be a strictly preconceived closed program. Flexibility will be needed to stay side by side of development as they arise.

11 Challenges and Future Trends in Using Nanomaterials in Humans

Nanotechnology-based production uses minimal human resources, land, maintenance, and it is cost-effective, high productivity with modest requirements of materials and energy. The extensively growing field offers scientists and engineers an excellent opportunity to manipulate or alter the nanoscale materials to yield benefit of enhanced material characteristics like increased strength, lightweight, higher electrical conductivity, and chemical activity in comparison to their large-scale counterparts. However, for biomedical applications, the toxicity evaluation of nanomaterials should be performed. Broadly, detailed physicochemical characterization of nanomaterial should be performed before and during any toxicity study. Essential properties can control nanomaterial-induced toxicity, including size and shape of the nanomaterials, coating, chemical composition, crystal growth, nanomaterials purity, structure, surface area, surface chemistry, surface charge, agglomeration, and solubility should also be taken care. Measurements should be performed in a sufficiently stable state of nanomaterials in the most suitable test medium, i.e., aggregation status and ion release from metallic nanomaterials. Various engineered materials should be tested for their multidisciplinary tiered toxicity using diverse models and experiments [165, 166]. Therefore, the first step in genotoxicity is an assessment of the physicochemical properties of nanomaterials. The validation of the proposed tiered approaches still waits for the future. The researchers are continuously trying to increase the relevant database with an increasing number of publications (papers, reviews, or even patents) every year [167], particularly the market share of the nanotechnology products is also growing up to thousands of billions of Euros [168]. Balanced use of the nanotechnologies/nanomaterials must be arranged to optimize the opportunities/risks factors.

Further studies related to the influence of size and shape, capping agents, receptors immobilization onto the metal nanoparticles are still necessary. Varying sizes can tune surface plasmon resonance, the shape of the nanomaterials and different surface functionalization of both silver and gold nanoparticles can reduce the toxicity and enhance a variety of biomedical applications in the future. For example, CNT toxicity can be reduced via functionalization, surface coating, and stimulation of the autophagic flux. The amino functionalization decreases the CNT toxicity to the cells [169] and albumin coating for SWCNTs [170]. We have summarized some comparative points about the advantages and disadvantages of nanotechnology discussed throughout our review in the form of the following Table 1.

12 Conclusions

Nanoparticles can enter and get distributed around the human body very easily. After entering into humans, it moves within the body and creates cellular toxicity. Then it attacks the respiratory system, cardiovascular system, brain, skin, gut, and other

Table 1 Comparative discussion about advantages and disadvantages of nanotechnology

Advantages	Disadvantages
Early-stage detection of some diseases	Still at its infancy stage
Reduction of the size of any material, machine or equipment	More research and developmental work need to be done
Reduction of the amount of energy and resource	Expensive technology till now
Helps to clean up the existing nano-pollution	Creates environmental nano pollution
Able to secure the economy once it can be fully implemented	It can create social-economic disruption in society
Applicable and implementable to most of the applications ever existed	The huge initial cost for implementation
Can alter the basis of technology for human, in its matured phase	Resistance from a culture perspective, activists, journalists and even within the government
Improvement of the therapeutic drug index by increasing efficacy and/or reducing toxicities	Knowledge limitation from many industries and misperception among many fields about its capabilities.
Targeted delivery of drugs in a tissue-, cell- or organelle-specific manner	The government does not regulate nanomaterials
Enabling sustained or stimulus-triggered drug release	Requirement of significant investment and research but yield is still a limiting factor
More sensitive cancer diagnosis and imaging	Some nanoparticles may be toxic to humans
Better pharmaceutical properties (i.e. stability, solubility, circulating half-life and tumor accumulation) of therapeutic molecules	Nanotechnology made weapons are more powerful and more destructive by increasing the explosion potential
Provision of new approaches for the development of synthetic vaccines	Lack of employment in the fields of traditional farming, manufacturing, and industrial sector

organs. Some nanomaterials kill harmful bacteria within the body, and some kill good bacteria and live-cells of the human body. Nanoparticles with different substances are used in SIM cards of cell phones or sunscreens. When these are used, free nanoparticles get released in the environment (air, water, or soil). Engineering fields like civil and electronics also create new occupational health risks, making new, potentially toxic nanomaterials. The toxicity of nanoparticles depends on their shape, size, and chemical composition. Centuries before, Paracelsus quoted, "everything is a poison, and nothing is a poison, it is only a matter of a dose." In regards to nanomaterials, the quotes hold value for both dose and particle size. The new interdisciplinary investigations explore the potentially harmful effects of these useful NPs and help in environmental preservation. Owing to a smaller size, the inhalation of nanomaterials imposes an adverse impact on human health. The inhalation causes severe injury to the lungs and can also become fatal. The deterioration of lungs can be observed even after the 60s of nanoparticle inhalation. Therefore, for sustainable nanotechnology development, it is mandatory to evaluate and spread knowledge about the short term and long term exposure benefits and hazards for nanomaterials.

To conclude, nanotechnology has the potential to impact society, both positively or negatively. Its consumers, producers, and dealers include all the community members and all stakeholders, so we should collectively raise the voice in its various growth and commercialization phases. Nanotechnology is currently in its infancy stage, with a significant lack of awareness about its effects on humans and the environment. As civilization moves forward, the vital query is: how should we manage the risks and uncertainties of this emergent technology? Is anyhow the COVID-19 pandemic situation human-made? If not, we can face such circumstances due to the careless application of nanotechnology in different fields.

References

1. Santra TS, Tseng F-G (Kevin), Barik TK (2015) Biosynthesis of silver and gold nanoparticles for potential biomedical applications—a brief review. J Nanopharmaceutics Drug Deliv 2:249–265. https://doi.org/10.1166/jnd.2014.1065
2. Kalantar-Zadeh K, Fry B (2008) Nanotechnology-enabled sensors. Springer, US
3. Bhushan B (2010) Introduction to nanotechnology. Springer handbook of nanotechnology. Springer, Berlin Heidelberg, Berlin, Heidelberg, pp 1–13
4. Jha AR (2008) MEMS and nanotechnology-based sensors and devices for communications, medical and aerospace applications. CRC Press
5. Bakhoum EG Micro- and nano-scale sensors and transducers
6. (2020) Handbook of single cell technology + reference. Springer
7. Microfluidics and BioMEMS: Devices and applications—1st Edition—Tu. https://www.routledge.com/Microfluidics-and-BioMEMS-Devices-and-Applications/Santra/p/book/9789814800853. Accessed 9 June 2020
8. Tseng F-G, Santra TS essentials of single-cell analysis: concepts, applications and future prospects
9. Tseng F-G, Santra TS (2015) Micro/nano fluidic devices for single cell analysis. MDPI AG Basel, Switz
10. Santra TS (2020) Bio-MEMS and bio-NEMS: devices and applications. Jenny Stanford Publisher Pvt. Ltd., Singapore
11. Shinde P, Kar S, Mohan L, Chang H-Y, Tseng F-G, Nagai M, Santra TS (2020) Infrared pulse laser activated highly efficient intracellular delivery using titanium micro-dish device. ACS Biomater Sci Eng. https://doi.org/10.1021/acsbiomaterials.0c00785
12. Santra TS, Kar S, Chang H-Y, Tseng F-G (2020) Nano-localized single-cell nano-electroporation. Lab Chip 17. https://doi.org/10.1039/d0lc00712a
13. Illath K, Narasimahan, AK, Nagai M, Wankhar S, Santra TS (2020) Microfluidic based metallic nanoparticle synthesis and applications. In: Microfluidics and bio-MEMS: devices and applications. Jenny Stanford Publisher Pte. Ltd
14. Medintz IL, Uyeda HT, Goldman ER, Mattoussi H (2005) Quantum dot bioconjugates for imaging, labelling and sensing. Nat. Mater
15. Oshida Y (2010) Bioscience and bioengineering of titanium materials. Elsevier
16. Mohan L, Anandan C, Grips VKW (2012) Corrosion behavior of titanium alloy Beta-21S coated with diamond like carbon in Hank's solution. Appl Surf Sci 258:6331–6340. https://doi.org/10.1016/j.apsusc.2012.03.032
17. Santra TS, Tseng F-G, Barik TK (2014) Green biosynthesis of gold nanoparticles and biomedical applications. Am J Nano Res Appl 2:5–12. https://doi.org/10.11648/j.nano.s.2014020602.12

18. Mohan L, Durgalakshmi D, Geetha M, Narayanan TSNS, Asokamani R (2012) Electrophoretic deposition of nanocomposite (HAp + TiO2) on titanium alloy for biomedical applications. Ceram Int 38:3435–3443
19. Mohan L, Anandan C (2013) Wear and corrosion behavior of oxygen implanted biomedical titanium alloy Ti-13Nb-13Zr. Appl Surf Sci 282:281–290. https://doi.org/10.1016/j.apsusc.2013.05.120
20. Mohan L, Anandan C, Rajendran N (2015) Electrochemical behaviour and bioactivity of self-organized TiO2 nanotube arrays on Ti-6Al-4 V in Hanks' solution for biomedical applications. Electrochim Acta 155:411–420. https://doi.org/10.1016/j.electacta.2014.12.032
21. Mohan L, Kar S, Nandhini B, Dhilip Kumar SS, Nagai M, Santra TS (2020) Formation of nanostructures on magnesium alloy by anodization for potential biomedical applications. Mater Today Commun: 101403. https://doi.org/10.1016/j.mtcomm.2020.101403
22. [PDF] A review of current research into the biogenic synthesis of metal and metal oxide nanoparticles via Marine Algae and Seagrasses|Semantic Scholar. https://www.semanticscholar.org/paper/A-Review-of-Current-Research-into-the-Biogenic-of-Fawcett-Verduin/04a57915cbbe6701fb08de0464260d5a3af7e3a7. Accessed 9 June 2020
23. Maynard AD (2007) Nanotechnology: the next big thing, or much ado about nothing? Ann Occup Hyg 51:1–12. https://doi.org/10.1093/annhyg/mel071
24. Cox S, Abu-Ghannam N, Gupta S (2010) An assessment of the antioxidant and antimicrobial activity of six species of edible Irish seaweeds. Int Food Res J 17:205–220. https://doi.org/10.21427/D7HC92
25. De Pádua M, Growoski Fontoura PS, Mathias AL (2004) Chemical composition of Ulvaria oxysperma (Kützing) bliding, Ulva lactuca (Linnaeus) and Ulva fascita (Delile). Brazilian Arch Biol Technol 47:49–55. https://doi.org/10.1590/s1516-89132004000100007
26. Kumari P, Kumar M, Gupta V, Reddy CRK, Jha B (2010) Tropical marine macroalgae as potential sources of nutritionally important PUFAs. Food Chem 120:749–757. https://doi.org/10.1016/j.foodchem.2009.11.006
27. Ravikumar S, Krishnakumar S, Inbaneson SJ, Gnanadesigan M (2010) Antagonistic activity of marine actinomycetes from Arabian Sea coast
28. Krishnakumar S, Premkumar J, Alexis RR, Ravikumar S (2011) Optimization of potential antibiotic production by salt-tolerant Actinomycetes streptomyces sp.-MSU29 isolated from marine sponge. Int J Appl Bio-Eng 5:12–18. https://doi.org/10.18000/ijabeg.10079
29. Manivasagan P, Venkatesan J, Kim S-K (2015) Marine algae: an important source of bioenergy production. In: Marine Bioenergy. CRC Press, pp 45–70
30. Ermakova S, Kusaykin M, Trincone A, Tatiana Z (2015) Are multifunctional marine polysaccharides a myth or reality? Front Chem 3:39. https://doi.org/10.3389/fchem.2015.00039
31. Venkatesan J, Bhatnagar I, Manivasagan P, Kang KH, Kim SK (2015) Alginate composites for bone tissue engineering: a review. Int J Biol Macromol 72:269–281
32. Fayaz AM, Balaji K, Girilal M, Yadav R, Kalaichelvan PT, Venketesan R (2010) Biogenic synthesis of silver nanoparticles and their synergistic effect with antibiotics: a study against gram-positive and gram-negative bacteria. Nanomed Nanotechnol Biol Med 6:103–109. https://doi.org/10.1016/j.nano.2009.04.006
33. Ra M, Gade A, Gaikwad S, Marcato PD, Durán N (2012) Biomedical applications of nanobiosensors: the state-of-the-art. J Braz Chem Soc 23:14–24
34. Kar S, Mohapatra DR, Freysz E, Sood AK (2014) Tuning photoinduced terahertz conductivity in monolayer graphene: optical-pump terahertz-probe spectroscopy. Phys Rev B Condens Matter Mater Phys 90:165420. https://doi.org/10.1103/PhysRevB.90.165420
35. Kar S, Nguyen VL, Mohapatra DR, Lee YH, Sood AK (2018) Ultrafast spectral photoresponse of bilayer graphene: optical pump-terahertz probe spectroscopy. ACS Nano 12:1785–1792. https://doi.org/10.1021/acsnano.7b08555
36. Kar S, Mohapatra DR, Sood AK (2018) Tunable terahertz photoconductivity of hydrogen functionalized graphene using optical pump-terahertz probe spectroscopy. Nanoscale 10:14321–14330. https://doi.org/10.1039/c8nr04154g

37. Kar S, Su Y, Nair RR, Sood AK (2015) Probing photoexcited carriers in a few-layer MoS2 laminate by time-resolved optical pump-terahertz probe spectroscopy. ACS Nano 9:12004–12010. https://doi.org/10.1021/acsnano.5b04804
38. Kar S, Jayanthi S, Freysz E, Sood AK (2014) Time resolved terahertz spectroscopy of low frequency electronic resonances and optical pump-induced terahertz photoconductivity in reduced graphene oxide membrane. Carbon N Y 80:762–770. https://doi.org/10.1016/j.carbon.2014.09.030
39. Li BL, Wang J, Zou HL, Garaj S, Lim CT, Xie J, Li NB, Leong DT (2016) Low-dimensional transition metal dichalcogenide nanostructures based sensors. Adv Funct Mater 26:7034–7056. https://doi.org/10.1002/adfm.201602136
40. Kar S, Sood AK (2019) Ultrafast terahertz photoresponse of single and double-walled carbon nanotubes: optical pump-terahertz probe spectroscopy. Carbon N Y 144:731–736. https://doi.org/10.1016/j.carbon.2018.12.081
41. Chiara R, Ciftci YO, Queloz VIE, Nazeeruddin MK, Grancini G, Malavasi L (2020) Green-emitting lead-free Cs4SnBr 6 zero-dimensional perovskite nanocrystals with improved air stability. J Phys Chem Lett 11:618–623. https://doi.org/10.1021/acs.jpclett.9b03685
42. Dubey SP, Lahtinen M, Sillanpää M (2010) Tansy fruit mediated greener synthesis of silver and gold nanoparticles. Process Biochem 45:1065–1071. https://doi.org/10.1016/j.procbio.2010.03.024
43. Prasad TNVKV, Elumalai EK (2011) Biofabrication of Ag nanoparticles using Moringa oleifera leaf extract and their antimicrobial activity. Asian Pac J Trop Biomed 1:439–442. https://doi.org/10.1016/S2221-1691(11)60096-8
44. Leela A, Vivekanandan M (2008) Tapping the unexploited plant resources for the synthesis of silver nanoparticles. African J Biotechnol 7:3162–3165. https://doi.org/10.5897/AJB08.425
45. Hedenborg M (1988) Titanium dioxide induced chemiluminescence of human polymorphonuclear leukocytes. Int Arch Occup Environ Health 61:1–6. https://doi.org/10.1007/BF00381600
46. Savithramma N, Linga Rao M, Ankanna S, Venkateswarlu P (2012) Screening of medicinal plants for effective biogenesis of silver nanoparticles and efficient antimicrobial activity. Int J Pharm Sci Res. https://ijpsr.com/bft-article/screening-of-medicinal-plants-for-effective-biogenesis-of-silver-nano-particles-and-efficient-anti-microbial-activity/. Accessed 9 June 2020
47. Chikramane PS, Suresh AK, Bellare JR, Kane SG (2010) Extreme homeopathic dilutions retain starting materials: a nanoparticulate perspective. Homeopathy 99:231–242. https://doi.org/10.1016/j.homp.2010.05.006
48. Aslan K, Leonenko Z, Lakowicz JR, Geddes CD (2005) Fast and slow deposition of silver nanorods on planar surfaces: application to metal-enhanced fluorescence. J Phys Chem B 109:3157–3162. https://doi.org/10.1021/jp045186t
49. Diao JJ, Cao Q (2011) Gold nanoparticle wire and integrated wire array for electronic detection of chemical and biological molecules. AIP Adv 1:012115. https://doi.org/10.1063/1.3568815
50. Lim JK, Imura K, Nagahara T, Kim SK, Okamoto H (2005) Imaging and dispersion relations of surface plasmon modes in silver nanorods by near-field spectroscopy. Chem Phys Lett 412:41–45. https://doi.org/10.1016/j.cplett.2005.06.094
51. Hutter E, Maysinger D (2011) Gold nanoparticles and quantum dots for bioimaging. Microsc Res Tech 74:592–604
52. Biological applications of colloidal nanocrystals—IOPscience. https://iopscience.iop.org/article/10.1088/0957-4484/14/7/201/pdf. Accessed 9 June 2020
53. Schider G, Krenn R, Hohenau A, Ditlbacher H, Leitner A, Aussenegg R, Schaich L, Puscasu I, Monacelli B, Boreman G (2003) Plasmon dispersion relation of Au and Ag nanowires. Phys Rev B Condens Matter Mater Phys 68:155427. https://doi.org/10.1103/PhysRevB.68.155427
54. Lou X, Zhang Y, Qin J, Li Z (2011) A highly sensitive and selective fluorescent probe for cyanide based on the dissolution of gold nanoparticles and its application in real samples. Chem A Eur J 17:9691–9696. https://doi.org/10.1002/chem.201100389
55. Pingarrón JM, Yáñez-Sedeño P, González-Cortés A (2008) Gold nanoparticle-based electrochemical biosensors. Electrochim Acta 53:5848–5866. https://doi.org/10.1016/j.electacta.2008.03.005

56. Geddes CD, Parfenov A, Gryczynski I, Lakowicz JR (2003) Luminescent blinking of gold nanoparticles. Chem Phys Lett 380:269–272. https://doi.org/10.1016/j.cplett.2003.07.029
57. Dey K, Kar S, Shinde P, Mohan L, Bajpai SK, Santra TS (2020) *Microfluidic electroporation and applications*. Jenny Stanford Publisher
58. Yu DG (2007) Formation of colloidal silver nanoparticles stabilized by Na+-poly(γ-glutamic acid)-silver nitrate complex via chemical reduction process. Colloids Surf B Biointerfaces 59:171–178. https://doi.org/10.1016/j.colsurfb.2007.05.007
59. Tan Y, Wang Y, Jiang L, Zhu D (2002) Thiosalicylic acid-functionalized silver nanoparticles synthesized in one-phase system. J Colloid Interface Sci 249:336–345. https://doi.org/10.1006/jcis.2001.8166
60. Petit C, Lixon P, Pileni MP (1993) In situ synthesis of silver nanocluster in AOT reverse micelles. J Phys Chem 97:12974–12983. https://doi.org/10.1021/j100151a054
61. Vorobyova SA, Lesnikovich AI, Sobal NS (1999) Preparation of silver nanoparticles by interphase reduction. Colloids Surf A Physicochem Eng Asp 152:375–379. https://doi.org/10.1016/S0927-7757(98)00861-9
62. Mallick K, Witcomb MJ, Scurrell MS (2005) Self-assembly of silver nanoparticles in a polymer solvent: formation of a nanochain through nanoscale soldering. Mater Chem Phys 90:221–224. https://doi.org/10.1016/j.matchemphys.2004.10.030
63. Kéki S, Török J, Deák G, Daróczi L, Zsuga M (2000) Silver nanoparticles by PAMAM-assisted photochemical reduction of Ag+. J Colloid Interface Sci 229:550–553. https://doi.org/10.1006/jcis.2000.7011
64. Pileni MP (2000) Fabrication and physical properties of self-organized silver nanocrystals. Pure Appl Chem, pp 53–65
65. Sun YP, Atorngitjawat P, Meziani MJ (2001) Preparation of silver nanoparticles via rapid expansion of water in carbon dioxide microemulsion into reductant solution. Langmuir 17:5707–5710. https://doi.org/10.1021/la0103057
66. Liu YC, Lin LH (2004) New pathway for the synthesis of ultrafine silver nanoparticles from bulk silver substrates in aqueous solutions by sonoelectrochemical methods. Electrochem Commun 6:1163–1168. https://doi.org/10.1016/j.elecom.2004.09.010
67. Sandmann G, Dietz H, Plieth W (2000) Preparation of silver nanoparticles on ITO surfaces by a double-pulse method. J Electroanal Chem 491:78–86. https://doi.org/10.1016/S0022-0728(00)00301-6
68. Bae CH, Nam SH, Park SM (2002) Formation of silver nanoparticles by laser ablation of a silver target in NaCl solution. In: Applied Surface Science. Elsevier, pp 628–634
69. Smetana AB, Klabunde KJ, Sorensen CM (2005) Synthesis of spherical silver nanoparticles by digestive ripening, stabilization with various agents, and their 3-D and 2-D superlattice formation. J Colloid Interface Sci 284:521–526. https://doi.org/10.1016/j.jcis.2004.10.038
70. Esumi K, Isono R, Yoshimura T (2004) Preparation of PAMAM- and PPI-Metal (Slver, Platinum, and Palladium) nanocomposites and their catalytic activities for reduction of 4-nitrophenol. Langmuir 20:237–243. https://doi.org/10.1021/la035440t
71. Murugadoss A, Khan A, Chattopadhyay A (2010) Stabilizer specific interaction of gold nanoparticles with a thermosensitive polymer hydrogel. J Nanoparticle Res 12:1331–1348. https://doi.org/10.1007/s11051-009-9668-0
72. Zhang H, Li X, Chen G (2009) Ionic liquid-facilitated synthesis and catalytic activity of highly dispersed Ag nanoclusters supported on TiO2. J Mater Chem 19:8223–8231. https://doi.org/10.1039/b910610c
73. Sun Y, Xia Y (2002) Shape-controlled synthesis of gold and silver nanoparticles. Science (80-) 298:2176–2179. https://doi.org/10.1126/science.1077229
74. Bhattacharya R, Mukherjee P (2008) Biological properties of "naked" metal nanoparticles. Adv Drug Deliv Rev 60:1289–1306
75. Raghunandan D, Borgaonkar PA, Bendegumble B, Bedre MD, Bhagawanraju M, Yalagatti MS, Huh DS, Abbaraju V (2011) Microwave-assisted rapid extracellular biosynthesis of silver nanoparticles using carom seed (Trachyspermum copticum) extract and in vitro studies. Am J Anal Chem 02:475–483. https://doi.org/10.4236/ajac.2011.24057

76. Kathiresan K, Manivannan S, Nabeel MA, Dhivya B (2009) Studies on silver nanoparticles synthesized by a marine fungus, Penicillium fellutanum isolated from coastal mangrove sediment. Colloids Surf B Biointerfaces 71:133–137. https://doi.org/10.1016/j.colsurfb.2009.01.016
77. Konishi Y, Tsukiyama T, Ohno K, Saitoh N, Nomura T, Nagamine S (2006) Intracellular recovery of gold by microbial reduction of AuCl 4-ions using the anaerobic bacterium Shewanella algae. Hydrometallurgy 81:24–29. https://doi.org/10.1016/j.hydromet.2005.09.006
78. (PDF) Biosynthesis of silver, gold and bimetallic nanoparticles using the filamentous fungus Neurospora crassa I Alfredo R Vilchis-Nestor—Academia.edu. https://www.academia.edu/22627587/Biosynthesis_of_silver_gold_and_bimetallic_nanoparticles_using_the_filamentous_fungus_Neurospora_crassa. Accessed 9 June 2020
79. Goodsell DS (2004) Bionanotechnology: lessons from nature. Wiley-Liss
80. Thakur NS, Dwivedee BP, Banerjee UC, Bhaumik J (2017) Bioinspired synthesis of silver nanoparticles: characterisation, mechanism and applications. In: Silver Nanoparticles for Antibacterial Devices. CRC Press, pp 3–36
81. Xie J, Lee JY, Wang DIC, Ting YP (2007) Silver nanoplates: from biological to biomimetic synthesis. ACS Nano 1:429–439. https://doi.org/10.1021/nn7000883
82. Mendiola JA, Rodriguez-Meizoso I, Señoráns FJ, Reglero G (2017) Advanced Microalgal Technologies for a Circular Economy. ALGATEC-CM View project. Project Allergy Car-Union European View project
83. Wenqiang G, Shufen L, Ruixiang Y, Yanfeng H (2006) Comparison of composition and antifungal activity of Artemisia argyi Lévl. et Vant inflorescence essential oil extracted by hydrodistillation and supercritical carbon dioxide. Nat Prod Res 20:992–998. https://doi.org/10.1080/14786410600921599
84. Santra TS, Bhattacharyya TK Diamond-Like Nanocomposite (DLN) Films for Microelectro-Mechanical System (MEMS). IJCA
85. Santra TS, Bhattacharyya TK, Tseng FG, Barik TK (2012) Influence of flow rate on different properties of diamond-like nanocomposite thin films grown by PECVD. AIP Adv 2:022132. https://doi.org/10.1063/1.4721654
86. Santra TS, Bhattacharyya TK, Patel P, Tseng FG, Barik TK (2011) Structural and tribological properties of diamond-like nanocomposite thin films. Surf Coatings Technol 206:228–233. https://doi.org/10.1016/j.surfcoat.2011.06.057
87. Santra TS, Liu CH, Bhattacharyya TK, Patel P, Barik TK (2010) Characterization of diamond-like nanocomposite thin films grown by plasma enhanced chemical vapor deposition. J Appl Phys 107:124320. https://doi.org/10.1063/1.3415548
88. IJCA—Diamond-like Nanocomposite (DLN) Films for Microelectro-Mechanical System (MEMS). http://sandbox.ijcaonline.org/proceedings/isdmisc/number6/3478-isdm132. Accessed 9 June 2020
89. Biomedical applications of diamond-like nanocomposite thin films: ingenta connect. https://www.ingentaconnect.com/content/asp/sam/2012/00000004/00000001/art00014. Accessed 9 June 2020
90. Das T, Ghosh D, Bhattacharyya TK, Maiti TK (2007) Biocompatibility of diamond-like nanocomposite thin films. J Mater Sci Mater Med 18:493–500. https://doi.org/10.1007/s10856-007-2009-x
91. Nanotechnology applications (With images) I Nanotechnology, Nano science, Nanotechnology art. https://in.pinterest.com/pin/574842339921055473/. Accessed 9 June 2020
92. Global Nanomaterials Market Worth USD 16.8 Billion by 2022. https://www.zionmarketresearch.com/news/nanomaterials-market. Accessed 9 June 2020
93. Nanotechnology global market value 2020 I Statista. https://www.statista.com/statistics/1073886/global-market-value-nanotechnology/. Accessed 9 Jun 2020
94. Global Nanotechnology Market To Reach $48.9 Billion In 2017. https://www.bccresearch.com/pressroom/nan/global-nanotechnology-market-reach-$48.9-billion-2017. Accessed 9 June 2020

95. Elsaesser A, Howard CV (2012) Toxicology of nanoparticles. Adv Drug Deliv Rev
96. Pridgen EM, Alexis F, Farokhzad OC (2015) Polymeric nanoparticle drug delivery technologies for oral delivery applications. Expert Opin Drug Deliv
97. Kafshgari M, Harding F, Voelcker N (2015) Insights into cellular uptake of nanoparticles. Curr Drug Deliv. https://doi.org/10.2174/1567201811666140821110631
98. De Jong WH, Borm PJA (2008) Drug delivery and nanoparticles: applications and hazards. Int J Nanomedicine 3:133–149
99. Papp T, Schiffmann D, Weiss D, Castranova V, Vallyathan V, Rahman Q (2008) Human health implications of nanomaterial exposure. Nanotoxicology 2:9–27. https://doi.org/10.1080/17435390701847935
100. Likus W, Bajor G, Siemianowicz K (2013) Nanosilver-does it have only one face?
101. Oberdörster G, Oberdörster E, Oberdörster J (2005) Nanotoxicology: an emerging discipline evolving from studies of ultrafine particles. Environ Health Perspect 113:823–839
102. Love SA, Maurer-Jones MA, Thompson JW, Lin Y-S, Haynes CL (2012) Assessing nanoparticle toxicity. Annu Rev Anal Chem. https://doi.org/10.1146/annurev-anchem-062011-143134
103. Ahamed M, AlSalhi MS, Siddiqui MKJ (2010) Silver nanoparticle applications and human health. Clin. Chim, Acta
104. Manke A, Wang L, Rojanasakul Y (2013) Mechanisms of nanoparticle-induced oxidative stress and toxicity. Biomed Res, Int
105. Beer C, Foldbjerg R, Hayashi Y, Sutherland DS, Autrup H (2012) Toxicity of silver nanoparticles-Nanoparticle or silver ion? Toxicol Lett. https://doi.org/10.1016/j.toxlet.2011.11.002
106. Liao MY, Liu HG (2012) Gene expression profiling of nephrotoxicity from copper nanoparticles in rats after repeated oral administration. Environ Toxicol Pharmacol 34:67–80. https://doi.org/10.1016/j.etap.2011.05.014
107. La Francesca S (2012) Nanotechnology and stem cell therapy for cardiovascular diseases: potential applications. Methodist Debakey Cardiovasc J 8:28–35
108. Singh R, Lillard JW (2009) Nanoparticle-based targeted drug delivery. Exp. Mol, Pathol
109. Wang AZ, Langer R, Farokhzad OC (2012) Nanoparticle delivery of cancer drugs. Annu Rev Med. https://doi.org/10.1146/annurev-med-040210-162544
110. Prow TW, Grice JE, Lin LL, Faye R, Butler M, Becker W, Wurm EMT, Yoong C, Robertson TA, Soyer HP, Roberts MS (2011) Nanoparticles and microparticles for skin drug delivery. Adv Drug Deliv, Rev
111. Xie J, Lee S, Chen X (2010) Nanoparticle-based theranostic agents. Adv Drug Deliv, Rev
112. Brannon-Peppas L, Blanchette JO (2004) Nanoparticle and targeted systems for cancer therapy. Adv Drug Deliv Rev 56:1649–1659
113. Gerloff K, Albrecht C, Boots AW, Frster I, Schins RPF (2009) Cytotoxicity and oxidative DNA damage by nanoparticles in human intestinal Caco-2 cells. Nanotoxicology 3:355–364. https://doi.org/10.3109/17435390903276933
114. Relating cytotoxicity, zinc ions, and reactive oxygen in ZnO nanoparticle-exposed human immune cells. | Semantic Scholar. https://www.semanticscholar.org/paper/Relating-cytotoxicity%2C-zinc-ions%2C-and-reactive-in-Shen-James/95df5b01a4a932d2c233884e7a4c9e8dcd942169. Accessed 9 June 2020
115. Yan G, Huang Y, Bu Q, Lv L, Deng P, Zhou J, Wang Y, Yang Y, Liu Q, Cen X, Zhao Y (2012) Zinc oxide nanoparticles cause nephrotoxicity and kidney metabolism alterations in rats. J Environ Sci Heal Part A 47:577–588. https://doi.org/10.1080/10934529.2012.650576
116. Sahu SC, Zheng J, Graham L, Chen L, Ihrie J, Yourick JJ, Sprando RL (2014) Comparative cytotoxicity of nanosilver in human liver HepG2 and colon Caco2 cells in culture. J Appl Toxicol 34:1155–1166. https://doi.org/10.1002/jat.2994
117. Semmler M, Seitz J, Erbe F, Mayer P, Heyder J, Oberdörster G, Kreyling WG (2004) Long-term clearance kinetics of inhaled ultrafine insoluble iridium particles from the rat lung, including transient translocation into secondary organs. In: Inhalation Toxicology, pp 453–459

118. Ferin J, Oberdorster G, Penney DP (1992) Pulmonary retention of ultrafine and fine particles in rats. Am J Respir Cell Mol Biol 6. https://doi.org/10.1165/AJRCMB/6.5.535
119. Jugan ML, Barillet S, Simon-Deckers A, Sauvaigo S, Douki T, Herlin N, Carrière M (2011) Cytotoxic and genotoxic impact of TiO2 nanoparticles on A549 cells. J Biomed Nanotechnol 7:22–23. https://doi.org/10.1166/jbn.2011.1181
120. Du Z, Zhao D, Jing L, Cui G, Jin M, Li Y, Liu X, Liu Y, Du H, Guo C, Zhou X, Sun Z (2013) Cardiovascular toxicity of different sizes amorphous silica nanoparticles in rats after intratracheal instillation. Cardiovasc Toxicol 13:194–207. https://doi.org/10.1007/s12012-013-9198-y
121. Shi H, Magaye R, Castranova V, Zhao J (2013) Titanium dioxide nanoparticles: a review of current toxicological data. Part Fibre Toxicol 10:1–33
122. Landsiedel R, Ma-Hock L, Hofmann T, Wiemann M, Strauss V, Treumann S, Wohlleben W, Gröters S, Wiench K, Van Ravenzwaay B (2014) Application of short-term inhalation studies to assess the inhalation toxicity of nanomaterials. Part Fibre Toxicol 11:16. https://doi.org/10.1186/1743-8977-11-16
123. Silver Nanoparticle induced blood-brain barrier inflammation and increased permeability in primary rat brain microvessel endothelial cells | Toxicological Sciences | Oxford Academic. https://academic.oup.com/toxsci/article-abstract/118/1/160/1664864?redirectedFrom=fulltext. Accessed 9 June 2020
124. Trickler WJ, Lantz-Mcpeak SM, Robinson BL, Paule MG, Slikker W, Biris AS, Schlager JJ, Hussain SM, Kanungo J, Gonzalez C, Ali SF (2014) Porcine brain microvessel endothelial cells show pro-inflammatory response to the size and composition of metallic nanoparticles. Drug Metab Rev 46:224–231
125. González C, Salazar-García S, Palestino G, Martínez-Cuevas PP, Ramírez-Lee MA, Jurado-Manzano BB, Rosas-Hernández H, Gaytán-Pacheco N, Martel G, Espinosa-Tanguma R, Biris AS, Ali SF (2011) Effect of 45 nm silver nanoparticles (AgNPs) upon the smooth muscle of rat trachea: Role of nitric oxide. Toxicol Lett 207:306–313. https://doi.org/10.1016/j.toxlet.2011.09.024
126. Genter MB, Newman NC, Shertzer HG, Ali SF, Bolon B (2012) Distribution and systemic effects of intranasally administered 25 nm silver nanoparticles in adult mice. Toxicol Pathol 40:1004–1013. https://doi.org/10.1177/0192623312444470
127. Tsuda H, Xu J, Sakai Y, Futakuchi M, Fukamachi K (2009) Toxicology of engineered nanomaterials—a review of carcinogenic potential. Asian Pac J Cancer Prev 10:975–980
128. Park EJ, Kim H, Kim Y, Yi J, Choi K, Park K (2010) Inflammatory responses may be induced by a single intratracheal instillation of iron nanoparticles in mice. Toxicology 275:65–71. https://doi.org/10.1016/j.tox.2010.06.002
129. Chen R, Zhang L, Ge C, Tseng MT, Bai R, Qu Y, Beer C, Autrup H, Chen C (2015) Subchronic toxicity and cardiovascular responses in spontaneously hypertensive rats after exposure to multiwalled carbon nanotubes by intratracheal instillation. Chem Res Toxicol 28:440–450. https://doi.org/10.1021/tx5004003
130. Maneewattanapinyo P, Banlunara W, Thammacharoen C, Ekgasit S, Kaewamatawong T (2011) An evaluation of acute toxicity of colloidal silver nanoparticles. J Vet Med Sci 73:1417–1423. https://doi.org/10.1292/jvms.11-0038
131. Clichici S, Filip A (2015) In vivo assessment of nanomaterials toxicity. In: Nanomaterials—toxicity and risk assessment. InTech
132. Shinde P, Kumar A, Kavitha, Dey K, Mohan L, Kar S, Barik TK, Sharifi-Rad J, Nagai M, Santra TS (2020) Physical approaches for drug delivery. In: Delivery of drugs. Elsevier, pp 161–190
133. Kumar A, Mohan L, Shinde P, Chang H, Nagai M, Santra TS Mechanoporation Toward single cell approaches
134. Santra TS, Chang H-Y, Wang P-C, Tseng F-G (2014) Impact of pulse duration on localized single-cell nano-electroporation. Analyst 139:6249–6258
135. Santra TS, Wang PC, Chang HY, Tseng FG (2013) Tuning nano electric field to affect restrictive membrane area on localized single cell nano-electroporation. Appl Phys Lett 103. https://doi.org/10.1063/1.4833535

136. Manoj H, Gupta P, Loganathan M, Nagai M, Wankhar S, Santra (2020) Microneedles: current trends & applications. In: Microfluidics and bio-MEMS: devices and applications. Jenny Stanford Publisher
137. Kar S, Shinde P, Nagai M, Santra TS (2020) Optical manipulation of cells. In: Microfluidics and bio-MEMS: devices and applications. Jenny Stanford Publisher
138. Santra TS, Wu TH, Chiou EPY (2016) Photothermal microfluidics. In: Optical MEMS for chemical analysis and biomedicine. Institution of Engineering and Technology, pp 289–323
139. Santra TS, Tseng F-G (2016) Electroporation for single-cell analysis, pp 55–83
140. Santra TS, Wang P-C, Tseng FG (2013) Electroporation based drug delivery and its applications. In: Advances in Micro/Nano Electromechanical Systems and Fabrication Technologies. InTech
141. Shanmugam MM, Santra TS (2016) Microinjection for single-cell analysis. Springer, Berlin, Heidelberg, pp 85–129
142. Santra TS, Kar S, Chen C-W, Borana J, Chen T-C, Lee M-C, Tseng F-G (2020) Near-infrared nanosecond-pulsed laser-activated high efficient intracellular delivery mediated by nano-corrugated mushroom-shaped gold-coated polystyrene nanoparticles. Nanoscale. https://doi.org/10.1039/d0nr01792b
143. Shinde P, Mohan L, Kumar A, Dey K, Maddi A, Patananan AN, Tseng F-G, Chang H-Y, Nagai M, Santra TS (2018) Current trends of microfluidic single-cell technologies. Int J Mol Sci 19. https://doi.org/10.3390/ijms19103143
144. Kar S, Loganathan M, Dey K, Shinde P, Chang H-Y, Nagai M, Santra TS (2018) Single-cell electroporation: current trends, applications and future prospects. J Micromechanics Microengineering 28:123002. https://doi.org/10.1088/1361-6439/aae5ae
145. Narasimhan AK, Lakshmi SB, Santra TS, Rao MSR, Krishnamurthi G (2017) Oxygenated graphene quantum dots (GQDs) synthesized using laser ablation for long-term real-time tracking and imaging. RSC Adv 7:53822–53829. https://doi.org/10.1039/c7ra10702a
146. Silvestre C, Duraccio D, Cimmino S (2011) Food packaging based on polymer nanomaterials. Prog. Polym, Sci
147. Mihindukulasuriya SDF, Lim LT (2014) Nanotechnology development in food packaging: a review. Trends Food Sci Technol. https://doi.org/10.1016/j.tifs.2014.09.009
148. Sekhon BS (2010) Food nanotechnology—an overview. Nanotechnol Sci Appl
149. Bajpai VK, Kamle M, Shukla S, Mahato DK, Chandra P, Hwang SK, Kumar P, Huh YS, Han YK (2018) Prospects of using nanotechnology for food preservation, safety, and security. J. Food Drug Anal
150. Chaudhry Q, Scotter M, Blackburn J, Ross B, Boxall A, Castle L, Aitken R, Watkins R (2008) Applications and implications of nanotechnologies for the food sector. Food Addit Contam Part A Chem Anal Control Expo Risk Assess
151. Sharma C, Dhiman R, Rokana N, Panwar H (2017) Nanotechnology: an untapped resource for food packaging. Front Microbiol
152. Ibrahim RK, Hayyan M, AlSaadi MA, Hayyan A, Ibrahim S (2016) Environmental application of nanotechnology: air, soil, and water. Environ Sci Pollut Res
153. Mehndiratta P, Jain A, Srivastava S, Gupta N (2013) Environmental pollution and nanotechnology. Environ Pollut. https://doi.org/10.5539/ep.v2n2p49
154. Zhang B, Misak H, Dhanasekaran PS, Kalla D, Asmatulu R (2011) Environmental impacts of nanotechnology and its products. Am Soc Eng Educ
155. Karn B, Kuiken T, Otto M (2009) Nanotechnology and in situ remediation: a review of the benefits and potential risks. Environ Health Perspect
156. Van Hee VC, Kaufman JD, Scott Budinger GR, Mutlu GM (2010) Update in environmental and occupational medicine 2009. Am J Respir Crit Care Med 181:1174–1180
157. Pope AC, Burnett RT, Krewski D, Jerrett M, Shi Y, Calle EE, Thun MJ (2009) Cardiovascular mortality and exposure to airborne fine particulate matter and cigarette smoke shape of the exposure-response relationship. Circulation 120:941–948. https://doi.org/10.1161/CIRCULATIONAHA.109.857888
158. Bernd N (2010) Pollution prevention and treatment using nanotechnology. In: Nanotechnology

159. Wood S, Geldart A, Jones R (2003) The social and economic challenges of nanotechnology. TATuP - Zeitschrift für Tech Theor und Prax. https://doi.org/10.14512/tatup.12.3-4.72
160. Roco MC, Bainbridge WS (2005) Societal implications of nanoscience and nanotechnology: maximizing human benefit. J Nanoparticle Res. https://doi.org/10.1007/s11051-004-2336-5
161. Khan A (2015) Ethical and social implications of nanotechnology. QScience Proc. https://doi.org/10.5339/qproc.2015.elc2014.57
162. Springer handbook of nanotechnology
163. (2012) Encyclopedia of nanotechnology
164. Arora S, Rajwade JM, Paknikar KM (2012) Nanotoxicology and in vitro studies: the need of the hour. Toxicol Appl Pharmacol 258:151–165. https://doi.org/10.1016/j.taap.2011.11.010
165. Savolainen K, Alenius H, Norppa H, Pylkkänen L, Tuomi T, Kasper G (2010) Risk assessment of engineered nanomaterials and nanotechnologies—a review. Toxicology 269:92–104
166. Kumar A, Dhawan A (2013) Genotoxic and carcinogenic potential of engineered nanoparticles: an update. Arch Toxicol 87:1883–1900
167. Scopus—Document search | Signed in. https://www.scopus.com/search/form.uri?display=basic. Accessed 9 June 2020
168. Nanomaterials: toxicity and risk assessment—Google Books. https://books.google.co.jp/books?id=ammQDwAAQBAJ&pg=PA18&lpg=PA18&dq=Rodgers,+P.,+Chun,+A.,+Cantrill,+S.,+Thomas,+J.+Editorial,+Small+is+different,+Natur+Nanotechnol.+1,+1,+2006.&source=bl&ots=NX2ybJWdwk&sig=ACfU3U0mIa1_5cq6sbgJJpAeR7SQxHn7Ig&hl=en&sa=X&ved=2ahUKEwjP8L2Kj_TpAhVrw4sBHQE4DLUQ6AEwAHoECAoQAg#v=onepage&q=Rodgers%2CP.%2CChun%2CA.%2CCantrill%2CS.%2CThomas%2CJ.Editorial%2CSmallisdifferent%2CNaturNanotechnol.1%2C1%2C2006.&f=false. Accessed 9 June 2020
169. Chen W, Xiong Q, Ren QX, Guo YK, Li G (2014) Can amino-functionalized carbon nanotubes carry functional nerve growth factor? Neural Regen Res 9:285–292. https://doi.org/10.4103/1673-5374.128225
170. Liu Y, Ren L, Yan D, Zhong W (2014) Mechanistic study on the reduction of SWCNT-induced cytotoxicity by albumin coating. Part Part Syst Charact 31:1244–1251. https://doi.org/10.1002/ppsc.201400145

Metallic Nanoparticles for Biomedical Applications

Kavitha Illath, Syrpailyne Wankhar, Loganathan Mohan, Moeto Nagai, and Tuhin Subhra Santra

Abstract Metallic nanoparticles have found various biomedical applications due to their intrinsic physicochemical properties. As the size decreases, the high surface area of particles gives rise to distinctive features, which are entirely different from that of a macro-sized structure. Several methods are involved in synthesizing metallic nanoparticles, and in general, it can be categorized into either bottom-up or top-down approaches. The top-down method consists of cutting down the bulk materials into nano-sized particles through physical, chemical, or mechanical treatments, whereas, in a bottom-up approach, nanoparticles are formed by joining individual atoms or molecules. The top-down approach produces metallic nanoparticles in naked form, which can further agglomerate and hence not suitable for biomedical applications. The bottom-up approach involves solid-state, liquid state, gas phase, biological, microfluidic-technology based, and other methods. Chemical reduction in the bottom-up approach is the most common method of metallic nanoparticle synthesis, which is flexible, simpler, inexpensive, and produces particles in homogenous form. Recently biological method of nanoparticle synthesis has become popular due to its toxic-free nature, inexpensiveness, sustainability, and eco-friendly. In this chapter, we describe the top-down and bottom-up approach and current trends in the synthesis of metallic nanoparticles for biomedical purposes. Further, it explains how the parameters can be tuned to get metallic nanoparticles with the desired shape, size, morphology, composition and crystallinity.

Keywords Metallic nanoparticles · Physicochemical properties · Top-down and bottom-up approaches

K. Illath · T. S. Santra (✉)
Department of Engineering Design, Indian Institute of Technology Madras, Chennai, India
e-mail: tuhin@iitm.ac.in

S. Wankhar
Department of Bioengineering, Christian Medical College, Vellore, India

L. Mohan · M. Nagai
Department of Mechanical Engineering, Toyohashi University of Technology, Toyohashi, Japan

T. S. Santra and L. Mohan (eds.), *Nanomaterials and Their Biomedical Applications*, Springer Series in Biomaterials Science and Engineering 16,
https://doi.org/10.1007/978-981-33-6252-9_2

1 Introduction

The history of nanoparticles (NPs) dates back to almost 4500 years ago; ancient humans utilized natural asbestos nanofibers in developing ceramic matrices. Four thousand years ago, Egyptians used ~5 nm diameter lead sulfide NPs for hair dye treatment, synthesized from a chemical synthetic process. 'Egyptian blue,' a multi-faceted mixture of cuprorivaite and silicon dioxide, is considered as the first synthetic material prepared from nanometer-sized glass and quartz, used by Egyptians around third century BC. Later in the thirteenth- and fourteenth-century BC, Egyptians and Mesopotamians started synthesizing glass from metallic nanoparticles (MNPs). During the Bronze age, surface plasmon excitation of cobalt nanoparticles (Co NPs) has been utilized in colouring the surface of red glass. Co NPs were also used in Celtic red enamels [1]. Another notable pioneering work in the MNPs era is the Lycurgus cup, used back to fourth century AD. The cup can change its colour depends on the location of the light source. Further analysis of this cup revealed that it consists of minimal quantities of ~70 nm gold and silver crystals with an appropriate molar ratio of 14:1, which results in its surprising characteristics [1, 2]. Later, gold and silver nanoparticles (Au & Ag NPs) in appropriate ratios have been utilized in red and yellow stained glass to decorate churches. Optical properties of Ag and copper nanoparticles (Cu NPs) were employed in glazed ceramics by Mesopotamians. It displays bright green and blue colours under particular specific radiation. Investigation revealed that ceramics has an outer double layer of Ag NPs and inner thinner layer with different sizes. Hence, light, when travels through the ceramics possess interference effect and scattered light from the second layer, has a phase shift due to the interference of scattered light from the first layer. This technique has been utilized in producing red glass, and Satsuma glass used in Japan during the mid-nineteenth century. Satsuma glass can be brightened by ruby colour by using the absorption properties of Cu NPs [1].

In 1857, Michael Faraday developed a systematic synthesis of colloidal gold solutions and studied optical and coagulation properties of colloidal gold [3]. This is considered the first scientific record on NP synthesis. The study gave an idea about the optical properties of a colloidal gold solution. This leads Mie to explain the specific colours of the colloidal metallic particle solution [1]. The birth of the concept of NP was originated in the pharmaceutic field, related to targeted drug delivery. Professor Peter-Paul Spieser did one of the pioneering works on NPs in this field at Eidgenössische Technische Hochschule Zurich. His research group investigated polyacrylic bead for oral therapy and started focusing on synthesizing microcapsules. Eventually, in the 1960s they developed first NPs for vaccination purpose and drug delivery approaches [4]. Later, it started to synthesize several NPs for various purposes.

In this book chapter, we review the different approaches in synthesizing MNPs for biomedical applications. The following sub-section describes the properties of NPs, the formation of NPs, and theories behind it. The second section briefly explains top-down methods in synthesizing MNPs and their limitations. The bottom-up approach is illustrated in the third section including solid-state, gas-phase, liquid-phase, biological, microfluidic-based and other techniques. Advantages, limitations and importance of each method is illustrated with key examples describing the controlled synthesis. Finally, it summarizes the synthesis method and discusses prospects.

1.1 Nanoparticles and Its Properties

NPs are the entities whose size ranges from 10 to 1000 nm. Due to its small size, it gives rise to a various exciting phenomenon which is not shown by their bulk counterpart [5]. As the size decreases, surface area to volume ratio increases exponentially and hence, make its surface and surroundings more reactive itself [6]. Apart from size, shape, and other physicochemical properties of NPs enables it to use in various fields such as biomedical and environmental to use it in catalysis, sensing, imaging, drug delivery, energy production, storage and water treatment [7–9]. The physical properties include size and size distribution, shape, specific surface area, aspect ratio, agglomeration, surface morphology, structure, and solubility. The chemical properties cover chemical composition, phase identity, surface chemistry, and hydrophilicity. The important physicochemical properties include large surface area, electronic and optic properties, magnetic properties, mechanical properties, and thermal properties. Optical and electronic properties are closely related. Noble MNPs show size-dependent optical properties and exhibit a strong UV-Vis absorption spectrum. The excitation band is due to the phenomenon of localized surface Plasmon resonance (LSPR) of conduction electrons of metals, which will discuss later. It has been found that the maximum wavelength of the absorption spectrum depends on the size of the NPs. Magnetic properties of NPs are originating from the unequal electronic distribution. It has been observed that when the size of the synthesized NP is less than the critical value, magnetic properties dominate. The mechanical properties of NPs have to be studied in detail based on their applications. NP solutions are expected to show thermal properties, because it usually synthesizes in solvents such as water, oil or ethylene glycols (EG). Also, higher surface area favours the heat transfer along the surface of particle. It has been reported that nanofluids containing copper oxide or aluminium oxide NPs in water or ethylene showed advanced thermal conductivity [5].

1.2 Theories on the Formation of Nanoparticles

The type of synthesis approach adopted dictates the size, shape, morphology, and uniformity of NPs, which in turn dictates the novel properties of NPs. Thus, a controllable synthesis and processing become the basis of nanotechnology to realize potential applications of NPs.

Synthesis of NPs can be classified into two; top-down and bottom-up approaches. In the top-down method, bulk materials are etched to produce NPs, whereas, in a bottom-up approach, NPs are formed from the atoms or molecules through nucleation, aggregation, growth, and crystallization processes. Nucleation is the process during which nuclei (seed) from the parent is formed to produce crystals. It may happen either homogeneously or heterogeneously in the solution. Several theories explain nanoparticle formation; classical nucleation theory (CNT), LaMer's nucleation and growth mechanism, two-step nucleation, and pre-nucleation cluster mechanism [12–16]. As per CNT, nucleation starts from the nuclei, which are formed from the monomers within the bulk liquid as a new phase, and they are considered as the building blocks of NPs. These nuclei aggregate together to form critical nucleation of a new phase. According to LaMer's concept, nucleation and growth processes can be separated in time, as shown in Fig. 1a. A prenucleation process happens within a short time to produce nuclei needed for monodispersity. Once the nuclei are formed, nucleation starts and continues until the concentration reaches critical limiting supersaturation. Soon after that, the growth process begins. During this process, nuclei monomers are consumed, and gradually, their concentration decreases. The growth process can be either a diffusion-limited or reaction-controlled growth process. Another theory explains NP formation through two-step nucleation and growth mechanism. The two-step nucleation mechanism starts before nucleation, generally in the existence of amorphous metastable clusters or disordered liquid in homogenous solution [17]. During the first step, nuclei are formed directly from the precursor as a result of a pseudo-first-order reaction. In the second step, autocatalytic growth of nuclei happens and results in the formation of the particle [17, 18]. Another theory is the prenucleation cluster mechanism, which says that prenuclei (small solutes in their lowest kinetic energy) are formed from the chemical species of atoms or molecules and are considered to be initiated during the phase transition [17]. Polte et al. studied the mechanism of Au NPs formation using in situ small-angle X-ray scattering at millisecond resolution [11]. Figure 1b illustrates the result of the study. Au NPs were synthesized using sodium borohydride ($NaBH_4$) as a reducing agent through the chemical reduction method (explain later). Initially, the gold precursor has reduced into nuclei Au (0), which is very fast, usually within 200 ms nuclei are formed. It is followed by the coalescence of nuclei into bigger particles, as illustrated.

1.3 Metallic Nanoparticles

NPs synthesized from the metallic precursors are known as MNPs [5]. Among MNPs, Au NPs have been studied in detail across the research sectors and have provided new insights, especially in biomedical applications. Gold is a material that is considered as a quintessential noble metal, and its NPs exhibit unique properties such as non-toxicity, ease of functionalization, physicochemical, catalytic, and biological

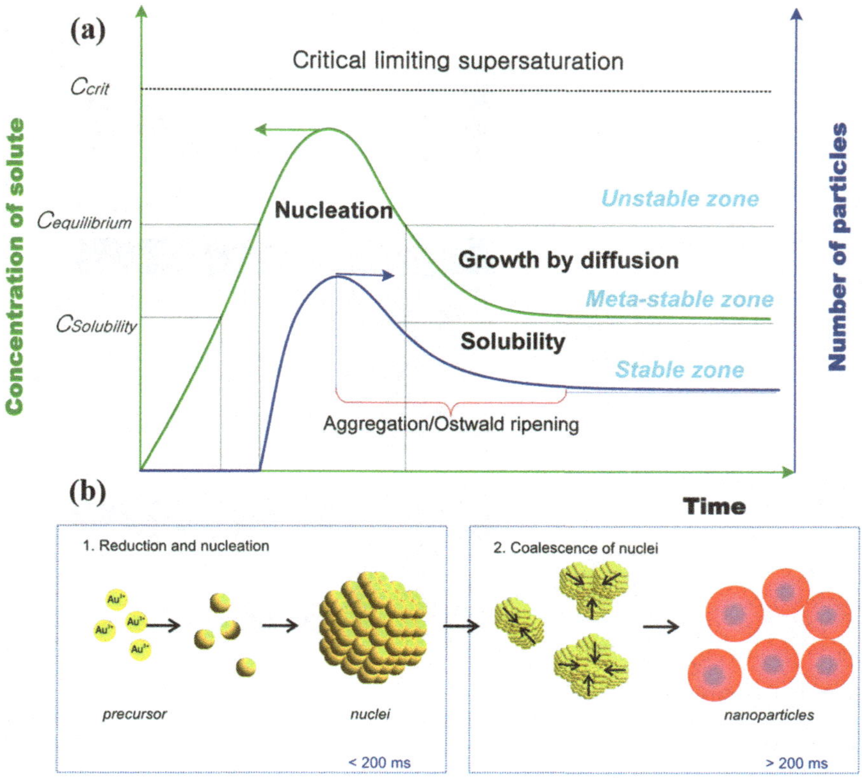

Fig. 1 **a** Schematic diagram used to represent nucleation and growth process in the formation of MNPs according to LaMer's concept. The concentration of solute and the number of particles over time are described. Republished with permission of the Royal Society of Chemistry, from Ref. [10]. **b** Schematic illustration of Au NPs formation using $NaBH_4$ reduction. It consists of reduction and nucleation within 200 ms, accompanied by coalescence of nuclei into bigger particles. Reprinted with permission from [11] Copyright (2010) American Chemical Society

properties. Due to these features, Au NPs are useful in applications like biomedical imaging, therapeutics, sensing, catalysis, analytical science, medical diagnosis, and drug delivery [19–26]. Based on size and shape, the colour of the yellow gold precursor turns red, violet, blue, green, and yellow [27]. Figure 2 depicts the colour dependency of Au NPs on shape and size for commonly employed particles in biomedical applications. MNPs have some exciting features which make them suitable in various fields such as thermal, electronics, photonics, catalysis, optoelectronics, and biomedical. Among these features, interesting behaviour is surface plasmon resonance, which happens when the size of the particle becomes comparable with the wavelength of incident light [28]. Usually, MNPs in size range of 10–100 nm possess variable surface plasmonic resonance in the visible region with

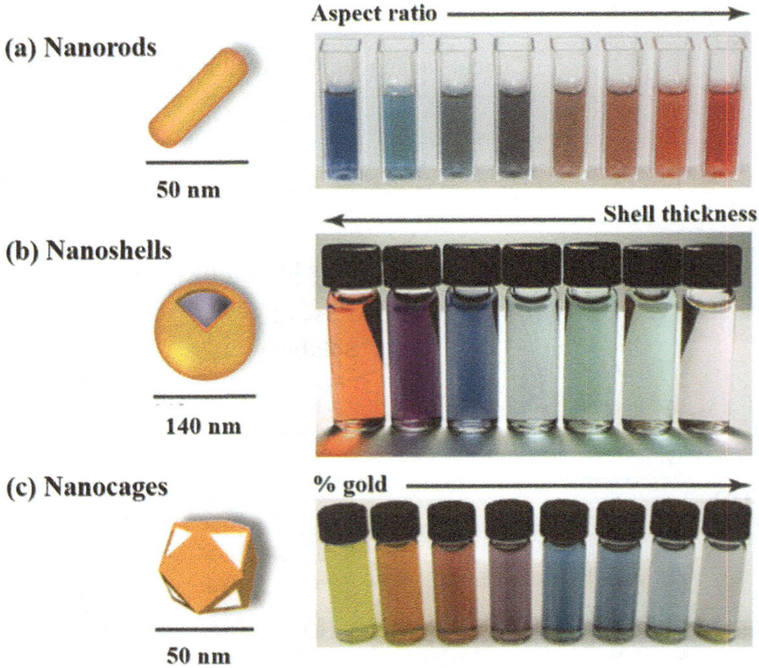

Fig. 2 Colour dependence of Au NPs on shape and size; **a** nanorods with increasing aspect ratio, **b** nanoshells with decreasing shell thickness, and **c** nanocages with increasing order of % gold. Republished with permission of the Royal Society of Chemistry, from Ref. [24]

unique optical properties based on their size and shape [27]. Surface plasmon resonance can be defined as the collective oscillation of electrons in MNPs, which are restricted locally due to small size. If the frequency of oscillation of electrons matches with that of the incident light, maximum absorption and scattering of light occur. It is known as localized surface plasmon resonance (LSPR). LSPR not only depends on size and shape but also the refractive index of the surrounding medium and complex dielectric constant. Au, Ag, and Cu NPs show LSPR in the visible region of the electromagnetic spectrum [28].

Au and Ag NPs possess shape-dependent optical properties. Isotropic Ag nanostructure like spherical and anisotropic nanostructures such as triangular and pentagonal shapes can display the colours blue, red, and green, respectively [29]. Ag NPs are found tremendous applications in agricultural, biological, and medical fields. Its properties, such as macroscopic quantum tunnelling, quantum size, surface, and volume effects, enable them to serves as chemical sensors, catalysts, chemical probes, antimicrobial material, DNA detection material, anti-cancer material, and drug delivery carriers. All these properties depend on size, shape, and particle distribution [30, 31]. Generally, MNPs possess a high surface-to-volume ratio, which enables them to interact with different functional moieties. Hence, MNPs can be functionalized with biomolecules like antibodies and drugs on the surface of MNPs easily, and they serve as specific biomarkers or drug carrier respectively. Absorption and fluorescent properties of these MNPs are utilized in fluorescence imaging and marker for the detection of proteins using surface-enhanced Raman spectroscopy (SERS) [27]. Apart from Au and Ag NPs, palladium (Pd), platinum (Pt), Cu, rhodium (Rh), iron (Fe), and nickel (Ni) NPs are also useful in biomedical applications [32, 33]. Pd is a major component in commercial devices and industrial processes due to its catalytic behaviour. Further, it has a strong affinity with hydrogen; hence it is useful for purification, detection, and storage purposes. Pt NPs are found its applications in electronic, chemical, automotive, photochemical, and pharmaceutical fields due to its features such as corrosion-resistant, electrical and catalytic properties [34]. Cu NPs have significant application in antimicrobial systems [28].

Synthesized MNPs can be characterized using various techniques; ultraviolet-visual (UV-vis) absorption spectrophotometry, transmission electron microscopy (TEM), high-resolution transmission electron microscopy (HR-TEM), high-resolution-scanning transmission electron microscopy (HR-STEM), selected area electron diffraction, infrared (IR) spectroscopy, atomic force microscopy, energy-dispersive X-ray spectroscopy, fibre spectrometry, dynamic light scattering (DLS) and scanning electron microscopy (SEM). Generally, for the size distribution, DLS, SEM, and TEM can be used. For the detailed morphological analysis, TEM, HR-TEM, and HR-STEM are the best choices. UV-Vis absorption spectrum is used to find the plasmonic peak of synthesized particles. Apart from these details, it also provides information about the dispersity of particle and morphology.

Varieties of synthesis method involved in fabricating MNPs. However, developing NPs at high volumes and with specific sizes, shapes, and crystalline nature is always challenging and requires more trials. As mentioned earlier, the synthesis of MNPs can be divided into two based on the starting material; top-down and bottom-up method. In the top-down method, the bulk metallic material serves as the starting material for nanoparticle formation. In contrast, in the bottom-up method, individual nuclei or atom forms the initial material.

2 Top-Down Methods

In this method, the bulk material is chiselled down into required nano-sized particles through physical, chemical, or mechanical treatments. According to the type of treatment, top-down approaches can be mechanical milling, laser ablation, or sputtering [34].

2.1 *Mechanical Milling*

Mechanical milling involves a reduction in particle size using high energy ball milling or mechanochemical milling. With ball milling, the bulk metallic powder is added in a container along with massive balls. By using a high-speed rotating ball, high mechanical energy is applied. Particle size is reduced in the desired form with the help of various high energy mills. The particles synthesized in this manner possess superior physical properties such as enhanced solubility, and some new features to the component are relying on the material and grain size. The requirement of high energy, more prolonged time consumption, contamination of powder, and synthesis of sensitive nanostructures are disadvantages of the ball milling method. In mechanochemical based milling, the initial bulk material is added stoichiometrically and milled. During milling, it undergoes repeated deformation, welding, and fracture of reactants mixture. Several chemical reactions are also evolved at the substrate and reagent interface. Even though the method is simple, synthesized NPs are highly sensitive to grinding conditions and require long term milling in the case of smaller particles [34].

2.2 *Laser Ablation*

In this method, laser irradiation is used to synthesize NPs from bulk material. Due to the application of laser energy, fragmentation of solid particles in the form of NPs are formed, and they remain in liquid which surrounds the bulk material known as a colloidal solution. The relative number of ablated atoms and generated NPs depends on the pulse duration and energy of the laser. Further, ablation efficiency and characteristics of formed particles rely on the wavelength of the laser, pulse duration, ablation time, and laser fluence. It has several advantages such as simplicity and effectiveness, properties can be changed by tuning the parameters, and particles can be formed without adding surfactant also. The major drawback is that exposure to laser for a long duration may reduce the overall ablation rate [34].

2.3 *Sputtering*

In sputtering, the bulk solid target material is vaporized with the help of inert gas ions. The whole procedure is carried out in a vacuum chamber, and sputtering gas is supplied and is kept under working pressure. By introducing high voltage into the cathode, free electrons are generated. It follows a spiral path using a magnetic system, where they collide with sputter gas atoms and leads to ionization of the gas. Positively charged ions move towards the target, where they continuously impinge. This process repeats, and if it occurs at an energy level higher than the surface binding energy of target, an atom can be removed. This scattering of atoms forms a diffused cloud in the form of NPs. This method is suitable for the synthesis of alloy NPs with better control [34].

All the top-down approaches produce MNPs in naked form, which can further agglomerate and hence not suitable for biomedical applications.

3 Bottom-Up Methods

In this method, the building block of NPs is atoms or molecules. NPs are formed from these building blocks through various phenomena and chemical reactions. The top-down method is compared with the bottom-up approach, as given in Fig. 3. Different approaches in bottom-up methods are solid-state methods, liquid state synthesis, gas phase, biological, microfluidic-based, and other methods. In all methods, building blocks are generated at first, followed by the assembling of building blocks to form NPs of desired size and shape [34].

3.1 Solid-State Methods

Here, the material is deposited on a surface in the form of thin-film or as a nanomaterial. Two types of deposition approaches exist; physical vapour deposition (PVD) and chemical vapour deposition (CVD). In PVD, techniques like laser ablation is used, which further causes the formation of plasma of ablated target, and they are deposited on a substrate to produce a thin film. MNPs are deposited on carbon nanotubes in this manner. Even though the method is simple, it is unable to provide a higher volume of material, and the procedure is expensive. The chemical reaction of gaseous molecules of atoms to form the thin film is utilized in depositing thin films on the substrate in CVD. Three types of CVD are used; thermally active chemical vapour deposition (TACVD), plasma-enhanced chemical vapour deposition (PECVD), and photoinitiated chemical vapour deposition (PICVD). In the form of volatile molecules, the target material is deposited on the substrate. In the case of temperature-sensitive material such as polymer substrate, TACVD is not applicable. For PECVD, plasma is generated inside the void chamber with the help of microwave and inductively induced electric current. Due to the requirement of specific operating conditions, PECVD suffers from scale-up issues. For PICVD, low energy treatment, as well as a

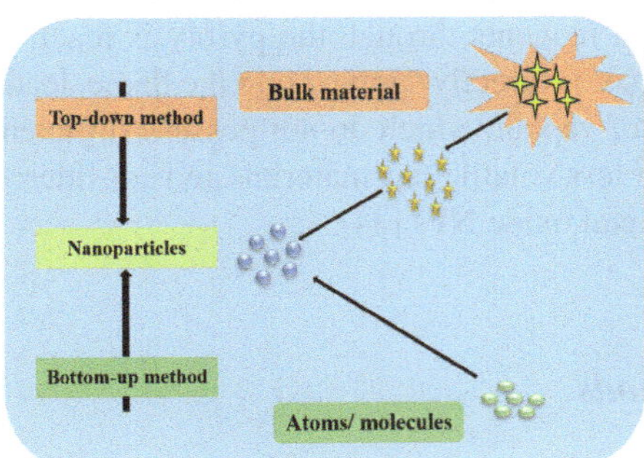

Top Down Method	Bottom Up Method
Starting material is smaller than the destination material	Starting material is larger than the destination material
Present certain amount of wasted material	No wasted materials present
Possess lower precision accuracy	Synthesized particles possess higher precision accuracy
Difficult to synthesize very small sized particles compared to bottom up	Able to synthesize small sized structures
Synthesized particles possess cavities or wrinkles	It can synthesize particles with perfect structures and edges

Fig. 3 Overview and comparison of synthesis of metallic nanoparticles; top-down method and bottom-up method. Redrawn from [34]

broad spectrum of variations, are possible. Compared to the other methods discussed above, crystal structure and surface morphology can be controlled in CVD. Further, the thin-film coating exhibits high durability, and it is easy to scale up [34, 35].

PVD and CVD methods are usually employed for forming thin films on a substrate. Later, corresponding NPs or composites can be synthesized, which are generally not preferred for large scale preparation of pure MNPs. It has been found that nanocomposites containing Ag NPs synthesized using solid-state methods are suitable in biomedical applications due to their antimicrobial activity and toxic-free nature [17].

3.2 *Gas-Phase Methods*

Gas-phase methods include spray pyrolysis, laser pyrolysis, and flame pyrolysis. Spray pyrolysis is done with the help of an apparatus that has three parts; one nebulizer to spray precursor in the vapour form into the hot reactor, a tubular reactor, which is kept at 200 °C to 120 °C (thermostatically controlled) and a precipitator for collecting the NPs. The apparatus can be modified by replacing the nebulizer with ultrasound and adopting atomized techniques like two fluids nozzle or air-assisted pumps or sprayers, spinning disk, and vibrating orifice. Out of these modified methods, ultrasound-based pyrolysis is commonly employed, where ultrasound of specific frequency is used to produce atomized droplets from the precursor. Atomized droplets are transported to reaction furnace by the carrier gas, and later NPs are collected in the precipitator. The advantages of spray pyrolysis are low-cost, simplicity, reproducibility, and easiness in controlling particle size. This is applicable for metal oxide and mixed metal oxide nanoparticle synthesis. In laser pyrolysis, laser energy is used to activate a homogeneous nucleation reaction by exposing laser on the precursor. Commonly employed laser energy heating is infrared carbon dioxide (CO_2) laser energy, which is easily absorbed by inert photosensitizers like sulfur hexafluoride. Once the sufficient amount of supersaturation of condensable product is reached in the vapour phase, CO_2 pyrolysis starts immediately. NPs of uniform size distribution is possible with this approach. Further, the size of the particles can be controlled by adjusting the flow rate of reagents through the pyrolysis reaction zone. In flame pyrolysis, the liquid precursor is directly sprayed into the flame, leads to the delivery of precursors in the form of vapour, which do not possess sufficient vapour pressure. This method is useful for less volatile raw materials and considered as a promising approach for producing metal oxide NPs [34].

3.3 *Liquid State Synthesis Methods*

Liquid state synthesis consists of sol-gel, chemical reduction, hydrothermal and solvothermal method. The sol-gel process involves the formation of networks of colloidal suspension (sol) and gelatine in the liquid phase. It can be performed in three ways; (1) formation of colloids of metal (oxide) and mix up with the sol consists of matrix-forming species, followed by the formation of a gel, (2) direct mixing of metal and metal oxide within a prehydrolysed silica sol, and (3) the formation of a complex compound of the metal with silone and subsequent reduction of metal before hydrolysis. Colloids are generally formed from the ions of metal alkoxides and alkoxysilanes. For example, silica gel is formed most commonly from tetramethoxysilane and tetraethoxysilane. The overall formation of sol-gel involves the following steps; hydrolysis, condensation, particle growth, and agglomeration of particles into NPs. It is possible to control the particle size and morphology by systematically monitoring the reaction entities. The method is generally used to synthesize thin films, zinc peroxide nanostructures, and metallic oxide NPs like nickel oxide, iron oxide, etc. [34, 36].

The hydrothermal method is due to the reaction between a solid material and aqueous solution vapour at high temperature and pressure. Cations precipitate in the form of polymeric hydroxide get dihydrate and accelerate the formation of metal oxide crystals. When a base is added to the salt, another cation is formed, which is essential in controlling particle formation by preventing complex hydroxide formation. Even though particles with high crystallinity can be synthesized with this method, processes are complicated to control. This method is suitable for synthesizing nanoparticles in the powdered form [34].

Solvothermal based synthesis involves the formation of NPs in liquid phase solvents such as water or organic compounds like methanol, polyol, and ethanol. These solvents are allowed to heat above their boiling temperature in a pressure vessel. It uses a metallic precursor, a reducing agent (reductant), and a solvent medium as essential reagents and sometimes ligands and facets specific capping agents for controlled synthesis. Narrow sized distribution of monodispersed crystals of NPs can be synthesized using this method. With the help of a microwave, kinetics of the crystallization step can be increased by the orders of one or two. MNPs such as Pt, Pd, Au, Ag, Rh, Ru, Co, and Ni NPs can be rapidly synthesized by this method with the help of methanol or polyethylene glycol (PEG) as reducing agent [34, 37].

Research analysis showed that the physicochemical properties of MNPs could change with morphology and size. Hence, controlling the synthesis to achieve the desired size and morphology is required, and is very challenging. The key parameters to tune the shape, size and crystallinity of MNPs by the solvothermal method are temperature and reaction time, type of solvent, ligand type and its concentration, facet specific capping agent, and reductant. Temperature and reaction time have a huge impact on controlling the synthesis of MNPs. Since the synthesis carries out at a high temperature above the boiling temperature of solvents, its properties change; viscosity decreases, mobility increases, and dielectric constant reduces. Metal precursor decomposes itself at high temperatures, or some can be carbonized. Further, at high temperature, the reactivity of reactant increases, hence the reduction rate of metal precursor increases, results in the formation of a lot of nuclei and a decrease in the size of particles, as shown in Fig. 4a–b. Here Pd/Pt NPs synthesized at 180 °C is smaller in size compared to those prepared at 120 °C [37, 38, 41]. High

Fig. 4 Effect of reaction temperature and time in solvothermal based MNP synthesis; TEM images of Pd/Pt NPs synthesized at **a** 180 °C and **b** 120 °C for 8 h, respectively [38]. TEM images of concave polyhedral Pd nanocrystals prepared with a reaction time of **c** 1 h, **d** 2 h, **e** 4 h, **f** 8 h, and **g** 15 h respectively, **h** a curve shows the dependency of the side length of nanocrystal with the reaction time. Reprinted with permission from [39]. Copyright (2009) American Chemical Society

temperature favours the formation of crystalline phases and alloy metal nanocrystals, as well as restructuring the metallic nanostructures by reducing the interaction between metal ions with ligands, metals, and facet specific capping agent. As explained in Sect. 1.2, the concentration of metal atoms increases as reaction progress without any nuclei. But once it starts to form nuclei, concentration reduces, and at the beginning of the growth stage, formed NPs are poorly shaped. As the reaction progress, it evolves into a regular shape, and with the help of shape-controlled capping agents, NPs with specific facets can be formed. If the temperature is high enough, it can reconstruct the MNPs and promote the Ostwald ripening process. Hence, the final products are usually morphologically evolved, and longer time promotes thermodynamic growth [37, 41]. Figure 4c–g shows the effect of reaction time on synthesized Pd nanocrystals. It can be easily observed the morphological evolution on synthesized particles; the crystals produced within 1 h has a side length of 16 nm, and it starts to increase to 24 nm at 2 h, 38 nm at 4 h and 45 nm at 8 h. Hence, the trend is increasing in the length of NPs as reaction time progresses. Further, it is noted that concave morphology is shaping as the size increases. At 2 h, concave morphology is seen, and it continues to maintain even after increasing the reaction time. After 8 h of reaction time, there is not much significant change on morphology is observed. Figure 4h shows the dependence of side length with reaction time. It is observed that beyond 8 h, there is no increment in length of NPs. Hence, in solvothermal based synthesis, reaction time can be adjusted to get a particle with specific size [39]. Similar to size, the shape also depends on reaction time in the solvothermal treatment of Ag NPs. In a typical solvothermal treatment in ethanol at 80 °C was performed with Ag nanocubes at different reaction times. Initial nanocubes and other structures evolved after 6 and 12 h of reaction time are shown in Fig. 5. From Fig. 5a–c it is clear that Ag nanocubes are transformed into nanocubes with concave facets and then into wavey edged NPs with 6 and 12 h of reaction time, respectively. Nanocubes obtained after 6 h have dark contrast in their {100} faces, and with 12 h of reaction time, NP yield is increased. The dramatic change in shape might be due to oxidative etching or Ostwald ripening. As reaction time increases to 6 h, oxidative etching happens due to the distribution of different surface energies of Ag NPs and sufficient availability of etching pair. But as reaction time is increased to 12 h, Ostwald ripening happens between in situ generated multiply twinned particles and plate-like seeds [40].

The effects of solvents on the synthesized NPs were studied in detail for various metals [43–45]. Reactants may react with solvents and form products that can significantly affect the reaction kinetics, equilibrium, and final products. Some solvents act as structure modulator, and some can influence synthesis if surfactants are used. Solvent properties such as viscosity, dielectric constant, and polarities are responsible for changes in particle synthesis. Further, mixtures of solvents can influence the synthesis greatly by reacting with the other reactants used. As an example, consider the synthesis of Ag NPs synthesized in various mixed solvent solutions by Rosa et al. They studied the effect of Ag nanograins synthesized in solvents including a mixture of ethanol + water, hexane + water, toluene + water, acetone + water, and ethanol alone. It has been found that grain size increases in the order of hexane + water, ethanol + water, toluene + water, followed by pure ethanol and acetone + water. With acetone + water, particles were present separately in a bigger size, whereas with hexane + water, grains are small in size present together in groups. Hence, it can be concluded that particles synthesized in immiscible solvents are bigger compared to the other solvents [37, 44].

The ligand, facet specific capping agents, reductant, and reactant molar ratio can also act as tuning parameters in controlling the MNP synthesis based on the

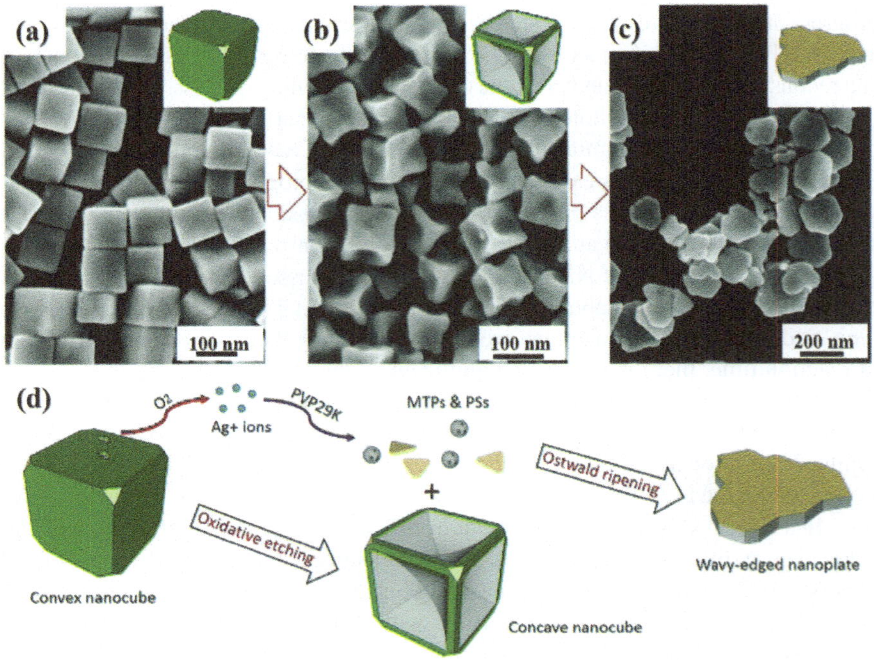

Fig. 5 Shape evolution with reaction time in the solvothermal process; TEM images of **a** initial Ag nanocubes, after a reaction time of **b** 6 h and **c** 12 h, **d** schematic illustration of a possible mechanism of shape evolution. Republished with permission of the Royal Society of Chemistry, from Ref. [40]

solvothermal approach [43, 46–54]. Ligands that bind with metal cations can reduce its redox potential and make their reduction reaction very difficult. Some ligands can change their redox potential and can effectively contribute to the formation of metallic nanocrystals. Depends on the reactions, ligands serve as a capping agent to stabilize some facets or etching agents. In solvothermal synthesis, facet specific capping agents are used to expose distinct facets. These capping agents are usually counter ions present in the ionic species, solvents, precursors, polymers, organic molecules, surfactants, thermal decomposition products, etc. [37]. Amine and glycine are commonly employed surface controller or facet specific capping agents in synthesizing Pt nanocrystals with high index facets. It has been found that the addition of 1-octylamine leads to multipod Pt NPs with dominant {211} surfaces, while formaldehyde gives concave cubes with {411} surface exposed. IR spectroscopic investigation of synthesized Pt NPs reveals that pure amine can adsorb on monoatomic step edges consisting of (111) terrace and (100) steps, results in the formation of {211} exposed surface. In the case of formaldehyde, by-product stabilizes (100) terrace, which enlarges it and leads to exposing {411} surface. Without formaldehyde, at the beginning of synthesis, formed nuclei clusters grow quickly along <111> direction with {211} facet exposed, which results in multipod nanoclusters. In the presence of formaldehyde, cuboctahedral Pt NPs were formed in the initial stage [49]. Figure 6 shows the effect of glycine in the morphology of synthesized Pt nanocrystals. It is found that as the number of glycine increases, the size of the particles increases. Nanocrystals formed without glycine consists of various shapes, but the majority of particles were in the form of nanowires, and they were assembled together. Mixed morphologies (cubes, polyhedral, and icosahedra) of particles started forming when the glycine amount is 30 mg. Concave nanocubes are present with 50 and 75 mg of glycine, but the sizes are different. As the amount reaches 150 mg, the concavity is easily visible, results in exposure to other different high index facets. Moreover, an increase in glycine amount produce particles with concavity, but less in number. The overall analysis shows that glycine and amine can control the size as well as the morphology of Pt nanocrystals [42]. Small organic molecule can also be used

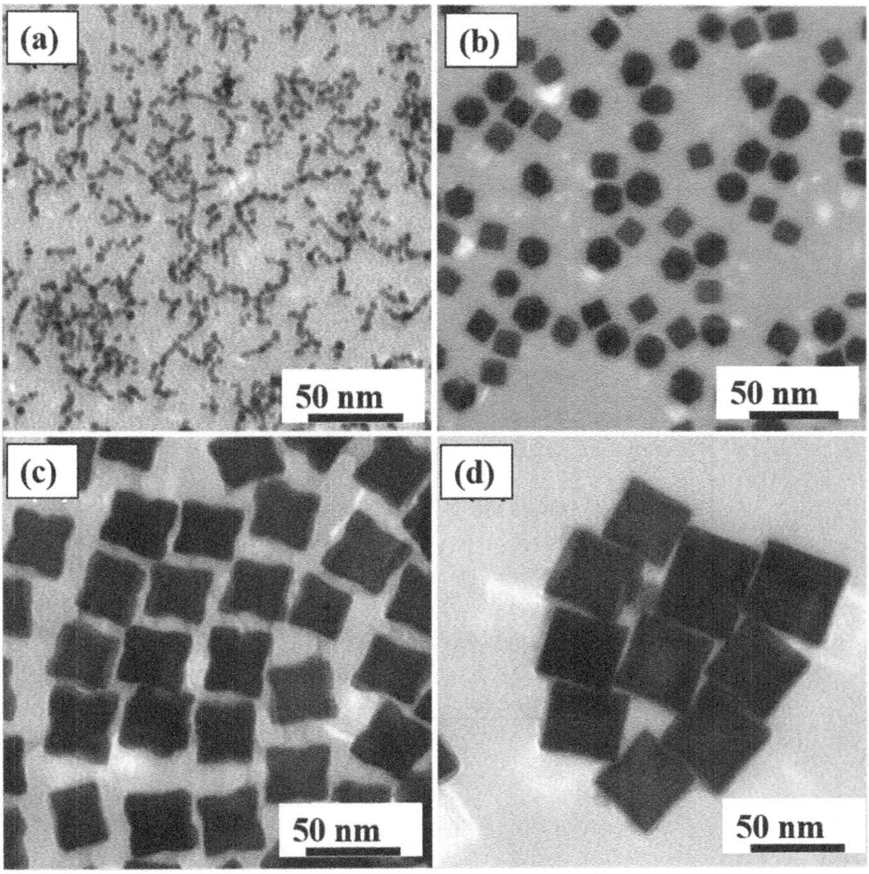

Fig. 6 Effect of the surface controller on the MNPs synthesis using the solvothermal method; TEM images of Pt nanocrystals synthesized using glycine of **a** 0, **b** 30, **c** 50, and **d** 150 mg respectively. Reprinted with permission from [42]. Copyright (2012) American Chemical Society

as facet specific capping agents. It can attract halide ions through electrostatic interactions of different strengths. Figure 7 shows the effect of small organic molecule acetonitrile (or acetone, 1,4-dioxane and 1,3,5-trioxane) and ethanol in forming facet specific morphological structures. Pd nanocrystals synthesized with acetonitrile are nanowires and with ethanol it is mainly bipyramids. Strong electrostatic interaction between acetonitrile and halide ion leads to oxidative etching of O_2/halide pair, controls twinned nanostructures. With ethanol, less electrostatic interaction decreases oxidative etching, results in single twinned structure [46].

Reducing agents can affect the synthesis procedure as follows; strong reducing agents such as $NaBH_4$ and hydrazine can reduce very fast, leads to the formation of nuclei within a short time. Hence, it is challenging to control the shape of the nanostructure. Organic phase-based synthesis uses mild reducing agents like polyol and amines, while aqueous phase uses glycine, ascorbic acid, formaldehyde, and glucose. Some reducing agents can generate capping agents also. Some additives like N, N-dimethylformamide (DMF) act as a weak reducing agent in solvothermal based MNP synthesis. As an example, while synthesizing Ag NPs using polyvinylpyrrolidone (PVP) as a reductant, particles were produced even in the absence of PVP. Investigation on the mechanism revealed that DMF acted as a weak reducing agent. Hence, Ag+ was easily reduced into Ag NPs [37]. Another important tuning parameter is the molar ratio (MR) of the reactant. It has been found that different shapes of Ag NPs can be synthesized by varying the MR of PVP (monomer unit) and $AgNO_3$, as

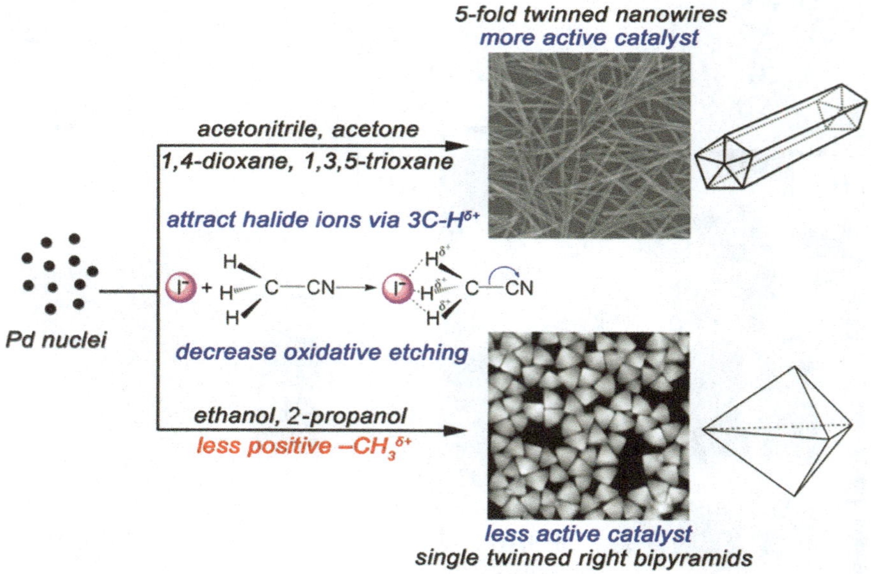

Fig. 7 Schematic illustration of the synthesis of fivefold twinned nanowires and single twinned right bipyramids using solvothermal synthesis in the presence of acetonitrile and ethanol, respectively. Reprinted with permission from [46]. Copyright (2009) American Chemical Society

depicted in Fig. 8. As the MR increases, evolution in shape, size, and crystallinity is also observed. With MR = 0.2, synthesized particles contain nanorods in excess with some spherical and triangular shape. Increasing the amount of PVP (MR = 0.9) produced more spherical shaped particles and less amount of nanorods. Further, the length of the nanorods is decreased. With MR = 5, only triangular-shaped NPs are produced, both spherical, and nanorods are disappeared. When MR increased to 12.1, synthesized particles are of hexagonal plates of various sizes and triangular plates. Bigger NPs are produced by increasing the MR and heating time. Big triangular prisms, hexagonal plates, and enneahedral plates were synthesized with MR = 16.6. It has been found experimentally that PVP can selectively adsorb on three facets of {100}, which results in the addition of Ag atoms to the {111} facets. Hence, three {100} of hexagonal plate becomes bigger, and {111} disappears, results in the

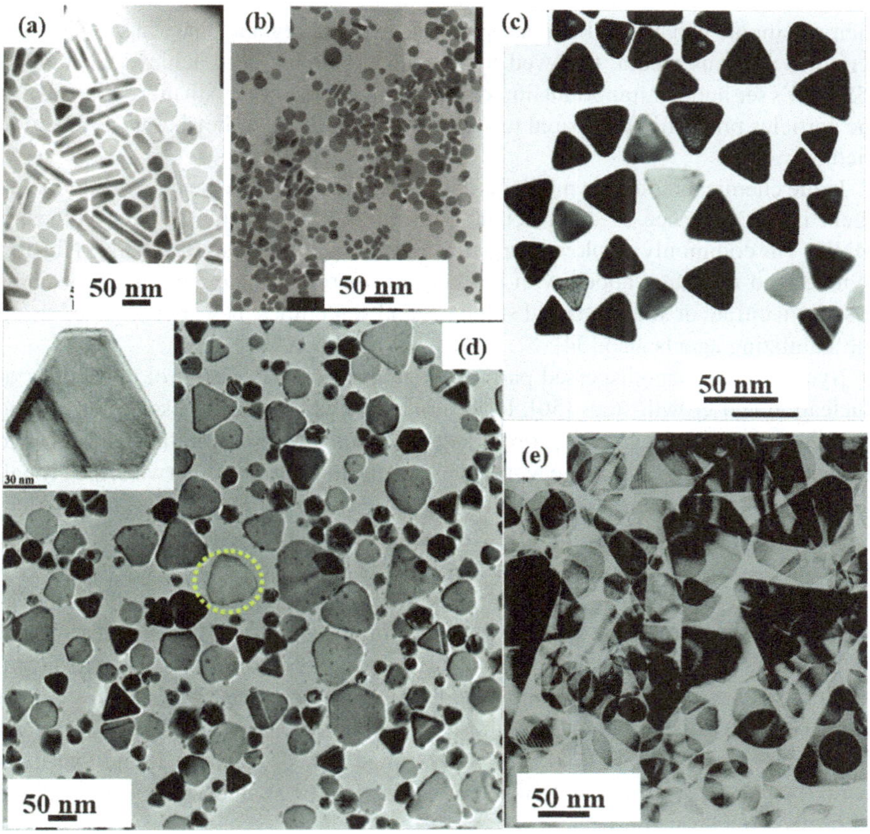

Fig. 8 Effect of molar ratio in the solvothermal synthesis of Ag NPs; TEM images of particles synthesized when the MR of PVP and $AgNO_3$ is **a** 0.2 **b** 0.9, **c** 5, **d** 12.1 and **e** 16.6. Reprinted with permission from [48]. Copyright (2007) American Chemical Society

formation of bigger triangular plates through Ostwald ripening. The resulted triangular plates are bounded by two {111} planes (top and bottom) and three {100} planes as side facets. Further increase in MR results in truncated cubic shape (not shown). When the MR becomes relatively high, NPs are of quasi-spherical polyhedrons in excess with few nanorods, triangular, and rhombic plates [48].

Synthesizing Cu NPs in aqueous media is very challenging compared to Au and Ag NPs, due to its oxidative nature during and after the synthesis. Hence, it always prepares in non-aqueous media under an inert atmosphere. The solvothermal based synthesis uses hydrazine, ascorbate, $NaBH_4$, polyol, and isopropyl alcohol as reducing agents with a surfactant like a hexadecyl trimethyl bromide (CTAB). The generally used metallic salt is cupric chloride [28]. Solvothermal based synthesis is usually preferred for preparing particles (Pt, Pd, Ag, and Au) with controlled facets which have applications mainly in catalysis. Few studies were reported that solvothermal based particles could be used in the biomedical application by tuning their plasmonic behaviour. Particles synthesized in the organic phase has to be transferred to aqueous media, followed by bio-functionalization to achieve biostability [55]. Let's see another important liquid state synthesis method, with more control over the particles properties compared to solvothermal, known as the chemical reduction method.

In the chemical reduction method, ionic salt is reduced with the help of a reducing agent in the presence of a surfactant. This is the simplest method to synthesize MNPs. The commonly employed reducing agent for the synthesis of MNPs is $NaBH_4$. Synthesized NPs are capped with the help of appropriate capping agents such as trisodium citrate or sodium lauryl sulphate, and sometimes they are stabilized using the stabilizing agents also [34].

Synthesizing monodispersed particles demand reasonable control of kinetics of nucleation and growth steps [56]. It, in turn, requires successful dissolution among these steps. Nucleation and growth steps happen almost simultaneously, hence reducing the concentration of reagents can limit the nucleation. Otherwise, the growth step has to be favoured by adjusting the temperature to limit nucleation. This problem can be solved with the help of weak reducing agents like trisodium citrate [57] or adopting a seed-mediated growth process. In the seed-mediated process, seed NPs prepared in advance form the seed for the synthesis of various-shaped particles. It was introduced first by Murphy et al. to synthesize bigger Au NPs [56]. The ratio of seed to the metal salt concentration can control the size of NPs (will discuss later) [56]. Compared to seeded growth synthesis, the unseeded method produces low aspect ratio NPs since it is sensitive to synthesis parameters [57]. By adding polymers and surfactants, the growth of MNPs can be confined [58]. It has been found that face-centred cubic (fcc) metals with different shapes can be synthesized easily with the seed-mediated growth process.

Evolution to seed from nuclei decides the final formation of nanocrystals. Seed can be single crystal, singly twinned, multiply twinned, and plate with stacking faults. Single crystal seed can synthesize polyhedron shapes such as octahedron, cuboctahedron, and cubes. Surface activation of these structures can yield anisotropic nanocrystals like octagonal rod and bar. Right bipyramids can be obtained from singly twinned seeds, and surface activation of bipyramids can produce beam-like structures. Multiply twinned seeds can grow into decahedron and icosahedron, while plate-like seed gives a hexagonal or triangular plate. The reason behind the formation of producing different shapes with the same seed is the different capping ability introduced by the polymer or ionic species present in the synthesis [59]. It has been found that La-Mer nucleation, followed by the Ostwald ripening mechanism, help to modify polydispersed particles into monodispersed [58]. By this, the lamellar structure can be generated and get adsorbed on the surface of a metal crystal plane. Then metal starts to grow in between layers into a 2D structure [57, 60].

Anisotropic Au NPs are found to have enormous applications in the biomedical field mainly due to its multiple plasmonic behaviours, and it can be tuned to the near-infrared region simply by varying the reaction concentration, seed parameters and seed to reagent ratio [61]. The shape evolution of Au NPs with a seeded growth method has been studied by Langille et al. [62]. They fabricated ten different shapes of Au nanostructures successfully and investigated the effect of reducing agents, halides, and the presence of additives on the shape evolution of Au NPs using a seeded growth method. Figure 9a–c shows how the concentration of weak reducing agent ascorbic acid can control the shape. Increasing the concentration of reducing agent increases the rate of gold ion reduction. Hence, it is preferred for the kinetically favourable growth of particle morphology. {111} faceted octahedra were formed with 0.5 mM ascorbic acid, while {100} faceted cubes were evolved with 2 mM ascorbic acid. Further increase in ascorbic acid concentration (10 mM) leads to the formation of ill-formed trisoctahedra with high index facets. Hence, the overall trend is that increased concentration of reducing agent leads to the formation of higher energy

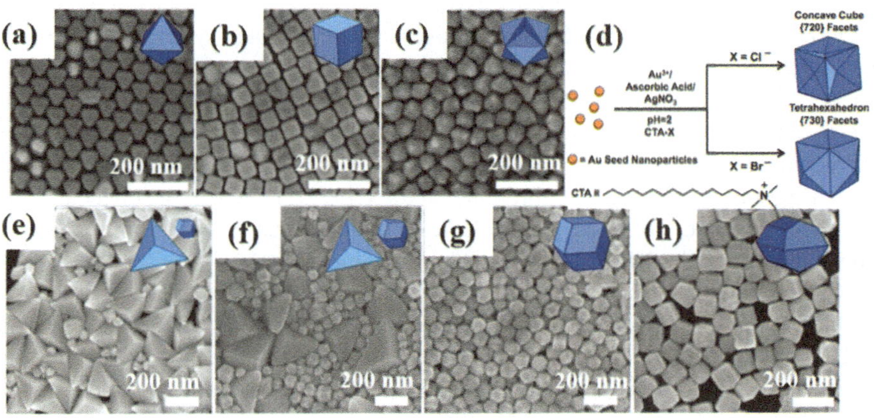

Fig. 9 Shape controlled synthesis of Au NPs using seed-mediated growth method under controlled pH environment; SEM images of NPs synthesized using various concentrations of reducing agent, 10 mM CTA-Br and **a** 0.5 **b** 2 and **c** 10 mM of ascorbic acid, **d** schematic representation of the effect of halide on seeded growth of Au NPs. Effect of halide in the presence of silver ions; SEM images of synthesized particles with 40 mM CTA-Cl, 10 μM $AgNO_3$ and **e** 0, **f** 10, **g** 20 and **h** 50 μM NaBr. Reprinted with permission from [62]. Copyright (2012) American Chemical Society

surface faces. Figure 9d shows the schematic illustration of silver assisted Au NPs synthesis using chloride and bromide ion. Addition of chloride ion leads to the formation of {720} faceted concave cubes, while the addition of bromide leads to the formation of a tetrahedron with {730} facets. Hence, the effect of halides has to be investigated in detail. It should be studied in the presence and absence of silver ions. The effect of halides in the absence of silver ions are further studied and found that {100} faceted cubes are formed with bromide ion. In contrast, the addition of iodide ion leads to the formation of {111} faceted octahedra and {111} faceted truncated bitetrahedra with 10 and 75 μM sodium iodide, respectively. Introduction of iodide or bromide ion in the presence of chloride ion reduces the rate of Au NPs formation by two mechanisms; by lowering the reduction potential and solubility of gold ion and inhibiting the growth by binding the halides strongly to the surface of gold particles. However, these two mechanisms are difficult to separate from obtaining specific shapes. Another important parameter in shape-controlled seeded growth of Au NPs is silver ions. Silver prefers to deposit epitaxially onto gold sites, where it has a high coordination number concerning gold. Hence, favourable growth of high index facets happens by stabilizing the facets through increasing the silver ions in the growth solution. This is due to the underpotential deposition (UPD) of silver onto the growing surface of gold. The effect of halides in the presence of silver ions is depicted in Fig. 9e–h. In the absence of bromide, a mixture of {110}-faceted bipyramids and {110}-faceted rhombic dodecahedra are formed. By increasing the amount of bromide, particle size, and yield of bipyramids decreased while that of rhombic dodecahedra increased. Comparison of the effects of halides with and without silver ions shows that surface effects are more influential with silver assisted synthesis, while kinetic effects dominate in the absence of silver. But increasing the amount of halide beyond some value inhibits the deposition of silver due to the destabilization of the Ag_{UPD} layer in silver assisted synthesis. This is not possible with iodide. Figure 10 shows the schematic diagram of the summary of the shape-controlled synthesis of Au NPs using halides in the presence or absence of silver ions. A higher concentration of halide ions are more favourable for surface effects and results in destabilization of the Ag_{UPD} layer. This effect is observed in increasing order with chloride, bromide, and iodide [62]. Hence, the concentration of reducing agent, silver ions, and selection of halides has to be optimized for the specific shape of Au NPs using a seeded growth method.

Critical parameters in controlling the size of MNP in the chemical reduction method without the involvement of seed are the type of reagent and its concentration [31, 63–69]. Other parameters like temperature, reaction time, reagent concentration, molar ratio, pH, and type of additives can also influence, as explained previously [63, 68–74]. Effect of reducing agents trisodium citrate and $NaBH_4$ on the formation of Au NPs and its absorbance is shown in Fig. 11. The observed trend is that as the concentration of trisodium citrate increases, particle distribution becomes monodispersed. The reverse trend is observed with $NaBH_4$ as a reducing agent. It might be because $NaBH_4$ is a strong reducing agent; hence as the concentration increases, the availability of reducing agents to reduce the metallic salt is higher. So, nucleation is favoured compared to the growth process. Since nucleation is chaotic than

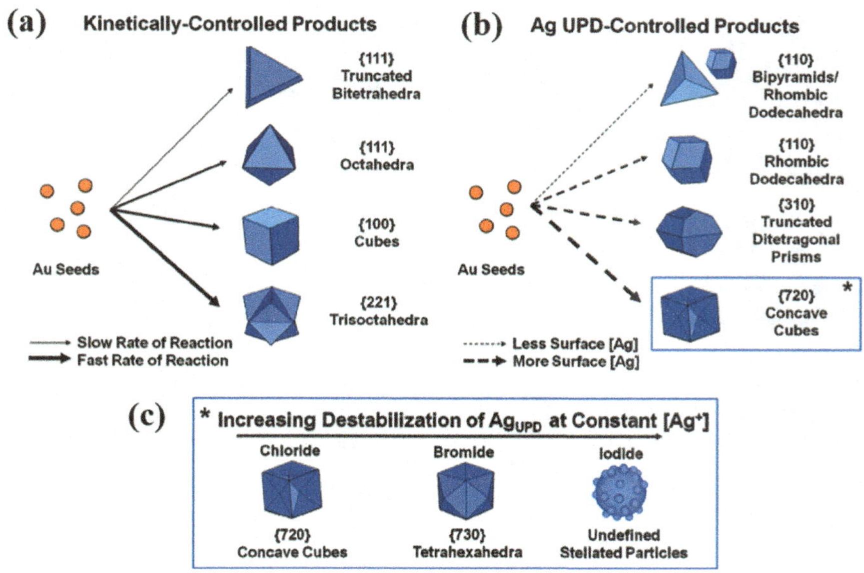

Fig. 10 Schematic illustration of shape control of Au NPs synthesized using seed-mediated growth method; **a** kinetically-controlled particles can be obtained by varying halides in the absence of silver ions, **b** surface-controlled synthesis of particles using silver ions by changing halides and **c** increasing order of destabilization of Ag $_{UPD}$ layer with halides. Reprinted with permission from [62]. Copyright (2012) American Chemical Society

growth, particle size distribution is random or polydispersed. Another finding is that the average size of particles synthesized with trisodium citrate is higher compared to $NaBH_4$ reduced particles. Also, the mean diameter of particles synthesized using trisodium citrate decreases as its concentration increases. Maximum absorbance is observed with a minimum amount of trisodium citrate, while with $NaBH_4$, maximum absorbance is observed for higher concentration. It is because maximum absorbance depends on the polydispersity of particles [69].

In the case of seed-mediated synthesis of MNPs, size of the seed, seed age and its concentration (or ratio) play an important role in shaping the morphology as well as the size of the NPs [31, 63–66, 75, 76]. Consider the synthesis of anisotropic Ag NPs by a chemical method using CTAB stabilized seed. It has been found experimentally that only by varying the seed ageing, the morphology of Ag NPs can be changed, as illustrated in Fig. 12. The UV-Vis absorption spectrum analysis reveals that as the seed age increases, it possess multiple plasmonic peaks. With 0 min seed age, the synthesized particles have a single peak at ~400 nm. As seed age increase, another peak starts to appear after this. With 5 min, the second peak is not much visible, indicates NPs with excess spherical and few anisotropic structures. At 15 min, the second peak is becoming sharper with a reduction in the first peak absorbance intensity. For 20 and 30 min of seed age, the second peak starts to dominate over the first peak and starts to appear the third peak before 400 nm. The appearance of the third peak indicates the

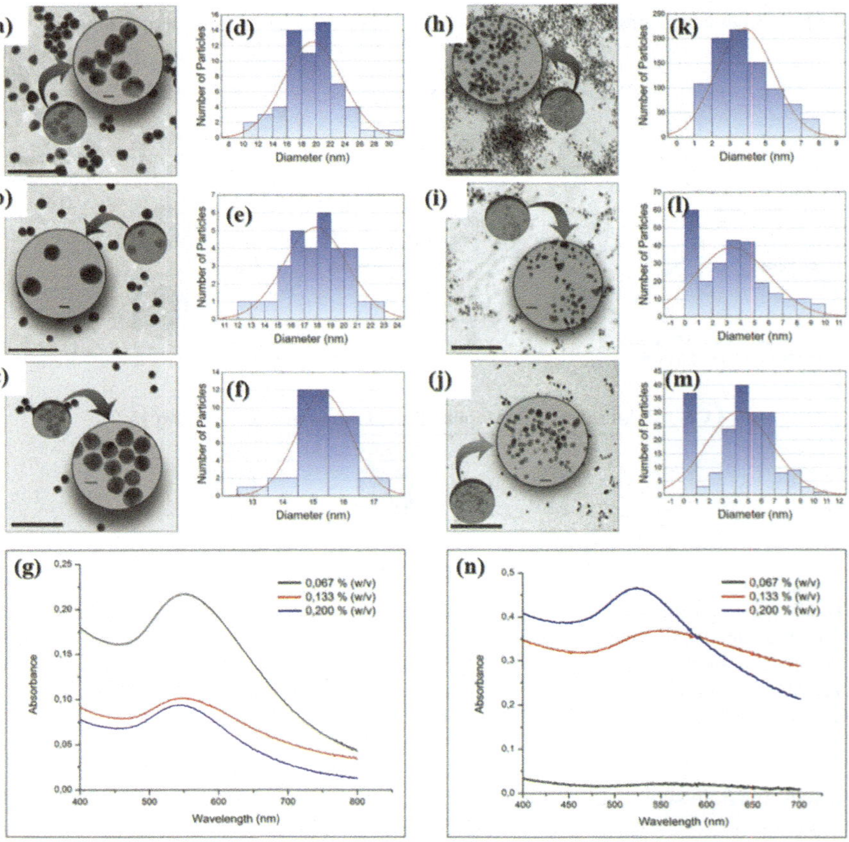

Fig. 11 Effect of reducing agent on size distribution of MNPs; TEM images of Au NPs synthesized with citrate concentration of **a** 0.067, **b** 0.133 and **c** 0.200% and its particle distribution in **d**, **e** and **f** respectively, **g** corresponding absorbance spectrum. TEM images of Au NPs synthesized with $NaBH_4$ concentration of **h** 0.067, **i** 0.133 and **j** 0.200% and its particle distribution in **k**, **l** and **m** respectively, **n** corresponding absorbance spectrum. All scale bar in TEM images is 100 nm and inset is 10 nm. Reproduced from [69] under a creative commons license deed (https://creativecommons.org/licenses/by-nc-nd/4.0/)

abundance of triangular NPs. Hence the tendency of the spectrum with increasing seed age is the occurrence of the second peak with redshift and increased intensity of absorption. The findings from the absorbance spectrum are further analysed with the TEM images, as shown in Fig. 12b–e. With 0 min seed age, synthesized NPs are spherical, and the particle solution has a yellow colour. With 15 min, anisotropic structures such as rods, triangles, and few spheroidal shapes are seen, and the particle solution is red. The corresponding histogram of shapes revealed that nanorods are abundant with 53.9%, spheres with 33.6%, and the remaining constitute triangle and other shapes. With 30 min seed age, triangular shapes are more compared to rods, spherical, and spheroids [31]. It has also been investigated the effect of seed ageing

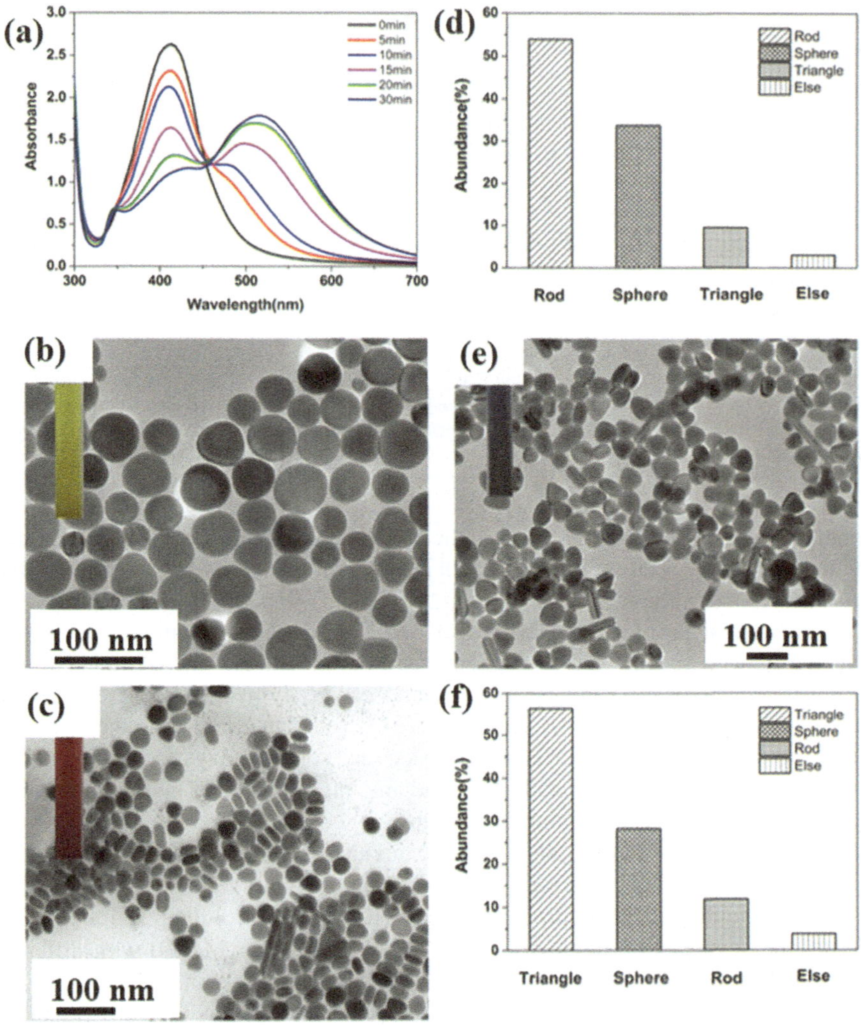

Fig. 12 Effect of seed ageing in the morphology of MNPs **a** UV-Vis absorption spectrum of synthesized Ag NPs at different seed ageing, TEM image of Ag NPs with seed age of **b** 0 min, **c** 15 min, **e** 30 min and shape distribution at seed age of **d** 15 min and **f** 30 min. Reproduced from [31] under a Creative Commons Attribution 4.0 International License (https://creativecommons.org/licenses/by/4.0/)

for a longer period, as depicted in Fig. 13. The absorption spectrum revealed that three peaks (350 nm, 420, and ~600 nm) are seen for all the seed ageing with a decrease in the intensity of absorption. This shows that there is a reduction happens in the synthesized nanoplates. The TEM image of Ag NPs obtained after 6 h seed age shows that majority of particles are truncated nanotriangle plates. It has been

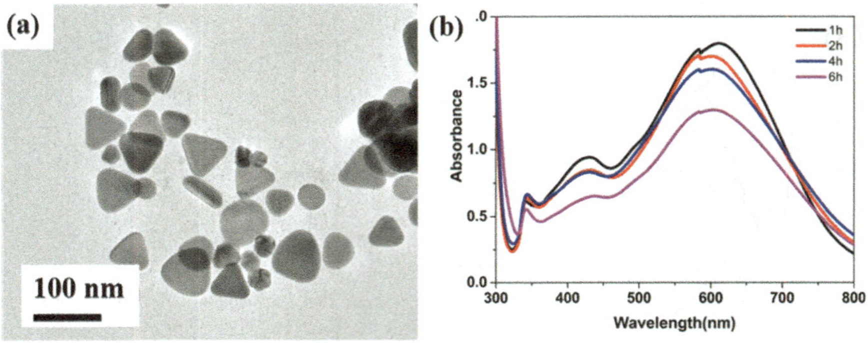

Fig. 13 Effect of more extended seed ageing **a** UV-Vis absorption spectrum of Ag NPs synthesized with different seed ageing and **b** TEM image of Ag NPs with seed age of 6 h. Reproduced from [31] under a Creative Commons Attribution 4.0 International License (https://creativecommons.org/licenses/by/4.0/)

found that crystal defects were seen in the particles prepared immediately after seed synthesis. Further, seed ageing beyond 5 h results in all-sided adsorption of surfactant on seed, results in the formation of complete crystalline particles with aggregation. The overall analysis of seed ageing revealed that there is a minimum seed age above which NPs with specific crystalline shape can be formed, and there is a maximum limit of seed age below which sharp structures can be formed. Hence, the optimum seed age in Ag NPs with CTAB stabilized seed is found as 2 h [31]. In the synthesis of Au nanorod with tuneable plasmonic peaks, seed solution plays an important role. Increasing the reaction time with the same amount of seed can give higher intensity plasmonic peaks. In addition, the volume of seed has a direct relationship with plasmonic peak, but there exists a specific amount of seed solution beyond which plasmonic peak cannot be further increased [66].

Another finding is that hydrochloric acid (HCl) can control the reaction kinetics during the synthesis of single-crystalline Pd NPs (or other MNPs) with different shapes. The summary of the mechanism is depicted in Fig. 14. Without HCl, Pd nuclei are formed rapidly due to a high reduction rate, and nanocubes with {100} facets are formed. As HCl is added, different shapes are evolved depending on the concentration of HCl. It is noted that with a small amount of HCl, slight truncation happens at the corner of cubic Pd. As the amount of HCl increases, truncation continuous with the further addition of atoms to {100} facets and finally results in the formation of octahedral bounded by {111} facets. The reason behind the shape evolution with HCl is that introduction of HCl can promote oxidative etching and depends on its concentration, etching rate varies. Also, it is found that a certain amount of HCl reduces the reduction rate, leads to reduce the number of seeds in the nucleation stage, results in the formation of bigger Pd NPs [74]. Further investigation is required to understand the role of HCl in other MNPs also.

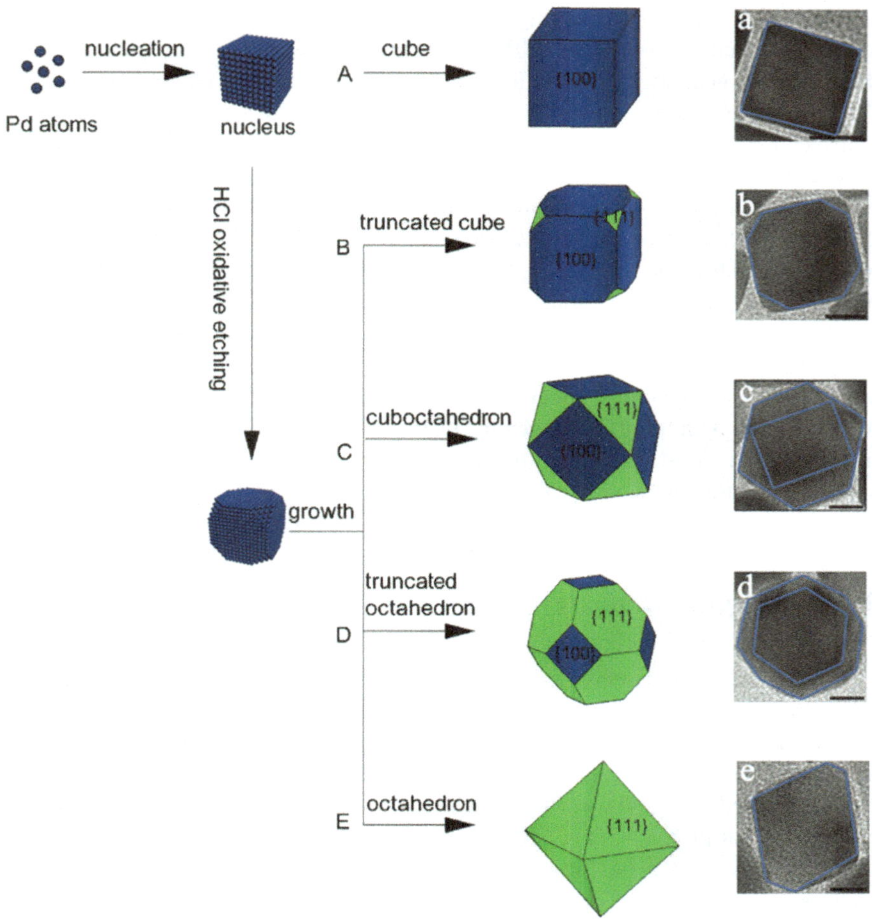

Fig. 14 Schematic diagram of shape evolution of Pd NPs with different shapes using HCl oxidative etching. **a–e** TEM images of different shapes. Reprinted with permission from [74]. Copyright (2014) American Chemical Society

A universal method has been reported to synthesize MNPs with the assistance of sugar using a chemical reduction method. Au, Ag, Pd, and Pt NPs were successfully synthesized using sugar like sucrose, glucose, fructose, etc. as reducing agents. $HAuCl_4$, $AgNO_3$, chloroplatinic acid (H_2PtCl_6), and palladium chloride ($PdCl_2$) were used as metallic salt and glucose, fructose, and sucrose were used as three separate reducing agents. A mixture of the metal salt and reducing agent solution were heated to 70–75 °C. The colour of the solution is an indication of the formation of NPs. Pink and yellow for Au and Ag NPs respectively, whereas the formation of black indicates Pd and Pt NPs. The analysis revealed that Ag and Au NPs possess plasmonic peak at 404 and 514 nm, respectively, while Pd and Pt NPs do not show any peak.

UV-Vis spectrum recorded during the synthesis revealed that as it progresses, the absorption band for Au and Ag NPs narrowed, absorbance intensity increased, and the plasmonic peak was shifted to low wavelength in the case of glucose and sucrose. Prolong heating results in adhering metal particles as a layer on the glass substrate, which causes a decrease in absorbance. After drying and sonicating, a blue shift was observed, and it was more prominent with sucrose and glucose. Fructose reduced NPs were distributed homogeneously concerning the size. Further, it was found that Ag and Pd NPs were unable to produce with sucrose. Generally, sucrose hydrolyses to glucose and fructose in the presence of an acid. In the case of $HAuCl_4$ and H_2PtCl_6, hydrolysed products take a longer time to reduce the metallic salt. But for Ag and Pd, even in the lower pH using sucrose as reductant could not produce particles. Compared to sucrose and fructose, glucose reduced NPs are smaller in size due to its strong reducing capability, but actual TEM analysis is contradictory; smallest NPs were obtained with fructose as reducing agent [77]. Table 1 lists the important works in the synthesis of MNPs with liquid state method along with synthesis type, shape, size, tuning parameters, and reaction condition.

3.4 Biological Methods

The biological method is an emerging approach to synthesize MNPs. It overcomes the difficulties associated with traditional methods such as reaction complications, safety issues, and high cost. Further, it does not utilize high pressure, temperature, energy, and toxic chemicals. It is also known as biomimetic method or green synthesis. The biological method of NPs synthesis involves the use of microorganisms and their enzymes, plant products, or its extracts. In general, the method can be categorized into two; bioreduction and biosorption. In the bioreduction method, with the help of microorganisms and their enzymes, metal ions are reduced chemically into a stable biological form. The formed MNPs can be safely separated from the sample, and they are found to be inert and stable. In biosorption, metal cations in aqueous media bind with the organism cell wall, and reaction between the cell wall and peptide further results in the formation of stable MNPs [34, 78]. A schematic illustration of the biological synthesis of MNPs is shown in Fig. 15. Biological species can be either prokaryotic like bacteria or eukaryotic such as fungi, plants or extracts. Extracts are mixed with precursor solution under controlled reaction condition, and the formation of NPs are confirmed with the colour.

3.4.1 Nanoparticle Synthesis Using Fungi

Here, metal salts are reduced with the help of protein and enzymes secreted by fungi. Generally used fungi for this purpose are *Fusarium oxysporum*, *Aspergillus fumigatus,* and *Trichoderma reesei.* Ag NPs can be easily prepared in this manner as it can bind with cytoplasmic membrane due to electrostatic interaction; hence Ag

Table 1 Overview of synthesis of MNPs using liquid state approach

S. No.	Metal	Shape	Size (nm)	Type	Reaction condition	Tuning parameters	References
1	Cu	Spherical	45 ± 8	Chemical reduction	Ambient temperature	Reducing agent concentration, reaction temperature and precursor injection rate	[47]
2	Cu	Cubes Polyhedral Spherical	100 ± 25	Chemical reduction	Stirring and heating at 140 °C	Molar ratio, temperature, reaction temperature and solvent	[79]
3	Cu	Spherical	<10	Chemical reduction	Ambient atmosphere	Reaction time, pH, and the ratio of copper sulphate to the surfactant	[80]
4	Pt	Cubic, tetrahedral and spherical	12.9–70.7	Chemical reduction	Room temperature	pH of protective agent	[81]
5	Ag	Sphere	10–30	Solvothermal	Shape dependent temperature and time	Temperature, concentration ratio of reagents, reaction time, amount of PVP and $AgNO_3$	[48]
		Rods	–				
		Triangular	50–150				
		Quasi-spherical	60–80				
		Polyhedron	–				
		Hexagonal	30–50				
		Nanocubes	50–80				
		Enneahedral	25–60				

(continued)

Table 1 (continued)

S. No.	Metal	Shape	Size (nm)	Type	Reaction condition	Tuning parameters	References
6	Pt	Nanowire	10 nm–10 μm	Solvothermal	Autoclave at 170 °C for 8 h, drying at 80 °C	Reducing agent and solvent	[43]
7	Pt	Multipod	–	Solvothermal	200 °C	Structure regulator	[49]
		Concave	–				
		Cube	–				
		Rod	30–100				
8	Pt	Cubes in excess Triangular, rod, hexagon, prism, etc. …	10 ± 3	Hydrothermal	heated at 140 °C for 6 h	Reaction time, reducing agent	[50]
9	Pt	Assembly of nanowires	20 (aspect ratio)	Solvothermal	–	Surfactant	[51]
10	Pt	Concave	~70 (apex to apex diameter)	Solvothermal	Autoclave at 160 °C for 11 h	Reaction time	[82]
11	Pt	Concave cubic	44.4 ± 3.2 (apex to apex diameter)	Solvothermal	200 °C for 6 h	Surface controller and reaction time	[42]
		Spherical	3.7 ± 0.5 Dozens of nm				
		Wire	–				
		Mixed structures (cubes, icosahedra, polyhedral, etc.)	–				

(continued)

Table 1 (continued)

S. No.	Metal	Shape	Size (nm)	Type	Reaction condition	Tuning parameters	References
12	Pd	Concave polyhedral	45 ± 2 and 35 ± 2	Solvothermal	heated at 100 °C for 8 h	Reaction time	[39]
13	Pd	Nanowires	9–15 (diameter)	Hydrothermal	At 200 °C for 2 h	Reaction time	[83]
		Rod	19 (diameter)				
		multiply twinned NPs	~57				
14	Pd	Cubes	12.5 ± 0.8	Solvothermal	heated at 150 °C for 8 h	Effect of halide, reaction time and temperature	[38]
		Multiply twinned NPs	50				
15	Pd	Cubes	7.6 ± 0.4	Solvothermal	180 °C	–	[84]
	Rh	Cubes	3–4				
16	Pd	Twinned wires	15.4 (diameter)	Solvothermal	140 °C for 4-12 h	Reaction time, the effect of reducing agent and small organic molecule	[46]
		Rods	14–15.4				
		Twinned right bipyramids	43.5 (short length)				
17	Ag	Cubes	55 ± 5	Solvothermal	At 120 °C	Molar ratio	[52]
18	Ag	Grain	5	Solvothermal	At 80 °C for 40 h	–	[85]
19	Ag	Various structures	4 ± 0.5 4 ± 0.7 (isolated particle)	Solvothermal	Temperature 180 and 200 ◦C	Temperature	[86]
20	Ag	Grain	8.219–59.36	Solvothermal	At 80 °C for 40 h	Solvent medium	[44]
21	Au	Spherical	7 and 15	Solvothermal	At 60 °C for 2 h	Reducing agent	[53]

(continued)

Table 1 (continued)

S. No.	Metal	Shape	Size (nm)	Type	Reaction condition	Tuning parameters	References
22	Au	Cages	7.2–9 (water soluble) 20.5 (nanocluster)	Solvothermal	60-210 °C	Reaction temperature and time	[87]
23	Au	Corolla	300	Solvothermal	Boiling for 10 min	–	[88]
24	Rh	Sheets	500–600	Solvothermal	At 180 °C for 8 h	–	[89]
25	Rh	Dendrites	18–25	Solvothermal	At 220 0C for 1.5-24 h	Reaction time, halogen anions, and the molar ratio	[54]
		Polyhedrons	8.5 ± 0.3				
		Grains	<2				
		Cubes	4.8 ± 0.3				
		Horned particles	3–11				
		Dendrites network- shaped wires	5.1 ± 0.3				
26	Ni	octahedral	40.0 ± 5.1	Solvothermal	200 °C for 12 h	Solvents	[45]
		truncated octahedral	25.2 ± 4.3		200 °C for 12 h		
27	Au	Star	39–104 (diameter) 50–187 (tip-tip length)	Chemical reduction—seed-mediated	Ascorbic acid reduction	Order of reagent addition, size and amount of seed, molar ratio, the addition of halides and pH	[63]

(continued)

Table 1 (continued)

S. No.	Metal	Shape	Size (nm)	Type	Reaction condition	Tuning parameters	References
28	Au	Star	78 ± 13 (tip-tip length)	Chemical reduction—seed-mediated	Ascorbic acid reduction	Ageing	[58]
29	Au	Rod	1.83–5.04 (aspect ratio)	Chemical reduction—seed-mediated	Ascorbic acid reduction	Concentration of $AgNO_3$	[70]
30	Au	Rod	8.9–22.6 (aspect ratio)	Chemical reduction—3 step seeding	Stirring at 25 °C for 16 h	Seed size	[64]
		Hexagonal	–				
		Triangular	–				
		Spherical	–				
31	Au	Spherical	9–120	Chemical reduction	1. Citrate reduction at 100 °C 2. After UV illumination at 25 °C 3. Ascorbate reduction at 25 °C	Concentration ratio	[71]
32	Au	Rods	3–10 (aspect ratio)	Chemical reduction—seed mediated	Gentle stirring at 25 °C	–	[90]
33	Au	Star	55–1000	Chemical reduction—seed-mediated	After synthesis keep at 25 °C overnight	Seed concentration	[65]

(continued)

Table 1 (continued)

S. No.	Metal	Shape	Size (nm)	Type	Reaction condition	Tuning parameters	References
34	Au	Star	–	Chemical reduction—seed-mediated	At room temperature	Amount of $AgNO_3$ and pH	[91]
35	Au	Rod	–	Chemical reduction—seed-mediated	–	Growth time, seed amount and concentration of $AgNO_3$	[66]
36	Au	Rod	3.4 (aspect ratio)	Chemical reduction—seed-mediated	Keep for 3 h for complete growth of particle	Reaction time, amount of $NaBH_4$, type of metal seed and amount of Au ion	[67]
37	Au	Spherical	30–60	Chemical reduction	Citrate reduction	Amount of silver acetate and trisodium citrate	[68]
		Rod	–	Chemical reduction—seed-mediated	Ascorbic acid reduction		
		Cubic	7–10	Photochemical method	X-ray irradiation		
38	Au	Spherical	–	Chemical reduction	Trisodium citrate dehydrate and NaBH4	The concentration of reducing agent and reaction time	[69]
39	Au	Spherical	3.2 ± 0.8	Chemical reduction	NaBH4 reduction	Reaction time	[72]
40	Ag	Spherical	41 ± 14.3	Chemical reduction -CTAB stabilized seed	NaOH reduction	Seed ageing time	[31]
		Rods	–				
		Triangular nanoplates	52.2 ± 10.3				

(continued)

Table 1 (continued)

S. No.	Metal	Shape	Size (nm)	Type	Reaction condition	Tuning parameters	References
41	Ag	Nanowires	30–40	Chemical reduction	PVP reaction medium, polyol process	Solution medium	[73]
42	Ag	Faceted pentagonal rods	62–430 (average length)	Photochemical- decahedral seeds growth	Ethylene glycol solvent medium, at 150-180 °C	Amount of silver	[75]
43	Ag	Wires and mixture of rods	30–40	Chemical reduction—Pt/Ag seeds	Ethylene glycol as a reducing agent	Reaction time and temperature, molar ratio and precursor concentration	[76]
44	Au	Octahedra Cubes Trisoctahedra truncated bitetrahedra Bipyramids rhombic dodecahedra truncated ditetragonal prisms Tetrahexahedra	–	Chemical reduction—seeded growth	Ascorbic acid reduction and silver as additives	Concentration of reducing agent, silver ions and halides	[62]
45	Ag	Cubes	–	Solvothermal	80 °C for 0, 6 and 12 h	Reaction time	[40]

(continued)

Table 1 (continued)

S. No.	Metal	Shape	Size (nm)	Type	Reaction condition	Tuning parameters	References
46	Pd	Cubes	17, 26 (average)	Chemical reduction—unseeded method	Temperature of 80 °C with ascorbic acid reduction	Molar ratio, reducing agent and halide ion	[74]
		Cuboctahedrons	–				
		Octahedrons	–				

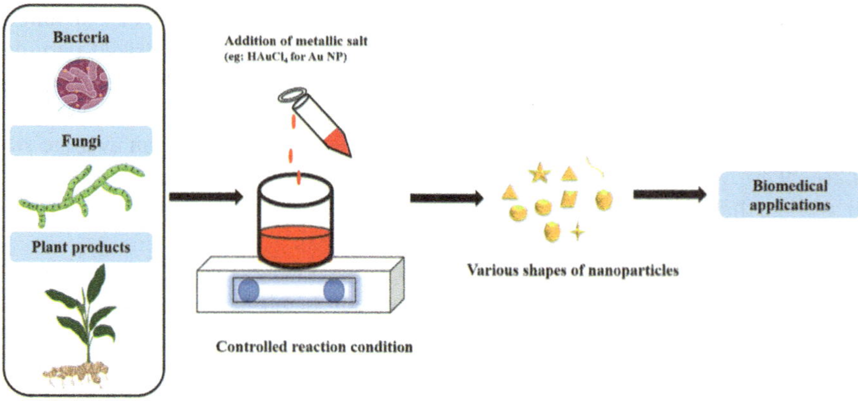

Fig. 15 Schematic illustration of the synthesis of MNPs using the biological method. It uses biological species such as bacteria, fungi, or plant extracts to reduce the metallic salt to form various shaped NPs under controlled reaction conditions such as temperature. Synthesized particles can be used for various biomedical applications due to its non-toxic behaviour

nuclei can be easily formed by reducing the metal ions, and they aggregate further to form NPs [34]. The location of synthesis can be either intracellular or extracellular. Pt NPs can be synthesized with the help of fungi such as *Neurospora crassa* and *Fusarium oxysporum*. *Neurospora crassa* based synthesis involves bioreduction of proteins in intracellular (cell wall, cytoplasm), which results in single-crystalline, quasi-spherical, and round nanoaggregates. Bioreduction of enzymes in *Fusarium oxysporum* leads to the Pt NPs of crystalline nature with spherical shape (5–30 nm). In contrast, the bioreduction of the hydrogenase enzyme leads to rectangular, triangular, and monodispersed spherical NPs. Hexagonal, pentagonal, circular, square, and spherical shaped Pt NPs can be easily synthesized through bioprecipitation of enzymes and peptides in intracellular or extracellular in *Fusarium oxysporum* [92]. Figure 16 shows the controlled synthesis of Au NPs with the help of *Aspergillus*

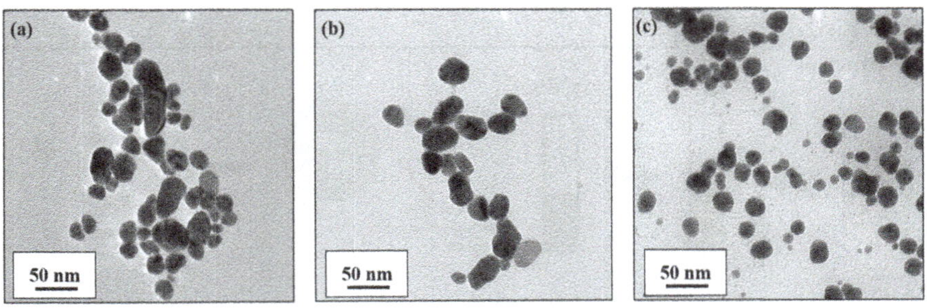

Fig. 16 Biological synthesis of Au NPs using *Aspergillus terreus*; TEM images of synthesized particles under ECF pH of **a** 5, **b** 8, and **c** 10, respectively. Reproduced from [93] under a Creative Commons Attribution 3.0 International License (https://creativecommons.org/licenses/by/3.0/)

terreus If0 in various pH conditions. Extracellular culture filtrate (ECF) of fungus served as the synthesis medium. When the pH of ECF is 5, synthesized particles contains multiple shapes such as triangular, rod, and elongated shapes, hence the shape distribution is not uniform. As pH is increased to 8, elongated shapes disappeared and dominated with rods and other asymmetric shapes with an average size of 20–29 nm. At a pH of 10, rod shape also disappeared, only spherical shape having an average diameter of 10–19 nm is present. Hence, the observed trend is reduction in shape distribution and size, with increasing pH [93].

Mushroom assisted MNPs synthesis is generally known as mycosynthesis or green chemistry, in which extracts from medicinal and edible mushrooms or fungi are used to form functional MNPs. Numata et al. synthesized NPs from mushrooms for the first time in 1994 [94]. They used *Schizophyllum commune* mushroom to extract purified polysaccharides to produce nanofibers. The species of *Pleurotus* (Oyster mushroom) such as *P. giganteus*, *P. pulmonarius*, *P. eous*, *P. platypus*, *P. salmoneostramineus*, *P. djamor*, *P. ostreatus*, *P. Florida*, *P. citrinopileatus*, *P. comucopiae var citrinopileatus*, *P. djamor var. roseus* and *P. sajor caju* are used to synthesize Ag NPs. *Pleurotus sp.* has several medicinal advantages like antimicrobial, antibacterial, antifungal, anti-pathogenic yeast, anti-oxidant, anti-cancer, anti-tumour, anti-viral, antidiabetic, antihypercholesterolemic, and anti-arthritic properties. The mechanism of synthesis of Ag NPs from *Pleurotus sp.* is depicted in Fig. 17. In intracellular based synthesis, electron released by nitrate reductase is used to reduce Ag^+ into Ag^0, while extracellular based synthesis involves myco-materials like amino acids, proteins or enzymes, polysaccharides, phenol, etc. Generally, extracts from *Pleurotus sp.* mix with mineral salt like $AgNO_3$. Produced NPs contain Ag NPs (66%) in the majority, and the remaining contains Au, Fe, Se, CdS, and ZnS NPs [94].

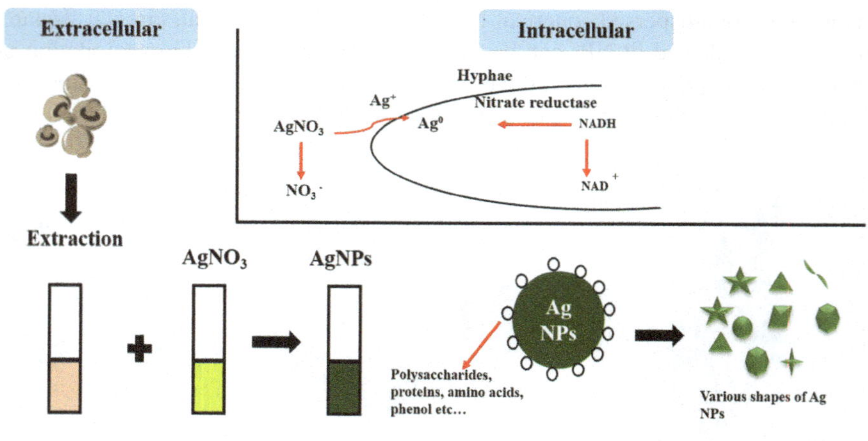

Fig. 17 Overview of synthesis (intracellular and extracellular) of Ag NPs from *Pleurotus sp.* using silver nitrate as mineral salt. Redrawn from [94]

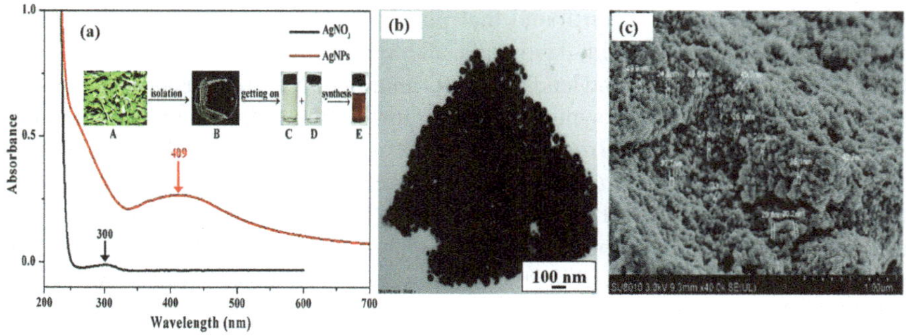

Fig. 18 Synthesis of Ag NPs using endophytic bacterium; **a** UV-Vis absorption spectrum along with its schematic illustration of synthesis procedure. **A** plant *Coriandrum sativum*, **B** extracting endophytic bacterium, **C** cell-free supernatant, **D** $AgNO_3$ and Ag NPs. The colour change in the solution confirms synthesis. **b** TEM and **c** SEM image of synthesized particles. Republished with permission of the Royal Society of Chemistry, from Ref. [95]

3.4.2 Nanoparticle Synthesis Using Bacteria

Ag, Au, Pd, and Pt NPs have been successfully synthesized with the help of bacteria. The commonly employed bacteria are *Lactobacillus species*, *Klebsiella pneumonia*, *Enterobacter coacoae,* and *Bacillus* species for producing Ag NPs, *Rhodococcus* species for Au, and *E. Coli.* for Pd, Pt, and Ag NPs. Due to the ability to adjust in the environment and its abundance in nature, prokaryotes are getting more attention in MNPs synthesis. Productivity is less compared to fungi since bacteria secrets less amount of protein compared to fungi [34]. Spherical shaped Ag NPs were synthesized from endophytic bacterium *Bacillus siamensis* strain C1 extracted from the plant *Coriandrum sativum.* Figure 18 shows the characterization of Ag NPs synthesized using endophytic bacterium. The absorbance peak of $AgNO_3$ is 300 nm, while Ag NPs possess a plasmonic peak at 409 nm after incubating NPs for 24 h at 30 °C. TEM and SEM analysis revealed that synthesized particles are spherical with an average diameter of 34 $\pm$ 3 nm [95]. For the bacteria-based synthesis of Pt NPs, the hydrogenase enzyme extracted from sulphate-reducing bacteria was used. It has been reported that spherical Pt NPs were synthesized by the bioreduction mechanism from *Desulfovibrio vulgaris.* At the same time, cuboidal shapes were obtained by enzymatic reduction of extracts from *Acinetobacter calcoaceticus* in intracellular based synthesis. Further, it has been employed as a reduction pathway to synthesize intracellular Pt NPs from the periplasmic membrane [96].

3.4.3 Nanoparticle Synthesis Using Plant Products

MNPs such as Au, Ag, Pt, Cu, and Zn can be easily synthesized with the help of plant extracts. Usually, in plants, primary and secondary metabolites are involved in redox reactions associated with plant metabolism. This property is utilized in the synthesis of NPs. It has been found that plant extracts are a source of secondary metabolites like phenolic acid, flavonoids, alkaloid, and terpenoid, which can selectively reduce metallic ions and leads to the formation of NPs, or can act as stabilizing or capping agent during the synthesis. Photoremediation of heavy metal is possible with some plants such as *Brassica juncea* (brown mustard), *Clethra barbinervis* (Japanese clethra), *Sesbania drummondi* (rattlebox) and *Acanthopanax scidophylloides*. After extracting the bio-reducing agents (bioreductant), it is mixed with an aqueous solution of metal precursor appropriately. Various spontaneous reactions occur at room temperature and lead to the formation of nanostructures. Sometimes, to hasten the synthesis process, Cd is added with stirring and heating. Plant-based synthesis is preferred over other biological methods due to its ease of availability, bulk production, and the products formed are not harmful to the environment [34, 97].

Plant parts like fruit, leaf extract, herb extract, tuber extract, dried leaves, and crusts can be used to synthesize Pt NPs of various shapes and sizes. Commonly employed plants are date palm (*Phoenix dactylifera*), pomegranate (*Punica granatum*), tulsi (*Ocimum sanctum*), vjradanti (*Barleria prionitis*), etc. Biofunctional groups involved in the synthesis of MNPs are shown in Fig. 19, along with synthesis procedure [92, 98].

Au NPs can be synthesized using plant parts like leaf, stem, fruit, bark, root, peel, flower, etc. These extracts are washed with distilled water, chopped finely and boil in distilled water to obtain the extracts. Derived extracts can be further purified by filtration and centrifugation, and later it can be mixed with Au salt solution. Parameters like the ratio of Au salt to the extracts, pH, and reaction temperature can control the synthesis by this method. Low pH leads to agglomeration and results in a smaller number of nucleation. The reaction pot can be incubated for some time, and NPs formation can be analysed by visually observing it. The advantage of this approach is that it does not require any stabilizing or capping agent. Synthesized Au

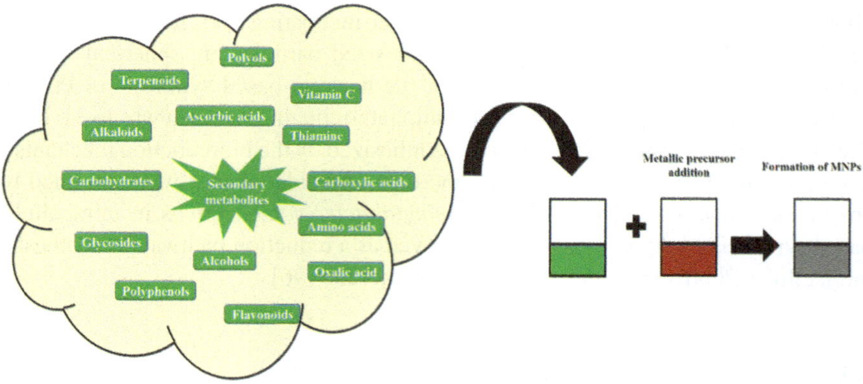

Fig. 19 Schematic illustration of bioactive functional groups involved in the plant-based synthesis of MNPs along with synthesis procedure. Plant extracts contain secondary metabolites that can selectively reduce the metallic salt, leads to the formation of MNPs identified by the colour change

NPs can be further purified by performing centrifuge at higher speed and washed with distilled water. *Lonicera japonica* flower extract has been utilized to synthesize Au NPs. Extracts contain alkane, amide, alcohols, and amino, which helps in reducing $HAuCl_4$ solution as well as act as a stabilizing agent. *Escherichia coli* K12 cells have been used to synthesize Au NPs with a mean diameter of 50 nm at room temperature [99, 100].

Figure 20 shows the synthesis of Au NPs using plant extracts. Figure 20a shows the TEM analysis of Au NPs synthesized from banana fruit waste extract. It is observed that the majority of particles are spherical, few are hexagonal, rods, and triangular. Au NPs synthesized using locust bean gum (LBG) is depicted in Fig. 20b. Particles are spherical with (111), (200), (220), and (311) planes exhibit fcc crystalline structure. The effect of reaction time on absorbance is illustrated in Fig. 20c; the intensity of

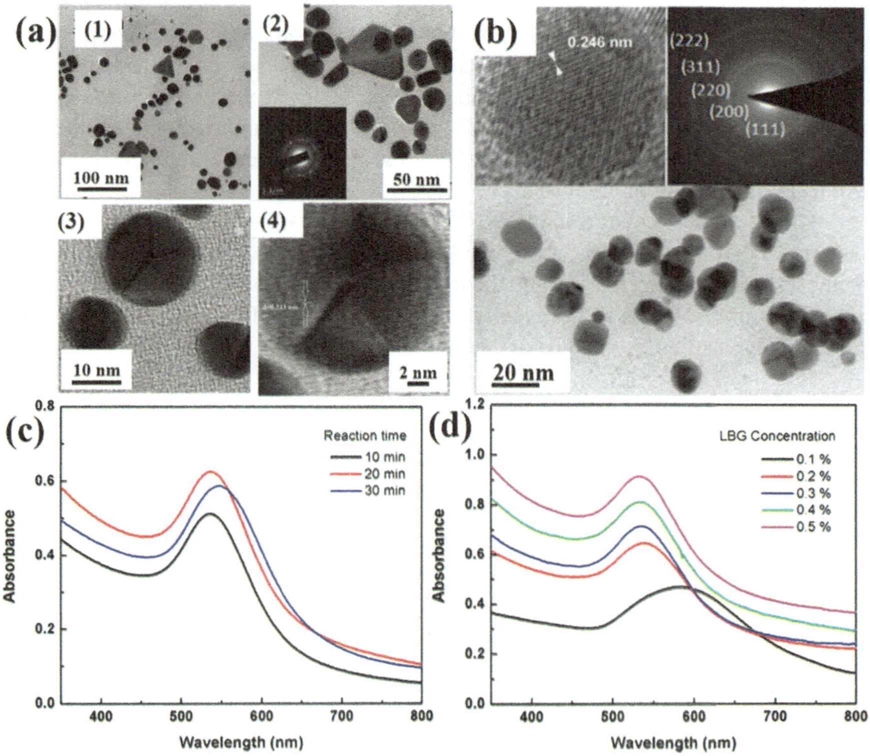

Fig. 20 Biosynthesis of Au NPs using plant extracts; **a** TEM images of synthesized Au NPs from banana fruit waste extract; **1** different shapes of Au NPs, **2** Au NPs with different morphology and SAED pattern, **3** HR-TEM image of spherical Au NPs and, **4** HR-TEM image of single spherical Au NP with lattice fringe. Republished with permission of the Royal Society of Chemistry, from Ref. [101], characterization of Au NPs synthesized using LBG, **b** TEM and HR-TEM image with clear lattice fringe, absorption spectrum with different, **c** reaction time and, **d** LBG concentration. Republished with permission of the Royal Society of Chemistry, from Ref. [102]

absorption corresponds to the plasmonic peak of 537 nm increased with reaction time 10 min to 20 min. Beyond 20 min, the intensity of absorption is decreased. Hence, the optimum reaction time of 20 min is needed in autoclaving the reaction mixture. Effect of concentration of bioreductant LBG is also investigated; An increase in the intensity of absorption is observed with increase in LBG. With 0.1% LBG, the amount of bioreductant is not sufficient to reduce all the gold ions. Hence the obtained spectrum is broad, and the peak is very weak (~579 nm). 0.2–0.5% of LBG is found to be the optimum amount for producing spherical nanoparticles with a narrow absorption band and sharp peak.

3.4.4 Miscellaneous Synthesis of MNPs

Apart from microorganisms and plant extracts, higher organisms can also be used to synthesize MNPs. It has been reported that products from higher organisms such as insects, birds, and mammals can be used to produce Pt NPs with controlled size and shapes. As an example, protein from sheep milk, honey from the bee, and egg yolk of quail were used to synthesize Pt NPs [92]. Table 2 summarizes the synthesis of MNPs using biological methods along with the type of microorganism, location of synthesis for specific metal, its shape, and size.

All the methods of synthesizing MNPs discussed above involve the use of conical flask in a laboratory set up, which are known as conventional batch reactor-based synthesis or macrosystems. Tuning the parameters to achieve the desired size, shape, and crystallinity of NPs is technically challenging and difficult to attain in a batch reactor. These difficulties can be overcome with the help of microfluidic technology.

3.5 Microfluidic Technology-Based Synthesis

Microfluidics is a part of micro-electro-mechanical system technology, which involves the flow of fluid of 10^{-9} to 10^{-18} litres through micron-sized channels. Since the dimension of the channel becomes very small, fluid flowing through a microfluidic channel shows some fascinating features, which are different from those of macrosystems. Flow-through microfluidics is laminar, electro-osmotic flow is observed for charged particles flow, and microfluidics has the ability to control water in channels, the dimensions of which match with the Debye layer [131]. The key feature of the microfluidic device is a high surface-area-to-volume ratio. It allows abrupt heat and mass transfer, results in rapid cooling and heating of reagents, and maintains an isothermal condition. Consumption of reagent in microfluidics is minimum, hence the cost of synthesis is low, making transportation and storage safer. It also helps to protect while handling toxic and inflammable solvents. The evolution of reactions in the spatial and temporal domain is possible in microfluidics. At any point of time, the reaction can be modified by controlling the reactant flow rate. Scaling up of products can be performed by conducting parallel experiments at the same time. Further, it is

Table 2 Overview of synthesis of MNPs using biological methods

S. No	Metal	Shape	Size (nm)	Type of microorganism/bioreductants	Location of synthesis	References
1	Au	Spherical, triangular, hexagonal, and rod	20–50	Leaf extracts of *Cymbopogon citratus*	Extracellular	[103]
2	Ag	Spherical	10–15	*Syzygium cumini* fruit extract	Extracellular	[104]
3	Au	Spherical	45–75	seed aqueous extract of *Abelmoschus esculentus*	Extracellular	[105]
4	Au	Spherical, cubical and hexagonal	37.7–79.9	saponin isolated from *T. decandra*	–	[106]
	Ag	Spherical	17.9–59.6	saponin isolated from *T. decandra*	–	[106]
5	Au	Hexagonal, spherical and triangular	15–25	*Cassia auriculata* aqueous leaf extract	Extracellular	[107]
6	Au	Spherical	6	*Micrococcus luteus*	Extracellular	[108]
7	Au	Spherical	70.90 ± 8.42	fruit peel extract of *Punica granutum*	Extracellular	[109]
8	Au	Spherical in excess	40 (average)	Streptomyces sp. MBRC-82.	Extracellular	[110]
9	Ag	Spherical	10–30	Chitosan immobilized on *Bacillus cereus* (Bc-C)	Extracellular	[111]
10	Au	Spherical in excess	5–50	*Basidiomycete Lentinus edodes*	Intracellular and extracellular	[112]
11	Ag	Spherical (dominant), parallelogram and trapezoidal	2–50	dried *Stevia rebaudiana* leaf extract	Extracellular	[113]
12	Au	Spherical in excess	20–25	grape skin, stalk and seeds	Extracellular	[114]

(continued)

Table 2 (continued)

S. No	Metal	Shape	Size (nm)	Type of microorganism/bioreductants	Location of synthesis	References
13	Au	Spherical and hexagonal	20	*Hovenia dulcis* fruit extract	Extracellular	[115]
14	Au, Ag	Spherical	5–65	anaerobic enriched mixed bacteria (AEMB)	Extracellular	[116]
15	Au	Spherical and triangular	5–30	*Scutellaria barbata* D. Don	Extracellular	[117]
16	Au	Spherical	4–24	*Psidium guajava* leaves extract	Extracellular	[118]
17	Au	Spherical	10–50	*Brevibacterium casei*		
18	Ag	Spherical	20–30	leaf extract of *Aca- lypha indica*	Extracellular	[119]
19	Au	Quasi-spherical	21	Apiin extracted from henna leaves	Extracellular	[120]
20	Ag	Spherical, triangular	39	Apiin extracted from henna leaves	Extracellular	[21]
21	Au, Ag & Ag-Au alloy	Spherical, triangular and hexagonal	5–35 & 50–100	*Azadirachta indica* leaf	Extracellular	[121]
22	Pd	–	3.2–6.0	broth of *Cinnamomum camphora* leaf	Extracellular	[122]
23	Pt Pd	–	–	*Desulfovibrio vulgaris*	Periplasmic space/cellular surface	[123]
24	Pt	Cuboidal	2–3	*Acinetobacter calcoaceticus* PUCM 1011	Intracellular	[124]
25	Pt	hexagons, pentagons, circles, squares and rectangles	10–100	*Fusarium oxysporum*	Intracellular and extracellular	[125]

(continued)

Table 2 (continued)

S. No	Metal	Shape	Size (nm)	Type of microorganism/bioreductants	Location of synthesis	References
26	Pt	Quasi-spherical, single-crystalline and round nano aggregates	2–3, 4–35, 7–76 and 20–110	*Neurospora crassa*	Intracellular	[126]
27	Pt	–	4	*Bidens tripartitus*	Extracellular	[127]
28	Pt	Irregular rod-shaped	–	dried leaf powder of *Anacardium occidentale*	Extracellular	[128]
29	Pt	Spherical	1.3–2.6	*Phoenix dactylifera* fruit extract	Extracellular	[129]
30	Pd	Spherical	20	aqueous extract of *Acacia concinna*	Extracellular	[130]
31	Ag	Spherical	25–50	*Bacillus siamensis* strain C1	Extracellular	[95]
32	Au	Spherical	6–18	Extract of banana fruit waste	Extracellular	[101]
		Hexagonal	–			
		Triangular	–			
		Rod	–			
		Other irregular shapes	–			

easy to add reagents downstream and easy to quench the reaction once the product is formed [132, 133]. NPs in microfluidics are synthesized using a single solvent (single-phase) or multiple miscible solvents (various phases). For the synthesis of MNPs, two flow types with different mixing geometries are used. Diffusion mixing of reagents happens in continuous flow microfluidic, whereas segmented flow uses either gas-liquid or liquid-liquid flow. Gas-phase is due to inlet air, or air evolved through reaction, whereas the liquid phase is due to the injection of immiscible liquid [134–136]. An important segmented flow device is based on droplet formation. Shear force and interfacial tension within two immiscible liquid generate droplets. It can be achieved with the geometries such as flow focusing, T-junction, and co-flow. Winding and distributed mixing modules are used in laminar flow mixing. T and Y-junction geometries are simple and easy to handle, where particle size can be controlled with channel width as well as flow rate. Geometry and flow condition determine particle size in cross channel geometry microfluidics [137].

Compared to other methods described above, particles prepared with microfluidics technology revealed superior performance in targeted cancer imaging, enhanced therapeutic efficiency, low sensing detection limit and higher catalytic activity [132, 138, 139]. For this, smart construction and design of the microfluidic device are required so that MNPs with desired shape, size, and crystallinity can be achieved. Parameters such as concentration ratio of reactants, flow rate, seed quantity, seed ageing time, residence time, temperature, heating time, and microchannel wall thickness influence the controlled synthesis of MNPs.

3.6 Other Methods of Synthesis

Other methods of synthesis include electrochemical deposition, microwave-assisted synthesis, supercritical fluid technology, and ultrasound technique. In an electrochemical deposition, baths containing metal salts are used. Baths can be either acidic or basic, and an electrode, namely cathode, is used, where metals are deposited. A three-terminal potentiostat and Ag/AgCl electrode (anode) is used, which acts as a reference electrode. A voltage is applied for a specific time for the deposition of metal. The advantages of this technique are simple, inexpensive, and particle synthesis with uniform morphology and size directly on the substrate. This method can quickly synthesize nanowires of Au, Co, Ni, and Pt [34, 140].

Recently, Ag NPs were successfully synthesized with the application of microwave of 300 MHz to 300 GHz. With this, polar molecules such as water are oriented with the electric field, creates friction and loss of energy in the form of heat. Usually, a metallic salt solution is irradiated with a reducing agent, and with or without a stabilizer. Sometimes this technique is employed along with plant-mediated synthesis. Compared to other methods, homogeneous heating in this method can speed up the reaction, leads to shorter crystallization time and homogeneous nucleation [34, 141]. Another technique is supercritical fluid technology, in which reaction happens in a reactor containing high-pressure pumps controlled by a pressure regulator. Precursor, along with different solvents, is injected into the reactor. As a result, the physicochemical property of the liquid phase can be modified into the gaseous phase, and formed products are nanomaterials. The method is suitable for synthesizing a few MNPs and a wide variety of metal oxides [142]. Ultrasound method uses either sonochemistry or ultrasonic spray pyrolysis to prepare unusual inorganic nanomaterials like carbonyl compounds [34].

4 Conclusions and Future Prospects

In this chapter, synthesis methods for metallic nanoparticles, which are essential for biomedical applications, are explained. The technique consists of two approaches; top-down and bottom-up. Particles synthesized with top-down methods are not suitable for biomedical applications due to their naked form. Important bottom-up approaches are solvothermal, chemical reduction, and biological methods. Solvothermal based synthesis involves the formation of nanoparticles in liquid phase solvents such as water or organic compounds. These solvents are allowed to heat above their boiling temperature in a pressure vessel. It uses a metallic precursor, a reducing agent (reductant), and sometimes ligands and facet specific capping agents. In the chemical reduction method, metallic salt is reduced with the help of reducing agent under particular conditions. Nanoparticles can be formed either through seeded-growth or unseeded method. With the seeded approach, desired nanostructures can be formed by changing seed size, type, amount, and ageing. Other tuning parameters are reaction time, temperature, solvent medium, reducing agent, ligands, capping agent, pH, and molar ratio. The biological method uses microorganisms and plant extracts to synthesize nanoparticles and can be carried out extracellular or intracellular. The concentration of bioproduct extracted, temperature, reaction time and pH can influence the particle synthesis. Compared to all other methods, biological method has proven so many advantages as discussed. If we can convert this biological method into the microfluidic platform, we can utilize the features offered by microfluidics for precise control of nanoparticles and eco-friendly synthesis. Further, the biomimicking materials can be used to fabricate the device, which will be beneficial for producing nanoparticles for biomedical applications. In future, it is expected the effective usage of biological method in microfluidics with more advanced fabrication procedures. Microfluidic-based metallic nanoparticle synthesis are well established. This technology is an emerging field, and more materials have to be invented for fabricating the microfluidic device as the majority of device material shows issues such as particle agglomeration and their fabrication requires clean room environment. Very few articles addressed these issues by simply connecting the glass capillary microreactors for metal hybrid core-shell nanoparticles. In addition, some researchers performed parameter tuning during synthesis, makes the overall process simpler for biomedical applications. More research has to be carried out in fabricating the new material, device fabrication and easy parameter scanning. The growth in material science and improvisation in method of metallic nanoparticle synthesis continue, and expect more research on new materials (metal combination) for biomedical applications. All the methods described in this chapter have their own advantages and limitations, hence more research is needed to synthesize metallic nanoparticles for specific biomedical application.

Acknowledgements The authors greatly appreciate the financial support from the DBT/Wellcome Trust India Alliance Fellowship under grant number IA/E/16/1/503062 and the Science and Engineering Research Board (SERB) under grant number ECR/2016/001945, Department of Science and Technology (DST), Government of India. We also acknowledge all authors and publishers who provided copyright permissions.

References

1. Jeevanandam J, Barhoum A, Chan YS, Dufresne A, Danquah MK (2018) Review on nanoparticles and nanostructured materials: history, sources, toxicity and regulations. Beilstein J Nanotechnol 9:1050–1074. https://doi.org/10.3762/bjnano.9.98
2. Barber DJ, Freestone IC (1990) An investigation of the origin of the colour of the lycurgus cup by analytical transmission electron microscopy. Archaeometry 32:33–45
3. Faraday M (1857) The bakerian lecture: experimental relations of gold (and other metals) to light. Philos Trans R Soc London 147:145–181. https://doi.org/10.1098/rstl.1857.0011
4. Kreuter J (2007) Nanoparticles-a historical perspective. Int J Pharm 331:1–10. https://doi.org/10.1016/j.ijpharm.2006.10.021
5. Khan I, Saeed K, Khan I (2019) Nanoparticles: properties, applications and toxicities. Arab J Chem 12:908–931. https://doi.org/10.1016/j.arabjc.2017.05.011
6. Gatoo MA, Naseem S, Arfat MY, Mahmood Dar A, Qasim K, Zubair S (2014) Physicochemical properties of nanomaterials: Implication in associated toxic manifestations. Biomed Res Int 2014: https://doi.org/10.1155/2014/498420
7. Cartaxo ALP (2018) Nanoparticles types and properties—understanding these promising devices in the biomedical area. Int J Nanomedicine: 1–8
8. Tiquia-arashiro S, Rodrigues DF (2016) Extremophiles: applications in nanotechnology. Springer Nature, Gewerbestrasse
9. Tiquia-Arashiro S, Rodrigues DF (2016) Application of nanoparticles. In: Extremophiles: applications in nanotechnology: biotechnological applications of extremophiles in nanotechnology, pp 163–193
10. Wu Z, Yang S, Wu W (2016) Shape control of inorganic nanoparticles from solution. Nanoscale 8:1237–1259. https://doi.org/10.1039/c5nr07681a
11. Polte J, Erler R, Thu AF, Sokolov S, Ahner TT, Rademann K, Emmerling F, Kraehnert R (2010) Nucleation and growth of gold nanoparticles studied via in situ Small angle X-ray scattering at millisecond time resolution. ACS Nano 4:1076–1082
12. Vekilov PG (2010) The two-step mechanism of nucleation of crystals in solution. Nanoscale 2:2346–2357. https://doi.org/10.1039/c0nr00628a
13. Gebauer D, Kellermeier M, Gale JD, Bergström L, Cölfen H (2014) Pre-nucleation clusters as solute precursors in crystallisation. Chem Soc Rev 43:2348–2371. https://doi.org/10.1039/c3cs60451a
14. LaMer VK, Dinegar RH (1950) Theory, production and mechanism of formation of monodispersed hydrosols. Am Chem Soc 72:4847–4854. https://doi.org/10.1016/s0033-3506(55)80003-0

15. Burda C, Chen X, Narayanan R, El-Sayed MA (2005) Chemistry and properties of nanocrystals of different shapes. Chem Rev 105:1025–1102. https://doi.org/10.1021/cr030063a
16. Polte J (2015) Fundamental growth principles of colloidal metal nanoparticles—a new perspective. CrystEngComm 17:6809–6830. https://doi.org/10.1039/c5ce01014d
17. Karatutlu A, Barhoum A, Sapelkin A (2018) Theories of nanoparticle and nanostructure formation in liquid phase. In: Emerging applications of nanoparticles and architectural nanostructures: current prospects and future trends. Elsevier Inc., pp 597–619
18. Amirjani A, Haghshenas DF (2019) Modified Finke-Watzky mechanisms for the two step nucleation and growth of silver nanoparticles. Nanotechnology: 0–22
19. Abadeer NS, Murphy CJ (2016) Recent progress in cancer thermal therapy using gold nanoparticles. J Phys Chem C 120:4691–4716. https://doi.org/10.1021/acs.jpcc.5b11232
20. Boies AM, Lei P, Calder S, Shin WG, Girshick SL (2011) Hot-wire synthesis of gold nanoparticles. Aerosol Sci Technol 45:654–663. https://doi.org/10.1080/02786826.2010.551145
21. Huang H, du Toit H, Besenhard MO, Ben-Jaber S, Dobson P, Parkin I, Gavriilidis A (2018) Continuous flow synthesis of ultrasmall gold nanoparticles in a microreactor using trisodium citrate and their SERS performance. Chem Eng Sci 189:422–430. https://doi.org/10.1016/j.ces.2018.06.050
22. Shinde P, Mohan L, Kumar A, Dey K, Maddi A, Patananan AN, Tseng FG, Chang HY, Nagai M, Santra TS (2018) Current trends of microfluidic single-cell technologies. Int J Mol, Sci, p 19
23. Watt J, Hance BG, Anderson RS, Huber DL (2015) Effect of seed age on gold nanorod formation: a microfluidic, real-time investigation. Chem Mater 27:6442–6449. https://doi.org/10.1021/acs.chemmater.5b02675
24. Dreaden EC, Alkilany AM, Huang X, Murphy CJ, El-Sayed MA (2012) The golden age: gold nanoparticles for biomedicine. Chem Soc Rev 41:2740–2779. https://doi.org/10.1039/c1cs15237h
25. Santra TS, Tseng F-G (2016) Pulse laser activated photoporation for high efficient intracellular delivery using nano-corrugated mushroom shape gold nanoparticles. In: International conference on Miniaturized systems for chemistry and life sciences, pp 1097–1098
26. Santra TS, Kar S, Chen C-W, Borana J, Chen T-C, Lee M-C, Tseng F-G (2020) Near-infrared nanosecond-pulsed laser-activated high efficient intracellular delivery mediated by nano-corrugated mushroom-shaped gold-coated polystyrene nanoparticles. Nanoscale. https://doi.org/10.1039/d0nr01792b
27. Johnson CJ, Dujardin E, Davis SA, Murphy CJ, Mann S (2002) Growth and form of gold nanorods prepared by seed-mediated, surfactant-directed synthesis. J Mater Chem 12:1765–1770. https://doi.org/10.1039/b200953f
28. Parveen F, Sannakki B, Mandke MV, Pathan HM (2016) Copper nanoparticles: synthesis methods and its light harvesting performance. Sol Energy Mater Sol Cells 144:371–382. https://doi.org/10.1016/j.solmat.2015.08.033
29. Chen S, Carroll DL (2002) Synthesis and characterization of truncated triangular silver nanoplates. Nano Lett 2:1003–1007. https://doi.org/10.1021/nl025674h
30. Liu G, Ma X, Sun X, Jia Y, Wang T (2018) Controllable synthesis of silver nanoparticles using three-phase flow pulsating mixing microfluidic chip. Adv Mater Sci Eng. https://doi.org/10.1155/2018/3758161
31. Jin W, Liang G, Zhong Y, Yuan Y, Jian Z, Wu Z, Zhang W (2019) The influence of CTAB-capped seeds and their aging time on the morphologies of silver nanoparticles. Nanoscale Res Lett 14. https://doi.org/10.1186/s11671-019-2898-x
32. Santra TS, Tseng F-G (2021) Handbook of single cell technology. Springer Nature
33. Tseng F-G, Santra TS (2016) Essentials of single-cell analysis. Springer Nature
34. Jamkhande PG, Ghule NW, Bamer AH, Kalaskar MG (2019) Metal nanoparticles synthesis: an overview on methods of preparation, advantages and disadvantages, and applications. J Drug Deliv Sci Technol 53. https://doi.org/10.1016/j.jddst.2019.101174

35. Piszczek P, Radtke A (2018) Silver nanoparticles fabricated using chemical vapor deposition and atomic layer deposition techniques: properties, applications and perspectives: review. In: Noble and precious metals—properties, nanoscale effects and applications, pp 187–213
36. Owens GJ, Singh RK, Foroutan F, Alqaysi M, Han CM, Mahapatra C, Kim HW, Knowles JC (2016) Sol-gel based materials for biomedical applications. Prog Mater Sci 77:1–79. https://doi.org/10.1016/j.pmatsci.2015.12.001
37. Lai J, Niu W, Luque R, Xu G (2015) Solvothermal synthesis of metal nanocrystals and their applications. Nano Today 10:240–267. https://doi.org/10.1016/j.nantod.2015.03.001
38. Huang X, Zhang H, Guo C, Zhou Z, Zheng N (2009) Simplifying the creation of hollow metallic nanostructures: one-pot synthesis of hollow palladium/platinum single-crystalline. Angew Chemie Int Ed 48:4808–4812. https://doi.org/10.1002/anie.200900199
39. Huang X, Tang S, Zhang H, Zhou Z, Zheng N (2009) Controlled formation of concave tetrahedral/Trigonal bipyramidal palladium. J Am Chem Soc 131:13916–13917. https://doi.org/10.1021/ja9059409
40. Zhang Q, Dong K, Wang C, Cheng Y (2015) Dramatic shape transformation of Ag nanoparticles with concave facets in a solvothermal process. CrystEngComm 17:7469–7472. https://doi.org/10.1039/c5ce01350j
41. Li J, Wu Q, Wu J (2015) Synthesis of nanoparticles via solvothermal and hydrothermal methods. In: Handbook of Nanoparticles. Springer Switzerland, pp 1–1426
42. Zhang ZC, Hui JF, Liu ZC, Zhang X, Zhuang J, Wang X (2012) Glycine-mediated syntheses of Pt concave nanocubes with high-index hk0 facets and their enhanced electrocatalytic activities. Langmuir 28:14845–14848. https://doi.org/10.1021/la302973r
43. Xia BY, Bin WuH, Yan Y, Lou XW, Wang X (2013) Ultrathin and ultralong single-crystal platinum nanowire assemblies with highly stable electrocatalytic activity. J Am Chem Soc 135:9480–9485. https://doi.org/10.1021/ja402955t
44. Rosa P de F, Cirqueira SSR, Aguiar ML, Bernardo A (2014) Solvothermal synthesis and characterization of silver nanoparticles. Mater Sci Forum 802:135–139. https://doi.org/10.4028/www.scientific.net/MSF.802.135
45. Wu Y, Cai S, Wang D, He W, Li Y (2012) Syntheses of water-soluble octahedral, truncated octahedral, and cubic Pt-Ni nanocrystals and their structure—activity study in model hydrogenation reactions. J Am Chem Soc 134:8975–8981. https://doi.org/10.1021/ja302606d
46. Lu N, Chen W, Fang G, Chen B, Yang K, Yang Y, Wang Z, Huang S, Li Y (2014) 5-fold twinned nanowires and single twinned right bipyramids of Pd: utilizing small organic molecules to tune the etching degree of O 2/halides. Am Chem Soc 26:2453–2459. https://doi.org/10.1021/cm4042204
47. Park BK, Jeong S, Kim D, Moon J, Lim S, Kim JS (2007) Synthesis and size control of monodisperse copper nanoparticles by polyol method. J Colloid Interface Sci 311:417–424. https://doi.org/10.1016/j.jcis.2007.03.039
48. Yang Y, Matsubara S, Xiong L, Hayakawa T, Nogami M (2007) Solvothermal synthesis of multiple shapes of silver nanoparticles and their SERS properties. J Phys Chem C 111:9095–9104. https://doi.org/10.1021/jp068859b
49. Zhang L, Chen D, Jiang Z, Zhang J, Xie S, Kuang Q, Xie Z, Zheng L (2012) Facile syntheses and enhanced electrocatalytic activities of Pt nanocrystals with hkk high-index surfaces. Nano Res 5:181–189. https://doi.org/10.1007/s12274-012-0198-1
50. Fu G, Wu K, Jiang X, Tao L, Chen Y, Lin J, Zhou Y, Wei S, Tang Y, Lu T, Xia X (2013) Polyallylamine-directed green synthesis of platinum nanocubes. Shape and electronic effect codependent enhanced electrocatalytic activity. Phys Chem Chem Phys 15:3793–3802. https://doi.org/10.1039/c3cp44191a
51. Xia BY, Ng WT, Bin WuH, Wang X, Lou XW (2012) Self-supported interconnected Pt nanoassemblies as highly stable electrocatalysts for low-temperature fuel cells. Angew Chemie Int Ed 51:7213–7216. https://doi.org/10.1002/anie.201201553
52. Yu D, Yam VWW (2004) Controlled synthesis of monodisperse silver nanocubes in water. J Am Chem Soc 126:13200–13201. https://doi.org/10.1021/ja046037r

53. Ahmad T, Wani IA, Lone IH, Ganguly A, Manzoor N, Ahmad A, Ahmed J, Al-Shihri AS (2013) Antifungal activity of gold nanoparticles prepared by solvothermal method. Mater Res Bull 48:12–20. https://doi.org/10.1016/j.materresbull.2012.09.069
54. Yuan Q, Zhou Z, Zhuang J, Wang X (2010) Tunable aqueous phase synthesis and shape-dependent electrochemical properties of rhodium nanostructures. Inorg Chem 49:5515–5521. https://doi.org/10.1021/ic100249t
55. Guerrini L, Alvarez-puebla RA, Pazos-perez N (2018) Surface modifications of nanoparticles for stability in biological fluids. Materials (Basel) 11:1–28. https://doi.org/10.3390/ma11071154
56. Jana NR, Gearheart L, Murphy CJ (2001) Seed-mediated growth approach for shape-controlled synthesis of spheroidal and rod-like gold nanoparticles using a surfactant template. Adv Mater 13:1389–1393
57. Uson L, Sebastian V, Arruebo M, Santamaria J (2016) Continuous microfluidic synthesis and functionalization of gold nanorods. Chem Eng J 285:286–292. https://doi.org/10.1016/j.cej.2015.09.103
58. Vega MM, Bonifacio A, Lughi V, Marsi S, Carrato S, Sergo V (2014) Long-term stability of surfactant-free gold nanostars long-term stability of surfactant-free gold nanostars. J Nanoparticle Res. https://doi.org/10.1007/s11051-014-2729-z
59. Skrabalak SE, Xia Y (2009) Pushing nanocrystal synthesis toward nanomanufacturing. ACS Nano 3:10–15. https://doi.org/10.1021/nn800875p
60. Tsuji M, Hashimoto M, Nishizawa Y, Kubokawa M, Tsuji T (2005) Microwave-assisted synthesis of metallic nanostructures in solution. Chem A Eur J 11:440–452. https://doi.org/10.1002/chem.200400417
61. Kohout C, Santi C, Polito L (2018) Anisotropic gold nanoparticles in biomedical applications. Int J Mol Sci 19. https://doi.org/10.3390/ijms19113385
62. Langille MR, Personick ML, Zhang J, Mirkin CA (2012) Defining rules for the shape evolution of gold nanoparticles. J Am Chem Soc 134:14542–14554. https://doi.org/10.1021/ja305245g
63. Indrasekara ASDS, Johnson SF, Odion RA, Vo-dinh T (2018) Manipulation of the geometry and modulation of the optical response of surfactant-free gold nanostars: a systematic bottom—up synthesis. Am Chem Soc Omega 3:2202–2210. https://doi.org/10.1021/acsomega.7b01700
64. Gole A, Murphy CJ (2004) Seed-mediated synthesis of gold nanorods: role of the size and nature of the seed. Chem Mater 16:3633–3640
65. Ahmed W, Khan HI, Khalid MU, Abdullah A, Ali A (2018) Enhanced Raman scattering and fluorescence quenching properties Facile synthesis of gold nanostars over a wide size range and their excellent surface enhanced Raman scattering and fluorescence quenching properties. J Vac Sci Technol B 36:03E101-1–6. https://doi.org/10.1116/1.4996541
66. Cui J, Fan J, Zhao T, Wang A, Drezek RA, Zhu M (2009) Real-time monitoring and scale-up synthesis of concentrated gold nanorods. J Biomed Nanotechnol 5:1–6. https://doi.org/10.1166/jbn.2009.1063
67. Samal AK, Sreenivasan ST, Thalappil P (2010) Investigation of the role of NaBH4 in the chemical synthesis of gold nanorods. J Nanoparticle Res 12:1777–1786. https://doi.org/10.1007/s11051-009-9733-8
68. Wang C, Wang T (2009) Synthesis and optical properties of colloidal gold nanoparticles synthesis and optical properties of colloidal gold nanoparticles. In: Journal of Physics: Conference Series
69. Oliveira JP, Prado AR, Juve W, Ribeiro RN, Pontes MJ, Nogueira BV, Guimara MCC (2017) A helpful method for controlled synthesis of monodisperse gold nanoparticles through response surface modeling. Arab J Chem. https://doi.org/10.1016/j.arabjc.2017.04.003
70. Feng L, Xuan Z, Ma J, Chen J, Cui D, Su C (2015) Preparation of gold nanorods with different aspect ratio and the optical response to solution refractive index. J Exp Nanosci 10:258–267. https://doi.org/10.1080/17458080.2013.824619
71. Kimling J, Maier M, Okenve B, Kotaidis V, Ballot H, Plech A (2006) Turkevich method for gold nanoparticle synthesis revisited. J Phys Chem B 110:15700–15707. https://doi.org/10.1021/jp061667w

72. Deraedt C, Salmon L, Gatard S, Ciganda R, Hernandez R, Ruiz J, Astruc D (2014) Sodium borohydride stabilizes very active gold nanoparticle catalysts†. Chem Commun 50:14194–14196. https://doi.org/10.1039/c4cc05946h
73. Sun Y, Gates B, Mayers B, Xia Y (2002) Crystalline silver nanowires by soft solution processing. Nano Lett 2:165–168. https://doi.org/10.1021/nl010093y
74. Zhang J, Feng C, Deng Y, Liu L, Wu Y, Shen B, Zhong C, Hu W (2014) Shape-controlled synthesis of palladium single-crystalline nanoparticles: the effect of HCl oxidative etching and facet-dependent catalytic properties. Chem Mater 26:1213–1218. https://doi.org/10.1021/cm403591g
75. Pietrobon B, McEachran M, Kitaev V (2009) Synthesis of size-controlled faceted pentagonal silver nanorods with tunable plasmonic properties and self-assembly of these nanorods. ACS Nano 3:21–26. https://doi.org/10.1021/nn800591y
76. Sun Y, Yin Y, Mayers BT, Herricks T, Xia Y (2002) Uniform silver nanowires synthesis by reducing AgNO3 with ethylene glycol in the presence of seeds and poly(vinyl pyrrolidone). Chem Mater 14:4736–4745. https://doi.org/10.1021/cm020587b
77. Panigrahi S, Kundu S, Ghosh SK, Nath S, Pal T (2004) General method of synthesis for metal nanoparticles. J Nanoparticle Res 6:411–414
78. Santra TS, Tseng F-G (Kevin), Barik TK (2015) Biosynthesis of silver and gold nanoparticles for potential biomedical applications—a brief review. J Nanopharmaceutics Drug Deliv 2:249–265. https://doi.org/10.1166/jnd.2014.1065
79. Wang Y, Chen P, Liu M (2006) Synthesis of well-defined copper nanocubes by a one-pot solution process. Nanotechnology 17:6000–6006. https://doi.org/10.1088/0957-4484/17/24/016
80. Dang TMD, Le TTT, Fribourg-Blanc E, Dang MC (2011) Synthesis and optical properties of copper nanoparticles prepared by a chemical reduction method. Adv Nat Sci Nanosci Nanotechnol 2. https://doi.org/10.1088/2043-6262/2/1/015009
81. Nagao H, Ichiji M, Hirasawa I (2017) Synthesis of platinum nanoparticles by reductive crystallization using polyethyleneimine. Chem Eng Technol 40:1242–1246. https://doi.org/10.1002/ceat.201600656
82. Huang X, Zhao Z, Fan J, Tan Y, Zheng N (2011) Amine-assisted synthesis of concave polyhedral platinum nanocrystals having 411 high-index facets. J Am Chem Soc 133:4718–4721. https://doi.org/10.1021/ja1117528
83. Huang X, Zheng N (2009) One-pot, high-yield synthesis of 5-fold twinned Pd nanowires and nanorods. J Am Chem Soc 131:4602–4603. https://doi.org/10.1021/ja9009343
84. Yuan Q, Zhuang J, Wang X (2009) Single-phase aqueous approach toward Pd sub-10 nm nanocubes and Pd–Pt heterostructured ultrathin nanowires w. Chem Commun: 6613–6615. https://doi.org/10.1039/b913974e
85. Wani IA, Khatoon S, Ganguly A, Ahmed J, Ganguli AK, Ahmad T (2010) Silver nanoparticles: large scale solvothermal synthesis and optical properties. Mater Res Bull 45:1033–1038. https://doi.org/10.1016/j.materresbull.2010.03.028
86. Rosemary MJ, Pradeep T (2003) Solvothermal synthesis of silver nanoparticles from thiolates. J Colloid Interface Sci 268:81–84. https://doi.org/10.1016/j.jcis.2003.08.009
87. Choi J, Park S, Stojanović Z, Han HS, Lee J, Seok HK, Uskoković D, Lee KH (2013) Facile solvothermal preparation of monodisperse gold nanoparticles and their engineered assembly of ferritin-gold nanoclusters. Langmuir 29:15698–15703. https://doi.org/10.1021/la403888f
88. Xu X, Zhang H, Liu B, Yang J (2020) One-pot synthesis of corolla-shaped gold nanostructures with (110) planes. RSC Adv 10:8286–8290. https://doi.org/10.1039/d0ra00715c
89. Duan H, Yan N, Yu R, Chang C, Zhou G, Hu H, Rong H, Niu Z, Mao J, Asakura H, Tanaka T, Dyson PJ, Li J, Li Y (2014) Ultrathin rhodium nanosheets. Nat Commun 1–8. https://doi.org/10.1038/ncomms4093
90. Sharma V, Park K, Srinivasarao M (2009) Shape separation of gold nanorods using centrifugation. PNAS 106:4981–4985
91. Li Y, Ma J, Ma Z (2013) Synthesis of gold nanostars with tunable morphology and their electrochemical application for hydrogen peroxide sensing. Electrochim Acta 108:435–440. https://doi.org/10.1016/j.electacta.2013.06.141

92. Puja P, Kumar P (2019) A perspective on biogenic synthesis of platinum nanoparticles and their biomedical applications. Spectrochim Acta PartA Mol Biomol Spectrosc 211:94–99
93. Priyadarshini E, Pradhan N, Sukla LB, Panda PK (2014) Controlled synthesis of gold nanoparticles using Aspergillus terreus if0 and its antibacterial potential against gram negative pathogenic bacteria. J Nanotechnol 2014. https://doi.org/10.1155/2014/653198
94. Owaid MN (2019) Green synthesis of silver nanoparticles by Pleurotus (oyster mushroom) and their bioactivity: review. Environ Nanotechnology, Monit Manag 12:100256. https://doi.org/10.1016/j.enmm.2019.100256
95. Ibrahim E, Fouad H, Zhang M, Zhang Y, Qiu W, Yan C, Li B, Mo J, Chen J (2019) Biosynthesis of silver nanoparticles using endophytic bacteria and their role in inhibition of rice pathogenic bacteria and plant growth promotion. RSC Adv 9:29293–29299. https://doi.org/10.1039/c9ra04246f
96. Ali J, Ali N, Wang L, Waseem H, Pan G (2019) Revisiting the mechanistic pathways for bacterial mediated synthesis of noble metal nanoparticles. J Microbiol Methods 159:18–25. https://doi.org/10.1016/j.mimet.2019.02.010
97. Iravani S (2011) Green synthesis of metal nanoparticles using plants. Green Chem 13:2638–2650. https://doi.org/10.1039/c1gc15386b
98. Nasrollahzadeh M, Sajjadi M, Dadashi J, Ghafuri H (2020) Pd-based nanoparticles: Plant-assisted biosynthesis, characterization, mechanism, stability, catalytic and antimicrobial activities. Adv Colloid Interface Sci 276:102103. https://doi.org/10.1016/j.cis.2020.102103
99. Ahmed S, Ikram S, S SY (2016) Biosynthesis of gold nanoparticles: a green approach. JPB 161:141–153. https://doi.org/10.1016/j.jphotobiol.2016.04.034
100. Santra TS, Tseng F-G, Barik (2014) Green biosynthesis of gold nanoparticles and biomedical applications. Am J nanoresearch Appl 2:5–12
101. Deokar GK, Ingale AG (2016) Green synthesis of gold nanoparticles (Elixir of Life) from banana fruit waste extract-an efficient multifunctional agent. RSC Adv 6:74620–74629. https://doi.org/10.1039/c6ra14567a
102. Tagad CK, Rajdeo KS, Kulkarni A, More P, Aiyer RC, Sabharwal S (2014) Green synthesis of polysaccharide stabilized gold nanoparticles: chemo catalytic and room temperature operable vapor sensing application. RSC Adv 4:24014–24019. https://doi.org/10.1039/c4ra02972k
103. Murugan K, Benelli G, Panneerselvam C, Subramaniam J, Jeyalalitha T, Dinesh D, Nicoletti M, Hwang JS, Suresh U, Madhiyazhagan P (2015) Cymbopogon citratus-synthesized gold nanoparticles boost the predation efficiency of copepod Mesocyclops aspericornis against malaria and dengue mosquitoes. Exp Parasitol 153:129–138. https://doi.org/10.1016/j.exppara.2015.03.017
104. Mittal AK, Bhaumik J, Kumar S, Banerjee UC (2014) Biosynthesis of silver nanoparticles: elucidation of prospective mechanism and therapeutic potential. J Colloid Interface Sci 415:39–47. https://doi.org/10.1016/j.jcis.2013.10.018
105. Jayaseelan C, Ramkumar R, Rahuman AA, Perumal P (2013) Green synthesis of gold nanoparticles using seed aqueous extract of Abelmoschus esculentus and its antifungal activity. Ind Crops Prod 45:423–429. https://doi.org/10.1016/j.indcrop.2012.12.019
106. Geethalakshmi R, Sarada DVL (2013) Characterization and antimicrobial activity of gold and silver nanoparticles synthesized using saponin isolated from Trianthema decandra L. Ind Crops Prod 51:107–115. https://doi.org/10.1016/j.indcrop.2013.08.055
107. Ganesh Kumar V, Dinesh Gokavarapu S, Rajeswari A, Stalin Dhas T, Karthick V, Kapadia Z, Shrestha T, Barathy IA, Roy A, Sinha S (2011) Facile green synthesis of gold nanoparticles using leaf extract of antidiabetic potent Cassia auriculata. Colloids Surf B Biointerfaces 87:159–163. https://doi.org/10.1016/j.colsurfb.2011.05.016
108. Arunkumar P, Thanalakshmi M, Kumar P, Premkumar K (2013) Micrococcus luteus mediated dual mode synthesis of gold nanoparticles: involvement of extracellular α-amylase and cell wall teichuronic acid. Colloids Surf B Biointerfaces 103:517–522. https://doi.org/10.1016/j.colsurfb.2012.10.051
109. Ganeshkumar M, Sathishkumar M, Ponrasu T, Dinesh MG, Suguna L (2013) Spontaneous ultra fast synthesis of gold nanoparticles using Punica granatum for cancer targeted drug

delivery. Colloids Surf B Biointerfaces 106:208–216. https://doi.org/10.1016/j.colsurfb.2013.01.035
110. Manivasagan P, Venkatesan J, Kang KH, Sivakumar K, Park SJ, Kim SK (2015) Production of α-amylase for the biosynthesis of gold nanoparticles using Streptomyces sp. MBRC-82. Int J Biol Macromol 72:71–78. https://doi.org/10.1016/j.ijbiomac.2014.07.045
111. Gupta VK, Atar N, Yola ML, Darcan C, Idil Ö, Üstündağ Z, Suhas (2013) Biosynthesis of silver nanoparticles using chitosan immobilized Bacillus cereus: nanocatalytic studies. J Mol Liq 188:81–88. https://doi.org/10.1016/j.molliq.2013.09.021
112. Vetchinkina EP, Loshchinina EA, Burov AM, Dykman LA, Nikitina VE (2014) Enzymatic formation of gold nanoparticles by submerged culture of the basidiomycete Lentinus edodes. J Biotechnol 182–183:37–45. https://doi.org/10.1016/j.jbiotec.2014.04.018
113. Yilmaz M, Turkdemir H, Kilic MA, Bayram E, Cicek A, Mete A, Ulug B (2011) Biosynthesis of silver nanoparticles using leaves of Stevia rebaudiana. Mater Chem Phys 130:1195–1202. https://doi.org/10.1016/j.matchemphys.2011.08.068
114. Krishnaswamy K, Vali H, Orsat V (2014) Value-adding to grape waste: Green synthesis of gold nanoparticles. J Food Eng 142:210–220. https://doi.org/10.1016/j.jfoodeng.2014.06.014
115. Basavegowda N, Idhayadhulla A, Lee YR (2014) Phyto-synthesis of gold nanoparticles using fruit extract of Hovenia dulcis and their biological activities. Ind Crops Prod 52:745–751. https://doi.org/10.1016/j.indcrop.2013.12.006
116. Siva Kumar K, Kumar G, Prokhorov E, Luna-Bárcenas G, Buitron G, Khanna VG, Sanchez IC (2014) Exploitation of anaerobic enriched mixed bacteria (AEMB) for the silver and gold nanoparticles synthesis. Colloids Surf A Physicochem Eng Asp 462:264–270. https://doi.org/10.1016/j.colsurfa.2014.09.021
117. Wang Y, He X, Wang K, Zhang X, Tan W (2009) Barbated Skullcup herb extract-mediated biosynthesis of gold nanoparticles and its primary application in electrochemistry. Colloids Surf B Biointerfaces 73:75–79. https://doi.org/10.1016/j.colsurfb.2009.04.027
118. Khaleel Basha S, Govindaraju K, Manikandan R, Ahn JS, Bae EY, Singaravelu G (2010) Phytochemical mediated gold nanoparticles and their PTP 1B inhibitory activity. Colloids Surf B Biointerfaces 75:405–409. https://doi.org/10.1016/j.colsurfb.2009.09.008
119. Krishnaraj C, Jagan EG, Rajasekar S, Selvakumar P, Kalaichelvan PT, Mohan N (2010) Synthesis of silver nanoparticles using Acalypha indica leaf extracts and its antibacterial activity against water borne pathogens. Colloids Surfaces B Biointerfaces 76:50–56. https://doi.org/10.1016/j.colsurfb.2009.10.008
120. Kasthuri J, Veerapandian S, Rajendiran N (2009) Biological synthesis of silver and gold nanoparticles using apiin as reducing agent. Colloids Surfaces B Biointerfaces 68:55–60. https://doi.org/10.1016/j.colsurfb.2008.09.021
121. Shankar SS, Rai A, Ahmad A, Sastry M (2004) Rapid synthesis of Au, Ag, and bimetallic Au core-Ag shell nanoparticles using Neem (Azadirachta indica) leaf broth. J Colloid Interface Sci 275:496–502. https://doi.org/10.1016/j.jcis.2004.03.003
122. Yang X, Li Q, Wang H, Huang J, Lin L, Wang W, Sun D, Su Y, Opiyo JB, Hong L, Wang Y, He N, Jia L (2010) Green synthesis of palladium nanoparticles using broth of Cinnamomum camphora leaf. J Nanoparticle Res 12:1589–1598. https://doi.org/10.1007/s11051-009-9675-1
123. Martins M, Mourato C, Sanches S, Noronha JP, Crespo MTB, Pereira IAC (2017) Biogenic platinum and palladium nanoparticles as new catalysts for the removal of pharmaceutical compounds. Water Res 108:160–168. https://doi.org/10.1016/j.watres.2016.10.071
124. Gaidhani SV, Yeshvekar RK, Shedbalkar UU, Bellare JH, Chopade BA (2014) Bio-reduction of hexachloroplatinic acid to platinum nanoparticles employing Acinetobacter calcoaceticus. Process Biochem 49:2313–2319. https://doi.org/10.1016/j.procbio.2014.10.002
125. Riddin TL, Gericke M, Whiteley CG (2006) Analysis of the inter- and extracellular formation of platinum nanoparticles by Fusarium oxysporum f. sp. lycopersici using response surface methodology. Nanotechnology 17:3482–3489. https://doi.org/10.1088/0957-4484/17/14/021
126. Castro-Longoria E, Moreno-Velásquez SD, Vilchis-Nestor AR, Arenas-Berumen E, Avalos-Borja M (2012) Production of platinum nanoparticles and nanoaggregates using Neurospora crassa. J Microbiol Biotechnol 22:1000–1004. https://doi.org/10.4014/jmb.1110.10085

127. Dobrucka R (2016) Synthesis and structural characteristic of platinum nanoparticles using herbal Bidens Tripartitus extract. J Inorg Organomet Polym Mater 26:219–225. https://doi.org/10.1007/s10904-015-0305-3
128. Sheny DS, Philip D, Mathew J (2013) Synthesis of platinum nanoparticles using dried anacardium occidentale leaf and its catalytic and thermal applications. Spectrochim Acta Part A Mol Biomol Spectrosc 114:267–271. https://doi.org/10.1016/j.saa.2013.05.028
129. Al-Radadi NS (2019) Green synthesis of platinum nanoparticles using Saudi's dates extract and their usage on the cancer cell treatment. Arab J Chem 12:330–349. https://doi.org/10.1016/j.arabjc.2018.05.008
130. Gaikwad DS, Undale KA, Kalel RA, Patil DB (2019) Acacia concinna pods: a natural and new bioreductant for palladium nanoparticles and its application to Suzuki-Miyaura coupling. J Iran Chem Soc 16:2135–2141. https://doi.org/10.1007/s13738-019-01682-7
131. Whitesides GM (2006) The origins and the future of microfluidics. Nature 442:368–373
132. Günther A, Khan SA, Thalmann M, Trachsel F, Jensen KF (2004) Transport and reaction in microscale segmented gas-liquid flow. Lab Chip 4:278–286. https://doi.org/10.1039/b403982c
133. Zhang X, Wiles C, Painter SL, Watts P, Haswell SJ (2006) Microreactors as tools for chemical research. Chim Oggi 24:43–45. https://doi.org/10.1007/978-3-642-56763-6_39
134. Casadevall I, Solvas X, Demello A (2011) Droplet microfluidics: recent developments and future applications. Chem Commun 47:1936–1942. https://doi.org/10.1039/c0cc02474k
135. Song Y, Cheng D, Zhao L (2018) Microfluidics fundamentals, devices, and applications
136. Ma J, Lee SMY, Yi C, Li CW (2017) Controllable synthesis of functional nanoparticles by microfluidic platforms for biomedical applications—a review. Lab Chip 17:209–226. https://doi.org/10.1039/C6LC01049K
137. Hao N, Nie Y, Zhang JXJ (2018) Microfluidic synthesis of functional inorganic micro-/nanoparticles and applications in biomedical engineering. Int Mater Rev 63:461–487. https://doi.org/10.1080/09506608.2018.1434452
138. Illath K, Narasimahan AK, Nagai M, Wankhar S, Santra TS (2020) Microfluidic based metallic nanoparticle synthesis and applications. In: Bio-MEMS and Bio-NEMS: devices and applications. Jenny Stanford Publishers
139. Santra TS (2020) Bio-MEMS and Bio-NEMS: devices and applications. Jenny Stanford Publisher, Singapore
140. Mohanty US (2011) Electrodeposition: a versatile and inexpensive tool for the synthesis of nanoparticles, nanorods, nanowires, and nanoclusters of metals. J Appl Electrochem 41:257–270. https://doi.org/10.1007/s10800-010-0234-3
141. Wang B, Zhuang X, Deng W, Cheng B (2010) Microwave-assisted synthesis of silver nanoparticles in alkalic carboxymethyl chitosan solution. Engineering 02:387–390. https://doi.org/10.4236/eng.2010.25050
142. Xu Y, Musumeci V, Aymonier C (2019) Chemistry in supercritical fluids for the synthesis of metal nanomaterials. React Chem Eng 4:2030–2054. https://doi.org/10.1039/c9re00290a

Size and Shape Selective Metal Oxide Nanomaterials: Preparation, Characterization and Prospective Biomedical Applications

Ananth Antony and Jin-Hyo Boo

Abstract Structure dependent (size and shape) properties are fundamental to nanoscience and technology. Metal oxide nanomaterials (MONMs) are considered suitable for a variety of chemical and biological applications due to their stability. Preparation of oxide NMs does not require energy-intensive methods to protect their surface nature as compared to metallic counterparts. Size and shape are controlled by alternations in the reaction parameters such as temperature, concentration, method of preparation, and utilization of small quantities of surfactant materials. Normal wet chemical approach is extremely cheaper, hydrothermal methods are excellent for preparing uniform shape NMs, and recent atmospheric pressure plasma assisted approaches are powerful for obtaining one-dimensional structures which are usually difficult by following other techniques. The MONMs are considered an excellent system for anti-bacterial and biomedical applications. This chapter deals with definitions of NMs, suggested growth and nucleation theories, different preparation methods for obtaining powder MONMs, characterization techniques, and MONMs usage in different biomedical applications. In addition, mechanism of interaction between NMs and internal structures of micro-organisms and various institutes working on the NMs standardization are discussed.

Keywords Metal oxide · Biomedical · Nanomaterials · Mechanism · Size and shape

---

A. Antony (✉) · J.-H. Boo
Department of Chemistry, Sungkyunkwan University, Suwon 16419, Republic of Korea
e-mail: sebastiananath@gmail.com

J.-H. Boo
e-mail: jhboo@skku.edu

T. S. Santra and L. Mohan (eds.), *Nanomaterials and Their Biomedical Applications*, Springer Series in Biomaterials Science and Engineering 16,
https://doi.org/10.1007/978-981-33-6252-9_3

1 Nanomaterials

Nanomaterials (NMs) are defined based on their size which is one-billionth of a meter (10^{-9} m) in any one of its dimensions. The joint research centre of the European Commission defines nanomaterials as "materials with any external dimension in the nanoscale or having an internal structure or surface structure in the nanoscale" [1]. In such an extremely small size, materials exhibit unique and spectacular performance due to an increment in surface to volume ratio. The NMs exhibit different physicochemical properties as compared to their bulk counterpart. Those are changes in the melting points, unique optical properties, increases in the specific surface area, improved mechanical strength, and magnetization properties. Generally, NMs are classified as zero-dimensional (e.g. nanoparticles), one dimensional (nanotubes, nanorods and nanowires), two-dimensional (nanofilms, nanolayers and coatings) and three dimensional (bundles of nanowires and multi-nanolayers) materials. They can exhibit amorphous or crystalline, single or polycrystalline nature [2]. Among different materials such as metallic, polymers and ceramics, metal oxide NMs are considered for many applications on account of their good stability in normal atmospheric working conditions. The chemical and physical properties of NMs are highly influenced by size and shape [3]. The optical properties are heavily dependent on size (shift in the surface plasmon resonance by increasing particle size) and catalytic or biomedical applications are generally influenced by shape [4].

2 Metal Oxide Nanomaterials (MONMs)

The transition metals naturally show affinity towards oxide compounds and result in metal oxides. When considering only about the material, distribution of electron cloud in the metal oxide valence shell determines the physical and chemical properties. Metal-oxide interactions depend on three factors such as (i) the band structure, base-acid characteristics, cationic nature (these determines the main physicochemical properties), (ii) the morphological characteristics coming out of kinks and exposed faces and (iii) surface defects such as ion vacancies and trapped electrons [5].

Particle size in MONMs influences three important basic properties. First one is related to the lattice symmetry and cell parameters. Particle size decrement increases surface free energy and stress which in turn affects mechanical or structural stability. Thus materials should exhibit low surface free energy (intrinsically and extrinsically) at the end of production (by suitable synthesis method and calcination temperature). Second one deals with electronic properties such as quantum confinement effect and the absence or limited Madelung field. The MONMs' size can affect the degree of ionicity or covalency in the metal-oxygen bond. The final one is directly related to size of NMs itself. The conductivity and chemical reactivity are strongly related to the bandgap which can be tuned by size control [6]. The surface morphology determines the degree of polarization and localization of light in different planes. The shape and crystalline properties are related to each other [7] and the MONMs differ from metals in terms of their intrinsic charge separation capacity.

In terms of biological systems, particle size, shape and surface area play vital roles for efficient interaction. If the NMs' size is <50 nm, then it is expected to transverse to all tissues without hindrance and delivers designated functions [8]. In addition, NMs' shape controls electron delocalization leading to enhanced electric fields which act as a hot spot thus bioactivity, catalytic and sensing efficiency are highly shape-dependent. For biological applications, the NMs with its surrounding medium should be studied. The colloidal behavior (net potential energy) was predicted by two accepted models. The Derjaguin-Landau-Verwey-Overbeek (DLVO) and the extended DLVO theories explain the aggregation or dispersion behaviors of NMs based on the van der Waal's force and electrostatic double layer (EDL). The NMs should exhibit the zeta potential (a measure of EDL) values $> \pm 30$ mV which is an indication of stability against coagulation in solution. This value is strongly dependent on the ionic strength (total ions present in the solution) of NMs and can be altered by changing the acidity or basicity of the colloid. At low ionic strength, the NMs exhibit good dispersion which is suggested for experiments [9].

3 Physicochemical Properties of Size and Shape

The spherical NMs (nanospheres) exhibit good thermodynamic stability and their optical properties are dependent on the diameter. Increment in size leads to formation of more valence and conduction band due to contribution of more atomic orbitals thus result in a decrease of electronic bandgap. Elongated structures such as rods, wires and tubes show electron delocalization in the lateral dimension thus their properties depend on the aspect ratio (r = length/width). Higher dimensions such as nanocubes and nanohexagons exhibit accumulation of dipoles due to multiple corners. The facet shows more number of edge and corner atoms which leads to different activities as compared to other shapes [7]. In biological application viewpoint, size and shape-dependent properties decide (i) the cell surface adsorption and transport across cell membranes, (ii) the light-induced redox activities and (iii) release of metal ions by dissolution [10]. Also, release of reactive nitrogen species (RNS) was high for NMs exhibiting small size [9]. Similarly, NMs shape was strongly corroborated with biological system since the aggregation was mainly determined by shape.

4 Preparation of Metal Oxide NMs

Nanomaterials powder can be prepared using different synthesis routes. Depending on the preparation method, size and shape-selective NMs are obtained. Several easy and cost-effective techniques such as wet chemical, green synthesis, hydrothermal and plasma-assisted methods with possible growth mechanisms are detailed in the next section. Every method shows its advantages and limitations based on bulk preparation, morphology tailoring, chemical purity and so on. For example, wet chemical and hydrothermal methods are convenient both in laboratory and industrial scales. Ease of handling, controlling reaction parameters and proposing the growth mechanisms are possible. The limitations of these methods can be related to the long time synthesis. Green methods (plant and micro-organisms based) rely mainly on the corresponding extracts for NMs synthesis thus it is environmentally benign. But lack of opportunity to tune size and morphology, surface chemistry and finding out the mechanism of NMs formation limits its widespread usage. Plasma assisted methods are excellent means to control surface morphology [by alternations in the plasma species or by electrical forces (power)] for the NMs which are not easily controlled structurally by other methods. At the same time bulk preparation is the limitation. This shortcoming can be addressed by using many arrays of plasma sources.

4.1 Wet Chemical Approach

Wet chemical method is a widely used and simple laboratory route to prepare NMs. The target NMs precursor solutions and an alkali solution such as NaOH, KOH, NH_3OH, etc. are prepared and reacted at room or higher temperature. The stirring speed, concentration of reactants, temperature and duration of reaction are considered as important parameters. Wet chemical approaches come under bottom-up approach in nanotechnology which enables good control over size, composition, and growth by tuning the kinetic and thermodynamic parameters. Nucleation and growth kinetics decides the size and shape of the NMs. In such cases, application of surfactants or stabilizer molecules (such as polymers or mild acids) are used to induce specific shapes (see Table 1 for some examples).

4.1.1 General Theory in Wet Chemistry

The theory of nucleation and growth was previously described by several models in colloidal synthesis. Those are LaMer burst nucleation, Ostwald ripening, Lifshitz-Slyozov-Wagner (LSW), and Watzky and Finke. Taken from the above models, a review by N.T.K. Thanh et al. [17] shortly describe that the surface free energy (γ) and the bulk free energy (ΔG_v) of the nanoparticle defines its total free energy (ΔG) in which the (ΔG_v) depends on the temperature (T), supersaturation of the solution (S) and the Boltzmann constant (k_B) [17]. The surface free energy calculation is done using indirect methods (such as contact angle measurement or light scattering methods [18]).

$$\Delta G = 4\pi r^2 + \frac{4}{3}\pi r^3 \Delta G_v \quad (1)$$

Table 1 Some examples of MONMs prepared using wet chemical approach with reaction parameters

No.	Metal oxide	Precursors	Methods	Observed shape	Ref.
1.	CuO	$CuNO_3$ + NaOH	Wet chemical and hydrothermal, PEG surfactant	Nano-needle, spherical, and sheet-like	[11]
2.	CuO	$CuNO_3$ + NaOH	Room temp. and cinnamic acid	Nano-rod, wires and belts	[12]
3.	ZnO	$ZnNO_3$ + NH_3	Room temp., cyclohexyl amine and KCN	Polyhedral	[13]
4.	ZnO	$ZnNO_3$ + 2-aminoethanol + NH_3	60 and 94 °C, ultra-sonication	Hexagonal	[14]
5.	Fe_2O_3	$FeCl_3$ + sodium oleate + ethanol + hexane	Room temp., and dry at 105 °C	Cubic, octahedral	[15]
6.	CeO_x	$Ce(NO)_3 \cdot 6H_2O$ + Glycine + $NaBH_4$	Chemical combustion	Spherical	[16]
7.	RuO_x	$RuCl_3$ + NaOH	75 °C	Spherical and rods	[4]

$$\Delta G_v = \frac{-\ln(S)}{v} k_B T \quad (2)$$

The nucleated seed should attain a critical radius (r_c) in order to avoid re-dissolution and should exhibit critical free energy to attain stability in solution.

$$r_c = \frac{-2\gamma}{\Delta G_v} \quad (3)$$

$$r_c = \frac{-2\gamma v}{-k_B T \ln(S)} \quad (4)$$

The surface free energy (γ) can be modified using surfactants thus different shape NMs are prepared by selecting suitable surfactants. Further growth of nanoparticles is strongly dependent on the surface reaction and monomer diffusion toward the surface. The monomers are considered to move or roll and attach or detach on to the seed surface. The seed or core controls the symmetric growth to certain extent and in latter stages, diffusional fluxes lead diffusing atoms to attach on the mostly curved surface in a non-equilibrium manner. It has been considered that initial seed formation is in a spherical shape and the experimental process determines the final shape of the NMs [19]. It should be noted that the above mechanism is applied to the simplest and generalized cases. At the same time, explicit theoretical nucleation and growth mechanism dealing with metal oxides need further developments in the above theory.

4.2 Green Methods-Bacteria-Mediated and Plant Extracts

In green methods, MONMs are prepared with reactions from the bacterial supernatant solution. The bacterial supernatant contains variety of energetic sources such as electrons, radicals and chemical molecules. The precursor for the required MONMs are made to react with such biological moieties and they are reduced to metallic or MONMs (in the presence of oxygen or hydroxyls) which are formed after certain incubation time. These are generally based on enzymatic reductions. For example, probable mechanism for the preparation of metallic nanoparticles using fungi extracts is due to an enzymatic reduction (reductase). It has been shown that two key aspects in *B. licheniformis* bacteria mediated biosynthesis of silver nanoparticles are based on NADH (nicotinamide adenine dinucleotide) and NADH-dependent nitrate reductase [20]. The intracellular formation of NMs by microorganisms are due to genetic and proteomic responses.

For synthesizing metal/metal oxide NMs using plant extr acts [21–23], organic molecules and phytochemicals such as ketones, flavones, aldehydes, amides, carboxylic acids, terpenoids, phenols, sugars, and ascorbic acid are believed to play vital roles in reducing metal salts into metal or metal oxide nanoparticles. For example, flavonoids show different functional groups that release reactive hydrogen atom due to tautomeric transformations from which enol-form is transformed into keto-form. This process is achieved by the reduction of metal ions into metal nanomaterials [24] using sweet basil (*Ocimum basilicum*) extracts for preparing silver NMs. The functional groups (for example, –C–O–C–, –C–O–, –C=C–, and –C=O–) act as capping ligands and stabilize the NMs during growth and control further aggregation. Alcoholic compounds such as hydroquinone and quinol act as reducing agents. At the same time, exact mechanisms for MONMs preparation using plant extracts is not fully understood which can be conceived as the presence of oxygen-containing compounds would facilitate the formation. It is stated that there are three steps of nanoparticle synthesis from plant extracts; (i) activation phase or nucleation process involving bio-reduction of metal ions/salts, (ii) growth phase as guided by the Ostwald ripening, and (iii) termination phase which defines the final shape of NMs [24].

4.3 Hydrothermal Method of Synthesis

Hydrothermal methods of MONMs synthesis are considered effective to prepare high uniformity bulk synthesis with an oriented growth. This method relies based on the chemical reactions and changes in solubility of a substance inside a sealed chamber (the heated solution kept above ambient temperature and pressure). It is perfectly suitable for preparing one and two-dimensional nanostructured powder and films (on a substrate). The synthetic strategy of hierarchical nanoarrays can be achieved by single-step self-template growth and multistep-graded growth methods. A multi-step-graded growth method is done by making secondary structure on an already obtained primary nanostructure and it is repeated many times. Applying changes in the reaction condition during secondary process, the size and shape can be controlled effectively [25]. On account of the variations in equilibrium with temperature, formed particles (at lower temperatures) re-dissolve and re-crystallize (at higher temperatures). During heating up or changing the aging period at a constant temperature the shape control is achieved. The dielectric constant of water is considered important for controlling factors of reaction equilibrium, rate of reaction and solubility. The chemical potential of ions is lowered by the usage of high dielectric constant medium, and dissociation of electrolytes produces reasonable stability [26].

The effects of temperature, concentration of reducing agents such as NaOH, reaction duration and addition of seed particles are known to control the final product formation such as flower or other particle shapes [27, 28]. The coordination numbers play a vital role in obtaining specific structures in the same compound. For example, in preparation of HfO_2, structures that exhibit coordination number 8 have shown excellent geometric symmetry and it forms easily. This property is also important in deciding room temperature stability. In normal temperature, the monoclinic-HfO_2 is very stable as compared to tetragonal crystals [29]. The thermodynamics and kinetics govern the reaction mechanism. If the kinetic stability is good, then it is expected to achieve better symmetry at higher alkaline condition. The concentration of alkalinity has been shown to decide the crystal structure (such as monoclinic or tetragonal) due to increase in the oxygen ions required for perfect coordination and symmetry. In terms of nucleation, overall Gibbs free energy change (ΔG) due to the changes from original particles and to the cluster formation (by new surface) is maximum at the critical size, r^*. This is actually the maximum free energy change required for nucleation. An excess or changes in this energy is further required for growth and achieving stability [29].

4.4 Plasma Assisted Methods of Synthesis

Plasma is a collection of freely moving charged particles consisting of electrons, ions, radicals and neutral species. The low temperature, non-equilibrium plasmas provides many advantages for synthesizing size and shape specific MONMs at fast bulk production mode. In this process, the reaction is carried out at room temperature, and no polymer surfactants or solution agitation are needed. There are different plasma methods such as dielectric barrier discharge (DBD), soft jet (APPJ), micro plasma and corona discharge can be used. The NMs precursor solution is mixed and exposed to the mentioned plasma source for an hour or more. The incoming plasma species create turbulence and the MONMs nucleate and grow in the presence of electric field established by plasma thus specific shapes can be achieved. For

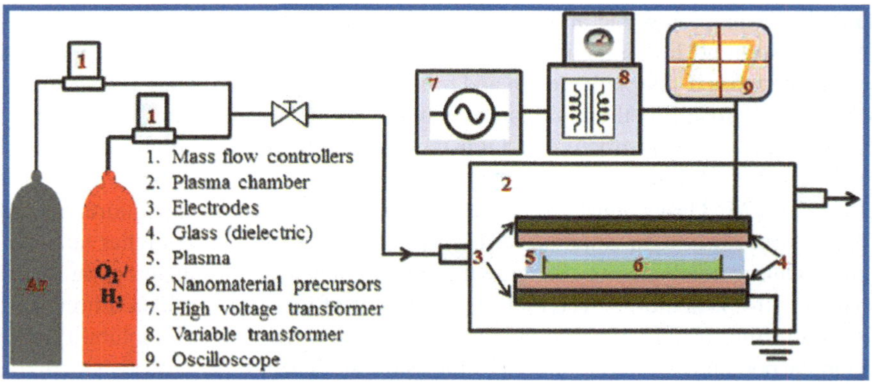

Fig. 1 The MONMs synthesis using DBD plasma reactor. The ruthenium and NaOH solution are exposed to plasma for an hour and nanorod type RuO_2 is obtained [31] (*Source* Reproduced with permission from IOP Publishing)

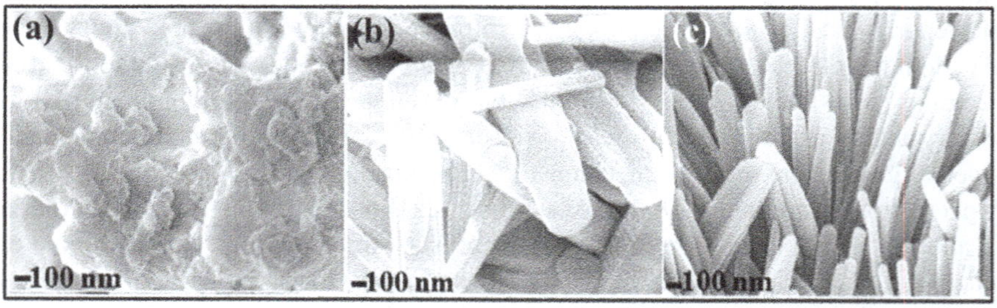

Fig. 2 The scanning electron microscopic images of RuO_2 prepared with DBD plasma assistance shows different surface morphological features such as aggregated (**a**), nano-pillars (**b**) and nanorods [31] by changing the plasma feed gas such as Ar + H_2, Ar + O_2 and Ar, respectively (*Source* Reproduced with permission from IOP Publishing)

example, depending on the concentration of the Zn and Ru reactants, the formed ZnO and RuO_2 exhibited different shapes such as nanorods, flower, nanolayers, etc. [30–32]. The schematics of preparation of MONMs using DBD and the observed surface morphology are given below (Figs. 1 and 2). At the same time, the feed gases used for plasma formation would be a deciding factor in terms of the product chemistry (i.e. oxidation state).

5 Surface Modification and Functionalization

When considering surface properties for biomedical applications, two kinds of stabilization processes should be considered. One is related to the stabilization of

MONMs itself with polymeric materials (MONMs/polymer) and the other one is the (MONMs/polymer) with biomolecules.

The prominent forces that reign NMs, in general, are electrostatic (double layer), polymeric (steric) and van der Waals interaction. In colloidal state, the NMs should exhibit small attractive forces to prevent thermal motion breaking process thus it can exhibit good stabilization. It is achieved by creating an electric double layer at the interface (solid-liquid) by the addition of ions or changing pH of the solution. If it is not sufficient then through by an addition/functionalization of polymer or surfactant stabilization, it is done. When polymers are introduced, its segments or chains enlarge and forms a protective adsorbed layer on the NMs. This is called polymer induced steric stabilization. The van der Waals force refers to (MONMs/polymer) with biomolecule interaction. This force arises due to the fluctuating or permanent diploe moments between biomolecule and MONMs in close interaction. This value of van der Waals force depends on the separation distance as explained by Hamaker and used in several review articles [33–35].

$$V_v\,dW = -\frac{Ar}{12s} \tag{5}$$

where A is the Hamaker constant (in the range of $1^{-10} \times 10^{-20}$ J for metal oxides in water); r—radius of two spheres (NMs) of interacting pair; s—separation distance.

By employing suitable surface modifications, the MONMs can be made to interact with biomolecules in a specific manner. The interaction between the MONMs and biomolecules would be a weak van der Waals or hydrogen bonding or strong covalent or ionic bonding. The presence of oxygen, hydrogen and hydrocarbons forming on metal oxide surface dissociate biomolecules through catalytic reactions. The protein, carbohydrates, cell membranes and other internal organelles are responsive to specific ions, radicals and chemical functionalities released either from MONMs or polymer-functionalized MONMs. For example, proteins denature easily when they interact with OH- from water molecules. Also the cells may induce inflammatory response when contacting with MONMs if the NMs surfaces are not functionalized or prepared as tissue-like material [36]. The polymers can provide charged, neutral or amphiphilic nature to NMs thus it can facilitate suitable cellular uptake. Some prominent polymers used for surface functionalization are polyethyleneglycol, polyvinylalcohol, polyacrylic acid, dextran, chitosan and its derivatives [19].

The surface charge due to functionalization is the result of partially saturated bonds present on the surface and partially filled metal atom's d-orbital and p-orbitals of oxygen atom [37]. This property is responsible for generation of reactive oxygen species which is important for producing toxicity (cyto and geno) to microorganisms.

6 Several Characterization Methods of MONMs in Biomedical Application Context

Metal oxide NMs does not produce any surface plasmon resonance (SPR) thus application of UV-visible spectroscopy is not much useful to identify size. The size exclusion chromatography is performed during the synthesis process at a periodic interval to study size and growth. After synthesis, other well-known methods such as scanning electron microscopy, transmission electron microscopy, X-ray diffraction and chemical analysis such as X-ray photoelectron spectroscopy etc. can be performed. Some specific methods routinely used with respect to MONMs for biomedical applications are detailed below.

6.1 Size Exclusion Chromatography (SEC)

This technique is used to analyze growth, size and shapes of MONMs during synthesis stage and to separate it. The SEC is a process working based on an entropy control. There is no interaction that occurs between the stationary phase and the analytes thus separation is done exclusively from the pores of the stationary phase. The size-specific separation is achieved based on the hydrodynamic volume of the nanoparticles. At the same time, the shape-specific nanoparticles are separated based on the combined effect of size-exclusion and adsorption. In both cases, large molecules elute initially as compared to the small nanoparticles through the column packing material [38]. For practical application, the elution time should be reproducible since it is an indication of the nanomaterial size with minimal standard deviation (Fig. 3).

6.2 Hydrophilicity or Hydrophobicity of Nanomaterials Surface

Hydrophilicity or hydrophobicity is a measurement to determine the interaction of NMs with bio-systems. For example, NMs' interaction with biological membranes, cellular uptake, immune response, protein adsorption and hemolytic effects are strongly dependent on the hydrophobic nature. In the case of NMs powder such measurement was considered impossible through direct water contact angle measurement (which is conventionally used for thin films and polymers) but some indirect methods are available recently. By measuring the surface energy through possibly four implicit methods it is performed. Those are (a) AFM based adhesion force measurement between the functionalized (hydrophilic/hydrophobic) tips and immobilized NPs on silicon surface [40], (b) the NMs surface adsorption assays based on rose bengal (RB) dye. In this method, amount of RB dye adsorbed on to the NMs is taken as a measure of hydrophobicity. The binding constant and the maximum amount

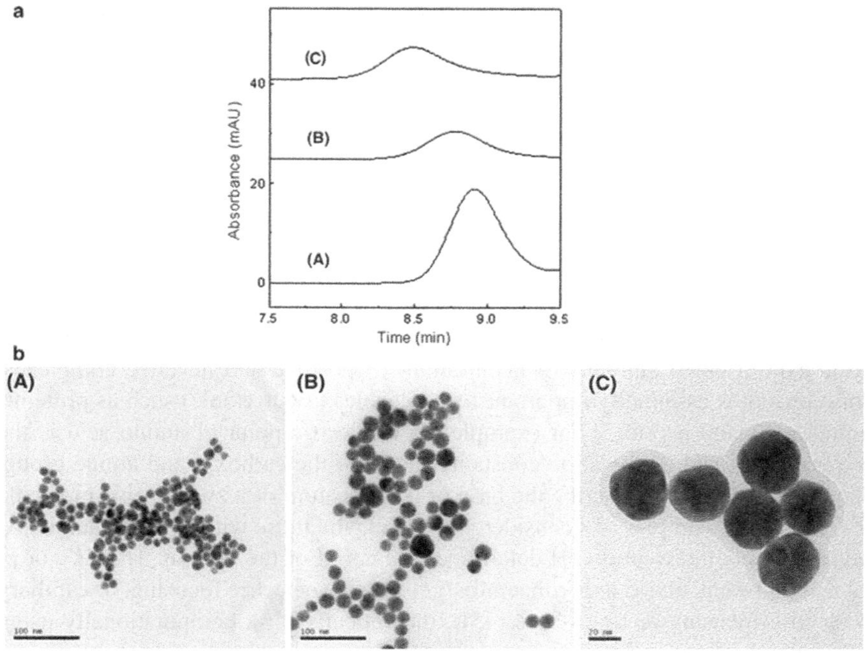

Fig. 3 Result of **a** chromatograms of Au nanoparticles exhibiting three different sizes such as 13.1, 21.3 and 40.1 nm (A–C) and **b** the corresponding TEM images of them, respectively [39] (*Source* Reproduced with permission from Springer Nature)

bound is calculated from Scatchard plots according to the following equation [41, 42]:

$$\frac{\mathrm{r}}{\mathrm{a}} = \mathrm{K} \cdot \mathrm{N} - \mathrm{K} \cdot \mathrm{r} \quad (6)$$

where, r-amount of RB adsorbed per mg of nanoparticles (μg/mg);

a - equilibrium concentration of rose bengal (μg/mL);
K - binding constant (mL/μg); and
N - maximum amount bound (μg/mg),

(c) Hydrophobic interaction chromatography and (d) direct quantification using different collectors [43, 44]. By using different collectors, nature of NMs is calculated by analyzing its adsorption kinetics (NMs binding affinity with respect to time) towards pre-calibrated surface free energy collectors (observed for different molecules or samples) in a dark field microscopy and comparing the result with extended DLVO theory. According to this theory, the total interaction energy G^{tot} between a flat surface and NMs can be represented as below:

$$G^{tot} = G^{el} + G^{AB} + G^{LW} \tag{7}$$

where G^{el}, G^{AB} and G^{LW} refers to the electrostatic, acid-base and Lifshitz-Van der Waals interactions, respectively.

6.3 Isoelectric Point for Biomolecules and MONMs

The isoelectric point (pI) or zero-point charge is referred to as the pH value at which a molecule can carry no electrical charge. At this point, the immersed solid oxide exhibit equal amounts (concentration) of positive and negative complexes. This concept is essentially important for molecules (zwitterionic) such as proteins, amino acids and peptides. For example, an isoelectric point of amino acid is the average pK_a (acid dissociation constant) values of the carboxyl and amine group. The pI can be used to describe the base or acidic nature of a zwitterionic molecule or compounds with pI > 7 is considered as basic, and those with pI < 7 is acidic. At the same time, the solution pH determines the net pI of the protein. The pK_a or pI values can be calculated experimentally (gel-based isoelectric focusing, or capillary isoelectric focusing or test assays [45]), theoretically [46], computationally using artificial neural networks, support vector machines or web based isoelectric point calculator programs [47]. For the MONMs the pI can be calculated using the zeta potential studies (Sect. 6.4). At the same time, at pI the NMs can easily aggregate due to the net-zero charge thus by changing the pH of the solution it can further be used for biomedical studies. For example, Fe_2O_3 exhibits pI at the pH 6 [48]. The value of pI for many oxides, hydroxides and complex systems has been listed by G.A. Parks [49]. The pI values are different for the same MONMs when it is in compound and dissolved state. For example, Fe_2O_3 shows pI values of 6.7 and 8.6 when they are in the powder and in dispersion state. When exposed to water, MONMs can attain hydration due to hydrogen bonding with surface oxide, formation of hydrate due to chemical sorption occurs and then convert to oxyhydroxide. Similarly, in the organic environment such as in biosystems, the MONMs can exhibit changes as follows [49]. (1) dissociation of surface functional groups such as –COOH, (2) specific adsorption of ionic molecules onto neutral metal oxide sites (natural tendencies toward formation of ion-pair and hydrogen bonding), (3) polar molecules such as waters' oriented adsorption with subsequent adsorption of ions and (4) weak adsorption of ions due to induced polarization of the underlying substrate [49], all of which affect pI.

6.4 Zeta Potential Studies

The zeta potential study is particularly useful for analyzing the net charges due to MONMs in solution. Similar to pI which is routinely used for protein zero charges at particular pH, zeta potential shows the NMs's surface charge which is a measure of stability against aggregation. In aqueous environment, the surface functional groups may dissociate or the ions may adsorb on to the NMs surface from solution (stern layer). Such adsorption promotes formation of a layer made of counter ions which is called diffuse layer. The potential that exists between the layer boundary is called as zeta potential [50]. Analysis of zeta potential is very much useful for number of applications such as electro-kinetic transport of particles, biocompatibility study, characterization of polymers used in biomedical fields. Mikolajczyk et al. [51] have studied the zeta potential of several MONMs in different experimental conditions and have proposed quantitative structure-property relationship for predicting agglomeration or aggregation phenomena. The zeta potential is the result of surface ionization, lattice ion dissolution, and ion adsorption. It was mentioned that the NMs size affects the zeta potential values which are size and concentration-dependent as per the simplified molecular line-entry systems model [52].

6.5 Biocompatibility Study

The biocompatibility of NMs should be considered an important criterion to use it in biological systems. Many MONMS are known for its photocatalytic performance thus it can be considered for anti-bacterial applications. Also, when MONMs are used with organic chemicals it can impart additional activity such as hydrophobicity (which is a measure of bio-adhesion). The biocompatibility studies can be performed using an enzyme metabolic assay [53] in the required cell cultures. Thus the MONMs should exhibit suitable non-toxic effect to the test physiological condition and at the same time it should render anti-bacterial properties. The MONMs can be modified or surface-functionalized with other biologically active materials in order to protect against unwanted aggregation, non-specific interaction, and toxicity.

7 MONMs-from Synthesis to Product Formation

The sequence of steps involved in commercial product formation are given below as per the review article by Stavis et al. [54].

a. *Synthesizing materials aiming at specific or useful properties*: Control over synthesis, physicochemical parameters, identification of the process variations to reduce the heterogeneity and removal of chemical contaminants should be considered while synthesizing the MONMs.
b. *Stabilization*: This allows internal storage (against Ostwald ripening and other aggregative ingredients) and for transfer between other liquids (in ionic liquids and further centrifugation before dispersing in the application medium). Storage of MONMs against thermal influence involves lyophilization or external coatings (such as polymers) which can increase its stability. Stabilization to sustain the electrosteric repulsion can be maintained by utilization of polyethyleneglycol.

c. *Functionalization*: This is done for improving interfacial interactions such as by following click chemistry [55, 56], which provides potential high-efficiency quantitative functionalization with less energy consumption. The click chemistry follows synthesis of surface functionalization of MONMs using the pre-designed ligands whose structures contain the following features: (a) a strong anchor (for example, phosphonic acid or carboxylic acid) that can easily bind with MONMs surfaces, (b) surface functional groups (contain any number of carbon chains) that act as spacers or branches from the anchors located on the metal oxide surface, and (c) attachment of a functional perimeter (for example azide or alkyne groups) to the spacers to induce orthogonal functionality.
d. *Purification-to minimize heterogeneity*: This involves removal of excess ligands, contaminants, and other products formed through the steps (a–c) mentioned above. Precipitation and re-suspension, density-gradient ultracentrifugation, gel and size exclusion chromatography, and electrophoresis (such as agarose gel and free flow) and filtration (such as dia-filtration and nanofiber membranes) are routinely used purification methods.
e. *Characterization*: It gives information for regulation or quality control. This is generally done to obtain sufficient information about a product at a reasonable cost. The characterization techniques are vast and need to be chosen based on the specific application needs.
f. *Integration-ultimate use of MONMs in products*: The integration of MONMs into materials or devices requires common criteria for spatial control and product complexity. For instance, colloidal nanoparticles require higher electronic mobility in solution. Similarly, each biomedical application demands specific integration methods.

8 Risk Assessment

The risk assessment for any MONMs is performed in three steps [37]. Those are (a) assessment based on exposure which includes frequency, magnitude, and population duration, (b) assessment based on hazard such as NMs dose response in tissues, cells, and organs and (c) quantification of final risk due to likelihood of hazard and exposure. The in vitro and in vivo assay tests using bacterial cultures and animal studies would provide the cytotoxic mechanism, redox potentials of appropriate intercellular reactions, levels of cell oxidation etc.

The heavy exposure of MONMs leads to deposition and agglomeration in the parts of the body and further distributes across blood brain barrier, gut epithelium or skin. The ultra-fine particles are known to induce cardiovascular effects, fibrosis, lung cancer, neurodegenerative and teratogenic effects. For example, high administration of TiO_2 NMs in mice has caused morphological changes in the cerebral cortex neurons and changes in gene expression. The copper and iron oxides also have shown induced changes in the neurotransmitter secretion and accumulation of NMs in the brain. The MONMs containing products in suspended liquids or as aerosols cause severe consumer exposure thus lead to health hazards. Some examples of the effects caused due to the risk of MONMs conducted on the mice and cellular assays could be found in the article by Dekkers et al. and Simko et al. [57, 58].

9 Common Antibacterial Activity Assessment Methods

9.1 Plate count

The concentration of viable bacterial cells are estimated by counting colony-forming units (CFU) before and after exposure to the MONMs. It is performed through serial dilution of the original bacterial culture in lysogeny broth (LB) medium, taking 10 micro-litre of the serially diluted culture, subsequent spreading using sterile glass rods onto an agar plate in triplicate and finally counting the available colonies after incubation [59, 60].

9.2 Flow Cytometry

Heterogeneous mixtures of bacterial cells are placed in a fluid suspension and passed through a window chamber or a small compartment in which laser scattering from the bacterial cells gives detailed information (in a fluorescence microscope). The light signals emitted from the suspension particles are collected and interpreted into details such as cell morphology, nature of surface, and expressions such as intracellular protein, gene and cellular physiology. The population of dead or live bacterial cells is correlated by fluorescence variation from the control groups [61].

9.3 Zone of Inhibition

This test is called as Kirby-Bauer test which is a qualitative method for measuring antibiotic resistance of MONMs solids and textiles against microbial growth [62, 63]. In this method, approximately one million bacterial cells are spread over an agar plate and then it is incubated in the presence of the antimicrobial materials (MONMs) loaded oxacillin disk. If the microbial strain is susceptible to the MONMs, then a zone of inhibition will appear on the agar plate. If the strain is resistant to the said antimicrobial agent, then no zone will appear. The thickness of the zone is measured (using a ruler or vernier caliper) at the back of agar plate in the presence of good illumination.

10 The MONMs-Biomolecular Interactions

When the nanoparticles are suspended in a medium containing biological culture, it encounters with different biological interfaces such as lipids, DNA, proteins, polysaccharides, flavonoids, etc. Understanding such interactions at the nano-bio interfaces would help in preferring a suitable approach for safe utilization of NMs in pharmaceutical, biomedical and clinical industries. Inside biological culture medium, interactions exist between functional groups of MONMs and biomolecules such as phospholipid, lipopolysaccharide, protein, and lipoteichoic acid present over the bacterial envelop which contribute for interaction at the interface. The functional groups present in biomolecules enhance adhesion to different surfaces. Similarly, different kinds of non-specific interactions such as dipole-dipole, electrostatic, hydrophobic, hydrogen-bond, and van der Waals interactions play important roles in adhesion of bacteria on any MONMs [64].

The NMs (such as polymers, biomarkers, metals and drugs) that interact with biological systems can be understood (in three ways) from the research done by different research groups around the world and compiled in a review article by Zhang et al. [65]. Those are (a) protein binding (b) ligand-mediated interactions and (c) interaction due to intracellular processing. Such understanding may further extend to oxide nanomaterials logically it seems.

The biological makeup such as blood, plasmic fluid and secretions (made of proteins and other biochemicals) that interact with protein-coated NMs are mainly size-dependent. The hydrodynamic size may increase due to functionalization and lead to aggregation. Thus the final size of NMs becomes an important criterion to expect specific-binding. The bimolecular proteins (such as albumin, fibrinogen, and apoliproteins) are reported to bind easily with MONMs (such as iron oxide) and carbon nanotubes [65]. The plasma protein corona recognizes NMs or foreign bodies firstly and responds immunologically which depends on the surface properties and size. The high or low-affinity binding of proteins is determined by their hydrophobicity character. In addition, if the protein's isoelectric point (pI) is greater than 5.5 then it can easily bind with NMs [66]. Ligand-mediated interactions are particularly useful for spatial localization, placing of NMs exactly at the diseased tissues, and to remove adverse effects caused by off-target moieties. For this mode, the MONMs are modified using a surface ligand to facilitate preferential interaction to a diseased tissue or binding with it. The ligands are usually small molecules, antibodies, proteins, and aptamers. Intra-cellular processes are related to the safe delivery of drugs loaded NMs with ease intra-cellular trafficking. Appropriate engineering and design of carriers would deliver therapeutics to the specified organelles. For inside cell delivery, polymer-MONMs conjugate using covalent bonding of chemo drugs with hydrophilic polymers [65] can be utilized.

11 MONMs as Antibacterial Products and Mechanism of Action

There are varieties of metal and MONMs have shown antibacterial property [67, 68]. The main external influencing factors are the particle size, shape, crystal structure, MONMs concentration, surface roughness, surface charge, etc. Internally, the metal (especially Ag, Au, Ga and composites) nanoparticles show (a) attachment to sulfur-containing proteins (or phospholipid bilayers) found on the bacterial membrane and cause pits or damages thus leading to the cell-lysis, (b) attachment to cysteine residues of NADH dehydrogenase which further inactivate other enzymes thus preventing respiratory processes, (c) production of reactive oxygen species (ROS) and subsequent DNA damage, (d) binding into cytokines, (e) attachment leading to changes in the membrane potential and decrement in the ATP level and (f) influencing cell functions for example cell adhesion, differentiation and spreading, all of which responsible for antibacterial activity.

Also, for MONMs (such as ZnO, MgO, CuO, Fe_2O_3 and TiO_2), (a) formation of ROS due to photocatalysis in the presence of UV light, (b) membrane phosphate group binding and subsequent permeation in dark condition, (c) electrochemical interaction between MONMs and cell wall leading to disruption and leakage of metabolites, (d) oxidation of proteins and DNA by ROS were shown (Figs. 4 and 5). In addition, the most popular antibacterial mechanisms are (a) dissolution of metal ions from the MONMs and its oxidative damage, (b) release of ROS species and its membrane damage and (c) the direct contact mechanism which is about rupturing the cell body due to shape [11, 69].

It has also been reported that in addition to the above mechanism, there are many factors which are responsible for the antibacterial action by MONMs. Those are bacterial cell wall interactions (such as Gram-positive or negative), surface charge of NMs, type, and shape of NMs, dispersion of NMs, size, and release of ions [71].

12 Nanoproducts Regulatory Authorities Around the World

There are several regulatory agencies or institutions around the globe (the following information are just examples, but the agencies involved are many) to assess the safety, company promises, potential risks and uncertainty of nanotech-based products. The Korea food and drug safety (KFDS) with many umbrella institutions [72, 73], The USA's food and drug administration [74], environmental protection agency (EPA) and national nanotechnology initiative (NNI), European Commission (Joint Research Centre [75, 76], and REACH [77]), Nano mission-managed by department of science and technology of India and various central organizations such as CDRI, ICMR, NIPER and CFTRI [78], China's National Nanotechnology Standardization Technical Committee (NSTC) and the corresponding authorization agencies [79],

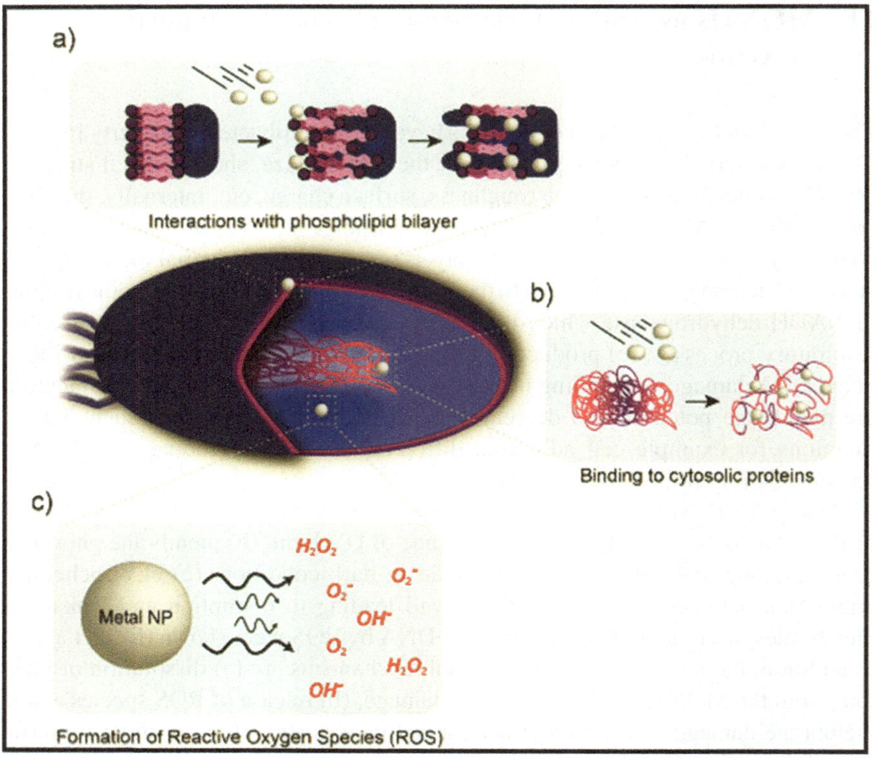

Fig. 4 Interaction of metal nanoparticles with prokaryotic cells. **a** The NMs destabilize the phospholipid bilayer and cause cell lysis, **b** binding of nanoparticles with DNA and leads to cell death, and **c** production of ROS which finally leading to oxidative stress and cell destruction (Gold et al.) [67] (*Source* Reproduced with permission from Wiley-VCH Verlag GmbH)

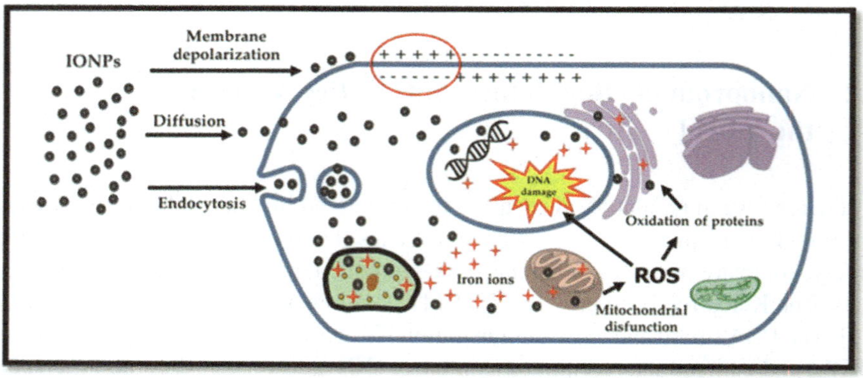

Fig. 5 The mechanism of antibacterial action due to iron oxide NMs as explained by L.S. Arias et al. in a review article [70]. Membrane depolarization, ROS generation, DNA damage and cell impairment were reported as responsible for antibacterial action (*Source* Reproduced with permission MDPI Publishing)

Therapeutic Goods Administration (TGA) in Australia [80] are some of the known organizations working for nanotechnology regulations in their country. Generally regulation on the food, biomedical and cosmetic products needs a pre-market notification before it is available to the commercial market. The common criteria for the assessment of nano-product include (a) the chemical identification of the MONMs, (b) amount of the materials present in exact unit, size and shape (c) physicochemical properties, (d) yearly production level, and (e) toxicology aspect of the NMs, safety and expected exposure effects. Other than MONMs, added ingredients, colorants, preservatives and other restricted substances should be indicated in the final product printed with precise units. The information should be continuously updated based on the analysis results from time to time [75] in the company's product file. Every country keeps watching on the regulations imposed by the competent authority in order to ensure its products shine on the different business markets.

13 Summary

This book chapter dealt with metal oxide nanomaterials and their usage in biomedical applications. Beginning from the definitions of NMs as suggested by various international organizations, theory behind the metal oxides' behaviors in different applications especially in biomedical field were discussed. The morphological evolutions such as size and shapes of NMs which are fundamental to many applications were discussed from the available theory. It was because of the importance of size and shapes for accessing particular target cells or tissues or organs in biomedicines. This chapter also discussed the simple preparation methods with possible theories corresponding to each synthesis method for easy understanding. Once the MONMs are synthesized, it should render preferential properties to interact with biological system internally and externally to maximize its expected performance. Those parameters are amphiphilicity, iso-electric points, surface charge, biocompatibility, size, shapes, concentration, etc. all of which play critical roles. At the same time, studies on the MONMs with biomolecular interactions call for further understanding and to elucidate the theoretical mechanism which is lacking still. The process of commercializing a product, prior risk analysis, different anti-bacterial assessment methods, and possible interaction mechanism were discussed in detail. This chapter (in addition to the already available sources) would provide a glance from synthesis to applications of MONMs to the readers, students, and research groups who are actively engaged in biomedical applications of nanomaterials.

Acknowledgements This work was supported by Basic Science Research Program through the NRF funded by the Ministry of Education (NRF-2018R1D1A1B07051012), Korea (to Sungkyunkwan University).

References

1. https://ec.europa.eu/jrc/sites/jrcsh/files/jrc_reference_report_201007_nanomaterials.pdf
2. https://www.ttu.ee/public/m/Mehaanikateaduskond/Instituudid/Materjalitehnika_instituut/MTX9100/Lecture5_NanomatFundamentals.pdf
3. https://application.wiley-ch.de/books/sample/3527331972_c01.pdf
4. Ananth A, Dharaneedharan S, Gandhi MS, Heo MS, Mok YS (2013) Novel RuO_2 nanosheets—facile synthesis, characterization and application. Chem Eng J 223:729–736
5. Chiesa M et al (2005) Nature of the chemical bond between metal atoms and oxide surfaces: new evidences from spin density studies of K atoms on alkaline earth oxides. J Am Chem Soc 127:16935–16944
6. https://www.bnl.godocuments/41042.pdf.v/isd/
7. Hunyadi Murph SE et al. (eds) Anisotropic and shape-selective nanomaterials. Nanostruct Sci Technol. https://doi.org/10.1007/978-3-319-59662-4_2
8. Gatoo MA, Naseem S, Arafat MY, Dar AM, Qasim K, Zubair S (2014) Physicochemical properties of nanomaterials: implication in associated toxic manifestations. Biomed Res Int 1–9. http://dx.doi.org/10.1155/2014/498420
9. Teske SS, Detweiler CS (2015) The biomechanisms of metal and metal-oxide nanoparticles' interactions with cells. Int J Environ Res Public Health 12:1112–1134
10. Yang Y, Zhang C, Hu Z (2013) Impact of metallic and metal oxide nanoparticles on wastewater treatment and anaerobic digestion. Environ Sci Proces Impacts 15:39–48
11. Ananth A, Dharaneedharan S, Heo MS, Mok YS (2015) Copper oxide nanomaterials: synthesis, characterization and structure-specific antibacterial performance. Chem Eng J 262:179–188
12. Bhosale MA, Karekar SC, Bhanage BM (2016) Room temperature synthesis of copper oxide nanoparticles: morphological evaluation and their catalytic applications for degradation of dyes and C-N bond formation reaction. Chem Select 1:6297–6307
13. Bagabas A, Alshammari A, Aboud MF, Kosslick H (2013) Room-temperature synthesis of zinc oxide nanoparticles in different media and their application in cyanide photodegradation. Nanoscale Res Lett 8:516–526
14. Naz T et al (2015) 2-Aminoethanol-mediated wet chemical synthesis of ZnO nanostructures. Appl Nanosci 5:425–433
15. Shavel A, Liz-Marzan LM (2009) Shape control of iron oxide nanoparticles. Phys Chem Chem Phys 11:3762–3766
16. Babu TS, Anandkumar M, Tsai TY, Kao TH, Inbaraj BS, Chen BH (2014) Cytotoxicity and antibacterial activity of gold-supported cerium oxide nanoparticles. Int J Nanomed 9:5515–5531
17. Thanh NTK, Maclean N, Mahiddine S (2014) Mechanisms of nucleation and growth of nanoparticles in solution. Chem Rev 114:7610–7630
18. Strey R, Wagner PE, Viisanen Y (1994) The problem of measuring homogeneous nucleation rates and the molecular contents of nuclei: progress in the form of nucleation pulse measurements. J Phys Chem 98:7748–7758
19. Matijevic E (2012) Fine particles in medicine and Pharmacy. Springer Publication, London. ISBN 978-1-4614-0378-4
20. Kalishwaralal K et al (2008) Extracellular biosynthesis of silver nanoparticles by the culture supernatant of Bacillus licheniformis. Mater Lett 62:4411–4413
21. Saif S, Tahir A, Chen Y (2016) Green synthesis of iron nanoparticles and their environmental applications and implications. Nanomater 6: 209 (1–26)
22. Charbgoo F, Ahmad MB, Darroudi M (2017) Cerium oxide nanoparticles: green synthesis and biological applications. Int J Nanomed 12:1401–1413
23. Mahamuni PP et al (2019) Synthesis and characterization of zinc oxide nanoparticles by using polyol chemistry for their antimicrobial and antibiofilm activity. Biochem Biophys Rep 17:71–80
24. Singh J, Dutta T, Kim KH, Rawat M, Kumar P (2018) 'Green' synthesis of metals and their oxide nanoparticles: applications for environmental remediation. J Nanobiotechnol 16:84–108

25. Yang Q et al (2013) Metal oxide and hydroxide nanoarrays: hydrothermal synthesis and applications as supercapacitors and nanocatalysts. Prog Nat Sci Mater Int 23:351–366
26. Adschiri T, Hakuta Y, Sue K, Arai K (2001) Hydrothermal synthesis of metal oxide nanoparticles at supercritical conditions. J Nanopart Res 3:227–235
27. Ramimoghadam D, Hussein MZB, Taufiq-Yap YH (2013) Hydrothermal synthesis of zinc oxide nanoparticles using rice as soft biotemplate. Chem Central J 7:136–235
28. Sagadevan S, Podder J, Das I (2016) Hydrothermal synthesis of zirconium oxide nanoparticles and its characterization. J Mater Sci Mater Electron 27:5622–5627
29. Wan Y, Zhou X (2017) Formation mechanism of hafnium oxide nanoparticles by a hydrothermal route. RSC Adv 7:7763–7773
30. Ananth A, Dharaneedharan S, Seo HJ, Heo MS, Boo JH (2017) Soft jet plasma-assisted synthesis of Zinc oxide nanomaterials: Morphology controls and antibacterial activity of ZnO. Chem Eng J 322:742–751
31. Ananth A, Gandhi MS, Mok YS (2013) A dielectric barrier discharge (DBD) plasma reactor: an efficient tool to prepare novel RuO_2 nanorods. J Phys D Appl Phys 46:155202–155209
32. Polonskyi O et al (2018) Plasma based formation and deposition of metal and metal oxide nanoparticles using a gas aggregation source. Eur Phys J D 72:93–106
33. Hamaker HC (1937) The London—van der Waals attraction between spherical particles. Physica 4:1058–1072
34. Faure B et al (2013) Dispersion and surface functionalization of oxide nanoparticles for transparent photocatalytic and UV-protecting coatings and sunscreens. Sci Technol Adv Mater 14:023001–023023
35. Leite FL et al (2012) Theoretical models for surface forces and adhesion and their measurement using atomic force microscopy. Int J Mol Sci 13:12773–12856
36. Kasemo B, Lausmaa J (1994) Material-tissue interfaces: the role of surface properties and processes. Environ Health Perspect 102:41–45
37. Escorihuela L, Martorell B, Rallo R, Fernandez A (2018) Toward computational and experimental characterisation for risk assessment of metal oxide nanoparticles. Environ Sci Nano 5:2241–2251
38. Pitkanen L, Striegel AM (2016) Size-exclusion chromatography of metal nanoparticles and quantum dots. Trends Anal Chem 80:311–320
39. Liu FK (2012) Using size-exclusion chromatography to monitor the stabilization of au nanoparticles in the presence of salt and organic solvent. Chromatographia 75:1099–1105
40. Fu W, Zhang W (2018) Measurement of the surface hydrophobicity of engineered nanoparticles using an atomic force microscope. Phys Chem Chem Phys 20:24434–24443
41. Scatchard G (1949) The attraction of proteins for small molecules and ions. Ann NY Acad Sci 51:660–672
42. Iohara D et al (2011) Formation of stable hydrophilic C60 nanoparticles by 2-hydroxypropyl-β-cyclodextrin. Mol Pharm 8:1276–1284
43. Valsesia A et al (2018) Direct quantification of nanoparticle surface hydrophobicity. Commun Chem 1:53–64
44. Desmet C et al (2017) Characterisation of nanomaterial hydrophobicity using engineered surfaces. J Nanopart Res 19:117–134
45. Pihlasalo S, Auranen L, Hanninen P, Harma H (2012) Method for estimation of protein isoelectric point. Anal Chem 84:8253–8258
46. Audain E, Ramos Y, Hermjakob H, Flower DR, Riverol YP (2016) Accurate estimation of isoelectric point of protein and peptide based on amino acid sequences. Bioinformatics 32:821–827
47. Kozlowski LP (2016) IPC—Isoelectric point calculator, kozlowski biol. Direct 11:55–71
48. Chang XC et al (2014) Synthesis of monodisperse iron oxide nanoparticles without surfactants. J Nanomater 1–5. http://dx.doi.org/10.1155/2014/740856
49. Parks GA (1965) The isoelectric points of solid oxides, solid hydroxides, and aqueous hydroxo complex systems. Chem Rev 65:177–198

50. Salgin S, Salgin U, Bahadir S (2012) Zeta potentials and isoelectric points of biomolecules: the effects of ion types and ionic strengths. Int J Electrochem Sci 7:12404–12414
51. Milolajczyk A et al (2015) Zeta potential for metal oxide nanoparticles: a predictive model developed by a nano-quantitative structure—property relationship approach. Chem Mater 27:2400–2407
52. Toropov AA et al (2018) Towards the development of global nano-quantitative structure-property relationship models: zeta potentials of metal oxide nanoparticles. Nanomater 8:243–257
53. Acosta LS et al (2011) Biocompatible metal-oxide nanoparticles: nanotechnology improvement of conventional prosthetic acrylic resins. J Nanomater: 1–8. http://dx.doi.org/10.1155/2011/941561
54. Stavis SM, Fagan JA, Stopa M, Liddle JA (2018) Nanoparticle manufacturing—heterogeneity through processes to products. ACS Appl Nanomater 1:4358–4385
55. White MA, Johnson JA, Koberstein JT, Turro NJ (2006) Toward the syntheses of universal ligands for metal oxide surfaces: controlling surface functionality through click chemistry. J Am Chem Soc 128:11356–11357
56. Yi G, Son J, Yoo J, Park C, Koo H (2018) Application of click chemistry in nanoparticle modification and its targeted delivery. Biomater. Res. 22:13–20. https://doi.org/10.1186/s40824-018-0123-0
57. Dekkers S et al (2016) Towards a nanospecific approach for risk assessment. Regul Toxicol Pharmacol 80:46–59
58. Simko M, Mattsson MO (2014) Interactions between nanosized materials and the brain. Curr Med Chem 21:4200–4214
59. Bankier C et al (2018) A comparison of methods to assess the antimicrobial activity of nanoparticle combinations on bacterial cells. PLOS ONE 13(2): 1 e0192093 (1–13). https://doi.org/10.1371/journal.pone.0192093
60. http://users.aber.ac.uk/hlr/mpbb/index_files/Page299.html
61. Ibrahim SF, Gvd Engh (2007) Flow cytometry and cell sorting. Adv Biochem Engin/Biotechnol 106:19–39
62. Barry AL, Coyle MB, Thornsberry C, Gerlach EH, Hawkinson RW (1979) Methods of measuring zones of inhibition with the bauer kirby disk susceptibility test. J Clinical Microbiol 10:885–889
63. http://microchemlab.com/test/zone-inhibition-test-antimicrobial-activity
64. Arakha M et al (2015) Antimicrobial activity of iron oxide nanoparticle upon modulation of nanoparticle-bacteria interface. Sci Rep 5:14813–14825
65. Zhang XQ et al (2012) Interactions of nanomaterials and biological systems: implications to personalized nanomedicine. Adv Drug Deliv Rev 64:1363–1384
66. Deng ZJ et al (2009) Differential plasma protein binding to metal oxide nanoparticles. Nanotechnol. 20:455101–455109
67. Gold K, Slay B, Knackstedt M, Gaharwar AK (2018) Antimicrobial activity of metal and metal-oxide based nanoparticles. Adv Therap 1:1700033–1700045
68. Dizaj SM, Lotfipour F, Barzegar-Jalali M, Zarrintan MH, Adibkia K (2014) Antimicrobial activity of the metals and metal oxide nanoparticles. Mater Sci Eng, C 44:278–284
69. Zunita M, Makertihartha IGBN, Saputra FA, Syaifi YS, Wenten IG (2018) Metal oxide based antibacterial membrane. Mater Sci Eng 395:012021–012029
70. Arias LS et al (2018) Iron oxide nanoparticles for biomedical applications: a perspective on synthesis, drugs, antimicrobial activity, and toxicity. Antibiotics 7: 46–78
71. Hoseinzadeh E et al (2017) A review on nano-antimicrobials: metal nanoparticles, methods and mechanisms. Curr Drug Metabol 18. https://doi.org/10.2174/1389200217666161201111146
72. http://www.mfds.go.kr/eng/index.do
73. http://www.egg2012.de/tl_files/pdf/Paper/EGG2012_C6_1_Park_presentation_slides.pdf
74. https://www.fda.gov/default.htm
75. https://ec.europa.eu/info/departments/joint-research-centre_en

76. Rauscher H, Rasmussen K, Sokull-Kluttgen B (2017) Regulatory aspects of nanomaterials in the EU. Chem Ing Tech 89:224–231
77. http://ec.europa.eu/environment/chemicals/reach/reach_en.htm
78. Srivastava N, Chowdhury N (2006) Nanotechnology and global sustainability. In: Maclurcan D, Radywyl N (eds) Nanotechnology regulation in India: a framework for exploring the risks and opportunities. CRC Press, pp 241–260
79. Jarvis DS, Richmond N (2011) Regulation and Governance of Nanotechnology in China: Regulatory Challenges and Effectiveness. Euro J Law Technol 2. http://ejlt.org/article/view/94/155
80. https://www.science.org.au/curious/technology-future/nanotechnology-regulation-australia

Nanofibers and Nanosurfaces

Pearlin Hameed, Vignesh K. Manivasagam, Magesh Sankar, Ketul C. Popat, and Geetha Manivasagam

Abstract Medical implants are developed to replace the diseased or fractured hard and soft tissues. Materials scientist and clinicians initially opted for materials which exhibited high strength, non-toxic behaviour and inert as they are not any rejection by the human body. However, with time, the failures of these materials were encountered as their properties were not close to that of human tissues (Sakka and Coulthard in Med Oral Patol Oral Cir Bucal 16:e42–e44, 2011). Developing the right environment for cells to grow with the necessary biomolecules is the major focus and to achieve this scaffolds are being developed with various materials ranging from polymers to ceramics using different processes. Advancements in characterization techniques has thrown more light on the structure of human tissues and bone. Studies have shown that the unique properties of bone were attributed to its micro/nanostructures formed by the nanostructured collagen and apatite crystals (Palmer et al. in Chem Rev 108:4754–4783, 2008; Perez et al. in J Tissue Eng Regen Med 7:353–361, 2013). In order to mimic the bone, materials were modified at micro/nano level both at the bulk and surface level. Studies have also revealed that materials with nanograins have superior osseointegration capability when compared to conventional micron materials (Thakral et al. in J Clin Diagn Res 8:ZE07–ZE10, 2014). Surface characteristics such as surface chemistry, topography, roughness, stiffness and surface charge influenced the biocompatibility [Ferrari et al. in Colloids Interfaces 3:48, 2019). This understanding led to development of several nanomaterials for biomedical applications. In the case of metals, several processing techniques were developed to form nano grained materials, whereas, in polymers, various nanofibers were prepared and tested for their bioactivity. This chapter presents some of the important surface properties and their influence on the biocompatibility and describes the effect of nano versus micron surfaces on the cellular attachment.

P. Hameed · V. K. Manivasagam · M. Sankar · G. Manivasagam (✉)
Centre for Biomaterials Cellular and Molecular Theranostics (CBCMT), Vellore Institute of Technology (VIT), Vellore 632 014, TN, India
e-mail: geethamanivasagam@vit.ac.in

K. C. Popat
Department of Mechanical Engineering, School of Biomedical Engineering, School of Advanced Materials Discovery, Colorado State University (CSU), Fort Collins, CO, USA

T. S. Santra and L. Mohan (eds.), *Nanomaterials and Their Biomedical Applications*,
Springer Series in Biomaterials Science and Engineering 16,
https://doi.org/10.1007/978-981-33-6252-9_4

Further, nanofibers development on polymers has enhanced the scope for fabricating scaffolds that can potentially mimic the architecture of natural human tissue at the nanometer scale (Vasita and Katti in Int J Nanomedicine 1:15–30, 2006). The biomimicry by nanofibers made from biocompatible polymers has opened novel and innovative therapeutic avenues in the area of tissue regeneration. In recent years, a plethora of bio-nanofibers has been fabricated to imitate the natural extracellular matrix for cartilage tissue engineering. Nanofibers have broadened their horizons in the past five years for potential biomedical applications from triaxial fibers to electrospinning of drug-loaded polymer and ceramic composites (Kiran et al. in Ceram Int 45:18710–18720, 2019). However, the effect of surface topology on stem cells for cartilage regeneration and the biophysics involved in the stress distribution of nanofibers needs to be elucidated. Apart from this, the fine-tuning of metallic implant surfaces for enhanced chondro-integration and antibacterial activity is a need of the hour and clear understanding towards the sensitivity of various surface characteristics is also needed. Various modifications on the surface of metallic implants are done using different chemical and physical modification techniques. The commonly used techniques are anodization, laser texturing, hydrothermal treatments and sand blasting (Vishnu et al. in Nanomedicine Nanotechnology, Biol Med 20, 2019; Manivasagam and Popat in ACS Omega 5:8108–8120, 2020). Studies have shown that these modification have enhanced suitability for the implant material for the chondrocyte conduction by providing the necessary roughness and surface chemistry in the nanoscale level. Hence, in this chapter, a detailed discussion about different synthesis routes associated with nanofibers' development for cartilage regeneration will be discussed. Along with this, the fabrication and effect of nanosurfaces on metallic implants for enhanced chondrocyte conductivity will also be highlighted.

Keywords Nanofibres · Nanosurfaces · Anodization · Cartilage regeneration · Chondrointegration · Scaffolds

1 Introduction

Nanobiomaterials have a unique advantage of bio mimicking the human tissue and thus considered as a suitable replacement for damaged tissues. The surface topography, chemical composition, surface mechanical properties and surface energy of a biomaterial dictate the cellular responses on the surface and, hence, biocompatibility. As the bone is made of apatite, which is around 2–5 nm in diameter and 50 nm in length [1], while the collagen is of 80–300 nm*1.5 nm [2]. The cells are more familiar with nano features around them, and hence it is of prime importance to develop nanomaterials or tune the surface of biomaterials in nano regime. Cells when introduced on a new surface, try to adhere with the help of focal adhesion proteins present on the cell membrane. Post protein adhesion the cytoskeleton of the cell undergoes re-arrangement for spreading onto the surface. The nearby cells also undergo similar process and begin interaction. These cell-cell interactions can be influenced by various nano topologies and surface chemistry. Nano topologies play a major role in the cell adhesion, cell proliferation, extracellular matrix (ECM) generation and cell metabolism [3]. The natural ECM in which fibrous proteins and proteoglycans whose size ranges from 50 to 500 nm consists mainly of collagen and found extensively in tissues such as bone, skin, and tendons [2]. Further, chemical signals and mechanical cues such as the elastic modulus of the surface are sensed by the cells [4]. These signals dictate how cells grow and differentiate. Hence, it is essential to understand the surface chemistry, mechanical properties and surface topology of any new surface developed. Surface related failure is the initial sign of understanding that alerts for an immediate repair action to be initiated [5, 4]. In case of biomedical devices and materials used both internally and externally, it needs much more caution as the materials to come in contact with human blood, blood is one of the most complex liquids and it compatibility is dictated by the cells, platelet and proteins in it. Hence, the surface of the implant should also be hemocompatible. Apart from this, the environment might also contains bacteria, and live cells that come in contact to the surface and the race between these two to adhere decides the fate of the implant. This is also dictated by the surface properties. The significance of surface is well established in the biomedical perception, and the dire need for analyzing this surface and the constant research on improvising the surface for the implant's longevity is the real global challenge. Several studies reported on enhanced cellular activity on nanofibers, nanotubes, nanophase oxides, and nano-HaP (Hydroxyapatite), which were of different shapes such as sphere, rosettes, needle, fibres etc. [6]. Thus, it is evident that the nanosurfaces' surface energy profoundly influenced cell activity through protein adhesion. Surface energy has two components polar and dispersive components. The dispersive components are mainly due to the London force between the solid and liquid, and the polar component is dictated by the surface charge, which leads to dipole interaction between the surface and the liquid. When there are nano architectures on the surface, these features can play a major role in making a surface hydrophobic or hydrophilic. Without surface features, it is impossible to attain a superhydrophobic configuration (apparent contact angle $> 150°$). For cardiovascular applications, we need blood repellent surfaces to reduce blood material interaction as that leads to blood clotting. Whereas, for bone applications, super hydrophilic (apparent contact angle $< 10°$) surfaces are preferred for cell adhesion and calcium deposition. Thus, tailoring the surfaces using nanotechnology has paved a path for the induction of the right pathways for biomaterials' cellular activity. The field of tissue engineering demands the need for the development of scaffolds for faster healing, and these scaffolds are developed using different processes like chemical synthesis, lyophilization, and spin coating. The most promising development of fibrous materials results in the path-breaking advancement in tissue engineering, as it helps develop very thin scaffolds, like the wound healing patches that can be loaded with anti-microbial drugs. The potency of utilizing the nanofibers for various applications in the biomedical arena is critical, and a lot of research is going on around the world in this aspect. Overall, in today's scientific advancement, it could be seen that the nanosurfaces and nanofibers modification processes could bring a paradigm shift to the biomedical industry.

This chapter is split into two parts comprising the nanofibers as the first part and nanosurfaces as a second. It helps explain how these processes can help in the overall osseointegration and antibacterial activity of the biomaterials developed.

2 Part I: Nanofibers

Nanofibers can steer cell alignment [7, 8], enhance cell-cell interaction [9] and provide topological cues for better adhesion, increase proliferation [9], induce differentiation [10] of progenitor cells. Nanofibers have demonstrated their ability to enhance the differentiation of progenitor cells into neuronal [11, 12], epithelial [13], myogenic [14] osteogenic [15], tendon [16] and chondrogenic [17] lineages as summarized in Fig. 1. In this, section the different synthesis routes other than electrospinning for cartilage regeneration, nanofibers for biomedical applications, and hybrid nanofibers will be discussed.

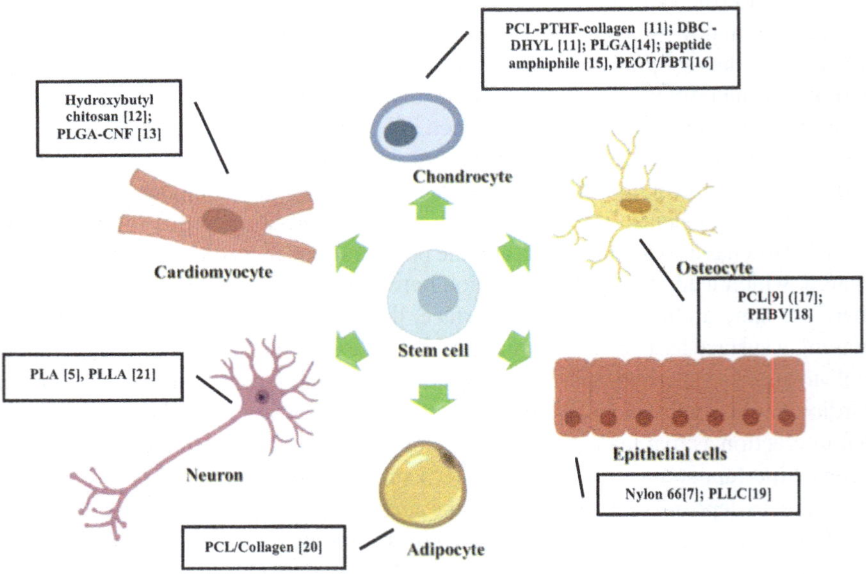

Fig. 1 Stem cell lineages and various nanofibers (Hydroxybutyl chitosan [18]; PLGA-CNF [19] PCL-PTHF-collagen [17];DBC-DHYL [17]; PLGA [20]; peptide amphiphile [21], PEOT/PBT [22]; PCL [15] ([23] Xue et al. 2017); PHBV [24] Nylon 66 [13]; PLLC [25]; PCL/Collagen [26]; PLA [11], PLLA [27])which enhanced their lineage markers when cultured

2.1 Cartilage Regeneration

Cartilage plays an integral role as connective tissue in vertebrates. As cartilage lacks a lymphatic and blood vessel system and has meagre percent of chondrocytes (approximately 1–5% of chondrocyte of a total matrix), regeneration of cartilage becomes a strenuous process [28, 29]. Due to confinement of chondrocytes within lacunae and low metabolic activity, reparation of cartilage to its natural form after the damage becomes unattainable. Surgical interventions to repair cartilage using microfracture and mosaicplasty have been the conventional line of treatment for chondral defects [30]. However, restoration of the damaged cartilage using a biomaterial scaffold, which epitomizes extracellular matrix (ECM) along with cells has been a more promising approach in cartilage repair than conventional methods. Apart from microarchitecture, studies show that progenitor cells' fate can be influenced by the pore size of the scaffold [31]. The broader mean pore size of 300 μm has unveiled enhanced cell proliferation, chondrogenic markers than lower mean pore sizes (94 & 130 μm) in Collagen-hyaluronic acid (CHyA) scaffolds [22]. Cells seeded in gradient scaffold showed enhanced chondrogenesis with higher expression of GAG and ECM deposition than non-gradient scaffold [22]. Consequently, pore size plays a crucial role in cartilage regeneration by enhancing and inducing chondrogenesis along with the orientation of the fibres [32, 33]. Chondrocytes cultured on aligned nanofibers tend to secrete more GAG and collagen type 2 then random nanofibers demonstrating that orientation played an essential role in enhancing chondrogenesis [32, 34].

Cartilage is predominantly made up of collagen, proteoglycan and water; Collagen type II accounts for about 80% of the total volume of cartilage. To mimic the natural structure of cartilage, researchers in the past have tried fabricating electrospun mats made up of collagen type 2 extracted from chicken cartilage [35]. However, there is a lack of clear understanding as there are scarce reports on the *in-vitro* and *in-vivo* assessment of collagen type II Mats. Jiang et al. demonstrated a growth factor free chondrogenesis of mesenchymal stem cells using composite electrospun fibres of poly(ε-caprolactone)/polytetrahydrofuran (PCL-PTHF urethane) and collagen I. Interestingly, in the absence of collagen 1, the PCL-PTHF nanofibers on its own failed to induce chondrogenesis. The difference in modulus of collagen free nanofiber may have attributed to this behaviour.

On the other hand, a softer and low modulus (4.3Mpa) PCL-PTHF-collagen nanofiber was able to block NF-kappa B signalling leading to chondrogenesis [36]. The nanofibers' microenvironment and modulus have shown to regulate chondrogenesis in progenitor cells [17]. A co-axial P(LLA-CL) & collagen type 1 loaded with rhTGF-β3 was evaluated for cartilage regeneration in trachea repair. The inner core comprised of rhTGF-β3, while the outer layer was a blend of P(LLA-CL) & collagen. On differentiating, stem cells on core-shell nanofiber, which was a blend of P(LLA-CL), an approximately 0.3-fold increase in the expression of collagen type II and a 0.1-fold increase in Sox 9 was observed. Implying that the expression of chondrogenic genes is upregulated on the addition of collagen [37].

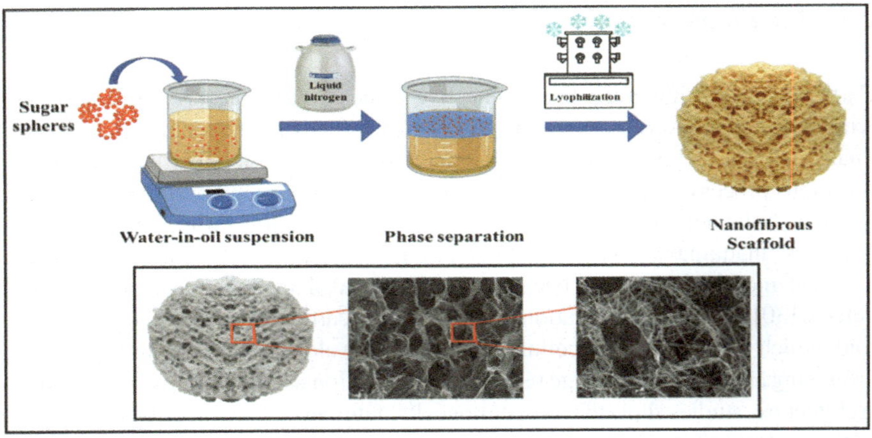

Fig. 2 Schematic representation of the preparation of nanofibrous scaffold and its animated microstructure

In a study by Wang et al., successful repair of cartilage tissue defects, using bone marrow-derived mesenchymal stem cells (BMSCs) grown in nanofibrous microcarriers (NF-MCs) made up of dialdehyde bacterial cellulose (DBC) with DL-allo-hydroxylysine (DHYL), and chitosan (CS) was achieved. A different technique produced fabrication of NF-MCs than electrospinning. The precursor was phase-separated in liquid nitrogen followed by lyophilization, as shown in Fig. 2. On *in-vivo* investigation, NF-MCs culture in a bioreactor with BMSCs showed a significant increase of approximately 15 and 20% in bone volume/tissue volume compared to NF-MCs without stem cells and control (fibrin glue) [17]. Another similar technique to fabricate nanofibrous scaffolds was developed by Wei et al. [38] and Liu et al. [39]. Briefly, to create micro and nanostructure, sugar spheres of the desired dimension were made and sieved, followed by phase separation and lyophilization. Liu et al. used this technique to develop poly (L-lactic acid) (PLLA) scaffold cultured with BMCs for the regeneration of articular cartilage in the rabbit. In addition to stem cells, matrilin-3, a non-collagenous, was also added to augment the cartilage regeneration and maintain the cartilage property. It was observed that on treatment with matrilin-3, there was a slight increase in gene expression of collagen-2 when compared to untreated PLLA and a significant increase in expression of Aggrecan, implying enhanced chondrogenesis. On the other hand, Collagen 1 expression was significantly less in matrilin-3 treated PLLA than untreated implying suppression of osteogenesis. Overall, untreated PLLA, in comparison to control, exhibited a significant increase in chondrogenic markers, implying PLLA without matrilin-3 also promotes chondrogenesis [39].

Hyaluronic acid has been shown to induce chondrogenesis in adipose-derived stem cells by activating CD 44 receptors [26], a cell surface glycoprotein that binds primarily with hyaluronic acid ligand [40]. Activation of the CD44 receptor by self-assembled supramolecular GAG- nanofibers for cartilage regeneration was studied by Yaylaci et al. [41]. Supramolecular GAG- nanofibers are hypothesized to emulate natural hyaluronic acid, thereby inducing chondrogenesis. Without chondrogenic media, the supramolecular GAG- nanofibers were able to display an approximate increase of 2.5-fold of collagen II expressions, a 3-fold increase in aggrecan and a 5-fold increase in Sox 9 expression in comparison to untreated control cells. In another study, a composite nanofiber composed of peptide amphiphile (PA) and hyaluronic acid (HA) was investigated for cartilage repair in the mouse model. The results showed PA-HA nanofiber was able to encapsulate chondrocyte encapsulation and displayed the lowest joint degradation score [42].

2.2 *Nanofibers in Neuronal Tissue Engineering*

For neuronal tissue engineering, autologous grafting of nerves from another site is considered the gold standard. However, autologous nerve grafts are restricted by limited availability, the possibility of permanent function loss at the harvest site, and mismatch of the diameter of an injured nerve [27, 43]. Due to these limitations, autologous nerve grafting has been substituted with artificial nerve guide conduit (NGC). Since nanofibers can manoeuvre the alignment of neural precursor cells, can guide Schwann cell migration [10], and can augment regenerate impaired neurons, they are considered as a promising alternative for autologous nerve grafting [8]. Electrospun NGC has shown the potential to direct axonal regeneration. A polycaprolactone (PCL) /poly-L-lactic acid (PLLA), conduit successfully bridged a gap of 10 mm in a rat sciatic nerve defect [27]. Similarly, a new U-shaped nerve guide conduit composed of Co-polymers; poly (lactic- co -glycolic acid) (PLGA) and polyurethane (PU) was fabricated for nerve regeneration using the electrospinning process. This inspiring technique made the central part of the tube double-coated with randomly oriented nanofibers over the aligned nanofibers, to strengthen the weak mechanical strength of the aligned nanofibers. Apart from new technical developments, biodegradability of polymers used for fabrication of nanofibers can increase the nerve restoration rate [43]. Schuh et al. made a novel self-shaping conduit with electrospun fibrin-PLGA nanofiber on fibrin gel, seeded with Schwann cell-like cells (SCLs) for peripheral nerve regeneration. Although the study lacked *in-vivo* assessment, invitro analysis showed that these aligned electrospun fibrin-PLGA fibres promoted the formation of Büngner-like structures of SCLs [44]. On the other hand, in a different study, a tubular conduit made of silk nanofiber and polyaniline nanoparticle was electrospun for peripheral nerve regeneration in a sciatic nerve injury model. Since polyaniline is a conducting polymer, it supports electrical conduction to repair lost nerve impulses. The normal nerve conduction velocity in a normal rat is 58 m/sec. On seeding with Schwann cells an excellent nerve conduction velocity of 50 ms-1 was achieved after 12 months of implantation, implying improved functional neuronal regeneration [45].

2.3 Nanofibers Based Sensor for Myocardial Infarction

Nanofibers have been highly touted for cardiac tissue repair as they guide cardiomyocytes' alignment [46]. Nanofibers assist in orienting cells, providing milieu, and controlling cell to cell interaction, making it ideal for multi-layered implantable patches than scaffolds [45]. Polymer/biopolymer-based nanofibers [41–48] and nanoparticle embedded in nanofibers [49, 50] have been used successfully as electrodes for various type of sensors such as biosensor [47], chemical sensor [51, 52], gas sensor [48], etc. Researchers in the past have measured the level of cardiac marker Troponin -I to detect Myocardial Infarction (MI) [53]. Troponin-I is a specific and sensitive biomarker for myocardial damage. The level of Troponin in a healthy person varies from 0.5–2 ng/ml, but in a patient suffering from MI, it increases to 20–550 ng/ml [54]. For accurate detection of troponin and rapid results, researchers like Periyakaruppan et al. [55] devised a label-free immunosensor for early diagnosis of MI using carbon nanofiber as electrode arrays. The immunosensor measured, Troponin I level as low as $\sim$ 0.2 ng/mL, which was 25 times more sensitive than conventional methods. On the similar ground, Kumar et al. devised a multiplexed immunosensor using carbon nanofiber (CNF) as electrode arrays to detect cardiac markers in real-time [56, 57]. In this sensor, apart from Troponin-I, C-reactive protein and myoglobin were also covalently bound to carbon nanofiber surface, and the device demonstrated a sensitivity of $\sim$ 0.2 ng/mL, alike to Periyakaruppan et al.'s. Rezaei et al. further advanced the idea by making a sandwich-type immunosensor made up of carboxylated multiwalled carbon nanotube-with whiskered nanofibers to measure Troponin level demonstrating an incomparable detection limit of ~ 0.04 ng/ml [54].

2.4 Nanofibers in Aneurysms

In 2014, Gonzalez et al. discovered that by growing silver nanofibers on the surface of the blood vessel, the shear stress experienced by blood vessels could be distributed to prevent the rupture of aneurysms. With silver nanofiber in place, the deformation force reduced up to 5 times. Metallic fillers in polymeric nanofibers can be used as a substitute for conventional endovascular coiling of platinum wire to prevent extravasation of blood in hemorrhage site [58]. Kim et al. fabricated platinum-coated Polyacrylonitrile (PAN) nanofibers as a substitute for endovascular coiling using platinum wire to fill the Aneurysmal cavity [59]. Metal embedded nanofibers can reduce the operational risk and cost of conventional endovascular coiling. For repair of carotid artery aneurysms, Wang et al. evaluated the potential of a biodegradable stent covered with PLA and PCL nanofibers in canines. Though the in-stent stenosis rate in the study was high, more than 87% successful occlusion of the aneurysm was achieved; indicating biodegradable stent coated nanofibers can be a viable attempt for stable aneurysm occlusion [60]. In 2018, Liu et al. developed a biodegradable, endovascular prosthesis, loaded with antibody eluting with electrospun PLGA and vancomycin nanofibers for reparation of mycotic aortic aneurysms [61]. A sustained release of vancomycin antibody was observed until 30 days in vitro study. Similarly, in *in-vivo* assessment, the protracting release of vancomycin in the aortic wall of rabbits for eight weeks was observed. The use of drug-eluting nanofibers as a coating on stents and prostheses can contribute to an increase in the success rate in the treatment of aneurysms.

2.5 Guided Bone Regeneration Using Nanofibers

Guided bone regeneration (GBR) uses a membrane with specific properties such as Osseo-conductivity, biocompatibility and mechanical properties as a barrier to allow separation of the defected bone from soft tissues [24, 62]. Kouhi et al. synthesized a nanofibrous membrane composed of poly (hybutyrate-co-3-hydroxyvalerate) (PHBV) -a degradable polyester, fibrogen and bredigite- a magnesium silicate bioceramic. PHBV/FG/BR nanofibers demonstrated a minimum 25% increase ALP activity of cells cultured on membrane compared to control PHBV and PHGV/FG membranes [24]. To incorporate antibacterial effect in the membrane for GBR, wang et al. developed metronidazole (MNA) and nano-hydroxyapatite (nHAs), based core-shell nanofibers. The outer layer of the shell was composed of polycaprolactone and nHA, while the nanofiber core was gelatin and metronidazole. The results indicated that core-shell nanofibers were able to release MNA slower than only MNA-loaded nanofibers, ie a slower MNA release profile was observed. In addition, core-shell nanofibers also enhanced bone formation and could prevent the colonization of anaerobic bacteria [63]. Correspondingly, Liam et al., a composite bilayer GBR membrane, the upper layer consisted of PLGA/gelatin nanofibers and silica nanoparticles loaded with dexamethasone, while the bottom layer comprised of PLGA nanofibers loaded with an antibiotic (doxycycline hyclate). Bone marrow stem cells (BMSCs) cultured on dexamethasone loaded mesoporous silica nanoparticles in PLGA/gelatin nanofibers (top layer) showed an upregulation of osteocalcin expression, highest ALP activity and mineralization than cells cultured on PLGA/gelatin nanofibers, PLGA/gelatin loaded with silica nanoparticles and plastic tissue culture plates. The bilayer membrane showed a 38.8% sustain release profile of dexamethasone loaded silica nanoparticles after 21 days and a total release of approximately 48% of antibiotic doxycycline hyclate over 21 days. Moreover, the bottom layer PLGA/loaded with doxycycline hyclate formed an inhibition zone 3 times and four times when tested with E.coli and S.aureus respectively indicating the effective broad-spectrum antibacterial property [20]. However, despite possessing good biocompatibility, antibacterial activity and osteoconduction/induction properties, these studies failed to assess the barrier properties of the GBR membrane. Tang et al. fabricated a core-shell GBR membrane comprising of PLGA/Hydroxyapatite (HAp) as core and collagen/amoxicillin as the shell. On culturing fibroblast on one side of the membrane, proliferated and distributed radially, however, no cells were found on the bottom of the membrane exhibiting the barrier property of PLGA/Hap [62].

Similarly, the result was achieved by Fu et al. they demonstrated 5wt% HAp in PLGA prevented penetration of fibroblast [64]. Apart from fibroblasts, researchers have also assessed the barrier function of the membrane against bacteria. Trobos et al. demonstrated the impermeability of streptococcus oralis against non-reabsorbed polytetrafluoroethylene (PTFE) membrane [65].

Periosteum, a thin lining that covers the bone's outer surface, plays an indispensable role in bone growth, repair, and supply of blood [66]. Periosteum can be divided into layers an outer fibrous layer and an inner osteogenic cambium [67], and can be mimicked using layers of electrospun fibre. Wang et al. constructed composite nanofiber sheets made up of PCL, collagen, and nano-hydroxyapatites seeded with bone marrow stem cell to mimic periosteum for reparation of a 4 mm segmental bone defect. The composite nanofiber sheets, when combining with allograft, were able to induce endochondral and intramembranous bone regeneration in mouse femur and completely repaired the defect. Compared to nanofiber sheets without stem cells, the composite nanofiber eliminated the fibrotic tissue capsule elicited by nanofiber sheets, leading to a marked improvement of osseointegration at the compromised periosteal site. Co-axial electrospinning using Platelet-rich plasma (PRP) and PVA has shown to release growth factors such as PDGF, TGF, IGF, VEGF for reparation of bone tissue. Cheng et al. incorporated PRP Silk fibroin, PCL, and PVA to make electrospun scaffolds. A sustained release of PRP growth factors was observed until four weeks; they also discovered increases in cell proliferation, cell migration, and level of collagen 2 in MSCs [68].

The ability of polymeric nanofibers to encapsulate drugs like dexamethasone and BMP-2 for reparation of a calvarial defect in the murine model was tested by Li et al. [69]. Nanofibers loaded with drugs demonstrate a stronger ability to induce differentiation toward osteoblasts in *in-vitro* study while in *In-vivo* osteogenesis studies revealed that the drug-loaded nanofiber scaffolds were significantly better in repairing than unloaded nanofibers. Yoa et al. explored a modern method of synthesizing PCL-PLA nanofibrous scaffold, using a thermally induced self-agglomeration (TISA) method, followed by freeze-drying using. Interestingly, improved mechanical property and osteogenic differentiation of stem cells were observed on 3D PCL-PLA nanofibrous scaffold. [70]. Nanofiber paste comprising of growth factors for reparation of cranial and spinal defects is an innovative therapeutic approach.

2.6 Drug Eluding Nanofiber

For the sustainable release of analgesics, various polymeric nanofibers have been deployed. PVA nanofibers loaded with neostigmine on the epidural application has resulted in extended analgesia in the murine model [71]. Whereas, subcutaneous implantation of PLGA nanofibers loaded with Ibuprofen demonstrated control release and slow degradation rate [72]. Similarly, PLGA nanofibers loaded with lidocaine and ketorolac [73]. PLGA nanofibers loaded with lidocaine- and ketorolac were electrospun on urethral stents for local drug delivery to alleviate the pain concomitant with stent implants [74]. Interestingly, sustained release of analgesics was observed until 30 days in an *in-vivo* rabbit model. Self-assembled biodegradable nanofibers made up of peptide amphiphiles and loaded with S-nitrosyl have shown to prevent restenosis and neointimal hyperplasia [75].

2.7 Hybrid Scaffolds Composed of Nanofiber and Gels

Aligned poly (ε-caprolactone-*co*-ethyl ethylene phosphate) (PCLEEP) nanofibers along with collagen type 1, and loaded with non-viral gene and drug was used by Nguyen et al. [76] for neural tissue engineering. This biodegradable hybrid scaffold successfully enhanced the regeneration of axons by aligning neutron and remyelination for treatment of spinal cord injury. Schuh et al. made a novel self-shaping conduit with electrospun fibrin-PLGA nanofiber on fibrin gel, seeded with Schwann cell-like cells (SCLs) for peripheral nerve regeneration. The *in vitro* assessment showed these aligned electrospun fibrin-PLGA fibres promoted the formation of Büngner-like structures of SCLs [44]. A recent study has reported the incorporation of carbon nanofibers in silk fibril scaffold leads to an increase in the compressive modulus to more than four times when compared to control silk fibril scaffold [77].

Moreover, Weng et al. fabricated a hybrid porous and nanofibrous scaffold composed of aerogel and bioactive glass fibres embedded with BMP-2 [78]. Overall, the percentage of new bone formation increased to twice in BMP-2 implanted hybrid scaffold than in no treatment control. A hybridization of 3D printed PLLA scaffold layered with Gelatin electrospun fibres on the surface has shown to improve cell adhesion and proliferation [79]. This technique can be very well used in the future for the patient and shape specific chondrogenic tissues such as nasal cartilage, ear, or epiglottis. The following section elaborates on the need for nanosurface development on the metallic materials for effective utilization as a useful implant material.

3 Part II: Nano Surfaces

3.1 Anodization—The Potential Nano Surface Modification Technique

Surface topography plays a significant role in governing the physiochemical, electrochemical and biological activity of a biomaterial. The development of nano features on the surface of the biomaterials is the key challenge, and this can be addressed with several surface modification processes. However, the most common, simple, and cost-effective surface modification, which results in nano features is anodization.

Anodization is a chemical surface modification process in which etching of the native oxide layer formed on the metal surface is performed, especially titanium which tends to form a strong oxide layer which can be tailored to develop nanotubular structures with the help of an electrolyte in the presence of an external electric field. This surface modification process is a straightforward and robust technique that helps develop nanotubes of different diameters and height by tailoring the following parameters: 1. Electrolyte composition, 2. Voltage and 3. exposure time. The setup consists of two electrodes: the working electrode being the anode and the platinum or graphite electrode being employed as a cathode. The electrolyte mostly contains solvents that have better dielectric potential and, at the same time, do not result in any residual impurities on the substrate. The most fascinating advantage of this method is not only the development of nanotubes but nanopillars, nanoflowers, and nanodots with resultant 3D morphology. The nanotubes are post-processed with the help of sintering.

As mentioned earlier, the concentration of electrolyte, voltage and anodization time [80, 81] that govern the nanotube growth can help in tailoring the tube diameter with the size range of 15–300 nm and different tube length can also be developed [80].

The development of self-organized nanotube layers has been attempted for the past two decades with the help of fluoride-containing electrolyte solutions [82, 83]. Apart from the fluoride-based electrolyte, an attempt to utilize organic solvents with low water content has been made in developing the nanotube structures by Wei et al. [84]. The amount of water utilized in the electrolyte is the key factor in determining uniform nanotube growth. This finding resulted in the development of a mechanism termed "pore-wall-splitting" in which the nanopores oxide layer forms as the first step, following which the tubes tend to originate from the ordered porous oxide [84].

The mechanisms involved in anodization and the process involved in developing different nanostructured oxides like nanotubes pores with hole morphology, nanochannels and microcones on the surface of titanium and titanium alloys has been critically reviewed by Roy et al. [81] and Kowalski et al. [85].

The nanotubes were formed on cp-Ti initially; however, this can be achieved in titanium alloys as well if the alloying elements are from the transition group. Such alloy combinations which tend to result with uniform nanotube formation are Ti-6Al-4 V [86], Ti-6Al-7Nb [86, 87], Ti-xZr alloy with varied Zr concentrations [88, 89] to name a few.

3.2 Nanotube Diameter the Key for the Cellular Response

Cellular attachment on nanotubular structures has been demonstrated in the in vitro level [90, 91], resulting in higher adhesion, proliferation, and ALP activity. Several researchers attempted to influence the cellular adhesion to the titania nanotubes, and

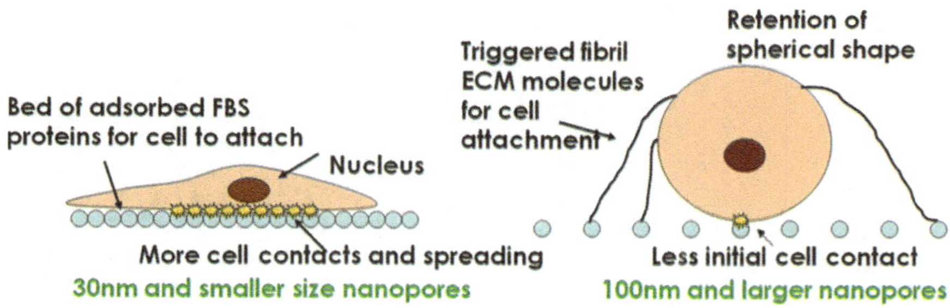

Fig. 3 Comparison of the cellular attachment on the nanotube surface with different tube diameters; Larger size nanopores result with poor adhesion [94]

the breakthrough happened when the nanotube diameter was tailored to a size range of 15–20 nm, which seemed to be an optimal size range for increased cell adhesion and proliferation [92].

Interestingly it was observed that the diameter range of 15–20 nm gives way for a larger surface area, and hence the cell adhesion is very high. On the contrary, when the diameter was increased to 100 nm, the cell adhesion decreased; however, the proliferation of the cells and the stretching of the filopodia and the networking of cells increased. This interaction of cells on different nanotube diameters is schematically depicted in Fig. 3. However, reports on cell apoptosis are also observed in 100 nm size nanotubes [93].

Mesenchymal stem cells, endothelial cells, and osteoclasts have been employed to study the size effect of nanotube and the adhesion behaviour [81] and it could be observed that the integrin cluster in the cell membrane with focal adhesion complex in size range of about 10 nm in diameter is perfectly fitting with the nanotube size range of 15 nm [92]. Toxicity was one concern on these TiO_2 nanotubes as TiO_2 nanopowder tends to be toxic; however, it has been verified that these nanotubes do not possess any in vitro cytotoxicity [95].

3.3 Nanotube—Protein Interaction: The First Line of Action

Any implant material that is placed in the human body encounters the blood and the protein complexes present in the blood. The adsorption of these proteins on the surface of the implant plays a crucial role in deciding the implants' early fate. Studies have shown that if the surface promotes fibrinogen adhesion that can improve the rate of blood clotting. Whereas, if the surface promotes albumin adhesion it can indirectly prevent blood clotting. However, few proteins can also influence the osteoblast cell adhesions. If the protein adhesion is successful, it paves the way for the cellular attachment. Gongadze et al. has proposed the mechanism involved in the cell attachment to nano rough titanium implant. [96, 97]. According to this study, the assumption made was that the negatively charged osteoblast would interact with the

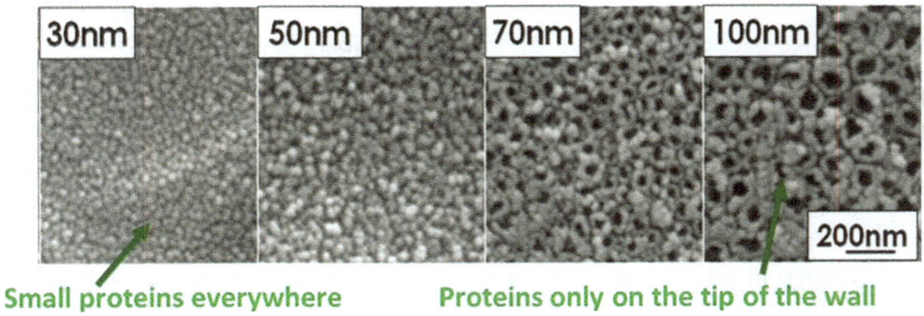

Fig. 4 SEM images showing the protein interaction on different nanotube diameter surfaces; smaller diameter substrates results with more protein adhesion compared to large diameter substrates [94]

negatively charged nanotubes with the help of the proteins present in the blood, which tend to pose a distinct quadrupolar internal charge distribution. Also, high surface charge plays a significant role in the attraction of fibronectin to the titanium surface, ending with an integrin-mediated osteoblast adhesion. When the negative surface charge density is high, it helps in the strong attachment of osteoblast to the convex edged nano rough morphology of the titanium surface. The surface with smaller diameter nanotube results with larger convex edges per unit area when compared to the larger ones, hence, favourable for improved osteoblast attachment and cellular response [96, 97]. The size dependency of nanotubes plays a key role in protein interaction as well, as shown in Fig. 4. The smaller diameter can accommodate more protein interaction due to the size of the proteins are also in the same nano regime, whereas the nanotubes of larger diameter tend to have a very minimal attachment of proteins on the tip of the wall [94].

3.4 Nanotubes on Orthopedic Implant Surfaces

As mentioned earlier, surface properties decides the performance of an implant, and hence, a modified surface with nano topographical features can result in improved osseointegration by serving as an attractive site for the protein interaction and cell attachment. To evaluate this, nanotubes were studied *in-vitro* [93] and *in-vivo* [93] condition, and it was reported that the osteoblast adhesion and proliferation was highly improved compared to unmodified titanium. These nanotubular surfaces improve the bone-bonding strength by nine-fold compared with conventional grit blasting surface, which results in micro-roughness [98]. In the *in-vivo* testing, it was observed that these nanotubes have better bone-implant contact and collagen expression. Several biomolecules are being utilized to promote cell differentiation of mesenchymal lineage via covalent immobilization to transform into the desired cells. One such attempt was the immobilization of epidermal growth factor on larger diameter (100 nm) nanotubes to reduce the cell apoptosis and enhance the proliferation

[99]. Yet another most prominently studied biomolecule is the bone morphogenic protein-2 (BMP-2), which has been loaded on the nanotubes with 30 nm diameter resulting in higher osteocalcin and osteopontin levels the in vitro analysis [100]. When this BMP-2 was immobilized by carbonyl diimidazole (CDI), the differentiation pathway was governed by the nanotubes' diameter, enabling improved osteogenic differentiation [101].

3.5 Antibacterial Efficacy of Nanotubes

Infection serves as a primary cause for implant failure, and the medical device industry is investing a lot of money in improving the antibacterial activity of its products to ensure safe implantation. The multidrug resistance bacteria and biofilm formations have resulted in the inability of traditional antibiotics to protect the implants from infection [102, 103]. The most common bacterial strains that result with implant failure due to biofilm formation are the *Staphylococcus aureus* (*S. aureus*) and *Staphylococcus epidermidis* (*S. epidermidis*). It was observed that nanotubes of controlled diameter showed a positive response compared to the pristine titanium [91, 102]. The size of nanotubes plays a key role not only for cellular attachment but also for bacterial attachment. In the case of bacterial attachment, the larger diameter tubes (100 nm) resulted in a decreased number of live bacteria compared to the smaller diameter ones (20 nm) [102, 104]. The antibacterial activity of these titania nanotubes can be attributed to three factors: the diameter of the tubes, alloys with a native anti-microbial based alloying element or adding nanoparticle with antimicrobial property to the nanotubes and finally by loading drugs into the tubular structures which can be released in a sustained manner thus maintaining the antibacterial activity for a prolonged duration. The most common nanoparticles that hold the anti-microbial property are Ag [105] and Zn [103] and in the case of alloying addition zirconium (Zr) tends to inhibit bacterial growth (esp. E. coli) with smaller diameter nanotubes [106]. In the case of drug loading, the nanotubes are loaded with infection-reducing drugs like penicillin/streptomycin or inflammation-reducing drugs like dexamethasone through physical adsorption process or deposition from simulated body fluids (SBF). Magnetic nanoparticles (Fe_3O_4) loaded into nanotubes have been demonstrated to be a successful process by Shrestha et al. [107] to have a guided delivery to the desired site using the photocatalytic activity to target the cancer cells. Development of amphiphilic nanostructures, i.e., coating the nanotubes with a hydrophobic cap like material to prevent the entry of body fluids and can be activated only with the help of photocatalytic activity for targeted release of the drugs [108, 109]. Degradable biopolymers approved by FDA are also into a study for sealing the drug-loaded nanotubes, which help in the programmed release of the drug with controlled degradation of the polymer [110]. With all these benefits and advantages, anodization seems to be a promising surface modification to promote osseointegration while imparting antibacterial efficacy to the implant material. However, this technique serves with a limitation of applicability only with titanium-based implants due to its oxide layer stability compared to other materials and hence, other surface modifications need to be explored for other metallic implant materials.

3.6 Laser Surface Modification

Laser surface modification is yet another field of research growing rapidly in supporting the development of safe and successful medical devices with micron, submicron and nano surface modifications. Apart from the development of laser-based bulk metallic materials for biomedical applications, laser is currently been extensively used in imparting structural modification on the surface of the implant, without compromising the properties of the bulk material. With the help of nano and femtosecond laser, the development of sub-micron features can be achieved on the implant surface, improving the surface energy, and modifiying protein interaction, and cellular attachment. Lithography technique has been used to imprint shark skin structure on the surface of titanium implant materials to enhance the antibacterial activity [111, 112], and the currently researchers are trying to develop this cost-effectively with the help of Femto and picosecond lasers. Several works [107–114] on the improved *in-vitro* and *in-vivo* characteristics of sub-micron featured laser-treated surfaces has been reported, and hence, laser treatment is another promising area to look up for nano surface modification.

3.7 Hydrothermal Treatment

Hydrothermal treatment is a very simple process in which the surface modification takes place in the presence of an alkaline or acidic solution under pressure and temperature (Fig. 5). In this process, an alkaline/acidic solution containing different salts that can help in the development of passive oxide films on titanium substrates. Substrates when subjected to pressure using a Teflon lined stainless steel autoclave and maintained in a temperature-controlled environment react with the solution and that leads to the formation of different morphology of oxides, namely nanoflowers [5], rods and branched nanostructures on the surface. These structures tend to have a more significant impact on cell proliferation and antibacterial activity than the untreated titanium substrate. Several governing parameters control the end produced nanostructures, namely concentration, pH of the solution, pressure, temperature, and exposure time [115, 116].

The nanosurface mentioned above modifications are widely researched, and apart from these, several other process routes are being studied to develop nano modified surface structures, which will help promote osteogenesis and antibacterial efficacy. Thus, nanofibers and nanosurfaces are the key research areas that can be used as the right tool to tailor the biomaterials and attain maximum efficiency.

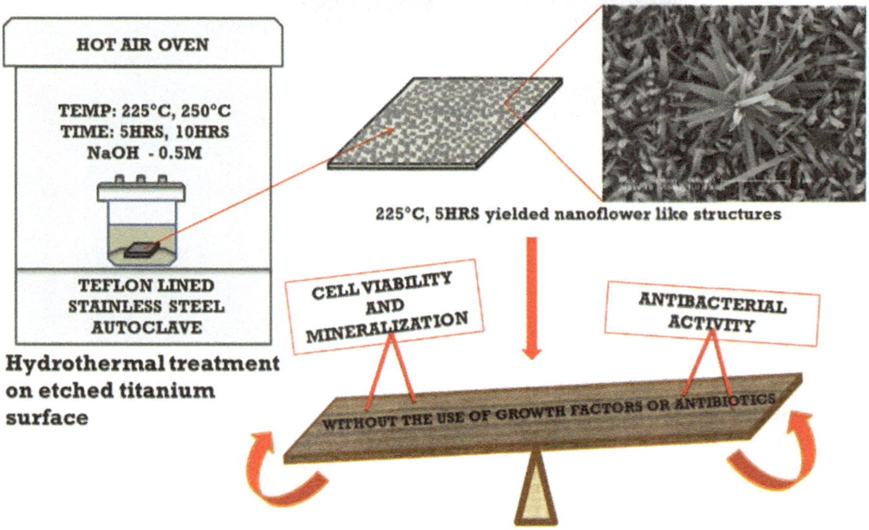

Fig. 5 Schematic representation of the hydrothermal treatment with nanoflower surface modification with balanced cell viability and antibacterial activity (adapted from [5])

4 Future Scope

The research focus on this chapter has been on developing nanofibers and nanosurfaces with different processes based on their application. Recently, researchers have focused on developing nanofibers on metallic implant surfaces, and this can enhance the behaviour of implants with the bulk properties, and at the same time, the fibres can enhance osteogenesis. Also, development of additively manufactured bulk materials with laser surface modified nanostructures are being studied to enhance the *in-vitro* and *in-vivo* behaviour of these implant materials. Even though the osseointegration and antibacterial activity of nano surface modified materials have been widely studied, the synergic behaviour needs to be evaluated in vitro to mimic the human body condition.

Abbreviations

BMP-2 – Bone Morphogenic Protein-2
CDI- Carbonyldiimidazol
CNF- Carbon Nanofiber
DBC- Dialdehyde Bacterial Cellulose
DHYL- DL-allo-hydroxylysine
ECM- Extra Cellular Matrix
NGC- nerve guide conduit (NGC)
PAN- Polyacrylonitrile
PHBV- Poly (hybutyrate-co-3-hydroxyvalerate)
PCL- Poly(ε-caprolactone)

PCLEEP- Poly (ε-caprolactone-*co*-ethyl ethylene phosphate)
PLLA- Poly (L-lactic acid)
PLGA- Poly (lactic- co -glycolic acid)
PTHF-Poly tetra hydrofuran
PU- Polyurethane
SBF- Simulated Body Fluid.

References

1. Cai Y, Liu Y, Yan W et al (2007) Role of hydroxyapatite nanoparticle size in bone cell proliferation. J Mater Chem 17:3780–3787. https://doi.org/10.1039/B705129H
2. Ushiki T (2002) Collagen fibers, reticular fibers and elastic fibers. A comprehensive understanding from a morphological viewpoint. Arch Histol Cytol 65:109–126. https://doi.org/10.1679/aohc.65.109
3. Decuzzi P, Ferrari M (2010) Modulating cellular adhesion through nanotopography. Biomater 31:173–179. https://doi.org/10.1016/j.biomaterials.2009.09.018
4. Ferrari M, Cirisano F, Morán MC (2019) Mammalian cell behavior on hydrophobic substrates: influence of surface properties. Colloids Interfaces 3:48. https://doi.org/10.3390/colloids3020048
5. Vishnu J, K Manivasagam V, Gopal V, et al (2019) Hydrothermal treatment of etched titanium: A potential surface nano-modification technique for enhanced biocompatibility. Nanomedicine Nanotechnology, Biol Med 20. https://doi.org/10.1016/j.nano.2019.102016
6. Jackson TC, Patani BO, Israel MB (2017) Nanomaterials and cell interactions: a review. J Biomater Nanobiotechnology 8:220–228. https://doi.org/10.4236/jbnb.2017.84015
7. Lim SH, Liu XY, Song H et al (2010) The effect of nanofiber-guided cell alignment on the preferential differentiation of neural stem cells. Biomater 31:9031–9039. https://doi.org/10.1016/j.biomaterials.2010.08.021
8. Hackelberg S, Tuck SJ, He L et al (2017) Nanofibrous scaffolds for the guidance of stem cell-derived neurons for auditory nerve regeneration. PLoS ONE 12:e0180427. https://doi.org/10.1371/journal.pone.0180427
9. Gertz CC, Leach MK, Birrell LK et al (2010) Accelerated neuritogenesis and maturation of primary spinal motor neurons in response to nanofibers. Dev Neurobiol 70:589–603. https://doi.org/10.1002/dneu.20792
10. Jeffries EM, Wang Y (2013) Incorporation of parallel electrospun fibers for improved topographical guidance in 3D nerve guides. Biofabrication 5:035015. https://doi.org/10.1088/1758-5082/5/3/035015
11. Luo B, Tian L, Chen N et al (2018) Electrospun nanofibers facilitate better alignment, differentiation, and long-term culture in an *in vitro* model of the neuromuscular junction (NMJ). Biomater Sci 6:3262–3272. https://doi.org/10.1039/C8BM00720A
12. Mahairaki V, Lim SH, Christopherson GT et al (2011) Nanofiber matrices promote the neuronal differentiation of human embryonic stem cell-derived neural precursors in vitro. Tissue Eng Part A 17:855–863. https://doi.org/10.1089/ten.TEA.2010.0377
13. Hosseinzadeh S, Soleimani M, Vossoughi M et al (2017) Study of epithelial differentiation and protein expression of keratinocyte-mesenchyme stem cell co-cultivation on electrospun nylon/B. vulgaris extract composite scaffold. Mater Sci Eng, C 75:653–662. https://doi.org/10.1016/J.MSEC.2017.02.101
14. Guan J, Wang F, Li Z et al (2011) The stimulation of the cardiac differentiation of mesenchymal stem cells in tissue constructs that mimic myocardium structure and biomechanics. Biomater 32:5568–5580. https://doi.org/10.1016/J.BIOMATERIALS.2011.04.038

15. Xue R, Qian Y, Li L et al (2017) Polycaprolactone nanofiber scaffold enhances the osteogenic differentiation potency of various human tissue-derived mesenchymal stem cells. Stem Cell Res Ther 8:148. https://doi.org/10.1186/s13287-017-0588-0
16. Rothrauff BB, Lauro BB, Yang G et al (2017) Braided and stacked electrospun nanofibrous scaffolds for tendon and ligament tissue engineering. Tissue Eng Part A 23:378–389. https://doi.org/10.1089/ten.TEA.2016.0319
17. Wang Y, Yuan X, Yu K et al (2018) Fabrication of nanofibrous microcarriers mimicking extracellular matrix for functional microtissue formation and cartilage regeneration. Biomater 171:118–132. https://doi.org/10.1016/J.BIOMATERIALS.2018.04.033
18. Dang JM, Leong KW (2007) Myogenic induction of aligned mesenchymal stem cell sheets by culture on thermally responsive electrospun nanofibers. Adv Mater 19:2775–2779. https://doi.org/10.1002/adma.200602159
19. Stout DA, Basu B, Webster TJ (2011) Poly(lactic–co-glycolic acid): carbon nanofiber composites for myocardial tissue engineering applications. Acta Biomater 7:3101–3112. https://doi.org/10.1016/J.ACTBIO.2011.04.028
20. Lian M, Sun B, Qiao Z et al (2019) Bi-layered electrospun nanofibrous membrane with osteogenic and antibacterial properties for guided bone regeneration. Colloids Surfaces B Biointerfaces 176:219–229. https://doi.org/10.1016/J.COLSURFB.2018.12.071
21. Shah RN, Shah NA, Del Rosario Lim MM et al (2010) Supramolecular design of self-assembling nanofibers for cartilage regeneration. Proc Natl Acad Sci U S A 107:3293–3298. https://doi.org/10.1073/pnas.0906501107
22. Di Luca A, Szlazak K, Lorenzo-Moldero I et al (2016) Influencing chondrogenic differentiation of human mesenchymal stromal cells in scaffolds displaying a structural gradient in pore size. Acta Biomater 36:210–219. https://doi.org/10.1016/J.ACTBIO.2016.03.014
23. Kumar G, Tison CK, Chatterjee K, et al (2011) The determination of stem cell fate by 3D scaffold structures through the control of cell shape. https://doi.org/10.1016/j.biomaterials.2011.08.054
24. Kouhi M, Jayarama Reddy V, Fathi M et al (2019) Poly (3-hydroxybutyrate-co-3-hydroxyvalerate)/fibrinogen/bredigite nanofibrous membranes and their integration with osteoblasts for guided bone regeneration. J Biomed Mater Res, Part A. https://doi.org/10.1002/jbm.a.36607
25. Zhu Y, Leong MF, Ong WF et al (2007) Esophageal epithelium regeneration on fibronectin grafted poly(l-lactide-co-caprolactone) (PLLC) nanofiber scaffold. Biomater 28:861–868. https://doi.org/10.1016/J.BIOMATERIALS.2006.09.051
26. Wu S-C, Chen C-H, Chang J-K et al (2013) Hyaluronan initiates chondrogenesis mainly via CD44 in human adipose-derived stem cells. J Appl Physiol 114:1610–1618. https://doi.org/10.1152/japplphysiol.01132.2012
27. Frost HK, Andersson T, Johansson S et al (2018) Electrospun nerve guide conduits have the potential to bridge peripheral nerve injuries in vivo. Sci Rep 8:16716. https://doi.org/10.1038/s41598-018-34699-8
28. Akkiraju H, Nohe A (2015) Role of chondrocytes in cartilage formation, progression of osteoarthritis and cartilage regeneration. J Dev Biol 3:177–192. https://doi.org/10.3390/jdb3040177
29. Rana D, Ratheesh G, Ramakrishna S, Ramalingam M (2017) Nanofiber composites in cartilage tissue engineering. Nanofiber Compos Biomed Appl 325–344. https://doi.org/10.1016/B978-0-08-100173-8.00013-2
30. Kazemnejad S, Khanmohammadi M, Baheiraei N, Arasteh S (2017) Current state of cartilage tissue engineering using nanofibrous scaffolds and stem cells. Avicenna J Med Biotechnol 9:50–65
31. Gupte MJ, Swanson WB, Hu J et al (2018) Pore size directs bone marrow stromal cell fate and tissue regeneration in nanofibrous macroporous scaffolds by mediating vascularization. Acta Biomater 82:1–11. https://doi.org/10.1016/J.ACTBIO.2018.10.016
32. Lin H-Y, Tsai W-C, Chang S-H (2017) Collagen-PVA aligned nanofiber on collagen sponge as bi-layered scaffold for surface cartilage repair. J Biomater Sci Polym Ed 28:664–678. https://doi.org/10.1080/09205063.2017.1295507

33. Krishna L, Dhamodaran K, Jayadev C, et al Nanostructured scaffold as a determinant of stem cell fate. https://doi.org/10.1186/s13287-016-0440-y
34. Wise JK, Yarin AL, Megaridis CM, Cho M (2009) Chondrogenic differentiation of human mesenchymal stem cells on oriented nanofibrous scaffolds: engineering the superficial zone of articular cartilage. Tissue Eng Part A 15:913–921. https://doi.org/10.1089/ten.tea.2008.0109
35. Matthews JA, Boland ED, Wnek GE et al (2003) Electrospinning of collagen type II: a feasibility study. J Bioact Compat Polym 18:125–134. https://doi.org/10.1177/088391150 3018002003
36. Jiang T, Kai D, Liu S et al (2018) Mechanically cartilage-mimicking poly(PCL-PTHF urethane)/collagen nanofibers induce chondrogenesis by blocking NF–kappa B signaling pathway. Biomaterials 178:281–292. https://doi.org/10.1016/J.BIOMATERIALS. 2018.06.023
37. Wang J, Sun B, Tian L et al (2017) Evaluation of the potential of rhTGF- β3 encapsulated P(LLA-CL)/collagen nanofibers for tracheal cartilage regeneration using mesenchymal stems cells derived from Wharton's jelly of human umbilical cord. Mater Sci Eng, C 70:637–645. https://doi.org/10.1016/j.msec.2016.09.044
38. Wei G, Ma PX (2006) Macroporous and nanofibrous polymer scaffolds and polymer/bone-like apatite composite scaffolds generated by sugar spheres. J Biomed Mater Res, Part A 78A:306–315. https://doi.org/10.1002/jbm.a.30704
39. Liu Q, Wang J, Chen Y et al (2018) Suppressing mesenchymal stem cell hypertrophy and endochondral ossification in 3D cartilage regeneration with nanofibrous poly(l-lactic acid) scaffold and matrilin-3. Acta Biomater 76:29–38. https://doi.org/10.1016/J.ACTBIO.2018. 06.027
40. Basakran NS (2015) CD44 as a potential diagnostic tumor marker. Saudi Med J 36:273–279. https://doi.org/10.15537/smj.2015.3.9622
41. Ustun Yaylaci S, Sardan Ekiz M, Arslan E et al (2016) Supramolecular GAG-like self-assembled glycopeptide nanofibers induce chondrogenesis and cartilage regeneration. Biomacromol 17:679–689. https://doi.org/10.1021/acs.biomac.5b01669
42. Arslan E, Sardan Ekiz M, Eren Cimenci C et al (2018) Protective therapeutic effects of peptide nanofiber and hyaluronic acid hybrid membrane in in vivo osteoarthritis model. Acta Biomater 73:263–274. https://doi.org/10.1016/J.ACTBIO.2018.04.015
43. Kim JI, Hwang TI, Aguilar LE et al (2016) A controlled design of aligned and random nanofibers for 3D bi-functionalized nerve conduits fabricated via a novel electrospinning set-up. Sci Rep 6:23761. https://doi.org/10.1038/srep23761
44. Schuh CMAP, Morton TJ, Banerjee A et al (2014) Activated schwann cell-like cells on aligned fibrin-poly(lactic-co-glycolic acid) structures: a novel construct for application in peripheral nerve regeneration. Cells Tissues Organs 200:287–299. https://doi.org/10.1159/000437091
45. Orlova Y, Magome N, Liu L et al (2011) Electrospun nanofibers as a tool for architecture control in engineered cardiac tissue. Biomaterials 32:5615–5624. https://doi.org/10.1016/j. biomaterials.2011.04.042
46. Kai D, Prabhakaran MP, Jin G, Ramakrishna S (2011) Guided orientation of cardiomyocytes on electrospun aligned nanofibers for cardiac tissue engineering. J Biomed Mater Res Part B Appl Biomater 98B:379–386. https://doi.org/10.1002/jbm.b.31862
47. Çetin MZ, Camurlu P (2018) An amperometric glucose biosensor based on PEDOT nanofibers. RSC Adv 8:19724–19731. https://doi.org/10.1039/C8RA01385C
48. Virji S, Huang J, Kaner RB, Weiller BH (2004) Polyaniline nanofiber gas sensors: examination of response mechanisms. Nano Lett 4(3):491–496. https://doi.org/10.1021/nl035122e
49. Marx S, Jose MV, Andersen JD, Russell AJ (2011) Electrospun gold nanofiber electrodes for biosensors. Biosens Bioelectron 26:2981–2986. https://doi.org/10.1016/J.BIOS.2010.11.050
50. Li D, Lv P, Zhu J et al (2015) NiCu Alloy nanoparticle-loaded carbon nanofibers for phenolic biosensor applications. Sensors 15:29419–29433. https://doi.org/10.3390/s151129419
51. Teepoo S, Dawan P, Barnthip N et al (2017) Electrospun chitosan-gelatin biopolymer composite nanofibers for horseradish peroxidase immobilization in a hydrogen peroxide biosensor. Biosensors 7:47. https://doi.org/10.3390/bios7040047

52. Apetrei IM, Diaconu C, Apetrei C, Georgescu C (2016) Electrochemical biosensor based on carbon nanofibers and diamine oxidase for detection of norepinephrine. Rom Biotechnol Lett 21(1):11092
53. Rosalki SB, Roberts R, Katus HA et al (2004) Cardiac biomarkers for detection of myocardial infarction: perspectives from past to present. Clin Chem 50:2205–2213. https://doi.org/10.1373/clinchem.2004.041749
54. Rezaei B, Shoushtari AM, Rabiee M et al (2018) An electrochemical immunosensor for cardiac Troponin I using electrospun carboxylated multi-walled carbon nanotube-whiskered nanofibres. Talanta 182:178–186. https://doi.org/10.1016/J.TALANTA.2018.01.046
55. Periyakaruppan A, Gandhiraman RP, Meyyappan M, Koehne JE (2013) Label-free detection of cardiac troponin-I using carbon nanofiber based nanoelectrode arrays. Anal Chem 85:3858–3863. https://doi.org/10.1021/ac302801z
56. Kumar R, Meyyappan M, Koehne JE (2018) Multiplexed electrochemical immunosensor for label-free detection of cardiac markers using carbon nanofiber array device. Meet Abstr MA2018-01:2502–2502
57. Gupta RK, Pandya R, Sieffert T et al (2016) Multiplexed electrochemical immunosensor for label-free detection of cardiac markers using a carbon nanofiber array chip. J Electroanal Chem 773:53–62. https://doi.org/10.1016/J.JELECHEM.2016.04.034
58. Gonzalez M, Rivera D, Marcelino A et al (2014) The effect of silver nanofibers on the deformation properties of blood vessels: towards the development of new nanotechnologies to prevent rupture of aneurysms. J Nanomater 2014:1–8. https://doi.org/10.1155/2014/853120
59. Kim M-W, An S, Kim K et al (2018) Packing of metalized polymer nanofibers for aneurysm embolization. Nanoscale 10:6589–6601. https://doi.org/10.1039/C7NR09645C
60. Wang J-B, Zhou B, Gu X-L et al (2013) Treatment of a canine carotid artery aneurysm model with a biodegradable nanofiber-covered stent: a prospective pilot study. Neurol India 61:282–287. https://doi.org/10.4103/0028-3886.115069
61. Liu K-S, Lee C-H, Lee D et al (2018) Sustained local delivery of high-concentration vancomycin from a hybrid biodegradable, antibiotic-eluting, nanofiber-loaded endovascular prosthesis for treatment of mycotic aortic aneurysms. J Vasc Surg 68:597–606. https://doi.org/10.1016/J.JVS.2017.07.142
62. Tang Y, Chen L, Zhao K et al (2016) Fabrication of PLGA/HA (core)-collagen/amoxicillin (shell) nanofiber membranes through coaxial electrospinning for guided tissue regeneration. Compos Sci Technol 125:100–107. https://doi.org/10.1016/J.COMPSCITECH.2016.02.005
63. Wang Y, Jiang Y, Zhang Y et al (2019) Dual functional electrospun core-shell nanofibers for anti-infective guided bone regeneration membranes. Mater Sci Eng, C 98:134–139. https://doi.org/10.1016/J.MSEC.2018.12.115
64. Fu L, Wang Z, Dong S et al (2017) Bilayer poly(lactic-co-glycolic acid)/nano-hydroxyapatite membrane with barrier function and osteogenesis promotion for guided bone regeneration. Materials (Basel) 10:257. https://doi.org/10.3390/ma10030257
65. Trobos M, Juhlin A, Shah FA et al (2018) *In vitro* evaluation of barrier function against oral bacteria of dense and expanded polytetrafluoroethylene (PTFE) membranes for guided bone regeneration. Clin Implant Dent Relat Res 20:738–748. https://doi.org/10.1111/cid.12629
66. Dwek JR (2010) The periosteum: what is it, where is it, and what mimics it in its absence? Skeletal Radiol 39:319–323. https://doi.org/10.1007/s00256-009-0849-9
67. Wang T, Zhai Y, Nuzzo M et al (2018) Layer-by-layer nanofiber-enabled engineering of biomimetic periosteum for bone repair and reconstruction. Biomaterials 182:279–288. https://doi.org/10.1016/J.BIOMATERIALS.2018.08.028
68. Cheng G, Ma X, Li J et al (2018) Incorporating platelet-rich plasma into coaxial electrospun nanofibers for bone tissue engineering. Int J Pharm 547:656–666. https://doi.org/10.1016/j.ijpharm.2018.06.020
69. Li L, Zhou G, Wang Y et al (2015) Controlled dual delivery of BMP-2 and dexamethasone by nanoparticle-embedded electrospun nanofibers for the efficient repair of critical-sized rat calvarial defect. Biomaterials 37:218–229. https://doi.org/10.1016/J.BIOMATERIALS.2014.10.015

70. Yao Q, Cosme JGL, Xu T et al (2017) Three dimensional electrospun PCL/PLA blend nanofibrous scaffolds with significantly improved stem cells osteogenic differentiation and cranial bone formation. Biomaterials 115:115–127. https://doi.org/10.1016/J.BIOMATERIALS.2016.11.018
71. Yosefifard M, Hassanpour-Ezatti M (2014) Epidural administration of neostigmine-loaded nanofibers provides extended analgesia in rats. Daru 22:73. https://doi.org/10.1186/s40199-014-0073-6
72. Riggin CN, Qu F, Kim DH et al (2017) Electrospun PLGA nanofiber scaffolds release Ibuprofen faster and degrade slower after In Vivo implantation. Ann Biomed Eng 45:2348–2359. https://doi.org/10.1007/s10439-017-1876-7
73. Kao C-W, Lee D, Wu M-H et al (2017) Lidocaine/ketorolac-loaded biodegradable nanofibrous anti-adhesive membranes that offer sustained pain relief for surgical wounds. Int J Nanomedicine 12:5893–5901. https://doi.org/10.2147/IJN.S140825
74. Lin Y-C, Liu K-S, Lee D et al (2018) In Vivo and In Vitro elution of analgesics from multi-layered poly(D, L)-lactide-co-glycolide nanofibers incorporated ureteral stents. J Nanomater 2018:1–7. https://doi.org/10.1155/2018/4943210
75. Bahnson ESM, Kassam HA, Moyer TJ et al (2016) Targeted nitric oxide delivery by supramolecular nanofibers for the prevention of restenosis after arterial injury. Antioxid Redox Signal 24:401–418. https://doi.org/10.1089/ars.2015.6363
76. Nguyen LH, Gao M, Lin J et al (2017) Three-dimensional aligned nanofibers-hydrogel scaffold for controlled non-viral drug/gene delivery to direct axon regeneration in spinal cord injury treatment. Sci Rep 7:42212. https://doi.org/10.1038/srep42212
77. Naskar D, Ghosh AK, Mandal M et al (2017) Dual growth factor loaded nonmulberry silk fibroin/carbon nanofiber composite 3D scaffolds for in vitro and in vivo bone regeneration. Biomaterials 136:67–85. https://doi.org/10.1016/J.BIOMATERIALS.2017.05.014
78. Weng L, Boda SK, Wang H et al (2018) Novel 3D hybrid nanofiber aerogels coupled with BMP-2 peptides for cranial bone regeneration. Adv Healthc Mater 7:1701415. https://doi.org/10.1002/adhm.201701415
79. Rajzer I, Kurowska A, Jabłoński A et al (2018) Layered gelatin/PLLA scaffolds fabricated by electrospinning and 3D printing-for nasal cartilages and subchondral bone reconstruction. Mater Des 155:297–306. https://doi.org/10.1016/J.MATDES.2018.06.012
80. Rohani S, Rohani S Synthesis of titania nanotube arrays by anodization. AIDIC CONF SER 2009 121–129
81. Roy P, Berger S, Schmuki P (2011) TiO2 nanotubes: synthesis and applications. Angew Chemie Int Ed 50:2904–2939. https://doi.org/10.1002/anie.201001374
82. Assefpour-Dezfuly M, Vlachos C, Andrews EH (1984) Oxide morphology and adhesive bonding on titanium surfaces. J Mater Sci 19:3626–3639. https://doi.org/10.1007/BF00552275
83. Albu SP, Ghicov A, Macak JM, Schmuki P (2007) 250 μm long anodic TiO2 nanotubes with hexagonal self-ordering. Phys status solidi—Rapid Res Lett 1:R65–R67. https://doi.org/10.1002/pssr.200600069
84. Wei W, Berger S, Hauser C et al (2010) Transition of TiO2 nanotubes to nanopores for electrolytes with very low water contents. Electrochem Commun 12:1184–1186. https://doi.org/10.1016/J.ELECOM.2010.06.014
85. Kowalski D, Kim D, Schmuki P (2013) TiO2 nanotubes, nanochannels and mesosponge: self-organized formation and applications. Nano Today 8:235–264. https://doi.org/10.1016/J.NANTOD.2013.04.010
86. Macak JM, Tsuchiya H, Ghicov A et al (2007) TiO2 nanotubes: self-organized electrochemical formation, properties and applications. Curr Opin Solid State Mater Sci 11:3–18. https://doi.org/10.1016/J.COSSMS.2007.08.004
87. Mazare A, Dilea M, Ionita D, Demetrescu I (2014) Electrochemical behavior in simulated body fluid of TiO2 nanotubes on TiAlNb alloy elaborated in various anodizing electrolyte. Surf Interface Anal 46:186–192. https://doi.org/10.1002/sia.5364

88. Kim W-G, Choe H-C, Ko Y-M, Brantley WA (2009) Nanotube morphology changes for Ti–Zr alloys as Zr content increases. Thin Solid Films 517:5033–5037. https://doi.org/10.1016/J.TSF.2009.03.165
89. Minagar S, Berndt CC, Gengenbach T, Wen C (2014) Fabrication and characterization of TiO_2–ZrO_2–$ZrTiO_4$ nanotubes on TiZr alloy manufactured via anodization. J Mater Chem B 2:71–83. https://doi.org/10.1039/C3TB21204A
90. Bakir M (2012) Haemocompatibility of titanium and its alloys. J Biomater Appl 27:3–15. https://doi.org/10.1177/0885328212439615
91. Puckett SD, Taylor E, Raimondo T, Webster TJ (2010) The relationship between the nanostructure of titanium surfaces and bacterial attachment. Biomaterials 31:706–713. https://doi.org/10.1016/J.BIOMATERIALS.2009.09.081
92. Park J, Bauer S, von der Mark K, Schmuki P (2007) Nanosize and vitality: TiO2 nanotube diameter directs cell fate. https://doi.org/10.1021/NL070678D
93. Bauer S, Park J, Faltenbacher J et al (2009) Size selective behavior of mesenchymal stem cells on ZrO2 and TiO2 nanotube arrays. Integr Biol 1:525. https://doi.org/10.1039/b908196h
94. Brammer KS, Oh S, Frandsen CJ, Jin S (2011) Biomaterials and biotechnology schemes utilizing TiO2 nanotube arrays. In: Biomaterials science and engineering. InTech, pp 193–210
95. Lucchini J-P, Aurelle J-L, Therin M et al (1996) A pilot study comparing screw-shaped implants: Surface analysis and histologic evaluation of bone healing. Clin Oral Implants Res 7:397–404. https://doi.org/10.1034/j.1600-0501.1996.070414.x
96. Gongadze E, Kabaso D, Bauer S et al (2011) Adhesion of osteoblasts to a nanorough titanium implant surface. Int J Nanomedicine 6:1801–1816. https://doi.org/10.2147/IJN.S21755
97. Gongadze E, Kabaso D, Bauer S et al (2013) Adhesion of osteoblasts to a vertically aligned TiO2 nanotube surface. Mini Rev Med Chem 13:194–200. https://doi.org/10.2174/138955713804805166
98. Zhao G, Schwartz Z, Wieland M et al (2005) High surface energy enhances cell response to titanium substrate microstructure. J Biomed Mater Res, Part A 74A:49–58. https://doi.org/10.1002/jbm.a.30320
99. Bauer S, Park J, Pittrof A et al (2011) Covalent functionalization of TiO2 nanotube arrays with EGF and BMP-2 for modified behavior towards mesenchymal stem cells. Integr Biol 3:927. https://doi.org/10.1039/c0ib00155d
100. Lai M, Cai K, Zhao L et al (2011) Surface functionalization of TiO_2 nanotubes with bone morphogenetic protein 2 and its synergistic effect on the differentiation of mesenchymal stem cells. Biomacromol 12:1097–1105. https://doi.org/10.1021/bm1014365
101. Park J, Bauer S, Pittrof A et al (2012) Synergistic control of mesenchymal stem cell differentiation by nanoscale surface geometry and immobilized growth factors on TiO2 nanotubes. Small 8:98–107. https://doi.org/10.1002/smll.201100790
102. Ercan B, Taylor E, Alpaslan E, Webster TJ (2011) Diameter of titanium nanotubes influences anti-bacterial efficacy. Nanotechnology 22:295102. https://doi.org/10.1088/0957-4484/22/29/295102
103. Huo K, Zhang X, Wang H et al (2013) Osteogenic activity and antibacterial effects on titanium surfaces modified with Zn-incorporated nanotube arrays. Biomaterials 34:3467–3478. https://doi.org/10.1016/J.BIOMATERIALS.2013.01.071
104. Kummer KM, Taylor EN, Durmas NG et al (2013) Effects of different sterilization techniques and varying anodized TiO_2 nanotube dimensions on bacteria growth. J Biomed Mater Res Part B Appl Biomater 101B:677–688. https://doi.org/10.1002/jbm.b.32870
105. Zhao L, Wang H, Huo K et al (2011) Antibacterial nano-structured titania coating incorporated with silver nanoparticles. Biomaterials 32:5706–5716. https://doi.org/10.1016/J.BIOMATERIALS.2011.04.040
106. Grigorescu S, Ungureanu C, Kirchgeorg R et al (2013) Various sized nanotubes on TiZr for antibacterial surfaces. Appl Surf Sci 270:190–196. https://doi.org/10.1016/J.APSUSC.2012.12.165
107. Shrestha NK, Macak JM, Schmidt-Stein F et al (2009) Magnetically guided titania nanotubes for site-selective photocatalysis and drug release. Angew Chemie Int Ed 48:969–972. https://doi.org/10.1002/anie.200804429

108. Song Y-Y, Schmidt-Stein F, Bauer S, Schmuki P (2009) Amphiphilic TiO_2 nanotube arrays: an actively controllable drug delivery system. J Am Chem Soc 131:4230–4232. https://doi.org/10.1021/ja810130h
109. Song Y-Y, Schmuki P (2010) Modulated TiO2 nanotube stacks and their use in interference sensors. Electrochem Commun 12:579–582. https://doi.org/10.1016/J.ELECOM.2010.02.004
110. Aw MS, Gulati K, Losic D (2011) Controlling drug release from titania nanotube arrays using polymer nanocarriers and biopolymer coating. J Biomater Nanobiotechnol 02:477–484. https://doi.org/10.4236/jbnb.2011.225058
111. K Webb H, Hasan J, K. Truong V, et al (2011) Nature inspired structured surfaces for biomedical applications. Curr Med Chem 18:3367–3375. https://doi.org/10.2174/092986711796504673
112. Dundar Arisoy F, Kolewe KW, Homyak B et al (2018) Bioinspired photocatalytic shark-skin surfaces with antibacterial and antifouling activity via nanoimprint lithography. ACS Appl Mater Interfaces 10:20055–20063. https://doi.org/10.1021/acsami.8b05066
113. Lee BEJ, Exir H, Weck A, Grandfield K (2018) Characterization and evaluation of femtosecond laser-induced sub-micron periodic structures generated on titanium to improve osseointegration of implants. Appl Surf Sci 441:1034–1042. https://doi.org/10.1016/J.APSUSC.2018.02.119
114. Menci G, Demir AG, Waugh DG et al (2019) Laser surface texturing of β-Ti alloy for orthopaedics: Effect of different wavelengths and pulse durations. Appl Surf Sci 489:175–186. https://doi.org/10.1016/J.APSUSC.2019.05.111
115. Jaggessar A, Mathew A, Wang H et al (2018) Mechanical, bactericidal and osteogenic behaviours of hydrothermally synthesised TiO2 nanowire arrays. J Mech Behav Biomed Mater 80:311–319. https://doi.org/10.1016/J.JMBBM.2018.02.011
116. Manivasagam V, Popat K (2020) In vitro investigation of hemocompatibility of hydrothermally treated titanium and titanium alloy surfaces. ACS Omega 5(14):8108–8120. https://doi.org/10.1021/acsomega.0c00281
117. Sakka S, Coulthard P (2011) Implant failure: etiology and complications. Med Oral Patol Oral Cir Bucal 16(1) (Jan 1):e42–e44. https://doi.org/10.4317/medoral. PMID: 20526267
118. Palmer LC, Newcomb CJ, Kaltz SR, Spoerke ED, Stupp SI (2008) Biomimetic systems for hydroxyapatite mineralization inspired by bone and enamel. Chem Rev 108(11):4754–4783. https://doi.org/10.1021/cr8004422
119. Perez RA, Altankov G, Jorge-Herrero E, Ginebra MP (2013) Micro- and nanostructured hydroxyapatite-collagen microcarriers for bone tissue-engineering applications. J Tissue Eng Regen Med 7(5) (May):353–361. https://doi.org/10.1002/term.530
120. Thakral G, Thakral R, Sharma N, Seth J, Vashisht P (2014) Nanosurface—the future of implants. J Clin Diagn Res 8(5):ZE07–ZE10. https://doi.org/10.7860/JCDR/2014/8764.4355
122. Vasita R, Katti DS (2006) Nanofibers and their applications in tissue engineering. Int J Nanomedicine 1(1):15–30. https://doi.org/10.2147/nano.2006.1.1.15
123. Kiran ASK, Kizhakeyil A, Ramalingam R, Verma NK, Lakshminarayanan R, Sampath Kumar TS, Doble M, Ramakrishna S (2019) Drug loaded electrospun polymer/ceramic composite nanofibrous coatings on titanium for implant related infections. Ceram Int 45(15):18710–18720. ISSN 0272-8842. https://doi.org/10.1016/j.ceramint.2019.06.097

Nanoceramics: Synthesis, Characterizations and Applications

S. Anne Pauline

Abstract Nanoceramics are ultrafine particles with particle size less than 100 nm and have greater advantages over macroscale ceramics which are brittle and rigid. They are inorganic, metallic and non-metallic compounds that have high heat resistance. Their small particle size offers them unique properties which have led to their widespread use in various fields. Their improved properties include bioactivity, dielectricity, ferromagnetism, piezoelectricity, magnetoresistance and superconductivity. Hardness and strength of ceramics are greatly improved by reducing their particle size to be in the nanoscale. Nanoceramics can be conveniently prepared by various physical and chemical methods in various sizes and shapes such as nanoparticles, nanorods, nanotubes, nanoribbons, nanosheets and nanofluids which determines their properties. Characterization of nanoceramics can be carried out by surface characterization methods such as X-ray diffraction analysis, Infrared spectroscopy, Scanning electron Microscopy, Transmission Electron Microscopy, Atomic Force Microscopy, etc. Nanoceramic particles can be used for bone repair, drug delivery, energy supply and storage, communication, transportation systems and construction. The current article discusses in detail the nanoceramics, their preparation methods, various characterization techniques, their unique properties and their application in the biomedical field arising due to their excellent properties.

Keywords Nanoparticle · Bioactivity · Drug delivery · Scaffolds

1 Introduction

Nanoceramics are essentially ceramic materials whose length scale lies in the range of 1–100 nm at least in one dimension. Ceramics are inorganic solid materials that are characterized by heat resistance and are made of metallic as well as nonmetallic compounds. On macro scale, ceramics are rigid and brittle that breaks upon impact

S. A. Pauline (✉)
Department of Chemistry, Central Institute of Plastics Engineering & Technology, Chennai 600032, Tamil Nadu, India
e-mail: lillyanne68@gmail.com

T. S. Santra and L. Mohan (eds.), *Nanomaterials and Their Biomedical Applications*, Springer Series in Biomaterials Science and Engineering 16,
https://doi.org/10.1007/978-981-33-6252-9_5

against other hard objects. However, in the nanoscale, their properties are unique, differ widely from that of macroscale ceramics and are of greater use in various fields. Ceramic nanoparticles have properties lying in between metals and non-metals. The improved properties include dielectricity, ferroelectricity, piezoelectricity, pyroelectricity, ferromagnetism, magnetoresistance and superconductivity [1]. Nanoceramics also have excellent mechanical, processing and surface properties such as biocompatibility, superplasticity, mechanical resistance, chemical resistance, strength and hardness at normal as well as high temperatures [2]. The properties of nanoceramics depend and vary based on the type of nanoceramic, its size and shape. The bonding between their constituent atoms defines their properties which are a combination of ionic and covalent bonds [3].

Nanoceramic materials were first discovered in the early 1980s using sol-gel method, a form of chemical solution deposition. Larger scale materials have flaws that make them brittle, but due to their small size, nanoceramics are flawless. Interest and research in nanoceramics are blooming since then owing to its varied properties and extensive applications. In the 2000s, synthesis methods evolved using heat and pressure in sintering process. The method of synthesis of nanoceramics plays a great role in determining the shape and size of the particles and hence its properties. During the past 20 years, research in nanoceramics has resulted in positive outcomes and the advanced materials prepared are being used in several industries such as sensors, batteries, capacitors, corrosion-resistant coatings, thermal barrier coatings, solid electrolytes, catalysts, cosmetics, automotive, optoelectronics, computers, electronics, biomaterials, etc. [4]. They have special applications in the field of medicine owing to their biocompatibility, bioactivity and hydrophilicity and hence are used as bioimplants, as drug delivery devices and also in cancer treatment as chemotherapy delivery vehicles [5].

Nanoceramics are characterized by large surface area due to their small particle size. This is especially useful in cases, where their surface properties play a role in its effective functioning. For example, when nanoceramics are used as catalysts, their small particle size increases the rate of reaction [6]. When they are used as bioimplants, nanoceramics facilitate faster bone-implant interface establishment and aids in faster healing. When used as drug delivery systems, the drug reaches the target site faster, does not get released in non-target sites, react with the target effectively and carry out the intended purpose.

The design and development of nanoceramics of different sizes and shapes by new techniques have garnered much attention as each of them come with unique properties. Till now, nanoceramics in different sizes and shapes such as nanoparticles, nanorods, nanoribbons, nanotubes, nanosheets and nanofluids have been synthesized [4]. Modification in various physical, chemical and biological properties can be bought about by modifying the nanoceramics size, shape, by doping different elements in their crystallographic sites as impurity or by creating defects using ion implantation method [7]. Advanced nanoceramics are of 3 major types: (i) oxides of alumina, zirconia, titania, ceria, beryllia, etc., (ii) non-oxides such as borides, carbides, nitrides, silicide, etc., and (iii) composite ceramics such as particulate reinforced, fiber reinforced combinations of oxides and non-oxides [8]. Advanced nanoceramics have profound surface properties as their surface to volume ratio is more than 10% when their particle size goes below 100 nm at least in one dimension [9].

2 Synthesis of Nanoceramics

Nanoceramics can be produced by various techniques that apply both wet and dry conditions. Most of the synthesis methods use liquid as media and factors such as type of suspension, packing of particles and their dispersion play a great role in their synthesis and outcome. The method of synthesis not only defines the size and shape of nanoceramics but also influences its characteristics such as crystal habits, specific surface area and state of agglomeration [10]. Further, size and shape of the nanoparticles can be molded by adsorbing organic molecules on the surface of growing nanoparticles [11]. Nanoparticles tend to agglomerate due to their high surface area and thermodynamic instability and the synthesis method adopted should involve strategies to overcome this for effective usage [12].

Synthesis of nanoceramics can be broadly classified into two approaches (i) Top-down approach where a large particle is broken down into nanosized particles and (ii) bottom-up approach where ions, atoms, molecules or nanoparticles are assembled in a controlled manner to form nanoparticles. There are various methods of synthesis which come under any of these two broad classifications. The synthesis method plays a great role in the functionality of the produced nanoceramics. The focus of this chapter is on nanoceramics and its application as biomaterials. Hence, we will limit our discussion to the different synthesis methods by which nanoceramics that are used as biomaterials are produced. The common methods of synthesis of nanoceramics include mechanochemical synthesis, co-precipitation, sol-gel method, spray pyrolysis, microemulsion method, physical vapour deposition, etc., which are discussed in detail below [13].

2.1 Physical Vapour Deposition

Physical vapour deposition is a high vacuum coating technique that comprises vacuum deposition methods to produce thin coatings and films of pure metal or alloy. The material to be coated is heated and the metal in the condensed phase is converted to vapour phase [14]. In the vapour phase, the material gets supersaturated in an inert atmosphere to condense the metal nanoparticles and then deposits as thin adherent film of the condensed phase on electrically conductive material [15]. It is possible to develop a much thinner layer of the nanoceramic material over the implant surface with high adhesion strength. The porosity of the developed coatings could be close to that of the cortical bone which enhances osseointegration when the orthopaedic implant comes into contact with the cortical bone [16].

Titania coatings of different thicknesses were developed on machined Ti Screw shaped implants. New bone formation on the 120 nm coating was observed to be higher at the bone-implant interface compared to the 1430 nm coating implying that thinner PVD coating enhances a higher degree of bone growth soon after implantation [17]. Titania/silver combined coating has been developed by PVD method and it is reported to provide increased antimicrobial potency against microbial strains without affecting its mechanical performance [18].

2.2 *Spray Pyrolysis*

Spray pyrolysis method is widely used for preparing metal and metal oxide powders [19]. It involves formation of solid metal oxide particles by first converting the reactants into micro-sized liquid droplets and spraying it onto a hot substrate in the furnace. The precursor gets decomposed and the desired nanostructure is produced. The size and shape of the nanostructure can be modified by controlling the reaction conditions such as spray energy, droplet size of the precursors, distance between the substrate and the spray gun, etc. [20]. Biocompatible nano calcium phosphate (CaP) was synthesized by the aerosol-derived flame spray pyrolysis method. The obtained nanoparticles have a particle size of 23 nm with increased crystallinity and specific surface area. It showed reduced cytotoxicity at 5–50 μg/ml and higher alkaline phosphatase (ALP) enzyme activity indicating that CaP synthesized by this method can be used in biomedical applications [21]. Nano-sized HAp was produced by ultrasonic spray pyrolysis using a salt-assisted decomposition method. The added $NaNO_3$ salt interrupts agglomeration and results in the formation of rod-type, single-phase, nano-sized particles with high crystallinity [22].

2.3 *Chemical Vapour Deposition (CVD)*

Chemical vapour deposition is used to produce high-quality solid materials with better performance characteristics. The deposit can be produced on the substrate by depositing volatile precursors on its surface and then making them react or undergo decomposition in the presence of heat, light or plasma. Many CVD processes have been developed such as plasma-assisted chemical vapour deposition (PACVD), low-pressure chemical vapour deposition (LPCVD), laser-enhanced chemical vapour deposition (LECVD), plasma-enhanced chemical vapour deposition (PECVD), etc. Of these methods, PACVD is found to enhance biocompatibility, chemical stability and also increase corrosion resistance [23]. Nanostructured titania coating was developed on pure Ti implants by a metal-organic chemical vapour deposition method.

The coating is reported to accelerate the osseointegration rate and bone mineralization at the bone-implant interface [24]. Silica films were deposited onto titanium by PECVD and then functionalized with amino groups using 3-aminopropyl triethoxysilane (APTES). The functionalized coating was observed to have greater stability under physiological conditions and hence can be used as functional biomaterial coatings [25].

2.4 Mechanochemical Synthesis

Mechanochemical synthesis is a high-energy milling technique which involves the formation of nanoscale composites. This method involves coupling of chemical and mechanical phenomena on the molecular scale to produce ceramic nanoparticles. The application of mechanical action through ball mill enables the reaction to happen at room temperature or temperatures lower than the traditional methods avoiding external heating. It can be performed in the presence or absence of solvents. The precursor powders are milled together by placing them in a high-energy mill. Mechanically induced chemical reactions take place under a controlled atmosphere of high load and strain conditions. By modifying the reaction conditions nanostructured compounds having ultrafine grains and homogeneous composition can be produced [26]. Nanoceramics as small as 5 nm with high crystallinity without agglomeration can be produced by this method [27]. Precursors are normally a combination of salt and a metal oxide and they react during milling followed by a heating process. Nanocomposites of oxide, non-oxide as well as mixed ceramics can be prepared by this method.

Calcium phosphate can be synthesized by mechanochemical method under both wet and dry conditions. It was observed that wet grinding slowed down the reaction rate and increased powder contamination due to erosion and hence it is reported that dry mechanochemical synthesis is preferred for biomedical applications [28]. By milling $AlCl_3$ with CaO_5, nanoparticles of Al_2O_3 of size 10–20 nm can be formed [29]. Equal concentrations of lanthanum and silicate substituted apatite was produced by mechanochemical synthesis. Single-phase products can be produced by this method and the synthesized apatite was deposited on Ti substrate by micro-arc oxidation method. The resultant coating exhibited high biocompatibility and no cytotoxic action on mesenchymal stem cells [30].

Silver nanoparticles were successfully synthesized by combining this method and green synthesis using egg shell membrane or Origanum vulgare L. plant as the reducing agent. Silver nitrate was used as the silver precursor. The Ag nanoparticles synthesized by co-milling with Origanum plant exhibited higher antibacterial activity than the former one [31]. Hydroxyapatite with 20% Ti nanocomposite was synthesized by combining mechanochemical process with solid-state method. The resultant nanopowders exhibited high crystallinity, smaller size of about 25 nm and high purity and have improved bioactivity compared to calcium phosphate nanocomposites [32]. Calcium-deficient hydroxyapatite (CDHA) is used to prepare calcium phosphate ceramics and the CDHA with desired Ca/P ratio can be synthesized by mechanochemical synthesis method by varying the reaction conditions. Nano size crystals of hydroxyapatite of size around 20 nm and controlled Ca/P ratio was reported to be produced by this method [33].

2.5 Microemulsion Method

Micelles are formed by self-assembly of surfactants or block copolymers when they are present above a critical concentration in air or aqueous solution. The synthesis of nanoceramics takes place inside the confined space in the micelle or microemulsion. A microemulsion is produced by dispersing an organic solution in an aqueous solution in the form of fine liquid droplets. The reactants are dissolved and are present in the organic droplets. The reaction happens at the interface between the organic and aqueous part [19]. The reaction will continue and nanoceramic will be produced till reactants are available inside the micelle. This method enables control of particle size, morphology, shape and surface area of the nanoceramic. Mesoporous HAp was prepared by microemulsion technique using hecadecyltrimethyl ammonium bromide (CTAB), cyclohexane and n-octyl alcohol. The HAp powders had well-ordered and uniform morphology with broader pore size distribution. The HAp powders had good biocompatibility at low concentrations and low toxicity at high concentrations and hence are suitable for bone tissue grafts, drug delivery and also for coating material [34]. Iron oxide nanoparticles are produced using the copolymer, poly(styrene-block-allyl alcohol) by microemulsion method. Hydrophobic nanoparticles are encapsulated into the iron oxide nanoparticles to obtain multi-functional nanocomposite. Fluorescent dye and anti-cancer drug molecules are loaded into them and it was observed that the nanocomposite enabled imaging-guided and magnetic targeted drug delivery [35].

2.6 Wet Chemical Deposition

Wet chemical deposition method represents chemical reactions occurring in the solution phase using appropriate precursors and experimental conditions. Many wet chemical synthesis methods are known such as solvothermal synthesis, template synthesis, self-assembly, hot-injection, metal-organic decomposition, etc., by which nanomaterials can be effectively produced [36]. HAp nanostructures were synthesized by wet chemical synthesis method at different pH values and sintering temperatures. The rod and flake-like HAp structures that have formed have an enhanced Ca/P ratio of 1.83 and an increased crystallite size from 20 to 56 nm. Stable and porous Hap powders were synthesized by this method [37]. Needle-like and plate-like Mg-substituted Hap particles containing different amounts of Mg were prepared by wet chemical precipitation method from $Mg(OH)_2/Ca(OH)_2$ and H_3PO_4. It was observed that the specific surface area of the Hap powder increased with an increase in concentration of Mg [38].

2.7 *Sol-Gel Process*

Sol-gel technique is used to produce high quality, homogeneous, and highly stoichiometric nanostructures. For the fabrication of nanoceramic oxides, the respective alkoxides are converted into a colloidal solution (sol) in the first step, which is then converted into an integrated network (gel) in the second step. The gel can be used as a precursor and can be coated on a substrate to form a film or can be subjected to other treatment methods to form nanosized powders. Nanoceramics can be prepared in different shapes such as nanospheres, nanorods, nanoflakes, nanotubes, nanoribbons, nanofibers and nanocoating's for different applications. Sol-gel prepared nanoceramics have varied applications in the biomaterial field such as fabrication of bioactive implant coatings and scaffolds for orthopaedic applications and drug delivery systems. Nanocomposite coatings with high homogeneity and purity can also be prepared by this method. Orthopaedic implant materials such as 316L SS and titanium and its alloys were coated with nanoporous nanoceramic oxides such as titania, niobia, silica, etc., and their ability to accelerate bone growth was studied. It was reported that these coatings enhance bioactivity as well as increase the corrosion resistance of the implant [39, 40].

The composite coating of tantalum oxide—carbon nanotubes was developed on Ti plates by sol-gel co-deposition method. The carbon nanotubes were loaded with bisphosphonic acid moieties. The composite coating material supported HAp growth as revealed by in vitro analysis [41]. Mesoporous silica-titania composites were obtained by sol-gel method assisted by hydrothermal method and the drug oxytetracycline was loaded onto the mesochannels. The drug release studies reveal that the drug was released slowly and steadily and the system exhibited good antimicrobial activity against Staphylococcus aureus [42].

2.8 *Template-Based Synthesis*

Template-based synthesis is a simple procedure by which metallic particles in the nano size can be developed with the help of a host which has preexisting ordered porous structure. The developed nanoparticles will be of reduced dimension and also will be having the orientational order of the host material [43]. Different shapes of nanomaterials such as nanorods, nanowires, nanobelts, etc., can be developed with the help of templates using various methods such as electrophoretic deposition, filling of templates by capillary force, chemical conversion, etc. [44]. Silica-based mesoporous materials having a unique silica network of a well-ordered arrangement of pore system and cavities are developed by using surfactants as templates [45]. Nanoporous calcium phosphate ceramics were synthesized by a hard-templating method using ordered mesoporous carbon as the templating material. The developed calcium phosphate had disordered 3-D interconnected nanopores of 20–30 nm size with increased surface area and pore volume and exhibited higher charging capacity for antibiotics [46].

2.9 Biomimetic Deposition Method

Biomimetic deposition method is a synthetic deposition method where a coating is developed on the implant surface by mimicking physiological conditions. It is done by immersing the implants in simulated body fluid (SBF) at a temperature of 37 °C. Biomimetic deposition happens in two steps. Nucleation followed by formation and growth of the coating. Such synthetic coatings can be developed on materials such as metals, bioactive glasses, glass-ceramics, and also on polymers [47]. Calcium phosphate coatings can be developed on the surface of metals, ceramics, or polymers by this method. These coatings improve load-bearing mechanical strength, bioactivity, and osteoconductivity. When biomolecules, proteins, or growth factors are incorporated along in these coatings it is found to improve osteoinductivity and also sustainably deliver these biomolecules, thus enhancing regeneration of bone tissue [48]. Octacalcium phosphate coating developed on zirconia oral implants by biomimetic method was found to have good reproducibility and improved tensile adhesion strength [49]. Hydroxyapatite coatings were developed on T-i6Al-4 V substrate by immersing the specimen in supersaturated Ca/P solution that has ionic composition similar to that of SBF. Homogeneous HAp coating was developed by this method in a few hours whereas it will require 14 days to develop similar coating from SBF. The deposited coating consists of HAp globular aggregates with fine lamellar structure than those deposited from SBF [50].

2.10 Electrophoretic Deposition

Electrophoretic deposition includes many industrial processes such as cathodic electrodeposition, anodic electrodeposition, electrophoretic coating method, etc. In this process, the coating material is suspended in the form of a colloid in the electrolytic solution and under the influence of an electric field, it migrates and gets deposited onto the electrode. Materials such as ceramics, metals, pigments, polymers and dyes can be deposited by this method. Electroplating is a type of electrophoretic deposition. Graphene oxide reinforced calcium phosphate coating was developed on anodized Ti by pulse electrodeposition method. The graphene reinforced coating exhibited improved nano hardness, adhesion strength, crystallinity and decreased Young's modulus mismatch of the coating with the Ti substrate and hence offered better protection of anodized titanium against corrosion [51]. Cobalt substituted calcium phosphate coatings were developed on Ti-22Nb-6Zr alloy by electrodeposition method. The coating was observed to be made up of a low crystalline apatite phase and it protected the alloy against the corrosive Hank's solution [52]. Pure hydroxyapatite and HAp-ZrO_2-TiO_2 nanocomposite coatings were developed by merging two different electroplating methods. The coatings exhibited bioactive behavior and also had improved corrosion resistance as observed from decreased corrosion current density (I_{corr}) values [53].

3 Nanoceramics Characterization

Nanoceramics are key components finding application in different fields ranging from energy generation to applications in the biomedical field. New advanced technologies are developed continuously with the help of nanoceramics. Characterization of nanoceramics is essential as it helps us to understand and control the synthesis as well as discover possible applications. Nanoceramics can be characterized based on the conventional characterization methods and surface analysis techniques used for bulk materials and specialized techniques are being developed to study them in nanoscale too. In this section, let us briefly study the various characterization techniques, its significance, and outcome.

3.1 X-Ray Diffraction Analysis (XRD)

X-ray diffraction analysis (XRD) is an important, accurate, and non-destructive experimental technique used to identify the crystal structure of solids, geometry, identify unknown samples, orientation of single crystals, etc. Various characteristics of a single crystal or polycrystalline material in the form of powder can be studied. In this, a beam of X-rays of wavelength 0.7–2 A° is made to fall on a small amount of the material. The X-ray gets diffracted at the crystalline phases in the material governed by the Bragg's law:

$$n\lambda = 2d \sin \theta$$

where, λ is the X-ray wavelength, d is the spacing between atomic planes in the crystal and θ is the diffraction angle. The main disadvantage of XRD is that it is a time-consuming process and it requires large volume of sample compared to new advanced techniques.

3.2 *Small Angle X-Ray Scattering*

Small angle X-ray scattering is a surface analytical technique that measures the intensity of scattered X-rays as a function of the scattering angle. A very narrow and highly intense incident X-ray beam is focused on the material under study and the behaviour of X-rays that have undergone elastic scattering is studied. Measurements are made at very small angles in the range of $< 5°$. Nanoparticles in the size range of 1–100 nm even up to 300 nm can be measured [54]. Structural features and properties of nanomaterials such as particle shape, specific surface area, nanoparticle size distribution, pore size distribution, agglomeration behavior of nanoparticles can be studied. Any type of nanomaterial sample such as liquid nanoparticle dispersions, nanopowders, nanocomposites, etc. can be examined by this method.

3.3 *Energy-Dispersive X-Ray Spectroscopy*

Energy-dispersive X-ray spectroscopy (EDS or EDX) is also known as energy dispersive X-ray analysis (EDXA) or energy dispersive X-ray microanalysis (EDXMA). It is a microanalysis technique for identifying and measuring elemental compositions on specific particles, morphologies, or isolated areas of the material on nano-scale [55, 56]. It records the X-rays that are emitted from the material under study by bombarding it by an electron beam. The EDS detector measures the emitted X-rays as a function of their energy. The energy of the X-ray is characteristic of the elemental composition of the chemical substance. From this, the various elements present and their quantities can be determined [57]. The technique requires a very small sample quantity for analysis and it has less or no sample preparation. This technique is usually used in conjunction with SEM.

3.4 *Thermal Analysis*

Thermal analysis is a low cost and high-speed analysis method useful for verifying the morphology and composition of nanoceramics. Many properties of nanoceramics can be studied with the help of thermal analysis. In this method, nanoceramics are heated to higher temperatures and the changes in the material are studied as a function of temperature in the temperature range of -150 to 1600 °C. Properties such as purity, composition, crystallization behavior, glass transition, melting, phase changes, reaction enthalpies, surface area analysis, kinetics of reactive processes, etc., can be studied by thermal analysis [50]. The important thermal analysis techniques that are used to study nanoceramics are: (i) thermogravimetric analysis (TGA)—measures weight loss of the material during heating, (ii) differential thermal analysis—measures relative change in the material's temperature during heating, (iii) differential scanning calorimetry—measures the amount of heat required to raise the temperature of the material with respect to temperature, (iv) Brunauer-Emmett-Teller method—measures the specific surface area of the material, etc. New techniques are being developed which require less sample quantity which will extend the applications of thermal analysis in the characterization of nanoceramics.

3.5 *Scanning Electron Microscopy (SEM)*

Scanning Electron Microscopy is used to study the surface features of the nanomaterials and nanostructured materials by focusing a narrow and high-intensity electron beam in the range of 5–100 keV over it [58]. The electron beam-sample interaction gives rise to a variety of signals including secondary electrons, backscattered electrons (BSE), photons, visible light etc. By collecting these signals with suitable detectors, high resolution and high-magnification images of the sample surface can be obtained. Secondary electrons give information about the morphology and the topography of the sample surface [59]. Backscattered electrons reveal information about the composition of multiphase samples whereas diffracted backscattered electrons can give information about the crystallographic orientation in the sample. SEM is a non-destructive analysis method requiring less or no sample preparation. Nonconductive samples can be studied by coating them with a thin layer of electrically conductive material such as carbon, gold, etc. in order to avoid or minimize negative charge accumulation from the incident electron beam [58].

3.6 *Transmission Electron Microscopy (TEM)*

Transmission Electron Microscopy is a powerful tool used for studying nanoscale materials. A high energy beam of electrons is focused on a very thin specimen of thickness less than 200 nm [19]. The electron penetrates the sample and the subsequent electron—atom interaction results in either deflected or undeflected electrons carrying information about the crystal structure, composition, and defects such as dislocations and grain boundaries. A high magnification from 50–10^6 can be obtained by this method revealing finest details of the material even as small as individual atoms. High resolution—Transmission Electron Microscopy (HRTEM) has a resolution of approximately 0.08 nm. The diffraction pattern of a small selected area of the sample can be recorded with Selected area electron diffraction (SAED) from which information about the structure and orientation of the material can be obtained [60].

3.7 *Scanning Probe Microscopy (SPM)*

Scanning Probe Microscopy is a quantitative measuring instrument to study the physical, chemical and surface properties such as topography and nanotribology at the nanoscale [61]. It uses a physical probe to scan the surface of the sample and it can image several interactions simultaneously. Three dimensional (3-D) topographical images of the sample surface with high atomic-scale resolution as well as chemical information can be generated by this method as the sample surface is felt and not just seen with electrons or light waves. This technique includes a group of instruments such as Scanning tunneling microscope (STM), Atomic force microscope (AFM), Lateral force microscope (LFM), magnetic force microscope (MFM), scanning thermal microscope (SThM), Electrical force microscope (EFM) and Near-field scanning optical microscope (NSOM). Different characteristics of the sample can be studied by changing the material, configuration of the probe and by modifying the detection scheme [62]. STM is the most powerful microscope and is used to study the surface of nanomaterials and biological samples related to microelectronics. STM can only scan electrically conductive samples.

AFM is a high-resolution type SPM with resolutions of the order of fractions of a nanometer. AFM is used to study smoothness, texture, presence and size of pores over the sample surface. Shape and topography of different nanoceramics can be investigated with AFM [63] AFM can scan any solid surface such as insulators, conductors, semiconductors, etc., and does not essentially require the sample surface to be conductive [19].

3.8 *Brunauer-Emmett-Teller (BET) Analysis—Physical Gas Adsorption*

Brunauer-Emmett-Teller analysis is an important analysis technique for measuring the specific surface area, size of particles and pore size distribution of nanomaterials. Gas molecules are made to adsorb on the sample surface and the physically adsorbed gases are removed by reducing the partial pressure. The amount of gas required to fill the pores is measured with respect to gas pressure and the plot is known as gas adsorption isotherm. Evaluation of the adsorption and desorption branches of these isotherms and the hysteresis between them reveal information about the size, volume, and area of the pores. Specific surface area and pore volume of mesoporous or microporous materials can be particularly determined by measuring physical adsorption of gases [19]. The surface area measurement helps in predicting the bioavailability of the loaded materials into the nanopores.

3.9 Infrared Spectroscopy (IR)

Infrared Spectroscopy is also known as vibrational spectroscopy. It is a powerful, sensitive and non-destructive tool for analyzing, characterizing and identifying both organic and inorganic molecules. It also helps in identifying functional groups present over a depth of about 1 μm [64]. When a sample is irradiated with light, the bonds present in the sample molecule absorbs light in the infrared region and start to vibrate. The absorbed light is characteristic of the bonds present in the molecule. IR spectrum is obtained by plotting the amount of light absorbed against wavelength in the region of 4000–400 cm^{-1}. This spectrum is known as 'molecular fingerprint' that helps to identify samples.

Fourier Transform—Infrared Spectroscopy (FT-IR) produces a spectrum with high spatial resolution, good signal-to-noise ratios and it also enables measuring a broad region of the spectrum in short duration [65]. Fourier Transform is a mathematical process carried out to convert the raw data into actual spectrum. Solid, liquid, or gaseous samples can be studied with FT-IR. Attenuated Total Reflection—Infrared Spectroscopy (ATR-IR) is used to study solid or liquid samples without further preparation. It uses the property of internal reflection to analyze the sample. The infrared radiation penetrates the samples to a depth between 0.5–2 mm. Rapid sampling and ease of handling are main advantages of this method.

3.10 X-Ray Photoelectron Spectroscopy (XPS)

X-ray Photoelectron Spectroscopy is also known as electron spectroscopy for chemical analysis (ESCA). It is a surface-sensitive spectroscopic technique that does a quantitative elemental analysis of the surface of the sample. The elemental composition, empirical formula, chemical and electronic state of the elements present in the material can be measured by this method [66]. The sample surface is irradiated with a beam of X-rays and the kinetic energy of the emitted electrons from the top 1–10 nm of the material surface is measured [67]. The photoelectron spectrum is recorded by plotting the number of ejected electrons with respect to a range of electron kinetic energies. Atoms emitting particular energy is recorded as peaks. Identification and quantification of the different elements present on the material surface can be done by studying the energies and intensities of the peaks. XPS is a useful technique as it not only enables identification of elements present on the surface but also the elements it is bonded to and its oxidation state.

4 Biomedical Applications of Nanoceramics

Nanoceramic materials find a great deal of applications in various fields like electronics, energy storage devices, biomedical fields, catalysts, etc. It is widely used in the biomedical field owing to their mechanical properties, greater strength to weight ratio, biocompatibility and bioactivity. Their widespread application can be credited to its ability of being available in various forms such as powders, granules, dense and porous blocks, coatings, porous scaffolds, etc. [68]. Nano ceramics can function along with the physiological system without eliciting negative host response and get integrated with the host. Hence, nanoceramics are used to develop new biomedical systems to meet the increasing need for medical devices with increased functionality and extended lifetime. The various biomedical applications of nanoceramics are depicted in Fig. 1.

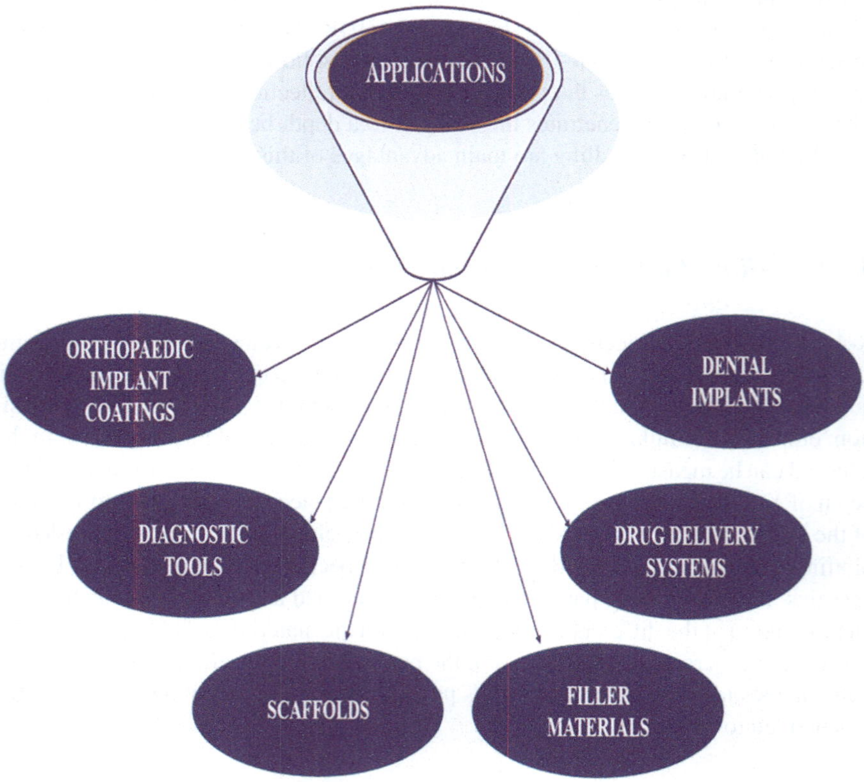

Fig. 1 Biomedical applications of nanoceramics

4.1 Nanoceramics as Orthopaedic Implant Coatings

Implants are used to replace and restore the functions of diseased and/or damaged hard tissues such as bone and teeth. A good artificial implant material should be able to mimic the physical structure, chemical composition and biological function of the natural bone. The increase in the life expectancy of people along with those undergoing treatments for musculoskeletal disorders is on the rise every year. The biological environment is harsh and it leads to loosening of the metal implants, corrosion in the physiological medium, inflammation, loosening, wear and/or tear debris, autoimmune reactions and ultimately failure of implants in patients with traditional metallic implants [69, 70]. Improving the bioactivity and corrosion resistance of the implant material as well as facilitating faster bone-implant interface establishment is crucial to avoid early implant failure and increase the lifetime of an implant.

Modifying the implant surface is a simple and cost-effective method to effect osseointegration and increase the bioactivity of the implant. This can be done by applying biocompatible coatings, modifying the surface topography or by removing material from the existing surface to create new topography [71, 72]. Considering the constraints in the physiological environment, nanoceramics is a suitable material for different bone system-related applications such as dental, periodontal, cranial, maxillofacial, spinal surgery etc. Applying nanoceramics in the form of coatings over the implant surface is an effective methodology.

Numerous nano bioceramics have been developed and categorized based on their activity. The first-generation bioceramics such as zirconia (ZrO_2), alumina (Al_2O_3), etc., had good mechanical properties but were bioinert [73]. Copper and silver incorporated ZrO_2 coatings were developed on pure Ti implant by magnetron sputtering. The nanostructured coatings changed the structure of crystalline zirconia coating and the incorporation of Cu and Ag improved the antibacterial resistance of the implant material [74]. A thin layer of dense Al_2O_3 was developed by micro-arc oxidation method. The Al_2O_3 layer exhibited good adhesion to the Ti implant, had high Vickers hardness and it is suggested to be a preferred material for load-bearing applications such as artificial hip joint [75].

The second-generation nano bioceramics were based on their chemical reactivity, i.e., they had bioactivity. Bioactive ceramics can bond with the living bone without having adverse reactions such as inflammation and toxicity [76]. The high reactivity of this class of materials is the main advantage for being used in periodontal repair and bone growth. Hydroxyapatite, silica-based bioactive glasses, etc., are the important second-generation bioceramics. Hydroxyapatite coatings were developed over NaOH treated and untreated Ti-6Al-4 V by electrodeposition method and compared with that of plasma-sprayed Hap coated Ti-6Al-4 V. The new bone area value for HAp electrodeposited coating on alkali-treated material was highest at 12 weeks indicating enhanced osseointegration in vivo [77]. A new family of glasses was formulated by partially substituting CaO by MgO and Na_2O by K_2O. The glass was deposited on the sample surface by a dip-coating method following by annealing. The glass coating's thickness was adjusted to be between 100–200 nm to have good adhesion and hence it does not crack or delaminate. The glass coating is highly bioactive with silica content of less than 60% [78]. It has been reported that 130 nm silica nanoparticles functionalized by amino group and silver nanoparticles aided the growth of human BMSCs [79].

The bioactivity of nanoceramics leads to the crystallization of hydroxyapatite (HAp) on the implant surface. HAp is similar to the bone tissue and this further facilitates the production of proteins and cell adhesion leading to a strong bond between the bone and implant. The third-generation bioceramics is based on its ability to activate genes that stimulate the regeneration of bone tissues. These can regenerate the bone tissue instead of acting as their substitutes [80, 81].

Silicon nitride (Si_3N_4) nano bio-ceramic exhibited less in vitro bacterial affinity than Ti. The ionic dissolution products contain Sr, Mg and Si ions and they increase the stimulatory effect for alkaline phosphatase activity. Increased bioactivity, no toxicity and reduced biofilm formation indicate that it can be applied as spinal fusion cages [82]. Magnesium based bioceramics are gaining a lot of attention owing to its ability to regulate ion channels, activate enzymes and stimulate cell growth and proliferation. Mg oxides, phosphates and silicates are employed in orthopaedic applications in the form of scaffolds, bone cements and also as implant coatings [83].

Nanostructured surfaces have been shown to elicit positive response from host, reduce inflammation and aid in faster bone-implant interface establishment [84, 85]. Growth factors, bioactive molecules, drugs, etc., can be loaded on to these nanoceramic coatings [86]. Ultrathin mesoporous TiO_2 coating was developed by evaporation-induced self-assembly method. Drugs like ibuprofen and vancomycin were loaded onto the pores in the coating. The coating exhibited excellent HAp growth and osteoblast adhesion indicating bioactivity while simultaneously eluting drugs from the coating. Thus, the coating exhibited improved therapeutic behavior for applications such as orthopaedic implants and drug delivery [87]. Recently, nanostructured composite materials have been developed in the form of coatings. Nanocomposites combine biodegradable or nonbiodegradable polymers or other compounds with nanoceramics to realize mechanical strength, effective biomineralization, and osseointegration [88]. Chitosan-bioactive glass nanocomposites with different concentrations of bioactive glass were developed over Ti-6Al-4 V by electrophoretic deposition method. Increasing the bioactive glass concentration leads to improvement in adhesion strength, roughness, wettability and also apatite growth. Good cell attachment and negligible cytotoxicity were observed during in vitro evaluation with osteoblast like MG 63 cell line confirming improved cellular performance of the nanocomposite coating [89]. Poly(3,4-ethylenedioxythiophene based nanocomposite coating with different concentrations of fluoro HAp nanoparticles) was developed on Ti-Nb0Zr alloy by an electrochemical deposition method. The uniformly distributed FHA nanoparticles lead to an increase in hardness and surface wettability. The coatings exhibited higher corrosion protection and increased cell adsorption and proliferation of MG 63 cells [90]. Silver particles are coated along with nano-titania on orthopaedic implants surface to prevent post-operative problems and infections [91].

4.2 *Nanoceramics as Dental Implants*

Ceramic materials have been used as dental implants due to their advantageous properties like compressive strength, wear resistance, radiopacity, color stability and biocompatibility. Many such ceramic dental implants have been developed in order to increase their durability and clinical lifetime. Dental implants get contaminated by settling of bacterial deposits especially the organic residues. Photocatalytic activity of TiO_2 can be used to degrade the organic residues and regain biocompatibility. A unique TiO_2 nanoceramic coating was developed on Ti implants by plasma electrolytic oxidation (PEO). The coating was able to decompose dyes like methylene blue, rhodamine B, and also pre-adsorbed lipopolysaccharide in the presence of visible light. The coating had good osteoconductivity than untreated Ti implant suggesting that it can be used in peri-implantitis treatments [92].

Feldspathic is an advanced ceramic material manufactured by high fusion and has excellent aesthetic properties, opacity and translucency. It is widely used in smile aesthetic recovery with the underlying tooth reinforcing the coating [91]. Feldspathic and alumina added Apatite-Wollastonite glass-ceramic were prepared by sintering MgO-CaO-SiO_2-P_2O_5-Al_2O_3 system at 1100 °C. The dental material produces an interface that is similar in characteristics to the commercially available dental material and hence is a suitable alternate [93]. Aluminized ceramics were prepared by incorporating metallic oxides. The recently developed glass-infiltrated aluminized ceramic with high alumina content has greater fracture resistance and they can be used for both anterior and posterior regions as prostheses [94, 95]. Zirconia based ceramics contain 69% aluminum oxide and 31% zirconium oxide. It's the best alternate for large metal-free fixed prostheses as it has good mechanical properties, clinical longevity and biocompatibility. Yttria-stabilized zirconia is developed by adding pure yttrium dioxide to zirconia. This material has high fracture toughness and it prevents crack propagation commonly observed in aluminized ceramics [90].

4.3 *Nanoceramics as Drug Delivery Systems*

Conventional drug delivery systems have a major limitation namely, limited drug solubility which leads to poor biodistribution, poor targeting, reduced efficacy, and serious side effects in non-target tissues. In some treatments, a definite amount of drug has to be maintained in the bloodstream over a stretch of time for effective treatment and faster recovery. With the conventional drug delivery method, this cannot be ensured as fluctuations in drug level is common. This leads to overdosage to achieve the result. Controlled and continuous in situ delivery of drugs is possible with biocompatible nanoceramics as they act as good drug delivery system (DDS) compared to the traditional ones such as lipids and polymers [96]. Their bioactive behavior along with their ability to control the rate and period of drug delivery and also target the release of drug in a specific area of the body makes them attractive for bone therapy purposes [97]. However, they suffer from certain disadvantages such as limited chemical stability, local inflammatory reactions, improper drug-release kinetics, etc. [71]. Bioactive calcium phosphates and bioactive glasses are widely used as matrices in drug delivery systems. Mesoporous silica-titania composites were developed to have ordered hexagonal array of pores. Antibiotic drug oxytetracycline was loaded onto the mesochannels by wetness impregnation method. Titania present on the silica surface enabled the composite to hold oxytetracycline through host-guest interaction and it avoided the initial burst release of the drug. It was observed that by controlling the distribution of titanium in the silica network it was possible to develop a drug delivery system with a controlled drug release profile [42]. Zeolites with different silica to alumina ratio was used as delivery vehicle for a chemotherapy drug, 5-fluorouracil. The molecular level interaction between the drug and zeolite was confirmed by FTIR spectroscopy. Alumina content played a key role in the drug release behavior and it was observed that greater the alumina content better the controlled drug release behavior [89]. Orthopaedic implants can also be coated with TiO_2 nanotubes that are loaded with anti-inflammatory drugs like Ibuprofen and antibiotics like gentamycin or antibacterial drugs like cefuroxime [99]. Thereby it is possible to control or prevent infection and inflammation without affecting the bioactive behavior of implant material which supports the adhesion of osteoblasts on to the implant surface [100].

4.4 Diagnostic Tools

The physicochemical properties of ceramic nanoparticles such as improved cellular adhesion, greater osteoblast proliferation and increased biomineralization allow them to be used as diagnostic tools in addition to effective drug delivery systems [99]. Nanoceramics such as porous silica nanoparticles, Fe_3O_4 nanoparticles, quantum dots, etc., are some examples that can be used in imaging, magnetic hyperthermia, or photothermal ablation in addition to drug delivery systems [67]. Mesoporous SiO_2 coated with europium hydroxide, $Eu(OH)_3$ core-shell microspheres were used as a fluorescent probe for biomedical applications. FETEM confirmed that SiO_2 nanospheres were covered evenly with luminescent $Eu(OH)_3$. The microsphere emitted strong red emission peak when irradiated with ultraviolet light [101]. Composite film of Fe_3O_4/cellulose was prepared by co-dispersing Fe_3O_4 particles and cellulose in an aqueous solvent and then regenerating them. The composite film was flexible, strong, had excellent mechanical property and thermal stability. The composite film can sense towards UV light and magnetic field and can be used to prepare sensors [102].

4.5 *Tissue Engineering Applications*

4.5.1 Scaffold

Bone tissue engineering involves usage of scaffolds in the oral cavity and craniofacial region and aims to restore alveolar bone after periodontal disease, peri-implantitis, and reconstructive surgery after trauma, after cancer, etc. Tissue engineering provides a suitable biochemical and physicochemical environment in which the osteoblasts can attach to the scaffolds providing mechanical support and also optimize cells osteogenic functions. With nanoceramics increased osteoblast adhesion and proliferation were observed on the material surface. Enhancement in their long-term function is also observed when their grain size is less than 100 nm [103, 84]. A good scaffold should be able to degrade in vivo at a specific rate combined with a controlled absorption rate that facilitates the formation of new bone in the space provided by the two processes. Nanoscale scaffold materials are preferred as they are porous, biodegradable and provide mechanical support during the process of bone repair [104]. Scaffolds for bone repair can be developed using ceramic, metal, polymer and composite materials. Nanoporous bioceramics have high mechanical strength, enhanced bioactivity and resorbability and hence are being used effectively in tissue engineering. Nanohydroxyapatite is now being clinically used on commercial scale. Biopolymers can be used along with HAp, bioactive glass, chitosan, etc., to modify the scaffolds properties such as porosity and growth factor delivering ability to have greater functionality [105]. Chitin, chitosan-based scaffolds, and those reinforced with nanoceramics such as hydroxyapatite (HAp), silicon dioxide (SiO_2), titanium oxide (TiO_2), etc., are being extensively used in bone tissue engineering applications [106]. Mesoporous silica nanocomposite scaffold loaded with BMP-7; enhanced differentiation of bone marrow-derived mesenchymal cells (BMSC) from osteocytes and initiated osteogenesis [107]. Three-dimensional (3D) periodic TiO_2 bio-ceramic scaffolds have finer feature size. The scaffold favoured cell growth and attachment for mouse osteoblastic cell line MC3T3-E1 indicating good biocompatibility of the scaffold. HAp scaffold with microporous structure and high interconnectivity is modified to have nanosheet, nanorod, or micro-nano-hybrids structure on the surface. The scaffold promotes cell adhesion, proliferation and osteogenic differentiation of adipose derived stem cells (ASCs) [108]. Nanoceramics such as HAp, β-tricalcium phosphate, and bioactive glass were combined with gelatin or chitosan to prepare composite scaffold material. The resultant scaffolds exhibited increased compressive strength, high bioactivity, osteoblast adhesion and proliferation and hence stimulated new bone regeneration [109]. Highly porous 3-D scaffolds of the Ag-bioactive glass system of type $58.6SiO_2$—$24.9CaO$—$7.2P_2O_5$—$4.2Al_2O_3$—$1.5Na_2O$—$1.5K_2O$—$2.1Ag_2O$ was found to have antibacterial property. The scaffold showed formation of Hap after 2 weeks of in vitro bioactivity study in SBF and had anti-methicillin-resistant Staphylococcus aureus (MRSA) effect on both direct and indirect exposure [110].

4.5.2 Filler Material

Bioactive glass-ceramic (BCG) is widely used as filler material for regenerating bone tissue as it can form strong interface between hard as well as soft tissue. Nano-bioglass ceramic particles doped with Calcarea phosphorica were formulated and their biological action in bone tissue engineering application was investigated [111]. Ca, Mg and Si-containing bioceramics such as calcium silicates have greater applications as they have better mechanical properties, controllable degradation rate, facilitate bone growth and aid healing [112]. Porous and non-porous calcium phosphate glass-ceramics were synthesized and used as injectable bone cement when added with xanthan gum for cell-based bone regeneration treatment. The possible damage of porous calcium phosphate during injection process is prevented by xanthan gum as a result of its viscoeleastic properties [113]. Surface pre-reacted glass is now increasingly used to fill tooth defects in dentistry. Various types of ions such as Al^{+3}, BO^{-3}, Na^{+}, $SiO3^{-2}$, F^{-}, Li^{+} etc., are released from these fillers and they exhibit high antibacterial activity and enhance osteoblast differentiation [114].

5 Summary

Nanoceramic materials are finding increased applications in the field of biomaterials owing to their biocompatibility, mechanical strength and greater surface area. They are increasingly used as implant coatings, scaffolds, bone grafts, drug delivery devices and biosensors. The various synthesis methods of nano bioceramics are discussed in this chapter. The characterization techniques used to study the various aspects of nanoceramics are briefly described. Nanoceramics are widely used as metal implant coatings to increase its functionality, bioactivity and resistance to corrosion and wear. Nanoceramic coatings play the dual role of in situ drug delivery system which deliver drugs directly in the implantation site and aid in faster healing and stronger bone-implant interface establishment. As nanoceramics have greater compressive strength and wear resistance they are successfully used as dental applications. The magnetic and radiopacity properties of nanoceramics are utilized in diagnostics and medical imaging. The nanoceramic's biodegradation property can be modified by converting them into nanocomposites with biopolymeric materials which are successfully used as scaffold materials. The application of scaffold as implant devices is increasing owing to its ability to biodegrade and aid natural bone growth in the defective site. Nanoceramics are being reinvented as filler materials in tissue engineering applications and the field is gaining momentum owing to its biocompatibility. Nanoceramics are continually implored in various capacities to bring out its advantageous properties to build functional biomaterials.

References

1. Khalil KA (2012) Advanced sintering of nano-ceramic materials. In: Ceramic materials-progress in modern ceramics, InTechOpen, London
2. Smith KT (2019) What are nanoceramics and their applications? Accessed March 3. https://azonano.com/article.aspx?ArticleID=5143
3. Thomas SC, Harshita, Mishra PK, Talegaonkar S (2015) Ceramic nanoparticles: fabrication methods and applications in drug delivery. Curr Pharm Des 21:6165–88
4. Virk HS, Poonam S (2010) Chemical route to nanotechnology. Int J Adv Eng Technol 1:114–129
5. Kiani A, Rahmani M, Manickam S, Tan B (2014) Nanoceramics: synthesis, characterization, and applications. J Nanomat 2014:1–2
6. Sharma RK, Sharma P, Maitra A (2003) Size-dependent catalytic behavior of platinum nanoparticles on the hexacyanoferrate(III)/thiosulfate redox reaction. J Colloid Interface Sci 265:134–140
7. Miyake H, Yuba Y, Gamo K, Namba S (1988) Defects induced by focused ion beam implantation in GaAs. J Vac Sci Technol B: Microelectron Process Phenom 6:1001
8. Ting HT, Hossein KA, Chua HB (2009) Review of micromachining of ceramics by etching. T Nonferr Metal Soc China 19:1–16
9. Wakamatsu MH, Salomão R (2010) Ceramic nanoparticles: what else do we have to know? InterCeram: Inter Ceram Rev 59:28–33
10. Hiemenz PC, Rajagopalan R (1997) Principles of colloidal and surface chemistry, revised and expanded, 3rd edn. CRC Press, New York
11. Wakamatsu MH, Salomão R (2011) (Unintentional) synthesis of ceramic nanoparticles. InterCeram: Inter Ceram Rev 60:364–369
12. Rao CNR, Müller A, Cheetham AK (2004) The chemistry of nanomaterials: synthesis, properties and applications, vol 1. Wiley-VCH Verlag, Weinheim
13. Vashist SK (2013) Magnetic nanoparticles-based biomedical and bioanalytical applications. J Nanomed Nanotechnol 4:1000–1130
14. Xie L, Abliz D, Li D (2014) Thin film coating for polymeric micro parts, Vol. 7 comprehensive materials processing, reference module in materials science and materials engineering, Elsevier Publications, London
15. Rane AV, Kanny K, Abitha VK, Thomas S (2018) Methods for synthesis of nanoparticles and fabrication of nanocomposites. In: Synthesis of inorganic nanomaterials, micro and nano technologies, Woodhead Publishing, Massachusetts.
16. Goharian A (2019) Porous osseoconductive layering for enhancement of osseointegration. In: Osseointegration of orthopaedic implants, Academic Press, Cambridge, England.
17. Ballo AM, Bjoorn D, Astrand M, Palmquist A, Lausmaa J, Thomsen P (2013) Bone response to physical-vapour-deposited titanium dioxide coatings on titanium implants. Clin Oral Implants Res 24(9):1009–1017
18. Bazaka K, Jacob MV, Crawford RJ, Ivanova EP (2012) Efficient surface modification of biomaterial to prevent biofilm formation and the attachment of microorganism's. Appl Microbiol Biotechnol 95:299–311
19. Cao G (2014) Nanostructures & Nanomaterials: synthesis, properties & applications. Imperial College Press, London
20. Kumar DS, Kumar BJ, Mahesh HM (2018) Chapter 3—Quantum Nanostructures'. In: Synthesis of inorganic nanomaterials, micro and nano technologies, Woodhead Publishing, Cambridge, England.
21. Ataol S, Tezcaner Duygulu O, Keskin D, Machin NE (2015) Synthesis and characterization of nanosized calcium phosphates by flame spray pyrolysis and their effect on osteogenic differentiation of stem cells. J Nanopart Res 17:1–14
22. An GH, Wang HJ, Kim BH et al (2014) Fabrication and characterization of a hydroxyapatite nanopowder by ultrasonic spray pyrolysis with salt-assisted decomposition. Mater Sci Eng, A 449–451:821–824

23. Kulkarni M, Mazare A, Schmuki P et al (2014) Biomaterial surface modification of titanium and titanium alloys for medical applications. In: Nanomedicine, UK Central Press, Cambridge
24. Giavaresi G, Ambrosio L, Battiston GA et al (2004) Histomorphometric, ultrastructural and microhardness evaluation of the osseointegration of a nanostructured titanium oxide coating by metal-organic chemical vapour deposition: an in vivo study. Biomater 25(25):5583–5591
25. Szili EJ, Kumar S, Smart RSC, Voelcker NH (2009) Generation of a stable surface concentration of amino groups on silica coated onto titanium substrates by the plasma enhanced chemical vapour deposition method. Appl Surf Sci 255(15):6846–6850
26. Gennari FC, Gamboa JJA (2018) A Systematic approach to the synthesis, thermal stability and hydrogen storage properties of rare-earth borohydrides. In: Emerging Materials from Energy Conversion and Storage, Elsevier Publications, London.
27. Tsuzuki T, McCormick PG (2004) Mechanochemical synthesis of nanoparticles. J Mater Sci 39:5143–5146
28. Benabdeslam HEB, Ginebra MP, Vert M (2008) Wet or dry mechanochemical synthesis of calcium phosphates? Influence of the water content on DCPD-CaO reaction kinetics. Acta Biomater 4(2):378–386
29. Singh Z (2018) Nanoceramics in bone tissue engineering: the future lies ahead. Trends J Sci Res 3:120–123
30. Bulina NV, Chaikina MV, Prosanov IY (2018) Lanthanum-silicate-substituted apatite synthesized by fast mechanochemical method: characterization of powders and biocoatings produced by micro-arc oxidation. Mater Sci Eng, C 92:435–446
31. Balaz M, Daneu N, Balazova L (2017) Bio-mechanochemical synthesis of silver nanoparticles with antibacterial activity. Adv Powder Technol 28(12):3307–3312
32. Fahami A, Kahrizsangi RE, Tabrizi BN (2011) Mechanochemical synthesis of hydroxyapatite/titanium nanocomposite. Solid State Sci 13(1):135–141
33. BenAbdeslam HEB, Mochales C, Ginebra MP et al (2003) Dry mechanochemical synthesis of hydroxyapatites from dicalcium phosphate dihydrate and calcium oxide: a kinetic study. J Biomed Mater Res A 67A(3):927–937
34. Huang A, Dai H, Wu X, Zhao Z et al (2019) Synthesis and characterization of mesoporous hydroxyapatite powder by microemulsion technique. J Mater Res Technol 8(3):3158–3166
35. Xu H, Cheng L, Wang C et al (2011) Polymer encapsulated upconversion nanoparticle/iron oxide nanocomposites for multimodal imaging and magnetic targeted drug delivery. Biomater 32(35):9364–9373
36. Pottathara YB, Grohens Y, Kokol V et al (2019) Synthesis and processing of emerging two-dimensional nanomaterials. In: Nanomaterials synthesis, design, fabrication and applications, micro and nano technologies, Elsevier Publications, London
37. Lugo VR, Karthik TVK, Anaya DM, Rosas ER (2018) Wet chemical synthesis of nanocrystalline hydroxyapatite flakes: effect of pH and sintering temperature on structural and morphological properties. R Soc Open Sci 5(8):180962
38. Stipniece L, Ancane KS, Borodajenko N (2014) Characterization of Mg-substituted hydroxyapatite synthesized by wet chemical method. Ceram Inter 40(2):3261–3267
39. Pauline SA, Mudali UK, Rajendran N (2013) Fabrication of nanoporous Sr-incorporated TiO_2 coating on 316L SS: evaluation of bioactivity and corrosion protection. Mater Chem Phy 142:27–36
40. Pauline SA, Rajendran N (2014) Effect of Sr on the bioactivity and corrosion resistance of nanoporous niobium oxide coating for orthopaedic applications. Mater Sci Eng, C 36:194–205
41. Maho A, Detriche S, Delhalle J et al (2013) Sol-gel synthesis of tantalum oxide and phosphonic acid-modified carbon nanotubes composite coatings on titanium surfaces. Mater Sci Eng, C 33(5):2686–2697
42. Georgescu D, Brezoiu AM, Mitran RA et al (2017) Mesostructured silica-titania composites for improved oxytetracycline delivery systems. C R Chim 20:1017–1025
43. Foss CA (2003) Optical properties of nanoparticle pair structures. In: Encyclopedia of materials: science and technology, Elsevier Publications, London

44. Cao G, Liu D (2008) Template-based synthesis of nanorod, nanowire, and nanotube arrays. In: Springer handbook of nanotechnology, Springer, New York. 136(1–2):45–64
45. Regi MV (2010) Evolution of bioceramics within the field of biomaterials. C R Chim 13(1–2):174–185
46. Fan J, Lei J, Yu C (2007) Hard-templating synthesis of a novel rod-like nanoporous calcium phosphate bioceramics and their capacity as antibiotic carriers. Mater Chem Phy 103(2–3):489–493
47. Wang M, Guo L, Sun H (2019) Manufacture of biomaterials. In: Encyclopedia of biomedical engineering, Elsevier Publications, London
48. Shin K, Acri T, Geary S et al (2017) Biomimetic mineralization of biomaterials using simulated body fluids for bone tissue engineering and regenerative medicine. Tissue Eng Part A 23(19–20):1169–1180
49. Stefanic M, Krnel K, Pribosic I et al (2012) Rapid biomimetic deposition of octacalcium phosphate coatings on zirconia ceramics (Y-TZP) for dental implant applications. Appl Surf Sci 258(10):4649–4656
50. Bigi A, Boanini E, Bracci B et al (2005) Nanocrystalline hydroxyapatite coatings on titanium: a new fast biomimetic method. Biomater 26:4085–4089
51. Fathyunes L, Khalil-Allafi J, Moosavifar M (2019) Development of graphene oxide/calcium phosphate coating by pulse electrodeposition on anodized titanium: Biocorrosion and mechanical behavior. J Mech Behav Biomed Mater 90:575–586
52. Drevet R, Zhuova Y, Dubinskiy S et al (2019) Electrodeposition of cobalt-substituted calcium phosphate coatings on Ti-22Nb-6Zr alloy for bone implant applications. J Alloys Compd 793:576–582
53. Poorraeisi M, Afshar A (2018) The study of electrodeposition of hydroxyapatite-ZrO_2-TiO_2 nanocomposite coatings on 316 stainless steel. Surf Coat Technol 339:199–207
54. Pontoni D, Narayanan T, Rennie AR (2002) Tr-saxs study of nucleation and growth of sillica colloids. Langmuir 18:56–59
55. Ebnesajid S (2014) Chapter 4—Surface and material characterization techniques. In: Surface treatment of materials for adhesive bonding. William Andrew Applied Science Publishers, New York
56. Ismail AF, Khulbe KC, Matsuura T (2019) Chapter 3—RO membrane characterization. In: Reverse osmosis. Elsevier Publications, London
57. Bergstrom J (2015) 2—Experimental characterization techniques, mechanics of solid polymers. In: Theory and computational modeling. William Andrew Applied Science Publishers, New York
58. Ratner BD (2013) Chapter I.1.5—Surface Properties and Surface Characterization of Biomaterials. In: Biomaterials science (Third Edition), an introduction to materials in medicine. Academic Press, Cambridge, England.
59. Ven ALVD, Mack A, Dunner Jr K et al (2012) Chapter one—preparation, characterization, and cellular associations of silicon logic-embedded vectors. In: Methods in enzymology, vol. 508. Elsevier Publications, London
60. Bajpai OP, Panja S, Chattopadhyay S et al (2015) Process-structure-property relationships in nanocomposites based on piezoelectric-polymer matrix and magnetic nanoparticles. In: Manufacturing of nanocomposites with engineering plastic. Elsevier Publications, London
61. Cuenat A, Leah R (2014) Chapter 7—Scanning probe and particle beam microscopy. In: Fundamental principles of engineering nanometrology (2nd edn), micro and nano technologies. William Andrew Applied Science Publishers, New York
62. Ramakrishna BL, Ong EW (2001) Surface evaluation by atomic force microscopy. In: Encyclopedia of materials: science and technology (2nd edn), Elsevier Publications, London
63. Shi D, Guo Z, Bedford N (2015) 2-Characterization and Analysis of Nanomaterials. In: Nanomaterials and Devices, Micro and Nano Technologies. William Andrew Applied Science Publishers, New York
64. Causserand C, Aimar P (2010) 1.15—Characterization of filtration membranes. In: Comprehensive membrane science and engineering, vol. 1. Elsevier Publications, London

65. McCluskey MD (2017) High-pressure IR. In: Encyclopedia of spectroscopy and spectrometry (3rd edn), reference module in chemistry, molecular sciences and chemical engineering. Elsevier Publications, London
66. Ohara S, Adschiri T, Ida T, Yashima M et al (2012) Chapter 5 - Characterization methods for nanostructure of materials. In: Nanoparticle technology handbook (2nd edn). Elsevier Publications, London
67. Mather RR (2009) 13—Surface modification of textiles by plasma treatments. In: Surface modification of textiles. Woodhead Publishing Series in Textiles, Cambridge
68. Arcos D, Regi MV (2013) Bioceramics for drug delivery. Acta Mater 61:890–911
69. Smith AJ, Dieppe P, Vernon K et al (2012) Failure rates of stemmed metal-on-metal hip replacements: analysis of data from the national joint registry of England and Wales. The Lancet 379:1199–1204
70. Wang CJ, Huang TW, Wang JW et al (2002) The often poor clinical outcome of infected total knee arthroplasty. J Arthroplasty 17:608–614
71. Simchi A, Eng D, Tamjid E (2011) Recent progress in inorganic and composite coatings with bactericidal capability for orthopaedic applications. Nanomedicine 7:22–39
72. Duan K, Wang R (2006) Surface modifications of bone implants through wet chemistry. J Mater Chem 16:2309–2321
73. In HSF, Hench LL, Editors WJ (1993) An introduction to bioceramics. World Scientific, Singapore
74. Huang HL, Chang YY, Weng JC (2013) Anti-bacterial performance of Zirconia coatings on Titanium implants. Thin Solid Films 528:51–156
75. Khanna R, Kokubo T, Matsushita T et al (2016) Fabrication of dense α-alumina layer on Ti-6Al-4 V alloy hybrid for bearing surfaces of artificial hip joint. Mater Sci Eng, C 69:1229–1239
76. Hench LL (1991) Bioceramics: from concept to clinic. J Am Ceram Soc 74:1487–1510
77. Lakstein D, Kopelovitch W, Barkay Z et al (2009) Enhanced osseointegration of grit-blasted, NaOH-treated and electrochemically hydroxyapatite-coated Ti-6Al-4 V implants in rabbits. Acta Biomater 5:2258–2269
78. Esteban SL, Saiz E, Fujino S et al (2003) Bioactive glass coatings for orthopedic metallic implants. J Eur Ceram Soc 23(15):2921–2930
79. Tarpani L, Morena F, Gambucci M (2016) The influence of modified silica nanomaterials on adult stem cell culture. Nanomaterials 6:104–114
80. Hench LL, Xynos ID, Polak JM (2004) Bioactive glasses for in situ tissue regeneration. J Biomater Sci. Polymer Edition 15:543–562
81. Habraken WJEM, Walke JGC, Jansen JA (2007) Ceramic composites as matrices and scaffolds for drug delivery in tissue engineering. Adv Drug Deliv Rev 59:234–248
82. Fu L, Xiong Y, Carlsson G et al (2018) Biodegradable Si_3N_4 bioceramic sintered with Sr, Mg and Si for spinal fusion: surface characterization and biological evaluation. Appl Mater Today 12:260–275
83. Nabiyouni M, Bruckner T, Zhou H et al (2018) Magnesium-based bioceramics in orthopedic applications. Acta Biomater 66:23–43
84. Liu H, Webster TJ (2007) Nanomedicine for implants: a review of studies and necessary experimental tools. Biomater 28:354–369
85. Ainslie KM, Tao SL, Popat KC et al (2008) In vitro inflammatory response of nanostructured titania, silicon oxide, and polycaprolactone. J Biomed Mater Res, Part A 91:647–655
86. Luginbuehl V, Meinel L, Merkle HP et al (2004) Localized delivery of growth factors for bone repair. Eur J Pharm 58:197–208
87. Chao CS, Liu KH, Tung WL et al (2012) Bioactive TiO2 ultrathin film with worm-like mesoporosity for controlled drug delivery. Micropor Mesopor Mat 152:58–63
88. Couto DS, Alves NM, Mano JF (2008) Nanostructured multilayer coatings combining chitosan with bioactive glass nanoparticles. J Nanosci Nanotechnol 8:1–8
89. Mahlooji E, Atapour M, Labbaf S (2019) Electrophoretic deposition of Bioactive glass—chitosan nanocomposite coatings on Ti-6Al-4 V for orthopaedic applications. Carbohydr Polym 226:115299

90. Kumar AM, Adesina AY, Hussein MA (2019) PEDOT/FHA nanocomposite coatings on newly developed Ti-Nb-Zr implants: biocompatibility and surface protection against corrosion and bacterial infections. Mater Sci Eng, C 98:482–495
91. Goncalves SEP, Bresciani E (2017) Reconstructions using alloys and ceramics. In: Material-tissue interfacial phenomena. Elsevier Publications, London.
92. Wu H, Xie L, He M et al (2019) A wear-resistant TiO_2 nanoceramic coating on titanium implants for visible-light photocatalytic removal of organic residues. Acta Biomater 97:597–607
93. Pekkan G, Pekkan K, Park J et al (2016) A study on microstructural characterization of the interface between apatite-wollastonite based glass ceramic and feldspathic dental porcelain. Ceram Inter 42(16):19245–19249
94. Donovan TE (2008) Factors essential for successful all-ceramic restorations. J Am Dent Assoc 139:14S–18S
95. Koutayas SO, Vagkopoulou T, Pelekanos S et al (2009) Zirconia in dentistry: part 2. Evidence-based clinical breakthrough. Eur J Esthet Dent 4(4):348–380
96. Regi MV, Balas F, Arcos D (2007) Mesoporous materials for drug delivery. Angew Chem Int Ed 46:7548–7558
97. Regi MV, Balas F, Colilla M et al (2008) Bone-regenrative bioceramic implants with drug and protein controlled delivery capability. Solid State Sci 1:163–191
98. Datt A, Burns EA, Dhuna NA et al (2013) Loading and release of 5-fluorouracil from HY zeolites with varying SiO2/Al2O3 ratios. Micropor Mesopor Mater 167:182–187
99. Ozdemir V, Glatt BWJSJ, Tsuang MT et al (2006) Shifting emphasis from pharmacogenomics to theragnostics. Nat Biotechnol 24:942–946
100. Yi H, Rehman FU, Zhao C (2016) Recent advances in nano scaffolds for bone repair. Bone Res 4:1–11
101. Ansari AA, Hasan TN, Syed et al (2013) *In-vitro* cyto-toxicity, geno-toxicity, and bio-imaging evaluation of one-pot synthesized luminescent functionalized mesoporous $SiO_2@Eu(OH)_3$ core-shell microspheres. Nanomedicine 9:1328–1335
102. Yang W, Tian H, Liao J (2020) Flexible and strong Fe_3O_4/cellulose composite film as magnetic and UV sensor. Appl Surf Sci 507:145092
103. Shi Z, Huang X, Cai Y et al (2009) Size effect of hydroxyapatite nanoparticles on proliferation and apoptosis of osteoblast-like cells. Acta Biomater 5:338–345
104. Khan Y, Yaszemski MJ, Mikos AG et al (2008) Tissue engineering of bone: material and matrix considerations. J Bone Jt Surg 90:36–42
105. Dziak R, Mohan K, Almaghrabi B et al (2020) Nanoceramics for bone regeneration in the oral and craniomaxillofacial complex. In: Nanobiomaterials in clinical dentistry. Elsevier Publications, London
106. Deepthi S, Venkatesan J, Kim SK et al (2016) An overview of chitin or chitosan/nano ceramic composite scaffolds for bone tissue engineering. Inter J Biol Macromol 93:1338–1353
107. Luo Z, Deng Y, Zhang R et al (2015) Peptide-laden mesoporous silica nanoparticles with promoted bioactivity and osteo-differentiation ability for bone tissue engineering. Colloids Surf B 131:73–82
108. Xia L, Lin K, Jiang X et al (2014) Effect of nano-structured bioceramic surface on osteogenic differentiation of adipose derived stem cells. Biomater 35:8514–8527
109. Chen P, Liu L, Pan J et al (2019) Biomimetic composite scaffold of hydroxyapatite/gelatin-chitosan core-shell nanofibers for bone tissue engineering. Mater Sci Eng C: Mater Biol Appl 97:325–335
110. Marsh AC, Mellott NP, Chamorro NP et al (2019) Fabrication and multiscale characterization of 3D silver containing bioactive glass-ceramic scaffolds. Bioact Mater 4:215–223
111. Kumar SD, Abudhahir KM, Selvamurugan N et al (2018) Formulation and biological actions of nano-bioglass ceramic particles doped with Calcarea phosphorica for bone tissue engineering. Mater Sci Eng C: Mater Biol Appl 83:202–209
112. Shokrollahi H, Salimi F, Doostmohammadi A (2017) The fabrication and characterization of barium titanate/akermanite nano-bio-ceramic with a suitable piezoelectric coefficient for bone defect recovery. J Mech Behav Biomed Mater 74:365–370

113. Veloza AM, Hossain KMZ, Scammell BE (2020) Formulating injectable pastes of porous calcium phosphate glass microspheres for bone regeneration applications. J Mech Behav Biomed Mater 102:103489
114. Ali M, Okamota M, Komichi S (2019) Lithium-containing surface pre-reacted glass fillers enhance hDPSC functions and induce reparative dentin formation in a rat pulp capping model through activation of Wnt/β-catenin signaling. Acta Biomate 96:594–604

Biomedical Applications of Carbon-Based Nanomaterials

Jyotsna, L. Stanley Abraham, Rathore Hanumant Singh, Ramesh C. Panda, and T. Senthilvelan

Abstract Nanomaterials are the solid colloidal particles in size ranging from 1 to 100 nm. Nanomaterials provide novel physicochemical properties with the increased surface to volume ratio from their bulk particles. Over the last two decades, several nanomaterials have been engineered precisely for their applications in diagnosis, targeted drug delivery, bio-imaging, biosensors, and also in therapeutics. Since the discovery of carbon allotropes, carbon-based nanomaterials have become important in the field of biomedicine due to their unique physio-chemical properties, high mechanical strength, and optical properties. In this chapter, carbon-based nanomaterials such as Nanodiamonds (NDs), Carbon nanotubes (CNTs), Buckminsterfullerene (C_{60}), Carbon quantum dots (CQDs), Carbon nanohorns (CNs) and its biomedical applications are described briefly. These nanoparticles allow the diagnosis and detection at the molecular scale with the addition of fluorescent probes. They provide more specificity and sensitivity in bio-imaging in drug delivery and cancer treatment.

Keywords Nanobiomaterials · Carbon nanotubes · Nanodiamonds · Fullerenes

Jyotsna · T. Senthilvelan (✉)
Centre for Ocean Research (DST-FIST Sponsored Centre), MoES–Earth Science and Technology Cell, Col. Dr. Jeppiaar Research Park, Sathyabama Institute of Science and Technology, Chennai, India
e-mail: senthiltlm02@gmail.com

L. Stanley Abraham · R. Hanumant Singh
Department of Biotechnology, School of Engineering & Technology, Nagaland University (A Central University Established by an Act of Parliament), D. C. Court Junction, Dimapur, Nagaland 797112, India

R. C. Panda
Chemical Engineering Department, CSIR-Central Leather Research Institute (CLRI), Chennai, India

T. S. Santra and L. Mohan (eds.), *Nanomaterials and Their Biomedical Applications*, Springer Series in Biomaterials Science and Engineering 16,
https://doi.org/10.1007/978-981-33-6252-9_6

1 Introduction

Nanoscience, together with biotechnology, is referred to as 'Nanobiotechnology,' which has shown remarkable applications in different fields of biological sciences. Nanomaterials are classified as materials having a particle size less than 100 nm in one dimension [1]. Nanomaterials with well-controlled size, surface distribution and surface modification using various chemicals and bioconjugate have applications in biomedical science. The nanomaterials derived from silica, noble metals, metal oxides, polymers are more attractive in various applications. Nanomaterials such as gold nanoparticles, quantum clusters, carbon nanotubes, nanodiamonds, fullerenes are most importantly used in fundamental research, device technology, and biomedical technology. They can incorporate multiple functional groups, targeting biomolecules, drugs, and genes, which offers the detection of subcellular organelles and biomolecules' structures and functioning.

Nanotechnology has found its applications in industrial sectors like communication, advanced materials, and biomedicine [2, 3]. Advanced nanomaterials with high performance, high optical, electronic, and structural properties were used to develop nanodevices that are used in nanomedicine applications [4]. These nanomedicines have been successfully applied in preclinical, translational, and clinical research, enabling them to diagnose and treat treatment methods in humans [5, 6]. Recently, a vast application of nanomaterials in drug delivery, biomarkers for diagnosis, biosensors, cancer chemotherapy, therapeutics, and molecular imaging has been studied. Carbon is a versatile element, and for over 6000 years, carbon is used to reduce metal oxides. The arrangement of atoms and forms of carbon provides enhanced functionalities in their structures and electronic properties [7–13]. A wide variety of carbon-based nanomaterials are available, such as carbon nanotubes, nanodiamonds, fullerenes, nanofibers, etc. used in biomedical applications, as shown in Fig. 1.

Recently, another attractive biomaterial invented from the carbon family is GQDs, a zero-dimensional graphene sheet with a dimension less than 100 nm with 3–10 layers [14, 15]. These GQDs have excellent quantum confinement, making them a potential tool for photoluminescence than other fluorescent dyes [16]. Considering the properties mentioned above, the current study aims to provide a broad snapshot of the recent trends in carbon-based nanomaterials' applications in the biomedical field.

2 Nanodiamonds

Nanodiamonds (NDs) or diamond nanoparticles are particles with a size less than 1 nm. Their structural, chemical, and biological properties make them suitable for high scientific and technological applications (Fig. 2). Based on the particle

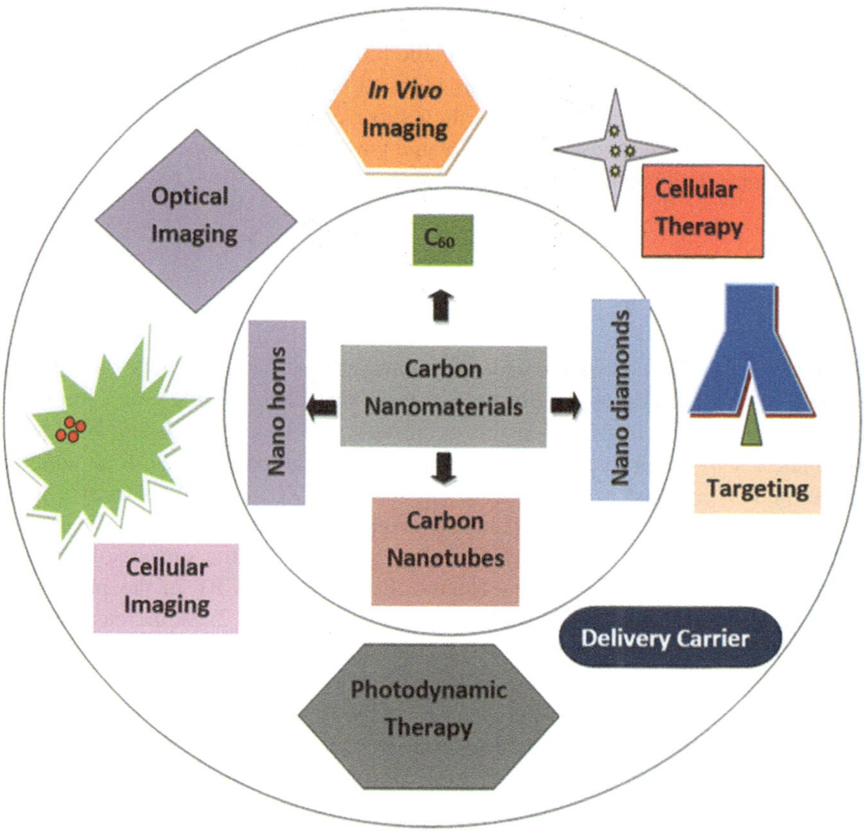

Fig. 1 Carbon-based nanomaterials and their diverse applications

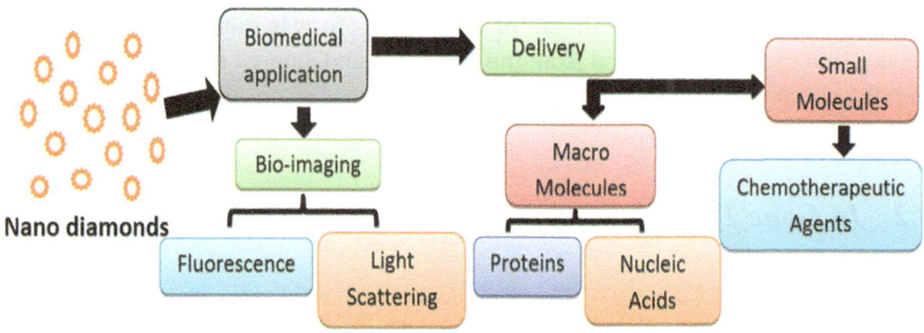

Fig. 2 Schematic diagram representing the biomedical application of Nanodiamonds

size, nanodiamonds are classified into nanocrystalline particles (1–150 nm), ultra-nanocrystalline particles (2–10 nm) and diamondoids (1–2 nm) [17]. These nanocrystalline and ultracrystalline diamonds are superior to silicon, platinum nanoparticles in biomedical implants coating due to their superior mechanical and wear-resistant properties. They possess low cytotoxicity and have high roughness, dense structure, and surface binding capacity [18, 19].

2.1 NDs as a Drug Delivery Agent

Nanodiamonds (NDs) are an ideal tool for conjugation of biomolecules via chemical bonding or physical adsorption due to their large surface to volume ratios, improved cellular delivery, the formation of loose clusters, and biocompatibility. NDs were conjugated with doxorubicin hydrochloride (DOX) via ionic forces in an aqueous medium and were used as a delivery agent. These DOX loaded NDs formed loose aggregates, which reduced the systemic adverse effects of plain DOX [20]. In another study, 10- hydroxycamptothecin (HCPT) was adsorbed on the surface of NDs via physical forces, which resulted in enhanced release of HCPT in the phosphate-buffered saline medium. The Nds-HCPT complex showed a 2.5% increase in cytotoxicity to HeLa cells than the plain chemotherapeutic agent HCPT [21]. Hence, it can be stated that drugs loaded on NDs improve the pharmaceutical properties of the drugs. Thus, the usage of NDs based delivery agent could overcome the use of a high dosage of chemotherapeutic drugs in cancer treatment and overcome the drawbacks of the chemotherapeutic drugs. In another study, conjugation of NDs with hydroxypolyethyleneglycol-4000 via covalent bond was performed, and later DOX was adsorbed on the surface of conjugated nanodiamond (ND-PEG-DOX). This formulation was used to treat human liver cancer cells, and the study reported efficient delivery of the formulation via clathrin-dependent pathways. The results also demonstrated that the half-life of this conjugate was double than free DOX uptake, and the DOX detached from the conjugate and entered into the nucleolus to stop the cell proliferation [22].

2.2 In Gene Therapy

Apart from acting as drug delivery agents, NDs efficiently deliver biological molecules such as DNA, RNA, and proteins. These nanodiamonds serve as a safe and effective method of gene therapy in comparison to viral gene therapy. It was reported that insulin when adsorbed on the surface of these nanodiamonds in a pH-dependent system at pH 10.5 via physical forces, the release of insulin was 20 times higher than the neutral pH system. This suggests that the aggregation properties of the insulin improved upon interacting with NDs [23, 24]. To the best of our knowledge, it is known that the biomolecules could not be delivered directly into the cells due to the larger size and excretion through glomerular filtration. Hence, to enhance the delivery of DNA and si-RNA into the cells, Plasmid DNA, along with NDs, were used to enhance the delivery properties via electrostatic forces. Similarly, anti-green fluorescent protein si-RNA was delivered to breast cancer cells using NDs and plasmid-based complex ND-PEI800 [25, 26].

2.3 Use as Bio-imaging Agents

Nds are emerging as efficient and safe candidates for cellular imaging due to their bright fluorescence, high photostability with a sufficiently long lifetime, and excellent biocompatibility. Nds were conjugated with transferrin through an amide linkage to form ND-transferrin conjugates. These conjugates were incubated with HeLa cells. The expression of transferrin was detected using a 514.5 nm laser source, and confocal fluorescence images were obtained to determine the uptake mechanism of transferrin in HeLa cells [27]. The uptake and interaction of NDs with cancer and non-cancer cells were analyzed via fluorescence imaging. The mechanism of NDs uptake by cells was analyzed by blocking the endocytosis pathways, mainly the clathrin-dependent pathways. It was observed that the cancer cells had significantly more amount of NDs than non-cancer cells.

3 Carbon Nanotubes (CNTs)

Carbon nanotubes consist of a rolled-up graphene sheet having a diameter of about 100 nm (Fig. 3). These carbon nanotubes (CNTs) are used in the nanomedicine and bioimaging areas due to their broad absorption spectra in UV-vis-NIR regions, large surface areas, and Raman bands. The Raman scattering features of the CNTs allow the evaluation of the signal intensities produced during the binding and conjugation of biomolecules to the CNT. This method allows the microarray-based biosensing of biomolecules with high sensitivity [28]. Functionalization of CNTs by hydrophilic molecules and reactive functional groups makes them suitable for biosensing, bioimaging, drug and gene delivery, and phototherapy. Chemical modifications of the CNTs allow the addition of various functional groups, probes, biomolecules, and antibodies to their surface [29]. Figure 4 shows the schematic of synthesis, characterization, and purification of CNTs.

3.1 CNTs as Biosensors

Properties like high conductivity, chemical stability and fast electron transfer rate make these CNTs an appropriate tool for biosensing applications [30]. Based on

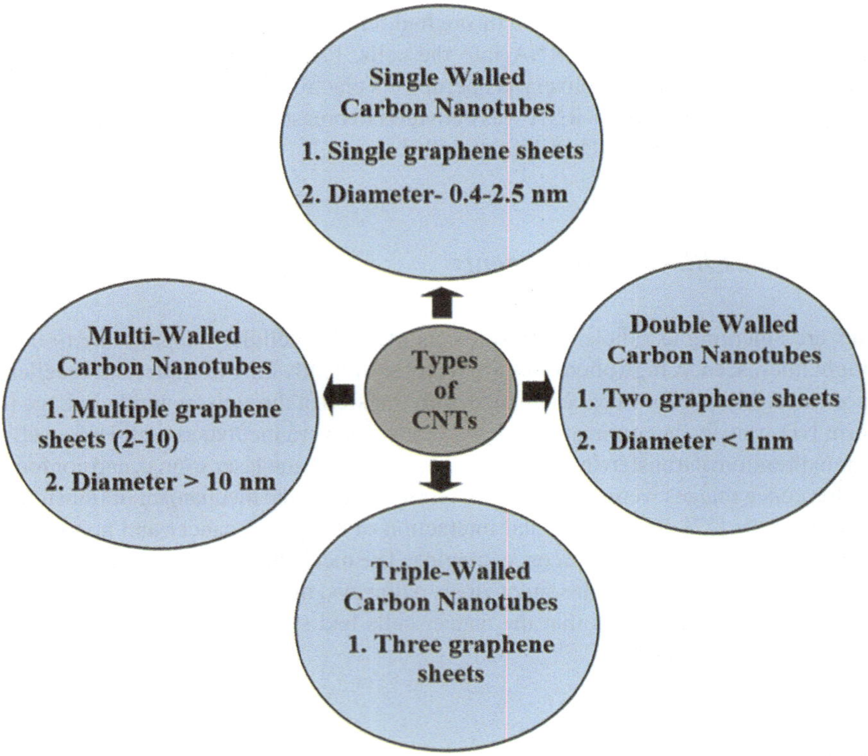

Fig. 3 Types of carbon nanotubes

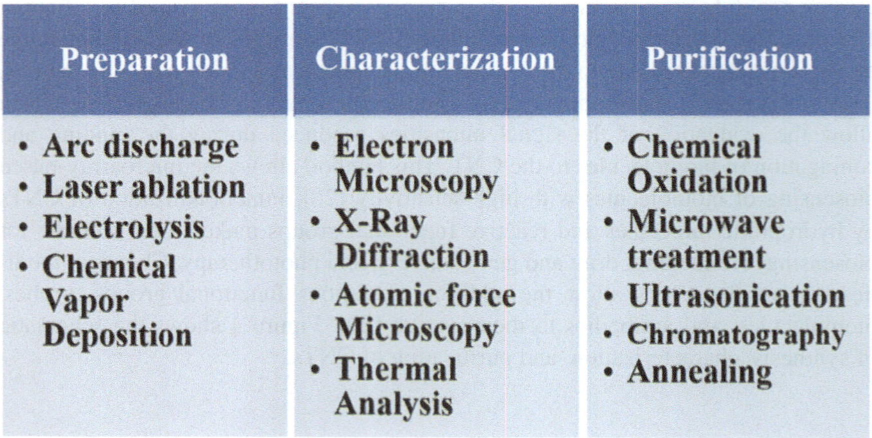

Fig. 4 Schematic showing the synthesis, characterization and purification of CNTs

the immobilization of biomolecules, their target recognition and the transduction process, they are classified into two categories viz., electrochemical and optical-based biosensors [31]. Different classes of CNT-based biosensors conjugated with DNA, antibodies, peptides, or enzymes have been developed to detect cancer biomarkers. CNT based optical biosensors have been developed to detect HIV in complex media, and an improved emission of response was detected in the presence of serum. The introduction of target microRNA-19 DNA and non-complementary control resulted in the analysis of the emission ability of biosensors. The electrical-based CNT-biosensors vertically functionalized with an amine group and conjugated with folic acid (FA-VCNT (Folic acid vertically aligned Carbon nanotubes)) were studied to detect cancer cells in normal MRC-5 and QUDM cancer cell lines isolated from lungs. The number of cancer cells entrapped on FA-VCNT improved as compared to Vertically aligned carbon nanotubes (VACNTs) [32]. Modifications in the structure, electrical conductivity, and other properties of the CNTs have been exploited in recent years. Electrochemically modified bioconjugated CNTs have shown potential for detecting molecular markers or immune sensors such as a carcinoembryonic antigen, interleukins for the early detection of cancers [33].

Enzyme biosensors, the tyrosinase biosensor, is widely used for the estimation of glucose level [34]. The reason behind this is the biomolecular conjugation is the electrocatalytic activity, greater surface area of MWCNTs. These CNTs, as biosensors, are involved in several fields like medical and forensic diagnosis. DNA is adsorbed on the CNTs and these CNT-DNA complexes are highly suitable in biosensing of several properties of the bioactive species. A cyclodextrin-CNT composite-based biosensor has been developed recently to detect cholesterol in rhodamine presence [35].

3.2 CNTs in Therapy

CNTs have the potential applications towards cancer bioimaging and gene and drug delivery. They can be used for therapeutic applications via two categories: (1) via oxidative stress and photothermal effect and (2) via drug and gene delivery. Anticancer drugs such as methotrexate, paclitaxel, and cisplatin can be conjugated to CNT in a high drug to CNT ratios. These bioconjugated CNTs can be delivered in specific cells by targeting the cancer molecules. Si-RNA can be delivered intracellularly using CNT and it can suppress the growth of tumour by activating the necrosis or inhibiting the hypoxia-inducible factors [36].

3.2.1 Cancer Therapy

Cancer is the uncontrolled proliferation of cells. CNTs have been used for cancer treatment due to their nanometric size, high drug loading efficiency and high surface functionality for the conjugation of bioactive molecules. It has been reported that

COOH functionalized cisplatin loaded MWCNTs were evaluated for their efficacy on MDA-MB-231 cells for breast cancer therapy. The developed formulations significantly decreased the cell viability after 48 h in contrast to the MWCNTs-COOH. This formulation inhibited the expression of caspase-3 activity and increased expression of procaspase-3 [37]. Doxorubicin loaded MWCNTs were evaluated for the cell inhibition study of HeLa-HFAR cells. This treatment significantly decreased tumour growth by 88.7% compared to the cells treated with free drug [38].

3.3 CNTs in Tissue Engineering

Carbon nanotubes have recently been developed for tissue engineering and regenerative medicines. For tissue engineering, the cells are encapsulated into biocompatible materials for the development of new tissues. They have been utilized as scaffolds for bone tissue engineering and for rejuvenating the neural system. It has to gain promising development in the field of neurosurgery, osteology and cardiology. These CNTs stimulate the cells electrically and aid in osteogenic differentiation of cells and their proliferation. Mostly the multiwalled CNTs were used as the substrate for the development of myoblast by electrical stimulation [39].

Tissue rejuvenation by CNTs takes place employing magnetic resonance and radiotracer contrast agents [40]. These CNTs improve the scaffold properties by enhancing their mechanical properties during bone and cartilage engineering. Scaffolds developed using polyester-CNTs were used for cardiac tissue engineering. The cells showed compaction of tissues around the scaffold [41]. These constructs revealed functional enhancement and did not show any cytotoxicity to the cells. The porous scaffold developed from collagen, MWCNTs, chitosan and hydroxyapatite was used for bone tissue engineering. This porous structure with high strength, greater in vitro activity, and pore volume showed more biocompatibility and enhanced biomineralization ability [42].

3.4 CNTS in Vaccine Delivery

CNTs have unique properties of low toxicity and in vivo stability and can attach multiple antigen copies, making them useful in vaccine delivery during cancer therapy and infectious disease. The hollow structure of CNTs allows the conjugation of multiple antigens at the same time without changing its structural conformation and it encourages antibody response with specificity [43]. MWCNTs were covalently attached with an immune–stimulatory drug (Lentinan) and the studies confirmed that this conjugate enhanced the accumulation of Lentinan in the dendritic cells and boosted humoral as well as cellular immunity [44]. Cytosine phosphate guanine oligodeoxynucleotide and anti-CD40 were conjugated on MWCNTs and antigen ovalbumin (OVA). This integration in the form of CpG-OVA improved OVA-specific immune response both in vitro and in vivo [45].

3.5 Toxicity Related to CNTs

Although nanotechnology has emerged with enhanced properties and is being extensively used in various fields, its chemical reactivity and biological activity are much more significant than larger particles. The mechanism behind CNTs toxicity is due to the formation of reactive oxygen species causing oxidative stress. The toxicity level is also influenced by impurities, shape, length, solubility, and routes of exposure [46]. The toxicity level of CNTs is described in Fig. 5.

4 Buckminsterfullerene (C_{60})

Fullerenes exist in various number forms in both natural and human-made sources and amongst all Buckminsterfullerene, (C_{60}) is the most common and abundant form. Functionalized fullerenes are mostly used in diagnostic, therapeutic applications.

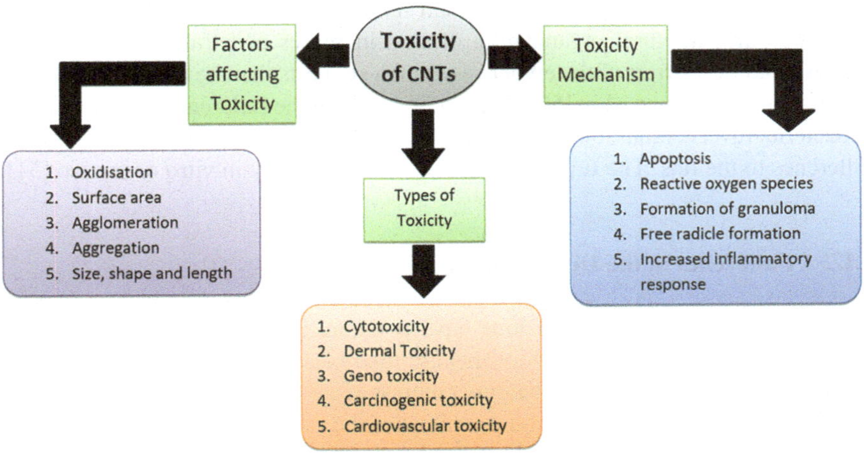

Fig. 5 The outlook of toxicity of CNTs

4.1 Biological Applications of Fullerenes

Along with their efficient distribution in blood and local organs, these fullerenes are extensively used in angiography, organ-specific imaging, oxidative stress reduction and gene/drug delivery. Fullerenes functionalized with iodine is a potential X-ray contrast agent with prolonged retention in the blood. It also provides excellent in vivo biocompatibility compared to other clinically available X-ray agents. Metallo-fullerenes are advantageous compared to other metal nanoparticles in terms of toxicity as they release fewer metals/ions due to the stable caging effect provided by C_{60} [47].

4.1.1 Fullerenes in Drug and Gene Delivery

The delivery of drugs/genes directly into the cells is a major challenge due to the three membrane barriers. Fullerenes are inorganic nanoparticles with a hydrophobic core. The functional groups attached to the core provide complexity to the behavior of the fullerene molecule. Attaching hydrophilic moieties makes them water-soluble and capable of carrying drugs and genes for delivery [48]. DNA functionalized fullerenes show better efficiency than commercially available lipid-based vectors in COS-1 cells, increasing the lifetime of DNA in endosomes by supporting their chromosomal incorporation [49]. The peptide-fullerene conjugate was evaluated for its ability to penetrate through flexed and unflexed skin. The mechanical flexion that alters the structural organization of skin increases penetration by conceding the permeability barrier of the epidermis, which suggests that the fullerenes can also penetrate through intact skin [50]. The genotoxicity of the fullerenes was determined in bacterial reverse mutation in hamster lung cells followed by oral administration of fullerenes to the rats. The fullerenes did not induce toxicity in vitro or in vivo [51].

4.1.2 Fullerenes in the Diagnostic Application

Fullerenes captured with metal ions have potential in the diagnostic application and are known as endofullerenes or metallofullerenes. These metallofullerenes are applied as magnetic resonance imaging contrast agent (MRI), X-ray imaging agents and radiopharmaceuticals. Holmium a metallofullerol molecule, could accumulate in the liver and can be detected in the bone. This shows that these fullerenes are targeted to tissues that are rich in macrophages. It could also be useful as a chemotherapeutic agent for the treatment of leukemia and bone cancer [52]. Another study revealed that when fullerene derivatives were used for carrying serum protein profiling, it helped identify proteins for pathologies and biomarker discovery using the material-enhanced laser desorption (MALDI) technique. Direct laser irradiation of the adsorbed protein was applied over the range m/z 2000–3000 for screening the proteins present in serum.

4.1.3 Fullerenes as a Photosensitizer

Photoexcitation of fullerenes is another important medical application. These can be excited from ground state to 1 C_{60} by photoirradiation. Whereas in the presence of molecular oxygen, its excitation can result in highly toxic species forming superoxide anion radical O_2^- [53]. In the presence of biological agents like guanosine, these exited fullerenes can be reduced, making them potential photosensitizers for their use in photodynamic therapy (PDT). C_{60} fullerenes were conjugated with oligonucleotide or acridine for their interaction with nucleic acids for increasing cytotoxicity in cancer cells. Dendritic monoadduct C_{60} was added to Jurkat cells, and on exposure to UV light, the cell count has decreased approximately by 19% within two weeks [54]. They have also been studied for biodistribution in tumour cells by radiotracer 125I-labeled C60 (OH)x [55]. The polyethylene glycol-conjugated fullerene containing $Gd3^+$ ions was used for PDT combined with MRI and the data signifies that photosensitizer significantly promoted the tumour targetability and MRI activity [56]. Fullerenes conjugated with hydrophilic cationic groups were used for PDT against mouse cancer cell lines, and they were highly effective photosensitizers for killing cancer cells by induction of apoptosis after illumination [57].

5 Carbon Quantum Dots (CQDs)

Carbon quantum dots (CQDs) are sometimes called carbon dots (CDs) are novel zero-dimensional carbon-based nanomaterials with size less than 10 nm [58]. They are synthesized using various techniques like chemical vapour deposition, laser ablation, microwave heating, pyrolysis, and chemical oxidation method [59]. They have been synthesized generally from organic materials like polymers and extracts of various fruits [60, 61]. These CQDs are smaller and have strong fluorescence characteristics, which are the inherited properties of traditional semiconductor quantum dots. They are water-soluble with high chemical stability and can be easily functionalized [62].

CQDs bioprobes with good stability have been engineered for their applications in live cells, tissues, and animals. During the last few years, their cytotoxicity evaluations were carried out on various cell lines without affecting the cell viability. The CQDs functionalized with various groups and polymers such as PEG, PEI, BPEI showed negligible cytotoxicity at lower concentrations when incubated for a short time [63].

5.1 CQDs as Biosensors

The fluorescent property of CDs allows the detection and identification of a wide range of analytes, drugs and macromolecules depending on high sensitivity and easy operation. CDs have been used as a biosensing device to identify biological molecules like amino acids, glucose, proteins, nucleic acids in the clinical sample, and early diagnosis of diseases. CQDs based biosensors have been used for visual monitoring of glucose, cellular ions, pH as well as nucleic acids. These CQDs based biosensors are highly soluble in water, non-toxic with good cell permeability and high stability. Mitochondrial H_2O_2 stress in the cells was detected and imaged using a multifunctional CQD-based fluorescence energy transfer (FRET) probe. These CQDs act as the donor and carrier of the sensing system that was used to detect exogenous H_2O_2 levels in L_92_9 cells and RAW 264.7 macrophage cells [64]. In a detailed study, N-doped CDs biosensor showed a high affinity towards DNA fragments from PC-3 cells, and melanoma tissues was demonstrated [65]. Bacterial contamination in tap water, human urine, and juice could also be detected using labeled CDs [66]. Graphene oxide derived N, and S doped Cds detected a wide range of biological species, including bacteria, viruses, and other molecules via fluorescent quenching effect [67].

5.2 In Drug and Gene Delivery

To serve as a drug delivery system (DDS), nanostructured materials should transport the drug to a specific target site and proper interaction of the drug and the target. These drug conjugated nanomaterials should improve the drug delivery system concerning absorption, distribution and elimination.

The application of nanotechnology in DDS results in site-specific delivery of the drug to the cells or tissue. Recently, CQDs have gained attention due to their superior properties of fluorescence imaging, cell permeability, low toxicity and water solubility. Recently, the drug 7-(3-bromopropoxy)-2-quinolylmethyl chlorambucil (Qucbl) was covalently attached to the CQDs and showed a controlled release of drug to the target cells through irradiation. Similarly, dopamine hydrochloride was conjugated on the surface of the CQDs to study its effect in vitro. The study showed that the half-life of the drug released by the DA-CQDs conjugate increased to 60 h when compared to the DA alone in Neuro 2A cells [68].

Recently, hollow nanostructures have gained importance in drug delivery due to their fluorescence imaging properties. Doxorubicin (DOX) was loaded onto the hollow CQDs for an in vitro experiment showed rapid delivery of drugs to the cells and using fluorescence microscopy, it was confirmed that the conjugate was internalized by the cells in the cytoplasm and the DOX was detected in the nucleus [69]. These CQDS have also been used to deliver a multifunctional theranostic agent prepared by conjugating oxaliplatin (oxa) on the surface of CQDs via amine linkage. This conjugate enhances biocompatibility, bioimaging function, and anticancer effects in vitro. The results showed the possibility to track the distribution of drugs via fluorescence signal provided by ox-CQDs, which helps in customizing the dosage of injection of the drug in vivo [70]. These CQDs provide multifunctional platforms for drug delivery, magnetic delivery, or simultaneous delivery of two or more drugs.

Gene therapy is the deletion/replacement of defective genes through the delivery and expression of the exogenous gene. CQDs have shown great potential in delivering genes with their electrostatic interaction between positively charged functionalized CDs and negatively charged nucleic acids. Plasmid DNA loaded-CQDs can be used in bioimaging, gene delivery and tissue engineering due to their superior capacity to condense plasmid DNA [71]. They are used as a gene vector for chondrogenesis from fibroblasts. By using CDs, the plasmid SOX9 was used to form nanoparticles with size ranging from 10 to 30 nm with excellent properties with low toxicity and high fluorescence [72].

The turn on-off theranostic application of CQDs were reported against hyaluronidase in cancer cells for self-targeted drug delivery. The CDs were coated with polyethyleneimine (PEI) via electrostatic interaction and functionalized with hyaluronic acid-Doxocubicin conjugate (p-CDs-HA-Dox). This conjugate was able to penetrate the cells with the target site on the cancer cells and successfully targeted bio-imaging and delivery vehicles for cancer therapy [73].

5.3 Bio-imaging

Carbon dots (CDs) with optical properties have been extensively used as fluorophores in the bio-imaging of cells and tissues for in vivo and in vitro applications. They have excellent biocompatibility, small size, broad absorption and narrow emission spectra with multicolored fluorescence properties suitable for biomedical applications. Several kinds of research have used CQDs over organic dyes and fluorescent proteins because of the imaging sensitivity, stability, and brightness. Earlier, semiconductor QDs have been used for in vitro and in vivo imaging; however, it caused aggregation in the tissues leading to toxicity. To overcome the toxicity of conventional QDs, these carbon dots have been used in live-cell imaging due to their small size with strong fluorescent and non-blinking activity compared to the commercial QDs. The smaller size of CDs provides efficient probing of biological structures and it reduces the in vivo injection volume [74]. Plant cells have a rigid cell wall, so it is more complicated than animal cells for in vivo imaging of cells. In a study, N and S heterodoped CDs were prepared for in vivo imaging of plant cells and the results indicated the successful penetration inside the cells, which suggested that CDs can be used for in vivo cell imaging for plants [70].

6 Carbon Nanohorns

Carbon nanohorns are one form of carbon nanomaterials closely related to nanotubes with narrowest opening and five pentagonal rings in its apex. Their average diameter size is 100 nm with a peculiar geometry similar to sponge with high porosity large surface area [71–75]. They have high affinity and absorbing capacity for organic compounds and other molecules. Although these nanohorns functions similarly to nanotubes, it is mandatory to functionalize these nanohorns to make them biocompatible.

Nanohorns have been loaded with various drugs to check its efficiency for drug delivery. Nanohorns functionalized with polyethyleneglycol chains, and doxorubicin was tested in vitro in lung cancer cells showed apoptosis of cells lower than the normal control drug. It could be possible that PEG-doxorubicin retained on the surface of the nanohorns thereby reducing its therapeutic effect. In another approach, cisplatin was trapped in the inner space of the nanohorns and used as an anticancer drug delivery vehicle for lung cancer cells. Following the administration of the drug and monitoring it for 48 h, it was observed that the conjugate showed comparable anti-cancer activity to the control drugs alone without any cytotoxicity [75]. Although this nanohorns did not show cytotoxicity, they formed aggregates which are a major concern for its in vivo application. Recently in a study, the cisplatin was encapsulated in the nanohorns followed by the PEG chain terminated with a peptide aptamer coating on their surface. This complex was able to induce cytotoxicity to the cancer cells. However, it might be possible that different molecules might induce other cellular problems which could not be neglected. Although these studies make nanohorns a vehicle for drug delivery, more clarity is needed to make a biocompatible nanohorns for paving their way towards biomedical applications [76].

7 Conclusions

Nanotechnology has undergone exponential growth in the field of nanomedicine from traditional strategies to modern applications. Carbon nanomaterials have shown great potential in biomaterials research and applications. The different forms of carbon-based nanomaterials such as nanotubes, fullerenes, nanohorns, and nanodiamonds, have proven to be successful candidates for carrying biological molecules due to their easy penetration into the cells. These nanomaterials can easily be functionalized under various physiological conditions to make them biocompatible to be administered in vitro or in vivo for targeted delivery. Although these materials have shown vital potential in therapeutics and biomedicines, a few safety concerns regarding the toxicity of these materials were demonstrated in the scientific literature. Hence, to eradicate the mild toxicity, many super-specific and excellent carbon-based materials are being studied for their immense potential in biomedicines.

Acknowledgements Dr. T. Senthilvelan would like to thank DST for providing a research fellowship through SERB National Post-Doctoral Fellowship (NPDF) (file no: PDF/2017/002894) for writing this book chapter.

References

1. Ambrosi A, Chua CK, Bonanni A, Pumera M (2014) Electrochemistry of graphene and related materials. Chem Rev 114:7150–7188
2. Nicolosi V, Chhowalla M, Kanatzidis MG, Strano MS, Coleman JN (2013) Liquid exfoliation of layered materials. Science 340:1226419
3. National Nanotechnology Initiative, What is nanotechnology? http://www.nano.gov/html/facts/whatIsNano.html
4. Thassu D, Deleers M, Pathak Y (2007) Nanoparticulate drug delivery systems. Informa Healthcare, New York, NY, USA
5. Kagan VE, Bayir H, Shvedova AA (2005) Nanomedicine and nanotoxicology: two sides of the same coin. Nanomed: Nanotechnol Biol Med 1:313–316
6. Ferrari M (2005) Cancer nanotechnology: opportunities and challenges. Nat Rev Cancer 5:161–171
7. Iijima S (1991) Helical microtubules of graphitic carbon. Nature 354:56–58
8. Aqel A, El-Nour KMMA, Ammar RAA, Al-Warthan A (2012) Carbon nanotubes, science and technology part (I) structure, synthesis and characterisation. Arab J Chem 5:1–23
9. Santra TS, Bhattacharyya TK, Patel P, Tseng F-G, Barik TK (2012) Diamond, diamond-like carbon (DLC) and diamond-like nanocomposite (DLN) thin films for films for MEMS applications. InTech, E.U. ISBN 978-953-307-905-9, pp 459–480
10. Santra TS, Bhattacharyya TK, Patel P, Tseng FG, Barik TK (2010) Characterization of diamond-like nanocomposite thin films grown by plasma enhanced chemical vapor deposition. J Appl Phys 10:124320
11. Santra TS, Bhattacharyya TK, Patel P, Tseng FG, Barik TK (2011) Structural and tribological properties of diamond-like nanocomposite thin films grown by PECVD. Surf Coat Technol 206:2–3
12. Santra TS, Bhattacharyya TK, Mishra P, Tseng FG, Barik TK (2011) Biomedical application of diamond like nanocomposite thin films. Sci Adv Mater 4:110–113
13. Santra TS, Bhattacharyya TK, Tseng FG, Barik TK (2012) Influence of flow rate on different properties of diamond like nanocomposite thin films grown by PECVD. AIP Adv 2:022132
14. Valcarcel M, Cardenas S, Simonet BM, Moliner-Martinez Y, Lucena R (2008) Carbon nanostructures as sorbent materials in analytical processes. TrAC Trends Anal Chem 27:34–43
15. Song L, Shi J, Lu JJ, Lu C (2015) Structure observation of graphene quantum dots by the single-layered formation in layered confinement space. Chem Sci 6:4846–4850
16. Wang J, Cao S, Ding Y, Ma F, Lu W, Sun M (2016) Theoretical investigations of optical origins of fluorescent graphene quantum dots. Sci Rep 6:24850
17. Shenderova OA, McGuire G (2006) Types of nanocrystalline diamond. Ultrananocrystalline diamond. New York, NY: USA, pp 79–114
18. Bajaj P, Akin D, Gupta A, Sherman D, Shi B, Auciello O, Bashir R (2007) Ultrananocrystalline diamond film as an optimal cell interface for biomedical applications. Biomed Microdevices 9:787–794
19. Shi B, Jin Q, Chen L, Auciello Q (2009) Fundamentals of ultra nanocrystalline diamond (UNCD) thin films as biomaterials for developmental biology: embryonic fibroblasts growth on the surface of (UNCD) films. Diam Relat Mater 18:596–600
20. Huang H, Pierstorff E, Osawa E, Ho D (2007) Active nanodiamond hydrogels for chemotherapeutic delivery. Nano Lett 7:3305–3314
21. Li J, Zhu Y, Li W, Zhang X, Peng Y, Huang Q (2010) Nanodiamonds as intracellular transporters of chemotherapeutic drug. Biomaterials 31:8410–8418
22. Chow EK, Zhang XQ, Chen M, Lam R, Robinson E, Huang H, Schaffer D, Osawa E, Goga A (2011) Nanodiamond therapeutic delivery agents mediate enhanced chemoresistant tumour treatment. Sci Transl Med 3:73ra21
23. Chen M, Pierstorff ED, Lam R, Li SY, Huang H, Osawa E, Ho D (2009) Nanodiamond-mediated delivery of water-insoluble therapeutics. ACS Nano 3:2016–2022

24. Shimkunas RA, Robinson E, Lam R, Lu S, Xu X, Zhang XQ, Huang H, Osawa E, Ho D (2009) Nanodiamond-insulin complexes as pH-dependent protein delivery vehicles. Biomaterials 30:5720–5728
25. Zhang XQ, Chen M, Lam R, Xu X, Osawa E, Ho D (2009) Polymer-functionalized nanodiamond platforms as vehicles for gene delivery. ACS Nano 3:2609–2616
26. Chen M, Zhang XQ, Man HB, Lam R, Chow EK, Ho D (2010) Nanodiamond vectors functionalized with polyethyleneimine for siRNA delivery. J Phys Chem Lett. 1:3167–3171
27. Weng MF, Chiang SY, Wang NS, Niu H (2009) Fluorescent nanodiamonds for specifically targeted bioimaging: application to the interaction of transferrin with transferrin receptor. Diam Relat Mater 18:587–591
28. Lin Y, Lu F, Tu Y, Ren Z (2004) Glucose biosensors based on carbon nanotube nanoelectrode ensembles. Nano Lett 4:191–195
29. Wang Y, Wei H, Lu Y, Wei S, Wujcik EK, Guo Z (2015) Multifunctional carbon nanostructures for advanced energy storage applications. Nanomater 5:755–777
30. Zanganeh S, Khodadade F, Rafizadeh Tafti S, Abdolahad M (2016) Folic acid functionalized vertically aligned carbon nanotube (FA-VACNT) electrodes for cancer sensing applications. J Mater Sci Technol 32:617–625
31. Gruner G (2006) Carbon nanotube transistors for biosensing applications. Anal Bioanal Chem 384:322–335
32. Asal M, Ozen O, Lu MSP (2018) Recent developments in enzyme, DNA and immune-based biosensors. Sensors 18:1–16
33. Gilbert S, Kyung H, Dowben PA (2017) Cyclodextrin-carbon nanotube composites for fluorescent detection of cholesterol. Chem Phys Lett 687:222–226
34. Liu Z, Tabakman S, Welsher K, Dai H (2010) Carbon nanotubes in biology and medicine: *In vitro* and *in vivo* detection, imaging and drug delivery. Nano Res 2:85–120
35. Badea M, Prodana M, Dinischiotu A, Crihana C, Ionita D, Balas M (2018) Cisplatin loaded multiwalled carbon nanotubes induce resistance in triple-negative breast cancer cells. Pharmaceutics 10:228
36. Yan Y, Wang R, Hu Y, Sun R, Song T, Shi X, Yin S (2018) Stacking of doxorubicin on folic acid-targeted multiwalled carbon nanotubes for *in vivo* chemotherapy of tumours. Drug Deliver 25:1607–1616
37. Tonelli FM, Santos AK, Gomes KN, Lorençon E (2012) Carbon nanotube interaction with extracellular matrix proteins producing scaffolds for tissue engineering. Int J Nanomed 7:4511–4529
38. Harrison BS, Atala A (2007) Carbon nanotube applications for tissue engineering. Biomaterials 28:344–353
39. Ahadian S, Davenport HL, Estili M, Yee B, Smith N, Xu Z, Sun Y, Radisic M (2017) Moldable elastomeric polyester-carbon nanotube scaffolds for cardiac tissue engineering. Acta Biomater 52:81–91
40. Turk S, Altinosy I, Celebi EG, Ipek M, Ozacar M, Bindal C (2018) 3D porous collagen/functionalized multiwalled carbon nanotube/chitosan/hydroxyapatite composite scaffolds for bone tissue engineering. Mater Sci Eng, C 92:757–768
41. Inhet Panhuis M (2003) Vaccine delivery by carbon nanotubes. Chem Biol 10:897–898
42. Xing J, Liu Z, Huang Y, Qin T, Bo R, Zheng S, Luo L, Huang Y, Nilu Y, Wang D (2016) Lentinan-modified carbon nanotubes as an antigen delivery system modulate the immune response *in vitro* and *in vivo*. AC Appl Mater Interfaces 8:19276–19283
43. Hassan HA, Smyth L, Wang JT, Costa PM, Ratnasothy K, Diebold SS, Lombardi G, Al-Jamal KT (2016) Dual stimulation of antigen-presenting cells using carbon nanotube-based vaccine delivery system for cancer immunotherapy. Biomaterials 104:310–322
44. Mohanta D, Patnaik S, Sood S, Das N (2019) Carbon nanotubes: evaluation of toxicity at biointerfaces. J Pharm Anal 9:1214
45. Partha R, Conyers JL (2009) Biomedical applications of functionalized fullerene-based nanomaterials. Int J Nanomed 4:261–275

46. Foley S, Crowley C, Smaihi M, Bonfils C, Erlanger BF, Seta P, Larroque C (2002) Cellular localisation of a water-soluble fullerene derivative. Biochem Biophys Res Commun 294:116–119
47. Isobe H, Tomita N, Jinno S, Okayama H, Nakamura E (2001) Synthesis and transfection capability of multi-functionalized fullerene polyamine. Chem Lett 12:1214–1215
48. Rouse JG, Yang J, Ryman-Rasmussen JP, Barron AR, Monteiro-Riviere NA (2007) Effects of mechanical flexion on the penetration of fullerene amino acid-derivatized peptide nanoparticles through the skin. Nano Lett 7:155–160
49. Mori T, Takada H, Ito S, Matsubayashi K, Miwa N, Sawaguchi T (2006) Preclinical studies on the safety of fullerene upon acute oral administration and evaluation for no mutagenesis. Toxicology 225:48–54
50. Thrash TP, Cagle DW, Alford JM, Wright K, Ehrhardt GJ, Mirzadeh S, Wilson LJ (1999) Toward fullerene-based radiopharmaceuticals: high-yield neutron activation of endohedral 165Ho metallofullerenes. Chem Phys Lett 308:329–336
51. Yamakoshi Y, Umezawa N, Ryu A, Arakane K, Miyata N, Goda Y, Masumizu T, Nagano T (2003) Active oxygen species generated from photoexcited fullerene (C_{60}) as potential medicines: O-.bul.2 versus 1O2. J Am Chem Soc 125:12803–12809
52. Rancan F, Rosan S, Boehm F, Cantrell A, Brellreich M, Schoenberger H, Hirsch A, Moussa F (2002) Cytotoxicity and photocytotoxicity of a dendritic C_{60} mono-adduct and a malonic acid C_{60} tris-adduct on Jurkat cells. J Photochem Photobiol, B 67:157–162
53. Ji ZQ, Sun H, Wang H, Xie Q, Liu Y, Wang Z (2006) Biodistribution and tumour uptake of C60(OH)x in mice. J Nanoparticle Res 8:53–63
54. Liu J, Ohta S, Sonoda A, Yamada M, Yamamoto M, Nitta N, Murata K, Tabata Y (2007) Preparation of PEG-conjugated fullerene containing Gd^{3+} ions for photodynamic therapy. J Controll Release 117:104–110
55. Mroz P, Pawlak A, Satti M, Lee H, Wharton T, Gali H, Sarna T, Hamblin MR (2007) Functionalized fullerenes mediate photodynamic killing of cancer cells: Type I versus Type II photochemical mechanism. Free Radical Biol Med 43:711–719
56. Semeniuk M, Yi Z, Poursorkhabi V, Tjong J, Jaffer S, Lu ZH, Sain M (2019) Future perspectives and review on organic carbon dots in electronic applications. ACS Nano 13:6224–6255
57. Wang Y, Meng Y, Wang S, Li C, Shi W, Chen J, Wang J, Huang R (2015) Direct solvent-derived polymer-coated nitrogen-doped carbon nanodots with high water solubility for targeted fluorescence imaging of glioma. Small 11:3575–3581
58. Xu Y, Chun-Jing T, Huang H, Chao-Qun S, Ya-Kun Z, Qun-Feng Y, Ai-Jun W (2014) Green synthesis of fluorescent carbon quantum dots for detection of Hg^{2+}. Chin J Anal Chem 42:1252–1258
59. De B, Karak N (2013) A green and facile approach for the synthesis of water-soluble fluorescent carbon dots from banana juice. RSC Adv 3:8286–8290
60. Yang X, Zhuo Y, Zhu S, Luo Y, Feng Y, Dou Y (2014) Novel and green synthesis of high-fluorescent carbon dots originated from honey for sensing and imaging. Biosens Bioelectron 60:292–298
61. Wang Y, Anilkumar P, Cao L, Liu JH, Luo PG, Tackett KN, Sahu S, Wang P, Wang X, Sun YP (2011) Carbon dots of different composition and surface functionalization: cytotoxicity issues relevant to fluorescence cell imaging. Exp Biol Med 236:1231–1238
62. Du F, Min Y, Zeng F, Yu C, Wu S (2014) A targeted and FRET-based ratiometric fluorescent nanoprobe for imaging mitochondrial hydrogen peroxide in living cells. Small 10:964–972
63. Weng CI, Chang HT, Lin CH, Shen YW, Unnikrishnan B, Li YJ, Hang CC (2015) One-step synthesis of biofunctional carbon quantum dots for bacterial labelling. Biosens Bioelectron 68:1–6
64. Chen L, Song L, Zhang Y, Wang P, Xiao Z, Guo Y, Cao F (2016) Nitrogen and sulfur codoped reduced graphene oxide as a general platform for rapid and sensitive fluorescent detection of biological species. ACS Appl Mater Interfaces 8:11255–11261
65. Kudr J, Richtera L, Xhaxhiu K, Hynek D, Heger Z, Zirka O, Adam V (2017) Carbon dots based FRET for the detection of DNA damage. Biosens Bioelectron 92:133–139

66. Khan MS, Pandey S, Talib A, Bhaisare ML, Wu HF (2015) Controlled delivery of dopamine hydrochloride using surface-modified carbon dots for neuro diseases. Colloids Surf B 134:140–146
67. Wang Q, Huang X, Long Y, Wang X, Rui Zhu HZ, Liang L, Teng P, Zheng H (2013) Hollow luminescent carbon dots for drug delivery. Carbon 59:192–199
68. Zhang M, Wang W, Cui Y, Zhou N, Shen J (2017) Near-infrared light-triggered phototherapy, in combination with chemotherapy using magneto fluorescent carbon quantum dots for effective cancer-treating. Carbon 118:752–764
69. Zhou J, Deng W, Wang Y, Cao X, Chen J, Wang Q, Xu W, Du P, Chen QYJ, Spector M, Yu J, Xu X (2016) Cationic carbon quantum dots derived from alginate for gene delivery: one-step synthesis and cellular uptake. Acta Biomater 42:209–219
70. Cao X, Wang J, Deng W, Chen J, Wang Y, Zhou J, Du P, Xu W, Wang Q, Yu Q, Spector M, Yu J, Xu X (2018) Photoluminescent cationic carbon dots as efficient non-viral delivery of plasmid SOX9 and chondrogenesis of fibroblasts. Sci Rep 8:7057
71. Gao N, Yang W, Nie H, Gong Y, Jing J, Gao L, Zhang X (2017) Turn-on theranostic fluorescent nanoprobe by electrostatic self-assembly of carbon dots with doxorubicin for targeted cancer cell imaging, *in vivo* hyaluronidase analysis, and targeted drug delivery. Biosens Bioelectron 96:300–307
72. Yang ST, Wang X, Wang H, Lu F, Luo PG, Cao L, Meziani MJ, Liu JH, Liu Y, Chen M, Huang Y, Sun YP (2009) Carbon dots as non-toxic and high-performance fluorescence imaging agents. J Phys Chem C 113:18110–18114
73. Xiao J, Liu P, Wang CX, Yang GW (2017) External field-assisted laser ablation in liquid: an efficient strategy for nanocrystal synthesis and nanostructure assembly. Prog Mater Sci 87:140–220
74. Jiménez-Soto JM, Moliner-Martínez Y, Cárdenas S, Valcárcel M (2010) Evaluation of the performance of single-walled carbon nanohorns in capillary electrophoresis. Electrophoresis 31:1681–1688
75. Ajima K, Yudasaka M, Murakami T, Maigné A, Shiba K, Iijima S (2005) Carbon nanohorns as anticancer drug carriers. Mol Pharm 2:475–480
76. Isobe H, Tanaka T, Maeda R, Noiri E, Solin N, Yudasaka M, Iijima S, Nakamura E (2006) Preparation, purification, characterization, and cytotoxicity assessment of water-soluble, transition-metal-free carbon nanotube aggregates. Angew Chem Int Ed 45:6676–6680

Solution Combustion Synthesis of Calcium Phosphate-Based Bioceramic Powders for Biomedical Applications

S. T. Aruna and M. Shilpa

Abstract Calcium phosphate (CaP)-based bioceramics are widely used in orthopedics and dentistry for bone regeneration due to their good biocompatibility, osseointegration and osteoconduction. Synthetic hydroxyapatite (HAp) is the most widely used bioceramic coating in biomedical implants as it has a chemical composition similar to that of the bone. The properties of the coating depend on the nature of powders and in turn on the source of the powders. Hydroxyapatite powders can be prepared by a variety of chemical routes like co-precipitation, hydrothermal, sol-gel, solution combustion, etc. Among these methods, the solution combustion method is promising as it is a single step, cost-effective and energy-efficient process and yields high purity ceramic powders compared to conventional multi-step wet chemical processes. This method can also yield powders in nano and micron size range and can be used for the fabrication of coatings using methods like plasma spraying, suspension plasma spraying, electrophoretic deposition, etc. This chapter gives an overview of the solution combustion synthesis of pure and doped hydroxyapatite powders and their characterization. The last section of this chapter discusses the solution combustion synthesis of plasma sprayable hydroxyapatite powder, fabrication of HAp coating and characterization of the developed plasma-sprayed coating.

1 Introduction

Ceramics are playing a pivotal role in all fields of science and technology. It is interesting to note that ceramics are also being used for a long time in the medical industry. Some of the important ceramic materials that are being used in biomedical applications are alumina (Al_2O_3), zirconia (ZrO_2), calcium aluminates, bio-glass and hydroxyapatite (Fig. 1) and they are referred to as bioceramics. Bioceramics are defined as a class of ceramics that are used for repairing and replacing diseased

S. T. Aruna (✉) · M. Shilpa
Surface Engineering Division, CSIR-National Aerospace Laboratories, Bengaluru 560017,, KA, India
e-mail: aruna_reddy@nal.res.in

T. S. Santra and L. Mohan (eds.), *Nanomaterials and Their Biomedical Applications*, Springer Series in Biomaterials Science and Engineering 16,
https://doi.org/10.1007/978-981-33-6252-9_7

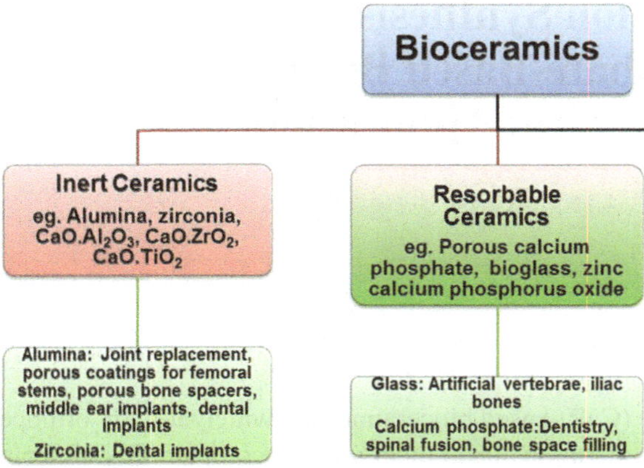

Fig. 1 Flowchart depicting the classification of bioceramics and their applications

and damaged parts of the animal musculoskeletal system [1]. Bioceramics are classified into (i) inert ceramics wherein the ceramics such as alumina and zirconia are chemically inert, do not exhibit any biological response and they also exhibit higher hardness and hence used as articulating surfaces in hip and knee joints, (ii) bioactive ceramics such as hydroxyapatite and bioactive glasses helps in promoting the integration with the body tissues through chemical bonding, and (iii) resorbable ceramics such as calcium phosphate cements, bioactive glass and glass-ceramics that gradually dissolve within the body (Fig. 1) [1–10].

Among the bioceramics, calcium phosphates are of astounding importance to sustain life [2–5]. The family of minerals containing calcium cations (Ca^{2+}) together with phosphate anions such as orthophosphate ($PO_3{}^{4-}$), metaphosphate ($PO_3{}^{-}$), or pyrophosphate ($P_2O_4{}^{7-}$) along with hydrogen (H^+) or hydroxide (OH^{-1}) ions is known as calcium phosphates (CaP). Calcium phosphate is the principal form of calcium found in milk, blood and bovine; about 60 and 90 wt% of bone and tooth enamel is made of CaP respectively. The CaP in bone is responsible for the mechanical durability, hardness, rigidity, and high resistance to compression. However, calcium phosphates, in general, do not possess sufficient hardness. The important calcium phosphate bioceramics are hydroxyapatite (HAp) with Ca/P = 1.67; α and β-tricalcium phosphate (TCP) with Ca/P = 1.5; tetracalcium phosphate with Ca/P = 2 and monocalcium phosphate (MCP) hydrate with Ca/P = 0.5 [5, 6]. Most of the calcium phosphates are resorbable and will dissolve when exposed to physiological environments. Unlike the other calcium phosphates, HAp is thermodynamically stable at physiological pH and does not break down under physiological conditions [5, 6].

1.1 Hydroxyapatite (HAp)

In the last several years, with the growth in medical science and advancements in surgical techniques, there is an increasing demand for artificial bone implants and synthetic hydroxyapatite has commanded extensive attention as a material for medical implants. Among the above described bioactive ceramics, from the view-point of biocompatibility, hydroxyapatite (HAp) is the most appropriate choice for hard tissue replacement implants. It directly binds to the bone and shows excellent biocompatibility with hard tissues and also with skin and muscle tissues and results in better osseointegration. However, it has inferior mechanical properties and hence it is applied in the form of a coating on metallic components such as titanium and its alloys, which bear the loads. HAp crystallizes in hexagonal structure and belongs to the apatite family wherein the compounds in this family possess a similar hexagonal structure but have different compositions. Calcium apatites have the general formula of $Ca_5(PO_4)_3X$ (X = Halide ion or OH group). The important calcium apatites are HAp, fluorapatite (FA) ($Ca_5[PO_4]_3F$) and chlorapatite (CA) ($Ca_5[PO_4]_3Cl$). The apatite structure can be substituted by ions such as Sr^{2+}, Pb^{2+}, Cd^{2+}, Mn^{2+}, etc., at Ca position which results in changes in the properties.

Hydroxyapatite coatings are applied by the atmospheric plasma spray (APS) method. However, due to the high processing plasma spray temperatures, careful control of plasma processing parameters is paramount to prevent the thermal decomposition of HAp into other soluble calcium phosphates. While dense as well as porous HAp blocks have found application as implant material in the field of dentistry and orthopedics, HAp powder and granules have been used for bone grafting and augmentation [10]. Hydroxyapatite is described as a non-stoichiometric, calcium-deficient carbonated apatite and is represented by the chemical formula $Ca_{10}(PO_4)_6(OH)_2$, wherein the molar ratio Ca/P is 1.67 [10]. Some of the important properties of hydroxyapatite are as follows: low thermal conductivity (0.013 W/cm.K), higher decomposition temperature (> 1000 °C), higher melting point (1614 °C), lower fracture toughness (0.7–1.2 $MPam^{1/2}$), lower Vickers hardness (3–7 GPa for dense HAp) and lower Young's modulus (35–130 GPa), besides higher biocompatibility, bioactivity, cellular compatibility and osteoconduction.

The different methods employed for the development of hydroxyapatite coatings include physical vapor deposition, chemical vapor deposition, sol-gel, electrophoretic deposition, plasma spray technique, etc., [5]. Among these methods, plasma spraying is the only method that has been approved by the Food and Drug Administration (FDA), USA [11]. The required key HAp coating properties include stable phase composition, phase purity, sufficient crystallinity (> 45%), suitable coating thickness, adequate porosity and roughness, high adhesive and cohesive strengths and devoid of residual coating stresses to avoid delamination of coatings during in vivo conditions. A thin (< 50 μm) HAp layer exhibited better adhesion and the porosity of plasma-sprayed HAp coatings was in the range of 3–20%. High porosity facilitates ingrowth of bone cells into the bioceramic coatings. However, very high porosity results in decreased tribological properties.

1.2 Atmospheric Plasma Spraying of Hydroxyapatite

The deposition of HAp coatings by APS is a matured well-researched technique to coat metallic hip endoprosthetic and dental root implants. The advantages of plasma-sprayed HAp coatings are as follows: macro-micro porosity, enhanced bio-adhesion, nanostructured surface topography, it provides osseointegration with adsorbed osteo-stimulating biological agents due to the chemical composition resembling that of the inorganic component of natural bone. Some of the drawbacks of the plasma spray technique are the thermal decomposition of the feedstock during spraying, line-of-sight process, difficulty to control pore sizes and porosity, and the inability to deposit coatings with lower thickness. A prudent design of plasma spray conditions and stringent quality management is required to develop HAp coatings with strong adhesion and provide bone regeneration over a long time. It has been established that a HAp coating must meet the minimum requirements described by the U.S. FDA and the ISO [12–16]. According to ISO 13779-2, some of the important desirable properties of plasma-sprayed hydroxyapatite coatings are as follows: hydroxyapatite content > 50%; Non-apatite CaP (%): > 50; Crystallinity > 45%; adhesion strength > 15 MPa; and Ca/P ratio 1.67–1.76 [17].

Plasma spraying is the most popular thermal spray technique [18]. In plasma spray technique, a plasma flame is produced by using argon as the primary gas and nitrogen/hydrogen is used as the secondary gas by striking an arc between the cathode and the anode. The electrical energy ionizes the gas and produces plasma. The plasma flame is characterized by relatively high temperatures (15,000 °C) and low velocity. The ceramic powder feed is fed into the plasma flame by using a carrier gas and the particles get melted in the plasma plume, traverses with high speed and impinges on the substrate forming a coating. The schematic of a plasma spray set-up is shown in Fig. 2.

Many of the methods employed for the synthesis of hydroxyapatite powder does not yield plasma sprayable grade flowable powders. Generally, HAp fine powders are prepared using methods such as coprecipitation, hydrothermal synthesis, solvothermal method, sol-gel method, hydrothermal method and solution combustion method. To obtain flowable powders, the powders are subjected to spray drying or fusing and drying. Thus obtained powders possess the required flowability for plasma spraying. Figure 3 shows the steps involved in the fabrication of the HAp coating starting from the fine powders to a coating on real component.

The review article by Mehdi et al. provides various methodologies employed in the synthesis of nanosized HAp powders [19]. The review by Surmenev et al. elaborates on the numerous techniques employed for enhancing the surface compatibility of implants with bone [5]. Generally, the hydroxyapatite fine powder is synthesized by solid-state, coprecipitation, sol-gel, flux, hydrothermal and solution combustion methods. The obtained powders generally do not possess the required flowability and hence are subjected to spray drying, agglomeration, crushing and fusing, etc. Thus it is an overarching challenge to prepare flowable hydroxyapatite powder suitable for plasma spraying without any agglomeration or spray drying process.

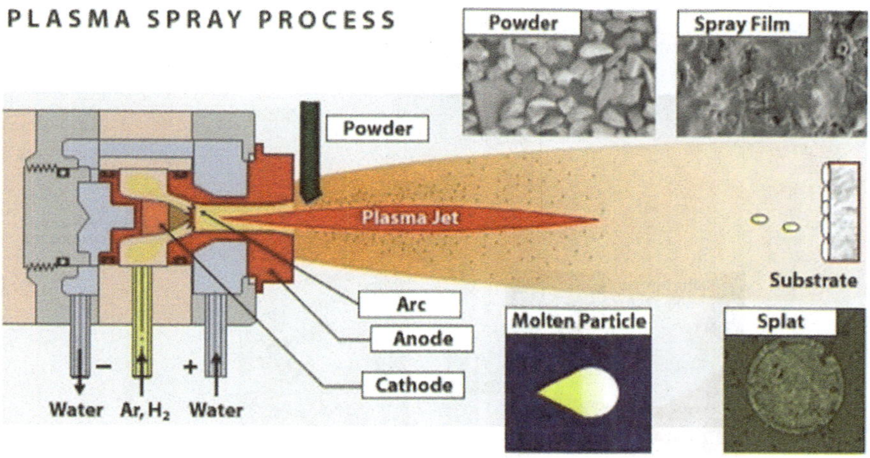

Fig. 2 Schematic representation of the formation of a coating by plasma spray process along with the main components of plasma torch [after ref. https://www.sciencelearn.org.nz/images/250-plasma-spray-process]

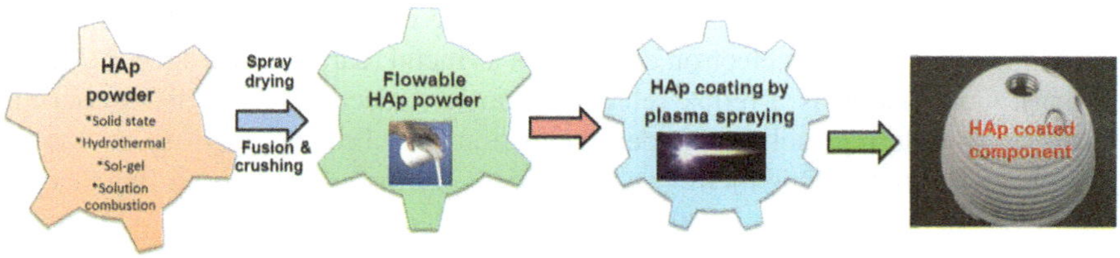

Fig. 3 Schematic showing the steps involved in the development of HAp coating

The main aim of this chapter is to provide a literature survey on the various aspects of the solution combustion synthesized hydroxyapatite and related oxide powders reported in literature. The chapter also focuses on the solution combustion synthesis of flowable plasma spray grade hydroxyapatite powder, fabrication of hydroxyapatite coating on Ti alloy and its characterization including the in vitro studies.

1.3 Solution Combustion Synthesis

Solution combustion synthesis is also known as flash combustion or fire synthesis is the most widely used wet chemical route for the synthesis of nano and micron-sized oxide powders [20–24]. This method is based on the concept of a fire triangle i.e., it requires an oxidizer, a fuel and an ignition source to create a fire and crystalline oxides are formed due to the high flame temperature. The steps involved in the solution combustion method are shown in Fig. 4. An oxidizer is usually an electron

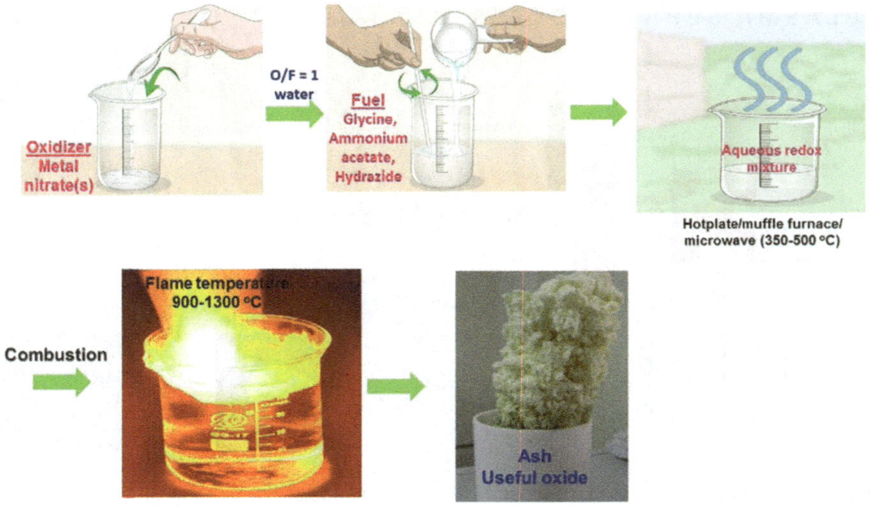

Fig. 4 Steps involved in the solution combustion synthesis of oxide materials

acceptor like non-metal or a compound containing oxidizing groups like nitrate or perchlorate and fuel is an electron donor such as metal (carbon, hydrogen, or hydrocarbon). Combustion is a redox (electron transfer) reaction which is exothermic. It can be smoldering (solid-state decomposition or solid-gas reaction) or flaming (gas phase reaction). This method was accidentally discovered at the Indian Institute of Science, Bangalore in 1988 by Patil and Kingsley during the synthesis of the high-temperature form of alumina by the combustion of a solution containing stoichiometric quantities of aluminum nitrate (oxidizer, 20 g) with urea (fuel, 8 g) [20]. The solution containing the above redox mixture when rapidly heated around 500 °C boiled underwent foaming followed by a flame to yield voluminous alumina in a few minutes. The solution combustion redox mixture is said to be stoichiometric when the oxidizer (O) to fuel (F) ratio, $\phi_c = 1$, fuel-lean when $\phi_c > 1$, and fuel-rich when $\phi_c < 1$. Stoichiometric mixtures produce maximum energy [20].

It has to be noted that the glass beakers will crack during solution combustion whenever the synthesized oxide material is not having sufficient insulating properties and hence alumina crucibles should be used in place of glass beakers while preparing non-insulating oxide powders. It is a simple, fast and economically attractive method to synthesize sinteractive fine ceramic oxide powders possessing high surface area. The process is very economical and does not require any sophisticated equipment. It is easy to dope or incorporate impurities in the atomic level in the oxide host matrix and prepare pigments, phosphors and catalytic oxides. This process can also be tailored to prepare various shapes of the oxide particles and micron-sized flowable powders suitable for plasma spraying.

2 Literature Survey on Solution Combustion Synthesis of Hydroxyapatite and Calcium Phosphate Powders and Coatings

The solution combustion method has been used for preparing calcium phosphate bioceramics. Bovand et al. have prepared β-tricalcium phosphate by microwave-assisted solution combustion method using three different fuels such as glycine, urea and citric acid. Citric acid as fuel facilitated the formation of high purity β-TCP powders and small amounts of hydroxyapatite (HAp), calcium pyrophosphate and calcium hydrogen phosphate were formed when glycine and urea were employed as fuels [25]. Calcium phosphates (CaP) with whisker-like structures were synthesized through microwave-induced and salt-assisted solution combustion synthesis employing urea as fuel and potassium chloride as an additive [26]. When potassium chloride was used, HAp and chlorapatite (ClP) were formed and in its absence calcium pyrophosphate was a major phase [26]. Single and biphasic CaP nanoparticles with a high aspect ratio (5–20) were formed with microwave-assisted solution combustion synthesis in presence of urea as fuel and the diameters and lengths were ranging from 250 to 500 nm and 2–10 μm respectively [27]. HAp has been synthesized by utilizing nitrate ions and nitric acid as oxidizers and various fuels such as citric acid, tartaric acid, sucrose, glycine and urea and a mixture of fuels [28, 29]. The mixture of fuels always resulted in the formation of TCP as a major product while single fuel resulted in carbonated hydroxyapatite. It is interesting to note that the solution combustion method was used to prepare single-phase HAp, β-tricalcium phosphate (β-TCP) and biphasic calcium phosphates (BCP) by regulating Ca/P ratio and annealing temperature [30].

Metal doped HAp is receiving a lot of attention for biomedical applications as it enhances the biocompatibility of HAp. Nanopowders of hydroxyapatite (HAp) and chromium (Cr^{3+}) doped HAp (Cr-HAp) have been synthesized using polyvinyl alcohol as fuel and the MTT (3-[4,5-dimethylthiazol-2-yl]-2,5-diphenyl tetrazolium bromide) assay studies confirmed the compatibility with human blood [31]. The effect of doping Al in HAp (Al-HAp) on the biocompatibility of HAp has been studied under the physiochemical conditions [32]. The Al-HAp nanoparticles synthesized by polyvinyl alcohol (PVA) assisted SCS was studied using L929 cell lines by MTT assays up to 24 h and they did not exhibit any toxic effects for further possible in vivo applications [32]. A hypothesis based on nucleation growth kinetics for the formation of one-dimensional strontium doped HAp (Sr-HAp) nanorods was proposed and as-synthesized powders exhibited better structural and chemical homogeneity with optimum crystallinity and thus satisfying the criteria for biomedical applications [33]. Solution combustion synthesized nanosized Sr-HAp powders were synthesized with structural and chemical nature matching the bone mineral and hence is a promising candidate material for hard tissue replacement and drug delivery systems [34]. Highly crystalline europium-doped calcium phosphate nanowhiskers with high aspect ratio suitable for multifunctional bioimaging probe applications were synthesized using a simple microwave-assisted SCS method [35]. They exhibited red and far- red wavelength fluorescence under UV excitation with emission at 696 nm and excitation at 350 nm. They are of promise for multicolor imaging applications as relatively narrow emission bands are observed in their fluorescence spectra [35]. Hydroxyapatite, fluorapatite (FAp) and chlorapatite (ClAp) were synthesized by the SCS method by employing reagent grade $Ca(NO_3)_2 \cdot 4H_2O$, (NH_4) H_2PO_4, NH_4F and NH_4Cl as sources of Ca, P, F and Cl, respectively. The powders synthesized by using citric acid and NH_4NO_3 respectively as oxidizer and accelerant were annealed at 800 °C after combustion [36]. Crystalline nano-sized powders of calcium hydroxyapatite and hydroxyapatite/tricalcium phosphate composites were produced by the SCS method using fuel-rich conditions. It was found that by varying the contents of calcium nitrate and diammonium hydrogen orthophosphate the phase composition of the composite products could be controlled. Considering the dependence of thermal stability of the HAp phase on the Ca/P atom ratio in the range of 1.45–1.67, it is proposed that the initially formed HAp transforms to TCP and results in the formation of composite material during solution combustion process [37].

The team at CSIR-CGCRI, India has carried out pioneering work on solution combustion synthesized calcium phosphates. Nanosized HAp with particle size in the range of 20–120 nm was synthesized by combustion method using urea and glycine as fuels and calcium nitrate–diammonium hydrogen orthophosphate as oxidizers [38]. The team has explored SCS method for synthesizing calcium phosphates and in particular HAp and carried out extensive studies on modeling the flame temperature during SCS and optimization of SCS parameter using Taguchi method [39], in vivo studies of calcium phosphate [40] and studies on the implantable delivery system [41]. The effectiveness of SCS synthesized porous calcium phosphates such as HAp and bi-phasic calcium phosphate (BCP) with predominantly β-tricalcium phosphate (β-TCP) impregnated with cefuroxime axetil for the treatment of experimental osteomyelitis was investigated and compared with parenteral treatment [41]. The study revealed that biphasic calcium phosphate with predominately β-TCP content was a potent carrier material for antibiotic compounds and also for refractory infections by S. aureus [41]. Porous struts were prepared using SCS synthesized pure hydroxyapatite (HAp) and a biphasic calcium phosphate-containing 90% β-TCP and 10% HAp along with bioglass. The struts were implanted to the lateral side of the radius bone of 24 black Bengal goats and to act as a control, a blank hole was left empty in a group of six specimens. β-TCP/bioglass-based implants exhibited higher bone in-growth and improved strength compared to HAp [40]. Solution combustion synthesized porous hydroxyapatite (HAp) powder was scrutinized as a bone substitute in healing bone defects in vivo and as determined by radiologic and histopathologic methods, oxytetracycline labeling, and angiogenic features. Porous HAp ceramic promoted bone formation over the defect, vouching for its biologic osteoconductive property [42].

The studies on solution combustion synthesized HAp powders were not limited to powder synthesis and evaluation of properties but also used for making scaffolds and coatings. Microporous HAp scaffolds were prepared by the slip casting method using the nanosized HAp powder synthesized by the SCS method using glycine as fuel [43]. The sintered (1250 °C for 1 h) product exhibited a bulk density in the range of 50–78% and the micropores were in the size range of 0.5–5.0 μm [43]. Efforts were also made by Liu et al. [44] to fabricate HAp coatings wherein solution combustion reaction was carried out by placing the Ti substrates in an aqueous redox mixture containing $Ca(NO_3)_2 \cdot 4H_2O$, $(NH_4)_2HPO_4$, HNO_3 (concentration of 85 vol.%), urea and $NaNO_3$ at 500 and 600 °C. This was followed by hydrothermal reaction in 1M NaOH solution at 80 °C for 10 h to obtain a coating of pure HAp and the addition of $NaNO_3$ resulted in improved interfacial bonding strength and coating thickness owing to the increase in the combustion temperature. Pure HAp coating was obtained when the Ca/P ratio was 1.75. Fully dense hydroxyapatite ceramic bodies with improved mechanical strength were synthesized from combustion synthesized nanostructured powders characterized by a high surface area to facilitate the sintering [45]. The phase evolution and sintering kinetics of HAp synthesized by solution combustion method have also been reported [46]. From the literature, it is evident that there are no reports on the solution combustion synthesis of flowable HAp powders and its utilization for plasma spray application. The efforts made on the synthesis of flowable HAp powder by SCS and plasma sprayed HAp coating is discussed in the next section.

3 Fabrication of Plasma-Sprayed HAp Coating Using Solution Combustion Synthesized HAp Powder

In the present study, the solution combustion method has been explored for the synthesis of plasma sprayable HAp powders using urea and also urea in combination with ammonium acetate as the fuels.

3.1 Synthesis of Plasma Sprayable Hydroxyapatite Powder by the Solution Combustion Process

Calcium carbonate ($CaCO_3$) was converted into calcium nitrate [$Ca(NO_3)_2$] using 70% HNO_3 which acted as an oxidizer and ammonium di-hydrogen phosphate ($NH_4H_2PO_4$) was used as the source for phosphate. Mixtures of fuels like urea and ammonium acetate were used and the ignition temperature was maintained at 400 °C. Varying amounts of urea and ammonium acetate were used to synthesize the HAp powder like (i) 100% urea, (ii) 100% ammonium acetate, (iii) 75% urea + 25% ammonium acetate, (iv) 50% urea + 50% ammonium acetate, and (v) 25% urea + 75% ammonium acetate corresponding to oxidizer to fuel ratio of 1. The redox mixture was placed in muffle furnace preheated to 400 °C. The solution combustion synthesized powder was white and was not very fluffy.

Table 1 Plasma spray parameters used for plasma spraying SCS HAp powder

Parameters	Value
Primary gas (bar)	5.1
Secondary gas (bar)	3.45
Carrier gas (bar)	4.0
Current (Amps)	420
Voltage (Volts)	60
Spray distance (cm)	10
Powder feed rate (g/min)	20

3.2 Plasma Spraying of HAp Powders

The Ti-6Al-4V substrate procured from TIMET, USA was cut to the following dimension 20 mm × 20 mm × 2 mm size. The obtained samples were grit-blasted using Al_2O_3 grits under 6 bar pressure for 30 seconds to obtain a rough surface. Grit blasted samples were cleaned in acetone. The plasma spraying of HAp powder (~250 g) on Ti-6Al-4V using air plasma spraying system (Sulzer Metco 9M) was carried out maintaining a stand-off distance of 10 cm from the flame and coatings with an average thickness of ~100 μm were coated. The parameters used for plasma spraying are listed in Table 1.

3.3 Characterization of HAp Powder and Plasma Sprayed HAp Coating

The particle size distribution of the synthesized HAp powder was obtained using a particle size analyzer (Mastersizer 2000, Malvern Instruments). X-ray diffraction analysis (Bruker-Advance D8) was carried out to analyze the phase of the HAp powder and coating using CuKα as the source. Due to the high temperature of plasma (7000 K–20,000 K), the hydroxyapatite powder gets transformed into phases like α-tricalcium phosphate, β-tricalcium phosphate and tetracalcium phosphate as a result of which the crystallinity of HAp reduces. It is very important to retain the crystallinity of the coatings to enhance the biocompatibility of the coatings. From the XRD data, the degree of crystallinity (Xc) of the HAp powder and coating were evaluated using the equation as reported in the literature [47].

Morphology and microstructural analysis of powders and coatings respectively were carried out using a field emission scanning electron microscope (FESEM, Carl Zeiss, Supra 40 VP model). Coated samples were cut with a diamond cutter and mounted in the epoxy resin for cross-section analysis. The mounted samples were ground using different grades of emery sheets (800, 1000, 1200 grade) and finally polished on a soft cloth using alumina grade 2 paste. The surface roughness of the plasma sprayed coatings was measured using surface roughness perthometer

(Taylor Hobson precision) and the average roughness (R_a) value of 5 such readings is reported. The Young's modulus of the substrate and coatings were obtained by nanoindentation (CSEM nano-hardness tester) technique by employing a Berkovich indenter at a load of 10 mN. The wear resistance of the coatings was measured using a reciprocating wear tester and more details are reported elsewhere [48]. For in vitro studies, HAp coating prepared from 600 °C calcined powders was used. More details of in vitro studies are reported elsewhere [48].

4 Results and Discussion

4.1 Effect of Fuel and Fuel Mixtures

The combination of the mixture of fuels used for the synthesis of the powder had affected the nature of the powder. The mixture of fuels (urea (U) + ammonium acetate [AA]) was found to be advantageous in getting flowable powders with average agglomerated particle size < 100 μm when compared to the use of a single fuel (U/AA). The powders obtained from different fuel combinations were sized in the range between 45 and 90 μm by sieving to get flowable powders. It was observed that with an increase in ammonium acetate concentration as the fuel, the particles became too fine (< 45 μm) and 100% urea gave very coarse particles (> 100 μm). The crystallite size of the powder depends on the flame temperature as well as the nature of the combustion reaction. It has been reported that when urea is used as the fuel it leads to the flaming type combustion and in course of this if the flame persists for a longer duration, the particles will have enough time and duration to sinter and yield larger particles [49]. Thus, with the use of a single fuel such as urea, larger-sized particles were obtained. To decrease the sinterability and reduce the exothermicity of the combustion reaction, the amount of urea (U) was reduced. Therefore, the stoichiometric amount of urea calculated was not used, instead varying concentrations of urea (0.75 U/0.5 U/0.25 U) were used and to compensate the fuel ratio, an extra fuel ammonium acetate (AA) was used. The reaction with 0.75U + 0.25AA was flaming and there was more frothing. With the increase in ammonium acetate concentration and decrease in the urea amount, the combustion reaction became smoldering type. The crystallite size of the particles decreases with the increase in the ammonium acetate concentration and yields amorphous particles. The ammonium acetate molecules decompose at lower temperatures and generate a large volume of gases that dissipates heat quickly from the system thereby releasing a large volume of gases. With a mixture of fuels (75% U + 25% AA), very less amount of fine powder was obtained and a large fraction of powder was in the range 45–90 μm size and hence this combination of the fuel was selected for the synthesis of flowable hydroxyapatite powder.

Table 2 Particle size distribution of solution combustion synthesized hydroxyapatite powder subjected to calcination at two different temperatures

Powder	$d_{0.1}$ (μm)	$d_{0.5}$ (μm)	$d_{0.9}$ (μm)
HAp calcined at 900 °C	3.435	59.531	124.533
HAp calcined at 600 °C	2.452	67.023	115.868

4.2 Effect of Calcination Temperature on the Flowability of the Powder

In order to increase the agglomeration of the particles and in turn, the flowability of powder, HAp powder synthesized from a mixture of fuels (75% U + 25% AA) was calcined at 600 °C and 900 °C for 3 hours. Kong et al. observed that when HAp powder was sintered at higher temperatures (1000 °C) the grain size of HAp increased [50]. In the present study, calcination of the powder was carried at 600 °C and 900 °C for 3 hours as the aim of the present study is different from that of Kong et al. [50]. The powder obtained after calcination at 900 °C was slightly crushed and sieved. About 50 g of HAp powder calcined at 900 °C was used for the flowability measurement using Hall flow meter (ASTM B213) and the flowability was found to be 15 s/50 g The powder obtained after calcination at 600 °C was crushed, sized between 45 and 90 μm by sieving and Hall flow rate was measured and the flow rate was 30 s/50 g.

4.3 Particle Size Analysis of the Powders

The particle size distribution of the powders was carried out and the results are tabulated in Table 2 and they exhibited a narrow size distribution. The $d_{0.9}$- of HAp powder calcined at 900 °C was slightly higher.

4.4 XRD Characterization of the Powder and Coating

The XRD patterns of HAp powders calcined at 900 °C and 600 °C (Figs. 5 and 6), showed the peaks corresponding to HAp and α- and β-tricalcium phosphates particularly in the 900 °C calcined powder and this was probably due to the use of higher calcination temperature (Fig. 5). The phases were confirmed from JCPDS card no 9–432, 9–348, and 9–169 respectively for hydroxyapatite, α-tricalcium phosphate and β-tricalcium phosphate. But the powder calcined at 600 °C did not exhibit the peaks corresponding to β tri-calcium phosphate (Fig. 6).

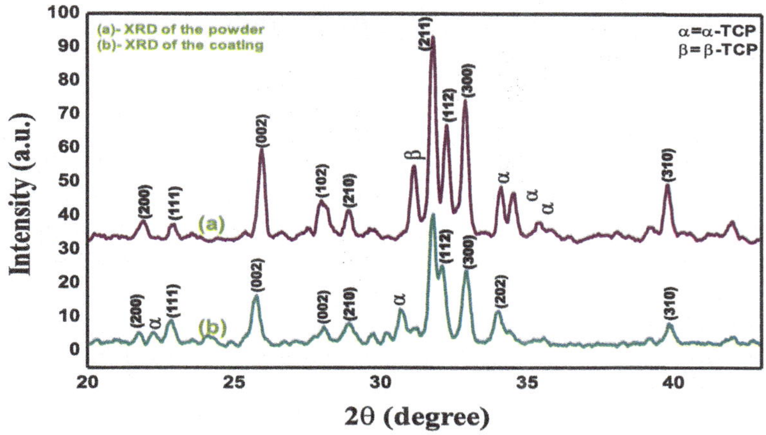

Fig. 5 XRD patterns of solution combustion synthesized HAp powder calcined at 900 °C and plasma sprayed coating

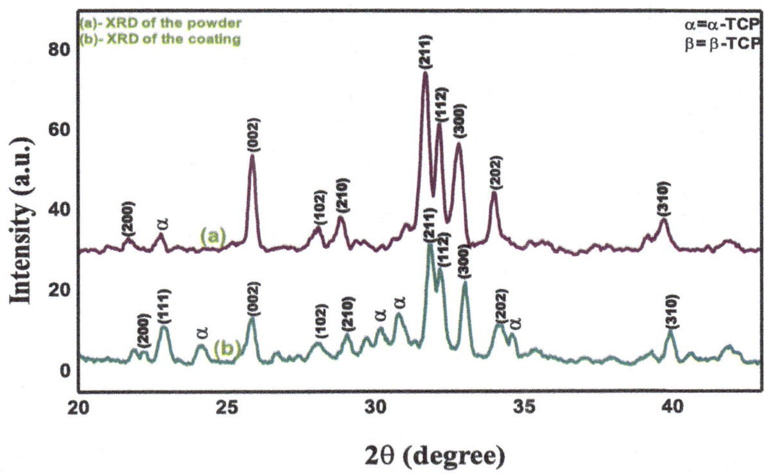

Fig. 6 XRD patterns of solution combustion synthesized HAp powder calcined at 600 °C and plasma sprayed coating

Solution combustion synthesized HAp powders calcined at 600 and 900 °C exhibited crystallinity of 67 and 82% respectively. However, plasma-sprayed HAp coatings obtained from 600 and 900 °C calcined powders exhibited a crystallinity of 74 and 90% respectively. Interestingly, there was an increase in the crystallinity of the coatings after plasma spraying.

4.5 FESEM Images of the Calcined Powders and Plasma-Sprayed Coatings

Figures 7 and 8 show the FESEM images of plasma sprayable HAp powder calcined at 900 °C and 600 °C at two different magnifications. The particles are blocky angular shaped. A higher magnification image shows each particle to be an agglomerate of smaller particles. The presence of a random distribution of voids and foamy agglomerated particles is seen. A large amount of gas evolved during the chemical reaction and fast quenching prevents further agglomeration or sintering of particles thereby resulting in a highly porous foam-like structure. During the combustion reaction, the release of extensive gaseous by-products and instantaneous reactions inhibit the sintering of particles and create the pores in the structure of resultant powder.

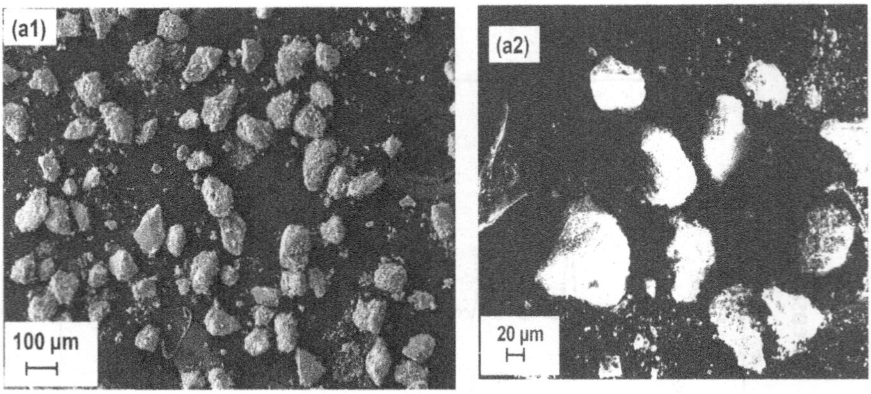

Fig. 7 FESEM images of HAp powder calcined at (a1) 900 °C and (a2) 600 °C

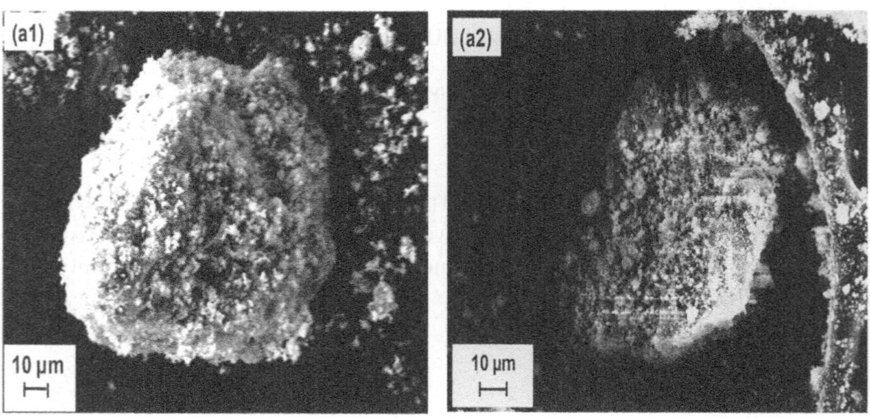

Fig. 8 FESEM images of HAp powder calcined at (a1) 900 °C and (a2) 600 °C

FESEM analyses of the plasma sprayed HAp coatings were carried out and it was observed that there was a good adhesion between the substrate and the coating (Fig. 9). Randomly distributed pores of different sizes and micro-cracks were also observed. During plasma spraying small size powder was completely melted and during fast cooling, large size powder was partially melted giving rise to bimodal microstructure. Besides there were a lot of defects inside the coatings like pores and these are frequently observed in plasma-sprayed HAp coatings. The coating showed a large area of fully melted regions. Coating thickness was ~120 μm and ~100 μm for coating (a1) and (a2) respectively. Coating (a1) showed more porosity compared to coating (a2) with a large number of bigger pores.

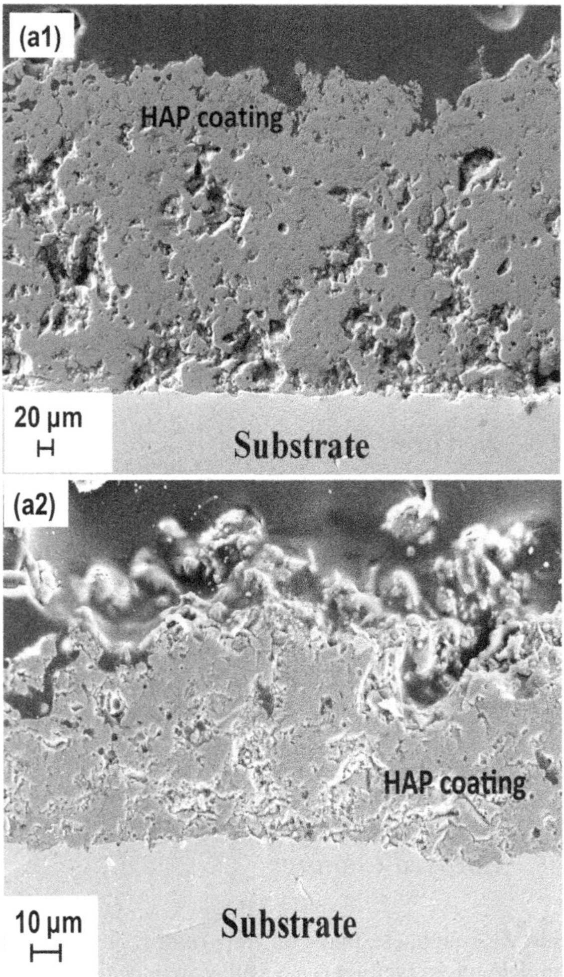

Fig. 9 FESEM images of HAp coatings obtained from calcined HAp powders (a1) 900 °C and (a2) 600 °C

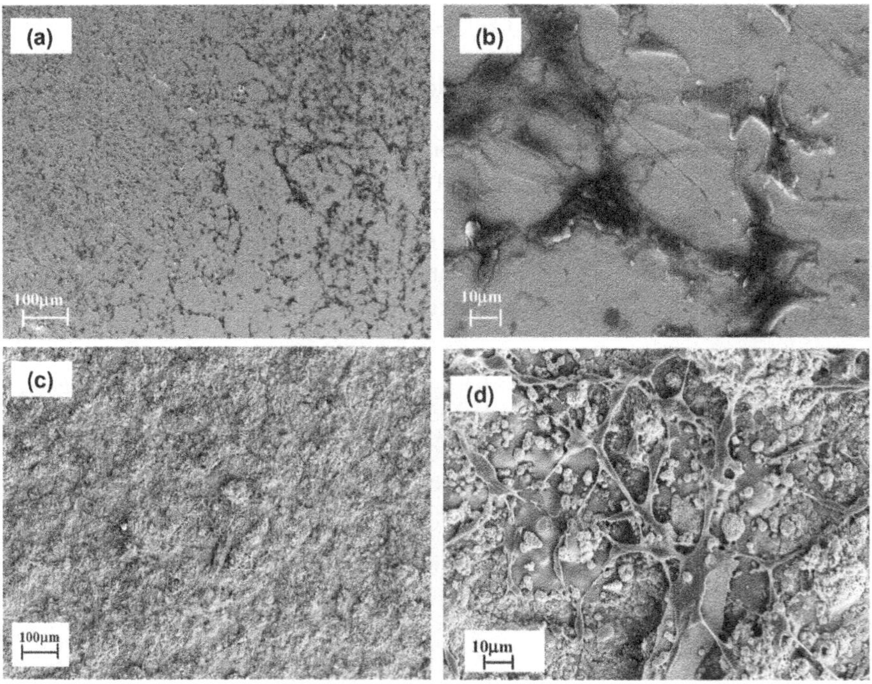

Fig. 10 FESEM images of **a**, **b** Ti-6Al-4V substrate and **c**, **d** HAp coating (SCS) after in vitro studies

4.6 *In Vitro Studies*

For in vitro studies, coating prepared from 900 °C calcined powders was utilized as the coating possessed higher crystallinity. The cell morphology of 3T3 fibroblast on the uncoated and HAp coated substrates were observed under FESEM (Fig. 10). Ti-6Al-4V substrate showed 100% cell viability and all the intended cells were viable and there was not much mat formation (Fig. 10a & b). Interestingly, on the HAp coating, a thick mat on the surface due to collagen synthesis was formed which was not observed in the case of the Ti-6Al-4V substrate (Fig. 10c & d).

4.7 *Surface Roughness and Young's Modulus of Plasma Sprayed Coatings*

Surface roughness plays an important role in the case of body implants. The bio implant has to facilitate easy cell growth once it is implanted in the body. The surface has to be rough for easy cell growth. The plasma-sprayed HAp coatings fabricated from powders calcined at 900 and 600 °C exhibited surface roughness values of 5.91

± 0.3 and 4.48 ± 0.2 μm respectively. The higher surface roughness values indicate more of partially melted regions and less of fully melted regions in the coating. The obtained surface roughness values of HAp coatings match well with the reported literature values (5.30–.49 μm) [51].

The Young's Modulus of HAp coatings were determined from indentation load-displacement data on the cross-section of the coatings using 10 mN. The substrate and plasma sprayed HAp coatings a1 and a2 exhibited Young's modulus of 133, 117 and 108 GPa respectively and these values are corroborating well with the literature values [48].

4.8 Wear Studies in Simulated Body Fluid (SBF) Medium

Reciprocating wear testing was carried out for Ti-6Al-4V substrate and plasma sprayed coatings in order to estimate the coefficient of friction and wear loss by employing loads of 1N, 2N, 5N and 7N. It was observed that the wear losses for the plasma sprayed coatings were found to be less at lower loads 1, 2 and 5N and at higher loads, the coating peeled off. The coating (a1) could withstand 7N load and then the coating peeled off whereas the coating (a2) could withstand 5N load and the coating peeled off at higher load (Fig. 11). This shows that the coating (a1) contained more of the fully melted region. The coefficient of friction (COF) was found to be lower for the coating obtained by the powder calcined at 900 °C at 1N load but with the increase in the load, the COF matched the COF of the substrate (Fig. 12). In general, the COF was lower for HAp coatings compared to the substrate at lower loads and at higher loads, the coating peeled off and thus resulting in COF equivalent to that of the substrate. The wear rate of coating a2 was higher than coating a1 (Table 3).

5 Conclusions

In the past decade, research efforts were focused on the development of hydroxyapatite powders and coatings exhibiting optimal osseointegration along with required coating thickness, phase composition, crystallinity, porosity, adhesion, residual stress, etc. It is evident that there is a need for careful engineering of hydroxyapatite coatings by optimizing the plasma spray parameters. The solution combustion method is a versatile and facile method for the synthesis of plasma sprayable grade hydroxyapatite powders. The present study has confirmed the expediency of preparing plasma sprayable powders by a single-step solution combustion method by employing the mixture of fuels and the biocompatibility of the developed plasma-sprayed HAp coating. However, there is a need for further optimizing the synthesis conditions of SCS to obtain phase pure HAp coatings with higher crystallinity and better phase purity.

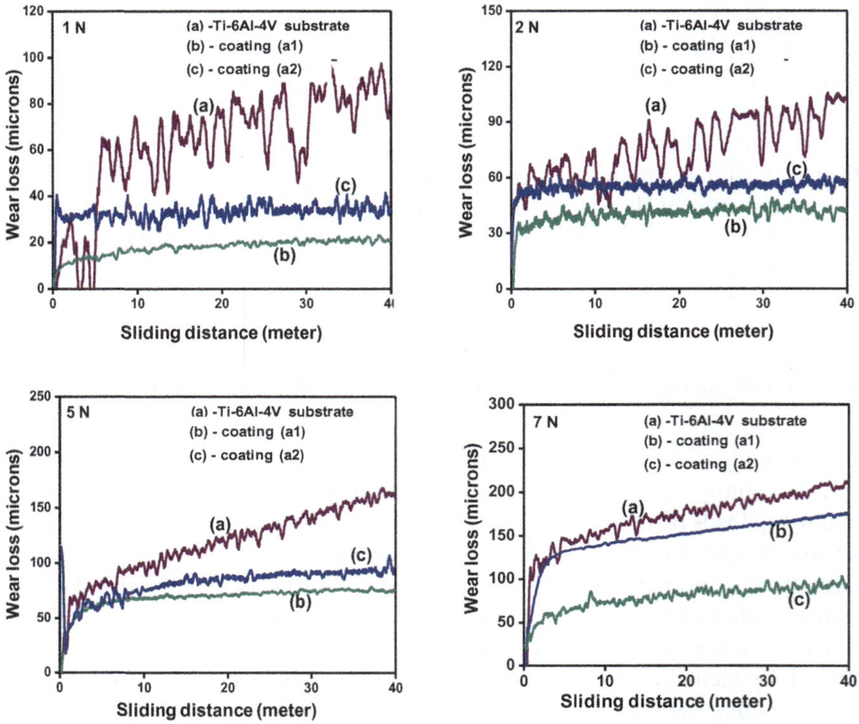

Fig. 11 Wear loss versus sliding distance for Ti-6Al-4V substrate and plasma sprayed coatings fabricated using SCS powder at 1N, 2N, 5N and 7N loads

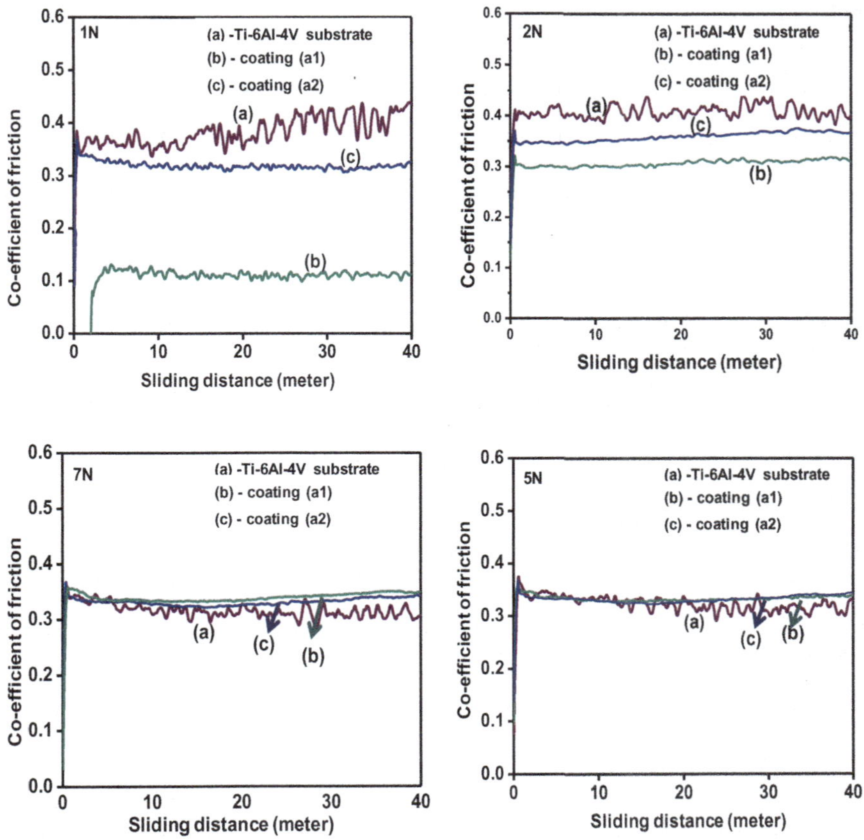

Fig. 12 Coefficient of friction versus sliding distance for substrate and plasma sprayed coatings fabricated using SCS powders at 1N, 2N, 5N and 7N loads

Table 3 Wear rates for Ti-6Al-4V substrate and plasma sprayed HAp coatings fabricated by SCS at varying loads

Sample	Wear rate (mm^3/m)			
	1N	2N	5N	7N
Ti-6Al-4V substrate	0.0004	0.0009	0.002	0.003
Coating (a1)	0.007	0.010	0.012	0.016
Coating (a2)	0.008	0.012	0.018	0.022

References

1. Heimann RB, Lehmann H (2015) Bioceramic coatings for medical implants: trends and techniques. Wiley VCH, Weinheim
2. Leon B, Jansen JA (2015) Thin calcium phosphate coatings for medical implants. Springer, New York
3. Heimann RB (2010) Classic and advanced ceramics from fundamentals to applications. Wiley-VCH, Weinheim
4. Heimann RB (2013) Structure, properties and biomedical performance of osteoconductive bioceramic coatings. Surf Coat Technol 233:27–38. https://doi.org/10.1016/j.surfcoat.2012.11.013
5. Surmenev RA, Surmeneva MA, Ivanova AA (2014) Significance of calcium phosphate coatings for the enhancement of new bone osteogenesis–a review. Acta Biomater 10:557–570. https://doi.org/10.1016/j.actbio.2013.10.036
6. Canillas M, Pena P, de Aza AH, Rodriguez MA (2017) Calcium phosphates for biomedical applications fosfatos de calcio para aplicaciones biomédicas. Boletín de la Sociedad Española de Cerámi cay Vidrio 56:91–112
7. Noam E, Noah M (2017) Calcium phosphate bioceramics: a review of their history, structure, properties, coating technologies and biomedical applications. Mater (Basel) 10(4):334. https://doi.org/10.3390/ma10040334
8. Rey C, Combes C, Drouet C, Glimcher MJ (2009) Bone mineral: update on chemical composition and structure. Osteoporos Int 20(6):1013–1021. https://doi.org/10.1007/s00198-009-0860-y
9. Chevalier J, Gremillar L (2009) Ceramics for medical applications: a picture for the next 20 years. J Euro Ceram Soc 29:1245–1255. https://doi.org/10.1016/j.jeurceramsoc.2008.08.025
10. https://en.wikipedia.org/wiki/Hydroxyapatite accessed 1 May 2020
11. Heimann RB (2016) Plasma-sprayed hydroxylapatite-based coatings: chemical, mechanical, microstructural, and biomedical properties. J Thermal Spray Technol 23:827–850. https://doi.org/10.1007/s11666-016-0421-9
12. ASTM F1185–03 (2009) Standard specification for composition of hydroxylapatite for surgical implants
13. ISO 13779-4 (2002) Implants for surgery—hydroxyapatite—part 4: determination of coating adhesion strength
14. ISO 13779-3 (2008) Implants for surgery—hydroxyapatite—part 3: chemical analysis and characterization of crystallinity and phase purity
15. ISO 13779-2 (2008) Implants for surgery—hydroxyapatite—part 2: coatings of hydroxyapatite
16. Food and drug administration (FDA) (1992) Washington, DC, p 1
17. ISO 13779-2 (2008) Implants for surgery-hydroxyapatite, part-2: coatings of hydroxylapatite, international organization for standardization, Geneva, Switzerland
18. Heimann RB (1996) Plasma-spray coating: principles and applications. Wiley-VCH, Weinheim
19. Mehdi S-S, Khorasani MT, Ehsan DK, Jamshidi A (2013) Synthesis methods for nanosized hydroxyapatite with diverse structures. Acta Biomater 9:7591–7621. https://doi.org/10.1016/j.actbio.2013.04.012
20. Patil KC, Hegde MS, Rattan Tanu, Aruna ST (2008) Chemistry of nanocrystalline oxide materials (Combustion synthesis, properties and applications). World Scientific, Singapore
21. Patil KC, Aruna ST, Mimani T (2002) Combustion synthesis: an update. Curr Opin Solid State Mater Sci 6(6):507–512. https://doi.org/10.1016/S1359-0286(02)00123-7
22. Aruna ST, Mukasyan AS (2008) Combustion synthesis and nanomaterials. Curr Opin Solid State Mater Sci 12(3–4):44–50. https://doi.org/10.1016/j.cossms.2008.12.002
23. Varma A, Mukasyan A, Rogachev SAS, Manukyan KV (2016) Solution combustion synthesis of nanoscale materials. Chem Rev 116(23):14493–14586. https://doi.org/10.1021/acs.chemrev.6b00279
24. Aruna ST (2010) Solution combustion synthesis-an overview. In: Lackner M (ed) Combustion synthesis: novel routes to novel materials. Bentham Publisher, Sharjah, p 206

25. Bovand D, Arabi AM, Bovand, M (2018) Microwave assisted solution combustion synthesis of β-tricalcium phosphate nano-powders. Boletin de la Sociedad Espanola de Ceram cay Vidrio 57(6):240–246. https://doi.org/10.1016/j.bsecv.2018.05.001
26. Lopera AA, Chavarriaga EA, Zuluaga B, Marin S, Giraldo GO, Estupiñan HA, Zapata V, Garcia CP (2017) Effect of salt concentration on the electrical and morphological properties of calcium phosphates obtained via microwave-induced combustion synthesis. Adv Powder Technol 28(10):2787–2795. https://doi.org/10.1016/j.apt.2017.08.007
27. Wagner DE, Lawrence J, Bhaduri SB (2013) Microwave-assisted solution combustion synthesis of high aspect ratio calcium phosphate nanoparticles. J Mater Res 28(22):3119–3129. https://doi.org/10.1557/jmr.2013.314
28. Sasikumar S, Vijayaraghavan R (2010) Synthesis and characterization of bioceramic calcium phosphates by rapid combustion synthesis. J Mater Sci Technol 26(12):1114–1118. https://doi.org/10.1016/S1005-0302(11)60010-8
29. Sasikumar S, Vijayaraghavan R (2008) Solution combustion synthesis of bioceramic calcium phosphates by single and mixed fuels-a comparative study. Ceram Int 34(6):1373–1379. https://doi.org/10.1016/j.ceramint.2007.03.009
30. Zhao J, Zhao J, Chen K, Wang X, Han Z, Li Y (2014) Rietveld refinement of hydroxyapatite, tricalcium phosphate and biphasic materials prepared by solution combustion method. Ceram Int 40(2):3379–3388. https://doi.org/10.1016/j.ceramint.2013.09.094
31. Bandgar SS, Yadav HM, Shirguppikar SS, Shinde MA, Shejawal RV, Kolekar TV, Bamane SR (2017) Enhanced hemolytic biocompatibility of hydroxyapatite by chromium (Cr^{3+}) doping in hydroxyapatite nanoparticles synthesized by solution combustion method. J Korean Ceram Soc 54(2):158–166. https://doi.org/10.4191/kcers.2017.54.2.11
32. Kolekar T, Thorat ND, Yadav HM, Magalad VT, Shinde AS, Bandgar SS, Kim JH, Agawane GL (2016) Nanocrystalline hydroxyapatite doped with aluminium: a potential carrier for biomedical applications. Ceram Int 42(4):5304–5311. https://doi.org/10.1016/j.ceramint.2015.12.060
33. Kavitha M, Subramanian R, Vinoth KS, Narayanan R, Venkatesh G, Esakkiraja N (2015) optimization of process parameters for solution combustion synthesis of strontium substituted hydroxyapatite nanocrystals using design of experiments approach. Powder Technol 271:167–181. https://doi.org/10.1016/j.powtec.2014.10.046
34. Kavitha M, Subramanian R, Narayanan R, Udhayabanu V (2014) Solution combustion synthesis and characterization of strontium substituted hydroxyapatite nanocrystals. Powder Technol 253:129–137. https://doi.org/10.1016/j.powtec.2013.10.045
35. Wagner DE, Eisenmann KM, Nestor-Kalinoski AL, Bhaduri SB (2013) A microwave-assisted solution combustion synthesis to produce europium-doped calcium phosphate nanowhiskers for bioimaging applications. Acta Biomater 9(9):8422–8432. https://doi.org/10.1016/j.actbio.2013.05.033
36. Zheo J, Dong X, Bian M, Zhao J, Zhang Y, Sun Y, Chen J, Wang X (2014) Solution combustion method for synthesis of nanostructured hydroxyapatite, fluorapatite and chlorapatite. Appl Surf Sci 314:1026–1033. https://doi.org/10.1016/j.apsusc.2014.06.075
37. Ghosh SK, Datta S, Roy SK (2004) Solution combustion synthesis of calcium hydroxyapatite nanoparticles. Trans Ind Ceram Soc 63(1):27 32. https://doi.org/10.1080/0371750X.2004.11012125
38. Ghosh SK, Roy SK, Kundu B, Datta S, Basu D (2011) Synthesis of nano-sized hydroxyapatite powders through solution combustion route under different reaction conditions. Mater Sci Eng B: Solid-State Mater Adv Technol 176(1):14–21. https://doi.org/10.1016/j.mseb.2010.08.006
39. Ghosh SK, Pal S, Roy SK, Pal SK, Basu D (2010) Modelling of flame temperature of solution combustion synthesis of nanocrystalline calcium hydroxyapatite material and its parametric optimization. Bullet Mater Sci 33(4):339–350. https://doi.org/10.1007/s12034-010-0052-6
40. Ghosh SK, Nandi SK, Kundu B, Datta S, De DK, Roy SK, Basu D 2008. In vivo response of porous hydroxyapatite and β-tricalcium phosphate prepared by aqueous solution combustion method and comparison with bioglass scaffolds. J Biomed Mate. Rese—Part B Appl Biomater 86(1):217–227. https://doi.org/10.1002/jbm.b.31009

41. Nandi SK, Kundu B, Ghosh SK, Mandal TK, Datta S, De DK, Basu D (2009) Cefuroxime-impregnated calcium phosphates as an implantable delivery system in experimental osteomyelitis. Ceram Inter 35(4):1367–1376. https://doi.org/10.1016/j.ceramint.2008.07.022
42. Nandi SK, Kundu B, Ghosh SK, De DK, Basu D (2008) Efficacy of nano-hydroxyapatite prepared by an aqueous solution combustion technique in healing bone defects of goat. J Vet Sci 9(2):183–191. https://doi.org/10.4142/jvs.2008.9.2.183
43. Rao RR, Roopa HN, Mariappan L (2016) Development of microporous scaffolds through slip casting of solution combustion derived nano hydroxyapatite. Inter Ceram: Inter Ceram Rev 65(3):100–105. https://doi.org/10.1007/BF03401159
44. Liu Y, Dan X, Du Y, Liu F (2004) Novel coating technique for preparation of hydroxyapatite/Ti composite. J Mater Sci 39(12):4031–4034. https://doi.org/10.1023/B:JMSC.0000031488.66341.ca
45. Canillas M, Rivero R, García-Carrodeguas R, Barba F, Rodríguez MA (2017) Processing of hydroxyapatite obtained by combustion synthesis. Boletin de la Sociedad Espanola de Ceramica Vidrio 56(5):237–242. https://doi.org/10.1016/j.bsecv.2017.05.002
46. Pratihar SK, Garg M, Mehra, S, Bhattacharyya S (2006) Phase evolution and sintering kinetics of hydroxyapatite synthesized by solution combustion technique. J Mater Sci: Mater in Med 17(6):501–507. https://doi.org/10.1007/s10856-006-8932-4
47. Landi E, Tampieri A, Celotti G, Sprio S (2000) Densification behaviour and mechanisms of synthetic hydroxyapatites. J Eur Ceram Soc 20:2377–2387. https://doi.org/10.1016/S0955-2219(00)00154-0
48. Aruna ST, Shilpa M, Lakshmi RV, Balaji N, Kavitha V, Gnanamani A (2018) Plasma sprayed hydroxyapatite bioceramic coatings from coprecipitation synthesized powder: preparation, characterization and in-vitro studies. Trans Ind Ceram Soc 77:90–99. https://doi.org/10.1080/0371750X.2018.1465358
49. Aruna ST, Rajam KS (2004) Mixture of fuels approach for the solution combustion synthesis of Al_2O_3–ZrO_2 nanocomposite. Mater Res Bull 39(2):157–167. https://doi.org/10.1016/jmaterresbull.2003.10.005
50. Kong LB, Ma J, Boey F (2002) Nanosized hydroxyapatite powders derived from coprecipitation process. J Mater Sci 37:1131–1134. https://doi.org/10.1023/A:1014355103125
51. Singh Gurbhinder, Singh Surendra, Prakash Satya (2011) Surface characterization of plasma sprayed pure and reinforced hydroxyapatite coating on Ti-6Al-4V alloy. Surf Coat Technol 205:4814–4820. https://doi.org/10.1016/j.surfcoat.2011.04.064

Nanomaterials in Medicine

Sam James

Abstract For decades, man has explored the cures for many diseases and illnesses. Scientists now believe that nanotechnology can perform that miracle. Nanomaterials in medicine are an attempt to limit or reverse pathological processes, and its advantages are due to their specific characteristics, such as the capacity to interact with biological systems with a high degree of specificity. The ultimate goal of nanomedicine is to identify and treat diseases as early as possible at the subcellular level. The application of nanotechnology in medicine ranges from diagnostics to therapeutics. In diagnostic imaging, nanomaterials are used to target a specific type of cancer cell, which would enable radiologists to visualize insignificant features at a better resolution possible. Moreover, nanomaterials are thought to stimulate and interact with target cells and tissues in controlled ways and to induce desired physiological responses with minimal side effects. Nanomedicine researchers developed an assay for early diagnosis of Alzheimer's disease with better accuracy and sensitivity than conventional methods. This new assay uses gold nanoparticles (NP) and magnetic microparticles (MMP) to bind to the biomarkers of Alzheimer's disease. Physicians use nanoparticles to target drugs at the source of the infection, thereby increasing the efficiency and minimizing the side effects. Nanoparticles are also used to stimulate the body's innate repair mechanisms by artificially activating and controlling the adult stem cells. To promote neuronal repair and regeneration, researchers use bio-reactive nanoscaffolding. In this chapter, we will explore how nanomaterials are used in medicine and its prospects.

Keywords Nanoparticles · Carbon nanotubes · Cancer · Diabetes · Liposomes

---

S. James (✉)
Director of Rehabilitation, Select Specialty Hospitals, Pontiac, MI 48342, USA
e-mail: samprasanna.j@gmail.com

Department of Industrial and Systems Engineering, Oakland University, Rochester Hills, MI 48309, USA

T. S. Santra and L. Mohan (eds.), *Nanomaterials and Their Biomedical Applications*, Springer Series in Biomaterials Science and Engineering 16,
https://doi.org/10.1007/978-981-33-6252-9_8

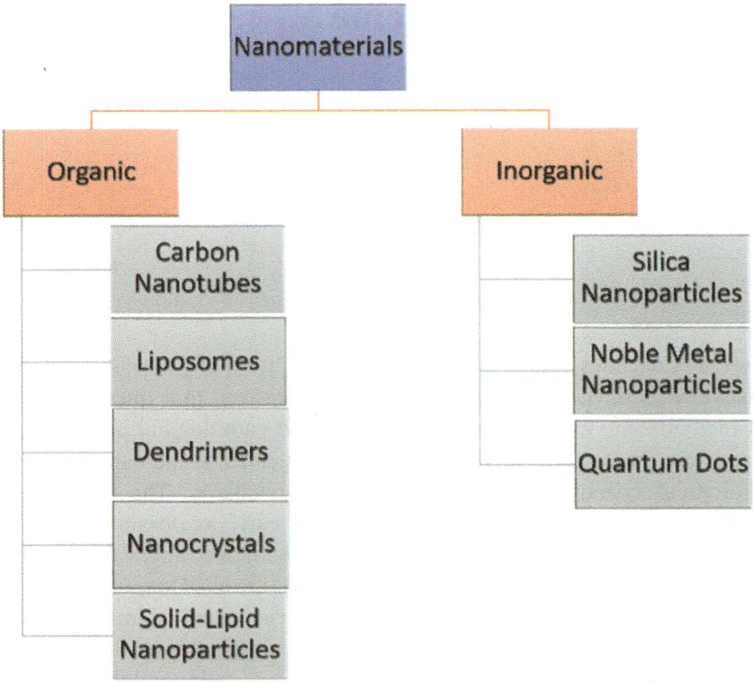

Fig. 8.1 Types of nanoparticles

Nanomedicine is the application of nanoscience and nanotechnology to diagnose, treat, monitor, and prevent diseases. Nanomedicine uses nanomaterials, nanodevices, nanosystems such as nanoparticles, nanorobots as drug carriers, sensors, and actuators.

1 Types of Nanomaterials in Medicine

Al-Kayiem et al. classify (Fig. 8.1) nanomaterials into organic nanoparticles, such as carbon nanotubes, liposomes, micelles, dendrimers, and inorganic nanoparticles such as metals, metal oxides, and quantum dots [1].

2 Carbon Nanotubes

Biomedical applications extensively use carbon nanotubes for their outstanding structural, mechanical, electrical, thermal, and optical properties [2]. Carbon nanotubes conduct heat and electricity very well [2]. They are not biodegradable and likely to form aggregates. The usage of carbon nanotubes in medical applications ranges from diagnostics to therapeutics. Examples include cancer therapeutics, gene therapy, infection control and antimicrobial therapy, tissue regeneration, treating neurodegenerative disorders, and NIR (Near-infrared) fluorescence microscopy.

Industries make carbon nanotubes with allographs of carbon—graphite by the process of chemical vapor deposition. This chemical manufacturing process arranges carbon atoms hexagonally in tube-like structures. There are two types of carbon nanotubes based on the number of layers—single-walled and multi-walled carbon nanotubes. Carbon nanotubes can easily cross the cell and nuclear membranes to deliver the drug in exact location without losing concentration. So scientists use carbon nanotubes as drug carriers to deliver exactly to the tumor cells. They have combined carbon nanotubes with many anticancer drugs and successfully tested both in vitro and in vivo, such as epirubicin, doxorubicin, and paclitaxel [2]

Studies show the usage of carbon nanotubes in antitumor immunotherapy [3]. They used a tumor vaccine or antigen in a carbon nanotube carrier to stimulate the body's immune system to attack tumor cells. They combined the tumor antigen with the carbon nanotube as a natural antigen-presenting cell to bring about the efficacy of the immune effector cells.

Carbon nanotubes, among other materials, are the best-suited candidate for tissue engineering. Carbon nanotubes are biocompatible, non-biodegradable, and functionalized with biomolecules for tissue regeneration. Since carbon nanotubes have strong mechanical properties, it reinforces tissue scaffolding and better conductivity with the host. Researchers use carbon nanotubes as labeling and imaging agents for their remarkable optical properties.

3 Liposomes

In 1970, Gregoriadis introduced liposomes as a drug delivery vesicle. A liposome is a small spherical vesicle filled with water, its size ranging from 30 nm to several micrometers. Liposomes are artificially made from cholesterol and phospholipids. Liposomes have several exploitable properties like size, hydrophobic, and hydrophilic walls, which make them suitable for drug delivery systems. Liposomes are made with either rigid or fluid vesicles based on the choice of the bilayer components; for example, unsaturated phospholipids from natural sources give more fluid and permeable bilayers.

In contrast, saturated phospholipids make it a rigid and impermeable bilayer [4]. Scientists use liposomal encapsulation technology to deliver drugs to the target body organs. They encapsulate active drug components inside these liposomes, which form a barrier around the drug. Liposome encapsulated drugs make the drug highly available in the body and protected from further chemical reactions like oxidation and degradation. Liposomes have low efficiency in holding hydrophilic drugs. Medications using liposomes as carriers are available as analgesics, anticancer, and antifungal drugs.

4 Dendrimers

Tomalia introduced the term "Dendrimer." Dendrimers are branched, tridimensional polymers resembling a sphere [5]. Dendrimer's structural and functional units are called dendrons, which branch out from the dendrimer's core. The dendrons have free functional groups that may be swapped for other substituents to modify the whole structure's chemical and physical properties [5].

5 Nanocrystals

Nanocrystals are crystalline nanoparticles smaller than one micrometer. Drug Nanocrystals are made of pure drugs, without any carriers, which improves pharmaco-dynamics and pharmaco-kinetics poorly soluble medications. Nanocrystals can be dispersed in aqueous and non-aqueous solutions that form nanosuspensions. Nanocrystals increase the bioavailability and solubility of the drug. According to the Noyes-Whitney equation, the decreased size makes nanocrystals dissolve at a faster rate, thus enhancing drug bioavailability. Drug solubility depends on the size of the particle. The smaller the particle, the more the soluble the drug. This small size increases the concentration gradient between the intestinal lumen and blood, thus increases the drug absorption passive diffusion [6]. During intravenous drug delivery, the size of these nanocrystals makes the drug 100% available for the target tissues without blocking the small capillaries. Nanocrystals make other drug delivery routes like pulmonary, ocular, dermal, and oral more bioavailable [7].

6 Solid Lipid Nanoparticles

Solid lipid nanoparticles are introduced as a carrier system for drugs with poor water dissolution. Solid lipid nanoparticle colloidal carrier systems have a high melting point lipid core. This nanomaterial consists of solid lipids core in an aqueous dispersion. To a certain extent, these nanoparticles resemble nano-emulsions that have inner liquid lipids replaced by solid lipids. Solid lipids contribute to improved control over drug release because the drug mobility is lower in solid lipids than in the oily phase. Aqueous biocompatible surfactants stabilize Solid Lipid Nanoparticles [8].

Solid Lipid Nanoparticles protect reactive and sensitive drugs from degradation during their passage through the intestines. These nanoparticles improve the bioavailability of highly lipophilic drug molecules. Using biodegradable and physiological lipids to produce nanoparticles and, with proper scaling, the lowers the cost of industrial production [8].

The properties of Solid Lipid Nanoparticles are diminutive size, vast surface zone, high medication stacking, and the communication of stages at the interface. They are appealing for their potential to enhance the execution of drugs. Their formulations for various application routes: parenteral, oral, dermal, visual, pulmonary, and rectal are available [9].

7 Inorganic Nanomaterials

7.1 Silica NPs

Silica Nanoparticles are considered as promising nanocarriers because of their excellent biocompatibility, exploitable structures, and high loading efficiency. The porous structure of silica nanoparticles makes them excellent nanocarriers for stem cells and stem-cell-related factors. Mesoporous silica nanoparticles have been found capable of loading neural growth factors to promote neural-cell and neurite growth, and even to control the fate of neural stem cells. Moreover, many contrast agents can be loaded into mesoporous silica nanoparticles to provide imaging modality further to track the transplanted stem cells in vitro or in vivo. It is worth noting that silicates are widely used as shell layers to encapsulate NPs for enhancing biocompatibility, loading efficiency, and stability of the original nanoparticles. These silica post modifications have been found helpful and useful in the neural stem cell-based therapy of neurological diseases [10].

7.2 Noble-Metal NPs

Noble-metal nanoparticles, such as gold and silver, are widely used in cancer theranostic applications [10]. These nanoparticles have unique optical properties like photoacoustics and photothermal response [10]. Researchers use gold and silver nanoparticles for bioimaging studies for their excellent localized surface plasmon resonance [11].

In particular, gold nanoparticles are easily modifiable with functional ligands, allowing the easy conjugation of biomolecules such as antibodies, proteins, and oligonucleotides. Silver nanoparticles have been renowned for their anti-inflammatory properties. Silver nanoparticles show a significant reduction of microglial toxicity toward dopaminergic neurons. They are capable of crossing the blood-brain barrier (BBB), which can cause neurotoxicity by accumulating in the brain. Further evidence indicates that silver nanoparticles can be cytotoxic to neurons that can cause neurodegeneration in vivo [12].

7.3 Quantum Dots

Quantum dots are 2–10 nanometer-sized artificial crystals that have peculiar properties. Quantum dots show size-dependent emissions, favors labeling and tracking of stem cells. [13]. Quantum dots exhibit high fluorescence stability, a nerve regeneration study in rats where stem cells labeled quantum dots are detected 35 days after transplantation [14]. Quantum dots provide comprehensive tracking property. In a study, injecting quantum dots with cell-penetrating lipopeptides, into the brains of embryonic chicks, made them track and identify neural stem cells [15].

8 Nanomaterials in Clinical Applications

8.1 Antibacterial /Sepsis Therapy

Sepsis is a life-threatening condition in response to the body fighting against severe infections, further leading to hemodynamic instability, failure of the vital organs, shock, and eventually death [16]. Several therapeutic protocols have implemented to prevent sepsis and septic shock [17]. However, there seem to be significant mortality rates secondary to sepsis. Recently researchers proposed nanoparticle-based diagnoses and treatments for sepsis [17]. An optimal size, surface, and composition have a longer circulation half-life than the free drug counterpart [18]. Hence, these drug nanoparticles can afford the time and exposure to neutralize pathogens and their molecular patterns in the blood and at the sites of infection [19]. When the nanoparticles were conjugated with poorly soluble drugs, it helped to disperse them in the aqueous medium and increased their bioavailability [17]. Sondi and Salopek-Sondi have explored Silver nanoparticles as an antimicrobial agent [20]. These silver nanoparticles induce toxicity and cause damage to the bacterial cell membrane by having a high affinity for the membrane [21]. Silver nanoparticles also serve as a repository of $Ag+$ ions, which form stable additions with cellular components to inactivate a broad spectrum of bacteria [21]. Pal and colleagues found that silver nanoparticles undergo shape-dependent interaction with the gram-negative bacterium E. coli [22]. The shape of the silver nanoparticles plays a significant role in their interactions with bacterial membranes [22]. The antibacterial activity varies not only with the shape but also by their size. Because of their smaller size, they affect the mobility of nanoparticles and the dissolution rate of $Ag+$ ions [22]. Silver is a metal known to have an insignificant risk in humans and eliminated by the liver and kidneys [23]. However, the long-term safety of systemically administered silver remains to be determined.

A study reports on a new antimicrobial agent SNAPP (structurally nanoengineered antimicrobial peptide polymers) [24]. SNAPPs demonstrate sub micromolecular activity against a broad range of pathogens due to the high local concentration of cationic peptides [24]. In a rat in vivo experiment, they induced peritonitis by colistin and A. baumannii, SNAPPs demonstrated significantly more favorable results than a conventional antibiotic imipenem, [24]. Overall, SNAPPs show great promise as low-cost and effective antimicrobial agents and may represent a weapon in combating the growing threat of multidrug-resistant Gram-negative bacteria [24].

9 Nanomaterial Applications in Common Diseases

9.1 Cancer

The goal of drug discovery is safety and efficacy. The limitations of the anticancer drugs are poor targeting, and exposure to healthy cells, which pose side effects [25]. The cell membrane blocks large and charged drug molecules, so transporting a drug into a target cell poses a critical challenge [25]. It is crucial to approach this issue by combining chemotherapeutics with cancer-targeting nanoparticles and other biomolecules [25].

Studies show that cell-penetrating peptides linked to active drugs facilitate drug delivery across the cell membrane [26]. Also, nanoparticles increase the pharmacokinetics of drugs [25]. Nanoparticles conjugated with cell-penetrating peptides increase cell penetration and drug delivery to intracellular targets [27].

Cell-penetrating peptides can virtually enter all kinds of cells so that the CPP-NP system can target the biomolecules in the membrane receptors of the cancer cells. Dos Santos Rodrigues et al. showed in a novel brain targeted gene delivery system that a CPP-NP could recognize transferrin receptors in the blood-brain barrier to enter the blood-brain barrier to transfect neuronal cells [28].

CytImmune is a Maryland based cancer nanomedicine company that has developed gold nanoparticle-based targeted chemotherapy and completed its Phase I clinical trial, CYT-6091: gold nanoparticles combined with thiolated PEG and tissue necrosis factor-alpha are better at targeting and lowering tumor toxicity [29]. Post-treatment tumor biopsies showed intracellular gold nanoparticles but not in healthy tissues [29]. Cerulean Pharma has developed another targeted chemotherapy that uses a nanoparticle: CRLX101 is a combination of cyclodextrin, a sugar molecule joined with camptothecin, a chemotherapy drug [30]. CriPec® from Cristal Therapeutics is a polymeric nanoparticle formulation containing the poorly water-soluble taxane docetaxel [31]. Researchers at MIT study a nanoparticle-based chemotherapeutic drug complex: 2 chemotherapeutic drugs along with liposomes coated on transferrin to pass through the brain barrier and target glioblastoma.

Additionally, transferrin accumulates at the tumor site by binding to proteins on the surface of tumor cells while avoiding healthy neurons [32]. Researchers at the University of Toronto designed manganese dioxide nanoparticles to accumulate in a tumor, altering the tumor microenvironment by generating oxygen to the efficiency of the chemotherapy drug doxorubicin [33]. See Table 8.1 for more nanomaterials for different cancers.

Table 8.1 Nanomaterials in cancer

Research Group	Nanoparticles	Chemotherapy Drug	Cancer	Physiological Effects	Reference
California NanoSystems Institute at UCLA and UCLA's Jonsson Comprehensive Cancer Center	LB-coated Mesoporous Silica	FOLFIRINOX (comprising *irinotecan*, 5-fluorouracil, oxaliplatin, and leucovorin)	Pancreatic ductal adenocarcinoma	reduces side effects; enhances drug's effectiveness; delivers the drug directly to the tumor site	Liu 2016
University of Georgia	sterically stabilized liposomes (SSL)	Inhibitor targeting PAK1 activation-3' (IPA-3) molecule; PAK1 (P21 activated kinases-1)	Prostate cancer	Quick absorption; reduced doses; improved drug effectiveness and retention	Al-Azayzih 2016
University of Texas—Southwestern Medical Center	modular degradable dendrimers	let-7 g microRNA	Hepatocellular carcinoma	high potency to tumors and low hepatotoxicity	Zhou 2016
University of Leicester, University College London, the Institute of Cytology of the Russian Academy of Sciences, and the Moscow Institute of Physics and Technology	molecularly imprinted polymer nanoparticles (nanoMIPs)	doxorubicin; anti-EGFR	General—Tumors and Cancers	cytotoxicity and apoptosis only in those cells that over-expressed EGFR	Canfarotta 2018
North Carolina State University, the University of North Carolina at Chapel Hill, and China Pharmaceutical University (CPU)	A cellular protease (furin)-mediated graphene	TRAIL (tumor-necrosis-factor-related apoptosis-inducing ligand,) and doxorubicin	human lung cancer tumor	bind directly to the surface of cancer cells	Jiang 2015

(continued)

Table 8.1 (continued)

Research Group	Nanoparticles	Chemotherapy Drug	Cancer	Physiological Effects	Reference
MIT	polymer nanoparticles	doxorubicin, camptothecin, and cisplatin	tested in ovarian cancer cells in the lab	Carry precise ratios of chemotherapy drugs	Liao 2014

9.2 Diabetes

Diabetes mellitus is a severe metabolic disease marked by fluctuations in blood sugar levels. A person with diabetes requires invasive methods to measure blood glucose levels by pricking a finger and injecting insulin if needed. These invasive methods cause pain and discomfort, which leads patients to less compliant [34]. Researchers have proposed non-invasive methods to measure blood glucose levels using nanoparticles as insulin carriers via less invasive pathways like oral and nasal routes, without injection [34]. Recent research has explored several nanoparticles for oral administration of insulin. Lipani et al. have proposed a proof-of-concept for a needle-free glucose monitoring system [35]. It is a non-invasive transdermal patch made up of graphene-based pixel-array of sensors, which measures the glucose levels in interstitial fluid [35]. These arrays extract the fluid through electro-osmosis to monitor the interstitial fluid-borne glucose [35].

9.3 Alzheimer's Disease

Brain complexity lies in its connections and communication pathways among neurons called synapses. These synapses make us think, talk, move, eat, and perform all our daily functions. A minor disruption in the communication pathway can cause brain disorders, which can cause problems [36]. During aging, the neurons shrink and eventually die, leading to loss of functions, skilled movements, memory, and thought processes. Neurological disorders have some common characteristics of neuronal loss, synaptic abnormalities, and deposits of misfolded proteins. The blood-brain barrier (BBB) prevents drug molecules from entering brain cells.

In the US, 5.5 million people have Alzheimer's Disease [37]. The pathology shows an accumulation of amyloid-beta plaques and neurofibrillary tangles [38]. The symptoms of Alzheimer's disease are memory loss, confusion, disorientation, and aphasia. Alzheimer's is diagnosed based on cognitive tests that do not detect the early onset of the disease [39]. Scientists studied several in vitro and in vivo experiments to remove the amyloid fibrils, improve drug permeability, improve drug retention, drug solubility, and drug stability. Nanomaterial based brain imaging studies can diagnose various neurological disorders using iron-oxide, gadolinium-based, and plasmonic nanoparticles. See Table 8.2 for more research.

9.4 Heart Diseases

Atherosclerosis is a gradual deposition of plaque in the arteries [40]. If this plaque is abundant in the coronary arteries, it can block blood flow, which is known as coronary artery disease (CAD). When the heart muscle receives less or no blood

Table 8.2 Nanomaterials in Alzheimer's Disease

Nanoparticles /Nanocarrier	Drug	Physiological effects	Reference
PEGylated nanoparticles	Aβ1-42 monoclonal antibody	• Removal of Amyloid Plaques • Improved memory	Carradori 2018
Dual-functional Nanoparticle (TQNP)	H102 peptide, a β-sheet breaker	• Removal of Amyloid Plaques • Improved memory	Zheng 2017
Polymeric nanocarriers	Curcumin	• Reduce oxidative stress • Reduce inflammation • Reduce plaque	Barbara 2017
Carbon nanotubes	Pittsburgh Compound B—an amyloid-beta binding molecule	• Remove the amyloid fibrils	Costa 2018
Chitosan nanoparticles	Saxagliptin	• Prevent premature release • Enhance targeting	Fernandes 2018
Polyamidoamine dendrimers	Carbamazepine	• Tripling the solubility of the drug; • Stability of the drug	Igartúa 2018
Micelles	Curcumin	• Neuroprotection; • Removal of Amyloid Plaques	Desai 2018
Plasmonic gold nanoparticles	Amyloidogenic β-amyloid binding peptides	• Increasing Blood-Brain Barrier permeability of the drug	Ruff 2017
Carbon nanotubes	Berberine	• Increasing the amount of drug	Lohan 2017

supply and damages the heart muscles, it is called myocardial ischemia [40]. Since these atherosclerotic changes start at a cellular level, only nanomedicine can prevent its progression at the cellular level [40]. Clinicians use invasive and non-invasive methods to treat CAD [40]. Lipoproteins carry cholesterol from the liver and deliver it to the other tissues. Lipoproteins oxidized and deposited on the vessel walls, cause endothelial injury, which sets off an inflammatory reaction, which sets off atherosclerosis [40]. Statins have anti-inflammatory and antioxidant effects, but high dose statins can have side-effects. Study shows that pravastatin loaded vesicles, functionalized with oligonucleotides, have a high affinity towards inflammatory macrophages, and reduces systemic toxicity [40].

Angiogenesis is a new blood vessel formation around the plaque sites. Fumagillin is an anti-angiogenic drug used in CAD that has an adverse neurocognitive impairment. When Fumagillin is delivered through integrin targeted nanoparticles, it targets atherosclerosis and reduces adverse side effects [40]. Clinicians use percutaneous coronary interventions to treat CAD. Clinicians will insert metal stents to support the arterial structural patency, but it causes arterial wall injuries. Animal experiments reveal that bioabsorbable stents, coated with antibodies, can reduce these potential side effects [40].

9.5 Future Prospects

The future of personalized medicine has its foundations on this nanotechnology. Nanotechnology had opened new avenues for drug delivery methods, which are more effective than the current drug delivery methods. Carbon nanotubes carry drugs into the target cells, which were previously not reachable. Overall, these nanomaterials in medicine can transform the therapeutic arena in the future and give hope for the treatment of many incurable diseases.

10 Summary

In this chapter, we briefly looked into commonly used nanomaterials in medicine, and their uses in diagnostics and therapeutics. We looked briefly into commonly occurring diseases such as cancer, diabetes, coronary artery disease. Research in nanomaterials in medicine has a prominent future in personalized medicine.

References

1. Al-Kayiem H, Lin S, Lukmon A (2013) Review on nanomaterials for thermal energy storage technologies. Nanosci Nanotechnology-Asia. https://doi.org/10.2174/22113525113119990011
2. He H, Pham-Huy LA, Dramou P, Xiao D, Zuo P, Pham-Huy C (2013) Carbon nanotubes: applications in pharmacy and medicine. Biomed Res Int 2013. https://doi.org/10.1155/2013/578290
3. Bhagath GP, Singh, Chandu Baburao, Vedayas Pispati H, Pathipati, Narashimha Muthy SP, BGR (2012) Carbon nanotubes—A novel drug delivery system. Int J Res Pharm Chem. https://doi.org/10.1007/s11060-011-0763-6
4. Akbarzadeh A, Rezaei-Sadabady R, Davaran S, Joo SW, Zarghami N, Hanifehpour Y, Samiei M, Kouhi M, Nejati-Koshki K (2013) Liposome: classification, preparation, and applications. Nanoscale Res Lett. https://doi.org/10.1186/1556-276X-8-102
5. Zdrojewicz Z, Waracki M, Bugaj B, Pypno D, Cabała K (2015) Medical applications of nanotechnology Zastosowanie nanotechnologii w medycynie. Postepy Hig Med Dosw 1196–1204
6. Junghanns JUAH, Müller RH (2008) Nanocrystal technology, drug delivery and clinical applications. Int J Nanomedicine 3(3):295–309
7. Gigliobianco MR, Casadidio C, Censi R, Di Martino P (2018) Nanocrystals of poorly soluble drugs: drug bioavailability and physicochemical stability. Pharm 10(3):134
8. Yadav N, Khatak S, Singh Sara UV (2013) Solid lipid nanoparticles—a review. Int J Appl Pharm 5(2):8–18

9. Mazzola L (2003) Commercializing nanotechnology. Nat Biotechnol 21:1137–1143. https://doi.org/10.1038/nbt1003-1137
10. Zhang B, Yan W, Zhu Y, Yang W, Le W, Chen B, Zhu R, Cheng L (2018) Nanomaterials in Neural-Stem-Cell-mediated regenerative medicine: imaging and treatment of neurological diseases. Adv Mater 30:1–23. https://doi.org/10.1002/adma.201705694
11. Zhao Y, Sultan D, Detering L, Cho S, Sun G, Pierce R, Wooley KL, Liu Y (2014) Copper-64-alloyed gold nanoparticles for cancer imaging: Improved radiolabel stability and diagnostic accuracy. Angew Chemie Int Ed. https://doi.org/10.1002/anie.201308494
12. Skalska J, Frontczak-Baniewicz M, Strużyńska L (2015) Synaptic degeneration in rat brain after prolonged oral exposure to silver nanoparticles. Neurotoxicol 46:145–154. https://doi.org/10.1016/j.neuro.2014.11.002
13. Rosen AB, Schuldt AJT, Kelly DJ, Potapova IA, Doronin SV., Brink PR, Gaudette GR, Cohen IS (2007) Quantitative analysis of hMSC-seeded biological scaffolds using quantum dot nanoparticles. In: Bioengineering, Proceedings of the Northeast Conference
14. Sanchez DNR, Bertanha M, Fernandes TD, Resende LA de L, Deffune E, Amorim RM (2017) Effects of canine and murine mesenchymal stromal cell transplantation on peripheral nerve regeneration. Int J Stem Cells. https://doi.org/10.15283/ijsc16037
15. Agarwal R, Domowicz MS, Schwartz NB, Henry J, Medintz I, Delehanty JB, Stewart MH, Susumu K, Huston AL, Deschamps JR, Dawson PE, Palomo V, Dawson G (2015) Delivery and tracking of quantum dot peptide bioconjugates in an intact developing avian brain. ACS Chem Neurosci. https://doi.org/10.1021/acschemneuro.5b00022
16. Singer M, Deutschman CS, Seymour C, Shankar-Hari M, Annane D, Bauer M, Bellomo R, Bernard GR, Chiche JD, Coopersmith CM, Hotchkiss RS, Levy MM, Marshall JC, Martin GS, Opal SM, Rubenfeld GD, Der Poll T, Vincent JL, Angus DC (2016) The third international consensus definitions for sepsis and septic shock (sepsis-3). JAMA J Am Med, Assoc
17. Yuk SA, Sanchez-Rodriguez DA, Tsifansky MD, Yeo Y (2018) Recent advances in nanomedicine for sepsis treatment. Ther Deliv 9(6):435–450
18. Ashton S, Song YH, Nolan J, Cadogan E, Murray J, Odedra R, Foster J, Hall PA, Low S, Taylor P, Ellston R, Polanska UM, Wilson J, Howes C, Smith A, Goodwin RJA, Swales JG, Strittmatter N, Takáts Z, Nilsson A, Andren P, Trueman D, Walker M, Reimer CL, Troiano G, Parsons D, De Witt D, Ashford M, Hrkach J, Zale S, Jewsbury PJ, Barry ST (2016) Aurora kinase inhibitor nanoparticles target tumors with favorable therapeutic index in vivo. Sci Transl Med. https://doi.org/10.1126/scitranslmed.aad2355
19. Ward NS, Levy MM (2017) Sepsis: definitions, Pathophysiology and the challenge of bedside management. Humana Press, Cham (Switzerland)
20. Sondi I, Salopek-Sondi B (2004) Silver nanoparticles as antimicrobial agent: a case study on E. coli as a model for Gram-negative bacteria. J Colloid Interface Sci. https://doi.org/10.1016/j.jcis.2004.02.012
21. Le Ouay B, Stellacci F (2015) Antibacterial activity of silver nanoparticles: a surface science insight. Nano Today
22. Pal S, Tak YK, Song JM (2007) Does the antibacterial activity of silver nanoparticles depend on the shape of the nanoparticle? A study of the gram-negative bacterium Escherichia coli. Appl Environ Microbiol. https://doi.org/10.1128/AEM.02218-06
23. Lansdown ABG (2006) Silver in health care: antimicrobial effects and safety in use. Curr Probl Dermatol 33:17–34
24. Lam SJ, O'Brien-Simpson NM, Pantarat N, Sulistio A, Wong EHH, Chen YY, Lenzo JC, Holden JA, Blencowe A, Reynolds EC, Qiao GG (2016) Combating multidrug-resistant gram-negative bacteria with structurally nanoengineered antimicrobial peptide polymers. Nat Microbiol. https://doi.org/10.1038/nmicrobiol.2016.162
25. Gessner I, Neundorf I (2020) Nanoparticles Modified with cell-penetrating peptides: conjugation mechanisms, physicochemical properties, and application in cancer diagnosis and therapy. Int J Mol Sci 21:1–21. https://doi.org/10.3390/ijms21072536
26. Silva S, Almeida AJ, Vale N (2019) Combination of cell-penetrating peptides with nanoparticles for therapeutic application: a review. Biomol

27. Wei G, Wang Y, Huang X, Hou H, Zhou S (2018) Peptide-based nanocarriers for cancer therapy. Small Methods 2:1700358. https://doi.org/10.1002/smtd.201700358
28. Dos Santos Rodrigues B, Lakkadwala S, Kanekiyo T, Singh J (2019) Development and screening of brain-targeted lipid-based nanoparticles with enhanced cell penetration and gene delivery properties. Int J Nanomedicine 14:6497–6517. https://doi.org/10.2147/IJN.S215941
29. Libutti SK, Paciotti GF, Myer L, Haynes R, Gannon W, Walker M, Seidel G, Byrnes A, Yuldasheva N, Tamarkin L (2009) Results of a completed phase I clinical trial of CYT-6091: a pegylated colloidal gold-TNF nanomedicine. J Clin Oncol
30. Chen G, Karzai F, Madan RA, Cordes LM, Bilusic M, Owens H, Hankin A, Williams M, Couvillon A, Gulley JL, Dahut WL, Thomas A (2018) CRLX101 plus olaparib in patients with metastatic castration-resistant prostate cancer. J Clin Oncol. https://doi.org/10.1200/jco.2018.36.15_suppl.tps5096
31. Atrafi F, Dumez H, Mathijssen RHJ, Menke CW, Costermans J, Rijcken CJF, Hanssen R, Eskens F, Schoffski P (2019) A phase I dose-finding and pharmacokinetics study of CPC634 (nanoparticle entrapped docetaxel) in patients with advanced solid tumors. J Clin Oncol. https://doi.org/10.1200/jco.2019.37.15_suppl.3026
32. Trafton A (2018) Tiny particles could help fight brain cancer. MIT News
33. Amini MA, Abbasi AZ, Cai P, Lip HY, Gordijo CR, Li J, Chen B, Zhang L, Rauth AM, Yu WuX (2019) Combining tumor microenvironment modulating nanoparticles with doxorubicin to enhance chemotherapeutic efficacy and boost antitumor immunity. J Natl Cancer Inst. https://doi.org/10.1093/jnci/djy131
34. Souto EB, Souto SB, Campos JR, Severino P, Pashirova TN, Zakharova LY, Silva AM, Durazzo A, Lucarini M, Izzo AA, Santini A (2019) Nanoparticle Delivery Systems in the treatment of diabetes complications. Mol 24:4209. https://doi.org/10.3390/molecules24234209
35. Lipani L, Dupont BGR, Doungmene F, Marken F, Tyrrell RM, Guy RH, Ilie A (2018) Non-invasive, transdermal, path-selective and specific glucose monitoring via a graphene-based platform. Nat Nanotechnol 13:504–511. https://doi.org/10.1038/s41565-018-0112-4
36. Ramanathan S, Archunan G, Sivakumar M, Tamil Selvan S, Fred AL, Kumar S, Gulyás B, Padmanabhan P (2018) Theranostic applications of nanoparticles in neurodegenerative disorders. Int J Nanomed 13:5561–5576. https://doi.org/10.2147/IJN.S149022
37. Alzheimer's facts and figures report
38. Kaur IP, Bhandari R, Bhandari S, Kakkar V (2008) Potential of solid lipid nanoparticles in brain targeting. J. Control Release
39. Costa PM, Wang JT-W, Morfin J-F, Khanum T, To W, Sosabowski J, Tóth E, Al-Jamal KT (2018) Functionalised carbon nanotubes enhance brain delivery of amyloid-targeting Pittsburgh Compound B (PiB)-derived ligands. Nanotheranostics 2:168–183. https://doi.org/10.7150/ntno.23125
40. Ambesh P, Campia U, Obiagwu C, Bansal R, Shetty V, Hollander G, Shani J (2017) Nanomedicine in coronary artery disease. Indian Heart J

Hydrogels: Biomaterials for Sustained and Localized Drug Delivery

Ganesan Keerthiga, Pallavi Gupta, and Tuhin Subhra Santra

Abstract Hydrogels are three dimensional (3D) cross-linked polymer networks capable of holding a large volume of water. The hydrophilic polymeric system sometimes exists as a colloidal gel inside water, i.e., dispersion medium. Hydrogels aim to mimic the 3D microenvironment of cells with the advantage of surpassing adverse gastrointestinal effects on the drug, therefore increasing patient compliance. This polymer-based hydrogel formulation has tunable properties such as porosity, tensile strength, drug loading capacity, and release kinetics that contribute towards better biocompatible hydrogel design. The monomeric units in hydrogels bind through physical and chemical forces such as hydrophobic interaction, hydrogen bonding, UV crosslinking, and many others. Albeit hydrogel is known for its water holding capacity and high biocompatibility, the cytotoxicity of hydrogel depends on the polymer selection. Deformable and injectable hydrogels that can alter its physical state in room and body temperature are in the research pipeline to avoid surgery for implantation. Further, environmental stimuli-responsive hydrogels like pH, temperature-sensitive hydrogels are evolving as 'Smart drug delivery' systems. This distinctive property of tunable hydrogel design and formulation finds its application in sustained and localized drug delivery. This chapter discusses the different classifications of the hydrogel, along with its crosslinking chemistry involved. We also have summarised various forms of hydrogel from lab scale to industrial level. Finally, this chapter also covers the synthesis, functionalization, tailoring mechanism of the hydrogel matrix, followed by in vitro, ex vivo, and in vivo characterization and drug loading/delivery efficiency.

Keywords Hydrogel · 3D polymer · Nanocomposites · Stimuli-responsive hydrogel · Controlled drug delivery

---

G. Keerthiga
Centre for Biotechnology, Anna University, Chennai 600032, Tamil Nadu, India

P. Gupta · T. S. Santra (✉)
Department of Engineering Design, Indian Institute of Technology Madras, Chennai 600032, Tamil Nadu, India
e-mail: tuhin@iitm.ac.in

T. S. Santra and L. Mohan (eds.), *Nanomaterials and Their Biomedical Applications*, Springer Series in Biomaterials Science and Engineering 16,
https://doi.org/10.1007/978-981-33-6252-9_9

1 Introduction

Hydrogels are three-dimensional cross-linked polymeric network capable of imbibing large volumes of water (90–99wt%) [145, 7]. These hydrophilic polymeric networks are formed by cross-linking monomers in addition to a physical or chemical cross-linking agent. Different polymeric sources (natural or synthetic) and varied crosslinking methodologies have led hydrogel into the limelight of research. Back in the early 1960s [215], a study on poly (2-hydroxyethyl methacrylate) comprehensively predicted a net repulsion between polymer network and a poor solvent causes phase transition, change in the degree of swelling and volume. Progress has been made to find a biocompatible and non-toxic formulation for in vivo delivery of drugs to overcome repeated dosing and drug loss during activation and transportation [19]. Significant research (publications over the years as shown in Fig. 1) and advances in polymer chemistry for understanding the underlying physiology has allowed researchers to consider hydrogel as a promising candidate for biomedical applications such as biosensors [157], micro-total analysis systems (μTAS) [204], molecular imprinting [20], contact lenses, targeted drug delivery vehicle for delivering biomolecule(s) of interest, mimicking extracellular matrix in tissue engineering applications.

Nanotechnology has extended its roots in various fields in the last two decades. Its fundamental property of a high surface to volume ratio of any material in the nanoscale dimension has found multiple applications in the fields of science and

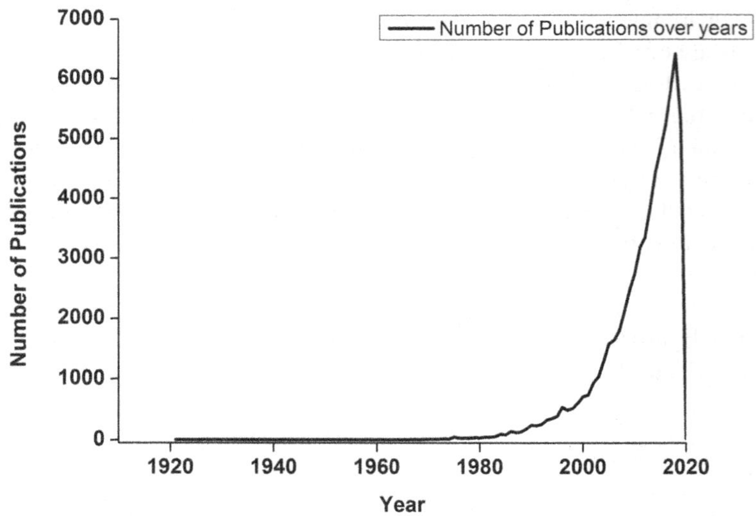

Fig. 1 Graph representing publications concerning hydrogel research over the years (Based on Scopus data)

engineering, including biomedical relevance [8–202]. Polymeric network encapsulated with nanoparticles that aim at targeted drug delivery has proven hydrogel a valuable biomaterial for pharmaceutics [42].

Nanocomposites take over the advantage of hydrophilicity of the polymers, versatility, biocompatibility as well as the increased surface to the volume aspect ratio of the nanoparticles. Diverse nanomaterials such as graphene, carbon nanotubes (CNTs) (see Sect. 4.2.1); polymeric/dendritic nanomaterial (see Sect. 4.2.2); inorganic/organic nanomaterials such as silicates, hydroxyapatite nanoparticle-based formulations (see Sect. 4.2.3); metallic nanoparticles such as gold, silver nanoparticles (see Sect. 4.2.4) are utilized in hydrogel nanocomposite formulation. This interdisciplinary field of research is known to have a significant effect on developing nanocomposites such as biodegradable polymeric nanoparticles, polymeric micelles, solid lipid nanoparticle (SLN), lipid drug conjugate (LDC), nanostructured lipid carriers (NLC), and quantum dots [12–160].

A combinatorial approach of nanotechnology and hydrogel for nanogel formulations produces three-dimensionally cross-linked submicron hydrophobic (or less soluble) particles for drug delivery. Nanogel increases the solubility of the drug, accumulation at the intended site of action, and stability of bioactive molecules in physiological conditions while reducing cytotoxicity. Temperature responsive, pH-temperature dual responsive nanogels are also constructed using copolymer blocks. Micro/nano-particles prepared by spray drying, microemulsion, phase separation procedures are used for mucoadhesive drug delivery systems (MDDS). Mucal adhesion occurs through the surface to surface contact in the form of micro/nanospheres or asymmetric patches. Stimuli responsive nanogels have gained considerable interest in research for its drug delivery capabilities and controlled drug release at the intended absorption site [19–69].

This chapter discusses the significant classifications of hydrogels based on their polymeric source and crosslinking techniques; different parameter constraints for hydrogel design; biomaterials for hydrogel nanocomposite formulation for biomedical applications; drug release mechanisms by diffusion to various stimuli in the physiological environment. Further, this chapter briefly emphasizes the current challenges and future research focus of hydrogel for clinical translations.

2 Classification of Hydrogel

Hydrogels are designed to encapsulate drug and release in a sustained mode in targeted location for prolonging the effect as well as minimizing repeated dosage. These are promising biomaterials for their unique physicochemical properties, which include molecular weight of the polymer, method of crosslinking and intermolecular bonding among monomer and between the monomer and crosslinking agent [37].

The physical nature of hydrogel plays a vital role in introducing the drug-loaded hydrogel into the target system. The 'sol-gel' characteristics of the prepared hydrogel decide the route of administration of the hydrogel into the host system [109].

Injectable hydrogel formulation is currently overpowering the matrix hydrogel for its capability to skip surgery for implantation [118]. Thus physical form significantly decides the route of hydrogel introduction, depending on the physical structure of the polymer, rigidity, covalent bonding and intermolecular interaction within the polymeric chains [37, 25–58].

On the other hand, chemical characteristics involve matrix density, the mechanical strength of the hydrogel formulation, biodegradability, and biocompatibility. These properties are constricted to the individual and collective properties of monomers and crosslinking agents used. Tailored hydrogel formulations are prepared by modifying the above parameters. Thus hydrogel formulation in itself contains various parametric choices such as polymeric source, natural or synthetic; polymeric composition (homo-polymeric, hetero-polymeric); their nature of gelation (physical, chemical crosslinking); their porosity as micro, macro or super porous; their degree of swelling and absorbance capacity [28–35]. The two main classifications depending upon the polymeric source and crosslinking (see Fig. 2) are discussed in this chapter as it encompasses various monomers involved in a hydrogel formulation.

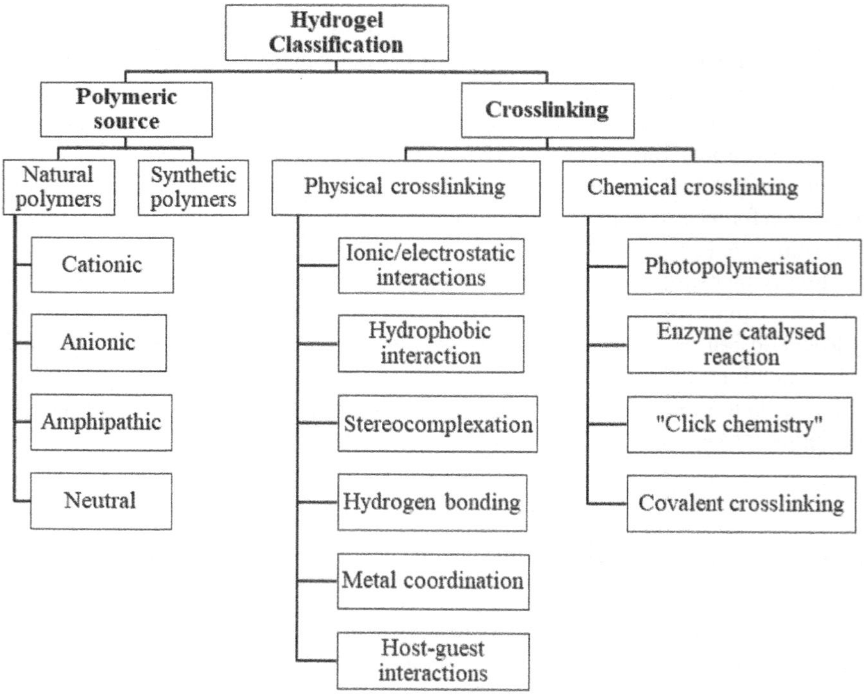

Fig. 2 Overall classification of hydrogel

2.1 Polymeric Source

The polymeric source can range from natural to synthetic monomers for hydrogel formulation depends upon the application intended.

2.1.1 Natural Polymers

Natural polymers are developing biomaterials that are the primary forms of renewable biomass. Extracted from a macromolecular matter of microbes, plants, and animals, they offer a range of properties such as biodegradability, biocompatibility, and non-toxicity. They are classified into cationic, anionic, neutral, and amphipathic, depending on their physicochemical property. Listed below in Table 1 are different types of naturally occurring polymeric sources (± crosslinkers) [73].

- **Chitin**, the next abundant polymer to cellulose is highly hydrophobic and insoluble in water and organic solvents. Deacetylation in alkaline conditions or by chitin deacetylase to yield chitosan is used in gel formulations at low pH. Chitosan-based hydrogels are mainly used to deliver drugs targeted to the mucosal layer [9, 33].
- **Hyaluronic acid (HA)**, an anionic copolymer of 2-acetamido-2-deoxy-D-glucose and D-glucuronic acid, allows us to form strong electrostatic interactions. Chemical reactions such as addition, condensation, or radical polymerization of the active moieties are carried out to obtain different physical formulations. Generally, HA is formulated as a firm hydrogel, viscoelastic liquid, mesh, and fiber that find its application majorly in nasal, pulmonary, ophthalmic, parenteral, and topical routes. HA-based hydrogels are also reported to show tissue healing, promotion of cell growth and proliferation, inflammatory response control, etc. [34–34].
- **Dextran** is a linear glucose polymer of 1,6-glycosidic bonds with some extent of branching through 1,3 linkages with many reactive moieties, makes it a great candidate for formulation. Dextran is known to have antithrombotic property. Its biodegradability and bio-compatibility have made it an efficient drug vehicle for carrying the drugs or bioactive agents of choice. Different forms of dextrans are hydroxyethyl methacrylate dextran, dextran urethane, glycidyl methacrylate dextran, and dextran hydroxyl-ethyl-methacrylate lactate [194], which are used for hydrogel formulations.

Table 1 Different types of naturally occurring polymers

Type	Polymer name
Cationic	Poly lysine, chitosan
Anionic	Pectin, Hyaluronic acid, dextran sulphate, alginic acid, carrageenan
Neutral	Dextran, agarose, pullulan
Amphipathic	Collagen (and gelatin), fibrin, carboxymethyl chitin

- **Collagen** is a triple helix protein found in the extracellular matrix (ECM) of fibroblasts and osteoblasts. Generally utilized form type I collagen is isolated from the body tissues, neutralized and broken down to single molecules called 'gelatin.' The final step determines the type of gelatin obtained. It is known to undergo a 'sol-gel' transition in response to the surrounding temperature, making it a suitable material for designing thermoresponsive hydrogel. Thus, collagen is an impending component for a 'smart drug delivery system' [52].

2.1.2 Synthetic Polymers

Natural polymeric sources are being entirely replaced by synthetic polymers, such as poly 2-hydroxylethyl methacrylate (pHEMA), polyvinyl alcohol (PVA), polyethylene glycol (PEG), etc. for increased of mechanical strength, acclimatized functionality, and degradability. The method of crosslinking effects hydrogel degradation that can be controlled in response to external stimuli such as pH, temperature, light, and electric field, resulting in making a 'smart drug delivery system' [51]. Some of the polymers, such as PEG, are amphiphilic, that aid in more natural absorption of hydrophobic drug molecules or intended cargos than conventional drug vehicles. Polyethylene glycol (PEG) is a synthetic polymeric unit that is a polymer or oligomer of ethylene oxide. Polyoxyethylene (POE), Polyethylene Oxide (PEO), polyvinyl alcohol (PVA), polypropylene oxide (PPO) are chemically synonymous and differentiated by their molecular weights. Polymers with molecular weight <100,000 are grouped under PEGs, and the higher molecular weight is arranged as PEOs. PEGs of low molecular weight (<1000) are generally colorless, viscous, and increased molecular weights are white color and waxy solids. All PEGs are soluble in both aqueous and organic solvents, such as chloroform, ethanol, toluene, methyl chloride [3]. It has unique properties such as biocompatibility, non-immunogenicity, and resistance to protein adsorption [135]. PEG is a Food and Drug Administration (FDA), USA approved polymer that has found application in various frontiers of biomedical applications, such as bone prostheses, wound healing and tissue engineering, and drug delivery [165]. PEG cannot be a standalone polymer for polymeric network formation. Additional functional groups such as acrylate, thiol, vinyl sulfone, amine, and carboxyl, determine together of its mechanical strength and degradability (both natural and stimuli-responsive) of the polymeric network. In situ gelling systems are designed on the principles of photosensitive systems that can be cross-linked upon ultraviolet (UV) or light irradiation and are self-assembled in vivo. The latter alters its physical state according to the physiological environment inside the host system [44–1].

Poloxamers are triblock polymers that contain a hydrophobic chain surrounded by hydrophilic units on either side to facilitate hydrophobic drug delivery [192]. Hydrophobic polymers cross-linked in aqueous environment exhibit sol-gel transition by reversible thermal gelation as shown in the scheme below (Fig. 3).

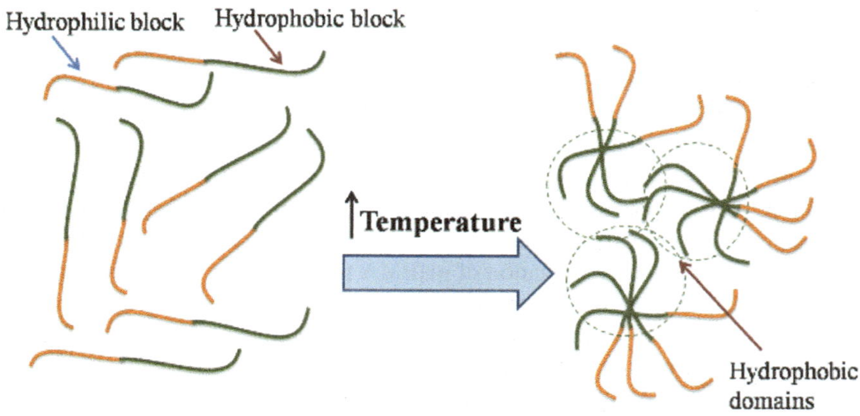

Fig. 3 Thermo-responsive physical gelation mechanism (Redrawn from [62])

Poloxamersare triblock polymers with two hydrophilic units and a hydrophobic entity flanked in between. They are noted as a potent drug delivery vehicle for delivering hydrophobic cargos (drugs or biomolecules) over conventional drug delivery vehicles. They do not entailthe natural absorption of hydrophobic molecules. Besides, thistri-block polymer offers various advantages such as biocompatibility, mechanical strength, and stability for targeted drug delivery [198, 182]. Pluronics were known to be an excellent drug vehicle until the polypropylene (PPO) unit in PEG-PPO-PEG triblock was non-degradable under physiological conditions. Modifications were made to replace the PPO by polypropylene glycol (PPG) using a template emulsion method to promote biodegradability. Other polymers with varying molecular weights can be used to control drug release purposes. Semi-hydrophilic triblock polymeric units such as PEO-PPO-PEO, hydrophobic polymer polycaprolactone (PCL) are also employed for site-specific drug delivery of poorly soluble drugs [201]. Pluronic/PCL blocks are known to change their size with respect to temperature precisely and for thermosensitive controlled drug release [93]. In addition to this, pluronic are also used in bioprinting for increasing tissue engineering augmentation. 3D printing limits the availability of the polymer substrate as a starting material. However, nanostructuring utilizing a mixture of polymers has resulted in better biocompatibility and cell stability. Acrylatedpluronic and pluronic F127 were cross-linked using UV, to preserve viability, enhance mechanical integrity, and cell adhesion [132].

Polyhydroxyethyl methacrylate (pHEMA) achieved considerable attention in the early 1960s. Popularly known for its high water absorbing capacity, biocompatibility, this polymer finds various applications in medical and biological fields. The mechanical and swelling properties of this polymer are tailored based on its use in mimicking bone substitute, tissue engineering, scaffold generation, synthetic biomaterial, and drug delivery purpose [123, 90]. The ability to release drugs or any bioactive agents of interest is directly proportional to its swelling property of the hydrogel pHEMA. This swelling property, in turn, depends upon the crosslinking agent employed for hydrogel formulation [69]. For tissue engineering scaffold generation, positively charged polymer scaffolds show a better attachment than the negatively charged polymer [186]. Implantable hydrogels for drug delivery/tissue engineering frequently face rejection due to the immune response of the host system. A thick collagenous of matter around the implant is formed, that favorably thwarts the implant- body interactions preventing implant rejection. pHEMA does not favor non-specific protein absorption, thus supporting its non-fouling property, which is highly desirable for an implant. However, pHEMA can only partially prevent surface capsule formation, thereby necessitates for modification of pHEMA formulations. Zwitterionic hydrogels of poly carboxy betaine methacrylate (PCBMA) has known to develop smaller surface capsule formation, thus causing lesser inflammation, ultimately leading to a lowered chance of implant rejection [223].

Polyacrylamide (PAAm) based hydrogels are synthesized from monomers of acrylamide. Different conditions of polymer synthesis, co-polymerization with other different monomers, to achieve for tailored chemical properties of the hydrogel. Hydrogels constructed with PAAm backbone are often exploited for stimuli-responsive drug delivery. Gastrointestinal systemic delivery with varied pH, ranging from highly acidic to alkaline (pH 1.0–8.2), chronic wound treatment, and treatment for cancer employed by using PAAm based hydrogels are reported in the literature. These formulations contain weak acidic (carboxylic or sulfonic) or primary (ammonium) functional groups bonded to the main polymer configurations, that become ionized in micro-environmental pH conditions either by accepting or donating protons, thereby affecting the swelling pattern of the hydrogel [58–203].

In addition to pH sensitivity, temperature-responsive PAAm polymers are also used for their distinct properties. Frequently used thermoresponsive modified PAAm polymers are polyN-isopropylacrylamide (PNIPAAm), polyN, N-dimethyl acrylamide (PDEAAm) [100]. One another polymer is employed for similar thermoresponsive properties is PNIPPAm, and the additional functionality of the pendant groups is poly2-carboxy isopropyl acrylamide (PCIPAAm) [100]. Similar yet distinctive polymers that are used for varied applications are as tabulated in Table 2 [119].

Table 2 Thermoresponsive polymers and their applications

Polymer	Applications
Poly(N-(1)-1-hydroxymethyl-propylmethacrylame) (poly(1-HMPMAAm)	pH responsive and photosensitive
Poly(N-acryloyl-N-alkylpiperazine)	pH and thermoresponsive
Poly (N,N-diethylacrylamide) (PDEAAm)	Thermoresponsive yet distinct structures
Poly(2-carboxyisopropylacrylamide) (PCIPAAm)	

2.2 Based on Crosslinking Methods

Different crosslinking strategies are employed for hydrophilic monomeric units in hydrogel fabrication to form stable polymeric networks [118, 54, 71]. Different physical and chemical crosslinking methods are listed in Fig. 2. Physical crosslinking include charge ion interactions, hydrogen bonding, crystallization/stereo-complex formation mechanisms, while synthetic strategies involve radical polymerization [64–102], photo-polymerization [11, 127], high energy irradiation, enzyme induced crosslink reaction, addition reaction of the polymeric units [184], Diels-Alder "click reaction," and Schiff-base formation. Depending upon the structure and desirable mechanical properties, nature of polymers, the crosslinking strategy is preferred for fabrication.

2.2.1 Physical Crosslinking

The physically cross-linked hydrogel, also known as a reversible hydrogel, is usually created by inter-molecular bonding through ionic interactions, polymerized entanglements, hydrophobic/hydrophilic bonding, etc. This reversible property is exploited in designing stimuli-responsive hydrogel with self-healing and injectable features for efficient drug delivery purposes. They are also referred to as the 'smart drug delivery systems' (SDDS).

i. Ionic/electrostatic interactions

Molecules with opposite charges tend to attract each other influencing formation of a hydrogel. Alginate a natural source of polysaccharide with mannuronic and glucuronic acid units, are cross-linked to achieve gel formation by divalent cations such as magnesium (Mg^{2+}), calcium (Ca^{2+}) and barium (Ba^{2+}) [22, 107]. It is widely employed in wound healing, drug delivery, tissue engineering. A similar mechanism occurs in macromolecules with opposite charges through electrostatic interactions to give polyelectrolyte complexes (PECs) [141]. For example, chitosan forms PECs by the electrostatic interactions between its amino group (cationic) and from other natural polyelectrolytes such as pectin, alginate, chondroitin sulphate or artificial polymers like polylactic acid (PLA), polyacrylic acid, and polyphosphoric acid (anionic) [9, 63]. These hydrogels fabrication using PECscan be modulated using the charge density of the polymer, amount of the polymer, the mix ratio, and the surrounding microenvironment of the polymer [210, 114]. When the net charge of the polymer is zero, the hydrogel complex precipitates in its microenvironment. The significant advantage of the preparation of the hydrogel through the ionic/electrostatic interactions is their deformability under high stress and ability to reform once the pressure is removed, making it a candidate for SDDS. However, limitations of ionic/electrostatic interaction-based hydrogel include the decreased mechanical strength due to the bonding involved [9, 107, 63, 114].

ii. Hydrophobic interaction

Water miscible monomers with hydrophobic tail groups on the side chains or hydrophobic monomers are subjected to thermal induction or ultrasonic treatment promoting sol-gel transition in hydrogel formation by hydrophobic interaction. Specific conditions in favor the gel transition are maintained for gel formation, while removal of the same leads to reversible physical state [62].

Thermal induction based on lower critical solution temperature(LCST) [76–39]/upper critical solution temperature (UCST) [12, 221] is used for hydrogel fabrication when thermally treated at a critical temperature, the sol-gel transition occurs. Graft copolymers or amphiphilic blocks can self-assemble to form an organized structure in an aqueous environment with a hydrophobic core such as micelles. Polymers such as PNIPAM and its derivatives blend with hydrophilic PEO with hydrophobic PPO/Poly glycolide/Polylactide/PCL are generally fabricated for their thermoresponsive property through LCST. These polymers below LCST are in solution form, while at higher temperature form an insoluble gel structure through hydrophobic interactions. Methylcellulose containing Calcium Phosphate (CaP) nanoparticles using one-pot reaction by LCST gelation, has been reported as an experimental hydrogel drug reservoir at physiological pH [144].

On the contrary, in UCST induced hydrogel formation, the cooling temperature of the polymer solution plays an important role. Here, the hydrogel state is achieved at the temperature below UCST. However, at UCST, the hydrogel disintegrates, making the hydrophobic water-soluble micelle cores. Thermo reversible linear triblock comprising with inner hydrophilic poly-polyethylene glycol methyl ether methacrylate (PPEGMMA) part and outer polyacrylamide-co-acrylonitrile (PAAm-co-AN) units as a cooling induced sol-gel transition. This physical transition can be modulated by the polymer concentration [45].

Ultrasonic induction [79, 209] induced crosslinking involves phase variation. Natural or synthetic polymers such as collagen and silk fibroins that possess complex secondary structures can be physically crosslinked to form interpenetrating networks (IPNs) with improved physical properties and tunable gelation properties when treated with physical shear, organic solvents, and heat treatment.

iii. Crosslinking by Crystallization

Crystallites of polymer unit work as sites of physical cross-linking for hydrogel formation. An aqueous solution of PVA is subjected to repeat freeze-thawing to yield hydrogel. Parameters such as molecular weight, polymer concentration, time, and the number of freeze-thawing cycles and freezing temperatures influence the hydrogel formation [66]. Similarly, exploiting the property of crystallization, hydrogen bonding in polymers PVA, and hydrophobic polyacrylamide (HAPAM), physical double network (PDN) was developed by one-potpolymerization followed by repeated freeze-thaw cycles (Fig. 4).

More compact hydrogel structures are obtained by stereo-complex interaction between the two enantiomeric polymers. Based on the racemic crystallite Poly L-Lactic acid (PLLA) and poly d-Lactic acid (PDLA), a unique gel-sol-gel transition occurs upon heating and stereo-complexation. Novel enantiomeric mixtures of

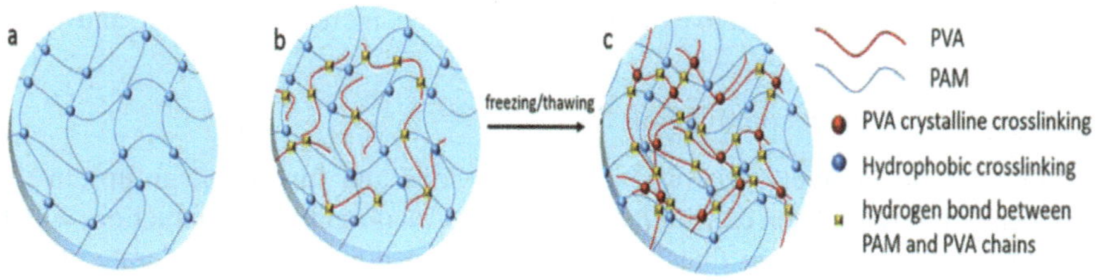

Fig. 4 Schematic illustration of Semi-IPN HAPAM/PVA hydrogel network structure fabricated via the freezing/thawing treatment: **a** HAPAM gel. **b** semi-IPN HAPAM/PVA gel. **c** PDN gel after subsequent freezing/thawing (Reproduced with permission from [80])

PDLA/PEG diblock polymeric systems and PLLA/PEG triblock copolymers are designed, where stereo-complex formation profoundly influence gel-sol-gel phase transitions [72].

iv. Crosslinking by Hydrogen Bonding

Hydrogen bonding is a crucial non-covalent interaction that widely influences by forming bonds within themselves, such as in hydroxyl, pyrrole, carboxylic acid, carbazole, or interact with electron donors such as imidazole and pyridine groups. Although a single hydrogen bond does not necessarily support healthy polymer network formation, multiple multivalent hydrogen bond interaction is found to influence the mechanical property of the hydrogel profoundly. Ureidopyrimidinone (Upy) dimer entrenched in hydrophobic domains of the PEG matrix assemble via 4-fold self-complementary hydrogen bonding, that reinforced networks (Fig. 5). This

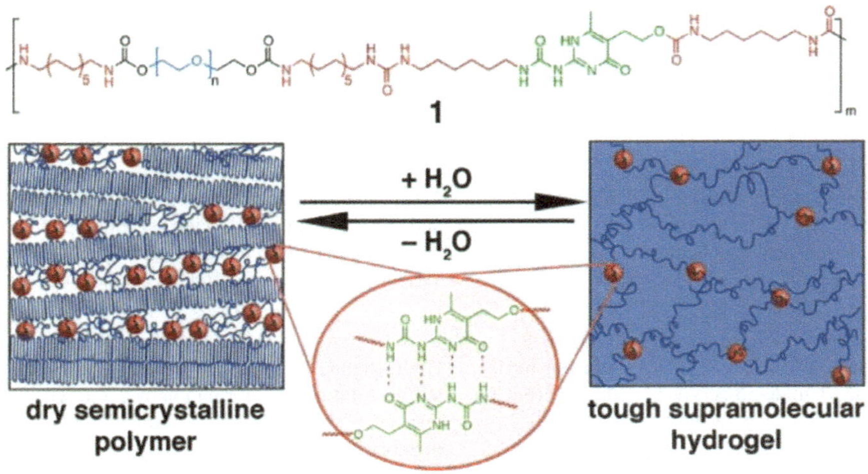

Fig. 5 PEG-UPy copolymers with multi-block architectures and self-complementary quadruple H-bonding interaction between 2 UPy segments. A graphical representation of drysemi-crystalline polymer and a reversible conversion to the hydrogel (Reproduced with permission from [59])

polymeric network is known to exhibit high resilience, increased strength, and shape memory behavior [59].

v. Crosslink by Metal coordination

Metal-ligand interaction between the metal ions and functional groups in the polymeric chains has been considered as a unique Lewis acid-base interaction, which is regarded as more potent than other non-covalent bonds. This moderate energy facilitates metal-ligand interaction occurrence or breakdown, which helps in self-healing property [111]. Mechanical properties of metal coordination based hydrogels such as hydrogel strength, toughness, the strength of the coordination bond, required tunable temperature-sensitive/pH-responsive/shape memory ability can be acquired accordingly to the application intended as they form sol-gel characteristics based on their metal coordination bonds [90–222]. Listed below in Table 3 are the polymers and their unique property achieved by tuning their metal coordination bond.

i. Crosslinking by host-guest interactions

'Host' refers to the molecule with a large cavity, such as cyclodextrins (CDs), calixarenes (CAs), cucurbiturils (CBs). At the same time, 'guest' molecule is complementary shapes (polymer) that form reversible interactions through various non-covalent binding such as hydrophobic interactions, hydrogen bonding, Van der Waals, electrostatic interactions, etc. Host–guest interactions occur by hydrophobic guest

Table 3 Different polymers and corresponding metal ion facilitating hydrogel formation

Polymer	Crosslinking metal ion	Property achieved	Mechanism involved
Poly (2-oxazoline) [29]	Fe (II), Co (III)	Stable at room temperature but disintegrates at 30 C	Intermolecular crosslink to intramolecular crosslink conversion
	Ru (II)	Hydrogel stable at boiling water	Dry gel recovers gelation in water after liquid evaporation
PEG functionalized with 3,4-dihydroxyphenylalanine (dopa) [95–76]	Fe (III)	Controlled gelation and cross-linking density by pH variation	Catechol ligand iron (Fe (III)) ion coordination to form catechol- Fe^{3+} complexes (mono- (below pH 5), bis- (about pH 8), tris- (above pH 8))
Poly (acrylamide-co-acrylic acid) (P(AAm-co-AAc)) [228]	Aqueous ferric chloride ($FeCl_3$) solution	High toughness, stiffness, fatigue resistance, shape memory, and processing ability filled hydrogel	Supramolecular network formation by carboxyl-Fe3+ coordination bond

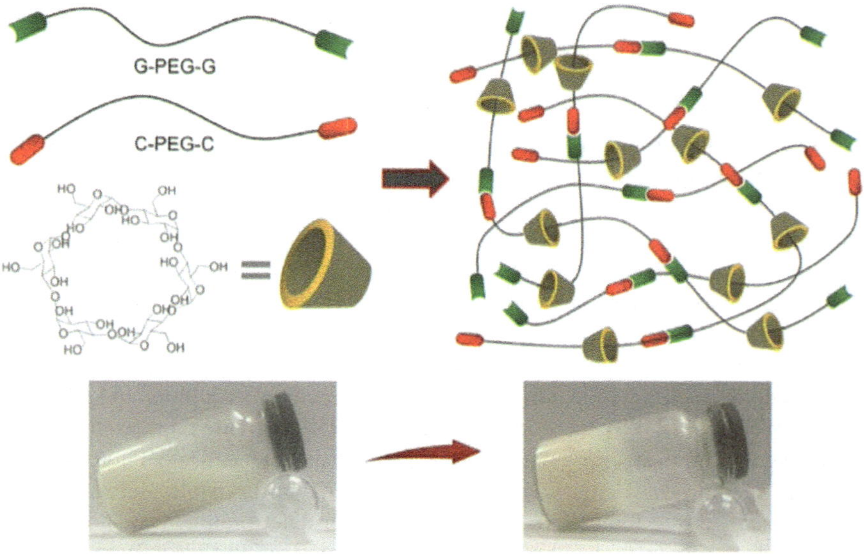

Fig. 6 Synthesis of supramolecular thermoresponsive hydrogels by host-guest inclusion between (G/C)-terminated PEG and α-CD. The G–C base pairing acts as additional network junctions and enhances the hydrogel mechanical properties (Reproduced from [80])

molecules encapsulation into the hydrophobic cavities in aqueous solution. Host–guest inclusion facilitates strong binding, fixed geometry, directionality, stimuli-responsive characteristics, making it useful for drug delivery [122]. One of the most universally used hydrophilic surfaces host molecule is cyclodextrin (CD). It is cross-linked with PEG to form injectable hydrogel for sustained drug release [109, 81]. The concentration of PEG and time required for gelation is dependent upon the molecular weight of PEG. Modified terminal ends of PEG with synthetic nucleobase guanine/cytosine (G/C) for base-pair formation can act as an additional network junction, as shown below in Fig. 6. This interaction enhances the storage module of PEG-CD inclusion complex increases mechanical integrity and cytocompatibility of the thermoresponsive controlled drug release [101–211].

2.2.2 Chemical Crosslinking

The chemically cross-linked hydrogel, also known as the permanent gel, is formed by covalent crosslinks in the polymeric chain with higher mechanical strength when compared to physically cross-linked hydrogels. Different chemical crosslinking strategies are employed such as free radical polymerization, enzyme-mediated crosslinking, Diels-Alder "click" reaction [136, 50], Michael addition reaction, Schiff base formation and oxime formation [131], etc. These hydrogels show higher stability under physiological conditions, enhanced mechanical property, and modifiable degradation behavior.

i. Crosslinking by photopolymerization

Light-activated cross-linking has been broadly used for hydrogel formation and therapeutics encapsulation and delivery [107–85]. Hydrogel formation is rapid at ambient temperature under placid conditions and alterable mechanical properties by controlling crosslinking reaction [135]. The choice of photo-initiated polymerization site is crucial as hydrogel formation occurs only in the irradiated light areas [220]. The composition of photo crosslinked biomedical hydrogels usually involves the presence of unsaturated groups that are incredibly reactive to favor free radical chain intensification polymerization when exposed to light. Also, cytocompatible-photoinitiators such as riboflavin phosphonate [61], camphorquinone [44], eosin Y [181], Irgacure 1173 [80], Irgacure 819, Irgacure 2959 [216] are required depending upon the wavelength absorption needed and given at the intended site. Photoinitiators absorb at specific wavelengths such as UV (250–370 nm), visible blue and purple (405–550 nm) or red light (750–810 nm) and either deteriorate (Type I) or acquire hydrogen from donor molecule (Type II) for polymerization [226]. Type II photoinitiators always require a co-initiator and some of which are listed below in Table 4 for visible light activated photo-polymerization.

UV crosslinking of γ-PGA and glycidyl methacrylate (GMS) to form methacrylate γ-PGA, showed ionic sensitivity and low cytotoxicity. A chitosan hydrogel prepared by UV crosslinking method with a cell loading pattern showed low cytotoxicity as well [15]. As light can penetrate the tissues, a prepolymer solution of the desired polymer and photo-initiator (Type I)/a photoinitiator and co-initiator (If Type II) were prepared and given as a hypodermic injection [181]. As discussed earlier, it is necessary to select a light penetrating site for photo-crosslinking, and as UV exposure risks the DNA damage, favoring visible light as an alternative [74]. Some of the commonly used visible light initiators are CQ, eosin Y, riboflavin [78], ruthenium [10], and lithium phenyl-2,4,6-trimethyl eosin benzoylphosphinate (LAP). Chemical modification of the photo-initiators is also made to improve crosslink density, higher photoactivity, and better mechanical properties. Only a few millimeters depth is achievable in photo-crosslinking and limiting the maximum attainable cure. Also, the homogeneity of the polymer structure and uniform tensile strength is uncertain. To overcome these limitations, polymerization of co-monomers with complementary reactive groups, that can facilitate homogenous hydrogel formation rapidly, and at

Table 4 List of the combination of Type II photoinitiators and co-initiators [16]

Type II photoinitiator	Co- initiator
Camphorquinone (CQ)	4- N,N- dimethylaminobenzoate (4EDMAB)
Isopropyl thioxanthone	Triethanolamine (TEA)
Eosin Y	TEA and N- vinyl pyrrolidone (NVP)

greater depths with lack of light intensity, and absence of photo-initiator has gained much attention [139]. Thiol and acrylate reaction is one of the extensively employed Michael addition reactions that can be coupled with photopolymerization, which is referred to as mixed-mode polymerization. It can yield tunable polymeric network formation and degradation by adjusting the thiol and acrylate ratio. Thiol groups also facilitate the post-polymerization of the hydrogel. Poly butadiene and poly (allyl methacrylates) functionalized with thiol can undergo photocatalytic redox reaction in visible light to form linear polymers by step-growth addition reaction [127, 85, 121].

ii. Crosslinking by Enzyme catalyzed reaction

Enzyme catalyzed crosslinking offers rapid in situ gelation with a possibility of tuning gel formation by controlling enzyme concentration under physiological conditions. Many kinds of enzymatic crosslinking methods are identified, such as amide linkage between carboxamide and amine groups in the presence of transglutaminase (TG); horseradish peroxidase (HRP) catalyzed coupling of aniline, phenol and its derived tyramine catalyzed by hydrogen peroxide. Enzymatically conjugated collagen with tyramine (Collagen- Ph) in the presence of HRP and H_2O_2 has been employed for constructing novel vascularized tissue [86]. This hydrogel encapsulated with bone marrow-derived mesenchymal stem cells (MSCs) and with human blood-derived endothelial colony-forming cells (ECFCs). The encapsulation improved prolonged MSC differentiation in the mouse as model organism after one month of implantation [123–87].

iii. Crosslinking by "Click chemistry."

Chemical reactions that favor high yield under mild conditions, lesser by-products, and increased specificity and selectivity are termed as "click" reactions. This reaction highly depends upon the functional groups of the polymeric materials for hydrogel fabrication. Classical click chemistry reactions are discussed as follows.

Diels Alder (DA) reaction is a one-step, highly selective, cycloaddition (4 + 2) between a dienophile (maleimide) and a diene (furan), that can occur in the absence of any initiators, catalysts and coupling agents. Polymers are customized accordingly with furan or furan derivatives to react with poly (ethylene glycol) dimaleimide for hydrogel formation [43]. Modifications are further being made to increase the DA reaction rate at physiological conditions, such as replacement of furan as methylfuran [189].

Schiff base formation is a condensation reaction between formyl or carbonyl-containing derivative and primary amines in the presence of catalysts like alkaline earth metal ions to form imine linkages. It has been extended to form injectable in situ gelling hydrogel as aldehyde end can adhere to tissues or organs [130]. Schiff linkages can be considered to be pseudo- covalent bonds that facilitate uncoupling and re-coupling of these linkages in polymeric networks result in self-healing capability. The amino group of the acrylamide modified chitin (AMC) and dialdehyde yield oxidized alginate for self-healable polymeric hydrogel. The molar ratio of the monomers and the microenvironment pH largely influence hydrogel formation [38].

Oxime crosslinking reaction requires an amino-oxy/hydroxylamine group and a functional aldehyde or ketone. These reactions are highly specific and occur even in the presence of other functional groups. It offers an advantage of crosslinking at the acidic condition and the only by-product being water. Self-healing, oxime cross-linked hydrogel can undergo reversible gel-sol conditions at acidic conditions [31].

Michael addition is a simple reaction between nucleophiles (donor) and activated electrophilic alkene or alkynes (acceptor). Often referring to thiol-containing polymers are added to the α and β unsaturated carbonyl polymers under necessary conditions that are well suited for the formulation of a cellular scaffold, gene transfection, and tissue replacements [124, 158].

iv. Covalent chemistry crosslinking

Reversibility of crosslinking helps to develop self-healing hydrogel that maintains strong integrity and internal annuity. This property of physical crosslinking is extended to covalent bonding. For example, boronate esters prepared from boronic acids and 1,2- and 1,3-diols that are pH-responsive, and it is designed based on its pKa (Hydrogel formation is favored when pH > pKa; remains as an aqueous solution when pH < pKa). Polymers such as PEG, polyphenols such as ellagic acid, epigallocatechin gallate (EGCG), and tannic acid (TA) were modified with linkers to provide boronate ester bonds for gelation at pH 7.4 that promote self-healing property at the physiological environment [77]. Native enzymes such as cytochrome-c can also be incorporated into hydrogel by structural integration to provide controlled stimuli-responsive hydrogel [178].

3 Hydrogel Design

Biomedical applications of the hydrogel are highly depended upon the bulk structure and parameters such as polymer volume in the swollen state ($\upsilon_{2,s}$), the molecular weight of the polymer chain between two crosslinks ($\overline{M}c$) and its corresponding mesh size (ξ). The measured quantity of water imbibed and retained by the hydrogel is given as polymer volume fraction. As random polymerization occurs, molecular weight between two crosslinks (physical/chemical) is calculated as average value (M_c) in the hydrogel. The correlation distance between adjacent crosslinks is also an average quantification that gives a measure of space existing between the polymer chains in the hydrogel. These parameters are related to each other, and dependent on the nature and charge of the polymer, type of crosslink, which measured both theoretically and experimentally [105, 104]. Two prominent theories widely used to explain configuration of the hydrogel are [149]:

1. Equilibrium—swelling theory.
2. Rubber—elasticity theory.

Hydrogels with no ionic moieties can be analyzed by Flory-Rehner theory, which is a combinatorial conjecture of thermodynamics and elasticity. This theory states that, when a cross-linked polymer gel plunged in a fluid, it is allowed to reach in equilibrium exerts, which is equal to the thermodynamic force of mixing and the retraction force of the polymer chains. Defining in terms of Gibbs free energy as Eq. 1,

$$\Delta G_{\text{total}} = \Delta G_{\text{mixing}} + \Delta G_{\text{elastic}} \tag{1}$$

Where, $\Delta G_{\text{elastic}}$ is the total refractive elastic force developed inside the gel; ΔG_{mixing} is the spontaneous mixing force of the polymer with the surrounding fluid. The above equation is differentiated with respect to the number of solvent molecules, the temperature and pressure as constants result in Eq. 2,

$$\mu_1 - \mu_{1,o} = \Delta\mu_{\text{elastic}} + \Delta\mu_{\text{mixing}} \tag{2}$$

Where μ_1 is the chemical potential of the solvent (fluid) in the polymer gel and $\mu_{1,o}$ is the chemical potential of the pure solvent.

At equilibrium, the difference between the chemical potential of the fluid inside and outside of the gel will be 0. Thus, making the chemical potential due to mixing and elastic forces are equal (Eq. 3).

$$\Delta\mu_{elastic} = \Delta\mu_{mixing} \tag{3}$$

Change in the chemical potential due to elastic forces of the polymer chain can be evaluated using rubber elasticity theory. The interaction between the polymer and the surrounding fluid is given as χ_1. Below, Eq. (4) is used to calculate molecular weight between adjacent crosslinks $\overline{M}_c$ of a neutral hydrogel in the absence of solvent.

$$\frac{1}{\overline{M}_c} = \frac{2}{\overline{M}_n} - \frac{\left(\frac{\overline{v}}{V_1}\right)\left[\ln\left(1 - v_{2,s}\right) + v_{2,s} + \chi_1 v_{2,s}^2\right]}{\left(v_{2,s}^{1/3} - \frac{v_{2,s}}{2}\right)} \tag{4}$$

Where M_n is the molecular weight of the polymer prepared identically in the absence of cross-linking agent; v is the polymer specific volume, and V_1 is the molar volume of water.

Ideal conditions are modified according to the real-time application and speculated to accord the need. The presence of water influences the elastic force, and polymer is subjected to influencing the chemical potential. Considering the volume fraction density of the neutral hydrogel, the molecular weight is calculated by Eq. (5),

$$\frac{1}{\overline{M}_c} = \frac{2}{\overline{M}_n} - \frac{\left(\frac{\overline{v}}{V_1}\right)\left[\ln\left(1 - v_{2,s}\right) + v_{2,s} + \chi_1 v_{2,s}^2\right]}{v_{2,r}\left[\left(\frac{v_{2,s}}{v_{2,r}}\right)^{1/3} - \left(\frac{v_{2,s}}{2v_{2,r}}\right)\right]} \tag{5}$$

Where, $\upsilon_{2,s}$ is the polymer volume fraction in a relaxed state (hydrogel immediately after crosslinking, but before swelling). The above equations are extended to calculate the molecular weight of cationic and anionic hydrogels in Eqs. (6) and (7), respectively.

$$\frac{V_1}{4IM_r}\left(\frac{\upsilon_{2,s}}{\overline{\upsilon}}\right)^2\left(\frac{K_b}{10^{-pH}-K_a}\right)^2 = \left[\ln(1-\upsilon_{2,s})+\upsilon_{2,s}+\chi_1\upsilon_{2,s}^2\right] + \left(\frac{V_1}{\overline{\upsilon}\overline{M}_c}\right)\left(1-\frac{2\overline{M}_c}{\overline{M}_n}\right)\upsilon_{2,r}\left[\left(\frac{\upsilon_{2,s}}{\upsilon_{2,r}}\right)^{1/3}-\left(\frac{\upsilon_{2,s}}{2\upsilon_{2,r}}\right)\right] \tag{6}$$

$$\frac{V_1}{4IM_r}\left(\frac{\upsilon_{2,s}}{\overline{\upsilon}}\right)^2\left(\frac{K_b}{10^{pH-14}-K_a}\right)^2 = \left[\ln(1-\upsilon_{2,s})+\upsilon_{2,s}+\chi_1\upsilon_{2,s}^2\right]+ \left(\frac{V_1}{\upsilon\overline{M}_c}\right)\left(1-\frac{2\overline{M}_c}{\overline{M}_n}\right)\upsilon_{2,r}\left[\left(\frac{\upsilon_{2,s}}{\upsilon_{2,r}}\right)^{1/3}-\left(\frac{\upsilon_{2,s}}{2\upsilon_{2,r}}\right)\right] \tag{7}$$

Where I is the ionic strength; K_a & K_b are the dissociation constants of acid and base, respectively; M_r is the molecular weight of the repeating entity. Hydrogel under stress can go through structural deformation. Rubber—elasticity theory is a means to scrutinize the structure of hydrogel in the presence of a solvent. Remarkably, it is used to analyze polymers formed by physical, chemical, and impermanent crosslinking processes. Equation (8) is used to calculate the stress that hydrogel can sustain theoretically.

$$\tau = \frac{\rho RT}{\overline{M}_C}\left(1-\frac{2\overline{M}_c}{\overline{M}_c}\right)\left(\alpha-\frac{1}{\alpha^2}\right)\left(\frac{\upsilon_{2,s}}{\upsilon_{2,r}}\right)^{1/3} \tag{8}$$

Here, τ is the stress applied to the hydrogel; ρ is the density of the polymer; R is the universal gas constant; T indicates an absolute temperature and $\overline{M}_c$ is the desired molecular weight between the crosslinks. To evaluate this theory, the hydrogel is subjected to a tensile testing system [147].

3.1 Hydrogel-Porosity

One vital structural parameter for hydrogel analysis is the pore or mesh size, which refers to the space present between the polymer chains. Different hydrogel classification with their pore size is given below in Table 5.

This size of the pore is calculated using correlation length (ξ) (Eq. 9),

$$\xi = \alpha\left(r_0^2\right)^{1/2} \tag{9}$$

Table 5 Hydrogel classification based on pore size

Hydrogel	Pore size
Macroporous	mm to cm
Microporous	5 μm- ~1 mm (<5 μm for oral/pulmonary DD)
Nanoporous	10–100 nm

Here, α is the elongation ratio of the polymer chain in any direction; $(r_0^2)^{1/2}$ is the root mean square of the collective, end to end distance of the polymer chains between two neighboring crosslinks. These parameters help much in tailoring the molecular structure of the hydrogel, therefore influencing its mechanical, responsive, and diffusive properties [149]. Thus, the hydrogel is an actively studied material for its application in biomedical science. This chapter focuses on the potential biomedical applications of hydrogel and is described in the next section.

4 Hydrogel Nanocomposite: Biomedical Perspective

Hydrogel formulated using various physical and chemical cross-linking methods has been significantly explored as a biomaterial for drug delivery [152]. Controlled drug delivery system (DDS) is a significant research area for more than a decade, for increasing patient compliance, avoidance of repeated dosage, overcoming peak valley effect of the drug and prolonged drug release at the desired site of action [40, 128]. Targeted, extended drug release by a biocompatible material significantly influences the conventional routes of drug delivery. Hydrogel offers a complete paradigm of satisfying as a biomaterial for drug delivery. This is a 3D polymeric network, can carry water and deliver drug molecules (cargo) into cells. There are high similarities with extracellular matrix aid in being a support material for tissue regeneration and DDS in drug payload [142–30]. Deformable, elastic hydrogel with augmented drug residence time and tissue permeability facilitates bioadhesive drug delivery. Hydrogels with shape conforming features eliminate the need for implantation surgery, making it more efficient. This DDS also poses challenges, such as not being able to carry and release hydrophobic drugs, by the addition of solubilizing moiety, cell-selective therapy by introducing targeted functional group, controlled drug release of low steric interference molecules by employing spacers or modification of polymer backbone. Thus favoring the alternative of polymeric nanoparticles (NPs) for drug encapsulation and release [145–60]. The schematic compilation of nanocomposite hydrogels in the biomedical application is shown below (Fig. 7).

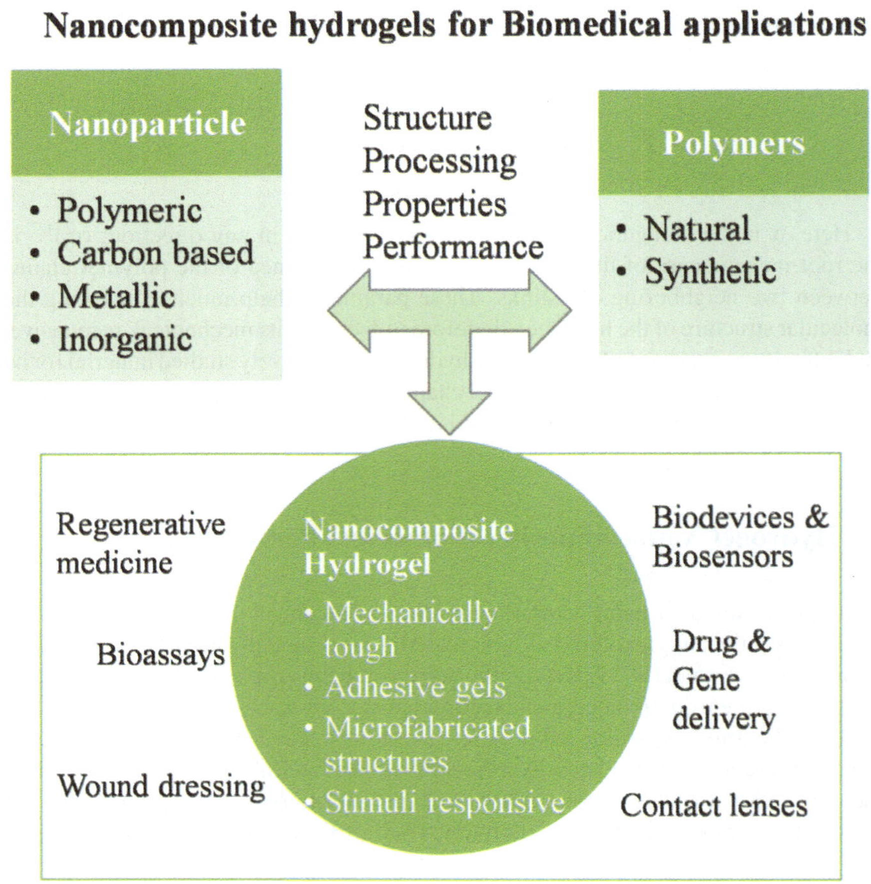

Fig. 7 Nanocomposite hydrogels for biomedical applications

4.1 Microfabrication Techniques for Hydrogel Formulation

In the last two decades, microfabrication approaches using polymeric materials to make hybrid materials are in research focus, to overcome the complexities in tissue engineering applications. Many such advanced techniques have been reported to locate different types of cells within the hydrogel and aid for active tissue formation. Fibrillar structures are necessary for intrinsic influence on cell morphology, differentiation, and polarity. Hydrogel offers fibrils of a wide range of diameters that can be prepared by various methods. Self-organization of the peptides with other hydrophilic and hydrophobic groups [173, 125], electrospinning method can provide fibers of nano to micrometer diameter with direct orientation and spacing [150–155], microfluidic co-flow of polymer and gelation solution in patterned devices are some of the techniques reported. Microfluidic devices are fabricated by well-established

Table 6 Comparative analysis of various microfabrication methods

Technique	Resolution	Description	Demerits
Bioprinting	~ 10 μm	Accurate deposition by the printer.	Expensive; Limited to specific hydrogel
Stereolithography	~ 1 μm	Photopolymerisable hydrogel block formation layer by layer with laser irradiation	Expensive; Limited to photopolymerizable polymers
Laser printing	~ 1 μm	Hydrogel block formation by laser ablation	Expensive; Time consuming; Non-homogeneous hydrogel network formation
Bioprinting	~ 100 nm	Hydrogel mold with master structure from microfabrication techniques	Highly sophisticated; trained personnel; multiple steps

micropatterning methodologies, such as photolithography, soft lithography [153–36], microcontact printing [151], bioprinting [83] and laser printing [57]. Comparative methods of microfabrication are listed below in Table 6 [205].

4.2 *Crosslinking Techniques for Hydrogel Nanocomposite (HNC) Formulation*

Hydrogel nanocomposite (HNC)/nanogels are fabricated by using physical/chemical crosslinking of polymers with different nanoscale features (see Fig. 8). It depends significantly on hydrogel mesh size, size of the nanoparticles (NPs), and nature of the NPs. The size of the NPs is crucial, as particles of diameter less than 10 nm are cleared by extravasation and renal filtration. With a diameter greater than 200 nm, HNCs can be seized by the spleen and eliminated eventually by the phagocytes. The optimal NPs diameter ranges from 70 to 200 nm for prolonged circulation, whereas 10–70 nm NPs can penetrate through tiny capillaries and immune system barriers [159–24]. Polymer crosslink formation can be inhomogeneous, thus necessitating novel NPs dispersion into the matrices. Different categories of nanomaterials and nanoparticles are involved in hydrogel nanocomposite formation are classified in Fig. 8.

4.2.1. Carbon-based nanomaterials have been well employed for various biomedical applications, such as CNTs, graphene/and its oxides, nanodiamonds, diamond-like carbon (DLC) [185] and diamond-like nanocomposite (DLN) [165–167], fullerene (C60) [163, 156], etc. Various studies are being carried out on CNTs due to its high electrical conductivity, mechanical strength, and optical properties [91].

Besides, these high aspect ratio cylindrical hollow tubes of carbon with sp^2 hybridization, Van der Waals force aid in the formation of ions of different molecular weight and charges. Therefore, carbon-based nanomaterials are a superior choice of

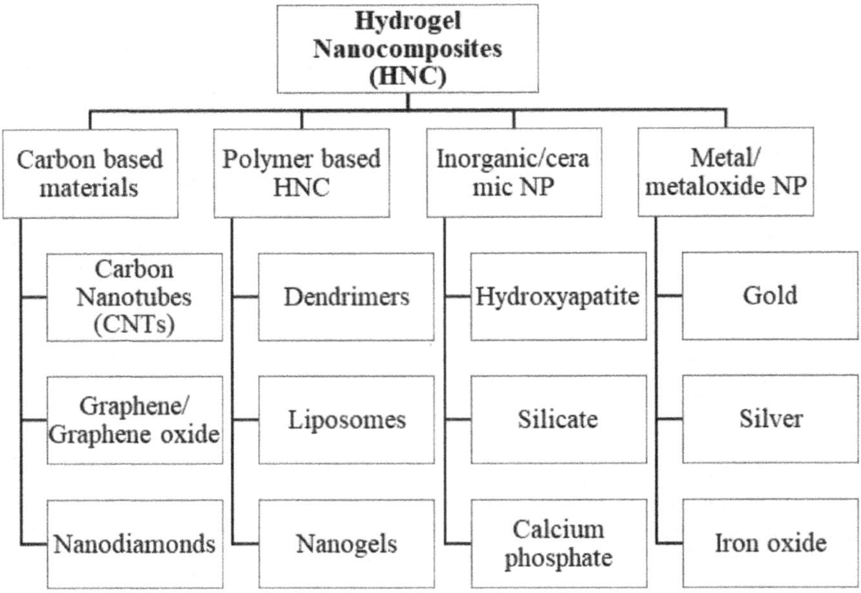

Fig. 8 Schematic classification of hydrogel nanocomposites

material for electrically conductive nervous, cardiac, and muscular tissues [99]. CNTs have limited interaction with the hydrophilic polymer, thus require surface modifications to enhance dispersion with polar groups, such as amines (NH_2), carboxyls (COOH), hydroxyls (OH) or polymeric inclusion. The diverse structures are being proposed with increased functionality for enhancing dispersion in the physiological conditions [173–225]. On the contrary, graphene is treated with strong oxidizers for surface attachment of oxygen, to get Graphene oxide (GO), one of the most explored 2D hydrogel actuators [208, 219]. The GO is less electro-conductive, but more hydrophilic, and making it inexplicable for site-specific genetic material delivery [146, 47]. Although carbon-based hydrogel offers various advantages, replacements to the body tissue require detailed cytotoxicity studies under in vitro and in vivo conditions [48].

4.2.2. Polymeric HNC comprised of monomers of similar or different nanostructured, that has gained enormous attention for its versatility like drug entrapment (hydrophobic/hydrophilic drugs, protein, genetic material, and other bioactive molecules) and stimuli responsiveness upon a change in temperature, light, concentration or pH [89, 174]. Dendrimers are hyperbranched polymers with a highly porous structure, and have multiple peripheral functional groups, that offering high reactivity and drug loading efficiency. The concentration of the dendrimer influence the stiffness of the hydrogel, degradation properties, hydration kinetics [184–224]. Nanocomposites containing dendrimers, demonstrate high stress absorbing capacity, and making it a viable candidate for cartilage tissue engineering. Disruption of cell membrane yields a hydrophilic core and hydrophobic shell yield phospholipid molecules, which offers a great advantage of loading all kinds of bioactive molecules. Drug entrapped liposomes can be conjugated with the degradable polymers by crosslinking to obtain HNCs for targeted drug delivery (see Fig. 9) [188].

Nanogel (nanosized gels) is yet another polymeric nanocomposites, that can trap the intended bioactive molecule into its nanoscale core. This preparation can be carried out by chemical (covalent bonds) or physical (hydrogen bonds) crosslinking methods. Amphiphilic block copolymer, self assembles in water as mono dispersive polymer micelles, with hydrophilic or hydrophobic chains in core or shell [170].

The size-controlled nanogels can be prepared through nano or microemulsion polymerization. This reaction occurs in the presence of surfactants, within the core of water-in-oil (w/o) or oil-in-water (o/w), using micro or nano-emulsions (Fig. 10). The different sizes of monodispersed nanogel fabrication have been reported using atom transfer radical polymerization (ATRP) water-in-oil nano-emulsion. Nanogels are hydrophilic (e.g., doxorubicin) and bioactive macromolecules (e.g., proteins) and potentially applicable for drug delivery purposes [190–171].

4.2.3. Inorganic/organic nanomaterials: Essential body minerals such as nano-hydroxyapatite (nHA), synthetic silicate nanoparticles (nanoclay), bioactive calcium phosphate, silica, glass, glass-ceramic, wollastonite are required for normal body functioning. These inorganic materials are widely used for hydrogel formulation and related bone applications, as they favor osseointegration through its chemical bonds as elaborated in Table 7,

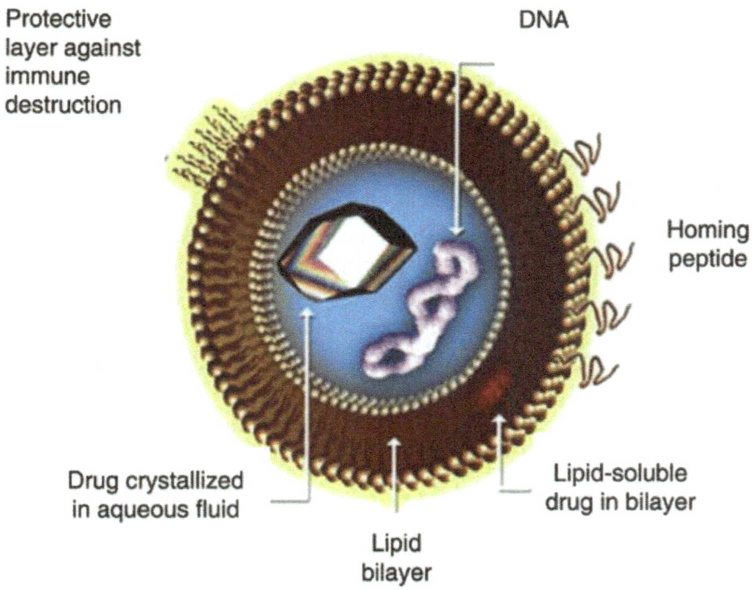

Fig. 9 Structure of Liposome and its drug delivery mechanism (Reprinted with permission from [154])

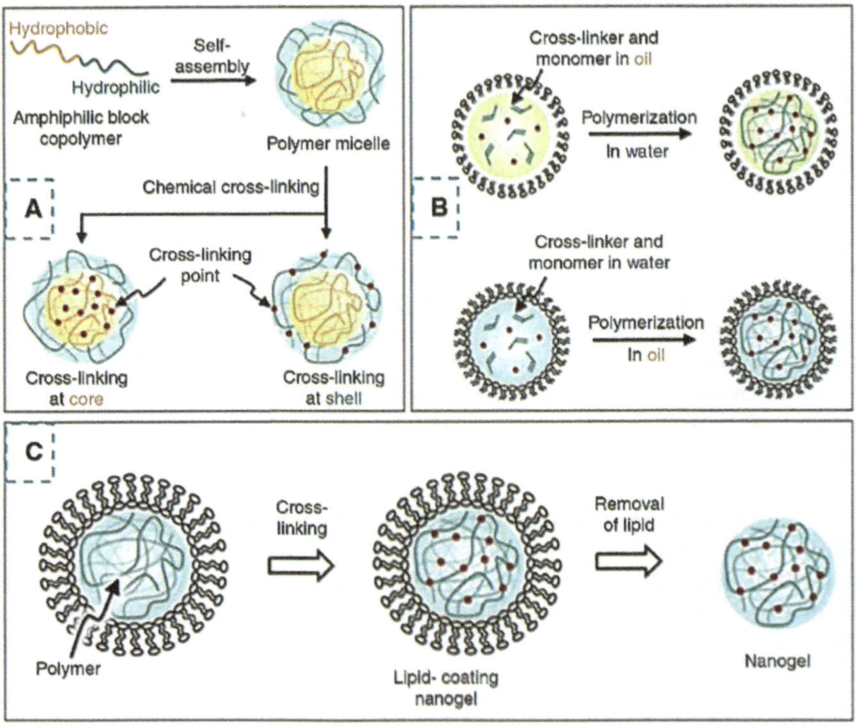

Fig. 10 Chemically prepared nanogel by **A** Crosslinking of amphiphilic block copolymer at core or shell of the polymer micelles in water. **B** Emulsion polymerization with/without emulsion. **C** Using liposome as a template (Reprinted with permission from [154])

Table 7 Interaction of various chemical bonds with body tissues

Chemical bond	Interaction with body tissues
Divalent covalent- ionic bond	PO_4^{3-} attached to the cation or oxygen ion on the ceramic surface and organic matter of the bone
Electrostatic bond	Involving positively charged amine (bone) and negatively charged oxygen (ceramic)
Hydrogen bond	Hydroxylated ceramic surface and the carboxyl group of amino acid
Van der Waals bonding	Negatively charged ceramics interacting with rigid hydrosphere of organic constituents

- Hydroxyapatite, filler for a bone substitution or implant coating, can be modified using silicon or methacrylate hyaluronic acid to lessen crack resistance and fatigue durability of the hydrogel in in vivo conditions. It also improves biodegradation properties, protein adsorption, and adhesion of the coating to the metal substrate [193–70].

- Nanoclays also show improved mechanical, high aspect ratio, and anisotropic morphology with tissue adhesive properties. When mixed with linear and branched polymers, it can form reversible structures, due to non-covalent interactions helping in absorption and desorption [195].

4.2.4. Metallic nanoparticles are classified as 0D hydrogel additives are used to prepare hydrogel nanocomposites [198–143]. In addition to the above, metallic NPs are widely used in creating conductive scaffolds, actuators, sensors, tracking agents in DDS. They can be functionalized to enhance polymer-NPs interactions.

5 Drug Release Mechanisms

Fabrication of drug-loaded polymeric network for the targeted and sustained release can follow any of the above mention methods. The drug release mechanisms largely depend upon the nature of the hydrogel. While in most cases, hydrogel follows diffusion controlled drug release (see Sects. 5.1 and 5.2). Stimuli-responsive smart polymer for drug release has been an incredible challenge. These polymeric systems respond to environmental stimuli (both physical and chemical) are called as environment-sensitive polymers (ESP)/stimuli-responsive polymer/intelligent polymers (refer Sect. 5.3).

5.1 Diffusion Controlled Drug Release

The hydrogel is one of the most explored biomaterials for designing both pulsatile and prolonged drug delivery systems (DDS). Diffusion controlled drug release kinetics is generally calculated as the amount of drug released concerning the given time interval. It is calculated by the 'Ritger- Peppas equation' where the mass fraction of the drug released follows by the power-law relationship Eq. (10)

$$\frac{M_t}{M_\infty} = kt^n \tag{10}$$

Where M_t is the amount of drug released at time 't'; M_∞ is the total amount of drug released; k is the rate constant, and 'n' is the diffusion constant that is between 0.5 and 1.

The following parameters can elucidate comparative drug release studies between two drug-laden hydrogel systems:

1. 'k' the rate constant measured over the given time. It can also indicate a burst release.
2. Half-life drug release, $t_{1/2}$ defined as the 50% mass fraction (M_t/M_∞) that signifies the sustained release in the given system

Different methods are used to control drug delivery in various hydrogel systems. Diffusion controlled hydrogel systems (given by the above equation) typically gives a half-life of 1 day and relatively higher burst release (k > 50%). To increase $t_{1/2}$ ~ 2–3 days, drug release is slow down by several mechanisms, such as the introduction of chemical bonds or polymer network degradation systems.

5.2 *Effect of Hydrogel Mesh Size in Diffusion Mediated Drug Release*

Diffusion is interrelated to the mesh size of the hydrogel, as it largely influences the drug-polymer interactions. When,$r_{mesh}/r_{drug} > 1$, drug release is governed by diffusion. Smaller drug molecules can freely move through the polymeric network, and diffusion is independent of the mesh size. In this case, Stokes-Einstein equation (Eq. 10) is used to measure the diffusivity (D), that is interdependent on the radius of the drug molecule (r_{drug}), usually ranges with its molecular weight and viscosity of the solution (η),

$$D = \frac{RT}{6\pi \eta r_{drug}} \tag{10}$$

Where R is the gas constant and T is the absolute temperature. In addition to this diffusion controlled drug release mechanisms, polymeric swelling, mechanical deformation, and network degradation are also considered, depending upon the nature of the polymer, crosslinking mechanism and intended application.

5.3 *Stimuli-Responsive Drug Release Mechanism*

HNCs, which are designed to respond to stimuli such as pH/temperature/concentration/light and release the drug act as controlled drug release centers. It is, therefore, necessary to understand the underlying principles of the same in drug delivery. Tabulated below in Table 8 are some of the physical/chemical/biological stimuli for ESP response [202–21].

Table 8 Different stimuli for ESP response

Physical stimuli	Chemical stimuli	Biological stimuli
• Light • Temperature • Ultrasound • Mechanical forces	• pH • Ionic strength • Solvent	• Enzyme • Glucose Concentration

One crucial feature of HNCs is its ability to surpass the immune system due to its nanoscale structures and, in some cases, of the blood-brain barrier itself. It is conceptualized by the Enhanced Permeability and Retention (EPR) mechanism using nanocarriers [2].

5.3.1 pH-Responsive Hydrogel for Drug Delivery

pH variation is observed in body tissues and cellular compartments. For instance, blood has pH in the range of 7.35–7.45; stomach 1.0–3.0; and duodenum 4.8–8.2 [6]. This difference in pH is exploited in formulating localized and sustained drug release in the intended areas in response to the specific substrates. The pH-sensitive hydrogel comprises of ionizable weak acids or a base is an attribute of the pendant groups of the polymeric unit. Hydrogel with numerous acidic groups is referred to as polyanions or polyacids, while the polymers with basic units are polybases or polycations [97]. Polycations, at basic pH, deprotonate and in the acidic environment, becomes positively charged. Combinational usages of nanocarriers that target and accumulate at the intended site and act as a localized drug storehouse embedded with drug are employed to design pH-sensitive hydrogel. It is the charge density of the acidic or basic polymer backbone of the polymer that induces pH-dependent swelling and deswelling of the hydrogel. Nanoparticles of poly β-amino ester (PbAE) were prepared to increase of biocompatibility and pH sensitivity, for paclitaxel drug delivery [180]. Another pH and temperature-sensitive HNC designed for treating breast cancer using poly ε-caprolactone (PCL) for tamoxifen drug delivery. It is channeled towards the estrogen receptor (ER) for targeted drug delivery [27]. The degree of a pH-responsive polymer depends upon the level of ionization, protonation, and deprotonation caused in response to its surroundings, as shown in Fig. 11 [129].

After completion of drug delivery, the degradation of the hydrogel is a mandatory requirement for clearance from the host system. It is achieved by cleaving the polymer backbone using hydrolysis or enzyme action [92]. Many drug delivery applications that involve drug entrapped pH-sensitive nanogels as a carrier have been reported, namely as anti-cancer doxorubicin, antibacterial tetracycline [218]. PLGA-Chitosan-based nanogel coated with eucalyptus oil is encapsulated with 5-fluorouracil (5-FU) by the 'solvent evaporation emulsification' process for prolonged drug release for skin cancer therapy [161].

5.3.2 Temperature Responsive Hydrogel for Drug Delivery

Polymers that exhibit sol-gel transition in response to the temperature alteration are one of the most widely studied physical stimuli responsive ESPs. Homogenous solutions of polymers that show temperature sensitivity at critical solution temperature (CST) are separated into polymer-rich and polymer-lean phase. When the polymer solution reaches CST, an alteration occurs between hydrophilic and hydrophobic chains of the polymer in an aqueous solvent. Phase diagram of the polymer/solvent

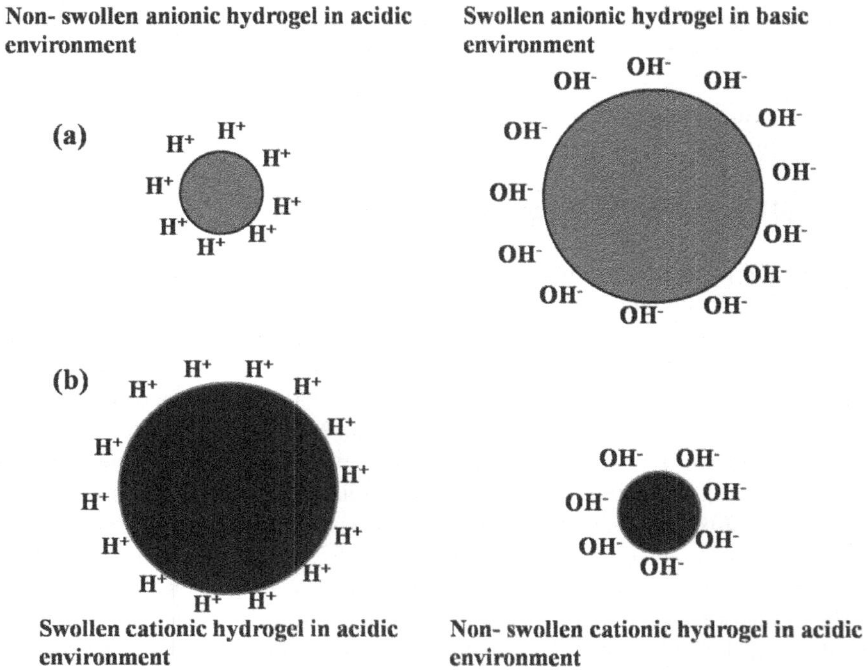

Fig. 11 pH-responsive swelling of **a** anionic. **b** cationic hydrogel (Redrawn from [6])

mixture against temperature, LCST, and UCST can be located. LCST is the lowest temperature value, at which the polymer remains soluble in the solvent and becomes biphasic on increasing the temperature. On the other hand, polymers that form gel below the CST and monophasic solution above CST are categorized as UCST polymer systems [213–108]. (Refer Sect. 2.2.1.)

Thermogelling polymer solutions find inexplicable applications for designing injectable biopolymers [112, 133]. In situ gelling of the physically cross-linked injectable polymer solution can deliver biomolecules such as transforming growth factor (TGF) - β1) [229], nanoparticles (super paramagnetic iron oxide nanoparticles (SPIONs)) [32], drugs (5-FU [4], daptomycin [23]) to the physiological systems are reported in the literature.

5.3.3 Enzyme Mediated Hydrogel Drug Release

Enzymes act as biocatalyst, that helps in controlled and sustained release of the drug payload from hydrogel [213]. One significant advantage of enzymes as stimuli is that it is highly selective and tightly regulated. Phase transition of enzyme sensitive hydrogel occurs by,

i. Ease of access and recognizable enzyme and substrate [26].

ii. Functional unit to regulate molecular interactions (Vander Waals forces, hydrogen bonding, hydrophobic interactions, electrostatic interaction, π-π interactions) that contribute to transitions in hydrogels.

Potential applications include PEG-based hydrogel microparticles for pulmonary drug delivery, and peptide delivery of PEG incorporated Gly-Leu-Lys (GLK) by overexpression of metalloproteinases [177].

One another major biological stimuli-responsive hydrogel is 'glucose responsive hydrogel' [101]. Insulin-dependent diabetes mellitus (IDDM), requires drug release in response to the changes in physiological blood glucose concentration [148]. Hydrogel designed to facilitate glucose-responsive mostly combines the use of saccharide-lectin complex as a crosslinker, incorporation of glucose binding site into a thermoresponsive polymer, utilizing enzymatic activity by the conjugating enzyme to pH-responsive polymer [175].

6 Current Trends and Future Prospects

Hydrogel though widely studied, has less translational efficiency. The reasons attributing to its limitation include reduced control over polymerization rate, weak mechanical property, stability, and inertness in the in vivo physiological conditions. Sterilization of mass production of the hydrogel, to withstand severe temperature cycles without losing its intended property is a significant challenge. Any biomaterial research is carried out to reach for clinical translation ultimately. Multiple problems arise for hydrogel formulation, storage, regulatory complexity, and cost are yet to be needed to overrun by researchers, to obtain robust data before clinical trials. As hydration is a typical mechanism involved, thus terminal sterilization is complicated to make it step-by-step validation from raw material to complete fabrication. Degradation or deterioration of the drug-polymer complex can occur during premature hydrolysis, calling out for optimum storage conditions. Regulatory approvals of the hydrogel drug systems also present a significant concern.

Stimuli-responsive polymers will have to stride through upfront challenges, such as availability of limited functional monomers, low binding capacity, template leakage, and small target intended delivery due to reduced recognition of particular site and cell surface receptors [153].

Current HNC development involves optimization of stimuli responsiveness for the improvement of mechanical strength and biodegradation. It mainly depends upon the embedded NPs in the polymeric networks for pH/enzyme/ion/temperature/magnetic/electric field responsiveness. HNCs also have embarked on its frontier in mimicking native tissues by multiphase combination. Concerning nanoparticle synthesis, biogenic sources such as plants, bacteria, yeast, fungi, and physical/chemical methods would be favorable. The organic, eco-friendly manner of NPs synthesis is yet to be achieved [154].

The functionalization of NPs-Polymer by covalent and non-covalent interactions is a crucial step to achieve HNC. The former involves unstable chemical bonds that can cause the loss of NP properties, while the latter is limited due to its weak interactions. Thus calling for alternative surface modifications that involve non-destructive approaches with more bond stability [113].

7 Conclusions

Despite several decades of research and development towards ideal hydrogel formation, aided by advances in material science, fabrication techniques, polymer chemistry, understanding of molecular and cell biology, and tissue engineering, till now, unmet challenges and need for sufficient clinical trials remain. The term 'ideal hydrogel' is expected to have an on-demand drug release with a predictable rate throughout its clearance from the system. The current research involves proof-of-concept studies for tissue repair and regeneration, which include multiple signaling molecules release and response. Advancement in hydrogel formulation is expected to withstand chemical and enzymatic reactions.

Theoretical modeling of the drug release profile remains challenging as drug release mechanisms can be combinatorial or vary in different hydrogel systems. Drug release systems with an integrated understanding of drug release and transport through a local tissue will be facilitating better hydrogel formulation.

Advancement of bioelectronics and biosensors research has widened prospects towards next-generation controlled drug delivery systems. Tailorable physical and chemical properties of hydrogel have extended its applications towards remote-controlled drug delivery purposes. Finally, the progress in the formulation, addressing the challenges, and clinical translation setbacks can make hydrogel an inevitable component for drug delivery systems.

Acknowledgements The authors greatly appreciate the financial support from the DBT/Wellcome Trust India Alliance Fellowship under grant number IA/E/16/1/503062. We also acknowledge all authors and publishers, who provided copyright permissions.

References

1. Abd Alla SG, Sen M, El-Naggar AWM (2012) Swelling and mechanical properties of superabsorbent hydrogels based on Tara gum/acrylic acid synthesized by gamma radiation. Carbohydr Polym. https://doi.org/10.1016/j.carbpol.2012.03.031
2. Aguilar M, Elvira C, Gallardo A, Vázquez B, Román J (2007) Smart polymers and their applications as biomaterials. Top Tissue Eng https://doi.org/10.1533/9780857097026.1.45
3. Akhtar MF, Hanif M, Ranjha NM (2016) Methods of synthesis of hydrogels … a review. Saudi Pharm J

4. Andrei M, Turturica G, Stanescu PO, Teodorescu M (2016) Thermosensitive injectable hydrogels from poly(N-isopropylacrylamide)–dextran aqueous solutions: thermogelation and drug release properties. Soft Mater. https://doi.org/10.1080/1539445X.2016.1172317
5. Annabi N, Nichol JW, Zhong X, Ji C, Koshy S, Khademhosseini A, Dehghani F (2010) Controlling the porosity and microarchitecture of hydrogels for tissue engineering. Tissue Eng Part B Rev
6. Bawa P, Pillay V, Choonara YE, Du Toit LC (2009) Stimuli-responsive polymers and their applications in drug delivery. Biomed Mater
7. Bawn CSH (1987) Encyclopedia of polymer science and engineering. Polymer (Guildf). https://doi.org/10.1016/0032-3861(87)90274-6
8. Berger J, Reist M, Mayer JM, Felt O, Peppas NA, Gurny R (2004) Structure and interactions in covalently and ionically crosslinked chitosan hydrogels for biomedical applications. Eur J Pharm Biopharm 57:19–34
9. Bhattarai N, Gunn J, Zhang M (2010) Chitosan-based hydrogels for controlled, localized drug delivery. Adv Drug Deliv Rev
10. Bjork JW, Johnson SL, Tranquillo RT (2011) Ruthenium-catalyzed photo cross-linking of fibrin-based engineered tissue. Biomaterials 32:2479–2488. https://doi.org/10.1016/j.biomaterials.2010.12.010
11. Bourke SL, Al-Khalili M, Briggs T, Michniak BB, Kohn J, Poole-Warren LA (2003) A photo-crosslinked poly(vinyl alcohol) hydrogel growth factor release vehicle for wound healing applications. AAPS PharmSci 5 https://doi.org/10.1208/ps050433
12. Boustta M, Colombo PE, Lenglet S, Poujol S, Vert M (2014) Versatile UCST-based thermoresponsive hydrogels for loco-regional sustained drug delivery. J Control Release 174:1–6. https://doi.org/10.1016/j.jconrel.2013.10.040
13. Brassinne J, Jochum FD, Fustin CA, Gohy JF (2015) Revealing the supramolecular nature of side-chain terpyridine-functionalized polymer networks. Int J Mol Sci 16:990–1007. https://doi.org/10.3390/ijms16010990
14. Brigger I, Dubernet C, Couvreur P (2002) Nanoparticles in cancer therapy and diagnosis. Adv Drug Deliv Rev
15. Bryant SJ, Nuttelman CR, Anseth KS (2000) Cytocompatibility of UV and visible light photoinitiating systems on cultured NIH/3T3 fibroblasts in vitro. J Biomater Sci Polym Ed 11:439–457. https://doi.org/10.1163/156856200743805
16. Bryant SJ, Nuttelman CR, Anseth KS (2012) Cytocompatibility of UV and visible light photoinitiating systems on cultured NIH/3T3 fibroblasts in vitro. http://dx.doi.org/101163/156856200743805. https://doi.org/10.1163/156856200743805
17. Burdick JA, Prestwich GD (2011) Hyaluronic acid hydrogels for biomedical applications. Adv Mater. https://doi.org/10.1002/adma.201003963
18. Burkert S, Schmidt T, Gohs U, Dorschner H, Arndt KF (2007) Cross-linking of poly(N-vinyl pyrrolidone) films by electron beam irradiation. Radiat Phys Chem. https://doi.org/10.1016/j.radphyschem.2007.02.024
19. Buwalda SJ, Boere KWM, Dijkstra PJ, Feijen J, Vermonden T, Hennink WE (2014) Hydrogels in a historical perspective: from simple networks to smart materials. J Control Release 190:254–273. https://doi.org/10.1016/j.jconrel.2014.03.052
20. Byrne ME, Park K, Peppas NA (2002) Molecular imprinting within hydrogels. Adv Drug Deliv Rev. https://doi.org/10.1016/S0169-409X(01)00246-0
21. Cabane E, Zhang X, Langowska K, Palivan CG, Meier W (2012) Stimuli-responsive polymers and their applications in nanomedicine. Biointerphases
22. Carré MC, Delestre C, Hubert P, Dellacherie E (1991) Covalent coupling of a short polyether on sodium alginate: synthesis and characterization of the resulting amphiphilic derivative. Carbohydr Polym 16:367–379. https://doi.org/10.1016/0144-8617(91)90055-H
23. Casadidio C, Butini ME, Trampuz A, Di Luca M, Censi R, Di Martino P (2018) Daptomycin-loaded biodegradable thermosensitive hydrogels enhance drug stability and foster bactericidal activity against Staphylococcus aureus. Eur J Pharm Biopharm. https://doi.org/10.1016/j.ejpb.2018.07.001

24. Centre AE, Science H (1996) Polymers in drug delivery Stanley S Davis * t +, Lisbeth lllum "! and Snjezana Stolnik *. Curr Opin Colloid Interface Sci 1:660–666. https://doi.org/10.1016/S1359-0294(96)80105-1
25. Chai Q, Jiao Y, Yu X (2017) Hydrogels for biomedical applications: their characteristics and the mechanisms behind them. Gels. https://doi.org/10.3390/gels3010006
26. Chandrawati R (2016) Enzyme-responsive polymer hydrogels for therapeutic delivery. Exp Biol Med. https://doi.org/10.1177/1535370216647186
27. Chawla JS, Amiji MM (2002) Biodegradable poly(ε-caprolactone) nanoparticles for tumor-targeted delivery of tamoxifen. Int J Pharm. https://doi.org/10.1016/S0378-5173(02)00483-0
28. Chembath M, Balaraju JN, Sujata M (2015) Surface characteristics, corrosion and bioactivity of chemically treated biomedical grade NiTi alloy. Mater Sci Eng, C. https://doi.org/10.1016/j.msec.2015.06.051
29. Chujo Y, Sada K, Saegusa T (1993) Synthesis of bipyridyl-branched polyoxazoline and its gelation by means of metal coordination. Polym J 25:599–608. https://doi.org/10.1295/polymj.25.599
30. Ciuffreda MC, Malpasso G, Chokoza C, Bezuidenhout D, Goetsch KP, Mura M, Pisano F, Davies NH, Gnecchi M (2018) Synthetic extracellular matrix mimic hydrogel improves efficacy of mesenchymal stromal cell therapy for ischemic cardiomyopathy. Acta Biomater. https://doi.org/10.1016/j.actbio.2018.01.005
31. Collins J, Xiao Z, Müllner M, Connal LA (2016) The emergence of oxime click chemistry and its utility in polymer science. Polym Chem. 7:3812–3826
32. Crippa F, Moore TL, Mortato M, Geers C, Haeni L, Hirt AM, Rothen-Rutishauser B, Petri-Fink A (2017) Dynamic and biocompatible thermo-responsive magnetic hydrogels that respond to an alternating magnetic field. J Magn Magn Mater. https://doi.org/10.1016/j.jmmm.2016.11.023
33. Croisier F, Jérôme C (2013) Chitosan-based biomaterials for tissue engineering. Eur Polym J
34. Dahiya P, Kamal R (2013) Hyaluronic acid: a boon in periodontal therapy. N Am J Med Sci
35. Das N (2013) Preparation methods and properties of hydrogel: a review. Int J Pharm Pharm Sci
36. Delamarche E, Bernard A, Schmid H, Bietsch A, Michel B, Biebuyck H (1998) Microfluidic networks for chemical patterning of substrates: design and application to bioassays. J Am Chem Soc. https://doi.org/10.1021/ja973071f
37. Deligkaris K, Tadele TS, Olthuis W, van den Berg A (2010) Hydrogel-based devices for biomedical applications. Sensors Actuators, B Chem
38. Ding F, Wu S, Wang S, Xiong Y, Li Y, Li B, Deng H, Du Y, Xiao L, Shi X (2015) A dynamic and self-crosslinked polysaccharide hydrogel with autonomous self-healing ability. Soft Matter 11:3971–3976. https://doi.org/10.1039/c5sm00587f
39. Dong LC, Hoffman AS (1986) Thermally reversible hydrogels: III. Immobilization of enzymes for feedback reaction control. J Control Release 4:223–227. https://doi.org/10.1016/0168-3659(86)90006-4
40. Euliss LE, DuPont JA, Gratton S, DeSimone J (2006) Imparting size, shape, and composition control of materials for nanomedicine. Chem Soc Rev 35:1095–1104
41. Evanoff DD, Chumanov G (2005) Synthesis and optical properties of silver nanoparticles and arrays. ChemPhysChem
42. Farooq MA, Aquib M, Farooq A, Haleem Khan D, Joelle Maviah MB, Sied Filli M, Kesse S, Boakye-Yiadom KO, Mavlyanova R, Parveen A, Wang B (2019) Recent progress in nanotechnology-based novel drug delivery systems in designing of cisplatin for cancer therapy: an overview. Artif Cells, Nanomedicine Biotechnol 47:1674–1692. https://doi.org/10.1080/21691401.2019.1604535
43. Feng J, Su W, Wang HF, Huang FW, Zhang XZ, Zhuo RX (2009) Facile fabrication of diblock methoxy poly(ethylene glycol)-poly (tetramethylene carbonate) and its self-assembled micelles as drug carriers. ACS Appl Mater Interfaces 1:2729–2737. https://doi.org/10.1021/am900452c

44. Fernandes-Cunha GM, Lee HJ, Kumar A, Kreymerman A, Heilshorn S, Myung D (2017) Immobilization of growth factors to collagen surfaces using pulsed visible light. Biomacromol 18:3185–3196. https://doi.org/10.1021/acs.biomac.7b00838
45. Fu W, Zhao B (2016) Thermoreversible physically crosslinked hydrogels from UCST-type thermosensitive ABA linear triblock copolymers. Polym Chem 7:6980–6991. https://doi.org/10.1039/c6py01517d
46. Fustin C-A, Guillet P, Schubert US, Gohy J-F (2007) Metallo-supramolecular block copolymers. Adv Mater 19:1665–1673. https://doi.org/10.1002/adma.200602170
47. Gaharwar AK, Patel A, Dolatshahi-Pirouz A, Zhang H, Rangarajan K, Iviglia G, Shin SR, Hussain MA, Khademhosseini A (2015) Elastomeric nanocomposite scaffolds made from poly(glycerol sebacate) chemically crosslinked with carbon nanotubes. Biomater Sci 3:46–58. https://doi.org/10.1039/c4bm00222a
48. Gaharwar AK, Peppas NA, Khademhosseini A (2014) Nanocomposite hydrogels for biomedical applications. Biotechnol Bioeng 111:441–453. https://doi.org/10.1002/bit.25160
49. Gandhi A, Paul A, Sen SO, Sen KK (2015) Studies on thermoresponsive polymers: phase behaviour, drug delivery and biomedical applications. Asian J Pharm Sci
50. Gandini A (2013) The furan/maleimide Diels-Alder reaction: a versatile click-unclick tool in macromolecular synthesis. Prog Polym Sci 38:1–29
51. Garnica-Palafox IM, Sánchez-Arévalo FM (2016) Influence of natural and synthetic crosslinking reagents on the structural and mechanical properties of chitosan-based hybrid hydrogels. Carbohydr Polym. https://doi.org/10.1016/j.carbpol.2016.06.036
52. Gasperini L, Mano JF, Reis RL (2014) Natural polymers for the microencapsulation of cells. J R Soc Interface
53. Geraldo S, Simon A, Elkhatib N, Louvard D, Fetler L, Vignjevic DM (2012) Do cancer cells have distinct adhesions in 3D collagen matrices and in vivo? Eur J Cell Biol. https://doi.org/10.1016/j.ejcb.2012.07.005
54. Gong C, Qian Z, Liu C, Huang M, Gu Y, Wen Y, Kan B, Wang K, Dai M, Li X, Gou M, Tu M, Wei Y (2007) A thermosensitive hydrogel based on biodegradable amphiphilic poly(ethylene glycol)-polycaprolactone-poly(ethylene glycol) block copolymers. Smart Mater Struct 16:927–933. https://doi.org/10.1088/0964-1726/16/3/043
55. Greco F, Mattoli V (2012) Introduction to active smart materials for biomedical applications
56. Gualdesi MS, Igarzabal CIA, Vara J, Ortiz CS (2016) Synthesis and physicochemical properties of polyacrylamide nanoparticles as photosensitizer carriers. Int J Pharm. https://doi.org/10.1016/j.ijpharm.2016.08.051
57. Guillotin B, Souquet A, Catros S, Duocastella M, Pippenger B, Bellance S, Bareille R, Rémy M, Bordenave L, Amédée J, Guillemot F (2010) Laser assisted bioprinting of engineered tissue with high cell density and microscale organization. Biomaterials. https://doi.org/10.1016/j.biomaterials.2010.05.055
58. Gulrez SKH, Al-Assaf SOG (2011) Hydrogels: methods of preparation, characterisation and applications. In: Progress in molecular and environmental bioengineering—from analysis and modeling to technology applications
59. Guo M, Pitet LM, Wyss HM, Vos M, Dankers PYW, Meijer EW (2014) Tough stimuli-responsive supramolecular hydrogels with hydrogen-bonding network junctions. J Am Chem Soc 136:6969–6977. https://doi.org/10.1021/ja500205v
60. Gurikov P, Smirnova I (2018) Non-conventional methods for gelation of alginate. Gels. https://doi.org/10.3390/gels4010014
61. Guvendiren M, Burdick JA (2010) The control of stem cell morphology and differentiation by hydrogel surface wrinkles. Biomaterials 31:6511–6518. https://doi.org/10.1016/j.biomaterials.2010.05.037
62. Gyles DA, Castro LD, Silva JOC, Ribeiro-Costa RM (2017) A review of the designs and prominent biomedical advances of natural and synthetic hydrogel formulations. Eur Polym J 88:373–392. https://doi.org/10.1016/j.eurpolymj.2017.01.027
63. Hamman JH (2010) Chitosan based polyelectrolyte complexes as potential carrier materials in drug delivery systems. Mar Drugs 8:1305–1322. https://doi.org/10.3390/md8041305

64. Haque MA, Kurokawa T, Gong JP (2012) Super tough double network hydrogels and their application as biomaterials. Polymer (Guildf)
65. Harrington MJ, Masic A, Holten-Andersen N, Waite JH, Fratzl P (2010) Iron-clad fibers: a metal-based biological strategy for hard flexible coatings. Science 328:216–220. https://doi.org/10.1126/science.1181044
66. Hassan CM, Peppas NA (2000) Structure and morphology of freeze/thawed PVA hydrogels. Macromolecules 33:2472–2479. https://doi.org/10.1021/ma9907587
67. Hauert R (2003) A review of modified DLC coatings for biological applications. Diam Relat Mater. https://doi.org/10.1016/S0925-9635(03)00081-5
68. Hawley AE, Davis SS, Illum L (1995) Targeting of colloids to lymph nodes: influence of lymphatic physiology and colloidal characteristics. Adv Drug Deliv Rev 17:129–148
69. He H, Guan J, Lee JL (2006) An oral delivery device based on self-folding hydrogels. J Control Release. https://doi.org/10.1016/j.jconrel.2005.10.017
70. Hench LL, Splinter RJ, Allen WC, Greenlee TK (1971) Bonding mechanisms at the interface of ceramic prosthetic materials. J Biomed Mater Res 5:117–141. https://doi.org/10.1002/jbm.820050611
71. Hennink WE, van Nostrum CF (2012) Novel crosslinking methods to design hydrogels. Adv Drug Deliv Rev 64:223–236
72. Hiemstra C, Zhou W, Zhong Z, Wouters M, Feijen J (2007) Rapidly in situ forming biodegradable robust hydrogels by combining stereocomplexation and photopolymerization. J Am Chem Soc 129:9918–9926. https://doi.org/10.1021/ja072113p
73. Hoare TR, Kohane DS (2008) Hydrogels in drug delivery: Progress and challenges, Polymer 49:8 1993–2007 https://doi.org/10.1016/j.polymer.2008.01.027
74. Hoeijmakers JHJ (2009) DNA damage, aging, and cancer. N Engl J Med 361:1475–1485
75. Hoffman AS (1987) Applications of thermally reversible polymers and hydrogels in therapeutics and diagnostics. J Control Release 6:297–305. https://doi.org/10.1016/0168-3659(87)90083-6
76. Holten-Andersen N, Harrington MJ, Birkedal H, Lee BP, Messersmith PB, Lee KYC, Waite JH (2011) pH-induced metal-ligand cross-links inspired by mussel yield self-healing polymer networks with near-covalent elastic moduli. Proc Natl Acad Sci 108:2651–2655. https://doi.org/10.1073/PNAS.1015862108
77. Hong SH, Kim S, Park JP, Shin M, Kim K, Ryu JH, Lee H (2018) Dynamic bonds between boronic acid and alginate: hydrogels with stretchable, self-healing, stimuli-responsive, remoldable, and adhesive properties. Biomacromol 19:2053–2061. https://doi.org/10.1021/acs.biomac.8b00144
78. Hu J, Hou Y, Park H, Choi B, Hou S, Chung A, Lee M (2012) Visible light crosslinkable chitosan hydrogels for tissue engineering. Acta Biomater 8:1730–1738. https://doi.org/10.1016/j.actbio.2012.01.029
79. Hu X, Lu Q, Sun L, Cebe P, Wang X, Zhang X, Kaplan DL (2010) Biomaterials from ultrasonication-induced silk fibroin-hyaluronic acid hydrogels. Biomacromol 11:3178–3188. https://doi.org/10.1021/bm1010504
80. Hu W, Wang Z, Xiao Y, Zhang S, Wang J (2019) Advances in crosslinking strategies of biomedical hydrogels. Biomater Sci 7:843–855
81. Huh KM, Ooya T, Lee WK, Sasaki S, Kwon IC, Jeong SY, Yui N (2001) Supramolecular-structured hydrogels showing a reversible phase transition by inclusion complexation between poly(ethylene glycol) grafted dextran and α-cyclodextrin. Macromolecules 34:8657–8662. https://doi.org/10.1021/ma0106649
82. Iizawa T, Taketa H, Maruta M, Ishido T, Gotoh T, Sakohara S (2007) Synthesis of porous poly(N-isopropylacrylamide) gel beads by sedimentation polymerization and their morphology. J Appl Polym Sci. https://doi.org/10.1002/app.25605
83. Ilkhanizadeh S, Teixeira AI, Hermanson O (2007) Inkjet printing of macromolecules on hydrogels to steer neural stem cell differentiation. Biomaterials. https://doi.org/10.1016/j.biomaterials.2007.05.018

84. Ishida O, Maruyama K, Sasaki K, Iwatsuru M (1999) Size-dependent extravasation and interstitial localization of polyethyleneglycol liposomes in solid tumor-bearing mice. Int J Pharm 190:49–56. https://doi.org/10.1016/S0378-5173(99)00256-2
85. Jeon O, Powell C, Solorio LD, Krebs MD, Alsberg E (2011) Affinity-based growth factor delivery using biodegradable, photocrosslinked heparin-alginate hydrogels. J Control Release 154:258–266. https://doi.org/10.1016/j.jconrel.2011.06.027
86. Jin R, Hiemstra C, Zhong Z, Feijen J (2007) Enzyme-mediated fast in situ formation of hydrogels from dextran-tyramine conjugates. Biomaterials 28:2791–2800. https://doi.org/10.1016/j.biomaterials.2007.02.032
87. Jin R, Moreira Teixeira LS, Dijkstra PJ, van Blitterswijk CA, Karperien M, Feijen J (2010) Enzymatically-crosslinked injectable hydrogels based on biomimetic dextran-hyaluronic acid conjugates for cartilage tissue engineering. Biomaterials 31:3103–3113. https://doi.org/10.1016/j.biomaterials.2010.01.013
88. Jin R, Moreira Teixeira LS, Dijkstra PJ, Karperien M, van Blitterswijk CA, Zhong ZY, Feijen J (2009) Injectable chitosan-based hydrogels for cartilage tissue engineering. Biomaterials 30:2544–2551. https://doi.org/10.1016/j.biomaterials.2009.01.020
89. Joshi N, Grinstaff M (2008) Applications of dendrimers in tissue engineering. Curr Top Med Chem 8:1225–1236. https://doi.org/10.2174/156802608785849067
90. Jung YP, Kim JH, Lee DS, Kim YH (2007) Preparation and properties of modified PHEMA hydrogel with sulfonated PEG graft. J Appl Polym Sci. https://doi.org/10.1002/app.25500
91. Kar S, Sood AK (2019) Ultrafast terahertz photoresponse of single and double-walled carbon nanotubes: optical pump-terahertz probe spectroscopy. Carbon N Y. https://doi.org/10.1016/j.carbon.2018.12.081
92. Khang G (2017) Handbook of intelligent scaffolds for tissue engineering and regenerative medicine, 2nd edn
93. Kim SY, Ha JC, Lee YM (2000) Poly(ethylene oxide)-poly(propylene oxide)-poly(ethylene oxide)/poly(ε- caprolactone) (PCL) amphiphilic block copolymeric nanospheres: II. Thermo-responsive drug release behaviors. J Control Release 65:345–358. https://doi.org/10.1016/S0168-3659(99)00207-2
94. Kim MH, Park H, Park WH (2018) Effect of pH and precursor salts on in situ formation of calcium phosphate nanoparticles in methylcellulose hydrogel. Carbohydr Polym 191:176–182. https://doi.org/10.1016/j.carbpol.2018.03.032
95. Kim JK, Kim HJ, Chung JY, Lee JH, Young SB, Kim YH (2014) Natural and synthetic biomaterials for controlled drug delivery. Arch Pharm Res
96. Kingsley JD, Dou H, Morehead J, Rabinow B, Gendelman HE, Destache CJ (2006) Nanotechnology: a focus on nanoparticles as a drug delivery system. J Neuroimmune Pharmacol
97. Kocak G, Tuncer C, Bütün V (2017) PH-responsive polymers. Polym Chem
98. Kong G, Braun RD, Dewhirst MW (2000) Hyperthermia enables tumor-specific nanoparticle delivery: effect of particle size. Cancer Res 60:4440–4445
99. Kumar S, Rani R, Dilbaghi N, Tankeshwar K, Kim KH (2017) Carbon nanotubes: a novel material for multifaceted applications in human healthcare. Chem Soc Rev 46:158–196
100. Kumar A, Srivastava A, Galaev IY, Mattiasson B (2007) Smart polymers: Physical forms and bioengineering applications. Prog Polym Sci
101. Kumarasamy D, Ghosh MK, Giri TK (2018) Polymer-based responsive hydrogel for drug delivery. Springer, Singapore, pp 1–25
102. Kuo CK, Ma PX (2001) Ionically crosslinked alginate hydrogels as scaffolds for tissue engineering: part 1. Structure, gelation rate and mechanical properties. Biomaterials 22:511–521
103. Kurisawa M, Chung JE, Yang YY, Gao SJ, Uyama H (2005) Injectable biodegradable hydrogels composed of hyaluronic acid-tyramine conjugates for drug delivery and tissue engineering. Chem Commun 4312–4314. https://doi.org/10.1039/b506989k
104. Langer R (2000) Biomaterials in drug delivery and tissue engineering: one laboratory's experience. Acc Chem Res 33:94–101. https://doi.org/10.1021/ar9800993

105. Langer R, Peppas NA (2003) Advances in biomaterials, drug delivery, and bionanotechnology. AIChE J 49:2990–3006. https://doi.org/10.1002/aic.690491202
106. Lee BP, Dalsin JL, Messersmith PB (2002) Synthesis and gelation of DOPA-modified poly(ethylene glycol) hydrogels. Biomacromol 3:1038–1047. https://doi.org/10.1021/bm025546n
107. Lee KY, Mooney DJ (2012) Alginate: properties and biomedical applications. Prog Polym Sci 37:106–126
108. Li S, Feng L, Lu H, Feng S (2017) From LCST to UCST: the phase separation behaviour of thermo-responsive polysiloxanes with the solubility parameters of solvents. New J Chem. https://doi.org/10.1039/c6nj03386e
109. Li J, Harada A, Kamachi M (1994) Sol-gel transition during inclusion complex formation between α-cyclodextrin and high molecular weight poly(ethylene glycol)s in aqueous solution. Polym J 26:1019–1026. https://doi.org/10.1295/polymj.26.1019
110. Li J, Li X, Ni X, Wang X, Li H, Leong KW (2006) Self-assembled supramolecular hydrogels formed by biodegradable PEO-PHB-PEO triblock copolymers and α-cyclodextrin for controlled drug delivery. Biomaterials 27:4132–4140. https://doi.org/10.1016/j.biomaterials.2006.03.025
111. Li H, Yang P, Pageni P, Tang C (2017) Recent advances in metal-containing polymer hydrogels. Macromol Rapid Commun 38:1700109. https://doi.org/10.1002/marc.201700109
112. Li Z, Guan J (2011) Thermosensitive hydrogels for drug delivery. Expert Opin Drug Deliv
113. Li J, Mooney DJ (2016) Designing hydrogels for controlled drug delivery. Nat Rev Mater
114. Lin H, Li Q, Du Q, Wang O, Wang Z, Akert L, Carlson MA, Zhang C, Subramanian A, Zhang C, Lunning M, Li M, Lei Y (2019) Integrated generation of induced pluripotent stem cells in a low-cost device. Biomaterials. https://doi.org/10.1016/j.biomaterials.2018.10.027
115. Lin CC, Anseth KS (2009) PEG hydrogels for the controlled release of biomolecules in regenerative medicine. Pharm Res
116. Liu L, Feng X, Pei Y, Wang J, Ding J, Chen L (2018) α-Cyclodextrin concentration-controlled thermo-sensitive supramolecular hydrogels. Mater Sci Eng, C 82:25–28. https://doi.org/10.1016/j.msec.2017.08.045
117. Liu L, Jiang S, Sun Y, Agarwal S (2016) Giving direction to motion and surface with ultra-fast speed using oriented hydrogel fibers. Adv Funct Mater. https://doi.org/10.1002/adfm.201503612
118. Liu M, Zeng X, Ma C, Yi H, Ali Z, Mou X, Li S, Deng Y, He N (2017) Injectable hydrogels for cartilage and bone tissue engineering. Bone Res 5:17014
119. Lu Y, Sun W, Gu Z (2014) Stimuli-responsive nanomaterials for therapeutic protein delivery. J Control Release
120. Lu AH, Salabas EL, Schüth F (2007) Magnetic nanoparticles: synthesis, protection, functionalization, and application. Angew Chemie Int Ed
121. Lyman MD, Melanson D, Sawhney AS (1996) Characterization of the formation of interfacially photopolymerized thin hydrogels in contact with arterial tissue. Biomaterials 17:359–364. https://doi.org/10.1016/0142-9612(96)85574-8
122. Ma X, Tian H (2014) Stimuli-responsive supramolecular polymers in aqueous solution. Acc Chem Res 47:1971–1981. https://doi.org/10.1021/ar500033n
123. Mabilleau G, Aguado E, Stancu IC, Cincu C, Baslé MF, Chappard D (2008) Effects of FGF-2 release from a hydrogel polymer on bone mass and microarchitecture. Biomaterials. https://doi.org/10.1016/j.biomaterials.2007.12.018
124. Mather BD, Viswanathan K, Miller KM, Long TE (2006) Michael addition reactions in macromolecular design for emerging technologies. Prog Polym Sci 31:487–531
125. Matson JB, Stupp SI (2012) Self-assembling peptide scaffolds for regenerative medicine. Chem Commun. https://doi.org/10.1039/c1cc15551b
126. Min KH, Park K, Kim YS, Bae SM, Lee S, Jo HG, Park RW, Kim IS, Jeong SY, Kim K, Kwon IC (2008) Hydrophobically modified glycol chitosan nanoparticles-encapsulated camptothecin enhance the drug stability and tumor targeting in cancer therapy. J Control Release. https://doi.org/10.1016/j.jconrel.2008.01.013

127. Mironi-Harpaz I, Wang DY, Venkatraman S, Seliktar D (2012) Photopolymerization of cell-encapsulating hydrogels: crosslinking efficiency versus cytotoxicity. Acta Biomater 8:1838–1848. https://doi.org/10.1016/j.actbio.2011.12.034
128. Mitragotri S, Lahann J (2009) Physical approaches to biomaterial design. Nat Mater 8:15–23. https://doi.org/10.1038/nmat2344
129. Moghimi SM, Hunter AC, Murray JC (2001) Long-circulating and target-specific nanoparticles: theory to practice. Pharmacol Rev
130. Morozowich NL, Nichol JL, Allcock HR (2016) Hydrogels based on schiff base formation between an amino-containing polyphosphazene and aldehyde functionalized-dextrans. J Polym Sci, Part A Polym Chem 54:2984–2991. https://doi.org/10.1002/pola.28184
131. Mukherjee S, Hill MR, Sumerlin BS (2015) Self-healing hydrogels containing reversible oxime crosslinks. Soft Matter 11:6152–6161. https://doi.org/10.1039/c5sm00865d
132. Müller M, Becher J, Schnabelrauch M, Zenobi-Wong M (2015) Nanostructured Pluronic hydrogels as bioinks for 3D bioprinting. Biofabrication. https://doi.org/10.1088/1758-5090/7/3/035006
133. Nadeem N, Sohail M, Bin Asad MHH, Minhas MU, Mudassir Shah SA (2018) Thermosensitive hydrogels: from bench to market. Curr, Sci
134. Nahar M, Dutta T, Murugesan S, Asthana A, Mishra D, Rajkumar V, Tare M, Saraf S, Jain NK (2006) Functional polymeric nanoparticles: an efficient and promising tool for active delivery of bioactives. Crit Rev Ther Drug Carrier Syst
135. Nguyen KT, West JL (2002) Photopolymerizable hydrogels for tissue engineering applications. Biomaterials 23:4307–4314. https://doi.org/10.1016/S0142-9612(02)00175-8
136. Nimmo CM, Owen SC, Shoichet MS (2011) Diels-alder click cross-linked hyaluronic acid hydrogels for tissue engineering. Biomacromol 12:824–830. https://doi.org/10.1021/bm101446k
137. Nolan A, Badminton J, Maguire J, Seymour RA (2009) The efficacy of topical hyaluronic acid in the management of oral lichen planus. J Oral Pathol Med. https://doi.org/10.1111/j.1600-0714.2008.00739.x
138. Nusgens BV (2010) Hyaluronic acid and extracellular matrix: a primitive molecule? Ann Dermatol Venereol. https://doi.org/10.1016/S0151-9638(10)70002-8
139. Nuttelman CR, Tripodi MC, Anseth KS (2005) Synthetic hydrogel niches that promote hMSC viability. Matrix Biol 24:208–218. https://doi.org/10.1016/j.matbio.2005.03.004
140. Oh JK, Drumright R, Siegwart DJ, Matyjaszewski K (2008) The development of microgels/nanogels for drug delivery applications. Prog Polym Sci 33:448–477
141. Ostrowska-Czubenko J, Gierszewska-Druzyńska M (2009) Effect of ionic crosslinking on the water state in hydrogel chitosan membranes. Carbohydr Polym 77:590–598. https://doi.org/10.1016/j.carbpol.2009.01.036
142. Panyam J, Labhasetwar V (2003) Biodegradable nanoparticles for drug and gene delivery to cells and tissue. Adv Drug Deliv Rev
143. Park J, An K, Hwang Y, Park JEG, Noh HJ, Kim JY, Park JH, Hwang NM, Hyeon T (2004) Ultra-large-scale syntheses of monodisperse nanocrystals. Nat Mater. https://doi.org/10.1038/nmat1251
144. Park H, Kim MH, Il Yoon Y, Park WH (2017) One-pot synthesis of injectable methylcellulose hydrogel containing calcium phosphate nanoparticles. Carbohydr Polym 157:775–783
145. Park H, Park K (1996) Hydrogels and biodegradable polymers for bioapplications. In: Hydrogels and biodegradable polymers for bioapplications
146. Paul A, Hasan A, Al Kindi H, Gaharwar AK, Rao VTS, Nikkhah M, Shin SR, Krafft D, Dokmeci MR, Shum-Tim D, Khademhosseini A (2014) Injectable graphene oxide/hydrogel-based angiogenic gene delivery system for vasculogenesis and cardiac repair. ACS Nano 8:8050–8062. https://doi.org/10.1021/nn5020787
147. Peppas NA, Brannon-Peppas L (1990) Hydrogels at critical conditions. Part 1. Thermodynamics and swelling behavior. J Memb Sci 48:281–290. https://doi.org/10.1016/0376-7388(90)85009-A

148. Peppas NA, Kim B (2006) Stimuli-sensitive protein delivery systems. J Drug Deliv Sci Technol 16:11–18
149. Peppas NA, Hilt JZ, Khademhosseini A, Langer R (2006) Hydrogels in biology and medicine: from molecular principles to bionanotechnology. Adv Mater
150. Qi C, Liu J, Jin Y, Xu L, Wang G, Wang Z, Wang L (2018) Photo-crosslinkable, injectable sericin hydrogel as 3D biomimetic extracellular matrix for minimally invasive repairing cartilage. Biomaterials. https://doi.org/10.1016/j.biomaterials.2018.02.016
151. Qin D, Xia Y, Whitesides GM (2010) Soft lithography for micro- and nanoscale patterning. Nat Protoc. https://doi.org/10.1038/nprot.2009.234
152. Qiu LY, Bae YH (2006) Polymer architecture and drug delivery. Pharm Res 23:1–30
153. Qureshi D, Nayak SK, Maji S, Anis A, Kim D, Pal K (2019) Environment sensitive hydrogels for drug delivery applications. Eur Polym J 120:109220. https://doi.org/10.1016/j.eurpolymj.2019.109220
154. Rafieian S, Mirzadeh H, Mahdavi H, Masoumi ME (2019) A review on nanocomposite hydrogels and their biomedical applications. IEEE J Sel Top Quantum Electron 26:154–174. https://doi.org/10.1515/secm-2017-0161
155. Raub CB, Suresh V, Krasieva T, Lyubovitsky J, Mih JD, Putnam AJ, Tromberg BJ, George SC (2007) Noninvasive assessment of collagen gel microstructure and mechanics using multiphoton microscopy. Biophys J. https://doi.org/10.1529/biophysj.106.097998
156. Rašović I (2017) Water-soluble fullerenes for medical applications. Mater Sci Technol (United Kingdom)
157. Richter A, Paschew G, Klatt S, Lienig J, Arndt KF, Adler HJP (2008) Review on hydrogel-based pH sensors and microsensors. Sensors
158. Rizzi SC, Hubbell JA (2005) Recombinant protein-co-PEG networks as cell-adhesive and proteolytically degradable hydrogel matrixes. Part I: Development and physicochemical characteristics. Biomacromol 6:1226–1238. https://doi.org/10.1021/bm049614c
159. Sadat-Shojai M, Khorasani M-T, Jamshidi A (2015) 3-Dimensional cell-laden nano-hydroxyapatite/protein hydrogels for bone regeneration applications. Mater Sci Eng, C 49:835–843. https://doi.org/10.1016/j.msec.2015.01.067
160. Sahoo SK, Labhasetwar V (2003) Nanotech approaches to drug delivery and imaging. Drug Discov Today
161. Sahu P, Kashaw SK, Jain S, Sau S, Iyer AK (2017) Assessment of penetration potential of pH responsive double walled biodegradable nanogels coated with eucalyptus oil for the controlled delivery of 5-fluorouracil: in vitro and ex vivo studies. J Control Release. https://doi.org/10.1016/j.jconrel.2017.03.023
162. Said HM, Alla SGA, El-Naggar AWM (2004) Synthesis and characterization of novel gels based on carboxymethyl cellulose/acrylic acid prepared by electron beam irradiation. React Funct Polym. https://doi.org/10.1016/j.reactfunctpolym.2004.07.002
163. Santra TS, Bhattacharyya TK, Mishra P, Tseng FG, Barik TK (2012) Biomedical applications of diamond-like nanocomposite thin films. Sci Adv Mater. https://doi.org/10.1166/sam.2012.1258
164. Santra TS, Bhattacharyya TK, Patel P, Tseng FG, Barik TK (2011) Structural and tribological properties of diamond-like nanocomposite thin films. Surf Coatings Technol. https://doi.org/10.1016/j.surfcoat.2011.06.057
165. Santra TS, Kar S, Chen C-W, Borana J, Chen T-C, Lee M-C, Tseng F-G (2020) Near-infrared nanosecond-pulsed laser-activated high efficient intracellular delivery mediated by nano-corrugated mushroom-shaped gold-coated polystyrene nanoparticles. Nanoscale. https://doi.org/10.1039/d0nr01792b
166. Santra TS, Liu CH, Bhattacharyya TK, Patel P, Barik TK (2010) Characterization of diamond-like nanocomposite thin films grown by plasma enhanced chemical vapor deposition. J Appl Phys 10(1063/1):3415548
167. Santra TS, Bhattacharyya TK, Patel P, Tseng FG, Barik TK (2012) Diamond, diamond-like carbon (DLC) and diamond-like nanocomposite (DLN) thin films for MEMS applications. In: Microelectromechanical systems and devices

168. Santra TS, Tseng F-G (2020) Handbook of single cell technologies. Springer, Singapore
169. Santra TS (2002) Microfluidics and bioMEMS applications. Springer US, Boston, MA
170. Sasaki Y, Akiyoshi K (2010) Nanogel engineering for new nanobiomaterials: from chaperoning engineering to biomedical applications. Chem Rec 10:366–376. https://doi.org/10.1002/tcr.201000008
171. Satarkar NS, Biswal D, Hilt JZ (2010) Hydrogel nanocomposites: a review of applications as remote controlled biomaterials. Soft Matter 6:2364–2371. https://doi.org/10.1039/b925218p
172. Sawhney AS, Pathak CP, Hubbell JA (1993) Interfacial photopolymerization of poly(ethylene glycol)-based hydrogels upon alginate-poly(l-lysine) microcapsules for enhanced biocompatibility. Biomaterials 14:1008–1016. https://doi.org/10.1016/0142-9612(93)90194-7
173. Schindler M, Nur-E-Kamal A, Ahmed I, Kamal J, Liu HY, Amor N, Ponery AS, Crockett DP, Grafe TH, Chung HY, Weik T, Jones E, Meiners S (2006) Living in three dimensions: 3D nanostructured environments for cell culture and regenerative medicine. Cell Biochem, Biophys
174. Schmaljohann D (2006) Thermo- and pH-responsive polymers in drug delivery. Adv Drug Deliv Rev 58:1655–1670
175. Schneider H-J (2015) Chemoresponsive materials: stimulation by chemical and biological signals
176. Schubert US, Eschbaumer C (2002) Macromolecules containing bipyridine and terpyridine metal complexes: towards metallosupramolecular polymers. Angew Chemie Int Ed 41:2892–2926. https://doi.org/10.1002/1521-3773(20020816)41:16%3c2892:AID-ANIE2892%3e3.0.CO;2-6
177. Secret E, Kelly SJ, Crannell KE, Andrew JS (2014) Enzyme-responsive hydrogel microparticles for pulmonary drug delivery. ACS Appl Mater Interfaces. https://doi.org/10.1021/am501754s
178. Seidler C, Ng DYW, Weil T (2017) Native protein hydrogels by dynamic boronic acid chemistry. Tetrahedron 73:4979–4987. https://doi.org/10.1016/j.tet.2017.06.066
179. Sharma VK, Yngard RA, Lin Y (2009) Silver nanoparticles: green synthesis and their antimicrobial activities. Adv Colloid Interface Sci
180. Shenoy D, Little S, Langer R, Amiji M (2005) Poly(ethylene oxide)-modified poly(β-amino ester) nanoparticles as a pH-sensitive system for tumor-targeted delivery of hydrophobic drugs. 1. in vitro evaluations. Mol Pharm 2:357–366. https://doi.org/10.1021/mp0500420
181. Shih H, Lin CC (2013) Visible-light-mediated thiol-ene hydrogelation using eosin-Y as the only photoinitiator. Macromol Rapid Commun 34:269–273. https://doi.org/10.1002/marc.201200605
182. Shin SJ, Lee JH, So J, Min K (2016) Anti-adhesive effect of poloxamer-based thermo-sensitive sol-gel in rabbit laminectomy model. J Mater Sci Mater Med. https://doi.org/10.1007/s10856-016-5773-7
183. Shin SR, Migliori B, Miccoli B, Li YC, Mostafalu P, Seo J, Mandla S, Enrico A, Antona S, Sabarish R, Zheng T, Pirrami L, Zhang K, Zhang YS, Wan KT, Demarchi D, Dokmeci MR, Khademhosseini A (2018) Electrically driven microengineered bioinspired soft robots. Adv Mater. https://doi.org/10.1002/adma.201704189
184. Shoda SI, Uyama H, Kadokawa JI, Kimura S, Kobayashi S (2016) Enzymes as green catalysts for precision macromolecular synthesis. Chem Rev 116:2307–2413
185. Simon J, Flahaut E, Golzio M (2019) Overview of carbon nanotubes for biomedical applications. Materials (Basel)
186. Sivashanmugam A, Arun Kumar R, Vishnu Priya M, Nair SV, Jayakumar R (2015) An overview of injectable polymeric hydrogels for tissue engineering. Eur Polym J. https://doi.org/10.1016/j.eurpolymj.2015.05.014
187. Slaughter B V, Khurshid SS, Fisher OZ, Khademhosseini A, Peppas NA (2009) Hydrogels in regenerative medicine. Adv Mater
188. Smeds KA, Grinstaff MW (2001) Photocrosslinkable polysaccharides for in situ hydrogel formation. J Biomed Mater Res. https://doi.org/10.1002/1097-4636(200101)54:1%3c115:AID-JBM14%3e3.0.CO;2-Q

189. Smith LJ, Taimoory SM, Tam RY, Baker AEG, Binth Mohammad N, Trant JF, Shoichet MS (2018) Diels-Alder click-cross-linked hydrogels with increased reactivity enable 3D cell encapsulation. Biomacromol 19:926–935. https://doi.org/10.1021/acs.biomac.7b01715
190. Sohail M, Mudassir Minhas MU, Khan S, Hussain Z, de Matas M, Shah SA, Khan S, Kousar M, Ullah K (2019) Natural and synthetic polymer-based smart biomaterials for management of ulcerative colitis: a review of recent developments and future prospects. Drug Deliv Transl Res 9:595–614. https://doi.org/10.1007/s13346-018-0512-x
191. Solouk A, Mirzadeh H, Shokrgozar MA, Solati-Hashjin M, Najarian S, Seifalian AM (2011) The study of collagen immobilization on a novel nanocomposite to enhance cell adhesion and growth. Iran Biomed J 15:6–14
192. De Souza Ferreira SB, Moço TD, Borghi-Pangoni FB, Junqueira MV, Bruschi ML (2016) Rheological, mucoadhesive and textural properties of thermoresponsive polymer blends for biomedical applications. J Mech Behav Biomed Mater. https://doi.org/10.1016/j.jmbbm.2015.10.026
193. Spang MT, Christman KL (2018) Extracellular matrix hydrogel therapies: in vivo applications and development. Acta Biomater
194. Stenekes RJH, Talsma H, Hennink WE (2001) Formation of dextran hydrogels by crystallization. Biomaterials. https://doi.org/10.1016/S0142-9612(00)00375-6
195. Syakir MI, Nurin NA, Zafirah N, Kassim MA, Khalil HA (2016) Nanoclay reinforced polymer composites. Springer, Singapore
196. Sydney Gladman A, Matsumoto EA, Nuzzo RG, Mahadevan L, Lewis JA (2016) Biomimetic 4D printing. Nat Mater. https://doi.org/10.1038/nmat4544
197. Söntjens SHM, Nettles DL, Carnahan MA, Setton LA, Grinstaff MW (2006) Biodendrimer-based hydrogel scaffolds for cartilage tissue repair. Biomacromol 7:310–316. https://doi.org/10.1021/bm050663e
198. Taheri A, Atyabi F, Dinarvnd R (2011) Temperature-responsive and biodegradable PVA: VP k30: poloxamer 407 hydrogel for controlled delivery of human growth hormone (hGH). J Pediatr Endocrinol Metab. https://doi.org/10.1515/JPEM.2011.079
199. Teotia AK, Sami H, Kumar A (2015) Thermo-responsive polymers: structure and design of smart materials. In: Switchable and responsive surfaces and materials for biomedical applications
200. Van Thienen TG, Lucas B, Flesch FM, Van Nostrum CF, Demeester J, De Smedt SC (2005) On the synthesis and characterization of biodegradable dextran nanogels with tunable degradation properties. Macromolecules 38:8503–8511. https://doi.org/10.1021/ma050822m
201. Torchilin VP (2004) Targeted polymeric micelles for delivery of poorly soluble drugs. Cell Mol Life Sci
202. Tseng F-G, Santra TS (2016) Essentials of single-cell analysis: concepts, applications and future prospects
203. Valade D, Wong LK, Jeon Y, Jia Z, Monteiro MJ (2013) Polyacrylamide hydrogel membranes with controlled pore sizes. J Polym Sci, Part A: Polym Chem. https://doi.org/10.1002/pola.26311
204. Van Der Linden HJ, Herber S, Olthuis W, Bergveld P (2003) Stimulus-sensitive hydrogels and their applications in chemical (micro)analysis. Analyst
205. Verhulsel M, Vignes M, Descroix S, Malaquin L, Vignjevic DM, Viovy JL (2014) A review of microfabrication and hydrogel engineering for micro-organs on chips. Biomaterials
206. Vinogradov SV, Bronich TK, Kabanov AV (2002) Nanosized cationic hydrogels for drug delivery: preparation, properties and interactions with cells. Adv Drug Deliv Rev. https://doi.org/10.1016/S0169-409X(01)00245-9
207. Voorhaar L, Hoogenboom R (2016) Supramolecular polymer networks: hydrogels and bulk materials. Chem Soc Rev 45:4013–4031
208. Wang E, Desai MS, Lee SW (2013) Light-controlled graphene-elastin composite hydrogel actuators. Nano Lett. https://doi.org/10.1021/nl401088b
209. Wang X, Kluge JA, Leisk GG, Kaplan DL (2008) Sonication-induced gelation of silk fibroin for cell encapsulation. Biomaterials 29:1054–1064. https://doi.org/10.1016/j.biomaterials.2007.11.003

210. Wang S, Liu K, Liu J, Yu ZT-F, Xu X, Zhao L, Lee T, Lee EK, Reiss J, Lee Y-K, Chung LWK, Huang J, Rettig M, Seligson D, Duraiswamy KN, Shen CK-F, Tseng H-R (2011) Highly efficient capture of circulating tumor cells by—supporting information. Angew Chemie 123:3140–3144. https://doi.org/10.1002/ange.201005853
211. Wang J, Williamson GS, Yang H (2018) Branched polyrotaxane hydrogels consisting of alpha-cyclodextrin and low-molecular-weight four-arm polyethylene glycol and the utility of their thixotropic property for controlled drug release. Colloids Surfaces B Biointerfaces 165:144–149. https://doi.org/10.1016/j.colsurfb.2018.02.032
212. Wang D, Jin Y, Zhu X, Yan D (2017) Synthesis and applications of stimuli-responsive hyperbranched polymers. Prog Polym Sci
213. Wang J, Zhang H, Wang F, Ai X, Huang D, Liu G, Mi P (2018) Enzyme-responsive polymers for drug delivery and molecular imaging. In: Stimuli responsive polymeric nanocarriers for drug delivery applications: volume 1: types and triggers
214. Whitesides GM, Ostuni E, Takayama S, Jiang X, Ingber DE (2001) Soft lithography in biology and biochemistry. Annu Rev Biomed Eng. https://doi.org/10.1146/annurev.bioeng.3.1.335
215. Wichterle O, Lím D (1960) Hydrophilic gels for biological use. Nature. https://doi.org/10.1038/185117a0
216. Williams CG, Malik AN, Kim TK, Manson PN, Elisseeff JH (2005) Variable cytocompatibility of six cell lines with photoinitiators used for polymerizing hydrogels and cell encapsulation. Biomaterials 26:1211–1218. https://doi.org/10.1016/j.biomaterials.2004.04.024
217. Wu XS, Hoffman AS, Yager P (1992) Synthesis and characterization of thermally reversible macroporous poly(N-isopropylacrylamide) hydrogels. J Polym Sci, Part A: Polym Chem 30:2121–2129. https://doi.org/10.1002/pola.1992.080301005
218. Xu L, Qiu L, Sheng Y, Sun Y, Deng L, Li X, Bradley M, Zhang R (2018) Biodegradable pH-responsive hydrogels for controlled dual-drug release. J Mater Chem B. https://doi.org/10.1039/c7tb01851g
219. Yang Y, Tan Y, Wang X, An W, Xu S, Liao W, Wang Y (2018) Photothermal nanocomposite hydrogel actuator with electric-field-induced gradient and oriented structure. ACS Appl Mater Interfaces. https://doi.org/10.1021/acsami.7b17907
220. Yao H, Wang J, Mi S (2017) Photo processing for biomedical hydrogels design and functionality: a review. Polymers (Basel) 10:11. https://doi.org/10.3390/polym10010011
221. Ye H, Owh C, Loh XJ (2015) A thixotropic polyglycerol sebacate-based supramolecular hydrogel showing UCST behavior. RSC Adv 5:48720–48728. https://doi.org/10.1039/c5ra08222f
222. Yuan J, Fang X, Zhang L, Hong G, Lin Y, Zheng Q, Xu Y, Ruan Y, Weng W, Xia H, Chen G (2012) Multi-responsive self-healing metallo-supramolecular gels based on "click" ligand. J Mater Chem 22:11515–11522. https://doi.org/10.1039/c2jm31347b
223. Zhang L, Cao Z, Bai T, Carr L, Ella-Menye JR, Irvin C, Ratner BD, Jiang S (2013) Zwitterionic hydrogels implanted in mice resist the foreign-body reaction. Nat Biotechnol. https://doi.org/10.1038/nbt.2580
224. Zhang H, Patel A, Gaharwar AK, Mihaila SM, Iviglia G, Mukundan S, Bae H, Yang H, Khademhosseini A (2013) Hyperbranched polyester hydrogels with controlled drug release and cell adhesion properties. Biomacromol 14:1299–1310. https://doi.org/10.1021/bm301825q
225. Zhang X, Pint CL, Lee MH, Schubert BE, Jamshidi A, Takei K, Ko H, Gillies A, Bardhan R, Urban JJ, Wu M, Fearing R, Javey A (2011) Optically- and thermally-responsive programmable materials based on carbon nanotube-hydrogel polymer composites. Nano Lett. https://doi.org/10.1021/nl201503e
226. Zhang X, Xi W, Huang S, Long K, Bowman CN (2017) Wavelength-selective sequential polymer network formation controlled with a two-color responsive initiation system. Macromolecules 50:5652–5660. https://doi.org/10.1021/acs.macromol.7b01117
227. Zhang H, Zhai Y, Wang J, Zhai G (2016) New progress and prospects: the application of nanogel in drug delivery. Mater Sci Eng C

228. Zheng SY, Ding H, Qian J, Yin J, Wu ZL, Song Y, Zheng Q (2016) Metal-coordination complexes mediated physical hydrogels with high toughness, stick-slip tearing behavior, and good processability. Macromolecules 49:9637–9646. https://doi.org/10.1021/acs.macromol.6b02150
229. Zhou T, Li X, Li G, Tian T, Lin S, Shi S, Liao J, Cai X, Lin Y (2017) Injectable and thermosensitive TGF-β1-loaded PCEC hydrogel system for in vivo cartilage repair. Sci Rep. https://doi.org/10.1038/s41598-017-11322-w

Nanomaterials: Versatile Drug Carriers for Nanomedicine

Ashwini S. Shinde, Pallavi S. Shinde, and Tuhin S. Santra

Abstract Certain medicines and therapies have emerged for the treatment of different ailments. But in many cases, they have poor solubility, lower bioavailability, inability to cross the blood-brain barrier (BBB), and drug resistance. It becomes essential to establish standard treatment systems for overcoming such challenges. In this connection, the nanomaterial application in medicine and pharmaceuticals has rapidly gained interest with revolutionary prospects. The idea of nano-carriers first observed in a biological system consisting of nanoparticles is committed to locomotory function and protein cargos like importin, exportin. This observation led to the development of biomimetic nanomaterials for drug delivery. The synthesized nanomaterials exhibit useful properties such as large surface area, maximum bioavailability, reduced toxicity, high specificity along with enhanced permeability and retention (EPR) effect. These properties contribute to enhancing the efficacy of drugs having short half-lives, monitoring drugs for sustained release, enhancing the rate of dissolution of drugs, and reducing required dosage volume. Thus, increased therapeutic action and fewer side effects are to improve the quality of human life. Currently, many nano-carriers such as niosomes, dendrimers, fullerene, polymer-based nanoparticles, micelle, liposomes, hydrogels, metallic, mesoporous silica, quantum dots, etc. show potential for better drug delivery systems. These help in carrying entities like drug molecules, DNA/RNA, proteins, viruses, cell receptor sites, lipid bilayers, and variable antibody region for drug delivery in therapeutics. Such nano-therapeutics and diagnostics will unfold the secrets of human longevity and help reduce human illness, including cardiovascular disease, genetic disorder, immunodeficiency, cancer, and even viral infections. This chapter highlights recent advancements and applications of nano-carriers for drug delivery in medicine, especially wound healing therapeutics. It also discusses different approaches to enhance drug cargo capacity, improve cell delivery efficiency, to avoid host immune systems, and to achieve specific cell targeting.

A. S. Shinde · P. S. Shinde · T. S. Santra (✉)
Department of Engineering Design, Indian Institute of Technology Madras, Chennai 600036, Tamil Nadu, India
e-mail: tuhin@iitm.ac.in

T. S. Santra and L. Mohan (eds.), *Nanomaterials and Their Biomedical Applications*, Springer Series in Biomaterials Science and Engineering 16,
https://doi.org/10.1007/978-981-33-6252-9_10

Keywords Nanomaterials · Drug delivery · Targeted nanomedicine · Immune-compatibility · Wound healing therapy

1 Introduction

The first-generation era made use of the simplest mechanisms of nanotechnology for the controlled release of the drug. Most of the formulations used for the drug delivery were either oral or transdermal administration. The stability and effectivity are found low in these delivery systems [1]. The first gold particle was synthesized in nanometre-scale by Michael Faraday, about more than 150 years ago [2]. These colloidal gold particles conjugated with antibodies for target-specific staining known to be as immune gold staining. The study of immune gold staining is pioneering for most of the recent gold particles application using nanotechnology. Polymer micelles and liposomes were first prepared in the 1960s with first-ever referred to as nanoparticles in 2000 [3]. In the 1970s, scientists synthesized for the first time the dendrimers and NPs without its known use in nanotechnological applications [4]. The 1980s period reported to successfully synthesize and develop micelles as drug delivery systems [5]. While in the 1990s, Kataoka invented block co-polymers of Polyethylene glycol (PEG)-polylysine [6]. Conventional formulations, suspension, or emulsion drug preparations have definite limitations tagged with them like instability, high dose-low availability, intolerance, first-pass effect, fluctuations in the levels of plasma drug, lacking in sustained impact, etc. Thus, there is a need for advanced drug delivery carrier systems to meet the ideal requirements. In the second generation, the era of modern nanotechnology, the use of nano-drug delivery systems began all with the launch of the Nation nanotechnology initiative by the United States [7]. It was the world's first program in the fields of nanotechnology. Various new methods were introduced as drug delivery systems among which nanotechnological approaches were found to be more efficient [1].

The nanomaterials have dimensions in nanoscale, which enables them to reach inaccessible areas like cancer cells or inflamed tissues. This property is achievable due to the high permeability and retention effect of the nanoparticles [8]. The nanomaterials should be safe, soluble, bioavailable as well as biocompatible. They should not rupture the blood vessels and be less invasive. The toxicity of nanomaterials associated with drug delivery should be very low so that in safe concentrations, they are usable in target-specific diseased tissues [9]. The nanomaterials protect drugs from hydrolytic and enzymatic degradation in the gastrointestinal tract and also helps to bypass from the metabolism in the liver. Nanoparticles coated with hydrophilic polymers remain in circulation for a longer time. Thus, they are also suitable for short half-lived drugs to enhance its efficacy and to maintain its sustained drug release formulations as well as for DNA delivery [10]. The rapid clearance and metabolism cause premature loss of drug, which is easy to prevent. The bio-adhesion increases retention of the nanoparticles used in drug delivery [11].

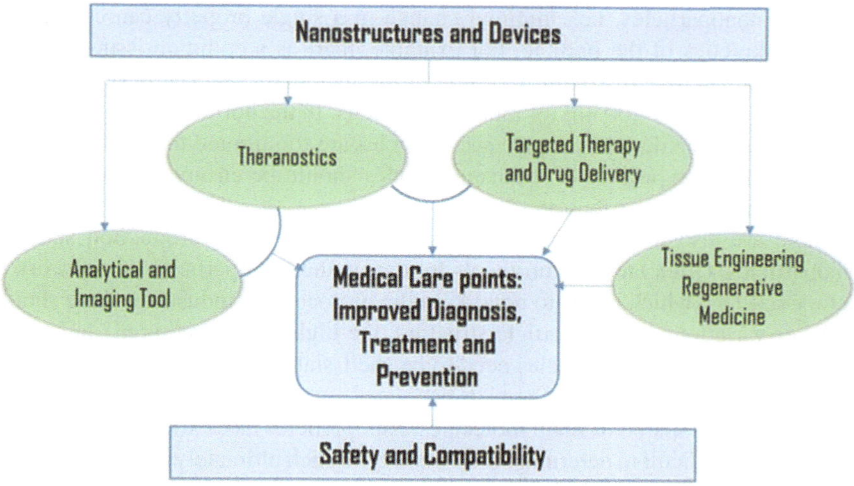

Fig. 1 Application and goals of nanomedicine in different spheres of biomedical research (Reproduced from Patra et al. [14] under a Creative Commons Attribution 4.0 International License. https://creativecommons.org/licenses/by/4.0/)

The acknowledgement of nanomedicines, for their nanostructures utilized for the targeted delivery agents for attached therapeutics drugs or encapsulated drugs with the controlled release [12]. Due to these fundamental advantages of the nanoparticles, there is a resultant reduction in drug quantity and toxicity of dosage, safe delivery of therapeutics drugs while toxic to healthy cells, thus protection from the undesired side effects [13]. Generally, the use of nanotechnology and nanocarriers-based drug delivery systems is to develop the effectiveness of therapeutic agents. The sub-micron sized colloidal nanoparticles are encapsulated with therapeutic agents of interest or dispersed within their matrix or conjugated or adsorbed onto the surface. It implicates the utilization of nano dimensional materials including nanosensors for diagnostics, nanorobots, delivery and actuate materials in live cells (Fig. 1).

In this chapter, we will discuss the different developments for drug delivery application using nanostructured carrier systems in medicine against different ailments and wound healing and also various approaches in the use of nanomaterials for more targeted delivery of nanomedicines along with the consideration of immune-compatibility of the nanomedicines.

2 Challenges in Nanoparticle Development

The establishment of nanomedicines technology involved many challenges in early and late stages of synthesis and development that were either non-nanoparticle-based therapy or currently non-existent. The primary issue observed is the nonuniform nature of nanoparticles, i.e., minimal change in a single property can change the pharmacokinetics of the particle. For example, there is a common issue of maintaining a narrow size distribution profile of the nanoparticles. The desired particle size should be below 200 nm for most applications. In the normal broad distribution of sizes, often, the desired average size particles are too limited to be useful. Thus, the manufacturing process of the nanoparticles should be ensured to maintain the desired narrow size distribution.

There are also many unique challenges faced during the production stage of nanoparticles. Often lab level protocols for the synthesis of particles do not work in factory settings, which needs to rework on the protocols. In industries, there should be fewer variations in nanoparticle structure, the higher yield required, and sterile synthesis is a must. The particles need to be shelf-stable, i.e., it should not degrade in solution and not clump over some time.

Finally, as compared to small molecules, nanoparticles face extra regulatory challenges as it is difficult to determine their toxicity. Which ultimately increases the cost and duration of clinical trials. All these challenges combined develops a more prudent attitude towards questions of reproducibility and scalability even at the earlier stage of development so as to prevent failure at a later stage.

3 Overview of Nanomaterials in Drug Delivery System

The most important part of the nano carrier-based drug delivery system is the sustained drug release from the nanoparticles. Polymer-coated nanoparticles, the drug is released either by controlled diffusion or erosion from the core of the carrier to the matrix or polymeric membrane. The polymeric layer around the nanoparticles acts as a physical barrier for the drug release. Thus, the regulating factor in drug release is solubility and diffusivity of a drug across the polymer membrane. Furthermore, due to the ionic interactions between auxiliary ingredients and the drug can affect its release rate. Supplemental components are co-polymers, which reduces the interaction of the drug with its matrix material due to electrostatic competitive interaction, helping to achieve increased drug release [15]. But when the drug is involved in an interaction with the auxiliary ingredients, they form less water-soluble complexes. Thus causes drug release very slow along with the no burst release effect [16]. Drug release and biodegradation of polymer are the most prime consideration factors in the development of a successful nanoparticulate system. Generally, the rate of drug release depends upon—drug solubility, desorption of the absorbed drug from the surface, diffusion through a matrix, degradation of the array, and combination of diffusion and erosion process [17]. While nanoparticles provide targeted drug delivery, thus enhancing the uptake of poorly soluble drug and bioavailability.

Nanoengineered devices mainly aim to lessen the side effects and increase the potency of treatment. Nanoparticles may be composed of a different substance such as inorganic nanoparticles, polymeric nanoparticles, protein nanoparticles, carbon-based nanoparticles, and lipid-based nanoparticles. These are useful as a carrier for large classes of therapeutic agents for cytotoxic activity, small interference RNA (siRNA), chemosensitizers, and antiangiogenic agents. A generalized overview of some nano-vehicles for release of drug which aims to progress the pharmacokinetics, pharmacodynamics, and bioavailability of drugs summarizes in Table 1.

Drug targeting, maintaining the therapeutic effect with a reduction in cytotoxicity and biocompatibility are the primary goals of drug delivery systems (DDS) [8]. The nanoscale carriers have upgraded the DDS. Thus comprises different biocompatible and biodegradable materials such as natural or synthetic materials [18]. At present, very few DDS has been approved by the FDA, and many are being studied in drug delivery. Here, we discuss the approaches to enhance the drug delivery systems based on drug targeting, controlled drug release, multidrug resistance, and multifunctional nano-carriers.

4 Drug Targeting

The synthesis of nanoparticles with an effective drug delivery system is one of the main challenges faced in the development process. Thus achieved with two critical functions by the DDS-evading host immune response and reaching their predetermined targets. Nanoparticles can be classified based on the potentiality to locate and drug delivery to the target area of the body as a passive and active target.

Intravenous drug delivery tends to disperse evenly in the body. Where the take-up ability of individual-sized particles by the tumor cells is to a greater extent than that of the healthy normal cells because of the combination of faulty particle screening and leaky tumor blood vessels, this phenomenon is known as enhanced permeability and retention (EPR) effect showing mechanism for passive targeting [19].

The properties of nanoparticles, such as particle size, shape, and surface charge, influences the EPR effect, which in return helps in circulation time, intracellular internalization, and penetration speed. For instance, phagocytic cells can uptake larger particles, while nonphagocytic cells prefer smaller particles for uptake [21]. Hence, due to the EPR effect, nanoparticles that encapsulate drugs get accumulated in the tumor at higher concentrations (5–10 times) as compared to healthy tissues within 1–2 days [22]. For example, FDA approved Doxil® (a PEG-coated liposomal system containing Doxorubicin delivery for cancer therapy), and Abraxane® (paclitaxel-coated with albumin nanoparticles for the metastatic breast cancer) tend to circulate in the body with longer (100 times) half-life with reducing systemic toxicity simultaneously than that of the free anticancer drugs [23]. The surface charge of the nanoparticles plays an essential role in blood circulation and the internalization of the cell. Negatively surface charged NPs will circulate in the blood for a more extended period while the NPs with a positive charge are more readily uptaken by the cancer cells (due to negative surface charge) [24].

EPR effect is not feasible for targeting cancer cells in all tumors, due to different porosity of tumor vessels and the degree of vascularization. Passively targeted NPs fails more than 95% to reach the cancer cells when injected intravenously, thus

Table 1 Examples of nanotechnology and nano-carrier-based drug delivery systems, compositions, aims, applications and route of administration for disease treatment

	Drug/active ingredients	System components	Aim	Major applications	Route of administration
Polymeric Nanoparticle	Amphotericin B	PLGA, PEG	To improve oral bioavailability	Increased oral bioavailability; decreased adverse effects (Nephrotoxicity)	Oral delivery
	Cisplatin	PLG-g-mPEG5K	To increase bioavailability	Increased oral bioavailability; decreased adverse effects	*In vitro* and *in vivo*
	Exenatide	PLGA-PEG-PLGA	To long-acting	Extended-release of drug	*In vitro*
	Ellipticine	PHAs-PHBVs	To improve bioavailability	Increased biocompatibility and bioavailability; increased cytotoxicity	*In vitro*
	Tamoxifen-Quercetin	PLGA	To increase bioavailability	Threefold increased oral bioavailability, reduced the therapy; higher cellular uptake	Oral delivery
	Topotecan-Tamoxifen	PLGA	To synergistic effect	Achieved synergism; reduced side effects; enhance permeation of cells; increased bioavailability	*In vitro* and *in vivo*
	Naringenin	PVA-Eudrgit(R) E 100	To improve oral bioavailability	Highly biocompatibility and high therapeutic efficacy	Oral delivery
	Curcumin	NIPAAM/VP/PEG-A	To improve the solubility of the drug	Increased oral bioavailability	Oral delivery

(continued)

Table 1 (continued)

	Drug/active ingredients	System components	Aim	Major applications	Route of administration
	Rivastigmine	Polysorbate-80	To prolong the drug release	Improved cognitive functions and memory	Oral delivery
Polymeric micelles	Doxorubicin	Styrene-Maleic acid	To improve the solubility of the drug	Decreased cardiac and bone marrow toxicity	*In vitro*
	NK012	PEG-P(Glu)	To reduce side-effects	Reduced gastrointestinal toxicity; reduced therapy	*In vitro*
	Cisplatin	PEG-P(Glu)	To reduce side-effects	Reduced nephrotoxicity, neurotoxicity; sustained release; readily accumulation with cancer cells	*In vitro*
	Docetaxel	PLA-PEG	To site specific	Targeted to prostate-specific membrane antigen; Prolonged tumor growth suppression	*In vitro*
	Daunomycin	SDS-Triton X-100	To hydrophobic contribution	Hydrophobic contribution to the free energy of DNA intercalation reaction	*In vitro*
	Paclitaxel	PGG	To improve the therapeutic index	Effectively inhibiting tumor growth	Oral delivery
Polymer conjugates	Doxorubicin	PEG	To improve the solubility of the drug	Decreased toxicity	*In vitro*

(continued)

Table 1 (continued)

	Drug/active ingredients	System components	Aim	Major applications	Route of administration
	Paclitaxel	P(GLU)	To improve the solubility of the drug	Improved therapeutics and reduced toxicity	*In vitro*
Mesoporous silica Nanoparticle	Cyclosporin A	Tetraethyl Orthosilicate	To improve the solubility of the drug	Increased solubility, bioavailability, and enhanced absorption	Oral delivery
	siRNA	Poly ethylenimine	To effective deliver siRNA and stable	Highly stable, highly effective siRNA delivery with optimal cellular integration and low toxicity	*In vitro*
	Doxorubicin	Cetyltrimethylammonium bromide (CTAB)	To sustained release	Increased biocompatibility and low toxicity	*In vitro*
	Docetaxel	Lactose	To targeted delivery	Increased bioavailability and site-specific delivery	*In vitro*
Chitosan Nanoparticle	Ethosuximide	Chitosan	To sustained release	Extended-release of drug and excellent cytotoxicity	*In vivo* and *in vitro*
	5-Fluorouracil	Chitosan	To prevent drug degradation	Preventing drug degradation at gastric pH and effective delivery of the drug in site-specific	Oral delivery
	Idebenone	Chitosan and N- carboxymethyl chitosan	To increase stability	Increased stability and reduced mucose membrane irritation	*In vitro*-topical or nasal

(continued)

Table 1 (continued)

	Drug/active ingredients	System components	Aim	Major applications	Route of administration
Albumin Nanoparticle	Doxorubicin	Human serum albumin	To targeted delivery	Increased bioavailability and site-specific delivery	*In vitro*
	Insulin	Calcium-Pectinate	To prepare oral insulin	Sustained release of insulin in the small intestine	Oral delivery
	Cisplatin	Bovine serum albumin	To increase the bioavailability	Sustained release of drug and decreased side effects (cytotoxicity, nephrotoxicity)	Intravenous infusion
Polymeric Liposomes	Sorafenib and Gadolinium	–	To improve bioavailability	Increased solubility and biodistribution	*In vitro* and *in vivo*
	Doxorubicin	MPEG-Phospholipid	To increase blood circulation time	Cross BBB and penetrate the vasculature of tumors	Intravenous
Gold Nanoparticles	*Azolla microphylla* extract	Gold (III) Chloride	To increase bioavailability	Increased bioavailability, biocompatibility, and reduced dose therapy	*In vitro* and *in vivo*
Gelatin drug conjugates	Methotrexate	Gelatin	To prepare controlled delivery	Faster drug release in gastric medium	*In vitro*
Silver Nanoparticle	Cauliflower leaf extract	Silver nitrate	To increase bioavailability	Increased bioavailability, biocompatibility, and reduced dose therapy	*In vitro*
	Elephantopus scaber extract	Silver nitrate	To increase bioavailability	Increased bioavailability, biocompatibility, and reduced dose therapy	*In vitro*

(continued)

Table 1 (continued)

	Drug/active ingredients	System components	Aim	Major applications	Route of administration
Fullerenes	Diethyl bromomalonate	C60	To neuroprotective agent	Ex-cytotoxic necrosis, protective effects of neuronal apoptotic neuronal deaths and increasing intercalation into brain membranes	*In vitro* and in vivo
Polymersome	Insulin	Dextran-b-poly(lactide-co-glycolide)	Oral delivery of insulin	Insulin release in gastric condition was negligible and released in the stimulated intestinal condition. Permeability increase compared with free insulin	*In vitro* and *in vivo*
Hybrid nanoparticles	Doxorubicin and Mitomycin C	Polymer-lipid	Synergistic activities	Increased biodistribution, pharmacokinetics, bioavailability and reduced toxicity	Implantable
Niosomes	Doxorubicin	Sorbitan Monostearate	To improve pharmacokinetics and tumoricidal activity	Increased bioavailability, pharmacokinetics and against multidrug resistance	*In vivo* and *in vitro*
	Doxorubicin	C16 triglyceryl ether with or without cholesterol	To distribution of drugs	*In vitro* drug release from cholesterol-containing niosomes is delayed, *in vivo*, there is little difference between the two preparations. More effective reduction in tumor growth	*In vitro* and *in vivo*

(continued)

Table 1 (continued)

	Drug/active ingredients	System components	Aim	Major applications	Route of administration
Carbon Nanotubes	Paclitaxel	PEG single-walled carbon nanotubes	To targeted delivery	Exhibit biocompatibility, excretion, and less toxicity.	*In vivo*

Reproduced from Chowdhury et al. [20]

limiting the passive targeting [25]. Thus attachment of target molecules at the surface of nanoparticles helped to overcome the limitation, also known as active targeting. The delivery system, when equipped with the 'homing device,' it guides the carrier towards the intended targets, is capable of recognizing and binding to the complementary receptors present on the surface of the cancer cells [22]. The ligands may also help in protecting the NPs from the enzymatic destruction increasing its effectiveness of the treatment and reduces the toxic effect on healthy tissues. For example, Kumar et al. synthesized functionalized gold nanoparticles (AuNPs) with PMI (p12), therapeutic peptide, and targeted peptide, CRGDK for binding to receptor neuropilin-1 which are overexpressed on cancer cells resulting in membrane receptor-mediated internalization [26]. In *in vitro* studies, after maximum binding interaction with the MDA-MB-321 cell surface, there is an increased rate of the delivery of therapeutic p12 peptide inside targeted cells [26]. FDA approved antibodies such as rituximab, ipilimumab, and trastuzumab used for clinical treatments, are widely studied ligands due to their high specificity and availability [27]. Dendrimer conjugated with antibody binds exclusively to human prostate adenocarcinoma cells [28].

Apart from the merits, antibodies are found difficult to conjugate with the NPs, resulting in shorter circulation time and thus expensive. Peptides have shown promising alternate solutions due to their smaller, stable, simple, and more comfortable to produce. Nucleic acid base aptamers consist of advantages of both peptides and antibodies, but on the other hand, they degrade quickly. Other small molecules can also be used as ligands on the NPs, such as folic acid for folate receptors [29]. Such units are small, stable, and easy to produce. Unfortunately, ligand detection for relevant substrates is challenging. Even with the proper binding of receptors and ligands, binding incompatibility can limit therapeutic efficiency. Multiple ligands with different charges can increase overall affinity for binding, but the limited binding ability and capacity of receptors will govern the quality and quantity of the binding. For instance, the overly strong coupling can reduce tumor penetration, hinder selectivity, and lead to an overdose of carriers [30].

Active targeting directs the nano-carriers to a specific cell, tissue, or organ, which remains at the site of disease for an extended period, thereby increasing the local accumulation of NPs. In comparison, passive targeting depends on the natural distribution of nano-carrier and the EPR effect. Both of these targeting processes depends on the location of initial drug delivery and blood circulation [31].

5 Controlled Drug Release

A flawless drug carrier should have the ability of controlled drug release along with the nature of a high load of the drug. Early release of drugs in blood circulation can harm healthy tissues and can show severe side effects. The evolving field of Nanotechnology can overcome these side effects with the use of nanomaterials for drug delivery. Many researchers use biodegradable materials such as polymeric nanoparticles, dendrimers, and liposomes as smart carriers for the release of pharmaceutical drugs in a controlled manner in the aqueous environment upon the structural degradation of the nano-carriers activated by different chemical factors, such as pH [32]. Though soft and structurally unstable materials are favorable for drug delivery, it is strenuous to attain the premature release of drugs with zero percent. In many scenarios, the drug molecules entrapped in the matrix would start to release from the biodegradable nano-carrier immediately after the system was injected *in vivo*. The premature release problem limits the usage of these DDS for effective disease treatment along with the significant challenge faced in oral administration for the site-selective delivery of protein and nucleotide-based drugs. These pharmaceutical or nutraceutical cargoes of enzymes, DNAs, and RNAs would be decomposed in the highly acidic environment of the stomach unless the nano-carriers could offer the essential protection [33].

Recent research has to concentrate on the development of structurally stable DDS that can help to deliver a large number of drug molecules with minimum premature release difficulty to the targeted tissues, cells, or even intracellular organelles. In 2003, the first designed mesoporous silica nanosphere (MSN) was based on a stimuli-responsive system using the advantages of various chemical structures used as "gatekeepers." Thus helped to modulate the encapsulation of the drug molecule and its release (Fig. 2) [34]. MSNs with cyclodextrin gatekeepers can be hydrolyzed by amylase enzyme, which helps to liberate the guest molecules from the porous

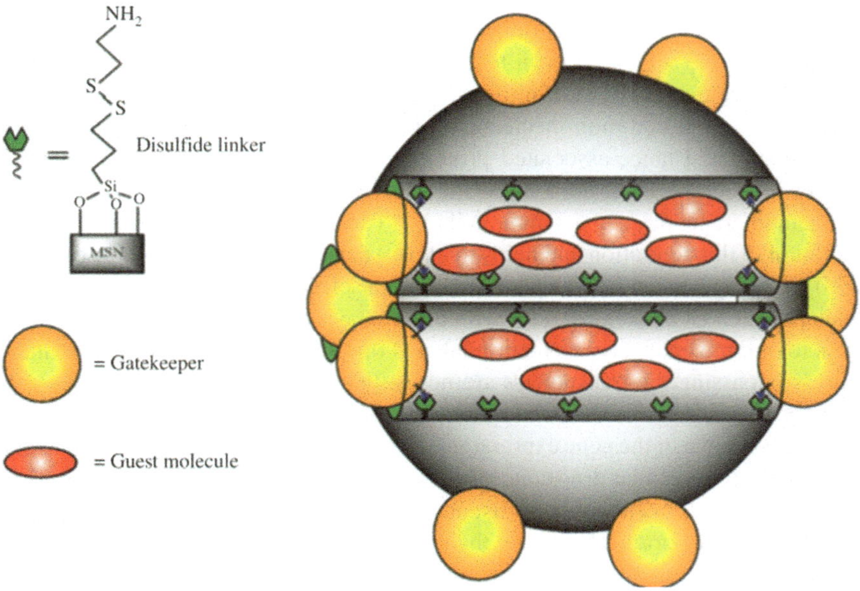

Fig. 2 Representation of an MSN loaded with the guest molecule and end-capped with a general gatekeeper (Reproduced from Chen et al. [34] under a Creative Commons Attribution 4.0 International License. https://creativecommons.org/licenses/by/4.0/)

reservoir. The cyclodextrin gatekeeper can be cleaved at the ester linkage present in the stalk part by lipase, releasing the guest molecules from the channel. Investigation of different categories, for the controlled DDS that uses various gate based on the surface of MSNs, such as AuNPs, AgNPs, dendrimer, and other gatekeepers [34]. Among the investigated DDS that are structurally stable, Bio-MSNs are suited to be the best candidate, because of its advantages in intracellular delivery, due to larger area of surface, tunable pore sizes, and volumes, and encapsulation of various drugs, proteins and biogenic molecules [35].

Also, to achieve the goal of the controlled release of the drug, nano-carriers been designed to have environment-sensitive delivery. Recently, with the help of the self-assembly approach, a pH/thermo-responsive nano-carriers were fabricated with dually responsive poly (N-isopropyl- acrylamide) (PNiPAM)-co-acrylic acid (AA) hydrogel core coated with silica. These core-shell particles are uniform sized with better dispersity as compared with MSNs. In *in vitro* studies, the nano-carriers are more biocompatible and less cytotoxic. Also, the controlled drug release results specify that PNiPAM/AA@SiO_2 particles are responsive in a pH-sensitive environment with the higher releasing efficiency as compared with the MSN particles, which is a desirable trait for cancer treatment [36].

6 Multidrug Resistance

In cancer treatment, drug-sensitive cells are killed, while the proportion of the drug-resistant cells are left out known to be multidrug resistance (MDR). The integral membrane protein also is known to be MDR transporters increases the drug efflux, which is the most common mechanism observed in MDR. The transporters such as multidrug resistance-associated proteins (MRPs) and p-glycoprotein (MDR1) reduces intracellular drug dose below the lethal threshold level by actively pumping the chemotherapeutic drug out of the cell [37]. The use of nanomaterials has provided a solution to this limitation and helps to bypass the MDR transporters. For example, SP1049C is an innovative micellar formulation of DOX, which manifests greater efficiency then DOX alone at odds with different drug-resistant tumor cell lines and helps to circumvent p-glycoprotein-mediated drug resistance [38]. However, nano-carriers bypassing the transporters are not sufficient to overcome drug efflux. The mechanism of drug efflux is due to the overexpression of transporters in the cancer cells. Knockdown of the gene expression for the transporters using siRNA could help reinstate the intracellular drug concentrations which are required for the induction of apoptosis and cell cytotoxicity. Thus, nano-based co-delivery systems are designed to target the transporters along with the delivery of the drug to increase the intracellular drug concentration. Meng et al. have reported MSNs to be functionalized with DOX along with the p-glycoprotein siRNA (to silence the expression of the efflux transporter) to an MDR cancer cell line synergistically [39].

The activation of anti-apoptotic pathways is another problem faced in MDR. Cells have developed anti-apoptotic pathways such as genetic and epigenetic modifications of cancer cells to avoid death. The alterations include overexpression of B-cell lymphoma 2 (Bcl-2), which is an anti-apoptosis regulator protein, and also mutation in p53 (tumor suppressor) gene [40]. The nano-carriers can also be used for the co-delivery of the cytotoxic drug along with the gene targeting apoptosis regulator protein such as transcription factor NF-κB, hypoxia-inducible factor-alpha (HIF-1α) [23] and Bcl-2 [41]. Such a synergistic effect can increase the therapeutic effect by intracellular drug accumulation. For example, the synthesis of nanoparticles attained by calcium carbonate ($CaCO_3$) co-precipitation method was used for co-delivery of DOX and p53 gene [42]. These nanoparticles have achieved higher cell inhibition and promoted tumor cell apoptosis with more efficiency.

7 Multifunctional Nano-Carriers

Multifunctional nano-carriers are used for the synergistic effects which are capable of delivering therapeutic drugs along with the image contrast enhancement agent at targeted body sites. Kim et al. synthesized chitosan-based nanoparticles, containing anti-cancer drug paclitaxel and near-infrared fluorescent (Cy5.5) [43]. C-NPs are stable in serum, deformability, and readily uptaken by cancer cells. A new era in cancer therapy is considered where early-stage cancer diagnostics, drug delivery and non-invasive real-time monitoring of therapeutic efficacy can be carried out simultaneously. Yang et al. have succeeded in nano-structure for multifunctional systems uniting chemotherapeutic drug DOX and magnetic nanocrystals with therapeutic antibodies [44]. These nanosystems have the ability of treatment, imaging, and targeting.

Nanomaterials with multifunctional abilities have been used for *in vivo* imaging, silencing of the tumor, and siRNA delivery. Magnetic nanoparticles are synthesized labeled with a near-infrared dye that is covalently linked to siRNA molecules that translocate to cytosol with high gene transduction efficiency [45]. Guo et al. synthesized a multifunctional nano-carrier using a graft-degradable cationic co-polymer named poly (ε -caprolactone)-graft-poly (2- (N, N-dimethylamino) ethyl methacrylate) (PCL-g-PDMAEMA) [46]. These nano-carriers were found to load DNA and entrap hydrophobic paclitaxel simultaneously. It was found that the nano-carriers are pH-sensitive, i.e., the drug is liberated faster in an acidic environment. While, as compared with Lipofectamine^{TM} 2000, PCL-g-PDMAEMA NPs exhibited high gene transfection efficiency. It was also found that the nano-carriers escapes the endosome and thus release the cargo in cytoplasm effectively. This has helped to achieve the synergistic effect of drug and gene therapy *in vivo*.

The application of nanomaterials as pharmaceutical nano-carriers, such as micelles, liposomes, and bio-MSN, have shown a heterogeneity of useful properties, such as longer blood circulations, increased intracellular penetration, and specific targeting. The continuous advancements in biomarkers and targeting ligands have been increasingly found suitable in the development of optimized nano-carriers as favorable candidates for clinical trials.

8 Nanomaterials in Medicine

Nanomedical approaches have been a major transforming factor in diagnosis and therapies for various ailments. There have been made significant efforts devoted to developing suitable nanotechnological platforms to improve drug delivery to the targeted area. Multiple nanoscale carriers such as inorganic, polymeric, carbon-based, lipid-based, and protein-based nanoparticles have been developed for this purpose. This section discusses various nanomaterials used as drug delivery systems towards infected areas.

9 Nanoparticle Targeting in Cancer Cells

Almost all anticancer drugs are the basis for injuring the normal cells, along with the cancer cells. During cancer chemotherapy, optimum dose and frequency are the prime factors. Now more focused efforts are attempted in killing the cancer cells using target specificity while sparing the normal cells. As compared to conventional cancer therapies, nanoparticulate drug delivery systems dispense better penetrations of therapeutic and diagnostic molecules inside the cancer tissue. Nanoparticles are formulated to take advantage of basic cancer cell morphology and to identify them based on their means of development, such as rapid cell proliferation, expression of antigen, leaky tumor vasculature, etc. It is crucial to develop nano-carriers for drug delivery in such a way that small doses of the chemotherapeutic agents could be delivered in an active form and controlled drug distribution within the body.

In cancer chemotherapy, nano-carriers are at an advantage over the free drug delivery. Here the drug is protected from premature degradation, prevention from early interaction with the biological environment, enhanced absorption into the solid tumor, improved intracellular penetration, controlling of drug kinetics and drug tissue distribution profile [47]. Nanoparticulate drug delivery systems are designed with the help of specific targeting agents for cancer cells. The targeting moieties are diverse ligands (such as peptides, aptamers, antibodies, and small molecules) that hold a high affinity towards distinctive signatures. They may also help in protecting the NPs from the enzymatic destruction and minimize the uptake of the drug by the healthy cells [48]. Hence, the enhanced retention and entry of the agent in tumor cells. Ligand molecules attached as a targeting agent on the surface of the nano-carriers will recognize and bind specifically to the receptor antigens on the cell surface. Even with the help of the antibody conjugated therapeutic agent, we can increase drug targeting efficacy.

There are three general categories for specific targeting in cancer—(i) uncontrolled cell proliferation targeting, (ii) tumor cell targeting, and (iii) angiogenesis-associated targeting. Various proangiogenic factors are responsible for the chemical stimulation of the angiogenesis, such as basic fibroblast growth factor (bFGF) and vascular endothelial growth factor (VEGF). These are the critical factors for a complex biological mechanism that is essential for life and also for the growth of solid tumors. Thus, it has become necessary to target angiogenesis in cancer therapeutics [49, 50]. Therefore, targeting angiogenic factors with the usage of drugs will cause tumors to cease from developing new blood vessels.

Cancer cells can also be targeted with the help of cell proliferation markers. The markers used are ordinary to the cancer cells. But specific cancer cells overexpress these markers, which are mainly transferrin receptors, folate receptors, and human epidermal receptors (HER). Nanoparticles are embedded with monoclonal antibodies to target overexpressed receptors for cell proliferation [51]. FDA has approved the treatment of monoclonal antibodies in some of the cancers. A humanized monoclonal antibody, Trastuzumab, is used against HER-2, overly expressed in positive breast cancer. A trastuzumab-conjugated nanoparticle can target specifically HER-2 positive cells, which have been demonstrated in several *in vitro* case studies using divergent cell lines [52] and also in some *in vivo* studies [53]. Trastuzumab-conjugated super magnetic iron oxide NPs serve as an MRI contrast agent to detect HER-2 positive tumors and can be easily identified by magnetic relaxometry [54]. This can be used for diagnostics of early breast cancer. Apart from this, trastuzumab is used as the delivery of chemotherapeutic drugs to cancer cells [52].

A serum glycoprotein, transferrin (TfR), transports iron into cells through the bloodstream via binding to the cell surface transferrin receptor. This results in receptor-mediated endocytosis and, thus, internalization of the iron [55]. The up-regulation of the transferrin receptors is found to be 100-folds higher in metastatic and drug-resistant malignant cells than that of the normal cells. Doxorubicin-loaded liposomal nano-carriers using TfR as targeting ligand is an intracellular drug delivery system [56]. The TfR-conjugated liposome NPs have resulted in being efficient drug delivery cargo into the neoplastic cells. These NPs bind specifically to the TfR receptors and thus internalized via a receptor-dependent endocytotic pathway, which helps it to bypass P-glycoprotein-mediated drug efflux to overcome the MDR [57].

Several studies have used transferrin to target cancer cells using paclitaxel-loaded polymeric PLGA NPs. TfR-conjugated PLGA NPs loaded with aromatase inhibitor were evaluated *in vitro* against SKBR-3 breast cancer cells. This resulted in enhanced activity of aromatase inhibitor of the TfR receptor-targeted NPs than that of the nontargeted NPs due to the TfR-mediated uptake [58]. This has achieved the highest accumulation of the nanoconjugates at the tumor sites with in significant reduction in weight of the tumor in the *in vivo* S180 murine sarcoma model [59].

A liposomal formulation of mifamurtide (MEPACT) is aimed to target the delivery of the drug to macrophages and monocytes, which are present in the spleen, liver, and lungs. Mifamurtide is a synthetic derivative of muramyl peptide used as a chemotherapeutic adjuvant. It majorly stimulated monocytes and macrophages to enhance tumoricidal effects. The liposomal encapsulation helps to enhance the tumoricidal capabilities of the monocytes/macrophages and to increase the biopersistence of the drugs in the lungs [60].

Abraxane, co-condensed with albumin and paclitaxel when conjugated with Lyp-1, which aims for actively targeting the cancer cells. Lyp-1 is known to be a cell-penetrating peptide. The following conjugate has been tested on MDA-MB-435 human cancer xenograft in mice. Abraxane-Lyp-1 conjugate showed enhanced drug delivery to the tumor with efficient tumor growth inhibition as compared with non-targeted NPs [61].

Also, in recent times many researchers have been reporting to achieve great success in the green synthesis of metal NPs using plant extracts against *in vitro* cancer cell treatments [62]. *Elephantopus scaber* L. hydro-methanolic extract mediated AgNPs showed good anti-proliferative activity with $GI_{50} < 10$ μg/ml against MCF-7, A-579, and SCC-40 malignant cell lines. Synergistic activity with standard Adriamycin was found to have better results on the cancer cell lines [63]. Another technique for the biosynthesis of AuNP was developed and studied for cancer treatment [64].

The majority of cancer nanodrugs are mainly based on liposomes and polymer nanoformulations, which help to lower the toxicity and enhances chemotherapeutic delivery via passive targeting. Nanoparticles consisting of receptor-based active targeting has also become a potential optimal delivery strategy. These nano vehicles are being used in early tumor diagnostics, therapy, and post-therapeutic follow-ups. Synergistic therapies and approaches can be used to increase the precision and accuracy of drug delivery for potent personalized treatment.

10 Nanotechnology for Brain Drug Delivery

The BBB prevents the transit of most circulating cells and molecules. The structure of BBB is organized by complex systems of pericytes, astroglia, endothelial cells, and perivascular mast cells. The nanodevices used as drug delivery systems should be between 1 and 100 nm, which works as a whole unit for transportation across BBB [65]. These delivery systems enhance the permeability of molecules across BBB and, at the same time, may also assist in the targeting ability of drug in the brain. This phenomenon depends entirely on the biomimetic features of the carrier and its physicochemical properties. The chemical composition of the drug inside the nanoparticles is the least of all concerns when it comes to cross BBB. The nanocarriers for drug delivery, which are unable to cross BBB, can be provided with stealth coating using some polymeric materials to avoid the reticuloendothelial system. This would help with longer circulation time, stability in blood, and functionalized to cross BBB successfully and target the brain.

The nanoplexes of AuRds-DARPP-32 siRNA have been synthesized to target and diminish the key protein expression (DARPP-32), protein phosphatase 1 (PP-1) and extracellular signal-regulated kinase (ERK) in the case of the dopaminergic signaling pathway for an ailment of drug addiction in the brain [66].

Transferrin receptor is found to be expressed on higher levels on malignant cells than that of the normal cells [67]. It is as well found to express on brain capillary endothelial cells [68]. BBB becomes an obstacle to the delivery of a chemotherapeutic drug. Thus, the transferrin receptors can be used for successful drug delivery beyond the BBB. Ulbrich et al. verified that loperamide-loaded human serum albumin NPs, surface-functionalized with TfR targeting mAbs (R17217 or OX26), could cross BBB as compared to loperamide alone [69]. Similarly, Gosk et al. have illustrated that OX26 (anti-transferrin receptor antibody) conjugated liposomes can selectively target brain capillary endothelial cells [70]. The formulation of liposome is used so that the liposomes may degrade in the brain capillary endothelial cells, which further allows liposomal cargo to travel into the brain.

Chitosan NPs have shown encouraging results in drug delivery to the brain via nasal route due to its excellent native properties such as good biocompatibility, low toxicity, high loading and entrapment efficiency, and having the potentiality to deliver hydrophilic molecules. Md et al. have produced chitosan NPs loaded with bromocriptine (BRC-CS NPs) using an ionic gelation technique with tripolyphosphate (TPP) as an anion [71]. Pharmacodynamic studies have exemplified the potential of BRC-CS NPs in neuroprotective by preventing the rise in cataleptic behavior, and akinesia score decreased sharply in haloperidol-induced swiss albino mice model of Parkinson's disease. The pharmacokinetic studies and biodistribution showed high brain/blood ratios of intranasal BRC-CS NPs, indicating direct transport of BRC from nose to brain along trigeminal nerve pathways or olfactory bypassing the BBB.

Nano neurotechnology is the application of nanotechnology in drug delivery for the treatment of Neurodegenerative disorders. There is a demand for the development of diverse therapeutic approaches to provide optimal treatments. The used surface engineered nano-carrier system for the delivery through passive and active targeting would be beneficial to promote progress in this field.

11 Nanotechnology in Gene Delivery

Various genetic disorders like cystic fibrosis, sickle cell anemia, diabetes mellitus, alpha 1 antitrypsin deficiency, etc. may be treated by the transfer of genetic material with the help of nano-carriers. Such systemic diseases are caused by the absence of enzymes, mainly due to missing or gene defects [72]. Apart from using gene therapy to treat genetic disorders, nowadays, these could be used as a carrier system that could be implanted to fight against other diseases like cancer, nervous conditions, cardiac ailments, etc. Nanoliposomes can be used for genetic material transfer into the cells. PEG and galactose incorporated nanoliposomes could target liver cells with efficient due to the prompt uptake by the Kupffer cells. Such type of therapy could be used to treat liver disorders like Wilson's and hereditary hemochromatosis. In another effective method, nanoliposomes can be administered using a ligand-receptor complex system where small protein EGF binds with receptor EGFR. In another approach, instead of encapsulating negatively charged plasmid DNA with nanoliposomes, they are mixed with the cationic lipids, which lead to the shaping of lipoplexes where the process is navigated by electrostatic interactions [73]. Such complexes can enter the infected cells by infusion with the endosome of the plasma membrane. For example, a plasmid, Allovectin-7 contains a gene for major histocompatibility complex (MHC) antigen HLA-B7 along with B2 microglobulin together formulated with cytofectin [74]. The unloading of the gene from nanoliposomes is decided by the nature lipid composition, which controls the means of release, Nanoliposome doping with neutral lipid-like 1,2-Dioleoyl-sn-glycero-3-phosphoethanolamine (DOPE) which is useful in endosomal membrane fusion. It helps to recognize and destabilize the phospholipids using the flip flop mechanism and paves the way for nanoliposomes to integrate with the membrane and release the nucleic acid into the cytoplasm [75].

12 Nanoparticle Therapy in Wound Healing

Wound healing is problematic for which wound management is necessary. When a biomaterial is embedded or incorporated into nanoparticles, it can be utilized as a potential wound dressing material. Nanotechnology propounds different new approaches for regenerative medicine. Nanomaterials help in burn treatments and enhance delayed wound healing. Nanoparticles help to deliver external substances that are persistently produced at the site of injury. Nitric acid (NO), is one of such endogenous molecules with the half-life in seconds. In diabetic wounds or ulcers, i.e., delayed or non-healing wounds, NO is generated in lower concentrations. Thus, the exogenous delivery of NO at the site of injury is promising therapy. But due to the lack of delivery material, its practical application is limited. The causes of delayed healing of the wound in diabetes as observed are wound dehiscence, prolonged inflammation, increased oxidative stress, impaired immune system, redox imbalance, growth factors degradation, reduced blood flow, sustained infection, and impaired macrophage activity. Blecher et al. found out that the NO-releasing nanoparticles (diazeniumdiolate) hastened the closure of the wound in diabetic mice along with few inflammatory cells, fibroblasts, increased blood vessels, and organized collagen content [76]. In another study, it was confirmed that NO-NPs treatment in mice increased wound healing and an increase in anti-inflammatory cytokines and various growth factors (mainly TGF-β). There was also increased collagen expression along with accelerated migration and proliferation of fibroblasts [77]. Topical application of PLGA containing recombinant human epidermal growth factor (rhEGF) NPs showed accelerated wound closure, the improved proliferation of fibroblast, and increased epithelization in diabetic mice *in vivo* [78].

Bioactive plant component-based NPs have been used due to its properties like anti-microbial and wound healing. Curcumin (the active compound of turmeric) has anti-inflammatory, anti-microbial, and anti-oxidant properties [79]. Curcumin, when topically applied, has resulted in enhanced re-epithelization, increased collagen deposition, fibroblast proliferation, and different growth factors in diabetic rats [80]. Due to its low water solubility and poor bioavailability, curcumin has limited its

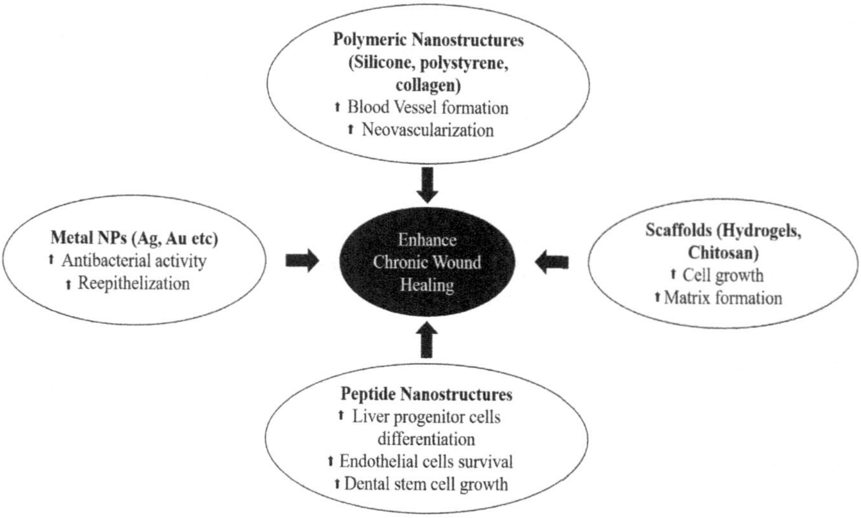

Fig. 3 Potential role of nanoparticles in accelerating wound healing (Redrawn from Rajendran et al. [84])

clinical uses. Curcumin loaded nanoparticulate system will consist of several benefits, including site-specific delivery, enhanced drug uptake, and improved solubility. For example, in one of the *in vivo* studies, it was shown that curcumin NPs have decreased microbial activity thus, reduced infection levels and enhanced wound healing in a methicillin-resistant *Staphylococcus aureus* infected mice [81].

Nanoparticles can be used for natural products, bioactive plant components, gaseous molecules, and growth factors directly at the site required. Nanoscaffold dressings can be used for a wound dressing, which may help lower the microbial infections and healing of wounds. In *in vitro* studies, the NPs showed strong antimicrobial activity against *Staphylococcus aureus*, *Pseudomonas aeruginosa*, and *Escherichia coli*. It also promoted the growth and proliferation of keratinocyte, thus a suitable environment for the growth of cells [82]. Many researchers have studied and confirmed the use of nano scaffolds as a substitute for damaged ECM. It also helps to increase fibroblast attachment, proliferation, and matrix formation. This provides solid grounds for enhanced healing activity [83]. Figure 3 illustrates the potential role of nanoparticles in expediting wound healing.

13 Silver Nanoparticles

Silver has bactericidal properties and commonly used for the treatment of wounds, burns, and a variety of ulcers. Silver nitrate has been used for chronic non-healing wound treatments. Nowadays, silver coatings in various forms are used in wound dressings, which are efficient in the even diffusion of the drug and play an integral part in the management of chronic wounds [85]. AgNPs, even used for a prolonged time, do not create any obstacles. The combination of AgNPs and collagen is a fitting component for wound dressing, giving it an intense antibacterial activity [86].

AgNPs show a wide range of anti-microbial activity, effectively disturbing the quorum sensing, which results in reduced formation of biofilms and detoxifying bacterial toxins [87]. Under *in vivo* conditions, AgNPs do not possess any of its biological functions. Thus the acidic environment facilitates the oxidization of AgNPs into Ag ions, which are then responsible for damaging cell wall and DNA and inhibits the synthesis of ATPs, thus resulting in ROS generation giving its antimicrobial properties [88].

In *in vitro* studies, the treatment of AgNPs using human dermal fibroblast and keratinocytes remarkably reduced the levels of oxidative stress, inflammatory cytokines, and thus promoted healing [89]. While in *in vivo* studies, mice with burn wounds was used to demonstrate that when AgNPs applied topically, there were reduced counts of neutrophils and Interleukin (IL)-6 levels. This also helped to elevate the levels of TGF-β, IL-10, interferon-gamma (IFN-γ), and vascular endothelial growth factor (VEGF) [90]. These findings have supported the prospective of AgNPs in anti-microbial activity and boosting the healing of the wound.

Wound dressings that are made up of biocompatible cellulose- AgNPs often results in secure attachments and proliferation of keratinocytes at the fringe of the wounds. AgNPs, when integrated with poly (dopamine methacrylamide-co-methyl methacrylate) (MADO) nanofiber and used in the dressing of the injuries, is found to have effectively inhibited the growth of microbes *in vitro*, such as *P. aeruginosa*, *E. coli*, and *S. aureus* [91]. When MADO-AgNPs were applied to *in vivo* studies to partial-thickness cutaneous wounds, the treatment resulted in complete wound healing within two weeks with an increase in epithelization as compared to the incomplete healing in the untreated group [92].

14 Gold Nanoparticles

AuNPs are used in targeted drug delivery, tissue engineering, and wound healing due to its biocompatibility properties. AuNPs, when embodied with various other biomolecules, can be effectively used in biomedical applications. AuNPs, when crosslinked with collagen, can be easily incorporated at the surface with different biomolecules such as peptized, polysaccharides, cell adhesion molecules, and growth factors. This modified version of AuNPs consists of properties like biodegradability and biocompatibility, which thus can be used in the healing of wounds [93].

Vancomycin (anti-microbial drug)-mediated AuNPs have increased activity against vancomycin-resistant enterococci (VRE) by 50-fold. It also showed remarkable activity against a gram-negative bacterium, *E. coli*, where vancomycin usually is ineffective [94]. AuNPs can be used for photothermal therapy [95] or photodynamic therapy [96], where they are conjugated with pathogen-specific antibodies and photosensitizing molecules, respectively to achieve the anti-microbial activity. Sherwani et al. synthesized AuNPs combined with photosensitizer showing anti-fungal activity in mice with *Candida albicans* infected wounds [82]. Topical application at the site of the injury helped to enhance wound healing distinguished by increased granulation tissue formation, re-epithelialization, collagen fiber content, and ECM deposition.

Chitosan-AuNPs showed increased free-radical scavenging activity by several folds with enhanced biocompatibility. In a rat surgical wound model study, chitosan-AuNPs substantially increased epithelial tissue formation, improved hemostasis with a faster rate of healing than that of chitosan dressing alone. Also, in the *in vitro* studies carried in rats, the cryopreserved human fibroblasts combined with AuNPs (CrHFC-AuNPs) were topically applied to burn wounds showed enhanced healing rate with increased deposition of collagen and reduced inflammatory phase [97].

15 Peptide Nanoparticles

Peptide nanoparticles can be synthesized through self-assembly approaches and molecular chemistry used in biomedicines for targeted drug delivery and also helpful in understanding cell signaling. Mimicking of the natural ECM is possible with the help of a self-assembled peptide scaffold with functionally modified for increased interaction with the other cells and tissues. A typical example is of peptide hydrogels, which are shown to have excessive cytocompatibility along with the biocompatibility in various biological systems [98].

Fibrils are made up of synthesized peptides by folding, thereby converting it into hydrogels, stimulated, and distributed evenly in cell cultures without affecting its viability [99]. In *in vitro* studies, the peptide hydrogels are shown to stimulate the ability of cell attachment, helps to differentiate liver progenitor cells into hepatocytes, thus significant contribution towards the regeneration of the liver tissue [100]. In many studies, peptide hydrogels help to smoothen the survival of endogenous endothelial cells, thus providing a better microenvironment [101].

Some tailor-made, peptide amphiphile (PA) systems have been synthesized expressing bioactive epitopes, helping in augmentation of the scaffolds for interesting cell types. For example, a PA was designed with specific cell adhesion epitope Arg-Gly-Asp-Ser (RGDS) aimed to alternately support the development of mononuclear cells of bone marrow and epithelial cells present in enamel. At the same time, the absence of the RGDS epitope in the control group resulted in a lowered number of mononuclear cells of bone marrow [102].

In a new strategy developed, a mixture of different hyaluronic acids was allowed to yield a self-sealing pouch. This was filled with specific liquids used to encapsulate human mesenchymal stem cells (MSCs) and to be used to target sites to deliver MSCs for tissue regeneration [103]. In recent studies, a self-assembled conjugated peptide was used in bone injury to promote regeneration, and the functional recovery of the chondrocytes [104]. The peptide hydrogels thus reveal to have significant benefits in tissue regeneration by providing a commendatory environment for cell growth and differentiation. Also, amino acids used for peptide nanostructures help them to retain maximum biodegradability, and biocompatibility and lastly, sequences can be tailor-made as required for peculiar target sites. Peptide nanomaterials, when used for wound treatments, it is free from graft rejection and do not show any immunogenic activity [105].

16 Polymeric Nanoparticles

Polymers are a long chain of organic molecules in nanostructures used in the development of surgical tools, vascular grafts, implantable devices, and medical device coatings. These are of low cost and easy to manufacture with the required flexibility for its use in several types of dressing materials. During the synthesis of polymers, its functional properties can be altered and modified as per the specific requirements of the biological applications. These synthetic polymers are utilized for the preparations of scaffolds, physiological barriers, surgical sutures, and vascular grafts. As compared to the biological scaffolds, these are inexpensive with longer shelf-life. In some studies, polyurethanes (PU) have been indirectly found to enhance cellular proliferation by effective stimulation of angiogenesis and re-epithelization in the wounds of rats [106].

Polymeric hydrogels have shown positive effects in boosting tissue repair. In the initial phase of healing, the hydrogel scaffolds help to stimulate the infiltration of inflammatory cells and successive stages of healing it facilitates the requirement of angiogenic cells, thus resulting in the progression of delayed healing. Further, the addition of growth factors or cytokines to the polymers helps to facilitate the formation of new blood vessels, neovascularization, and enhanced microenvironment around the area of wound [107]. These compositions are effective in treating both normal as well as delayed infectious wounds. In some illustrations, the cross-linked polymer linkages might swell in an aqueous environment from its original state. This reveals the remarkable role of polymers contributing to the supplementing healing of the wound.

Gelatin is used in the fabrication of biocompatible and biodegradable materials for wound dressing. Powell and Boyce demonstrated skin repairs with the use of gelatin scaffolds, mainly dependent on its porosity and inter-fiber distance [108]. Bilgic et al. showed the topical application of gelatin scaffolds to the rat wounds resulting in enhanced wound closure and healing [109]. Fibrin is polymerized from fibrinogen in the presence of the enzyme thrombin, possessing properties that include an increase in immunological response, reduction of inflammation and cell adhesion which is widely used in wound healing and tissue engineering [110].

Hydrogels are soft that retain water easily, thus preventing tissue dehydration and hence used in bandage preparations and dressing of acute/chronic wounds, burns, and diabetic foot ulcers. On the other hand, superior mechanical support is provided by the scaffolds along with the delivery of the growth factors at the targeted sites;

a trait required the most for tissue regeneration. The scaffolds are to be embedded at the site of a target restricting its use [111]. According to another point of view, the usage of hydrogels is more popular as they can be injected at the site of injury and achieve the shape of the defect with extraordinarily elastic and flexible due to its magnitude to retain water. In recent clinical trials, biodegradable gelatine hydrogel was combined with rhFGF-2, injected into the knee of osteoarthritic patients during the surgery resulting in safe and successful [112]. In another clinical trial, FGF was used in fibrin hydrogel and was used in the treatment of cervical spinal cord injury, revealing strong properties of nerve regeneration [113].

For the second-degree burns, the use of granulocyte-macrophage colony-stimulating factor (GM-CSF) along with hydrogel was found to be active and safe [114]. Even the dispensation of PDGF in a hydrogel with a periodontal defected patient resulted in remarkable advancement in none regeneration [115]. The FDA has approved the use of carboxymethylcellulose-PDGF based topical gel for the treatment of diabetic neuropathic ulcers. Hydrogels used as drug delivery vehicles have been studied in various preclinical researches showing positive outcomes in the regeneration of several injured tissues. These hydrogels are loaded with growth factors for implantations or as injectable solutions [116].

In some cases, a mixture of hydrogels containing growth factors is placed over a scaffold and embedded at the site of injury. For example, hyaluronic acid/fibrin hydrogel conjugated with bone morphogenetic protein-2 (BMP-2) exhibited an activity of robust bone regeneration. Similarly, BMP-2 was delivered using heparin hydrogel coated over the poly-lactide caprolactone scaffold, leading to active and improved bone formation and mineralization [117]. Researchers used new covalent binding strategies to avoid rupture of growth factors from the hydrogels, resulting in improved properties of hydrogel consisting of sustained release of growth factor at the targeted site.

17 Advanced Targeted Nanomedicine

The heterogeneity and adaptability of different ailments are strenuous to overcome, which creates problem to have an ordinary cure based on passive or active drug targeting. Therefore, it is foremost to consider the use of external physical stimuli as a targeting strategy. The infected area varies from the healthy tissues in various physical properties such as lack of oxygen, temperature, and pH, which can be exploited for targeted strategies. The principle behind the physical targeting is the use of nanoparticles along with a physical stimulus applied externally, or it is originated at the infected site. This prompts either a physical change in the nanoparticle structure, causing target eradication or modulating the rate of release of the drug. Electromagnetic radiation, ultra-sound, mechanical forces, lack of oxygen, temperature, and pH are various stimuli used for physical targeting.

18 Magnetic-Sensitive Systems

Magnetic sensitive nanoparticles are primarily fabricated in such a way that therapeutic agents are encapsulated inside the magnetic core or attached to the shell, thus functionalizing the surface. These functionalized NPs are injected into the bloodstream, positioned near the target site. Magnetic fields are applied over the target site with the use of powerful rare earth magnets with a high gradient affecting the particles. Here, the magnetic field captures and releases the nanoparticles at the target site. The clinical trials by Koda et al., deliver doxorubicin hydrochloride nanoparticles, showing success for hepatocellular carcinoma [118]. This method is helpful when the target is near the body's surface. The strength of the magnetic field falls off swiftly as the site deepens within the body, becomes tougher to target. Several researchers proposed to overcome this hurdle by focusing the infected sites with the help of implanted magnets inside the body [119]. But as the suggested method is invasive and not systematic, its application may restrict for clinical use. Synthesis of smaller NPs is essential for superparamagnetism to tackle the problem of magnetic agglomeration. As the size decreases, the magnetic field may fail to guide the particles and keep them at target proximity remarkably while resisting the drag of blood flow.

19 Ultra Sound-Sensitive Systems

The homogenous distribution of nanoparticles in the solid tumor becomes challenging as they cannot travel far away from the blood vessels to the extracellular matrix. Thus, nanoparticles can reach and release the cytotoxic drug to the cancer cells that are closely located to the blood vessels; hence, a minimized therapeutic index. These nanoparticles can target and damage healthy cells due to their ability to stay for more extended periods. Researchers have explored the use of ultrasound, which helps to increase the permeability of the capillary walls and to press them through the extracellular matrix; thus, enhanced delivery of drugs from nano-carriers and improved cellular uptake. Some studies have shown 2D and 3D models where ultrasound is used to help the nanoparticles to penetrate the solid tumor along with the controlled release of the drug into the intracellular spaces without any alterations in its chemical properties [120].

The main advantage of the ultrasound sensitive system is its absence of ionizing radiation, non-invasiveness, and smooth regulation of tissue penetration depth by tuning duty cycles, frequency, and exposure time. Ultrasound destabilizes the cell membrane that helps to release the drug into the cytosol; thus, bypassing the degenerative endocytotic pathway. Also, low-frequency ultrasounds can be utilized to deliver liposomal-based drugs past the skin, which helps to obstruct the progression of melanocytic abrasions [121].

20 Temperature-Sensitive Systems

Temperature-sensitive method relays on the intrinsic properties or the microenvironment found in diseased tissue. It was perceived that in some malignant cancers (such as prostate, bladder, etc.) and also some other ailment conditions appear to show some local temperature difference than that of the normal tissues [122]. This difference in temperature at the pathological sites helped to synthesize temperature-sensitive drug delivery systems. The two fundamental types of thermo-sensitive materials based on its potentiality to swell and de-swell to depend on the temperature, are—

1. Positive temperature-sensitive hydrogels—They are swollen as well as hydrated at higher temperatures, and when cooled at upper critical solution temperature (UCST), they contract.
2. Negative temperature-sensitive hydrogels—They swell at lower temperatures, and as they are heated above the lower critical solution temperature (LCST), they contract.

The nano-carriers made up of UCST/LCST polymers can respond to the changes in temperature present in their microenvironment where the nano-systems alter their characteristics such as solubility, conformation, and hydrophobic/hydrophilic balance to deliver the therapeutic agents at the targeted sites [123]. Thermosensitive liposomes are made up of DPPC/DSPC, the most common lipids. A temperature-dependent fusion of liposomes can be accomplished with the right composition of lipids at a mild hyperthermia region (39–42 °C), which helps to enhance levels of drug release at the infectious site [124]. In early *in vivo* work done in various tumor models, including sarcomas, carcinomas, and lymph nodes metastasis, with a range of drugs such as methotrexate, bleomycin, cisplatin, and adriamycin exhibited positive results of temperature-sensitive liposomes with mild hyperthermia [125, 126]. With an extensive study done, this gives a wide variety so as choose the design of nano-carriers of desired architecture.

21 Targeting Hypoxia

Hypoxia is a condition when the concentration of oxygen is low. This plays a crucial role in numerous tumors, which contributes to radio-resistance, chemoresistance, vasculogenesis, angiogenesis, reluctant to cell death, invasiveness, genomic instability, and altered metabolism. In solid tumors, hypoxia lies in surrounding areas of necrosis. A transcriptional factor, hypoxia-inducible factor (HIF), acts as a hypoxia sensor in a cell. Overexpression of the regulatory subunit, HIF-α is a measure of escalated severity of tumor grade, vascular density, collapsing of conventional treatment, and prognosis. Therefore, the use of HIF as a direct or indirect therapeutic agent has become attractive over the years [127]. Thambi et al. reported hypoxia-responsive self-assembled nanoparticles, which under hypoxia conditions release hydrophobic drugs [128]. The *in vivo* studies have shown the accumulation of nanoparticles in tumors at the hypoxia regions as differentiated to healthy cells.

The brain commands high amounts of oxygen supply. It is at risk of hypoxia, causing irregularity in the flow of blood induced either by traumatic brain injury, brain cancer, cardiac failure, or subcortical vascular disease. Hypoxia can also cause neurodegenerative diseases due to neuronal dysfunction and cell death. HIF based nanoparticles have the superiority of regulating the expressions of various genes, thereby treating the disease [129]. Hypoxia targeting NPs have a potential advantage to treat inaccessible brain tumors, which are difficult to achieve through conventional surgeries.

22 Targeting Acidity

In the presence of oxygen, cancer cells undergo glycolysis, which is significantly vital for survival and proliferation, resulting in the high production of lactic and carbonic acids. These chemicals are intensively pumped out of the cells to keep the intracellular pH nearly neutral. The acidification of the tumor microenvironment promotes cancer aggressiveness and invasiveness. Thus, the acidic microenvironment around cancer cells may provide a marker for detecting and targeting diseased tissues.

The pH-sensitive nanoparticles consist of lipids, polymers, or peptides such as pHLIPs (pH Low Insertion Peptides). The fundamental principle of synthesizing pH-sensitive fusogenic liposomes is to select such a lipid which is steady at physiological pH and destabilized in acidic conditions ensuing cellular internalization thus, the delivery of drugs within the nanoparticles into the cytosol [130]. The highly pH-sensitive NPs are developed to target acidity, help for both treatment and diagnosis of a wide array of cancers, and also many other pathological diseases.

Emmetiere et al. synthesized pHLIP using the dual-delivery method to tether liposomes to cancer cells [131]. First, Tetrazine (Tz) was coupled to pHLIP, which is a bioorthogonal reactive small molecule to form a pHLIP-Tz conjugate. This conjugate was injected into mice that label the surface of the cancer cells. Then, liposomes containing ^{18}F PET isotope was coated with bioorthogonal reactive trans-cyclooctenes and was given a second injection. After circulating for an extended period, the radiolabelled liposomes due to the EPR effect accumulated at the tumor site. This was then followed by a click reaction between trans-cyclooctenes and pHLIP-Tz in liposome coat resulting in tethering and covalent conjugation of liposome to the cancer cell surface. This *in vivo* clicks reaction permits to attain higher accumulation of radiolabelled liposomes in the tumor site, high signal/noise ratio, low non-specific binding, and reduced toxicity to bone marrow and kidneys.

In another study, pHLIP-liposome construction is used to deliver nanopores and induce apoptosis in cancer cells [132]. For the cell to function usually, the proper balance of ion between extracellular and intracellular is crucial. Any slightest changes in the conditions may lead to the death of the cell. A pore formation gramicidin A was delivered to the membrane of the cell using pHLIP-coated DOPC liposome, which would help to change cellular ion balance. When pHLIP-liposomes target the cancer cells, they are either fused to the plasma membrane or up-taken by endocytosis and then merged with a lysosomal layer. Here, the gramicidin A is delivered, and pores with a diameter of 4–5 °A are formed in the cell membranes of cancer cells [133]. The intracellular space is acidified due to the nanopores, thus eliminating the vital NA+/K+ ion balance. Along with the treatment strategy to solid acidic tumors, it also opens new opportunities to deliver various membrane proteins and peptides to the cells widening the application of nanomedicine and biotechnology.

Yao et al. showed AuNPs (1.4 nm diameter) were coupled to pHLIP at N-terminus, which successfully targeted cancer cells at low pH [134]. *In vivo* studies were done using mice-tumor models, showing high uptake of NPs by tumors in both intratumoral and intravenous administration when compared with non-functionalized AuNPs. The pHLIP-AuNPs provided the capability of specific targeting, increased homogenous local concentration in a solid tumor, allowing NPs to remain inside of a cell for several days, thus helping in radiation therapy and imaging.

Davies et al. used pHLIP luminescent europium coated nanoparticles into platelets [135]. The AuNPs (13 nm) were co-coated with pHLIP and luminescent europium (EuL) to give pHLIP-EuL-Au NPs. Human platelets are vulnerable to microinjection or transfection. But with the help of this method, NPs consisting of approximately 640 lanthanides per particle were delivered. This resulted that NPs could internalize into platelets only at low pH. This research signifies that pHLIP helps to translocate NPs in a pH-dependent manner.

Han et al. reported the use of pHLIP in delivering pDNA to cancer cells [136]. The surface of dendrigraft poly-l-lysines (DGLs) of generation 3 with 123 amino groups per molecules was conjugated to N-terminus of pHLIP. This electrostatically interacted between positively charged DGL head group of DGL-PEG-pHLIP and negatively charged pDNA to create DGL-PEG-pHLIP/pDNA NPs. The *in vitro* results revealed the uptake of NPs at pH 6.0. Thus, the NPs entered the cells by absorptive endocytosis. The same effects were observed in *in vivo* studies. This clearly showed the enhanced pH-controlled NPs localization by pHLIP in tumors.

pHLIP peptide could target tumors in acidic micro-environment along with the controlled release of NPs in the intracellular spaces [137]. The researchers chose mesoporous silica nanoparticles (especially MCN-41) to load with doxorubicin. MSN has high homogenous porosity, biocompatibility, payload capacity, inertness, and could quickly surface functionalized. MSN particles connected the pHLIP at C-terminus by a disulfide bond. At pH < 6.5, doxorubicin-loaded pHLIPs-MSN were pressed into the cell membrane and translocated into cytosol. The doxorubicin was released by cleaving the disulfide bond into the cytoplasm.

There are various challenges faced by the nano-drug delivery system and yet to overcome. Every nano-drug platform should be assessed experimentally, which can be found to be strenuous. Here the physical targeting of nanomedicine, especially the pHLIP nanotechnology, alone or combined with other approaches, has been found to have the inherent capability to address most of the difficulties and challenges with new formulations translatable to clinical trials.

23 Microfluidic Devices for Drug Delivery

Microfluidics has been developed over the years to address the challenges faced in developing an efficient DDS. Some of the key advantages in microfluidics are the ability to precisely control individual cells, controlled and targeted delivery with high efficiency and high cell viability [138, 139]. This has developed the concept of single-cell technology, which allows manipulation and study of an individual cell [140, 141]. The cell membranes are temporarily permeabilized to enable the passage of targeted cargo molecules into the cell. Based on the source of energy used for permeabilization the techniques can classify as electroporation (electric field) [142], photoporation (laser) [143] and mechanoporation (shear stress force) [144]. Once the membrane is permeabilized, the drug is pumped inside the cell either by an active or passive mechanism. The use of physical energy with a microfluidics platform enables localized permeabilization of the cell membrane by application of the nano field [145]. These cell membrane pores reseal within few seconds to few minutes based on the density and size of the pores created [146]. In a study, opsin encoded genes were transfected to mouse retina to repair the eyesight using optoporation [147]. The shape and size of the nanoparticles could be designed to achieve higher drug delivery efficiency with higher cell viability [148].

Further, optical tools have been developed that can insert individual drug molecules into the cell [149]. Microfluidics is also being explored for making microreactors to synthesize nanoparticles to, reduce the reagents required, and control the size distribution curve [150]. The development of microfluidic platforms for biomedical applications has increased the functionality of tools to explore hidden aspects of the biological study [151].

24 Immune-Compatibility of Biomedical Nanoparticles

Nanoparticle-based drugs, when administered intravenously, interact immediately with the blood components (e.g., serum proteins). This causes a change in their characteristics, which consequently changes the featured interactions with tissues and cells. The most important communication is between nanomedicine and the immune system, which may cause recognition and elimination of the nanoparticles considered as a dangerous foreign agent. Thus, nanomedicine must be able to avoid identification by immune cells to reach their therapeutic target and cause its effect. The defense mechanism should not be triggered, which may cause damage to the body tissues by complement activation or inflammation. The nanomedicine should also not obstruct the immunocompetent cells to avoid immune-related diseases.

The characteristics of nanomedicines, such as size, shape, chemical composition, surface charge, and route of administration, will help to prolong it's *in vivo* persistence and efficacy. Three immunological issues must be considered during the fabrication of drug-loaded nanoparticles and their administration into human patients.

1. Immune-mediated rejection or destruction—to avoid nanomedicine as a dangerous foreign entity, leading to initiate immune defense reaction and further elimination of the drug.
2. Immunotoxicity—to make sure nanomedicine do not cause unwanted inflammatory or immune responses, that may induce damage to the body.
3. Immunosafety—to design nanomedicine so that it won't interfere with the normal immune responses.

Nanoparticulate drug delivery system, when inoculated, immune system provokes a series of reactions to protect the body from possible harm. The immune system may trigger various defense mechanisms such as platelet activation, production of antibodies, and inflammation. These reactions may further cause pathological consequences such as thrombosis, hemolysis, hypersensitivity reactions, which may lead to the elimination of the nanomedicine from the system [152]. The reactions occurred may be either past complement activation or direct NPs interaction. Although, during the synthesis of nanomedicines, the bioactive contaminates, if eliminated carefully and biocompatible materials are used, then the danger of its destruction may be reduced.

Reticuloendothelial system (RES) helps in the capture and engulfment of the particulate material at the tissue level. This includes tissue leukocytes, macrophages, dendritic cells, epithelial and endothelial cells present in different organs (e.g., liver and spleen). Some studies have revealed that positively charged particles can voluntarily get attached to the surface of the phagocyte membrane (negatively charged) and are frequently more cytotoxic than anionic and neutral ones, thus abundantly internalized. The process of endocytosis can be receptor-independent or -dependent according to the active functional characteristics present of the size, shape, and NP surface. Such sub micrometric particles are engulfed mostly via a receptor-mediated mechanism, by the professional phagocytes (like monocytes and macrophages). Here, the particle surface characteristics act as a critical element to determine the receptor-mediated uptake. For example, if the NPs are opsonized with the IgG or C3ib (one among the activated fragment of complement system), it can be recognized by specific receptors for Fcγ or C3ib on the phagocyte membrane and internalized. At the same time, the adsorption of serum albumin can sharply decrease the recognition and uptake of the particle [153, 154]. After the nanomedicine is administered intravenously, its composition of the "bioshell" (preferred than that of 'protein corona,' as it is not entirely made up of proteins) that covers NPs is not an easy task to determine. The bioshells differ with time because of the on/off rate of blood molecules that are adsorbed on the surface of the NPs, time (prolonged incubation in 100% plasma results in more stable shell forms), temperature (feverish of patients) and blood composition (like age, sex and disease dependent). This causes NPs-blood interaction as more of a dynamic process, where bioshell influences the NPs surface characteristics and its degradation [155]. Many soluble defense or scavenging molecules (clotting components, inflammation complement fragments, adhesion molecules, collectins, etc.) are get adsorbed on the surface of the NPs. Sometimes, NPs may be able to adsorb anti-inflammatory factors (such as albumin, serpins, vitronectin) or factor H (complement inhibitor). This protects from the complement attack. This method is used by some bacteria and cancer cells, which helps to escape complement-mediating lysis [156].

Surface coating and derivatization with molecules that help to alter the recognition and engulfment by the phagocytes is the technique that makes the NPs 'invisible' to the recognition tools of the immune cells. This 'Trojan horse' strategy helps nanomedicine to escape uptake and destruction [157]. This invisibility of NPs fails to recognize it as dangerous and thus do not induce the polymorphonuclear phagocytes (PMNs) and inflammation reactions. For example, polysialic acids (negatively charged) is a capsule to some pathogenic bacteria (like *Neisseria meningitidis*) escapes recognition by the immune system. It is similar to that of the host polysaccharides like those existing on integrins molecules. Therefore, this can be utilized for the synthesis of nanomedicine to evade the immune defense system [158].

PEG is a block copolymer when used for the coating of the nanoparticles can help to decrease the opsonin adsorption and recognition by immune cells. It also significantly reduces the uptake and destruction by RES (*in vitro* and *in vivo*) along with the other procedures to manage NPs size and shape [159]. However, PEGylated liposomes act as T-cell-independent antigen (type II) and induce antibody-dependent (IgM) rapid clearance of its second dose of circulation, known to be accelerated blood clearance (ABC) phenomenon. This consequence lacks the therapeutic effect [160]. The rate of ABC phenomenon depends on the NPs size, shape, surface charge, liposomal composition, PEG density, dosage, and the interval between administered doses. Thus, while giving attention to these criteria while synthesizing the nanomedicine will help to avoid immune recognition and antibody responses. At the same time, the host immunological integrity will be spared.

Dendrimers are micellar nanostructures, branched symmetric structure polymeric macromolecules containing highly dense functional groups at their periphery. Significant applications of dendrimer nanocarriers hold promise to facilitate diagnostic imaging, gene transfection, non-viral gene transfer, drug delivery, detection, and therapeutics for cancer and other diseases [161]. With different surface modifications, glycodendrimers, have been designed with anti-viral or anti-microbial effect as anti-infective ligands. For example, a polysulfonate G4 polyamidoamine (PAMAM) dendrimer can block HIV-1 and HIV-2 activity *in vitro* [162]. The infection is prevented by the mechanism related to structural entrapment of the virus with help of the branched structure. The dendrimer-mediated anti-microbial activity may be due to the involvement of immune-modulation. For instance, the subcutaneous melanoma mice model, when administered with N-acetyl-glucosamine-coated G1 PAMAMs, decreases tumor growth, and thus the survival of mice in increased. These results were accompanied along with the increased number of CD69+ cells in tumor tissues and spleen. The upregulation of the IFN-g, IL-1 h, IL-2, and TNF-a was also observed [163]. Such a phenomenon was indicated with the over-expression of pro-inflammatory chemokines and cytokines, such as MIL-1 h, MIP-1a, IL-8, IL-1 h, IL-6, and TNF-a induced in macrophages and human dendritic cells by glucosamine-modified G3.5 PAMAMs [164]. Thus, many synthesized dendrimers must be studied using primary immune cells and then further animal models to recognize and categorize different immune responses.

Chitosan is the naturally occurring polymer with the chemical formula of a (1-4)-2-amino-2-deoxy b-D-glucan. It is substantially used in biomedical applications due to its characteristics, including biodegradability, rapid blood clotting, non-toxicity, high charge density, and muco-adhesivity. Thus, it can be used in various pharmaceutical applications like entrapment or coating of drugs, biochemicals, and antigenic molecules. Chitosan microspheres have a promising carrier for mucosal vaccination (nasal and oral) to enhance immune responses [165]. Chitosan can be coated over inorganic as well as organic nanostructures. It has been shown to improve immune cell functions such as macrophages and neutrophils, which helps to enhance the production of growth factors and cytokines [166].

Sodium alginate is a water-soluble, hydrophilic and biocompatible polysaccharide extracted from marine brown algae. It is often used for the coating of magnetic [167] and polymeric nanoparticles [168]. Alginate polymers, when digested enzymatically, form alginate oligomers that can secrete high levels of inflammatory chemokines and cytokines, like RANTES, MCP-1 *in vitro*, and *in vivo* [169]. Combinations of PLGA and alginate are used to develop vaccines. In immunization studies, incorporation of alginate by intradermal route in Balb/c mice showed enhanced cellular and humoral responses [170]. Non-specific immune response, when enhanced *in vivo*, suggests that NPs coated with alginate have directed at immune stimulation either for diagnostic purposes or adjuvant vaccine properties to reveal a deficiency in innate immune reactions.

Silicon dioxide NPs (SiO_2 NPs) have been used in various applications in the field of biomedical research. As previously described, phagocytes respond more to micro-sized particles rather than nano-size particles, due to which cell damages are decreased [171]. An interaction between endothelial cells and amorphous silica nanoparticles was studied, which promoted the up-regulation of the endothelial adhesion molecules expression. Due to which the adhesion of the monocytes to endothelial cells is enhanced. This is a typical early response to the inflammatory process [172]. Morishige et al. demonstrated that the chemically modified surface of silica NPs decreased the inflammasome activation and IL-1β release in the human THP-1 macrophage/monocyte cell line [173].

For the preparation of bimetallic particles, gold coatings are generally used on the magnetic core for functionalization of the surface, mainly designed for drug delivery [174] and hyperthermia treatments in cancer [175, 176]. Titanium dioxide (TiO_2) is also used as a coating for inorganic nanoparticles to escalate their cytocompatibility. It was illustrated that TiO_2 coating on ZnNPs helped to reduce its release of zinc ions in the human lung epithelial cell line [177]. Some recent studies mentioned the use of iron—titanium NPs can be used as a carrier for doxorubicin against cancer cells [178]. TiO_2 particles are inert and are unable to pass the damaged skin, thus used in cosmetics and sunscreens as an efficient U.V. light filter [179].

Precise organic or inorganic NPs coatings design to specifically interact or avoid the immune system could only be achieved if the binding sites are known. NPs show diverse reactions on their surface, depending on the crystal or atomic structures. Thus, testing of nanomedicine on the normal development of inflammatory responses and innate immunity is the key while accessing the immune safety profile of nanodrug. With all the complexity, different *in vitro* and *in vivo* models should be studied thoroughly for its immune system reactions. For this, representative and specific assays are needed to be designed. But the laboratory models provide limited knowledge and information for potential immune response in human subjects. In several cases, there is no molecular correlation found between immune reactions in human versus mice, where alternate molecular pathways have been used [180]. For example, cell lines are mostly transformed or tumor-derived. Thus, the NPs cytotoxicity on highly proliferating macrophage cell lines may not reflect what has its toxic effect on noncycling primary resident macrophages cells.

On the other hand, NPs can induce TNF-α and yield similar results in continuous versus primary macrophages. Thus, cell lines are beneficial for rapid and reproducible testing only for validating biomarkers, which reflect normal human cell responses. *In vivo* experimental models are also required to understand and study the systemic immune reactions. Genetically modified mice for the immunity system deliver straightforward answers for potential activation of the immune system by the NPs. The target delivery for protected organs like CNS, in vivo models, are mandatory. Blood circulation is the ideal way to deliver NPs to the CNS. Polymeric nanoparticles or coating with PEG have been used to cross BBB [181]. The nanomaterials used as biomedicine can be synthesized to target or avoid the immune system. Thus, there is a need for the evolution of new nanomaterial-mediated approaches and assays to help to justify the results obtained from different methods.

25 Conclusions

In the last few years, the scientist has developed several new technologies for the diagnostics and treatment of various diseases. The use of nanotechnology in the development of nanocarriers as DDS has shown high potential in research. Nano-DDS presents unique advantages which have proven to have enhanced intracellular uptake as compared to the other conventional drug forms. Ligands such as antibodies that are conjugated to nanocarriers are favored in a targeted therapeutic approach because of the better-controlled therapy where delivery of active drugs is achievable in smaller amounts at the required sites in the body. Building upon these advances towards human clinical trials, much work is needed to be done in over the next 5–10 years to more fully realize the promise of nanomedicine. First, the logical design of nanomaterials and the development of tools is needed to design a detailed understanding of biomedical processes required. For example, the targeting strategies for drug carrier design may vary in type, developmental stage, and location of the disease. Second, more complex DDS such as multi-functional NPs that are simultaneously are capable of targeting, imaging, and treatment therapy, are the motif of future research. Third, the main concern about these DDS is regarding their immune toxicity issues, which is essential for the fundamental research that is needed to be carried out to address them for the successful and efficient application of these technologies.

Besides, a significant challenge faced by the new generation in this field is its highly interdisciplinary nature, the best course will be to train an of truly interdisciplinary investigators having firm multidisciplinary grounding. The US FDA has approved some nanoparticle-based products, and several others are currently under development process and clinical assessment. In summary, although nanomedicine is still in its infancy, these practical applications demonstrate its enormous potential.

References

1. Park K (2013) Facing the truth about nanotechnology in drug delivery. ACS Nano 7:7442–7447
2. Edwards PP, Thomas JM (2007) Gold in a metallic divided state—from faraday to present-day nanoscience. Angew Chemie Int Ed 46:5480–5486
3. Wissing SA, Kayser O, Müller RH (2004) Solid lipid nanoparticles for parenteral drug delivery. Adv Drug Deliv Rev 56:1257–1272
4. Pachón LD, Rothenberg G (2008) Transition-metal nanoparticles: synthesis, stability and the leaching issue. Appl Organomet Chem 22:288–299
5. Hoffman AS (2008) The origins and evolution of "controlled" drug delivery systems. J Control release 132:153–163
6. Zhang Y, Chan HF, Leong KW (2013) Advanced materials and processing for drug delivery: the past and the future. Adv Drug Deliv Rev 65:104–120
7. Lane N, Kalil T (2005) The national nanotechnology initiative: present at the creation. Issues Sci Technol 21:49–54
8. Jong JW, Borm PJA (2008) Drug delivery and nanoparticles: applications and hazards. Int J Nanomed 3:133–149. https://doi.org/10.2147/ijn.s596
9. Webster DM, Sundaram P, Byrne ME (2013) Injectable nanomaterials for drug delivery: carriers, targeting moieties, and therapeutics. Eur J Pharm Biopharm 84:1–20
10. Chakarvarty G, Seth N, Sharma V (2013) Nanoparticles and nanotechnology: clinical, toxicological, social, regulatory and other aspects of nanotechnology. J Drug Deliv Ther 3:138–141
11. Sahoo SK, Parveen S, Panda JJ (2007) The present and future of nanotechnology in human health care. Nanomed Nanotechnol Biol Med 3:20–31. https://doi.org/10.1016/j.nano.2006.11.008
12. Jahangirian H, Lemraski EG, Webster TJ, Rafiee-Moghaddam R, Abdollahi Y (2017) A review of drug delivery systems based on nanotechnology and green chemistry: green nanomedicine. Int J Nanomedicine 12:2957
13. Irving B (2007) Nanoparticle drug delivery systems. Innov Pharm Technol 58–62. https://doi.org/10.1385/1-59259-427-1:117
14. Patra JK, Das G, Fraceto LF, Campos EVR, Rodriguez-Torres MDP, Acosta-Torres LS, Diaz-Torres LA, Grillo R, Swamy MK, Sharma S, Habtemariam S, Shin HS (2018) Nano based drug delivery systems: recent developments and future prospects. J Nanobiotechnology 16:1–33. https://doi.org/10.1186/s12951-018-0392-8
15. Calvo P, Remuñan-López C, Vila-Jato JL, Alonso MJ (1997) Chitosan and chitosan/ethylene oxide-propylene oxide block copolymer nanoparticles as novel carriers for proteins and vaccines. Pharm Res 14:1431–1436. https://doi.org/10.1023/A:1012128907225
16. Chen Y, McCulloch R, Gray B (1994) Synthesis of albumin-dextran sulfate microspheres possessing favourable loading and release characteristics for the anticancer drug doxorubicin. J Control release 31:49–54

17. Mohanraj VJ, Chen Y (2006) Nanoparticles—a review
18. Suri SS, Fenniri H, Singh B (2007) Nanotechnology-based drug delivery systems. J Occup Med Toxicol 2:16–22. https://doi.org/10.1186/1745-6673-2-16
19. Matsumura Y, Maeda H (1986) A new concept for macromolecular therapeutics in cancer chemotherapy: mechanism of tumoritropic accumulation of proteins and the antitumor agent smancs. Cancer Res 46:6387–6392
20. Chowdhury A, Kunjiappan S, Panneerselvam T, Somasundaram B, Bhattacharjee C (2017) Nanotechnology and nanocarrier-based approaches on treatment of degenerative diseases. Int Nano Lett 7:91–122. https://doi.org/10.1007/s40089-017-0208-0
21. He C, Hu Y, Yin L, Tang C, Yin C (2010) Effects of particle size and surface charge on cellular uptake and biodistribution of polymeric nanoparticles. Biomaterials 31:3657–3666. https://doi.org/10.1016/j.biomaterials.2010.01.065
22. Barreto JA, O'Malley W, Kubeil M, Graham B, Stephan H, Spiccia L (2011) Nanomaterials: applications in cancer imaging and therapy. Adv Mater 23:H18–H40. https://doi.org/10.1002/adma.201100140
23. Abdollahi A, Folkman J (2010) Evading tumor evasion: current concepts and perspectives of anti-angiogenic cancer therapy. Drug Resist Updat 13:16–28. https://doi.org/10.1016/j.drup.2009.12.001
24. Guo X, Shi C, Yang G, Wang J, Cai Z, Zhou S (2014) Dual-responsive polymer micelles for target-cell-specific anticancer drug delivery. Chem Mater 26:4405–4418. https://doi.org/10.1021/cm5012718
25. Bae YH, Park K (2011) Targeted drug delivery to tumors: myths, reality and possibility. J Control Release 153:198–205. https://doi.org/10.1016/j.jconrel.2011.06.001
26. Kumar A, Ma H, Zhang X, Huang K, Jin S, Liu J, Wei T, Cao W, Zou G, Liang XJ (2012) Gold nanoparticles functionalized with therapeutic and targeted peptides for cancer treatment. Biomaterials 33:1180–1189. https://doi.org/10.1016/j.biomaterials.2011.10.058
27. Gabizon AA (2001) Pegylated liposomal doxorubicin: metamorphosis of an old drug into a new form of chemotherapy. Cancer Invest 19:424–436
28. Patri AK, Myc A, Beals J, Thomas TP, Bander NH, Baker JR (2004) Synthesis and in vitro testing of J591 antibody-dendrimer conjugates for targeted prostate cancer therapy. Bioconjug Chem 15:1174–1181. https://doi.org/10.1021/bc0499127
29. Low PS, Henne WA, Doorneweerd DD (2008) Discovery and development of folic-acid-based receptor targeting for imaging and therapy of cancer and inflammatory diseases. Acc Chem Res 41:120–129. https://doi.org/10.1021/ar7000815
30. Cheng Z, Al Zaki A, Hui J, Muzykantov VR, Tsourkas A (2012) Multifunctional nanoparticles: cost versus benefit of adding targeting and imaging capabilities. Science 338:903–910
31. Meacham CE, Morrison SJ (2013) Tumour heterogeneity and cancer cell plasticity. Nature 501:328–337
32. Davis ME (2009) The first targeted delivery of sirna in humans via a self-assembling, cyclodextrin polymer-based nanoparticle: from concept to clinic. Mol Pharm 6:659–668. https://doi.org/10.1021/mp900015y
33. Slowing II, Vivero-Escoto JL, Wu CW, Lin VSY (2008) Mesoporous silica nanoparticles as controlled release drug delivery and gene transfection carriers. Adv Drug Deliv Rev 60:1278–1288
34. Chen S, Zhang Q, Hou Y, Zhang J, Liang XJ (2013) Nanomaterials in medicine and pharmaceuticals: nanoscale materials developed with less toxicity and more efficacy. Eur J Nanomed 5:61–79. https://doi.org/10.1515/ejnm-2013-0003
35. Tang F, Li L, Chen D (2012) Mesoporous silica nanoparticles: synthesis, biocompatibility and drug delivery. Adv Mater 24:1504–1534. https://doi.org/10.1002/adma.201104763
36. Hu X, Hao X, Wu Y, Zhang J, Zhang X, Wang PC, Zou G, Liang X (2013) Multifunctional hybrid silica nanoparticles for controlled doxorubicin loading and release with thermal and pH dual response. J Mater Chem B 1:1109–1118
37. Peer D, Margalit R (2006) Fluoxetine and reversal of multidrug resistance. Cancer Lett 237:180–187

38. Danson S, Ferry D, Alakhov V, Margison J, Kerr D, Jowle D, Brampton M, Halbert G, Ranson M (2004) Phase I dose escalation and pharmacokinetic study of pluronic polymer-bound doxorubicin (SP1049C) in patients with advanced cancer. Br J Cancer 90:2085–2091
39. Meng H, Liong M, Xia T, Li Z, Ji Z, Zink JI, Nel AE (2010) Engineered design of mesoporous silica nanoparticles to deliver doxorubicin and p-glycoprotein siRNA to overcome drug resistance in a cancer cell line. ACS Nano 4:4539–4550. https://doi.org/10.1021/nn100690m
40. Adams JM, Cory S (2007) Bcl-2-regulated apoptosis: mechanism and therapeutic potential. Curr Opin Immunol 19:488–496
41. Chen AM, Zhang M, Wei D, Stueber D, Taratula O, Minko T, He H (2009) Co-delivery of Doxorubicin and Bcl-2 siRNA by mesoporous silica nanoparticles enhances the efficacy of chemotherapy in multidrug-resistant cancer cells. Small 5:2673–2677. https://doi.org/10.1002/smll.200900621
42. Chen S, Zhao D, Li F, Zhuo RX, Cheng SX (2012) Co-delivery of genes and drugs with nanostructured calcium carbonate for cancer therapy. RSC Adv 2:1820–1826
43. Kim K, Kim JH, Park H, Kim YS, Park K, Nam H, Lee S, Park JH, Park RW, Kim IS, Choi K (2010) Tumor-homing multifunctional nanoparticles for cancer theragnosis: simultaneous diagnosis, drug delivery, and therapeutic monitoring. J Control Release 146:219–227
44. Yang J, Lee CH, Ko HJ, Suh JS, Yoon HG, Lee K, Huh YM, Haam S (2007) Multifunctional magneto-polymeric nanohybrids for targeted detection and synergistic therapeutic effects on breast cancer. Angew Chemie Int Ed 46:8836–8839. https://doi.org/10.1002/anie.200703554
45. Medarova Z, Pham W, Farrar C, Petkova V, Moore A (2007) In vivo imaging of siRNA delivery and silencing in tumors. Nat Med 13:372–377
46. Guo S, Qiao Y, Wang W, He H, Deng L, Xing J, Xu J, Liang XJ, Dong A (2010) Poly (ε-caprolactone)-graft-poly (2-(N, N-dimethylamino) ethyl methacrylate) nanoparticles: pH dependent thermo-sensitive multifunctional carriers for gene and drug. J Mater Chem 20:6935–6941
47. Ruoslahti E, Bhatia S, Sailor M (2010) Targeting of drugs and nanoparticles to tumors. J Cell Biol 188:759–768
48. Byrne JD, Betancourt T, Brannon-Peppas L (2008) Active targeting schemes for nanoparticle systems in cancer therapeutics. Adv Drug Deliv Rev 60:1615–1626
49. Chung AS, Ferrara N (2011) Developmental and pathological angiogenesis. Annu Rev Cell Dev Biol 27:563–584. https://doi.org/10.1146/annurev-cellbio-092910-154002
50. Seaman S, Stevens J, Yang MY, Logsdon D, Graff-Cherry C, Croix BS (2007) Genes that distinguish physiological and pathological angiogenesis. Cancer Cell 11:539–554
51. Gullotti E, Yeo Y (2009) Extracellularly activated nanocarriers: a new paradigm of tumor targeted drug delivery. Mol Pharm 6:1041–1051
52. Dilnawaz F, Singh A, Mohanty C, Sahoo SK (2010) Dual drug loaded superparamagnetic iron oxide nanoparticles for targeted cancer therapy. Biomaterials 31:3694–3706
53. Chattopadhyay N, Fonge H, Cai Z, Scollard D, Lechtman E, Done SJ, Pignol JP, Reilly RM (2012) Role of antibody-mediated tumor targeting and route of administration in nanoparticle tumor accumulation in vivo. Mol Pharm 9:2168–2179
54. Adolphi NL, Butler KS, Lovato DM, Tessier TE, Trujillo JE, Hathaway HJ, Fegan DL, Monson TC, Stevens TE, Huber DL, Ramu J, Milne ML, Altobelli SA, Bryant HC, Larson RS, Flynn ER (2012) Imaging of Her2-targeted magnetic nanoparticles for breast cancer detection: comparison of SQUID-detected magnetic relaxometry and MRI. Contrast Media Mol Imaging 7:308–319. https://doi.org/10.1002/cmmi.499
55. Qian ZM, Tang PL (1995) Mechanisms of iron uptake by mammalian cells. Biochim Biophys Acta (BBA)-Molecular Cell Res 1269:205–214
56. Li X, Ding L, Xu Y, Wang Y, Ping Q (2009) Targeted delivery of doxorubicin using stealth liposomes modified with transferrin. Int J Pharm 373:116–123
57. Fonseca C, Moreira JN, Ciudad CJ, de Lima MCP, Simoes S (2005) Targeting of sterically stabilised pH-sensitive liposomes to human T-leukaemia cells. Eur J Pharm Biopharm 59:359–366

58. Zheng Y, Yu B, Weecharangsan W, Piao L, Darby M, Mao Y, Koynova R, Yang X, Li H, Xu S, Lee LJ (2010) Transferrin-conjugated lipid-coated PLGA nanoparticles for targeted delivery of aromatase inhibitor 7α-APTADD to breast cancer cells. Int J Pharm 390:234–241
59. Hong M, Zhu S, Jiang Y, Tang G, Sun C, Fang C, Shi B, Pei Y (2010) Novel anti-tumor strategy: PEG-hydroxycamptothecin conjugate loaded transferrin-PEG-nanoparticles. J Control Release 141:22–29
60. Ando K, Mori K, Corradini N, Redini F, Heymann D (2011) Mifamurtide for the treatment of nonmetastatic osteosarcoma. Expert Opin Pharmacother 12:285–292. https://doi.org/10.1517/14656566.2011.543129
61. Karmali PP, Kotamraju VR, Kastantin M, Black M, Missirlis D, Tirrell M, Ruoslahti E (2009) Targeting of albumin-embedded paclitaxel nanoparticles to tumors. Nanomed Nanotechnol Biol Med 5:73–82
62. Santra TS, Tseng F-G, Barik TK (2014) Biosynthesize silver and gold nanoparticles for potential biomedical applications: a brief review. J Nanopharmaceutics Drug Deliv 2:1–17
63. Shinde AS, Mendhulkar VD (2020) Antiproliferative activity of Elephantopus scaber mediated silver nanoparticles against MCF-7, A-549, SCC-40 and COLO-205 human cancer cell lines. Asian J Pharm Clin Res 13:163–67. https://doi.org/10.22159/ajpcr.2020.v13i2.36497
64. Santra TS, Tseng F-G, Barik TK (2015) Green biosynthesis of gold nanoparticles and biomedical applications. Am J Nano Res Appl 2:5–12
65. Rabanel JM, Aoun V, Elkin I, Mokhtar M, Hildgen P (2012) Drug-loaded nanocarriers: passive targeting and crossing of biological barriers. Curr Med Chem 19:3070–3102
66. Bonoiu AC, Mahajan SD, Ding H, Roy I, Yong KT, Kumar R, Hu R, Bergey EJ, Schwartz SA, Prasad PN, Bonoiu AC, Mahajan SD, Ding H, Roy I, Yong KT, Kumar R, Hu R, Bergey EJ, Schwartz SA, Prasad PN (2009) Nanotechnology approach for drug addiction therapy: gene silencing using delivery of gold nanorod-siRNA nanoplex in dopaminergic neurons. Proc Natl Acad Sci 106:5546–5550
67. Daniels TR, Bernabeu E, Rodríguez JA, Patel S, Kozman M, Chiappetta DA, Holler E, Ljubimova JY, Helguera G, Penichet ML (2012) The transferrin receptor and the targeted delivery of therapeutic agents against cancer. Biochim Biophys Acta (BBA)-General Subj 1820:291–317
68. Jefferies WA, Brandon MR, Hunt SV, Williams AF, Gatter KC, Mason DY (1984) Transferrin receptor on endothelium of brain capillaries. Nature 312:162–163
69. Ulbrich K, Hekmatara T, Herbert E, Kreuter J (2009) Transferrin-and transferrin-receptor-antibody-modified nanoparticles enable drug delivery across the blood–brain barrier (BBB). Biopharmaceutics 71:251–256
70. Gosk S, Vermehren C, Storm G, Moos T (2004) Targeting anti-transferrin receptor antibody (OX26) and OX26-conjugated liposomes to brain capillary endothelial cells using in situ perfusion. J Cereb Blood Flow Metab 24:1193–1204. https://doi.org/10.1097/01.WCB.0000135592.28823.47
71. Md S, Khan RA, Mustafa G, Chuttani K, Baboota S, Sahni JK, Ali J (2013) Bromocriptine loaded chitosan nanoparticles intended for direct nose to brain delivery: pharmacodynamic, pharmacokinetic and scintigraphy study in mice. Eur J Pharm Sci 48:393–405
72. Davis PB, Cooper MJ (2007) Vectors for airway gene delivery. AAPS J 9:E11–E17
73. Buñuales M, Düzgüne N, Zalba S, Garrido MJ, Tros De Ilarduya C (2011) Efficient gene delivery by EGF-lipoplexes in vitro and in vivo. Nanomedicine 6:89–98. https://doi.org/10.2217/nnm.10.100
74. Felgner P, Tsai Y, Felgner J (1996) Advances in the design and application of cytofectin formulations
75. Alex SM, Sharma CP (2013) Nanomedicine for gene therapy. Drug Deliv Transl Res 3:437–445
76. Blecher K, Martinez LR, Tuckman-Vernon C, Nacharaju P, Schairer D, Chouake J, Friedman JM, Alfieri A, Guha C, Nosanchuk JD, Friedman AJ (2012) Nitric oxide-releasing nanoparticles accelerate wound healing in NOD-SCID mice. Nanomed Nanotechnol Biol Med 8:1364–1371

77. Han G, Nguyen LN, Macherla C, Chi Y, Friedman JM, Nosanchuk JD, Martinez LR (2012) Nitric oxide–releasing nanoparticles accelerate wound healing by promoting fibroblast migration and collagen deposition. Am J Pathol 180:1465–1473
78. Azuma K, Izumi R, Osaki T, Ifuku S, Morimoto M, Saimoto H, Minami S, Okamoto Y (2018) Functional biomaterials chitin, chitosan, and its derivatives for wound healing: old and new materials. J Funct Biomater 6:104–1426. https://doi.org/10.3390/jfb6010104
79. Hussain Z, Thu HE, Ng SF, Khan S, Katas H (2017) Nanoencapsulation, an efficient and promising approach to maximize wound healing efficacy of curcumin: a review of new trends and state-of-the-art. Colloids Surfaces B Biointerfaces 150:223–241
80. Kant V, Gopal A, Pathak NN, Kumar P, Tandan SK, Kumar D (2014) Antioxidant and anti-inflammatory potential of curcumin accelerated the cutaneous wound healing in streptozotocin-induced diabetic rats. Elsevier 20:322–330
81. Yallapu MM, Nagesh PKB, Jaggi M, Chauhan SC (2015) Therapeutic applications of curcumin nanoformulations. AAPS J 17:1341–1356. https://doi.org/10.1208/s12248-015-9811-z
82. Sherwani MA, Tufail S, Khan AA, Owais M (2015) Gold nanoparticle-photosensitizer conjugate based photodynamic inactivation of biofilm producing cells: potential for treatment of C. albicans infection in. PLoS One 10
83. Yates CC, Hebda P, Wells A (2012) Skin wound healing and scarring: fetal wounds and regenerative restitution. Birth Defects Res Part C Embryo Today Rev 96:325–333
84. Rajendran NK, Sundar S, Kumar D, Houreld NN, Abrahamse H (2018) A review on nanoparticle based treatment for wound healing. J Drug Deliv Sci Technol 44:421–430. https://doi.org/10.1016/j.jddst.2018.01.009
85. Zhang XF, Liu ZG, Shen W, Gurunathan S (2016) Silver nanoparticles: synthesis, characterization, properties, applications, and therapeutic approaches. Int J Mol Sci 17:1534–1568
86. Sarhan WA, Azzazy HME, El-Sherbiny IM (2016) Honey/chitosan nanofiber wound dressing enriched with Allium sativum and Cleome droserifolia: enhanced antimicrobial and wound healing activity. ACS Appl Mater Interfaces 8:6379–6390. https://doi.org/10.1021/acsami.6b00739
87. Lambadi PR, Sharma TK, Kumar P, Vasnani P, Thalluri SM, Bisht N, Pathania R, Navani NK (2015) Facile biofunctionalization of silver nanoparticles for enhanced antibacterial properties, endotoxin removal, and biofilm control. Int J Nanomedicine 10:2155
88. Butler KS, Peeler DJ, Casey BJ, Dair BJ, Elespuru RK (2015) Silver nanoparticles: correlating nanoparticle size and cellular uptake with genotoxicity. Mutagenesis 30:577–591
89. Franková J, Pivodová V, Vágnerová H, Juráňová J, Ulrichová J (2016) Effects of silver nanoparticles on primary cell cultures of fibroblasts and keratinocytes in a wound-healing model. J Appl Biomater Funct Mater 14:e137–e142. https://doi.org/10.5301/jabfm.5000268
90. Tian J, Wong KKY, Ho C-M, Lok C-N, Yu W-Y, Che C-M, Chiu J-F, Tam PKH (2007) Topical delivery of silver nanoparticles promotes wound healing. ChemMedChem Chem Enabling Drug Discov 2:129–136. https://doi.org/10.1002/cmdc.200600171
91. GhavamiNejad A, Rajan Unnithan A, Ramachandra Kurup Sasikala A, Samarikhalaj M, Thomas RG, Jeong YY, Nasseri S, Murugesan P, Wu D, Hee Park C, Kim CS (2015) Mussel-inspired electrospun nanofibers functionalized with size-controlled silver nanoparticles for wound dressing application. ACS Appl Mater Interfaces 7:12176–12183. https://doi.org/10.1021/acsami.5b02542
92. Liu J, Sonshine DA, Shervani S, Hurt RH (2010) Controlled release of biologically active silver from nanosilver surfaces. ACS Nano 4:6903–6913. https://doi.org/10.1021/nn102272n
93. Akturk O, Kismet K, Yasti AC, Kuru S, Duymus ME, Kaya F, Caydere M, Hucumenoglu S, Keskin D (2016) Collagen/gold nanoparticle nanocomposites: a potential skin wound healing biomaterial. J Biomater Appl 31:283–301. https://doi.org/10.1177/0885328216644536
94. Gu H, Ho PL, Tong E, Wang L, Xu B (2003) Presenting vancomycin on nanoparticles to enhance antimicrobial activities. Nano Lett 3:1261–1263. https://doi.org/10.1021/nl034396z
95. Norman RS, Stone JW, Gole A, Murphy CJ, Sabo-Attwood TL (2008) Targeted photothermal lysis of the pathogenic bacteria, pseudomonas aeruginosa, with gold nanorods. Nano Lett 8:302–306. https://doi.org/10.1021/nl0727056

96. Gil-Tomás JJ, Dekker L, Nair S, Gil-Tomás J, Tubby S, Parkin IP, Narband N, Nair SP, Wilson M, Street C (2007) Lethal photosensitisation of Staphylococcus aureus using a toluidine blue O-tiopronin-gold nanoparticle conjugate Lethal photosensitisation of Staphylococcus aureus using a toluidine blue O-tiopronin-gold nanoparticle conjugate. J Mater Chem 17:3739–3746. https://doi.org/10.1039/b706615e
97. Volkova N, Yukhta M, Pavlovich O, Goltsev A (2016) Application of cryopreserved fibroblast culture with Au nanoparticles to treat burns. Nanoscale Res Lett 11:1–6. https://doi.org/10.1186/s11671-016-1242-y
98. Michael GC, Darrin JP, Joel SP (2011) Controlled biodegradation of self-assembling β-hairpin peptide hydrogels by proteolysis with matrix metalloproteinase-13. Biomaterials 32:6471–6477
99. Haines-Butterick L, Rajagopal K, Branco M, Salick D, Rughani R, Pilarz M, Lamm MS, Pochan DJ, Schneider JP (2007) Controlling hydrogelation kinetics by peptide design for three-dimensional encapsulation and injectable delivery of cells. Proc Natl Acad Sci 104:7791–7796
100. Wang S, Nagrath D, Chen PC, Berthiaume O, Yarmush ML (2008) Three-dimensional primary hepatocyte culture in synthetic self-assembling peptide hydrogel. Tissue Eng Part A 14:227–236. https://doi.org/10.1089/tea.2007.0143
101. Webber MJ, Tongers J, Renault MA, Roncalli JG, Losordo DW, Stupp SI (2010) Development of bioactive peptide amphiphiles for therapeutic cell delivery. Acta Biomater 6:3–11
102. Webber MJ, Kessler JA, Stupp SI (2010) Emerging peptide nanomedicine to regenerate tissues and organs. J Intern Med 71–88
103. Jayawarna V, Smith A, Gough JE, Ulijn RV (2007) Three-dimensional cell culture of chondrocytes on modified di-phenylalanine scaffolds. Biochem Soc Trans 35:535–537
104. Smith AM, Williams RJ, Tang C, Coppo P, Collins RF, Turner ML, Saiani A, Ulijn RV (2008) Fmoc-diphenylalanine self assembles to a hydrogel via a novel architecture based on π-π interlocked β-sheets. Adv Mater 20:37–41. https://doi.org/10.1002/adma.200701221
105. Mohamed A, Xing MM (2012) Nanomaterials and nanotechnology for skin tissue engineering. Int J Burns Trauma 2:29
106. Heit YI, Dastouri P, Helm DL, Pietramaggiori G, Younan G, Erba P, Münster S, Orgill DP, Scherer SS (2012) Foam pore size is a critical interface parameter of suction-based wound healing devices. Plast Reconstr Surg 129:589–597. https://doi.org/10.1097/PRS.0b013e3182402c89
107. Jiang B, Larson JC, Drapala PW, Pérez-Luna VH, Kang-Mieler JJ, Brey EM (2012) Investigation of lysine acrylate containing poly(N-isopropylacrylamide) hydrogels as wound dressings in normal and infected wounds. J Biomed Mater Res Part B Appl Biomater 100:668–676. https://doi.org/10.1002/jbm.b.31991
108. Powell HM, Boyce ST (2008) Fiber density of electrospun gelatin scaffolds regulates morphogenesis of dermal-epidermal skin substitutes. J Biomed Mater Res Part A 84:1078–1086. https://doi.org/10.1002/jbm.a.31498
109. Bilgic H, Demiriz M, Ozler M, Ide T, Dogan N, Gumus S, Kiziltay A, Endogan T, Hasirci V, Hasirci N (2013) Gelatin based scaffolds and effect of EGF dose on wound healing. J Biomater Tissue Eng 3:205–211
110. Patel H, Bonde M, Srinivasan G (2011) Biodegradable polymer scaffold for tissue engineering
111. Ehrlich HP, Hunt TK (2012) Collagen organization critical role in wound contraction. Adv Wound Care 1:3–9. https://doi.org/10.1089/wound.2011.0311
112. Kawaguchi H, Jingushi S, Izumi T, Fukunaga M, Matsushita T, Nakamura T, Mizuno K, Nakamura T, Nakamura K (2007) Local application of recombinant human fibroblast growth factor-2 on bone repair: a dose-escalation prospective trial on patients with osteotomy. J Orthop Res 25:480–487. https://doi.org/10.1002/jor.20315
113. Wu JC, Huang WC, Chen YC, Tu TH, Tsai YA, Huang SF, Huang HC, Cheng H (2011) Acidic fibroblast growth factor for repair of human spinal cord injury: a clinical trial. J Neurosurg Spine 15:216–227

114. Zhang L, Chen J, Han C (2009) A multicenter clinical trial of recombinant human GM-CSF hydrogel for the treatment of deep second-degree burns. Wound Repair Regen 17:685–689. https://doi.org/10.1111/j.1524-475X.2009.00526.x
115. Devi R, Dixit J (2016) Clinical evaluation of insulin like growth factor-I and vascular endothelial growth factor with alloplastic bone graft material in the management of human two. J Clin diagnostic Res JCDR 10:ZC41
116. Kishimoto Y, Hirano S, Kitani Y, Suehiro A, Umeda H, Tateya I, Kanemaru SI, Tabata Y, Ito J (2010) Chronic vocal fold scar restoration with hepatocyte growth factor hydrogel. Laryngoscope 120:108–113. https://doi.org/10.1002/lary.20642
117. Peeters M, Detiger SEL, Karfeld-Sulzer LS, Smit TH, Yayon A, Weber FE, Helder MN (2015) BMP-2 and BMP-2/7 heterodimers conjugated to a fibrin/hyaluronic acid hydrogel in a large animal model of mild intervertebral disc degeneration. Biores Open Access 4:398–406. https://doi.org/10.1089/biores.2015.0025
118. Koda J, Venook A, Walser E, Goodwin S (2002) I/II trial of hepatic intra-arterial delivery of doxorubicin hydrochloride adsorbed to magnetic targeted carriers in patients with hepatocellular carcinoma. Eur J Cancer 38:S18–S18
119. Yellen BB, Forbes ZG, Halverson DS, Fridman G, Barbee KA, Chorny M, Levy R, Friedman G (2005) Targeted drug delivery to magnetic implants for therapeutic applications. J Magn Magn Mater 293:647–654
120. Escoffre JM, Novell A, de Smet M, Bouakaz A (2013) Focused ultrasound mediated drug delivery from temperature-sensitive liposomes: in-vitro characterization and validation. Phys Med Biol 58:8135
121. Tran MA, Gowda R, Sharma A, Park E-J, Adair J, Kester M, Smith NB, Robertson GP (2008) Targeting V600E B-Raf and Akt3 using nanoliposomal-small interfering RNA inhibits cutaneous melanocytic lesion development. Cancer Res 68:7638–7649. https://doi.org/10.1158/0008-5472.CAN-07-6614
122. Stefanadis C, Chrysochoou C, Markou D, Petraki K, Panagiotakos DB, Fasoulakis C, Kyriakidis A, Papadimitriou C, Toutouzas PK (2001) Increased temperature of malignant urinary bladder tumors in vivo: the application of a new method based on a catheter technique. J Clin Oncol 19:676–681. https://doi.org/10.1200/JCO.2001.19.3.676
123. Fitzpatrick SD, Fitzpatrick LE, Thakur A, Mazumder MAJ, Sheardown H (2012) Temperature-sensitive polymers for drug delivery. Expert Rev Med Devices 9:339–351. https://doi.org/10.1586/erd.12.24
124. Ta T, Porter TM (2013) Thermosensitive liposomes for localized delivery and triggered release of chemotherapy. J Control Release 169:112–125
125. Iga K, Hamaguchi N, Igari Y, Ogawa Y, Gotoh K, Ootsu K, Toguchi H, Shimamoto T (1991) Enhanced antitumor activity in mice after administration of thermosensitive liposome encapsulating cisplatin with hyperthermia. J Pharmacol Exp Ther 257:1203–1207
126. Nishimura Y, Ono K, Hiraoka M, Masunaga S, Jo S, Shibamoto Y, Sasai K, Abe M, Iga K, Ogawa Y (1990) Treatment of murine SCC VII tumors with localized hyperthermia and temperature-sensitive liposomes containing cisplatin. Radiat Res 122:161. https://doi.org/10.2307/3577601
127. Poon E, Harris AL, Ashcroft M (2009) Targeting the hypoxia-inducible factor (HIF) pathway in cancer. Expert Rev Mol Med 11. https://doi.org/10.1017/S1462399409001173
128. Thambi T, Deepagan VG, Yoon HY, Han HS, Kim SH, Son S, Jo DG, Ahn CH, Suh YD, Kim K, Kwon IC (2014) Hypoxia-responsive polymeric nanoparticles for tumor-targeted drug delivery. Biomaterials 35:1735–1743
129. Rapisarda A, Melillo G (2012) Overcoming disappointing results with antiangiogenic therapy by targeting hypoxia. Nat Rev Clin Oncol 9:378
130. Chiche J, Brahimi-Horn MC, Pouysségur J (2010) Tumour hypoxia induces a metabolic shift causing acidosis: a common feature in cancer. J Cell Mol Med 14:771–794. https://doi.org/10.1111/j.1582-4934.2009.00994.x
131. Emmetiere F, Irwin C, Viola-Villegas NT, Longo V, Cheal SM, Zanzonico P, Pillarsetty NVK, Weber WA, Lewis JS, Reiner T (2013) 18F-labeled-bioorthogonal liposomes for in vivo targeting. Bioconjug Chem 24:1784–1789. https://doi.org/10.1021/bc400322h

132. Wijesinghe D, Arachchige MC, Lu A, Reshetnyak YK, Andreev OA (2013) pH dependent transfer of nano-pores into membrane of cancer cells to induce apoptosis. Sci Rep 3:3560
133. Hladky SB, Haydon DA (1972) Ion transfer across lipid membranes in the presence of gramicidin A: I. Studies of the unit conductance channel. Biochim Biophys Acta (BBA)-Biomembr 274:294–312
134. Yao L, Daniels J, Moshnikova A, Kuznetsov S, Ahmed A, Engelman DM, Reshetnyak YK, Andreev OA (2013) pHLIP peptide targets nanogold particles to tumors. Proc Natl Acad Sci 110:465–470. https://doi.org/10.1073/pnas.1219665110
135. Davies A, Lewis DJ, Watson SP, Thomas SG, Pikramenou Z (2012) pH-controlled delivery of luminescent europium coated nanoparticles into platelets. Proc Natl Acad Sci 109:1862–1867. https://doi.org/10.1073/pnas.1112132109/-/DCSupplemental
136. Han L, Ma H, Guo Y, Kuang Y, He X, Jiang C (2013) PH-controlled delivery of nanoparticles into tumor cells. Adv Healthc Mater 2:1435–1439. https://doi.org/10.1002/adhm.201300013
137. Zhao Z, Meng H, Wang N, Donovan MJ, Fu T, You M, Chen Z, Zhang X, Tan W (2013) A controlled-release nanocarrier with extracellular pH value driven tumor targeting and translocation for drug delivery. Angew Chemie Int Ed 52:7487–7491. https://doi.org/10.1002/anie.201302557
138. Shinde P, Kumar A, Illath K, Dey K, Mohan L, Barik TK, Rad J, Nagai M, Santra TS (2020) Physical approaches for drug delivery-an overview. Elsevier
139. Shinde P, Mohan L, Maddi A, Dey K, Chang H-Y, Nagai M, Santra TS (2018) Current and emerging trends of single cell technology. Int J Mol Sci 19:3143
140. Tseng F, Santra T (2016) Essentials of single-cell analysis. Springer, Berlin Heidelberg
141. Santra TS, Tseng F-G (2020) Handbook of single cell technologies. Springer Nature Publisher
142. Kar S, Mohan L, Dey K, Shinde P, Chang H-Y, Nagai M, Santra TS (2018) Single-cell electroporation: current trends, applications and future prospects. J Micromech Microeng 28:123002
143. Santra TS, Chiou PY, Wu T-H (2016) Optical MEMS for chemical analysis and biomedicine. In: Photothermal microfluidics. The Institute of Engineering and Technology (IET), USA Inc
144. Kumar A, Mohan L, Shinde P, Chang H-Y, Nagai M, Santra TS (2018) Mechanoporation: towards single cell approaches. In: Handbook of single cell technologies. Springer
145. Santra TS, Chen S-C, Chang C-J, Chen T-J, Wang P-C, Tseng F-G (2012) Delivery of molecules into cells using localized single cell electroporation by ITO micro-electrode based transparent chip. Biomed Microdevices 14:811–817
146. Santra TS, Chang HY, Wang PC, Tseng FG (2014) Impact of pulse duration on localized single-cell nano-electroporation. Analyst 139:6249–6258. https://doi.org/10.1039/c4an01050g
147. Dhakal K, Batabyal S, Wright W, Kim YT, Mohanty S (2015) Optical delivery of multiple opsin-encoding genes leads to targeted expression and white-light activation. Light Sci Appl 4:e352–e352
148. Santra TS, Kar S, Chen T-C, Chen C-W, Chang H-W, Lee M-C, Tseng F-G (2020) Near-infrared nanosecond-pulsed laser-activated high efficient intracellular delivery mediated by nano-corrugated mushroom-shaped gold-coated polystyrene nanoparticles. RSC Nanoscale 12:12057–12067. https://doi.org/10.1039/D0NR01792B
149. Kar S, Shinde P, Nagai M, Santra TS (2020) Optical manipulation of cells. In: Microfluidics and bio-mems: devices and applications. Jenny Stanford Publisher Pte. Ltd
150. Illath K, Narasimahan AK, Nagai M, Wankhar S, Santra TS (2020) Microfluidic based metallic nanoparticle synthesis and applications. In: Microfluidics and bio-mems: devices and applications. Jenny Stanford Publisher Pte. Ltd
151. Kumar A, Shinde P, Mohan L, Mhapatra PS, Santra TS (2020) Microfluidic technologies for cell manipulation, therapeutics and analysis. In: Microfluidics and bio-mems: devices and applications. Jenny Stanford Publisher Pte. Ltd
152. Jiskoot W, Van Schie RMF, Carstens MG, Schellekens H (2009) Immunological risk of injectable drug delivery systems. Pharm Res 26:1303–1314
153. Thiele L, Rothen-Rutishauser B, Jilek S, Wunderli-Allenspach H, Merkle HP, Walter E (2001) Evaluation of particle uptake in human blood monocyte-derived cells in vitro. Does

phagocytosis activity of dendritic cells measure up with macrophages? J Control Release 76:59–71
154. Thiele L, Diederichs JE, Reszka R, Merkle HP, Walter E (2003) Competitive adsorption of serum proteins at microparticles affects phagocytosis by dendritic cells. Biomaterials 24:1409–1418
155. Lundqvist M, Stigler J, Elia G, Lynch I, Cedervall T, Dawson KA (2008) Nanoparticle size and surface properties determine the protein corona with possible implications for biological impacts. Proc Natl Acad Sci 105:14265–14270
156. Ferreira VP, Pangburn MK, Cortés C (2010) Complement control protein factor H: the good, the bad, and the inadequate. Mol Immunol 47:2187–2197
157. Limbach LK, Wick P, Manser P, Grass RN, Bruinink A, Stark WJ (2007) Exposure of engineered nanoparticles to human lung epithelial cells: influence of chemical composition and catalytic activity on oxidative stress. Environ Sci Technol 41:4158–4163. https://doi.org/10.1021/es062629t
158. Byrne B, Donohoe GG, O'Kennedy R (2007) Sialic acids: carbohydrate moieties that influence the biological and physical properties of biopharmaceutical proteins and living cells. Drug Discov Today 12:319–326
159. Kah JCY, Wong KY, Neoh KG, Song JH, Fu JWP, Mhaisalkar S, Olivo M, Sheppard CJR (2008) Critical parameters in the pegylation of gold nanoshells for biomedical applications: an in vitro macrophage study porphyryins and porphyrin monolayers to enhance graphene electronics view project Mueller matrix microscopy view project. J Drug Target 17:181–193. https://doi.org/10.1080/10611860802582442
160. Koide H, Asai T, Hatanaka K, Akai S, Ishii T, Kenjo E, Ishida T, Kiwada H, Tsukada H, Oku N (2010) T cell-independent B cell response is responsible for ABC phenomenon induced by repeated injection of PEGylated liposomes. Int J Pharm 392:218–223
161. Lamanna G, Kueny-Stotz M, Mamlouk-Chaouachi H, Ghobril C, Basly B, Bertin A, Miladi I, Billotey C, Pourroy G, Begin-Colin S, Felder-Flesch D (2011) Dendronized iron oxide nanoparticles for multimodal imaging. Biomaterials 32:8562–8573
162. Witvrouw M, Fikkert V, Pluymers W, Matthews B, Mardel K, Schols D, Raff J, Debyser Z, De Clercq E, Holan G, Pannecouque C (2000) Polyanionic (i.e., polysulfonate) dendrimers can inhibit the replication of human immunodeficiency virus by interfering with both virus adsorption and later steps (reverse transcriptase/integrase) in the virus replicative cycle. Mol Pharmacol 58:1100–1108
163. Vannucci L, Fiserová A, Sadalapure K, Lindhorst TK, Kuldová M, Rossmann P, Horváth O, Kren V, Krist P, Bezouska K, Luptovcová M (2014) Effects of N-acetyl-glucosamine-coated glycodendrimers as biological modulators in the B16F10 melanoma model in vivo. Int J Oncol 44:1410–1410
164. Shaunak S, Thomas S, Gianasi E, Godwin A, Jones E, Teo I, Mireskandari K, Luthert P, Duncan R, Patterson S, Khaw P (2004) Polyvalent dendrimer glucosamine conjugates prevent scar tissue formation. Nat Biotechnol 22:977–984
165. Islam MA, Firdous J, Choi YJ, Yun CH, Cho CS (2012) Design and application of chitosan microspheres as oral and nasal vaccine carriers: an updated review. Int J Nanomedicine 7:6077
166. Porporatto C, Bianco ID, Correa SG (2005) Local and systemic activity of the polysaccharide chitosan at lymphoid tissues after oral administration. J Leukoc Biol 78:62–69. https://doi.org/10.1189/jlb.0904541
167. Chen KL, Mylon SE, Elimelech M (2007) Enhanced aggregation of alginate-coated iron oxide (Hematite) nanoparticles in the presence of calcium, strontium, and barium cations. Langmuir 23:5920–5928. https://doi.org/10.1021/la063744k
168. Démoulins T, Bassi I, Thomann-Harwood L, Jandus C, Kaeuper P, Simon HU, von Gunten S, McCullough KC (2013) Alginate-coated chitosan nanogel capacity to modulate the effect of TLR ligands on blood dendritic cells. Nanomed Nanotechnol Biol Med 9:806–817
169. Yamamoto Y, Kurachi M, Yamaguchi K, Oda T (2014) Bioscience, biotechnology, and biochemistry induction of multiple cytokine secretion from RAW264.7 cells by alginate oligosaccharides. Biosci Biotechnol Biochem 71:238–241. https://doi.org/10.1271/bbb.60416

170. Salvador A, Igartua M, Hernández RM, Pedraz JL (2012) Combination of immune stimulating adjuvants with poly (lactide-co-glycolide) microspheres enhances the immune response of vaccines. Vaccine 30:589–596
171. Dobrovolskaia M, McNeil S (2013) Handbook of immunological properties of engineered nanomaterials, vol 1. World Scientific Publishing, Singapore
172. Napierska D, Quarck R, Thomassen LC, Lison D, Martens JA, Del-croix M, Nemery B, Hoet PH (2012) Amorphous silica nanoparticles promote monocyte adhesion to human endothelial cells: size-dependent effect. Small 9:430–438. https://doi.org/10.1002/smll.201201033
173. Morishige T, Yoshioka Y, Inakura H, Tanabe A, Yao X, Narimatsu S, Monobe Y, Imazawa T, Tsunoda SI, Tsutsumi Y, Mukai Y (2010) The effect of surface modification of amorphous silica particles on NLRP3 inflammasome mediated IL-1β production, ROS production and endosomal rupture. Biomaterials 31:6833–6842
174. Boraschi D, Italiani P, National I (2012) Interaction of nanoparticles with immunocompetent cells: nanosafety considerations. Nanomedicine 7:121–131. https://doi.org/10.2217/nnm.11.169
175. Elsherbini AA, Saber M, Aggag M, El-Shahawy A, Shokier HA (2011) Laser and radiofrequency-induced hyperthermia treatment via gold-coated magnetic nanocomposites. Int J Nanomedicine 6:2155
176. Elsherbini AA, Saber M, Aggag M, El-Shahawy A, Shokier HA (2011) Magnetic nanoparticle-induced hyperthermia treatment under magnetic resonance imaging. Magn Reson Imaging 29:272–280
177. Hsiao I-L, Huang Y-J (2011) Titanium oxide shell coatings decrease the cytotoxicity of ZnO nanoparticles. Chem Res Toxicol 24:303–313. https://doi.org/10.1021/tx1001892
178. Arora HC, Jensen MP, Yuan Y, Wu A, Vogt S, Paunesku T, Woloschak GE (2012) Nanocarriers enhance doxorubicin uptake in drug-resistant ovarian cancer cells. Cancer Res 72:769–778. https://doi.org/10.1158/0008-5472.CAN-11-2890
179. Nohynek GJ, Dufour EK (2012) Nano-sized cosmetic formulations or solid nanoparticles in sunscreens: a risk to human health? Arch Toxicol 86:1063–1075. https://doi.org/10.1007/s00204-012-0831-5
180. Davis MM (2008) A prescription for human immunology. Immunity 29:835–838
181. Gagliardi M, Bardi G, Bifone A (2012) Polymeric nanocarriers for controlled and enhanced delivery of therapeutic agents to the CNS. Ther Deliv 3:875–887

Nanomaterials for Medical Implants

Y. Sasikumar, A. Srinivasan, and E. Hrishikesan

Abstract Nanotechnology has spread worldwide because of its wide applications, in particular with its novel and unique properties. Various kinds of nanomaterials have been investigated as their properties are mainly dependent on the size, shape, and composition of the materials, widely applied in the biomedical field. Various biomedical devices like dental implants made of Ti, Mg, Co, etc. surgically placed in the jawbone where the teeth are absent. The hip joint implant is carried out by surgical procedures through a replacement of a prosthetic implant. Cardiovascular implants include pacemakers of artificial heart valves or prosthetic implants, stents implanted inside the body etc. Furthermore, in this chapter, we will discuss more on the general consideration of using nanomaterials in implantable devices, dental implants/prosthodontics, spinal, orthopaedic implants, hip and knee replacements, cardiovascular implants and others—phakic intraocular lens and cosmetic implants.

Keywords Nanostructure · Biomedical implant · Surface coatings · Cosmetic implants · Prosthodontics

Y. Sasikumar (✉)
School of Materials Science and Engineering, Tianjin University of Technology, Xiqing District, Tianjin 300384, China
e-mail: sasikumar.phd@gmail.com

A. Srinivasan
School of Mechanical Engineering and Automation, Beihang University, 37 Xueyuan Road, Beijing 100191, China

E. Hrishikesan
School of Chemical Engineering and Technology, Beiyang Yuan Campus, Tianjin University, Jinnan District, Tianjin 300350, China

T. S. Santra and L. Mohan (eds.), *Nanomaterials and Their Biomedical Applications*, Springer Series in Biomaterials Science and Engineering 16,
https://doi.org/10.1007/978-981-33-6252-9_11

1 General Consideration of Usage of Nanomaterials as Implant Devices

Nanotechnology is considered in various fields of science, integrating engineering technology with biology, physics, and chemistry. Nanotechnology can be considered for the application of science that "steps across the limit" of miniaturization. In general, if the dimensions of a piece of solid material become small, then its chemical and physical properties will alter and become entirely different from the same material on a large scale as bulk form. This research area can be described, which limits on its new properties, and strategies to develop the properties with controlled size [1]. Nanoscience is the study of materials at the atomic, molecular, and macromolecular scales. Further, the properties of nanomaterials can be significantly different from that of the larger-scale materials. Nanotechnologies refer to the application of systems, devices, and structures with proper design, production, and characterization with controlled shape and size at their nanometer-scale. Generally, the nanotechnology market was categorized into three sections, viz., (a) materials, (b) devices, and (c) tools.

Nanomaterials can be described as the materials that consist of one or more components in the range of 1–100 nm of at least with one dimension, which includes nanoparticles, nanofibers, nanotubes, nanocomposite materials, and nanostructured surfaces. Nanoparticles (NP) are considered as single particles of less than 100 nm in its diameter. Nanofibres are classified as a subclass of nanoparticles with two dimensions of less than (< 100 nm). However, their third dimension (axis) can be compared with these particles [1, 2]. In this chapter, we discuss various nanomaterials, and their technologies widely used in bio implantable devices.

1.1 Nanostructured Materials Used for Implant Devices

In the past decades, it can be seen that nanomaterials were given importance in several implant applications, promoting the healing process, treating the fractures, etc. It has been reported that about 7.7 million patients visit each year in the U.S for orthopedic replacement surgery. Further, 600,000 joint replacement surgeries are performed each year at the cost of 3 billion dollars. [3].

However, it can be seen that all the surgeries have not been successful, as under certain conditions, the implants may fail, even well before their expected average life span of about 10–15 years. The major reasons for the implant failures are due to the stress-shielding effect, infection, tribo-corrosion, and wear debris (which lead to the loosened implant material due to the osteolysis in the peri-implant region). In addition to that, the implant material surface will not provide proper or suitable characteristics that are essential for bone cell recruitment and its differentiation. Thus, the failed implant materials will face several challenges resulting from revised surgeries. Eventually, there is a drastic change in the cost, as well as for the recovery time. It significantly impacts the surgery cost as the overall population (> 65 years) percentage expected to increase from 12.4% to 23% between the years 2000 and 2100 [3, 4].

1.2 Nano Surface Coatings Used for Implant Devices

The global medical device industries have shown significant growth in particular for the past five years. The nanotechnology field has become widely unexplored in biomedical devices for implant applications as the implant surface needs to be carefully designed. Unfortunately, many biomedical implant devices triggered a series of unwanted reactions during the biological processes, including inflammation, thrombosis, fibrosis, and various infections leading to implant rejection. At first, the biocompatibility decided upon the hydrophilic coatings, but the adverse effects did not subside significantly.

Abbott Cardiovascular Systems, Inc.'s invention pertains to the incorporation of one or multiple nanobeads or nanocarriers in the same or separate layer of slurry coated onto a medical device. The coating also serves for the customizable release of therapeutic agents with nanobeads that construct a layer at outermost spheres with the least thickness. Also, the innermost spheres will have a greater thickness to provide a controlled or graduated release. The nanobead's size would be approximately in the range of 1–1000 nm [5].

Another category of nanocoatings on the biomedical devices is nanoporous structures, on which the matrix either holds the drug or, in some cases, it will be the nanoparticles. Generally, the need for an additional surface area will meet the generating or coating nanostructures to medical devices. This may serve the desired amount of therapeutic to incorporate and impart some additional properties for biomedical devices [5].

1.3 Surface Treatments to Produce Nanostructures on Medical Implant

Nanomaterials with more complex structures such as nanodots, nanotubes, nanowires, nanorods, etc. can be fabricated through self-assembly and/or one-dimensional concepts. The nanomaterials can be classified as nanoparticles, nanofibers, nanocrystals, nanocoatings, etc. based on their form, structure, etc. Various methods' viz., physical, chemical, and mechanical methods are employed to fabricate nanoscale features on the implant surfaces. Physical methods include creating bioactive nano surface over the biocompatible implant surfaces with self-assembled monolayers of nanoparticle compaction. Plasma spraying, sputtering, and ion implantation are also commonly used physical methods.

1.3.1 Plasma Spraying

The plasma spraying method is employed to create nano-engineered surfaces having a dimension of less than 100 nm. The process includes the deposition of various charged metallic ions using a vacuum or through plasma guided kinetic energy over the implant material surface. Different types of coating materials such as Ti, Ag, Au were coated on metals, polymers, and ceramics. This method is generally adopted, and its main usage was the deposition of hydroxyapatite (HA) or calcium phosphate coatings to accelerate the bone integration process during service, which is essential for the implant materials. Hydroxyapatite coated implants were employed and led to a high percentage of bone-implant contact. This method also has several limitations, such as variations in the coating composition, lack of long-term adherence, and the non-uniform thickness of the deposited layer over the substrate materials, resulting in health hazards [6].

1.3.2 Sputtering

Sputtering is a common process in which a thin film of the ceramic layer is deposited from the ejected molecules or atoms due to the bombardment of high energy ions. This method is commonly used to achieve higher adhesion between the coating and substrate. Surface properties such as corrosion, wear resistance, biocompatibility can be effectively improved using this method. Nevertheless, the main drawback of this technique is a slow process with a lower deposition rate. The slow deposition rate can be improved using magnetron and radiofrequency sputtering [6].

- *Magnetron sputtering*:

  Magnetron sputtering is mainly carried out to deposit the viable thin film coating. This method can preserve the mechanical properties of a material, and also maintain its bioactivity. The magnetron sputtering process can be carried out at ambient temperature as well as at various temperatures. Further, the sputtering chambers can be designed based on the experimental condition for the coating deposition [6].

- *Radiofrequency sputtering*:

  Radiofrequency sputtering is generally used for thin-film calcium phosphate coating deposition over the surface of implants. The advantage of this method was the superior adhesion compared with other coating methods. This was mainly due to the deposition of material with high energy penetrating to the surface of the substrates. However, the main disadvantage in this method i.e., the crystalline nature of the coating surface, strong adhesion to the coating surface, and the alteration of CaP ratio easily.

1.3.3 Ion Implantation

The ion implantation process involves the arrangement of atoms that enable injecting the element to the near-surface region, which results in hydroxyapatite layer formation over the metal surfaces. It has been synthesized by the implantation of calcium and phosphate ions. This method mainly uses a high energy beam (10 keV) ions over the metal surface, which falls over the vacuum chamber. The incident ions come to rest due to their energy loss in the near-surface region because of the collision between substrate and incident ions [6–11]. The major advantages of this method are as follows:

a. High purity layers with ultra-clean process
b. Easy control and determination with concentration and depth of impurities
c. Excellent adhesion of the coating to an implanted surface
d. Effect free properties (bulk) of the substrate because of its low substrate level
e. Easily controllable and reproducible

Chemical methods are used to modify the surface of implant material in the nanoscale range. Anodic oxidation, combined anodization with chemical etching, etching, acid treatment, alkali treatment, hydrogen peroxide treatment, nanoparticle deposition methods (sol-gel process), chemical deposition, combined chemical vapor deposition are the common chemical methods widely used to fabricate the micro/nano surfaces [6].

1.3.4 Anodic Oxidation

The anodic oxidation process is generally carried out on smooth Ti or Mg implant surfaces to fabricate nanotubular structures (< 100 nm in diameter). Physicochemical surface properties, spacing, and diameter of nanotubes can be controlled with varying current density, voltage, electrolyte composition, and temperature. The anodization process leads to the formation of nanostructures with a pillar form of large nanotube arrays of tunable size with 10 μm on multi-walled nanotubes, nano HAp coatings (15–25 nm) are deposited on Ti or Mg surfaces to improve bioactivity, which is essential for implant materials [6].

1.3.5 Combined Anodization with Chemical Etching

This method is a combination of both anodization and chemical etching combined to form the reaction with metal or polymers. Anodized nanotubular Ti or Mg surfaces are coated by treating with sodium hydroxide (NaOH) with nanoporous poly (lactic-co-glycolic acid) to stimulate cell activity. However, it can be seen that there is no significant difference when compared with the anodized surfaces. Hence, the combination of hydrothermal treatment (temperature, time duration, and tuning concentration) with NaOH has been used with unique nanostructures, which include nanorods, nanoneedles, nanoflowers, and mesoporous nano scaffolds [6].

1.3.6 Acid Treatment

Acid treatment is widely used to clean the surface contamination and produce a uniform oxide layer deposition over the implant surface. The acids normally used were nitric acid (HNO_3), hydrochloric acid (HCl), sulfuric acid (H_2SO_4), and hydrofluoric acid (HF). These acid treatments generally increase surface roughness and increase the surface area, enhancing bone-implant contact. Takeuchi et al. [12] have evaluated the influences of three acids, namely HCl, H_2SO_4, HF, and their efficiency of the decontamination of Ti implant surfaces for the implant surface modification method. The results showed that the HCl treatment possesses a strong influence compared to other acids. A combination of strong bases or acids along with oxidants should produce nano pit networks of pit diameter from 20–100 nm effectively on Ti, Cr-Co-Mo, Ta alloys. Various parameters like surface topography, wettability, roughness, and thickness of the protective oxide layer formation, are easily controlled through the etching solution composition, temperature modulation, and length of exposure.

1.3.7 Alkali Treatment

Alkali treatment is a method in which the Ti implant is immersed in either NaOH or KOH followed by heat treatment in the range (600–800 °C) for 0.5–1 h and further rinsed with distilled water. This treatment results in a nanostructured growth of bioactive sodium titanate layer on implant Ti surfaces. Further, the surfaces are immersed in SBF (simulated body fluid) solution to become bioactive and calcium phosphate formation. The sodium ions are released by the formation of Ti-OH through the ion exchange process. However, the negatively charged Ti-OH reacts with positively charged Ca^{2+} ions from the SBF solution for the formation of the calcium titanate. Thus, the apatite formation occurs from calcium titanate solution through calcium and phosphate ions. This provides favorable conditions for osseointegration and cell differentiation [6].

1.3.8 Hydrogen Peroxide Treatment

Hydrogen peroxide treatment is a method in which the Ti implant surface undergoes oxidation and chemical dissolution. Ti peroxy gel formation occurs because of the chemical reaction on the Ti surface and hydrogen peroxide solution. The immersion of these surfaces with titania gel in the SBF solution leads to the development of thick layers, which is beneficial for apatite crystal deposition [6].

1.3.9 Chemical Vapor Deposition

Chemical vapor deposition is a method where the reaction occurs between the implant surface and the presence of chemicals in the gas phases [7–11]. Due to the chemical reactions, non-volatile compound deposition takes place over the surface of the implant. In this method, the metallic surface properties at their nanoscale level can be modified. The nanoscale modification will alter the topography/chemistry over the surface of the implant. Various methods are available for imparting these nanoscale structures to the implant's surface [6].

1.3.10 Sol-Gel Method

The sol-gel process is a common method practiced widely for the deposition of CaP, TiO_2, TiO-CaP composites, silica-based coatings over Ti implant surfaces. The main process carried out in this method is the sol formation of submicroscopic oxide particles with uniform suspension liquid by controlled hydrolysis and condensation. Different factors like sintering temperature, and chemical pre-treatment, surface roughness will determine the adhesion of TiO_2 sol-gel coatings over the Ti implant surface. The dual acid etching process on Ti implant surfaces (on the rat) leads to bone healing with the enhanced mechanical interlocking of the bone of $CaPO_4$ nanoparticles deposition with 20–40 nm [6, 12, 13].

1.3.11 Combined Chemical Vapor Deposition with Sol-Gel Method

This is a combination of two methods and is widely used to improve the metallic surface and its properties. Niobium oxide and carbon nano topography of diamond-like structures were deposited by this method. This method is mainly used for the improvement of bioactivity for the implant materials [6].

Mechanical methods are widely used to modify the metal's surface at its nanoscale level. Various available mechanical methods are mechanical alloying, powder metallurgy, and plastic deformation. These methods are mainly employed to increase surface roughness and surface morphological modifications on the surface of the implant for HA formation. The functionalized coatings are widely used with a combined synergistic effect of surface morphological changes with biomimetic CaP coatings. The main goal was to achieve improved wear and corrosion resistance with enhanced biocompatibility for the metals or alloys for bio-implant applications. Hence, by altering these changes over the implant's surface, there will be an interaction with the ions, cell tissues, and biomolecules. Thus, these interactions mainly favor the influence of cellular and molecular activities for the osseointegration process. Nanoscale surface leads to an increase in the initial adsorption of protein and possesses excellent and high surface energy, which is essential to the implant surface for the regulation of the cellular interactions. The adhesion's impact will also have surface properties with their charge distribution with the chemistry of the material [14, 15].

Powder Metallurgy (PM) is defined as "the art and science of the production of fine metal powders, and finished or semi-finished objects from a mixed or alloyed powder of an individual with/without inclusion with non-metallic constituents." Powder metallurgy is a technique in which the production of metal powders was involved and the powders' conversion into useful engineering structures by compacting the mixture in a rigid die. Due to the weak bond formation and friction, the powders are densified through deformation, and the particles are kept under pressure, and the subsequent heating of the compacted powder bonds the particles together. The process is simple because of the operations required to produce a component is relatively easy to understand. However, the component of the characteristics has become more difficult to work when it becomes more precise. The control of mechanical and physical properties increases in complexity, depending on the required characteristics. Powder Metallurgy process has a special advantage as the material wastage is minimum compared to the conventional melting route. To produce complex metal shapes to exact dimensions at high rates and economical prices, the PM process is widely used [14, 15].

This technique is frequently used in powder metallurgy for making an immiscible system of alloys, nanocrystalline phases, composite materials, and ceramic materials from elemental powders are termed as "mechanical alloying (MA)." MA is a technique that consists of continuous welding, repeated fracturing with rewelding, and cold fracturing of powder particles mixed with fabricating extremely fine microstructure. It is an energy-and time-consuming process and is used to alloy the elements that are normally difficult to combine by conventional melting techniques. Mechanical alloying mainly refers to a mechanically induced solid-state chemical reactions. These chemical reactions are mainly controlled by the diffusion rates of reactants, and the initial reactant geometry through the barrier products.

MA can overcome many limitations experienced by other conventional methods such as the processing of materials with widely varying densities. This technique is generally employed for the systems, where the equilibrium solid solubility is limited to its solid solubility extension. This process is mainly used for powder production, which has a fine microstructural scale for alloying incompatible materials. During MA, the solid-state reaction proceeds across the interface of different components under intense mechanical deformation process. This results in the formation of several metastable phases, including quasicrystals. For reasonable routes, the elevated temperature is normally employed in this technique. In MA, a high energy ball mill is placed with a powder charged particles of suitable and proper grinder medium. The mills are usually one of the three configurations: vertical ball mill, vibratory ball mill, and conventional ball mill. In the MA process, the powder particles are plastically deformed by a periodic treatment with colliding balls. Thus, lattice defects with a huge number of dislocations occur during this process. Furthermore, the collision of balls will cause the fracture and continuous cold welding of the particles in the atomic scale range. The increase of milling time further leads to the increase of the elementary components area's particle size, which will decrease the length from the micrometer to the sub-micrometer scale. The alloying process mainly involves repeated continuous fracturing, welding of powder particles rewelding with a high energy ball mill under or/not by the protective atmosphere. Process controlling agents like toluene, stearic acid, and methanol, etc. are usually employed for excessive cold welding and rewelding during the milling process. This process is widely used to control and achieve from alloy powders to nanocrystalline for various materials with a wide range. [14, 15].

2 Various Applications of Nanostructured Biomedical Implant Devices

2.1 Dental Implants /Prosthodontics

Dental implants are widely used to provide support as dental prostheses, crowns, bridges, or dentures with a supported implant and an anchorage in orthodontic tooth movement. These dental implants are frequently used as single implants in any situation for the replacement of a missing tooth and all cases, including total and partial edentulism. Metallic implants, in particular, Ti alloys, are used widely in dental implants as they have higher mechanical strength and chemical stabilities. However, bone-bonding abilities, antibacterial activity, and cell adhesion performances on polished Ti implants during service still lack. Acceleration of possible ways is the osseointegration, i.e., the direct bone bonding of living tissues with the surgical implants is to fabricate the surface with micro-nano surface roughness called micro/nanotextured surfaces. The nano surfaces fabricated on the implant could enlarge the contact area on the implant with living cells, thus enhancing the osseointegration. Interestingly, the nano surface with enhanced wettability further enhances the bone mineralization and osteoblast differentiation. Several surface coating methodologies are reported to modify the metallic surfaces with micro/nano morphologies, accelerating the osseointegration. Electrochemical oxidation is the most commonly used technique which is widely applied for Ti alloys to fabricate the nano surface with pores and tubes with different morphological features with varying chemical composition. The MAO (micro-arc oxidation) process is used to fabricate the nanoscale porous structure (ϕ10–300 nm) and micro-sized slots (3–7 μm) with nano-sized pores (ϕ10–300 nm) with duty cycles and durations of 9%–2 min and 11%–13 min, respectively in 0.1 mol/l $Li_2B_4O_7$ [16].

The coatings are composed of rutile and anatase phases; further, the increase of coating duration increased the rutile phase due to the higher voltage and the reaction temperature. Further, the static CA (contact angle) value of the 13 min coated alloy was about 4 and 13 times lower than the 2 min coated and polished samples. The polygonal shape MG63 cell adhesion was observed on 13 min coated Ti sample after 2 h culture, and cell proliferation was also increased in 1 and 3 days. Moreover, the MG63 cell adhesion and its osteogenic differentiation results have shown that the 13 min coated samples promoted osteogenic differentiation. The extracellular matrix proteins, particularly its early-stage BMP2 expression, were higher on the 13 min coated alloy, attributed to the tight cell adhesion and pile-up, therefore enhancing the osteoblast differentiation the micro/nano morphologies produced from MAO process. Anodized porous titania amorphous tricalcium phosphate (ATP) coatings incorporating Ca (calcium) and K (potassium) were developed on to Cp-Ti of grade 4 implants from phosphoric acid aqueous solution. Coatings with pore size of about 0.3 μm and the spacing of ~ 1 μm were achieved, and the homogeneous distribution of Ca and K metal cations with P in the coating was noticed [17]. The Mg, Ca, and P incorporated Ta_2O_5 porous nanostructures were developed on to high purity tantalum sheet using reverse polarization and anodic oxidation process [18]. In the anodization process (Ta anode), the (ATP: $Ca_3(PO_4)_2$) was deposited on the surface, while in the reverse polarization approach (Ta cathode), the incorporation of Ca^{2+}, Mg^{2+}, and Na^+ ions was achieved. The increase of surface hydrophilicity and its roughness suggested its candidature for the biomineralization process.

2.2 *Spinal Orthopaedic Implants*

The antibacterial activity of the implant is an important criterion to control the formation of biofilm over the implant's surface. The bacterial infection was the main reason and limitation of the failure of the implant. Therefore, the surface coatings having antibacterial properties are developed over the implant surface by incorporating the antibacterial agents. In a recent study, pure Ti was coated with TiO_2 containing Ag nanoparticles through the MAO process. The coating consists of microporous morphology, and Ag was characterized with nanoparticles in the cluster form, and few stand-alone nanoparticles were also observed in the TiO_2 matrix [19]. Furthermore, doping of Ag nanoparticles will not affect a crystalline phase of TiO_2 and the wettability of the coated surface was increased (~ 20°) and only minor influence of Ag doping on the wettability was reported. It was found that the S.aureus and E.coli bacteria adhesion were controlled, and the amount of Ag will also influence antibacterial activity studies. The samples with higher Ag contents (0.5 and 1.0 g/l) showed a complete reduction of E.coli and reduction of 6-log S.aureus. The enhancement in the antibacterial activity was due to the contact of Ag with the bacteria, and the release of Ag^+ ions could damage the membrane and lead to cell death. Nano-Ceria coatings with different shapes (rod, cube, and octahedron) developed on the titanium surface showed variation in the Streptococcus sanguinis adhesion anti-inflammatory response [20]. The antibacterial and anti-inflammatory effects were better for the nanooctahedron and nanocubes of CeO_2 coated Ti compared to the nanorod CeO_2

coated sample. Furthermore, the nano-octahedron CeO_2 coated Ti had the best properties compared with nanorod and nanocubes. The improved properties of nano-octahedron CeO_2 were mainly due to the smallest size with more crystalline planes exposure exhibited the highest Ce^{3+} value and strongest ROS scavenging capacity. The ZnO and HAp nanoparticle coated grade 23 Ti discs were found to be effective in resisting bacterial attachment and enhanced the osseointegration [21].

Ceramic dental implant market is consistently growing since ceramics possess superior mechanical properties; in particular, zirconia is a good choice of implant materials that can be attributed to its high excellent strength and fracture toughness. The surface modification of zirconia implants through various techniques are carried out to increase the surface roughness and wettability through the nanoporous surface formation. A selective infiltration etching (SIE) technique is used to fabricate the nanoporous surface on zirconia implants. The hydroxyapatite derived from the bones of the femoral head of white rabbits (V Spain) were prepared in the nanoscale (100 nm) and coated with zirconia by immersion coating. Platelet-rich plasma (PRP) coatings were also prepared onto porous zirconia by immersion technique [22]. The SIE technique results in the porous surface formation with an average of 136.43 ± 2.76 nm of pore diameter, and the porosity was reduced about 9 and 4% for the HA, and PRP coated surfaces respectively. The hybrid surfaces enhanced the new bone formation of about 79.8% and 71% for the HA and PRP coated Zr, respectively, in comparison with the uncoated surface (49%). The bone-implant contact (BIC) of HA-coated zirconia was higher than the PRP coated one, and this was attributed to the production of bone cells with attachment, and its proliferation matrix was better than the PRP which has only limited time of effectiveness.

2.3 *Hip and Knee Replacements*

Hard tissues are damaged due to the cause of sudden accidents. In general, the usual method adopted was the replacement of bone and tissues. The surgical substitution of hard tissues with damaged artificial joint replacements is the common method widely employed. Different endoprosthetic materials with different requirements widely-used depend upon the place and regions where the implants were inserted. Hip replacement generally refers to a surgical procedure, where a replacement of a hip joint with a prosthetic or artificial implant is carried out. The replacement surgery is performed by total or hemi (half) replacements depending on the patient's requirements. As part of the hip replacement, this joint replacement surgery is mainly conducted to remove the arthritis pain or fix various physical joint damage. Knee and hip joint diseases in patients as the average life expectancy have increased. One of the major problems which have raised significantly was the hip revision surgeries have also been increased. The metallic, polymeric, and composite implants are available and frequently used for knee and hip replacement surgeries as they possess a promising load-bearing capacity. It should be noted that the metallic implants often undergo wear and corrosion in the physiological environment was important. Therefore, the metallic debris release, corrosion products to the surroundings resulting in deleterious biological responses, and implant failure take place. To overcome these issues and various problems, several strategies are adopted to control the microstructure, (while developing the alloys) and severe plastic deformation. Also, the process to refine the microstructure/second phase distribution, and surface coatings on implants. Surface modification methods over the surface of implants using nanocoating could improve wear and corrosion properties and simultaneously enhance the biological responses. Hydrogen-free a-C/a-C: Ti multilayer coating was deposited to Ti-6Al-4 V alloy using the magnetron sputtering technique and revealed good biocompatibility and the femoral coated head showed better wear properties, reduced metal ions release the longevity of prostheses for prolong time [23–28].

The lubrication on the biomedical devices is an important factor in reducing the implant failure because of its friction and wear loss. The hydrophilic coating comprised of hyaluronic acid (HA)-silica nanohybrid hydrogels deposited on stainless steel 316 L (SUS) substrate revealed a lower coefficient of friction values than the only hyaluronic acid-coated sample. Hydroxyapatite binds physically with SiO_2, and due to the electrostatic force between the charged polymers, HAp stretches away from the substrate. As a result, a thick hydration layer formed, enhancing the lubrication in an aqueous environment [29]. The bio-corrosion of Cp-Ti, grade 4 was significantly controlled by spin coating, a thin film of ZnO functionalized with organic bi-functional molecules. The PEG functionalized surface exhibited the CA value of 18°, which is significantly lower than the other organic molecules and pristine ZnO coated surfaces. The enhanced wettability was due to the increase of C-O/CH_2 ratio and the existence of oxygen in -(CH_2CH_2O-)n backbone [30]. A noble shift of about 300 mV in corrosion potential (E_{corr}) for the coated samples was compared to the uncoated Ti. The 3-mercaptopropionic acid (MPA) and polyethylene glycol (PEG) had a slight decrease in E_{corr} compared to ZnO, which can be attributed to the presence of defects water or oxygen. Electrochemical impedance spectroscopic (EIS) results revealed that all the coated Ti except ZnO-PEG, exhibited R_{ct} values of magnitude with three orders which are higher than the uncoated Ti. ZnO with organic molecules will significantly reduce the charge transfer at their electrolyte/substrate interface and reduce the metal ions release, thus enhancing the functionalized surface's corrosion resistance.

Prosthetic implant mainly used in the hip replacement surgery contains different components viz. femoral component, acetabular cup, and articular interface. In general, total hip arthroplasty (total hip replacement) refers to the replacement of both femoral head and acetabular component, whereas the half (hemiarthroplasty) will replace only the femoral component. The component acetabular cup was placed into the acetabulum (hip socket). Bone and cartilage are removed from the acetabulum. Further, the acetabular cup is filled and attached with cement by friction. Some acetabulum is a single piece, and others are of modular type pieces. One-piece of shells are usually made up of either metal or polyethylene with articular surface machined over the inner surface of the cup and do not rely upon locking mechanism to hold with a liner. Polyethylene cup is cemented in place, along with a metal cup where the metal coating is held outside the cup. On the other hand, modular cups are usually made up of two cups, a liner, and a shell. The liner is designed inside with a locking mechanism, while the shell is made up of metal with a porous coating outside. Also, a fixation is achieved for bone growth over the porous coating. Further, the screws are used for the lag of the shell to the bone for providing more fixation. Hence, Cp-Ti and its alloys can be used for hard tissue replacements, artificial bones, and hip joints.

2.4 Cardiovascular Implants

Cardiovascular implants are stents that are mostly used for the treatment of cardiovascular diseases for most of the frequent cases. Stents used for cardiovascular applications consist of several metallic materials like stainless steel, Ti, Co-Cr, Au alloys, etc. for the permanent implants and Mg-Zn alloys are widely used as degradable implants. It has recently received considerable attention on shape memory alloy (NITINOL and Co-Cr alloys) as intravascular devices mainly used for stents and occlusion coils. Ti alloys possess a strong, inert, non-magnetic nature, which has got the main advantage for cardiovascular applications. A powerful diagnostic tool like MRI (magnetic resonance imaging) has produced a few artifacts was not sufficient due to radio-opaque in finer structures. The materials used for both the blood-contacting surface and the mechanical pump components are the artificial heart and cardiovascular circulatory assist devices. Generally, the metallic stents are coated with drug-loaded polymers as a local drug reservoir to fabricate the drug-eluting stents (DES). The usage of DES reduces the rate of in-stent restenosis by introducing local anti-inflammatory and anti-proliferative therapeutics [31–37].

A versatile top-down approach develops biodegradable fibrous nanoplatforms as a coating formation through DES through electrospinning deposition method, which is very cost-effective. As a biodegradable material, it has been proven that polylactic acid (PLA) is a good potential stent material for the systems with drug loading. To combat this DES late thrombosis, a mechanism was suggested for this polymer coating, which can act as a storage release after the implantation with inflammation by a controlled manner are well associated with DES [38, 39]. Due to its anti-thrombotic effect, the pharmaceutical agent, like dipyridamole (DPM) is frequently used along with the combination of salicylic acid repeatedly for the patients who are in the dangerous condition (secondary stroke). It can be seen that the results of DPM exhibit the formation of a thrombus and platelet segregation in humans. The major role of DPM is the resultant inhibition along with vascular cells of equilibrative nucleotide transporters, which uptake the extracellular adenosine into platelets. Therefore, it can inhibit the proliferation of vascular cells with smooth muscle cells from preventing the restenosis of vascular grafts. It activates the surface of the cell with the adenosine receptors, which will increase the intracellular concentration (cyclic adenosine monophosphate) with anti-inflammatory properties [40]. Electrospinning deposition is fabricated using anti-platelet of drug-loaded biodegradable polymer nanoplatforms, which helps perform with artery thrombosis. Also, the drug-eluting coating into cardiovascular stents will often occur after the stent implantation. This evinced that the drug is entirely inserted throughout the polymer with a homogeneous release for the drug-loaded scaffolds. Therefore, the studies revealed a sustainable control release of DPM drug, within the line degradation of overtime by fibrous matrices for the drug-release kinetics [40–44].

3 Others—Phakic Intraocular Lens and Cosmetic Implants

3.1 Phakic Intraocular Lens

Phakic intraocular or phakic lens (PLs) consists of silicone or plastic material. These PLs can be directly used to implant the human eye permanently, such that person can avoid using glass or contact lenses. Generally, the term 'phakic' commonly refers to the intraocular lens, which is implanted to the human eye with or without removing the natural human eye's lens. Intraocular lenses (IOLs) are usually made up of small plastic struts known as haptics. This haptics are mainly used to hold lens inside the human eye where there is a place of capsular bag present in it. Usually, the materials used in the manufacture of these IOLs implants are silicone, PMMA, collamers, hydrophilic acrylate, and hydrophobic acrylate. However, among these materials, PMMA was found to be successful. Therefore, IOLs are frequently made up of inflexible PMMA and can be implanted in the human eye for the common myopia, or cataracts treatment. The commonly used IOLs are of pseudophakic type. This type IOLs are used for the treatment for cataract surgery over the cloudy natural lens to remove it. It also provides the focus of the same light to that of the natural crystalline lens of the human eye [45].

Phakic lenses (IOLs) consist of 3 types viz., angle-supported anterior chamber, posterior chamber, and iris-claw anterior chamber. In these, each type has its specific features, selection criteria, different surgical treatments, and also with complications. During the last decades of the early 1980s and 1990s, reasonable and technological progress has been made by IOLs manufacturers with various surgical techniques that have been developed. At first, the PMMA angle-supported anterior chamber was initially used. But all of these were phased out of the market, and subsequently vanished due to their unacceptably high complication rates. This includes of glare and halos, corneal cell loss, chronic anterior uveitis, and pupil ovalization. Any kind of the refractive procedure of preoperative workup on pIOLs implantation, which includes manifest refraction, cycloplegic refraction, applanation tonometry, corneal topography, pupillometry, pachymetry, Snellen uncorrected distance visual acuity, corrected distance visual acuity, central endothelial with cell count and a fundus examination are usually employed. Hence, the focus on PMMA angle supported pIOLs was no longer available [45].

Currently, they focus on surgical techniques with available pIOLs like Kelman Duet pIOLs, which are being widely used. These pIOLs are not foldable. However, it consists of two components, viz: optic and haptic. To facilitate and manipulate the components, cornea visions of two, with one mm clear, are created at various angles. Sequentially a small incision was inserted and assembled over the anterior chamber. Initially, the haptic is first inserted by forceps with one incision to reposition at a certain angle. Further, the optic is being injected over the anterior chamber with a 3 mm incision. Implantable Collamer Lenses (ICLs) are widely used as post chamber pIOLs. These ICLs are mainly incorporated with a material known as 'collamers,' which has increased biocompatibility. Collamers mainly consist of 60% hydroxyethyl methacrylate copolymer and 0.2% collagen, respectively. This material will inhibit the aqueous protein binding and the deposition of the monolayer, being attracted to fibronectin of IOL surfaces [44, 45].

3.2 Cosmetic Implants

Cosmetic implants are often referred to as prosthetic devices on which an attempt is made with an acceptable aesthetic norm to bring some portion of the body back to its position. Follow-up mastectomy (due to breast cancer) is commonly used for correcting out the forms of disfigurement, and it's modifying aspects present in the body. In order to improve the appearance, numerous plastic surgery techniques are developed for various types of cosmetic implants and restore its original function inside the body. Examples include the pectoral implant, injectable gluteal fillers, ocular prosthesis, nose prosthesis, and breast implants. Recently various types of cosmetic implants have been developed with various types of plastic surgery techniques for the appearance or restoring the function on the human body with further additional techniques. Pectoral implants (Male chest enhancement) are increasingly performed with men. Most of the patients are unable to do proper exercises to build up their pectoral muscles. So, it is generally performed with fewer numbers of patients initially with their chest wall of acquired or congenital deformities. E.g. pectoral denervation because of plexus injury, absence of unilateral or congenital pectoral muscles, post-surgical chest walls deformities, muscle-related injuries due to sports events, etc. For a specific patient, customized pectoral implants are typically made by various manufactures. Depending upon the deformity of defects, these pectoral implants are placed either in an oblique or transverse position. Generally, these pectoral implants are usually made up of silicone rubber of soft pliable consistency, which can be customized easily. Also, it is typically of a variable with CT appearance with denser soft tissue less dense than that of bone.

Gluteal augmentation (gluteal fillers) was found in 1965 and one of the very fastest-growing plastic surgery techniques. It mainly involves with prostheses insertion in the parts of breast and buttocks. Various techniques are performed, which include autogenous fat augmentation, solid silicone implant replacement, and filler injection. It can be seen that for the buttock augmentation, FDA has approved the solid silicone type of elastomer implants as the sole implants since it has got the proven potential to increase congenital, asymmetry, gluteal tone with ptosis for the patients.

Recently, autogenous augmentation has also been found to be a widely increase and the fastest-growing technique due to its natural appearance with ease of fabrication and availability. In this method, the procedure adopted is the fat is suctioned from lower parts of the body (E.g. lower back, flanks, and trochanteric regions of the human body). Employing multiple small incisions, the fat collected from the body is transferred, and sand injected into buttocks by a syringe. The excessive fat from the flank and lower back regions will help to contour the buttocks. Various surgeons are practicing and performing lipofilling and liposuction on buttocks. However, various complications of the fat augmentation include donor-site ischemia or fluid accumulation, seroma, fat embolism, cellulitis infection, partial reabsorption of fat grafts, transient sciatic paresthesias etc.

Losing the teeth in our day today-life is a significant problem in our living life. Cosmetic implants are widely used to get back our teeth, artificially. Hence, the artificial implants that have been affixed will possess the same strength, and their characteristic features should be quite reasonable to that of natural teeth. These artificial teeth not only have equal strength, but they help for the preservation of bones and teeth gums more effectively. Therefore, these cosmetic implants would be a welcome procedure whoever loses their teeth due to trauma or some accidents. These cosmetic implants can be applied through the process, where a titanium implant is inserted into teeth gums. After the implantation, once the bone starts to grow around, it completely looks like the natural teeth and matches up. However, the cosmetic procedure would take about a period from 3-6 months before the cosmetic implant becomes fully functional and looks good and similar to natural teeth. Though these cosmetic implants are recommended too by the dentists, the patients' age and health play a major role, and the patient needs to be in good health condition and able to support it completely [46–51].

4 Summary

Available research data on the nanoscale surface modification methods of implants were discussed. Changes with various surface modifications, along with the chemical-biological reaction of bio-implant materials, were also discussed. Nanostructured materials and nanocoatings of the implant materials were elaborated in detail. Surface treatments for the nanostructures were explained, and various methods like physical, chemical, mechanical methods have also been summarized. Various applications used for nano implant applications are dental implants, spine orthopedic implants, hip and knee replacements, cardiovascular artificial stents and valves. Cosmetic implants exhibit an important role concerning the eye lens or contact lens. The phakic intraocular lens (pIOLs) was employed. Pectoral implant, injectable gluteal fillers, breast implants, ocular, and nose prostheses are widely used in cosmetic plastic surgeries. In particular, the augmentation of the nose, teeth, buttocks, thigh, chest was discussed. Artificial teeth or cosmetic implants with equal strength help the gums, and bones for the preservation of the teeth more effectively. The change in interactions with the bio implant surfaces will influence the biomolecular activities for altering the osseointegration. Furthermore, the exploration of the change in behavior for the nano surface modifications is still indeed in the implant applications for their long-term usage.

References

1. Lehn JM (2002) Toward self-organization and complex matter. Sci 295:2400
2. Schmidt G, Decker M, Ernst H, Fuchs H, Grunwald W, Grunwald A, et al (2003) Small dimensions and material properties. Europaische Akademie Graue Reihe. In: A definition of nanotechnology (ed) Bad Neuenahr, p 134
3. Christenson EM, Anseth KS, Beucken JJ, Van Den JP et al (2007) Nanobiomaterial applications in Orthopedics. J Orthop Res 25:11
4. Balasundaram G, Webster TJ (2006) Nanotechnology and biomaterials for orthopedic medical applications. Nanomed (Lond) 1:169
5. Arsiwala Ammar M, Raval Ankur J, Patravale Vandana B (2013) Nanocoatings on implantable medical devices. Pharm Pat Analyst 2:499
6. Preeti P, Lakshmana Rao B, Rajashekar S (2014) Techniques for dental implant nanosurface modifications. J Adv Prosthodont 6:498
7. Santra TS, Bhattacharyya TK, Patel P, Tseng F-G, Barik TK (2012) Diamond, Diamond-Like Carbon (DLC) and Diamond-Like Nanocomposite (DLN) thin films for films for MEMS applications. InTech, E.U. ISBN 978-953-307-905-9, p 459–480
8. Santra TS, Bhattacharyya TK, Patel P, Tseng FG, Barik TK (2010) Characterization of diamond-like nanocomposite thin films grown by plasma enhanced chemical vapor deposition. J Appl Phys 10:124320
9. Santra TS, Bhattacharyya TK, Patel P, Tseng FG, Barik TK (2011) Structural and tribological properties of diamond-like nanocomposite thin films grown by PECVD. Surf Coat Technol 206:2–3
10. Santra TS, Bhattacharyya TK, Mishra P, Tseng FG, Barik TK (2011) Biomedical application of Diamond Like nanocomposite thin films. Sci Adv Mater 4:110–113
11. Santra TS, Bhattacharyya TK, Tseng FG, Barik TK (2012) Influence of flow rate on different properties of diamond like nanocomposite thin films grown by PECVD. AIP Adv 2:022132
12. Takeuchi M, Abe Y, Yoshida Y, Nakayama Y, Okazaki M, Akagawa Y (2003) Acid pretreatment of titanium implants. Biomaterials 24:1821
13. Nishimura I, Huang Y, Butz F, Ogawa T, Lin A, Jake Wang C (2007) Discrete deposition of hydroxyapatite nanoparticles on a titanium implant with predisposing substrate microtopography accelerated osseointegration. Nanotechnol 18:245101
14. McHenry ME, Willard MA, Iwanabe H, Sutton RA, Turgut Z, Hsiao A, Laughlin DE (1999) Nanocrystalline materials for high temperature soft magnetic applications: A current prospectus. Bull Mater Sci 22:495
15. McHenry ME, Johnson F, Okumura H, Ohkubo T, Ramanan VRV, Laughlin DE (2003) The kinetics of nanocrystallization and microstructural observations in FINEMET, NANOPERM and HITPERM nanocomposite magnetic materials. Script Mater 48:881
16. Pan X, Li Y, Abdullah AO, Wang W, Qi M, Liu Y (2019) Micro/nano-hierarchical structured TiO_2 coating on titanium by micro-arc oxidation enhances osteoblast adhesion and differentiation. R Soc Open Sci 6:182031

17. Marenzi G, Spagnuolo G, Sammartino JC, Gasparro R, Rebaudi A, Salerno M (2019) Micro-Scale Surface patterning of titanium dental implants by anodization in the presence of modifying salts. Materials 12:1753
18. Fialho L, Carvalho S (2019) Surface engineering of nanostructured Ta surface with incorporation of osteoconductive elements by anodization. Appl Surf Sci 143:573
19. Thukkaram M, Cools P, Nikiforov A, Rigole P, Coenye T, Van Der Voort P et al (2019) Antibacterial activity of a porous silver doped TiO_2 coating on titanium substrates synthesized by plasma electrolytic oxidation. Appl Surf Sci 144:235
20. Li X, Qi M, Sun X, Weir MD, Tay FR, Oates TW et al (2019) Surface treatments on titanium implants via nanostructured ceria for antibacterial and anti-inflammatory capabilities. Acta Biomater 94:627
21. Abdulkareem EH, Memarzadeh K, Allaker RP, Huang J, Pratten J, Spratt D (2015) Anti-biofilm activity of zinc oxide and hydroxyapatite nanoparticles as dental implant coating materials. J Dent 43:1462
22. Mostafa D, Aboushelib M (2018) Bioactive-hybrid-zirconia implant surface for enhancing osseointegration: an in vivo study. Int J Imp Dent 4:20
23. Li J, Li Z, Tu J, Jin G, Li L, Wang K et al (2019) In vitro and in vivo investigations of a-C/a-C: Ti nanomultilayer coated Ti6Al4V alloy as artificial femoral head. Mater Sci Engg C 99:816
24. Hwang C, Min Y, Seong Y-J, Kim D-E, Kim H-E, Jeong S-H (2019) Enhanced biolubrication on biomedical devices using hyaluronic acid-silica nanohybrid hydrogels. Coll Surf B Biointer 184:110503
25. Trino LD, Dias LFG, Albano LGS, Bronze-Uhle ES, Rangel EC, Graeff CFO et al (2018) Zinc oxide surface functionalization and related effects on corrosion resistance of titanium implants. Ceram Int 44:4000
26. Garza-Maldonado AV, Hernandez-Rodriguez MAL, Alvarez-Vera M, Ortega-Saenz JA, Perez-Unzueta A, Cue-Sampedro R (2017) Biotribological study of multi-nano-layers as a coating for total hip prostheses. Wear 376–377:243
27. Amaral M, Maru MM, Rodrigues SP, Gouvêa CP, Trommer RM, Oliveira FJ et al (2015) Extremely low wear rates in hip joint bearings coated with nanocrystalline diamond. Tribo Inter 89:72
28. Frandsen CJ, Noh K, Brammer KS, Johnston G, Jin S (2013) Hybrid micro/nano-topography of a TiO_2 nanotube-coated commercial zirconia femoral knee implant promotes bone cell adhesion in vitro. Mater Sci Engg C 33:2752
29. Hwang C, Min Y, Seong Y-J, Kim D-E, Kim H-E, Jeong S-H (2019) Enhanced biolubrication on biomedical devices using hyaluronic acid-silica nanohybrid hydrogels. Colloids Surf B 184:110503
30. Trino LD, Dias LFG, Albano LGS, Bronze-Uhle ES, Rangel EC, Graeff CFO et al (2018) Zinc oxide surface functionalization and related effects on corrosion resistance of titanium implants. Ceram Int 44:4000–4008
31. Soliman AM, Tolba SA, Sharafeldin IM, Gepreel MA-H, Allam NK (2019) Ni-free, built-in nanotubular drug eluting stents: experimental and theoretical insights. Mater Sci Engg C 103:109750
32. Fan Y, Zhang Y, Zhao Q, Xie Y, Luo R, Yang P et al (2019) Immobilization of nano Cu-MOFs with polydopamine coating for adaptable gasotransmitter generation and copper ion delivery on cardiovascular stents. Biomaterials 204:36
33. Bedair TM, Min IJ, Park W, Joung YK, Han DK (2018) Sustained drug release using cobalt oxide nanowires for the preparation of polymer-free drug-eluting stents. J Biomater Appl 33:352
34. Tang J, Liu Y, Zhu B, Su Y, Zhu X (2017) Preparation of paclitaxel/chitosan co-assembled core-shell nanofibers for drug-eluting stent. Appl Surf Sci 393:299
35. Yang J, Lee I-S, Cui F (2016) Sirolimus-loaded CaP coating on Co-Cr alloy for drug-eluting stent. Regen Biomater 3:167
36. Kushwaha M, Anderson JM, Bosworth CA, Andukuri A, Minor WP, Lancaster JR et al (2010) A nitric oxide releasing, self assembled peptide amphiphile matrix that mimics native endothelium for coating implantable cardiovascular devices. Biomaterials 31:1502

37. Rintoul TC, Bulter KC, Thomas DC, Carriker JW, Maher TR, Kiraly RJ, Massiello A, Himely SC, Chen JF, Fukamachi K (1993) Continuing development of the C level and clinic-nimbus total artificial heart. ASAIO J 39:168
38. Uurto I et al (2005) Drug-eluting biodegradable poly-D/L-lactic acid vascular stents: an experimental pilot study. J Endovasc Ther 12:371
39. Karagkiozaki VC et al (2010) Nanomedicine for the reduction of the thrombogenicity of stent coatings. Int J Nanomed 5:239
40. Bakola V, Karagkiozaki V, Tsiapla AR, Pappa F, Moutsios I, Pavlidou E, Logothetidis S (2018) Dipyridamole-loaded biodegradable PLA nanoplatforms as coatings for cardiovascular stents. Nanotechnol 29:275101
41. Repanas A et al (2016) The effect of dipyridamole embedded in a drug delivery system made by electrospun nanofibers on aortic endothelial cells. J Drug Deliv Sci Technol 35:343
42. Singh JP et al (1994) Dipyridamole directly inhibits vascular smooth muscle cell proliferation in vitro and in vivo: implications in the treatment of restenosis after angioplasty. J Am Coll Cardiol 23:665
43. Brown DG, Wilkerson EC, Love WE (2015) A review of traditional and novel oral anticoagulant and antiplatelet therapy for dermatologists and dermatologic surgeons. J Am Acad Dermatol 72:524
44. Zhuplatov SB et al (2006) Mechanism of dipyridamole's action in inhibition of venous and arterial smooth muscle cell proliferation. Basic Clin Pharmacol Toxicol 99:431
45. Güell Jose Luis, Morral Merce, Kook Daniel, Kohnen Thomas (2010) Phakic intraocular lenses Part 1: historical overview, current models, selection criteria, and surgical techniques. J Catar Refract Surg 36:1976
46. Yahyavi-Firouz-Abadi Noushin, Menias Christine O, Bhalla Sanjeev, Siegel Cary, Gayer Gabriela, Katz Douglas S (2015) Imaging of cosmetic plastic procedures and implants in the body and their potential complications. AJR 204:707
47. Flores-Lima G, Eppley BL, Dimas JR, Navarro DE (2013) Surgical pocket location for gluteal implants: a systematic review. Aesthet Plast Surg 37:240
48. Gonzalez R (2010) Gluteal implants: the "XYZ" intramuscular method. Aesthet Surg J 30:256
49. Darouiche RO (2004) Treatment of infections associated with surgical implants. N Engl J Med 350:1422
50. Ali A (2011) Contouring of the gluteal region in women: enhancement and augmentation. Ann Plast Surg 67:209
51. Bruner TW, Roberts TL 3rd, Nguyen K (2006) Complications of buttocks augmentation: diagnosis, management, and prevention. Clin Plast Surg 33:449

Fabrication of Nanostructured Scaffolds for Tissue Engineering Applications

Govindaraj Perumal and Mukesh Doble

Abstract The field of tissue engineering is rapidly growing nowadays, and the development of the bio-functional scaffolds to mimic the natural environment of the body tissue to replace the damaged or diseased tissues. Use of nanomaterials to develop nanostructured biomaterial scaffolds showing enhanced biological functions than microstructured materials. Nanostructured materials have higher surface area and biomimetic nature, hence allow better cell adhesion than conventional materials, which is needed for materials used for tissue engineering applications. Unlike conventional materials, nanomaterials mimic resembling natural extracellular matrix (ECM) to support better protein adsorption and also stimulate enhanced tissue regeneration. Here, we address different categories of biomaterials used to fabricating nanostructured scaffolds for tissue regeneration applications. The desired properties required for various tissue engineering scaffolds and its methods of fabrication, current development, and future directions of these methodologies are discussed. In addition to that, the special emphasis is given on bone tissue engineering and three dimensional (3D printing) technologies to manufacture tissue engineering scaffolds using various nanomaterials are also discussed in this chapter.

Keywords Nanostructured materials · Electrospinning · 3D printing · Tissue engineering · Regenerative medicine

1 Introduction

The nanostructure scaffold denotes that structures or components are in the nanoscale dimension and which makes them a suitable candidate to stimulate various cell functionalities ultimately required for tissue regeneration applications [1]. In other words, a material is considered to be nanostructured if the internal pores or surface structure

G. Perumal (✉) · M. Doble
Department of Biotechnology, Bhupat and Jyoti Mehta School of Biosciences, Indian Institute of Technology Madras, Chennai 600036, India
e-mail: govindvit07@gmail.com

T. S. Santra and L. Mohan (eds.), *Nanomaterials and Their Biomedical Applications*, Springer Series in Biomaterials Science and Engineering 16,
https://doi.org/10.1007/978-981-33-6252-9_12

is in the nanoscale [2]. In few cases reported the toxicity of the nanostructured interactions and posed serious problems, and this should be addressed for the effective use of such materials. Recently, nanotechnology-based scaffold fabrication is attracted to tissue engineering applications due to its ultimate biomimetic nature similar to ECM structure, which leads to better functionalities of the nanostructured scaffolds. Examples such as electrospun nanofibers, nanotubes, nanoparticles, and hydrogel preparations have shown successfully in tissue engineering applications both in vitro and in vivo for enhanced repair of damaged tissue over conventional scaffolds having micron-sized materials. One of the most significant advantages of nanostructured materials as previously mentioned, due to the nano dimension of native tissue structures, can efficiently interact with cells to stimulate nanostructured ECM, which is excellent biomimetic nature, which leads to enhanced guided tissue regeneration. To mimic the natural ECM nanostructure, the biomaterial scaffolds developed into various forms (examples mentioned above) for useful tissue regeneration applications [3]. In addition to that, natural tissues or organs resemble nanometer dimensions that allow nanostructured scaffolds with many improved cellular functions and tissue regeneration capacity [4].

Recently, numerous studies have reported that the effectiveness of the nanostructured scaffolds by incorporating various cells such as stem cells, chondrocytes, and osteoblasts for tissue regeneration applications [5]. The current attention is towards nanostructured materials due to their improved mechanical properties, porosity in nanoscale, biocompatibility, and bioactivity by better cell adhesion and proliferation, resulting in remarkable integration with the adjacent tissues [6, 7]. Nanomaterials including nanoparticles, nanofibers, nanocomposites, nanoclusters, nanocrystals, nanotubes, nanowires, nanorods, and nanofilms etc., currently, many top-down and bottom-up approaches existing to fabricate nanomaterials or nanostructured materials with well-organized or random nano topographies are namely electrospinning, phase separation, self-assembly processes, thin film deposition, chemical vapor deposition, chemical etching, nano-imprinting, photolithography, and electron beam or nanosphere lithographies [8–12].

2 Problems Associated with Conventional Biomaterials

Conventional or micron structured materials lack desired cytocompatibility, mechanical, optical, catalytic, and magnetic properties [13]. The autografts and allografts are the most widely used tissue substitutes for damaged tissue repair and which have limitations such as limited tissue availability, donor site morbidity, and high cost, which cannot meet the improved tissue regeneration [14, 15]. However, nanostructured material possesses higher surface area, allowing better adhesion of cells, proteins, or active ingredients than the former. Unlike conventional materials, nanomaterials biomimetic nature (shown in Fig. 1) similar to native bone tissue and promoting enhanced protein adsorption than the micron-sized materials, which has potential in stimulating more new bone regeneration [13].

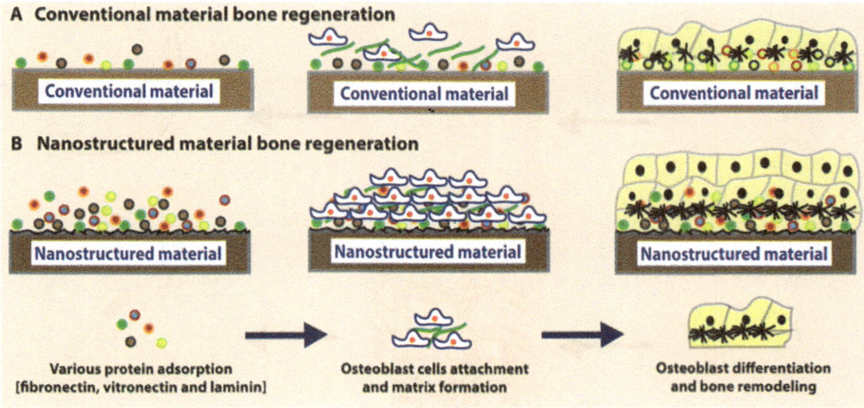

Fig. 1 Schematic illustration of the mechanism of interaction, **a** Conventional material and **b** Nanostructured materials mimicking the natural bone formation with more specific protein interactions and proliferation of osteoblast cells than conventional materials leading to the formation of extracellular matrix (ECM) to stimulate efficient new bone regeneration. Redrawn from Ref. [13]

The scaffold (micron or nano) architecture plays a significant role in cell attachment or binding. The response of the cells on the micron-sized scaffolds flat with limited interactions with adsorbed proteins. Whereas, the nanostructured scaffolds have a larger surface area for protein adsorption with higher binding capacity with receptors present in the cell membrane (shown in Fig. 2). This improved cell behavior with proteins and new binding sites can provide a better cellular response and mimic the ECM structure, which influences the enhanced tissue regeneration compared to micron-sized scaffolds [16].

3 Ideal Requirements of Materials for Tissue Regeneration

The design of the scaffold is more critical for tissue regeneration applications. The ultimate requirements of materials used for the development of tissue engineering scaffold should be biocompatible, biodegradable, and no or less immunogenic. The mismatched or non-biocompatible biomaterial causes the inflammatory response or foreign-body reaction as a result of rejection and/or necrosis or implant failure [17]. The scaffold surface should support cell adhesion and proliferation, provide interconnected porous structure for tissue ingrowth. Electrospun nanofibrous polymeric scaffolds, nanostructured composite scaffolds, and nanoparticles incorporated in bioactive molecule-delivering scaffolds significantly enhance cell responses than conventional materials used in tissue engineering applications. Therefore, fabricating biomaterials in nanoscale structures can overcome the problems of poor cytocompatibility, mechanical properties, and the absence of interconnected pores. An ideal tissue engineering scaffold should provide sufficient mechanical support during the

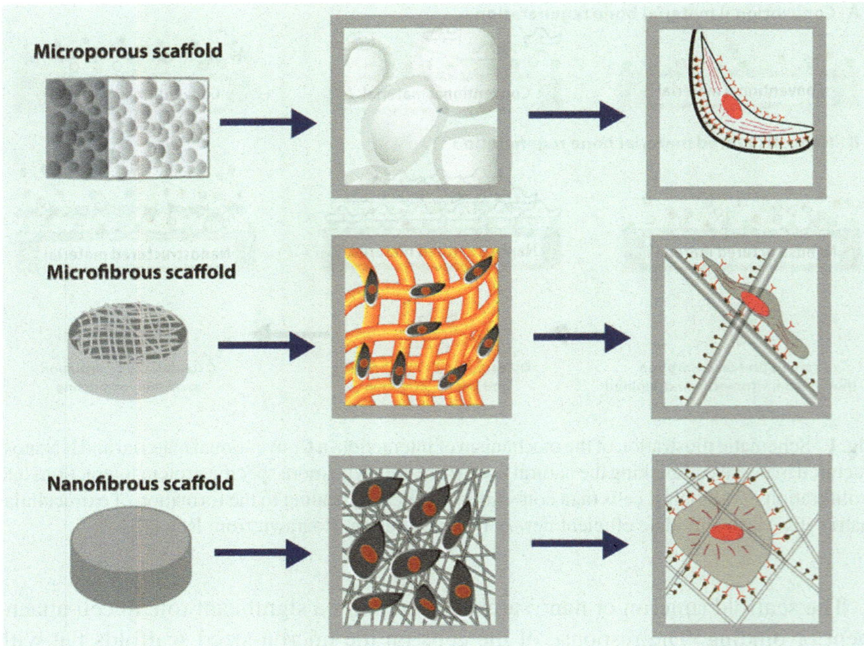

Fig. 2 Cell attachment and proliferation of micro and nanoarchitecture scaffolds. Redrawn from Ref. [16]

initial stage of the tissue regeneration/healing. The biomaterial scaffold should facilitate the three dimensional (3D) architecture; this could guide the new tissue formation [18]. These properties can be achieved by fabricating the biomaterial as nanofibers and nanocomposite, which can be used for tissue and bone regeneration applications. When compared to conventional biomaterials, the significant advantage of the nanofibrous scaffold with high surface to volume ratio, which considerably improves the biomimetic ECM environment leading design more complex tissues or organs [19]. Hence, the tissue-engineered scaffolds are typically fabricated to mimic the ECM to provide better functionality.

4 Biomedical Applications of Nanostructured Materials

Nanostructured materials have many biomedical applications [20] as listed below

i. Tissue Engineering—regeneration of injured or diseased tissues such as the skin, bones, cartilage, lymph nodes, blood vessels, muscle, and other tissues.

ii. Drug Delivery—controlled drug delivery incorporated drug or bioactive molecules can be effectively delivered from nanostructured materials by degradation or diffusion mechanism.
iii. Scaffolds –used to support improved cell adhesion, proliferation with larger surface area, and porous nature with interconnected pores, biocompatibility, and biodegradability.
iv. Wound Healing—novel wound healing scaffolds developed with various biopolymers and active compounds for enhanced tissue healing.

Several in vitro studies of nanofibrous wound healing bandages, scaffolds and drug delivery carriers have shown that nanostructured materials outperform their micro or macro architecture scaffold counterparts with similar material composition. Hence, nanofibrous assemblies can be tailor-made to distinct tissue engineering applications.

5 Nanostructured Scaffolds for Tissue and Bone Regeneration Applications

The tissue engineering market is valued at around USD 5 billion in 2016 and is expected to grow at a healthy rate. According to the Centers for Medicare and Medicaid, more than 900,000 surgeries are performed each year in the United States for bone reconstruction or replacement. Currently, nanotechnology-based products are few and are only available for tissue regeneration such as Acticoat™ Flex 3 and 7 (Smith & Nephew., UK) nano silver-containing wound dressing, Polymem® silver (Ferris Mfg. Corp., USA) and Regrenex® gel (Smith & Nephew., UK) [21]. When compared to topical application, scaffolds developed in nanostructured architecture potentially mimic the ECM structure. Currently, electrospinning technology widely used to fabricate nanofibrous scaffolds with improved cell adhesion and proliferation for wound healing and bone tissue regeneration applications.

6 Need for Biodegradable Materials

The scaffolds developed for drug delivery and tissue engineering applications need to be biocompatible and biodegradable [22]. The degradability rate must be equal to the healing rate of the tissue. Biocompatibility is an additional key feature for any scaffold used for tissue engineering applications, which should be devoid of toxicity, favor cell adhesion and proliferation, allow the exchange of gas and nutrients from the blood, and provide adequate mechanical strength [23]. The biomaterials used for tissue engineering include polymers, ceramics, composites, metals, and alloys [24]. These different classes of biomaterials were shown in Fig. 3. The degradation rate of a tissue-engineered scaffold is controlled and gradual, which leads to the transfer of the mechanical load to the surrounding tissue and bone, which avoids the

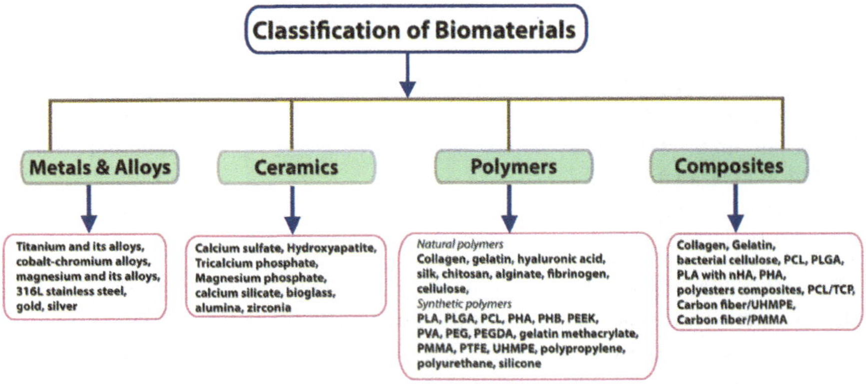

Fig. 3 Classification of various biomaterials used for tissue regeneration applications

stress shielding effect [24, 25]. Biodegradable polymers are the most commonly used biomaterials in tissue engineering and drug delivery applications. The degradation rate of the polymer can be altered with a different combination of materials and scaffold development techniques. The ideal biodegradable scaffold must not produce any adverse effects on the surrounding tissues with devoid of toxicity and cancer-causing effects. Additionally, degraded products from the scaffolds should be biocompatible without affecting the healing tissue [24].

7 Biomaterials Used for Scaffold Preparation

Biomaterials defined as any substance or combination of substances other than food and drugs and can be used for any period and biocompatible with living tissue, which replaces partly or totally any tissue organ or function of the body [26]. Among many materials, polymers are widely used in biomaterials applications for fabricating scaffolds due to their functional properties and design flexibility. The polymers used for tissue engineering applications can be classified into two groups, such as (i) natural polymers are biodegradable in nature and (ii) synthetic polymers (biodegradable and non-biodegradable in nature). The biodegradable polymers are a vital choice biomaterial for the development of scaffolds for tissue engineering applications compared to non-biodegradable polymers [27]. Metals and alloys, ceramics, polymers, and composites are the most commonly utilized biomaterials in tissue engineering or bone regeneration applications [28]. Utmost, the cells are scaffold dependent; therefore, the choice of biomaterials used for tissue engineering application requires significant consideration for successful biological functions.

8 Scaffold Requirements

For tissue engineering scaffold fabrication, the selection of materials with functional requirements for the preferred application is essential. The ideal tissue engineering scaffold should be a biocompatible, biodegradable, non-toxic, non-immunogenic, higher porous structure with interconnected pores, reproducible, higher surface to volume ratio, suitable for surface modification, acceptable mechanical strength, promote ECM development and able to deliver biomolecule signals [6, 29–33].

The ultimate aim of any tissue engineering scaffold should satisfy by providing a three-dimensional environment for cell attachment and better tissue ingrowth on it. The first biodegradable tissue scaffold fabricated and reported by Hutmacher et al. in 2002 using fused deposition modeling (FDM). The ideal qualities of the tissue regeneration scaffold are following; (i) micro (<20 μm) and macro (>100 μm) pore size respectively, (ii) high porosity with interconnected open pores for in vivo tissue in-growth; (iii) should provide adequate mechanical strength and well-controlled degradation rate (iv) easy and harmless to handle for sterilization, packaging, transportation to surgery and (v) should be sterile for seeding the cells [34]. Moreover, the difference among the required qualities for tissue regeneration and bone graft substitutes is still imprecise. However, in general agreement, the bone tissue regeneration scaffolds should have more open pores (>40–60%) to allow the easy diffusion dissolved gases and nutrients which permit more cell migration [35]. The pore sizes should be in the range of 50–1000 μm to attain the required porous structure needed for tissue regeneration [36, 37]. However, the ideal tissue engineering scaffold requires diverse configurations for suitable tissue healing or regeneration applications.

8.1 Polymers

Polymers are widely used materials in tissue engineering and drug delivery applications. As mentioned earlier, polymers can be either naturally derived or synthetic origin or both the natural polymers such as polysaccharides, including alginate, chitin/chitosan, hyaluronic acid and its derivatives similarly, proteins such as collagen, elastin, gelatin, fibrin gels, and silk. The natural polymers are beneficial due to their high biocompatibility [28]. Since the 1960s, synthetic biodegradable polymers play a significant role in tissue regeneration scaffold fabrication. They have many advantages compared to natural polymers by which they can easily custom-made to offer a wide range of functional properties and the significant benefit of being devoid of immunogenicity. In addition to that, they can be easy for processing, surface modification, and sterilized by different techniques. A limited number of synthetic biodegradable polymers are only used for tissue engineering applications and approved by the United States Food and Drug Authority (USFDA) for use in specific biomedical applications. The synthetic polymers such as poly(lactic acid) (PLA), poly(glycolic acid) (PGA), poly(lactic-co-glycolic acid) (PLGA) and polycaprolactone (PCL) are widely used in tissue engineering and drug delivery applications [27].

It has been proven that the efficiency of both natural and synthetic polymers was significantly enhanced through the utilization of nanotechnology, unlike metals and ceramics. As mentioned earlier, polymers possess high biocompatibility, controllable degradation with nanoarchitecture the polymers are important for cell attachment [38]. It is essential to consider that polymeric nanofibers are amongst better scaffolds for various tissue engineering applications. Hence, they replicate the structural characteristics similar to that of native ECM subsequently, serving an ideal platform for cellular functionalities required for specific tissue regeneration [28].

8.2 Biocomposites

Recently, composite biomaterials have been attracted to the development of bone tissue regeneration scaffolds. The composite materials have distinct advantages due to the combined properties of the different materials than the original documents. Since native bone is a nanocomposite of organic and inorganic materials of collagen matrix embedded nano-hydroxyapatite crystals, therefore, use of nanocomposite scaffolds could be a preferred biomaterial to mimic the natural bone composition and its effective regeneration [39]. Numerous studies have been designated the development and utilization of the ceramic/metal or ceramic/polymer composites as ideal bone replacements for bone tissue regeneration.

Ceramics are the ultimate choice for bone tissue regeneration owing to their very good osteoconductivity, bioactivity, and biocompatibility. Additionally, they have shown improved osteogenic cell adhesion, proliferation, and differentiation. The most widely used ceramics for bone tissue regeneration include calcium orthophosphates such as nano-hydroxyapatite (nHA) and Tricalcium phosphate (TCP) due to their native bone mineral composition [28]. However, they have a limitation of poor mechanical properties under weight-bearing applications. In contrast, metals and alloys are having higher mechanical strength and suitable for load-bearing applications. However, they have poor biocompatibility when compared to ceramics. Studies have shown that combining HA coating on titanium scaffolds for effective load-bearing applications [40]. Similarly, the plasma spray coating of HA on titanium is the most suitable method to improve the implants' bioactivity for orthopedic applications [41, 42]. However, this coating method encountered problems of non-uniform deposition of different phases of calcium phosphate and failure of the implant due to peeling off or degradation of the HA [43, 44].

Another category of scaffolds commonly used for bone tissue engineering includes polymers/ceramics composite due to their biomimetic nature and similar structure of ECM. When individual materials such as ceramics and polymers and their unique properties are very good osteoconductive and osteointegration ability, non-toxic and degradable nature in vivo are combined and used for intended bone regeneration applications [45, 46]. The following polymer-ceramic composite showed improved biomineralization with PLGA/HA nanofibers [47] and osteogenesis with collagen/HA composite nanofibrous scaffold [48] for bone tissue engineering applications. Similarly, gelatin is also used instead of collagen, which is devoid of immunogenicity and disease transmission problems of the later, which present in most tissues, including bone, tendon, ligament, and connective tissues [49]. While fabricating scaffold using gelatin and HA and electrospun nanofibers [50] showed improvement in cell adhesion having integrin-binding sites and calcium ions from HA bound with -COOH groups [51].

When it comes to bone tissue regeneration applications, the scaffold should mimic the native bone structure, at the same time, should support the tissues mechanically for cell attachment and differentiation and initial load-bearing with controlled degradation over the period. From a biological standpoint, the native bone structure of the human is composed of organic (collagen polymer)/inorganic (hydroxyapatite crystals) composite [52]. Hence, the researcher should consider these aspects before fabricating bone tissue engineering scaffold to mimic the nature bone architecture better. In addition to that, the suitable blend of the different materials can modify the mechanical properties and degradability of the scaffolds; thereby, composite biomaterials came into the picture for the development of biomimetic scaffolds [53]. Consequently, many researchers are attempted in this way and developed the bone tissue engineering scaffolds with the combination of nHA and various natural (collagen, gelatin, silk) and synthetic biodegradable polymers (PLA, PLGA, PCL [54–57] and poly(propylene fumarate) (PPF)) [1]. However, HA possess poor sintering capabilities and inherent hardness, brittleness, and absence of flexibility lead to show reduced mechanical properties, which limit its usage in weight-bearing applications. A recent study showed that poor sintering of HA could overcome by developing into nanostructured (nHA) materials [58], which can enhance the sinterability because of higher surface free energy as a result of improvement in mechanical properties can be achieved [59]. Currently, resorbable biomaterials used for bone-related applications majorly include bi-phasic calcium phosphates such as HA and TCP, or their combination can be resorbed in the biological system over the period. Principally, HA is the abundant inorganic mineral of Calcium phosphate (CaP) and present in the natural bone tissues and teeth. Synthetic nHA and its closest chemical structure to the native bone, which made them ideal candidate bone substitute used in various bone tissue regeneration applications [60, 61].To take advantage of its excellent biocompatibility and osteoconduction ability at the same time overcome the limitations of the HA, can be combined with a different type of biodegradable polymers and fabricate as bio-nanocomposite materials for bone defect repair applications [36, 62, 63].

Notably, to demonstrate the advantages of the nanostructured scaffolds, Shi et al. developed the nano and micron-sized glass substrate, and the study results concluded that the former nanostructured material showed greater MG63 cell attachment when compared to latter. This improved cellular response with nano-sized materials because of amorphous or lower crystalline nature and higher surface area for cell adhesion and differentiation [64]. Likewise, another study reported by Heo et al., with nHA incorporated with PCL polymer and fabricated as three dimensional (3D) scaffold have shown improved mesenchymal stem cells (MSCs) adhesion, proliferation alkaline phosphatase (ALP) activity and mineralization capacity [65]. Similarly, nanostructured scaffold fabricated with nHA/collagen/PLLA (Poly-L-lactic acid) composite reinforced with chitin fibers have showed improved mechanical strength and bioactivity [66].

9 Structural Characteristics of the Bone

Natural bone is a perfect example of composite material with various structural organizations. The structural arrangement of bone at various levels. Bone is composed of calcified hard cancellous or compact bone with varying levels of Haversian canals (osteons) at micron dimensions. Nanostructured linearly arranged collagen matrix embedded with nano-hydroxyapatite crystals acting as biofunctional tissues for various cellular functions. Bone tissue mainly composed of compact cortical bone and porous or trabecular bone (spongy bone). The combined hard shell with complex integrated cell structure of trabecular bone alignments, which makes them excellent resistance in bending forces at the same time, provide light-weight architecture [1]. At the nanostructured level, bone is mainly composed of inorganic calcium phosphate (hydroxyapatite crystals chemical formula of $Ca_{10}(PO_4)_6(OH)_2$) 70% and collagen fiber matrix 20–30%embedded some portion of water. The mechanical properties of the individual components as follows Collagen: Elastic Moduli (EM) = 1-2 gigapascal (GPa), ultimate tensile strength (UTS) between 50 and 1000 megapascal (MPa). Similarly, hydroxyapatite crystals: EM of 130 GPa and UTS of 100 MPa. These combined properties of both materials as composite provide bone to higher mechanical strength with ductility with excellent load-bearing capacity [67]. This composite structure at the macro level provides superior mechanical strength to the cortical and cancellous bone (compressive strength: 100–230 and 2–12 MPa, Young's modulus: 7–30 and 0.5–0.005 GPa respectively [68]. However, micron-level arrangements having osteons, Haversian canals, and nanoarchitecture (collagen matrix embedded nHA) levels still need to be addressed [69, 70].

10 Nanofibrous Biomaterials

Nanofibrous materials are one of the significant categories of widely used biomaterials for tissue engineering applications. Numerous studies have been reported for nanostructured scaffolds made into electrospun fibers using electrospinning technique with utilization of various naturally derived, synthetic biodegradable polymers or combination of both by encapsulating cells, growth factors, proteins, and ceramics, etc., used for various tissue regeneration applications. Especially nanofibrous scaffolds developed with biodegradable polymers are primarily employed for wound healing [71], nerve tissue regeneration, blood vessels, and cardiac tissue regeneration applications. Similarly, composite nanofibrous scaffolds also fabricated by incorporating ceramic particles such as nHA or TCP for bone tissue regeneration applications. This category of biomaterials was biomimetic nature with nanofibrous architecture of ECM. Nanofibrous collagen showed improvement in cell attachment and enhanced cell-matrix interactions due to its main constituent of ECM structure of varying tissues and studied its suitability as a scaffold for cell attachment, proliferation, and differentiation [27]. ECM mimicking nanofiber architecture can be developed using techniques such as electrospinning, phase separation, and self-assembly. The various diameter and orientation of the fibers were produced by these methods and shown its useful implications in various tissue regeneration applications. Several studies have reported the use of nanofibrous biomaterials for tissue engineering applications. Hence, the electrospinning technique is an important tool to fabricate nanostructured fibers utilized to engineer various biomaterials for wider tissue engineering applications [72].

11 Different Fabrication Techniques for Bone Tissue Engineering Scaffolds

11.1 Electrospinning

It is a simple, cost-effective, rapid method for the fabrication of different materials into nanostructured architecture by using electrostatic forces for producing polymeric fibers with a micro and nanometer scale for various applications. Zelency first presented this technique in 1914, and this process was patented by Formhal's for the production of micron fibers. However, at the early time, this method was not used for fabricating scaffolds for tissue engineering applications and meant for other industrial applications. Recently this technique attracted many researchers for successful applications for developing various tissue engineering scaffolds with micron (>1 mm) or nano-sized fibers (<1000 nm) [27].

During electrospinning, a high electric voltage is applied to a homogeneous polymer/composite solution or melt, which overcomes the surface tension of the solution and forms a charged jet. This charged polymer solution or melt is ejected from the syringe; eventually, it's dried and deposited on the grounded collector plate. The expelled polymer solutions repel each other while traveling towards collector plate and then which lead to the formation of fibers with different orientation subsequent solvent evaporation. By optimizing the electrospinning parameters such as flow rate, applied electric voltage, the distance between collector plate and needle tip, and collector type (plate or drum), which results in various diameters of the fibers range from about 0.02 to about 20 mm [72]. The distinctiveness of this method over other traditional methods is, can be able to electrospun continuously, we can obtain nanostructured scaffolds with large surface volume ratio and interconnected porous architecture with three-dimensional orientation. These factors are highly important for enhanced cell functions in vitro and in vivo.

The biomaterials employed in electrospinning were naturally occurring macromolecules, namely collagen, chitosan, silk fibroin, and biodegradable polymers synthesized via chemical routes such as PGA, PLGA, PLLA, and PCL and blends of both polymers. Furthermore, several constituents, such as proteins, growth factors, and hydroxyapatite, can also be combined and developed as nanofibrous materials. Based on these nano-features, the electrospun scaffolds are capable of supporting cell adhesion, proliferation, and differentiation in a three-dimensional (3D) manner. Additionally, the cells grown on these structures lead to biomimetic support for enhanced cellular functions [27]. However, considerable limitations persist with this technique for developing intricate 3D scaffold structures, and porous structures are confining its use for diversified tissue engineering applications. Nanofiber scaffolds with different arrangements for bone tissue regeneration and wound healing applications were shown in Fig. 4.

Different types of electrospinning techniques include basic setup, co-axial type, parallel type and multi-jet type electrospinning. The various types of spinning set up results, a variety of dimensions of the fibers (micron to nanofibers) fabricated in the

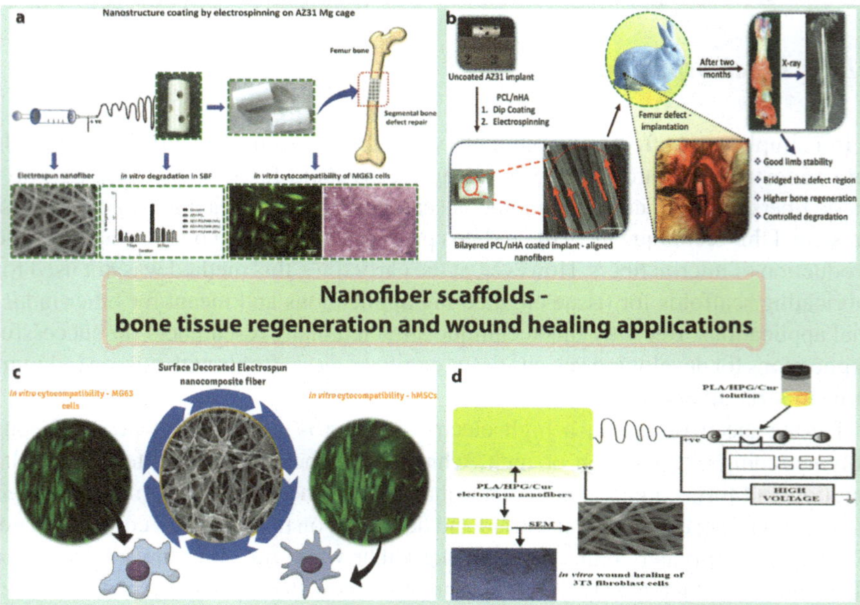

Fig. 4 Different types of nanofiber scaffolds for bone tissue engineering and wound healing applications—**a** Nanostructure coating of AZ31 magnesium implants, **b** Nanostructure coated AZ31 implants for critical segmental bone defect repair, **c** Surface decorated PCL/HPG/nHA/nMP nanocomposite electrospun fibers and **d** PLA/HPG/Cur nanofiber scaffolds for wound healing applications. The different arrangements of collectors used for obtaining woven and aligned fibers. [Reprint permission obtained from Refs. [54–56, 71] respectively]

form of non-woven, woven, uniformly aligned, and film. The collector plates such as metal plate (stainless steel or copper plate) wrapped with aluminum foil results aligned fibers with drum collectors and woven fibers were collected using counter electrodes.

11.2 3D Printing

Three dimensional (3D) printing is an additive fabrication technique used to construct tailor-made or patient specific biomaterial scaffolds. Unlike the conventional method, 3D Printing allows the design and fabrication of scaffolds using clinical images obtained with computed tomography (CT) and magnetic resonance imaging (MRI). Recently, 3D Printing emerged as an essential technique in medicine for manufacturing various tissue engineering scaffolds such as skeletal tissues—(bone, skin, tendon, ligament etc.), dental implants, different tissue models for drug screening, and disease/disorders [73].

Various biomaterials or hydrogel bioinks are used to fabricate 3D scaffolds, includes natural polymers majorly collagen, gelatin, hyaluronic acid, alginate, chitosan, and silk. Synthetic polymer includes Poly(lactic acid) (PLA), Poly(lactide-co-glycolide) (PLGA), Poly(caprolactone) PCL, Poly(propylene fumarate) (PPF), Polyether Ether Ketone (PEEK), polyvinyl alcohol (PVA), polyethylene glycol (PEG), polyethylene glycol diacrylate (PEGDA), pluronic etc. Additionally, various composite bioinks are used to fabricate the tissue engineering scaffolds, including hydroxyapatite, gold (AuNPs), silver (AgNPs), and magnetic iron oxide nanoparticles combined with either natural and synthetic polymers. The ideal requirements of biomaterials used for fabricating 3D printed scaffolds include good printability, biocompatibility, biodegradability, biomimetic in nature, adequate or tunable mechanical properties, reproducibility with limited batch to variation, sterilizability and good handling characteristics [74].

The 3D printing methods of tissue engineering scaffolds majorly based on the following four groups. (i) Extrusion method, (ii) Particle fusion method, (iii) Light-induced, or photo-polymerization method and (iv) Inkjet 3D printing.

i. *Extrusion type 3D printing*—this also includes fused deposition modeling (FDM) and direct ink writing (DIW), which are widely used for fabricating tissue engineering scaffolds. This technique allows the flow of bioink(s) through the nozzle by either extrusion of filament melted by heat by FDM and using a pneumatic pump in DIW with predefined computer model 3D object is created layer by layer manner.
ii. *Particle fusion type 3D printing*—this method includes selective laser sintering (SLS) and particle binding (PB) or indirect SLS. This type of fabrication used in industry for prototype development using materials such as polymers, ceramics, metals, and composites. SLS 3D printing involves using laser CO_2 to melt the polymers or metal particles and fuse to create 3D objects. Similarly, PB 3D

printing requires the use of a liquid binding solution to combine the particles collectively in each layer, and finally, fabricated 3D objects are sintered at high temperatures.

iii. *Light or photopolymerized 3D printing*—this fabrication technique also called stereolithography (SLA). This method involves creating 3D objects by patterning with ultra violet (UV) or laser beam of light in the bath containing photo-polymeriazable liquid, creating a hard polymerized layer by layer manner.

iv. *Inkjet 3D Printing*—this technique is used to deliver picoliters volume (1–100) of small droplets and the creation of 3D objects after solidification. Two subtypes of this technique are continuous inkjet (CIJ)—produce a stream of 100-micrometer diameter drops continuously and drop-on-demand (DOD) inkjet—produces 25–50 μm in diameter individual drops on demand, which is commonly used to fabricate tissue engineering scaffolds [73].

Recently, bone tissue engineering was involved using 3D printing and electrospinning techniques and used to create 3D objects and coating of the implants/scaffolds. The combination of 3D printing and melt electrospinning writing offers fabrication of the tissue engineering scaffolds with controlled architecture. From the input of the clinical image data, we can create the 3D model, which can be printed accurately to match the patient defect region. This method facilitates the controlling architecture of the scaffolds and also mimics the natural environment [75]. This biomimetic scaffold architecture can improve bone healing as well as improved patient compliance. The essential techniques used for fabricating bone tissue engineering scaffold includes the combination of 3D printing and electric field-based techniques such as electrospinning/writing, electrospraying, and electrophoretic deposition [75]. This methodology may be adopted for fabricating tissue engineering scaffolds with nanomaterials in the form of composite for skeletal tissue healing or repair.

12 Summary

In this chapter, we have discussed the nanostructured materials and their requirements for tissue engineering scaffolds and the advantages of nanostructured scaffolds over conventional materials. The various biomaterials or bioinks used, and fabrication techniques, notably electrospinning and 3D Printing, are used for developing tissue engineering scaffolds with nanomaterials. The utilization of nanostructured materials in tissue engineering and regenerative medicine has enormous potential. The selection of appropriate biomaterials and fabrication techniques will decide the success of the tissue engineering scaffolds.

References

1. Li X, Wang L, Fan Y, Feng Q, Cui FZ, Watari F (2013) Nanostructured scaffolds for bone tissue engineering. J Biomed Mater Res, Part A 101(8):2424–2435
2. Ramsden J (2011) Nanotechnology: an introduction. William Andrew. https://doi.org/10.1016/B978-0-08-096447-8.00006-5
3. McMahon RE, Wang L, Skoracki R, Mathur AB (2013) Development of nanomaterials for bone repair and regeneration. J Biomed Mater Res B Appl Biomater 101(2):387–397
4. Chen CS, Mrksich M, Huang S, Whitesides GM, Ingber DE (1997) Geometric control of cell life and death. Science 276(5317):1425–1428
5. Liu H, Yazici H, Ergun C, Webster TJ, Bermek H (2008) An in vitro evaluation of the Ca/P ratio for the cytocompatibility of nano-to-micron particulate calcium phosphates for bone regeneration. Acta Biomater 4(5):1472–1479
6. Pişkin E (1997) Biomaterials in different forms for tissue engineering: an overview. In: Materials science forum. Trans Tech Publ, pp 1–14
7. Weber JN, White EW (1972) Carbon-metal graded composites for permanent osseous attachment of non-porous metals. Mater Res Bull 7(9):1005–1016
8. Freeman JW, Wright LD, Laurencin CT, Bhattacharyya S (2008) Nanofabrication techniques. Biomedical nanostructures. Wiley, Hoboken, NJ, pp 3–24
9. Chen VJ, Ma PX (2004) Nano-fibrous poly (L-lactic acid) scaffolds with interconnected spherical macropores. Biomaterials 25(11):2065–2073
10. Sun S, Zeng H, Robinson DB, Raoux S, Rice PM, Wang SX, Li G (2004) Monodisperse mfe2o4 (m = fe, co, mn) nanoparticles. J Am Chem Soc 126(1):273–279
11. Hata K, Futaba DN, Mizuno K, Namai T, Yumura M, Iijima S (2004) Water-assisted highly efficient synthesis of impurity-free single-walled carbon nanotubes. Science 306(5700):1362–1364
12. Venugopal JR, Low S, Choon AT, Kumar AB, Ramakrishna S (2008) Nanobioengineered electrospun composite nanofibers and osteoblasts for bone regeneration. Artif Organs 32(5):388–397
13. Zhang L, Webster TJ (2009) Nanotechnology and nanomaterials: promises for improved tissue regeneration. Nano today 4(1):66–80
14. Komaki H, Tanaka T, Chazono M, Kikuchi T (2006) Repair of segmental bone defects in rabbit tibiae using a complex of β-tricalcium phosphate, type I collagen, and fibroblast growth factor-2. Biomaterials 27(29):5118–5126
15. Gamradt SC, Lieberman JR (2004) Genetic modification of stem cells to enhance bone repair. Ann Biomed Eng 32(1):136–147
16. Agarwal S, Wendorff JH, Greiner A (2008) Use of electrospinning technique for biomedical applications. Polymer 49(26):5603–5621
17. Atala A (2008) Bioengineered tissues for urogenital repair in children. Pediatr Res 63(5):569–575
18. Kim B-S, Mooney DJ (1998) Development of biocompatible synthetic extracellular matrices for tissue engineering. Trends Biotechnol 16(5):224–230
19. Kim N, Lee S, Atala A (2013) Biomedical nanomaterials in tissue engineering. In: Nanomaterials in tissue engineering. Elsevier, pp 1–25e
20. Namdari M, Eatemadi A (2016) Nanofibrous bioengineered heart valve—application in paediatric medicine. Biomed Pharmacother 84:1179–1188
21. Ashtikar M, Wacker MG (2018) Nanopharmaceuticals for wound healing–lost in translation? Adv Drug Deliv Rev 129:194–218
22. Goldberg M, Langer R, Jia X (2007) Nanostructured materials for applications in drug delivery and tissue engineering. J Biomater Sci Polym Ed 18(3):241–268
23. Smith I, Liu X, Smith L, Ma P (2009) Nanostructured polymer scaffolds for tissue engineering and regenerative medicine. Wiley Interdiscip Rev: Nanomed Nanobiotechnol 1(2):226–236
24. Sheikh Z, Najeeb S, Khurshid Z, Verma V, Rashid H, Glogauer M (2015) Biodegradable materials for bone repair and tissue engineering applications. Materials 8(9):5744–5794

25. Hutmacher D, Hürzeler MB, Schliephake H (1996) A review of material properties of biodegradable and bioresorbable polymers and devices for GTR and GBR applications. Int J Oral Maxillofac Implants 11(5)
26. Peppas NA, Langer R (1994) New challenges in biomaterials. Science 263(5154):1715–1720
27. Murugan R, Ramakrishna S (2006) Nano-featured scaffolds for tissue engineering: a review of spinning methodologies. Tissue Eng 12(3):435–447
28. Chiara G, Letizia F, Lorenzo F, Edoardo S, Diego S, Stefano S, Eriberto B, Barbara Z (2012) Nanostructured biomaterials for tissue engineered bone tissue reconstruction. Int J Mol Sci 13(1):737–757
29. Chaignaud BE, Langer R, Vacanti JP (1997) The history of tissue engineering using synthetic biodegradable polymer scaffolds and cells. In: Synthetic biodegradable polymer scaffolds. Springer, pp 1–14
30. Freed LE, Vunjak-Novakovic G, Biron RJ, Eagles DB, Lesnoy DC, Barlow SK, Langer R (1994) Biodegradable polymer scaffolds for tissue engineering. Nat Biotechnol 12(7):689
31. Hubbell JA (1995) Biomaterials in tissue engineering. Nat Biotechnol 13(6):565
32. Laurencin CT, Ambrosio A, Borden M, Cooper J Jr (1999) Tissue engineering: orthopedic applications. Annu Rev Biomed Eng 1(1):19–46
33. Widmer MS, Mikos AG (1998) Fabrication of biodegradable polymer scaffolds for tissue engineering. In: Frontiers in tissue engineering. Elsevier, pp 107–120
34. Bose S, Roy M, Bandyopadhyay A (2012) Recent advances in bone tissue engineering scaffolds. Trends Biotechnol 30(10):546–554
35. Will J, Melcher R, Treul C, Travitzky N, Kneser U, Polykandriotis E, Horch R, Greil P (2008) Porous ceramic bone scaffolds for vascularized bone tissue regeneration. J Mater Sci—Mater Med 19(8):2781–2790
36. Bohner M, Van Lenthe G, Grünenfelder S, Hirsiger W, Evison R, Müller R (2005) Synthesis and characterization of porous β-tricalcium phosphate blocks. Biomaterials 26(31):6099–6105
37. Karageorgiou V, Kaplan D (2005) Porosity of 3D biomaterial scaffolds and osteogenesis. Biomaterials 26(27):5474–5491
38. Tran N, Webster TJ (2009) Nanotechnology for bone materials. Wiley Interdiscip Rev: Nanomed Nanobiotechnol 1(3):336–351
39. Liu H, Slamovich EB, Webster TJ (2006) Increased osteoblast functions among nanophase titania/poly (lactide-co-glycolide) composites of the highest nanometer surface roughness. J Biomed Mater Res, Part A 78(4):798–807
40. Ning C, Zhou Y (2008) Correlations between the in vitro and in vivo bioactivity of the Ti/HA composites fabricated by a powder metallurgy method. Acta Biomater 4(6):1944–1952
41. De Groot K, Geesink R, Klein C, Serekian P (1987) Plasma sprayed coatings of hydroxylapatite. J Biomed Mater Res 21(12):1375–1381
42. Yang Y-C (2007) Influence of residual stress on bonding strength of the plasma-sprayed hydroxyapatite coating after the vacuum heat treatment. Surf Coat Technol 201(16–17):7187–7193
43. Radin S, Ducheyne P (1992) Plasma spraying induced changes of calcium phosphate ceramic characteristics and the effect onin vitro stability. J Mater Sci—Mater Med 3(1):33–42
44. Lee JJ, Rouhfar L, Beirne OR (2000) Survival of hydroxyapatite-coated implants: a meta-analytic review. J Oral Maxillofac Surg 58(12):1372–1379
45. Wei G, Ma PX (2004) Structure and properties of nano-hydroxyapatite/polymer composite scaffolds for bone tissue engineering. Biomaterials 25(19):4749–4757
46. Jose MV, Thomas V, Xu Y, Bellis S, Nyairo E, Dean D (2010) Aligned bioactive multi-component nanofibrous nanocomposite scaffolds for bone tissue engineering. Macromol Biosci 10(4):433–444
47. Lao L, Wang Y, Zhu Y, Zhang Y, Gao C (2011) Poly (lactide-co-glycolide)/hydroxyapatite nanofibrous scaffolds fabricated by electrospinning for bone tissue engineering. J Mater Sci—Mater Med 22(8):1873–1884
48. Xie J, Baumann MJ, McCabe LR (2004) Osteoblasts respond to hydroxyapatite surfaces with immediate changes in gene expression. J Biomed Mater Res, Part A 71(1):108–117

49. Harrington WF, Von Hippel PH (1962) The structure of collagen and gelatin. In: Advances in protein chemistry, vol 16. Elsevier, pp 1–138
50. Francis L, Venugopal J, Prabhakaran MP, Thavasi V, Marsano E, Ramakrishna S (2010) Simultaneous electrospin–electrosprayed biocomposite nanofibrous scaffolds for bone tissue regeneration. Acta Biomater 6(10):4100–4109
51. Jakobsen R, Brown L, Hutson T, Fink D, Veis A (1983) Intermolecular interactions in collagen self-assembly as revealed by Fourier transform infrared spectroscopy. Science 220(4603):1288–1290
52. Chu PK, Liu X (2008) Biomaterials fabrication and processing handbook. CRC press
53. Ma PX (2008) Biomimetic materials for tissue engineering. Adv Drug Deliv Rev 60(2):184–198
54. Perumal G, Sivakumar PM, Nandkumar AM, Doble M (2020) Synthesis of magnesium phosphate nanoflakes and its PCL composite electrospun nanofiber scaffolds for bone tissue regeneration. Mater Sci Eng, C 109:110527
55. Perumal G, Ramasamy B, Maya Nandkumar A, Sivaraman D, Selvaraj R, Doble M (2020) Bilayered nanostructure coating on AZ31 magnesium alloy implants for the healing of critical-sized rabbit femoral segmental bone defects. Nanomed: Nanotechnol, Biol Med:102232. https://doi.org/10.1016/j.nano.2020.102232
56. Perumal G, Ramasamy B, Doble M (2018) Nanostructure coated AZ31 magnesium cylindrical mesh cage for potential long bone segmental defect repair applications. Colloids Surf, B 172:690–698
57. Hanas T, Kumar TS, Perumal G, Doble M (2016) Tailoring degradation of AZ31 alloy by surface pre-treatment and electrospun PCL fibrous coating. Mater Sci Eng, C 65:43–50
58. Legeros RZ (1993) Biodegradation and bioresorption of calcium phosphate ceramics. Clin Mater 14(1):65–88
59. Kingery WB (1976) HK, and UHLMANN, DR introduction to ceramics. Wiley
60. Foroughi MR, Karbasi S, Ebrahimi-Kahrizsangi R (2012) Physical and mechanical properties of a poly-3-hydroxybutyrate-coated nanocrystalline hydroxyapatite scaffold for bone tissue engineering. J Porous Mater 19(5):667–675
61. Habraken W, Wolke J, Jansen J (2007) Ceramic composites as matrices and scaffolds for drug delivery in tissue engineering. Adv Drug Deliv Rev 59(4–5):234–248
62. Sun F, Zhou H, Lee J (2011) Various preparation methods of highly porous hydroxyapatite/polymer nanoscale biocomposites for bone regeneration. Acta Biomater 7(11):3813–3828
63. Perumal G, Ramasamy B, Nandkumar AM, Doble M (2018) Influence of magnesium particles and pluronic F127 on compressive strength and cytocompatibility of nanocomposite injectable and moldable beads for bone regeneration. J Mech Behav Biomed Mater 88:453–462
64. Shi Z, Huang X, Cai Y, Tang R, Yang D (2009) Size effect of hydroxyapatite nanoparticles on proliferation and apoptosis of osteoblast-like cells. Acta Biomater 5(1):338–345. https://doi.org/10.1016/j.actbio.2008.07.023
65. Heo S-J, Kim S-E, Wei J, Kim DH, Hyun Y-T, Yun H-S, Kim HK, Yoon TR, Kim S-H, Park S-A, Shin JW, Shin J-W (2009) In vitro and animal study of novel nano-hydroxyapatite/poly(ε-caprolactone) composite scaffolds fabricated by layer manufacturing process. Tissue Eng Part A 15(5):977–989. https://doi.org/10.1089/ten.tea.2008.0190
66. Li X, Feng Q, Cui F (2006) In vitro degradation of porous nano-hydroxyapatite/collagen/PLLA scaffold reinforced by chitin fibres. Mater Sci Eng, C 26(4):716–720. https://doi.org/10.1016/j.msec.2005.06.062
67. Wegst UGK, Ashby MF (2004) The mechanical efficiency of natural materials. Phil Mag 84(21):2167–2186. https://doi.org/10.1080/14786430410001680935
68. Hutmacher DW, Schantz JT, Lam CXF, Tan KC, Lim TC (2007) State of the art and future directions of scaffold-based bone engineering from a biomaterials perspective. J Tissue Eng Regen Med 1(4):245–260. https://doi.org/10.1002/term.24
69. Rho J-Y, Kuhn-Spearing L, Zioupos P (1998) Mechanical properties and the hierarchical structure of bone. Med Eng Phys 20(2):92–102. https://doi.org/10.1016/S1350-4533(98)00007-1

70. Schneider P, Voide R, Stampanoni M, Müller R (2009) Post-processing technique for improved assessment of hard tissues in the submicrometer domain using local synchrotron radiation-based computed tomography. Biomed Eng 54. http://doi.org/10.1515/BMT.2009.007
71. Perumal G, Pappuru S, Chakraborty D, Nandkumar AM, Chand DK, Doble M (2017) Synthesis and characterization of curcumin loaded PLA—hyperbranched polyglycerol electrospun blend for wound dressing applications. Mater Sci Eng, C 76:1196–1204
72. Wei G, Ma PX (2008) Nanostructured biomaterials for regeneration. Adv Func Mater 18(22):3568–3582. https://doi.org/10.1002/adfm.200800662
73. Guvendiren M, Molde J, Soares RM, Kohn J (2016) Designing biomaterials for 3D printing. ACS Biomater Sci Eng 2(10):1679–1693
74. Gungor-Ozkerim PS, Inci I, Zhang YS, Khademhosseini A, Dokmeci MR (2018) Bioinks for 3D bioprinting: an overview. Biomater Sci 6(5):915–946
75. Koons GL, Diba M, Mikos AG (2020) Materials design for bone-tissue engineering. Nature Rev Mater:1–20

Nanomaterials for Medical Imaging and In Vivo Sensing

N. Ashwin Kumar, B. S. Suresh Anand, and Ganapathy Krishnamurthy

Abstract Pre-clinical imaging is a technique that could help in investigating deep inside the rodents to obtain information regarding disease site and drug development process using a non-invasive approach. Diverse FDA approved contrast agents have been implemented since the evolution of these imaging technologies. The current limitations of these contrast agents include faster clearance and photo instability and could be unsuitable for multi-modal and hybrid imaging. These impediments can be overcome with the aid of developing new nanotechnology-based contrast agents. This opens up a new paradigm for researchers to visualize cancer to obtain nanomolecular information. During the past two decades, nano-based contrast agents have revolutionized pre-clinical imaging science, which offers to detect cancers early, rapidly, and effectively. Despite the fact, the concept and technology of imaging are old, the way we look at the disease using nanomaterials in a different perspective. Additionally, the imaging techniques combined with nanotechnology-based contrast agents can be used to investigate the interaction of drugs at a pre-clinical stage and the cellular level. Pre-clinical imaging is performed with two different strategies. The former techniques give anatomical information, which includes computed tomography, magnetic resonance imaging, and ultrasound. While the latter presents molecular information using optical techniques, photon-acoustic imaging, and positron emission tomography. Recent developments in nanotechnology-based contrast agents have opened new avenues to alter and improve the current imaging modalities resulting in hybrid and multi-modal imaging approaches. Here, we intend to provide fundamental knowledge and general considerations of using nanomaterials in pre-clinical imaging modalities. This chapter provides extensive information about the advancements in nanomaterials for pre-clinical imaging applications. In the later

N. Ashwin Kumar (✉)
Department of Biomedical Engineering, SRM Institute of Science and Technology, Kattankulathur, Chennai, India

B. S. Suresh Anand
Department of Biomedical Engineering, Rajalakshmi Engineering College, Chennai, India

G. Krishnamurthy
Department of Engineering Design, Indian Institute of Technology-Madras, Chennai, India

T. S. Santra and L. Mohan (eds.), *Nanomaterials and Their Biomedical Applications*, Springer Series in Biomaterials Science and Engineering 16,
https://doi.org/10.1007/978-981-33-6252-9_13

part, we discuss the science behind individual nanomaterials with different imaging systems and its improvements in pre-clinical imaging. Also, the advantages and drawbacks associated with nanomaterials are presented. Finally, we discuss the application of different nano-based contrast agents and their application in biomedical imaging.

Keywords Contrast agents · Imaging · Nanomaterials · Anatomical imaging · Molecular imaging · Biomedical applications · Pre-clinical imaging

1 Introduction

Drug development process (DDP) involves major checkpoints, including the selection of disease, drug discovery, characterization of drug, in vitro studies (both in cellular and non-cellular environment), in vivo or pre-clinical research, and finally to different phases of clinical trials [1]. However, in the process of clinical validation of novel drugs, several experiments showed a negative result that stops further commercialization [2]. For this reason, animal models that can explicitly mimic the human body's complexity are used for drug development studies [3]. Different animal models have developed based on specific requirements of the disease. The repeated investigation of a drug might jump between the phase-I clinical trials and pre-clinical experiments for a consecutive response. In this process, small animals or in vivo studies have an impact on DDP to avoid human experimentation when translated to the clinic [4]. This helps in understanding the effects of a drug in a real biological environment that comprises multidisciplinary information about the drug, as shown in Fig. 1.

On the other hand, the experimental time that occurs in the clinical phase can be reduced drastically by performing more optimization in the preclinical studies

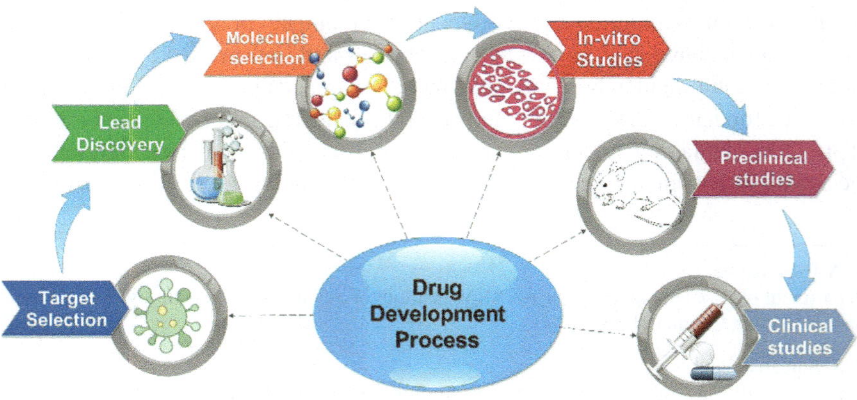

Fig. 1 A schematic diagram represents the flow of bringing up new drug molecules into the clinical practice. The process of drug development phases of preclinical studies is an important step for the next pace for clinical role

[1]. Also, knowledge of the in vitro cell and drug interaction behavior is not as similar to in vivo. To have a better understanding of the drug-disease interaction via a non-invasive method, pre-clinical imaging tools are essential [2].

2 Medical Imaging

Medical imaging is a science, aid to look deep inside the body non-invasively to obtain prognostic information about the disease region. Non-invasive internal structure imaging requires a specific region of electromagnetic (EM) spectrum based on location in the human body and disease type. Different energy in EM would aid in understanding the structures at different depths, either at a high or low resolution [3]. For instance, visible light has limited penetrating power when compared to ultrasound and X-rays, but the resolution of the optical imaging is in the range of nanometers (nm). Visible lights are generally used in skin, endoscopic, microscopic, and live-cell imaging applicable for clinics and research. Direct visualization, observation, and qualitative information are essential in understanding the clinical problems and research proofs [4]. Medical imaging technology is divided into two different categories majorly as anatomical or structural, molecular and hybrid, or multi-modal imaging. X-rays, computed tomography, ultrasound, and magnetic resonance imaging are applicable for structural imaging. Whereas, positron emission tomography and optical imaging are used as tools to obtain the molecular information at nanometer resolution [5]. Pre-clinical imaging is an effective technique to provide valid information on the new drug developed at a laboratory level for further commercialization. It is essential to diagnose cancer in its early stages, atrioventricular block in heart, and immunological diseases to understand the molecular information of the diseased site. Recent progress in the preclinical non-invasive imaging has led to the development of high-resolution systems to obtain qualitative and quantitative information, drug treatment response, and to study the changes in anatomical and molecular levels [6]. Three significant fields that widely use non-invasive imaging modalities are oncology, neurosciences, cardiology, and cardiovascular diseases. Various imaging modalities like micro-computed tomography (mCT), micro-magnetic resonance imaging (mMRI), micro-optical imaging (mOI), micro-ultrasound (mUS), and micro-positron emission tomography (mPET) are specifically designed for small animal imaging and currently in use as shown in Fig. 2. The letter prefix ‘micro (m)’ of the above imaging systems differentiates between clinical and pre-clinical modalities. This ‘m’ stands for ‘micro’, which represents that imaging system were specifically designed for small animals and not for clinical use [7]. The above-mentioned imaging modalities are chosen based on the locations and the type of disease. The fundamentals of the above imaging systems will be discussed in the later sections of this chapter.

Several aspects of imaging would help to understand the anatomic and molecular details of different clinical departments, and few are listed below. First, non-invasive imaging in oncology involves estimating the structural changes in the tumor

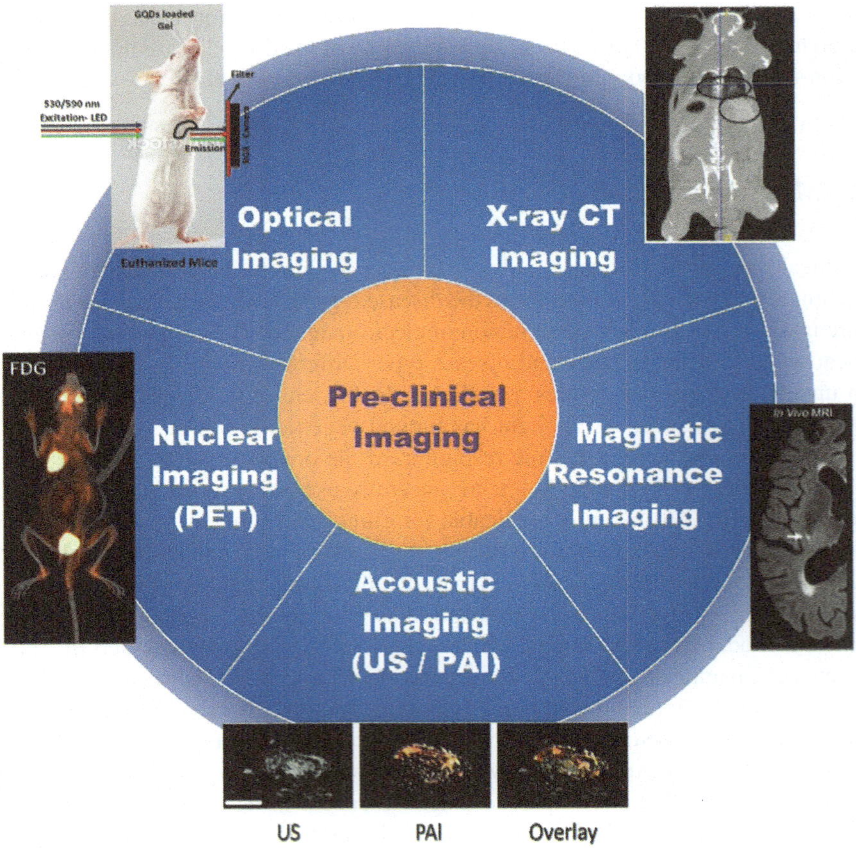

Fig. 2 Different types of imaging modalities for preclinical imaging applications. Each in-vivo image gives an example to the specific imaging system. For instance, molecular imaging can be acquired using optical, nuclear imaging modalities. Structural information can be obtained from X-ray CT, MRI, and Acoustic imaging system. A combination of systems, along with the specialized probe, can produce hybrid/multi-modal imaging with features of structural and molecular information

volume, location, and molecular identities like angiogenesis, expression of markers, and genetic information [8]. Second, the early risk of plaque formation and detection of blockage in heart vessels, therapeutic outcomes, and information about risk assessment can be evaluated for cardiovascular diseases [9]. Third, neurological disorders like brain injury, congenital disorders stroke, the flow of cerebrospinal fluid, spinal cord injury, and other causes related to head can be imaged via a non-invasive method [10]. Finally, in vivo optical imaging is widely used as a functional imaging tool to determine disease progression, target efficacy, and other molecular mechanistic effects of the disease. Compared to other in vivo imaging modalities, optical imaging is still in the infant stage [11].

3 Nanotechnology

Over the years, direct visualization of the diseased region using different imaging modalities have shown conflicts between resolution and penetration depth. Thus, an exogenous contrast agent that is biologically compatible with less toxicity is required to improve current imaging systems. Conventionally used agents approved by Food Drug Administration (FDA) are Iodine for X-ray imaging, fluor-deoxy-glucose (FDG) for PET, and visible and NIR fluorophores for OI are used widely in these systems [12]. Present contrast agents used in existing imaging systems have many limitations; for instance, fast clearance & photo instability and are inappropriate for multi-modal and hybrid imaging. Later, chemists worked with a clinical specialist to produce some novel materials to handle critical issues to ease the diagnosis at a faster rate. Imaging with nanomaterials compared to traditionally used agents can provide enhanced sensitivity and specificity of the diseased region.

Recent improvements in nano-based contrast agents have triggered to modify and enhance the current imaging techniques to produce hybrid and multi-modal imaging systems. During the past two decades, nano-based medicines have revolutionized medical imaging science. Early detection of cancer has become more rapid and efficient. Although the concept and technology of imaging are old, the way we look at the disease using nanomaterials is new. Imaging techniques, along with the nano-based contrast agents were used as a tool to characterize the interaction of drugs at both cellular and pre-clinical level [13]. The Fig. 3 summarizes different nanoparticles used in preclinical imaging modalities.

Inherent properties of elements that can be converted into nanomaterial would act as nanoprobe to either specific imaging systems. Nanoprobes show better performance compared to conventional contrast agents used for disease diagnosis. For instance, gadolinium and iron oxide (IO) NPs have been used as a probe in an MRI system as a T_1/T_2 contrast. Compared to the gadolinium complex, the contrast of IO NPs shows higher due to its superparamagnetic properties when converted into nano form [14]. Likewise, other materials have also been explored and compared with the conventional contrast used in specific imaging modalities. For example, gold nanoparticles (GNPs) (mCT and photoacoustic tomography (PAT)), quantum dots (mOI), and nanobubbles (mUS) show a better contrast compared to conventional agents [6]. Nano based contrast agents for pre-clinical imaging applications are not entirely replaced with the conventional contrast agents. Figure 4 shows the general requirements in designing nanomaterials for in vivo imaging as follows [15]: (i) high surface area due to their size, (ii) ease of functionalization, (iii) modulation in their physicochemical properties, (iv) high drug loading efficiency, (v) multi-modal imaging, (vi) optimization in image quality.

The size of the nanoparticles is an essential factor to be considered due to change in contrast for CT. For instance, Dong et al. showed that different sized GNPs ranging from 4 to 150 nm were studied in both clinical and pre-clinical modalities. Particles with size less than 15 nm have higher X-ray attenuation and more extended blood

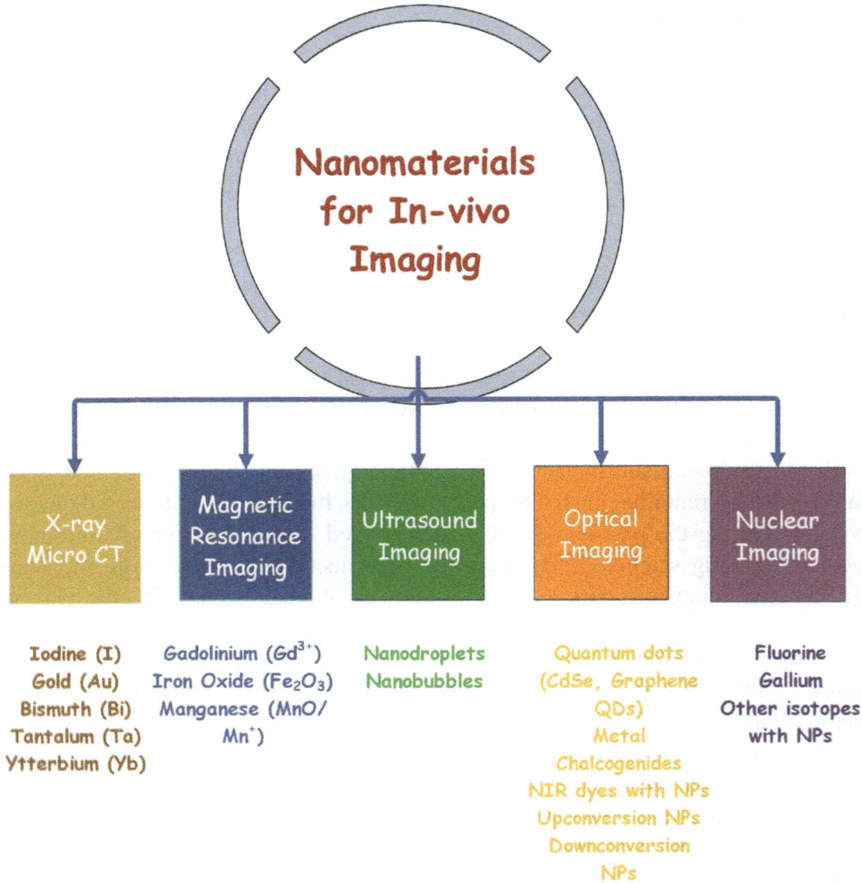

Fig. 3 Key elements in the periodic table are converted into nano-sized particles with their relevant imaging systems. Due to their innate nature of the elements and the nano regime provides an intrinsic property to the nanoparticles

circulation compared to larger particles. More than 50 nm has a significant attenuation in the liver and spleen due to the higher metabolic rate enabling them to retain in these organs [16]. Similarly, Chen et al. developed surfactant coated IO NPs via thermal decomposition for MR imaging applications studied in vivo size ranging from 14 to 26 nm. Poly-ethylene glycol (PEG) coated IO nanoparticles have shown better magnetic properties concerning the size. But the results provided that the optimum size of IO NPs for the in vivo imaging application is 22 nm is best for T2 contrast [17]. Marashdeh et al. also developed nanoparticles via the sol-gel method obtained a higher contrast at 30 nm compared to 22 nm particles [18]. These changes in MR contrast were not evident, and it can be concluded due to variation in synthesis procedure, size due to surfactant, and poly-dispersity. Also, the nanoparticle

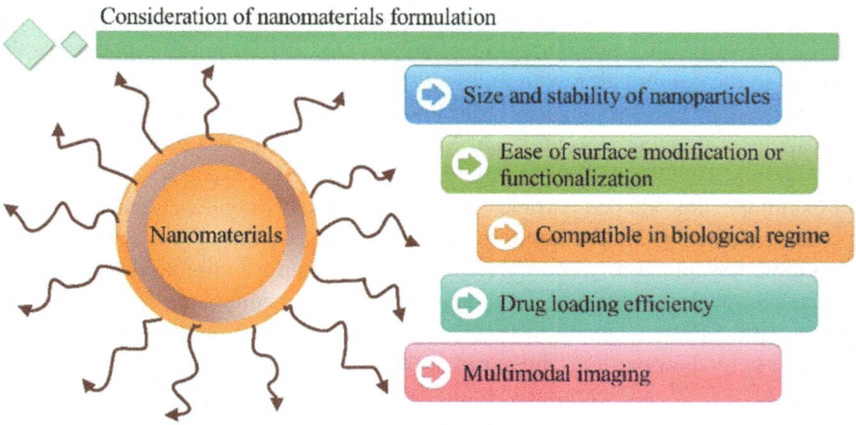

Fig. 4 Schematic representation of nanomaterial properties considered for medical imaging systems applicable for in-vivo

size factor that can have the ability to circulate systemically for a more substantial time compared to conventional agents. Different sizes of nanoparticles enable us to increases the surface area to volume ratio, which can enhance the functionalization of the different moieties.

The coating of nanoparticles with different surfactants, antibodies, and drugs enables them to work as a multi-functional agent [19]. Nanoparticles' functionalization is vital during the clinical translation and also to study the interaction between them to overcome biological constraints [20]. Reports suggested that immunological response can be reduced and level of bioavailability can be increased by coating bioresorbable polymers like polyethylene glycol, Poly-L-glycolic acid (PLGA), polyvinyl alcohol (PVA), chitin, cellulose, and dextran [21]. But, the usage of such polymeric coatings would increase the size of particles that in turn, change the biological property. Nanomaterials functionalized with suitable agents would increase the compatibility in a biological environment, improve targeting ability, and better cell uptake to specific tumor cells. Designing the appropriate changes onto the surface of nanoparticles would impact higher on the current treatment and diagnosis [20, 22]. Physicochemical properties of nanomaterials like selection and composition of materials (dopants), atomic arrangement, surface roughness, and shape of particles are the major considerations for better design. For instance, anisotropic nanostructures are well studied for various applications such as imaging, sensing, catalysis, optics, and therapy [23]. Inparticular, nanotechnology finds applications in the personal medicine. Nanoparticles are applicable in deliver drugs at the intracellular level [24]. This provides a larger understanding of the interaction and behavior of drugs with single cell analysis [25, 26]. Electroporation, optoporation, and mechanoporation are applied to the cells with anisotropic nanoparticles like gold for the enhanced cell delivery [27]. Also other types of magnesium, diamond and titanium based intracellular delivery using optoporation techniques [28, 29]. Also these materials also involved in the combination with imaging applications.

In particular, anisotropic nano gold and iron oxide are excellent materials for photoacoustic imaging (PAI)/CT and MR Imaging aided for cancer hyperthermia [30, 31]. In MRI, T1 and T2 relaxivity are affected based on the shape of the IO NPs. For example, cubic shaped particles showed higher in T2 contrast agents compared to rods or clusters [32]. The addition of rare earth dopant into nanomaterials might be suitable for optical as well as CT/MRI imaging. Some combinations change the physicochemical properties of a nanomaterial, e.g. by conjugating fluorescent dye, doping with different rare earth metals in CT/MR contrast, etc. Bimetallic (gold-iron) and core-shell particles are some of the best ways of using nanomaterials for imaging [5, 12, 33–35]. Similarly, bioresorbable nanomaterials like calcium phosphate, hydroxyapatite, and ceramic composites are used as a host material for loading drugs and dopants. Also, radiolabeling nanoparticles enables them to be used for hybrid or multi-modal imaging. Liposomes provide a multifunction capability that can hold imaging agents for MR/CT contrast with high payload drug molecules [12].

The next level of visualizing disease is through multi-modal imaging techniques that offer combinatorial structure and functional information about the disease. Multi-modal imaging is defined as the fusion of images from different modalities to form one single image [36]. Multi-modal imaging systems like PET-CT, CT-Optical PET-MRI are currently in clinical use to overcome the drawbacks associated with individual modality [37]. But probes that are required to use in these type of imaging systems is challenging. This core problem is addressed by nanotechnology, wherein one single probe can be either used for one or more modality. Nano based contrast agents provide more room for doping various elements into host material like Gadolinium, Iodine, loaded with fluorescent dyes [38]. The addition of these moieties mentioned above would enable a nano host material to act as a multifunctional probe for in vivo and in vitro cell imaging [39, 40]. Interconnection between imaging and therapy are close to each other to increase the usefulness of multi-modal imaging systems. For instance, image-guided therapy combines CT and echocardiography into a single image for heart valve replacement and catheter locations and to deliver contrast agents [41, 42]. Imaging parameters that are involved in increasing sensitivity, contrast, resolution, and in reducing artifacts, noise, and distortions. Conventional agents like Iodine, gadolinium, and fluorophore dyes have these better imaging properties in their respective imaging systems. But the usefulness of these agents inside biological systems showed limitations in terms of toxicity, circulation time, fast clearance, biological artifacts (i.e., autofluorescence), and photo instability [43]. Nano researchers need to consider these parameters and limitations to design nanoprobes for imaging systems. Also, it is mandatory to compromise the balance between image quality and concentration of nanoparticles. Many such agents developed were not useful for in vivo imaging due to their toxicity at a concentration to obtain an excellent image. In the case of a multi-modal imaging nanoprobe, understanding the optimum concentration to acquire images in this system would be affected [44, 45].

This book chapter covers the significant aspects and design principles of various types of nanoparticles (NPs) specific to pre-clinical imaging techniques. Sections of this chapter are well organized that explain the nanomaterials in research as well as in the drug development process. Also, the chapter provides a multidisciplinary approach for students, industries, the scientific community, and healthcare professions to enhance the knowledge on understanding the bond between nanoprobes and pre-clinical imaging.

4 Nanomaterials for CT Imaging

After the discovery of X-rays in 1895 by W.C Roentgen, medical imaging has evolved drastically for non-invasive clinical diagnosis. X-rays are used to determine the location of implants, bone deformities, dental issues, calcifications, arthritis, and determine small objects swallowed by infants [46]. X-rays radiographic projections (2D images) are currently used to visualize anatomical structures of bone in a post and pre-surgical treatment. Compared to other imaging systems, X-rays have been used as an initial screening for several diseases due to their low cost, high resolution, less toxic to tissues, and scans at short duration [47]. Medical image processing techniques in radiographic X-ray images convert 2D images to 3D that help physicians for precise diagnosis of disease. Computed tomography (CT), a collection of 2D images are acquired by rotating source and detector from the different axis and post-processing these images to visualize in 3D.

Further advancements in image processing tools provide high-resolution images with fewer artifacts [48]. Readers can get a more detailed review of the principles of X-ray CT reconstruction is provided in [49]. Contrast agents like Iodine and bismuth are used to understand the structural modification in the tissues. These exogenous contrast materials help in improving the resolution and sensitivity of the diseased location. Figure 5 shows the family of different contrast agents used in preclinical X-ray CT imaging [50].

5 Conventional CT Contrast for In Vivo Imaging

Clinically, used iodinated contrast agents (1,3,5-triiodo benzene) are injected intravenously, typically for enhancing X-ray CT images. These iodinated contrast agents are also used in fluoroscopy, angiography, and sometimes in plain radiography. Iodine is commonly used as a contrast medium for CT because its K-edge (33.2 keV) lies in the average diagnostic energy. There are two types of iodinated contrast agents, namely, ionic and non–ionic based on water solubility. Iodinated contrast agents have several disadvantages, (i) molecular weight of Iodine is less thereby clears rapidly through kidneys; (ii) high concentration of Iodine is required for better signals but shows toxicity (iii) Iodinated contrast agents cannot be targeted to a particular region

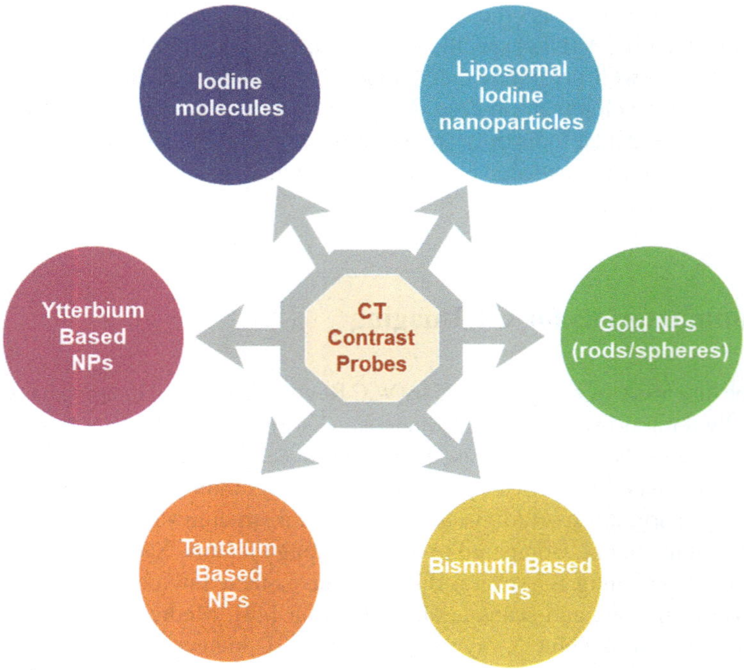

Fig. 5 Development of X-ray contrast agents for pre-clinical imaging applications to obtain the structural information tumor and blood vasculature

of interest. Conventionally, non-ionic iodinated contrast agents (water-soluble) are used by conjugating carboxylic acids and amine groups to the base molecule increase the solubility in water [51].

Over the decade, several attempts were made to increase the payload of iodine molecules in nanoparticle form to overcome issues (Fig. 6). For example, poly(iohexol) nano-based contrast agents exhibited substantially increased the retention time within the tumor sites and 36-fold enhancement at an equivalent dose of Iohexol. Since these particles are of a size of more than 10 nm, they are eliminated by the liver/ fecal route and not by renal clearance [52]. The choice of surfactant is vital for Iodine loaded nanoparticles to enable them to increase circulation time and better uptake in tumor tissues. Furthermore, fabricating these nanomaterials for the multi-modal platform would enhance better tumor imaging at both the structural and molecular levels. For instance, Polyethylene glycol (PEG) and 5-Amino-2,4,6-triiodoisophthalicacid adhered to up-conversion nanoparticles coated with silica improving the stability and circulation time. Another method of loading iodine using self-assembly nano micelles formed above critical micelle concentration with size range from 10 to 200 nm. A block copolymer of poly(butadiene)-b-poly (ethylene glycol) and pluronic F-127 were used to develop into nanoemulsion. This type of nano-emulsion showed high stability with a loading efficiency of 130 mgI/ml.

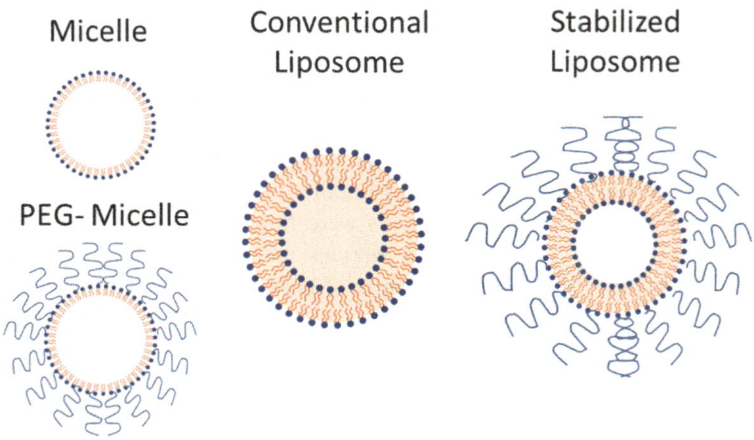

Fig. 6 Strategies to load Iodine molecules in nanoparticles as X-ray contrast agents

The half-life of these emulsion in vivo blood circulation was about 3 h [53]. Kim et al. developed a bimodal emulsion of Iodine (CT) and indocyanine green (optical imaging) using Tween-80 and Span-85 surfactants. This bimodal imaging contrast extract information about pre and intraoperative sentinel lymph node imaging in both anatomical and molecular functions. The emulsion showed optimum retention of a probe in lymph nodes enabling physicians for better treatment planning [54]. Hainfield et al. developed polymeric iodine nanoparticles with size 20 nm revealed longer circulation time of 40 h. Results showed less toxicity at 4 gI/kg via intravenous injection and cleared 50% if Iodine from the liver [55]. Zou et al. developed Iodine (I_2) conjugated with monomers of polypyrrole (PPy) to form nanoparticles using an oxidizing agent ($FeCl_3$). The monomers of PPy to form a polymer and capped with polyvinyl alcohol (PVA) surfactant for CT guided photothermal therapy [56].

6 Nano Agents for CT Imaging

Considering parameters such as toxicity at high concentrations, low signal to noise ratio at low concentrations, inability to target pathology and absorption of low energy photons at the K-edge, it is concluded that Iodine is not optimal for functional or molecular imaging at low concentrations [49]. Therefore, there is a need to design new contrast that can overcome some of the drawbacks of conventional X-ray CT contrast agents as well as address some drawbacks in existing nano-based contrast agents, for instance, toxicity. Prospective materials are several elements with a higher atomic number such as Gold (Au), Bismuth (Bi), Ytterbium (Yb) and Tantalum (Ta) that can be produced in nano form for a better contrast [49].

6.1 Gold-Based Nanomaterials

In early history, Gold has been used as a purple color coating in the paint glasses described by Faraday in 1857. Characterizing colloidal gold revealed high biocompatibility with better surface modification properties enabling them for applications like imaging, sensing, drug delivery, and photothermal therapy (see Fig. 7) [57]. Due to its high surface to volume ratio, GNPs were functionalized with moieties like monoclonal antibodies, cell-penetrating peptides, and inhibitors for target theranostics applications [58]. Also, different shapes of gold, such as rods, cubes, triangles, clusters, and stars, were formulated with a size range of 1–250 nm. Different shapes and sizes of nanomaterials possess' variable plasmonic resonance which can include various biomedical applications due to optical characteristics [59]. Gold has excellent X-ray absorbing material due to its high atomic number excited at K-edge 80 keV. Absorption of electrons in the biological tissues through the photoelectric mechanism and Compton scattering effects. Studies showed that the contrast enhancement between the Iodine and GNPs, colloidal gold produces 2.5 times higher than the conventional agent [60, 61].

Hanfield et al., first to study that nano-sized gold can be used as X-ray contrast agents and also studied the effect of radiotherapy in vivo [62]. Later, conjugating probes on the surface of GNPs enabled them to use for multi-modal or hybrid imaging. Compared to Iodine, X-ray attenuation of GNPs dispersed in water was not reduced when compared to that of Iodine in water. The contrast to noise ratio is high, with an average of 115% when excited at 140 keV. These results suggested that GNPs

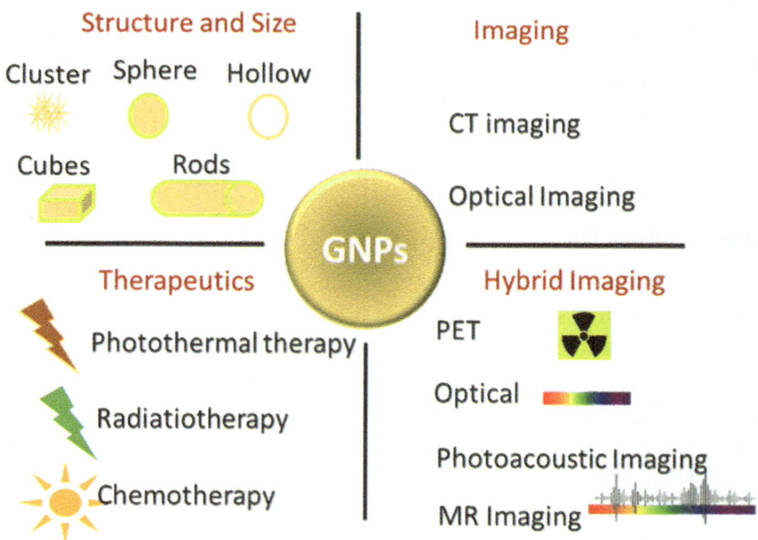

Fig. 7 Nano-sized gold can be formed in different sizes and shapes. They can be used multifunctional particles with both therapy and diagnostic applications

are efficient enough to be used as a CT contrast compared to conventional iodine conjugated nanoparticles [62]. GNPs with sizes in the range of 25 nm are retained in the vasculature providing excellent contrast for vessel imaging. GNPs with different sizes were also studied to use as blood pool contrast agents for pre-clinical imaging. But the study suggested by Ross et al. in 2014 suggested that X-ray attenuation of gold linearly varies with the concentration of gold chloride and independent of gold particles [63]. In 2007, Kattumuri et al. used biocompatible surfactant called gum-arabic onto GNPs to understand in vivo pharmacokinetics and potential contrast compared to Iodine in clinical CT. Later in 2010, more detailed experiments in juvenile swine were performed by the same group to determine the changes in CT Hounsfield units (HU) about the uptake concentration in pre and post-injection scans at different X-ray levels. The results revealed that the uptake is high spleen compared to the liver. At different X-ray voltages, the uptake of GNPs is consistently high at 80 kVp (22 HU) and 140 kVp (27 HU), respectively [64]. Cormode groups have studied the effect of GNP size from 4 to 152 nm coated with m-PEG in clinical and pre-clinical CT systems. Results indicated that GNP size of less than 15 nm had shown more extended circulation compared to larger particles. While the concentration of GNPs in ex vivo and in vivo determined using ICP OES analysis demonstrated that larger size particles rapidly accumulate in the liver and spleen [16].

6.2 Surfactant Functionalized GNPs for CT Imaging

Most research works developed GNPs using citrate as reducing agents (Turkevich method), but the toxicity of the citrate molecules in vivo showed enhanced toxicity. Researchers found that usage of the surfactants may eradicate the issues related to the toxicity in vivo. Surfactant polyethylene glycol (PEG) with a methoxy group was found biocompatible and stable in a biological environment. PEG enabled GNPs to enhance the contrast characteristics when exposed to X-rays. A first attempt in 2007 by Kim et al. about the capping of PEG after the synthesis of GNPs via chemical method [65]. The average GNPs size was around 30 nm, showed 5.7 times higher contrast with circulation time of 4 h compared to Ultravist (commercially available iodine contrast) having less than 10 min. GNPs based imaging can delineate the hepatoma region in rats exhibited twofold contrast compared to normal healthy cells with less toxicity.

Further, less sized GNPs with an average size of 2 and 8 nm were evaluated the changes in the contrast and behavior of the particles [66, 67]. In 2011, Zhang et al. studied the complete in vivo toxicity of nanoparticles coated with surfactant PEG [68]. Results concluded that the toxicity of the GNPs could not be determined based on the particle size. The influence of the 10 and 60 nm toxicity is relatively higher compared to the lesser sized GNPs like 5 and 30 nm due to the increased metabolic rate in the liver. In vivo analysis of 5 and 10 nm are loaded in the liver and 30 nm collected in the spleen. While 60 nm GNPs revealed the reduction of white blood cells and increased the transamination levels of alanine and aspartate.

Later, the same group studied the toxicological effects of GNPs in vivo by different administration routes using 13 nm sized particles. Parameters considered for toxicity involves concentration, organ index, hematology, weight, and longevity of the animal. GNPs concentration used range from 137.5 to 2200 μg/kg for 14–28 days. Data obtained from the experiments revealed that lower concentration of GNPs has less toxic, but at higher concentrations, GNPs have a substantial effect over the organ index. Also, comparing the administration routes, they suggested that administering targeted GNPs via intravenously is more promising for imaging applications [69]. Size-dependent radiosensitization have studied intensively, suggested that PEG-coated GNPs size range between 12.1 and 27.3 nm are considered to be effective with high accumulation in tumor region compared to other sized NPs [70].

Development of particular types of GNPs was used to improve the CT contrast like mixing of metals, incorporation of lanthanide materials, having dendrimers as a template, and site-specific functionalization over the particles. In particular, bimetallic nanoparticles have been more useful compared to the single metallic particles with the allied conjugations [71]. A combination of gold and silver with 2 nm size coated with bovine serum albumin via green synthesis. These nanoparticles were explicitly targeted to lysosome with high viability and reduction in the cost of materials. X-ray attenuation behavior was studied by varying the different concentrations of Au-Ag combinations (Au: 0, 40, 50, 60, 90, and 100) having Iodine as a control sample. In comparison, Au:60 showed increased X-ray attenuation and decreased at Au:90, and 100 for 80 mM could be due to an increase in the concentration of silver [72]. Dendrimers have also used a template to synthesize uniform particles with a size less than gold and silver particles as a potential CT contrast [73, 74]. The advantage of using these dendrimers is that they can modify the terminal groups in the Gold-dendrimer complex. Similar to GNPs, Gold nanorods also pose higher attenuation due to the increase in size and can also act as therapeutic agents. Gold nanorods possess strong absorption in the NIR region makes it suitable for photothermal therapy. Due to some controversial aspects of GNRs, toxicity, and pharmacodynamic studies have to be studied in more detail within-Vivo X-ray CT imaging applications compared to GNPs [75].

6.3 Targeted GNPs for CT Imaging

Conjugation of GNPs with different targeting agents has been deployed for early detection of cancer and plaque detection in valves. Localization of targeted nanoparticles in the diseased region to enhance the contrast between the normal and abnormal tissues. Various techniques involved in conjugating antibodies on the surface of GNPs are physical interaction and chemical interactions [76]. Physical interactions such as ionic interaction, hydrophobic interaction, and dative binding similarly thiol conjugation, linkers binding, and adapter molecules for chemical interactions [77]. Commonly used targeting moiety is vitamin B9, i.e., folate receptor binds with folic acid for intracellular delivery [39, 40, 78]. Wherein many cancers like breast, colon, glioblastoma, and leukemia would overexpress similar markers in humans and in vivo [79, 80]. In this context, a simple approach made by Zhou et al. conjugating GNPs with folic acid using a branched poly ethylenimine and coated with PEG [81]. They produced water-soluble GNPs with a size around 2.1 nm, which is non-toxic, and pre-clinical CT imaging was performed at different time points. Due to PEG functionalized on GNPs, they were not considered as foreign bodies and started to accumulate in heart, kidney, tumor, and other regions. Further circulation for 72 h in vivo, there is an increased GNPs accumulation compared to a previous time point. Eck et al., in 2010, conjugated GNPs with Anti-CD4 for peripheral lymph node CT contrast. Whole-body CT imaging before and after injection of targeted GNPs, as shown in Fig. 8.

Recently Khademi et al. in 2019, synthesized folic acid-GNPs via cysteamine with an average size of 13 nm specifically to enable molecular CT imaging of nasopharyngeal cancer. These findings showed a clear understanding between the targeted and non-targeted cancer cells in nude mice. In comparison to passive targeting GNPs, CT images show contrast two times higher in folic acid targeted nanoparticles to determine nasopharyngeal cancer in nude mice [84]. Annexin V targeted GNPs capped with ^{99m}Tc, a dual-modal imaging probe, to position and assessed the effects of atherosclerotic plaques. Synthesized nanoparticles showed monodispersed particles with a 98.9% labeling rate (see Fig. 8). They evaluated the uptake of these particles in apoptotic macrophages with and without the Annexin V target. Incubation of targeted GNPs with 30% apoptotic cells for 2 h showed enhanced uptake of 3.52%. In vivo studies revealed that the introduction of targeting moiety enhances the accumulation of these probes for vulnerable plaques for high-resolution SPECT/CT imaging [83,

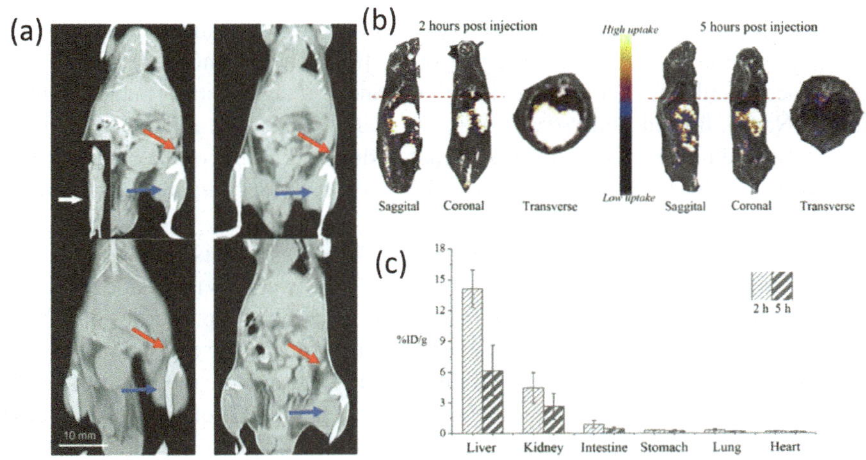

Fig. 8 Targeted GNPs: X-ray CT images of intravenous injected GNPs without conjugating immunoglobin (IgG) and after conjugating with IgG. Targeted images of GNPs show a clear demarcation of the region compared to non-specific nanoparticles. Uptake of the nanoparticles in different organs post-injection of 2 and 5 h. Adapted from reference [82, 83]

84]. Zhu et al. developed GNPs via citrate reduction method functionalized with an anti-collagen-I antibody for determining the severity of the renal fibrosis. This anti-collagen-I antibody binds with GTP protein regulates the morphology of the cell [85].

6.4 Gold Nanoclusters for CT Imaging

Nanoclusters are produced using different protocols, among which citrate reduction technique is the most popular in the literature. Proteins and small molecules such as bovine serum albumin (BSA), reduced glutathione (GSH), cysteine, cystidine, pepsin, cyclodextrin, chicken protein, and ovalbumin were used to act as reducing and capping agent [86]. Pre-clinical CT imaging was studied using 1.9 nm particles because of the smaller particle size and can excrete from the body like Iodine. Also, the clearance of such particles is slower and reduces circulation in the bloodstream. Injecting nanoclusters intravenously enables high-resolution imaging of tumor volume and development of blood vasculature in tumors [87]. Other properties of GNCs are applicable as an adjuvant with image-guided therapy along with radiotherapy and hyperthermia. Another type of particle that has few atoms of gold called gold nanoclusters (GNCs) with a size range of ~1–2 nm. These GNCs not only exhibit X-ray attenuation properties, but it also produces emission in red region around 610 nm when excited at 365 nm using a UV lamp. Glutathione capped gold nanoclusters have shown effective in vivo renal clearance studied with 3D CT images. Compared to cysteine capped GNCs, glutathione capped GNCs showed 50% clearance via urine with 3.7% accumulation in the liver (see Fig. 9) [88]. Liu et al. developed microwave-assisted multi-modal GNCs using lysozyme as a capping and reducing agent as a dual-modal imaging probe [89]. These nanoclusters exhibited good biocompatibility characteristics with enhanced imaging performance in both CT and Near infra-red fluorescence. Histologically lysozyme capped GNCs (Lyz-GNCs) showed no necrosis and had a similar body trend as that of the control mice. THE coronal CT view of Lyz-GNCs showed positive contrast compared to the control Kunming mouse as shown in Fig. 9 [89].

GNC does not only help in CT imaging, but it also acts as promising radiosensitizer material that can absorb and remit the energy [90]. However, this is outside the scope of this chapter, but it is significant to mention the radiotherapy properties of GNCs. Both radiotherapy and CT imaging are synchronous for the pre and post-treatment of cancer [91]. Zhang et al. developed ultra-small glutathione capped GNCs as X-ray contrast agents as well as radiotherapy to boost the killing efficacy [92]. These GNCs have shown an increase in radiotherapy. U14 tumor models were imaged using a micro CT scan for 10 min and reconstructed to know the tumor growth. Several combinations of GNCs have been developed to enhance the bioavailability of the particles. One such study performed by Domg Ma and his coworkers developed non-thiolate clusters using chitosan-graft-poly-(L)-Lysine in one step based on simple azide-alkyne cyclization reaction [93]. X-ray attenuation of gold concentration versus

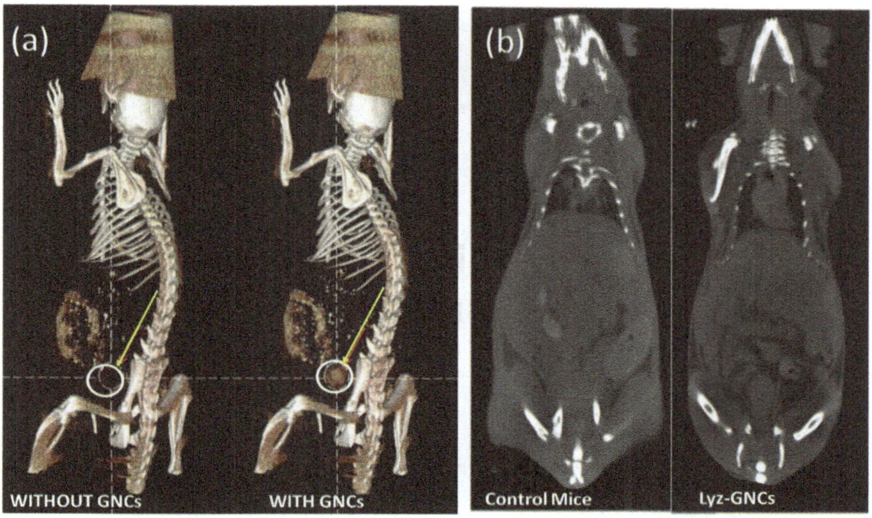

Fig. 9 GNCs with and without cysteine capped with enhanced performance in-vivo. Similarly, lysozyme capped GNCs with positive contrast compared to non-injected particles. Adapted from reference [88, 89]

Hounsfield units (HU) is linear and, at specifically around 160 mM concentration, showed 450 HU. GNCs is efficient in gene therapy for RGD targeted intracellular delivery of matrix metalloproteins-9 plasmid sh-RNA. Post-treatment efficacy of the tumor was imaged using GNCs acquired via CT. Though few kinds of literature are mentioned here, there is immense opportunity to explore more about the GNCs for imaging (CT and optical) and therapeutic (Radiotherapy and photothermal therapy) applications.

6.5 *Bismuth*

Similar to Gold, Bismuth is another promising material for pre-clinical research, which is less expensive, posses large X-ray attenuation properties, and highly biocompatible for the use in medical imaging [94]. In general, several forms of Bismuth such as metallic Bismuth (Bi), Bismuth Sulphide (Bi_2S_3), Bismuth Oxide (Bi_2O_3), in combination with other elements like Bismuth-Carbon, Bismuth Tungstate, Bismuth selenide are used for several biological applications (see Fig. 10) [95, 96]. In 2014, Cormode group reported ultra-high payload of metallic Bi nanoparticles coated with glucose and used as intravascular X-ray contrast agents. These synthesized bismuth nanoparticles were eventually decomposed to Bismuth ions and excreted through the renal route in a controlled fashion [97]. Wang et al. developed one step pH assisted BSA capped biocompatible Bismuth and Bi_2S_3 NPs for

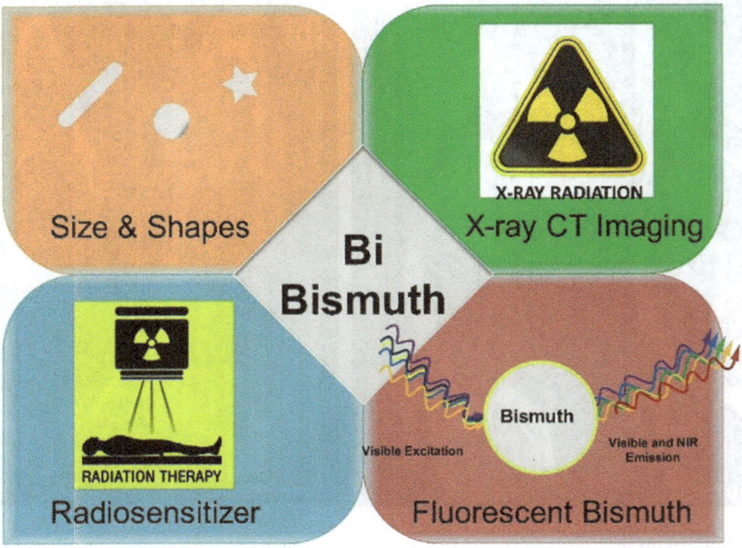

Fig. 10 Schematic representation of Bismuth based nanomaterials for in-vivo imaging applications. Bismuth possesses similar characteristics of GNPs, enabling them for X-ray imaging, radiosensitizer, and optical imaging applications

in vivo imaging applications. BSA-Bi nanomaterials were synthesized using acidic conditions facilitates the coordination of Bi ions and BSA. Further changing the pH from acidic to basic condition (pH = 12), the nucleation state of the BSA-Bi ions converts to BSA capped Bismuth nanoparticles with a size less than 10 nm. These nanoparticles possess a direct bandgap of 1.3 eV, which exhibits absorbance at NIR that aids in photo-thermal effects for therapeutic applications. BSA-Bi NPs were injected intravenously into the tumor region via EPR effect, as shown in Fig. 11 [95].

Compared to metallic Bismuth, bismuth chalcogenide possesses enhanced CT and therapeutic effects [96]. In 2009, Lanza's group developed soft bismuth encapsulated polymeric nanoparticles with high metallic content and provided enhanced CT signal in vivo. Another work reports Bi_2S_3 synthesized hard bismuth nanoparticles [98, 99]. Monodispersed Bi nanoparticles are produced using oleyl amine as a reducing agent but showed weak interaction during the washing step. However, the bismuth-based CT contrast was not well studied and is still in the primitive stage due to the challenges in loading bismuth, controlling the size and surface modification of the particles. Polyvinyl pyrrolidone coated cubic shaped bismuth sulphide nanocrystals were developed for pre-clinical lymph node imaging. These nanocrystals showed a five-fold increase in CT contrast compared to regular iodine-based contrast agents with longer circulation time and having a similar toxicity trend of Iodine. Compared to bismuth sulphosalicylate ionic form (LD50 at 8 mM) is more toxic than bismuth nanocrystals (LD50 at 100 mM) [98]. Ultrasmall sized Bismuth Sulphide nanoparticles (1–2 nm) are efficient compared to larger particles due to short circulation time,

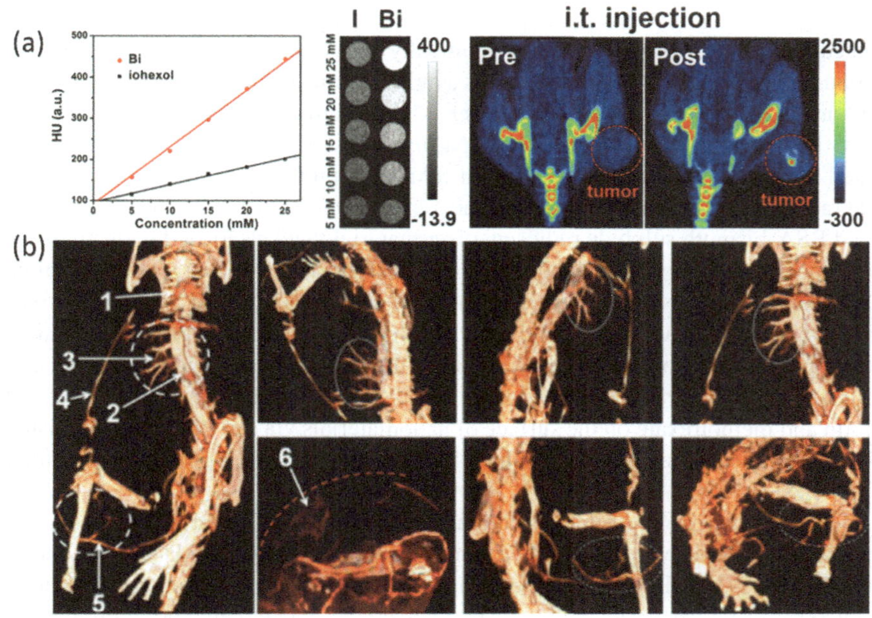

Fig. 11 Bismuth based nanoparticles for in-vivo imaging. Ultra-small semi-metal shows that the CT Hounsfield unit is high compared to conventional Iohexol at various concentrations. Intravenous tail vein injection depicts the location of the tumor (**a**). At the various time point of injected Bismuth sulphide nanoparticles showed an improved signal intensity than the pre-injected section (**b**). Adapted from reference [101, 102]

unique optical and magnetic properties with less toxicity, and high x-ray attenuation. In 2011 Ai et al. synthesized large-scale oleyl amine coated Bi_2S_3 nanodots, which are cost-effective than gold particles with unique CT contrast for in vivo imaging. Polyvinyl pyrrolidone is coated over nanodots for better dispersibility in an aqueous medium, having 80% viability in 3 mg Bi/mL with higher uptake in HeLa cells. Histopathological changes of organs after 1 month showed no tissue damage when administered intravenously and safe for biological applications [100].

Recent studies with Bismuth based X-ray contrast agents coated with biotin, Folic acid, hyaluronic acid, and epidermal growth factors are used as targeting moieties [103]. Other than surfactant coated bismuth nanoparticles, targeted nanoparticles have shown great potential as in vivo cancer imaging. Labeled particles that are targeted to specific cancer tissues as an exogenous contrast agent to detect at an early stage with enhanced sensitivity and specificity. Coating of particles can would able to protect from the lymph node clearance and protein adsorption. Amino acid peptide-functionalized Bi_2S_3 NPs for targeted breast cancer cells due to the elevated levels of protein in them. Likewise, LyP-1 linked with Bi_2S_3 NPs increases the level of particles in conjugation with p32 cellular receptors in the tumor region. Also, clearance of these nanoparticles is via fecal route and provides sufficiently high contrast in the X-ray imaging as shown in Fig. 11 [104]. Bismuth nanoparticles coated with trastuzumab linked via polyvinylpyrrolidone and chemically encapsulated on a single layer mesoporous silica [101, 105]. Bismuth acts as a radiosensitizer, conjugation of antibodies, and drugs for enhanced therapy of cancer. Accumulation of these bismuth nanorods is 16 times higher than the non-targeted particles. Bismuth nanoparticles were conjugated with RBC cells and targeted with folate receptors for simultaneous imaging and radiotherapy treatment of breast cancer [106]. Danafar and his coworkers developed BSA-bismuth sulphide radiosensitizer material that was loaded with curcumin anticancer agent and surface functionalized with folic acid to the target tumor region. Drug loaded Bi_2S_3 NPs were injected intravenously at 100 $\mu g\ ml^{-1}$ concentrations in aid with X-rays radiation 2, and 6 Gy showed higher toxicity compared to unexposed particles. Also, the survival rate of all the mice exposed with nanomaterials, drugs along with X-ray radiation was 60 days. Similarly, the same group conjugated biotin and methotrexate on the surface of nanomaterials via carbodiimide chemistry. Biocompatible studies with blood and HEK-293 revealed that possess high viability [107, 108].

Multi-modal imaging offers structural and molecular information of the disease diagnosis, early detection, assessing tumor growth, real-time guidance, and therapeutic effects. Hybrid nanoparticles which aid in multi-modal imaging is another critical factor for early cancer diagnosis. Zhou et al. folate targeted perfluoro hexane loaded with Bi_2S_3 NPs as a dual-modal imaging contrast agent using ultrasound and CT, respectively. This targeted theranostic contrast agents would allow non-invasive high intensity focused ultrasound ablates cervical cancer at high efficacy [109]. Similarly, Yu and coworkers developed a peptide labeled ultrasmall semimetal bismuth as a multifunctional nanomaterial that serves in CT/Photoacoustic imaging in combination with thermoradiotherapy. Accumulation of the targeted bismuth nanomaterials injected in the tumor region is 1.7-fold (1193.5 HU at 6 mg mL^{-1}) higher compared to pegylated particles. Bismuth (concentration of 3 mg mL^{-1}) has strong absorption in the NIR II region (1064 nm) along with radiotherapy (4 Gy) showed a synergetic reduction in the tumor volume. These were further evaluated using hematoxylin and eosin (H&E) stained images, which revealed the reduction in angiogenesis was observed [102]. Targeted nanoparticles were cleared over 30 days via fecal and renal clearance with no toxicity for the animal. Bismuth sulphide nanorods were designed for optical coherence tomography, and CT guided therapy via photothermal properties of nanorods. The change in temperature varies linearly with an increase in concentration compared to control water, and they are compatible up to 300 $\mu g/mL$ [101].

6.6 *Tantalum Based CT Contrast*

Tantalum (Ta), another promising material, has potential applications as a CT contrast agent. Conventionally Ta was used in biomedical implants along with a coating agent because of its biocompatibility, radio-opacity, and chemical inertness. For instance, the efficiency of Ta is 1.4 times higher than Iodine [110]. Both metallic and oxide form of Ta was used as contrast agents as well in dentistry, orthopedics, and regenerative medicine [111]. Tantalum is still considered as CT agent compared to gold, which has similar X-ray attenuation properties like Bismuth. In 2012, Marino's group developed Zwitterionic polymer-coated Tantalum oxide (TaO) nanoparticles as an X-ray contrast agent. Administering these particles at a dose of 1500 mg Ta kg^{-1} had no significant impact on rodent physiology [112]. They were able to hypothesize that the particles with size 2–3 nm would significantly accumulate in kidneys and liver, thereby reducing circulation time. The evaluation of TaO nanoparticles as single or multi-modal by conjugating a fluorophore or iron oxide nanoparticles has been investigated in both in vitro and in vivo imaging (see Fig. 12) [110, 113]. Recently reported work showed an accurate determination of Tantalum and TaO NPs bio-distribution in different organs by detecting the trace of Tantalum in vivo using ICP analysis [114]. TaO NPs can be easily modified with silane compounds on further conjugated with Rhodamine, and these are used as a bi-modal imaging probe. To make it a more multi-modal imaging probe, ultrasmall iron-oxide based nanoparticles were loaded into TaO NPs and characterized and showed as a versatile contrast agent

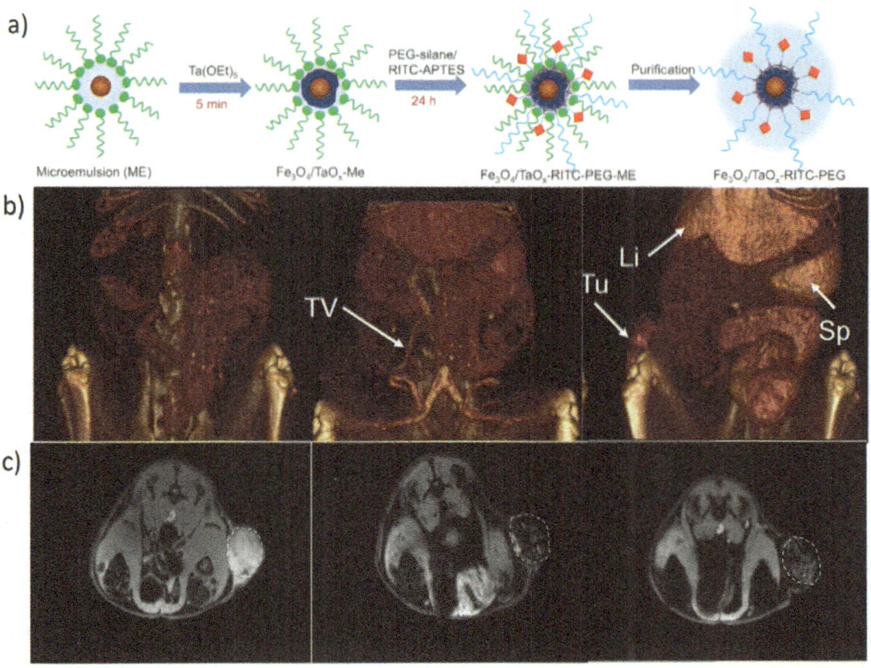

Fig. 12 Multimodal Iron oxide NPs into TaOx NPs: Incorporating SPIONs while synthesizing TaOx NPs via microemulsion technique as an effective contrast agent for MR and CT imaging (**a**). 3D reconstructed CT images depicts that accumulation of nanomaterials in liver, spleen, and tumor area Cross-sectional MR images represent negative contrast of SPIONs. Adapted from reference [113]

for MR/CT/Optical imaging systems (see Fig. 12). Also, General Electric (GE) has developed TaO nanoparticles coated with a zwitterionic agent serves as CT contrast with improved bioavailability characteristics [112].

On further extension to the above work, Freedman et al. in 2014 acquired 3D articular cartilage, composed of collagen and glucosaminoglycans (GAGs), in cadaver models of both mouse and human. TaO nanoparticles were functionalized with different charges of surfactants like phosphate, ammonium, and carboxylate. Compared to other negative and neutral charged nanoparticles, cartilage models with positively charged particles possess the highest affinity with visualizable defects [115]. The biodistribution and exposure of quantifying TaOx NPs with different organs and body fluids parameter to understand the toxicity levels. The quantitative information of Tantalum metal was at picomolar concentration was determined using ICP OES by 3 methods of tissue digestion, such as microwave, open beaker, and dry ash. Injected TaOx NPs in vivo were determined by these methods, the limit of detection and quantification varied at 6 ng to 6 μg and 0.02–20 μg Ta per gram [116]. TaO was shown some positive radiotherapy results when showing in vivo. For example, solid tumors exposed to radiation therapy have higher resistance and show negligible effects due to the hypoxic condition of the microenvironment. This is due to the higher production of H_2O_2, while decomposing into oxygen would reduce the resistance for enhanced radiotherapy. To overcome this, Song et al. developed PEG-TaOx nanoparticles were encapsulated with catalase, decomposing H_2O_2, as core-shell particles [117]. In continuation of the above work, drug-loaded mesoporous nanoparticles were developed to reduce the side effects produced from the free-drug. Liu and his coworkers together developed doxorubicin loaded into radiosensitizer mesoporous TaOx NPs for synergistic effect using radiotherapy and chemotherapy [118]. Later in 2017, Jin et al. developed TaOx as a multi-modal imaging probe by loading doxorubicin and polypyrrole in the core, and NIR dye conjugated on the shell of nanoparticles [119]. Other similar works with radiotherapy wherein TaOx NPs were used as triple sensitization [120] and molybdenum disulphide loaded TaOx NPs for photothermal therapy [121].

Shapiro and his coworkers developed the highest loaded Ta coated with hydrophilic and hydrophobic layers labeled with Rhodamine dye. At 100 mM dosage of TaO NPs they circulated for an extended period by having negligible cellular toxicity at a concentration of 2.4 mg mL^{-1}. In vivo imaging of ductal trees in rodents showed efficient imaging by injecting Ta in the ductal region of mammary pads. Further conjugation with PLGA and Mesoporous silica nanoparticles as a shell over the TaO NPs were synthesized with high cell viability at 1.2 mg Ta mL^{-1}. Encapsulation of Ta with different FDA approved capping agents like PLGA and silica loaded highest Ta materials was not attempted previously [123]. Blending of core up-conversion nanoparticle and shells as TaO NPs to form core-shell particles [122]. Wherein, Core particles were synthesized in a combination of $Yb^{3+}/Er^{3+}/Tm^{3+}$ doped $NaYF_4$ nanocrystals and further epitaxially $NaGdF_4$ grown on them. TaO was decorated over the core by simple water in oil microemulsion method functionalized rhodamine and coated with surfactant PEG silane. This unique

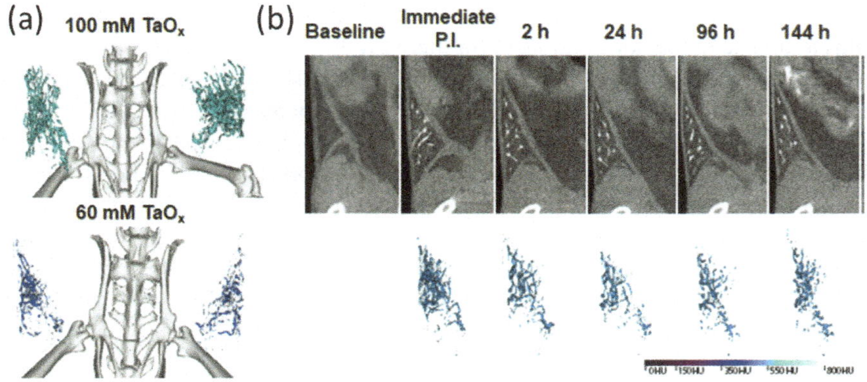

Fig. 13 Combination of up-conversion particles loaded into TaOx NPs. The unique combination of particles depicts localization in the lymph node exposed with two different concentrations (**a**). A gradual decrease in the contrast over time up to 144 h along with 3D reconstructed images (**b**). Adapted from reference [122]

combination of nanoformulation showed excellent in vivo imaging in CT/MRI/Up-conversion particle from TaO/Gd^{3+}/$NaYF_4$: Yb^{3+}/Er^{3+}/Tm^{3+}, respectively. Distribution of nanoprobes was imaged using CT by tail vein injection at a concentration of 64 mg mL^{-1} were able to localize and map lymph nodes at different time points. Within 10 min, CT imaging showed increased accumulation in the liver and spleen and enhanced contrast after 3 h of injection (see Fig. 13).

6.7 *Yttrium Based CT Agents*

The above elements, such as Au, Ta, Bi, were used as CT contrast agents either in its metallic or oxide or sulphide states [124]. In line with this, the ionic form of Ytterbium (Yb) is another promising material like Iodine but has a high X-ray attenuation at 62 keV. Most importantly, a chelated form of Ytterbium is less toxicity in rodents [125] and the highest availability in earth crust [126], which enables them to produce CT contrast agents in large amounts. Also, Yb has a significant absorption in the NIR range that can be used for the synthesis of multimodal imaging probes and also for up-conversion nanomaterials [127]. The first experiment was performed in 1986 by Unger et al. by chelating Yb using DTPA as a potential intravascular CT contrast [128]. Compared to Iodine, Yb and Gd showed much higher radiopacity when injected in different dogs showed high SNR in imaging pulmonary artery and its branches when excited at 125 kVp [129, 130]. Though there are several remarkable studies shown in the literature that Yb based chelating agents have some disadvantages when injected in vivo. 1. The large amount of Yb has to be injected to obtain valuable diagnostic information. 2. They have minimal time in the blood circulation and excreted within seconds. 3. The ionic form of Yb is less toxic, and it tends to uptake in the bone region. 4. Similar to Iodine, hight concentration of Yb is required for proper CT, thereby viscosity increases [124]. Later studies revealed that incorporating chelated agents would over the issues mentioned above [131]. Nano based contrast agents can load a large amount of Yb chelates/ions forms and can improve CT imaging. In 2011, Lehui Lu and his coworkers were the first to report homogenous Yb nanoparticle as a CT contrast agent co-doped with Gadolinium and Erbium as up-conversion nanomaterial [132]. $NaYbF_4$ nanoparticles were initially coat with oleyl amine to avoid aggregation and functionalized using DSPE-PEG-2000 to improve the bioavailability and reduce the toxicity in vivo (see Fig. 14). CT contrast was higher compared to other metallic nanoparticles and iodine contrast when excited at 120kVp.

Angiography played an important role in imaging chambers and vasculature system of preclinical systems. $BaYbF_4@SiO_2@PEG$ nanomaterials helped in prolonged circulation up to 2 h in the bloodstream due to increased stealth properties when coated with silica and PEG. Intravenous injection in the rabbit has enhanced imaging, even small blood vessels shown in Fig. 14. The main factors are the combined effect of Yb and Ba, and these materials have better X-ray attenuation at 120 kVp and, finally, the surfactant of nanomaterial [133]. Liu and coworkers synthesized Yb_2O_3 nanoparticles doped with Erbium and Gadolinium as an excellent X-ray contrast agent for in vivo imaging [134, 135]. A comparison of YbO based nanoparticles doped with Gd^{3+} and Er^{3+} were shown in Fig. 15a, b was performed by them. These nanomaterials are functionalized with PEG with a high content of Yb, and doped materials serve both as X-ray CT and up-conversion dual-modal imaging. Erbium-doped nanomaterials were subcutaneously injected at 1 mg ml^{-1} concentration compared with Iodine exposed at 120 kVp, 300 mA current. In vivo CT imaging was examined in different organs and distributed in the liver, heart, spleen, and kidney. In early 30 min, the maximum contrast was observed and slowly decreased over time. The retention of the nanoparticles in the bloodstream was up to 24 h can be visualized with 3D CT images (see Fig. 15). Erbium (5%) doped Yb_2O_3 NPs showed up-conversion fluorescence revealed red emission when excited using NIR laser (980 nm). The same group has also exposed to gadolinium doped Yb_2O_3 NPs as dual-modal probes such as X-ray CT and MR imaging. Figure 15 showed ICP analysis of nanoparticle accumulation was higher in liver and spleen compared to other organs post 30 min of injected nanomaterials clear contrast is visualized. Later several works were published with Yb as dopant materials. For instance, PEG-$BaGdF_4$: Yb^{3+}/Er^{3+}, nano colloids incorporated Yb^{3+}, $NaYF_4$:$Yb^{3+}/Er^{3+}@NaGdY_4$, Gd_2O_3:Yb^{3+}/Er^{3+},NaLuF4:Yb^{3+},$Tm^{3+}@SiO_2$-Gd-DTPA and BaYbF4@SiO_2@PEG showed evident results that nano-based Yb as CT contrast agent [133, 136–140].

Multispectral or dual-energy CT uses dual X-ray energies are exposed to sample having multiple detectors to classify tissues with different X-ray attenuation properties [141]. Multicolor spectral CT imaging was performed using Yb doped nano colloids helps in understanding the atherosclerotic plaques in coronary artery disease. Yb nanoclusters (1 mg mL^{-1}) were intravenously injected and exposed at 130 kVp,

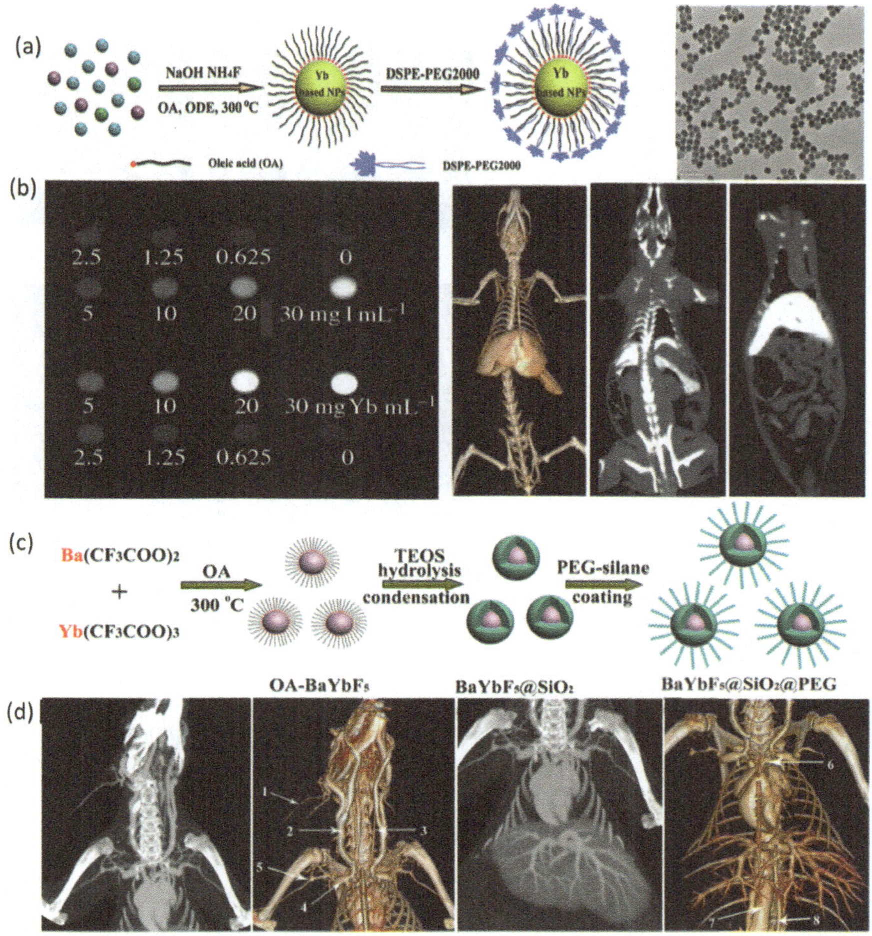

Fig. 14 High-performance CT imaging: Schematic representation of the synthesis of $BaYbF_4@SiO_2@PEG$ (**c**) & Gd/Er/YbO based nanoparticles with its TEM images (**a**). Concentration-dependent contrast enhancement with CT images (**b**) [132]. CT and 3D reconstructed images were obtained after injection of $BaYbF_4@SiO_2@PEG$ NPs for in vivo imaging (**d**) [133]

$I = 50$ uA at 6 varying threshold levels having 900 views in 72 s. These images were acquired in euthanized mice after 2 min of injection of nanoclusters. This was a significant milestone in Yb based nanoparticles in preclinical to image for the first time in spectral CT to localize atherosclerosis in blood vessels (see Fig. 16). Biodistribution and pharmacodynamics were studied post administration of 2, 24 h, and 7 days results in high uptake in the liver compared to other organs and clearance through RES [137]. Due to the limited understanding of X-ray attenuation factors of Yb loaded nanoparticles, they were not much studied in the literature. The mechanism of Yb as CT contrast has to be evaluated. But the doping of Yb^{3+} and

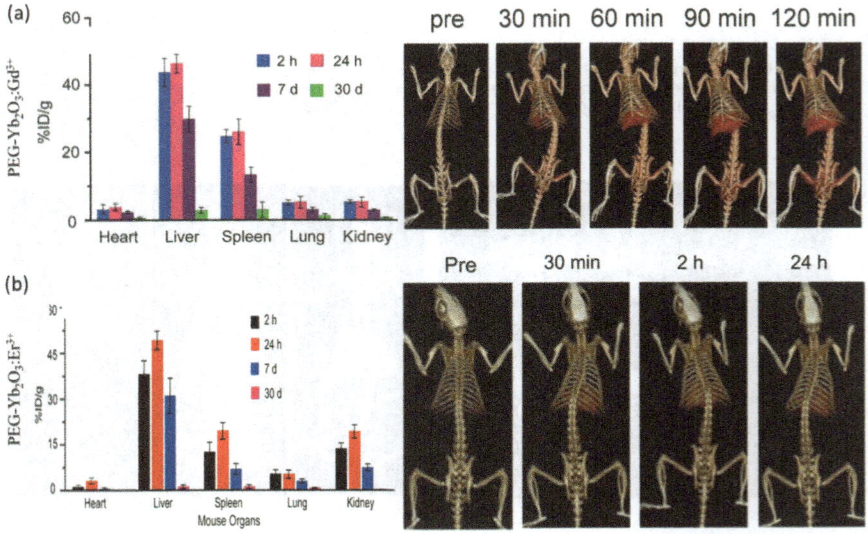

Fig. 15 PEGylated Gadolinium (**a**) and Erbium (**b**) doped Yb based oxide NPs for dual-modal imaging (X-ray CT and NIR fluorescence imaging). The percentage of injected dose per gram was determined using ICP analysis, which clearly showed that accumulation in liver and spleen are ~42 and ~30 after 24 h of post-injection [134, 135]

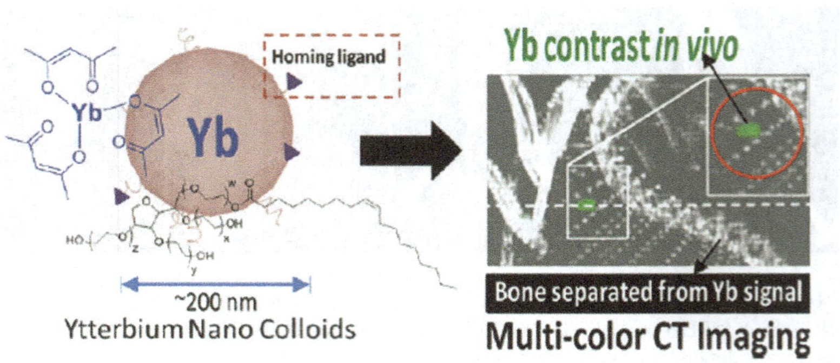

Fig. 16 Multispectral or dual-energy CT imaging using Yb based oxide NPs: Yb nano colloids are used as a contrast to determine plaque formation using multi-spectral CT imaging. Adapted from reference [137]

different forms of Yb were synthesized mainly as CT contrast and up-conversion hybrid nanomaterials.

Despite a long history of nanoparticles-based contrast agent development for X-ray imaging applications, there is still potential to improve in both pre-clinical and clinical practice. Improving properties such as prolonged circulation and the ability to target the particles would accelerate the innovation in CT imaging. This eventually can lead to detecting various disease diagnoses at an early stage. Since X-ray is commercially available at even primary health centers at a lower cost, the disease diagnosis in developing countries would be effective. This would also reduce the failure rate and fasten the clinical trials by improved results of the preclinical studies. Incorporating nanomaterials in preclinical studies would help in determining the dynamics and the fate of drugs. Nanoparticles delivered in the tumor-targeted region can eventually improve the sensitivity of the CT system for personalized diagnosis. The performance of different materials (Bi, Ta, Au, Yb, Eu, and Iodine) has no standard protocol to evaluate its performance in CT imaging. Also, these nanoparticles used in preclinical imaging with different concentrations. Also, a thorough examination of these materials on compatibility aspects is needed for a clinical translation. Since most of the materials are of high atomic number, their metabolism, degradation, and excretion have to be studied extensively. The cost of the new contrast agents should be addressed because it requires a high concentration for better imaging [49]. Even though the attention towards the nanoparticle for CT imaging is less when compared to MR and optical imaging, considerable research interest in developing hybrid nanoparticles would help combine the advantage of each modality.

7 Nanomaterials for MRI Imaging

Analogous with CT, information obtained from Magnetic resonance imaging (MRI) provides structural information with more detailed features in the tissues [142]. Images are acquired by magnetic behavioral changes of tissues when placed in a large magnetic field. A magnetic field is generated exposed to the protons or atoms in the tissues that align with a magnetic dipole. When a radio frequency (RF) signal is a resonance with Larmor frequency of the protons, then they relax to its normal state for a short duration of time known as T1 and T2 relaxation [3]. Magnetic relaxation of protons in tissues produces a RF signal that varies with the tissue type and density. For more details about the relaxation properties of MRI, readers can refer to this article [3, 143]. The complete mechanism of longitudinal and transverse (T1 and T2) relaxation of MRI concerning on/off RF signals.

MRI is a non-invasive technique that generates 3D images of tissues. Unlike CT, MRI does not involve any type of hazardous radiation, and it is entirely safe for repeated measurements [144]. Clinical MRI is not suitable for small animal imaging and is associated with many problems and limitations [145]. Clinical MRI has a large bore, less magnetic field up to 2 T) produces a low signal to noise ratio. Therefore, a particular type of MRI requires a high magnetic field up to 10 T is required to study the developmental biology of animals is called micro-MRI [146]. Micro-MRI produces high-resolution anatomical, molecular and functional imaging for drug discovery applications. The main disadvantages of micro-MRI are expensive due to uniform high magnetic field generation. Animal models have to be anesthetized for detection, localization, volume of tumors, and blood flow characteristics. Also, time taken to acquire MRI images takes longer compared to micro-CT [147]. Another major disadvantage of MRI is the contrast between the tissues within the organs is negligible. For instance, the delineation of the tumor region and the normal tissues show a poor diagnosis for physicians. This separation between the normal and abnormal regions requires a localized exogenous contrast to detect via micro-MRI is necessary [148]. Exogenous contrast agents are an excellent supplement for the current medical imaging systems to enhance the diagnosis. New types of nanomaterials are being developed to improve the contrast between the tissues in both single as well as multi-modal imaging aspects [149]. Paramagnetic (T1 contrast agents) and superparamagnetic (T2 iron oxide nanoparticles) properties were used in MRI to acquire the images and to enhance the contrast between the tissues [150].

7.1 Gadolinium Chelates Based Contrast Agents

Species of Gadolinium ions were chelated with different compounds that are used as a clinical contrast agent for MR imaging, as shown in Fig. 17. Young et al. was first to develop exogenous contrast for MRI in 1981 using ferric chloride as a bowel labeling agent. Later the same group reported that gadolinium-based complexes (Gadolinium Diethylene Triamine Penta-acetic Acid—Gd^{3+}-DTPA) as a contrast to determine cerebral tumor intravenously at a concentration of 0.1 mM/kg [151, 152]. Most of the clinical and preclinical studies were performed by injection of contrast agents and is growing in a comprehensive manner. After the early development of Gd^{3+}-DTPA, several lanthanide elements as core metal ions such as Mn^{2+}, Dy^{3+}, and Eu^{2+}/Eu^{3+} were used as paramagnetic contrast agents [153]. The MRI contrast agents should possess basic chemical and biological properties like high relaxivity,

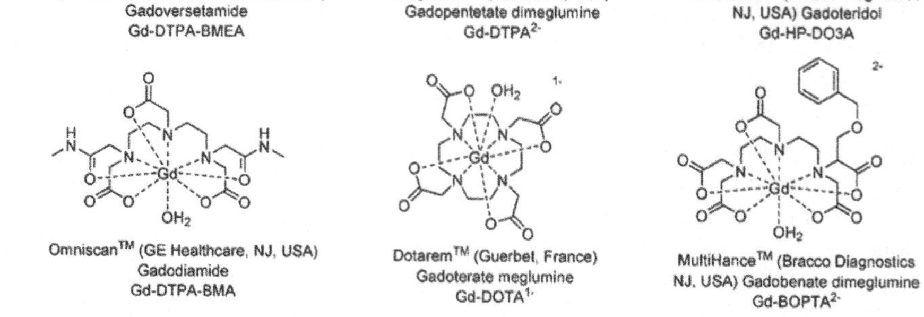

Fig. 17 Different types of FDA approved Gd chelates along with the brand name. Adapted from reference [157]

stability, biodistribution, rapid clearance, low osmolality, and viscosity, and should be biocompatible. The major problem associated with gadolinium contrast agents (GCA) is the leakage of highly positive Gd^{3+} ions compared to neutral or negative charges. The ions of Gd^{3+} competes with Ca^{2+}, thereby altering the cellular functions that require Ca^{2+} [154]. Over three decades after the development of Gd-DTPA, several chelates were developed to load Gd^{3+} ions to avoid the leakage to prevent toxicity [148, 155]. In preclinical MR imaging, Gd^{3+} complexes were immediately distributed in the bloodstream and cleared via kidneys and liver. The clearance time of these complexes would vary from minutes to a few hours based on the functions of a kidney that might cause nephrotoxicity called nephrogenic systemic fibrosis (NSF) [156]. With emerging trends in nanotechnology and its advancements, incorporation these Gd^{3+} complexes into various nanostructures have been investigated. Let us look some of the key examples of gadolinium-based nanostructures used for in vivo imaging applications.

7.2 *Liposomal Gd-Chelates Nanomaterials*

Liposomes are types of drug carriers synthesized using phospholipids and are promising drug delivery vehicles [158]. Liposomes can encapsulate hydrophilic, hydrophobic drugs, and can be functionalized with the targeted antibody on the surface [159]. Yang and his coworkers developed liposomes loaded with Magnevist (Gd^{3+}-DTPA) conjugated with interleukins-13 (IL-13) for the detection of glioma. Glioma cells are overexpressed with IL-13 receptors alpha compared to normal brain cells. Chelates of contrast agent loaded liposome conjugated with IL-13 antibodies aid to cross the blood-brain barrier (BBB). The result of in vivo imaging shows that IL-13-Liposomes-GD-DTPA possess enhanced contrast in the brain for early detection of glioma [160]. Also, similar paramagnetic liposomes with single and dual integrin functionalized liposomes loaded with Gd-DTPA were also potential candidates for early diagnosis of cancer. Zhoug and his coworkers reported a lipopeptide coupled with RGD peptide (as head) and palmitic acid. In vivo lung tumor model has been developed at a volume of 50–100 mm^3. Comparing non-targeted and targeted liposomes, targeted liposomes have long circulation in the blood and slowly accumulated in tumor regions with an increase in MR intensity over time. Also, the localization of targeted liposomes was high in the tumor than the normal tissues [161]. The same group further developed dual-targeted liposomes, in vivo MR signal intensity was higher than the Gd-DTPA, and RGD targeted contrast agents [162].

Imaging the inflammation in vivo for disease diagnosis, to understand the pathology and targeting the immune system [163]. The inflammation region would have an increased level of macrophages, wherein targeting these cells with mannose enables them for better imaging. Tian et al. developed mannose coated gadolinium nano-sized liposomes to determine the severity of acute pancreatitis inflammatory conditions [164]. These liposomes are with a size range of 100 nm with an encapsulation efficiency of 85% Gd-DTPA. Two types of liposomes with and without targeting moiety mannose compared with control Gd-DTPA. T_1 MR imaging with mannose Gd-DTPA showed increase macrophages uptake both in vitro and in vivo with less toxicity in organs. Discrimination of severity of the inflammation was revealed with T_1 MR images using these Gd loaded nanoliposomes. Detecting the absence of specific tumor markers in cancer cells is challenging until now, especially for triple-negative markers in breast cancer cells. Without these markers such as HER-2, estrogen, and progesterone hormones, and it is challenging to treat metastatic cells [165]. To address this issue, Schroeder and his coworkers developed targeted nanoliposomes (100 nm) to detect the remaining metastatic breast cancer cells post-surgery [166]. Triple-negative metastatic breast cancer cells were injected via tail vein (4T1 cells) and then treated with multi-modal imaging contrast to image the small lesion. Gadolinium loaded liposomes detect the premetastatic niche in 9 and 15 days and micrometastasis in the later stages. The small lesion images from MR were further confirmed by H&E staining of lungs.

Thermosensitive based liposomes (TSL) are promising drug carrier particles for drug delivery when the heat is generated by external sources of localized hyperthermia [24]. In 2015, Kuijten et al. synthesized biotin TSL loaded with imaging molecules and drug moieties in the hydrophilic regions. This work was developed to increase the relaxivity of MR agents rather than increasing the concentration of Gd^{3+} agents. Gd-DOTA linked with pegylated lipid functionalized with biotin and loaded with rhodamine (NLP) compared with conventional liposomes Gd-DTPA-BSA with pegylated lipid (CLP) [167]. The heat generated from TSL causes a change in the behavior of water molecules into NLP that changes the relaxivity of Gd-DOTA. MR/Optical based image-guided TSL would help using external sources like hyperthermia and ultrasound-mediated drug delivery for solid tumors. Also, biomagnetic properties like Gd^{3+} chelates (paramagnetic) and iron oxide nanoparticles (super-paramagnetic) was loaded into TSL for MR guided imaging. This TSL was exposed to highly focused ultrasound for increased relaxivity, as mentioned earlier. Co-labeled magnetic properties help physicians to select suitable MR sequences to study the accumulation and image-guided drug delivery of liposomes [168]. Delivery of membrane-impermeable drugs like gemcitabine (GEM) into solid tumors is challenging. Affram et al. synthesized TSL based particles loaded with GEM and Magnevist for effective image-guided drug delivery triggered via localized heat sources on an ex vivo model [169].

Apart from chelates of Gd^{3+} used as MR contrast, oxides of Gadolinium in nanometer-sized particles were synthesized and developed as MR contrast agents. Udayal developed the first synthesis of Gd_2O_3 nanoparticle with a size of ~ 5 nm, and his coworkers showed high T_1 relaxivity. Furthermore, pegylated-Gd_2O_3 nanoparticles were characterized by MR imaging having 4 times higher than the Gd-DTPA contrast [170, 171]. Glioblastoma cell labeled ultrasmall diethylene glycol capped Gd_2O_3 NPs (2–3 nm) were implanted in chicken embryo and studied the tumor growth in clinical MRI [172]. Agglomeration of Gd_2O_3 NPs has an impact on the change in the relaxation properties in NMR signals. Particles showed enhanced signals in non-agglomerated particles compared to agglomerated ones when imaged through MR 1.5 and 7 T [173]. Park et al. suggested that depending upon the particle size, the interaction behavior of MR signals produces high relaxivity properties [174]. Paramagnetic gadolinium oxide nanoclusters with size ~1 nm showed a larger relaxivity of 9.9 s^{-1} mM^{-1}, after injection of these nanoclusters with high positive contrast enhancement at various time points. This clearly shows that the injected particles with more positive contrast enhancement of nanoparticles.

The high longitudinal relaxivity of paramagnetic Gd_2O_3 NPs size of 1 nm was observed when refluxing the precursors of Gadolinium ions under an oxygen environment. The relaxivity of the 1 nm Gd_2O_3 NPs of water was around 9.9 s^{-1} mM^{-1}. The large relaxivity of these NPs depends on the direct coupling effect and interaction with Gd ions and the size of nanomaterials. From the above literature, we could narrow down that particle size with less than ~5 nm possesses high longitudinal relaxivity, and are suitable for in vivo imaging. But the biodistribution and compatibility have to study to understand the use of Gd based NPs as clinical contrast. Recent studies have shown the biodistribution of nanosized Gd_2O_3 NPs has shown less toxicity similar to chelated gadolinium. Tilement and his coworkers studied the in vivo biodistribution analysis of Gd_2O_3 NPs of a size range of 3–5 nm encapsulated with polysiloxane shell conjugated with cyanine dye. Intravenous injection of these NPs can be studied with optical imaging techniques. Initially, a small number of particles was uptake through RES from the systemic circulation with less retention time in RES. They suggested that the significant excretion was through the renal system, which was evident from both the optical and MR imaging methods. Large chelates of the Gd ions were accumulated from the liver and spleen. Hybrid imaging with Gd_2O_3 NPs is possible with the results given by the same group and further developments of contrast agents with other imaging modalities [175].

Despite Gd_2O_3 NPs, different nano compositions of Gadolinium have reported modifying MR relaxivity of the contrast agents, combined therapeutic effects, and image-guided surgery [176–178]. Capping and targeting agents of Gd_2O_3 NPs are essential to improvise the T_1 relaxivity in MRI. Polyacrylic acid capped Gd_2O_3 NPs were studied by Miao et al. with high positive contrast enhancement in liver, kidneys, and bladder observed in 10 min duration. On increasing the time up to 2 h, the gradual decrease in the contrast in the MR signals was due to the clearance from the bladder [179]. Mekuria et al. encapsulated Gd_2O_3 NPs with PAMAM dendrimer (G4.5-Gd_2O_3 NPs) and capped with polyethylene glycol. T_1 and T_2 images were acquired by injecting these NPs in vivo at a concentration of 0.5 uM per kg. In general, NPs with size, more than 20 nm would be uptake by RES in the body. But they were able to circulate for a while and accumulated in the organs with high positive T_1 contrast MR signals. These signals were observed in the kidney after 30 min of post-injection and decrease after an hour. This is due to the circulation in enterohepatic circulation. A similar effect of these G4.5-Gd_2O_3 NPs with higher negative contrast in T_2 or dark region in kidneys [180]. Navon and group showed a higher accumulation of PET agents in tumor regions when linked with glucosamine compared to glucose molecules [181]. Glucosamine linked with poly-cyclodextrin capped with Gd_2O_3 NPs were used as a better targeting moiety for early diagnosis of cancer. Compared to control Gd-DOTA ($T_1 = 2.79$ mM^{-1} s^{-1}), glucosamine functionalized nanoparticles show better contrast with a relaxivity with $T_1 = 4.86$ mM^{-1} s^{-1}. The metastasis breast cancer cells were injected in vivo for comparison of the accumulation of nanoparticles in rodents. After 1 h of post-injection, the accumulation was higher compared to normal chelates, and slow signals reduce due to the clearance mechanism. On the other hand, the injected chelates of Gd does not any change in tumor contrast enhancement and they are least significant [182]. Beta cyclodextrin capped Gd_2O_3 NPs and linked with folic acid to target early cancer diagnosis. Contrast to noise ration of targeted particles showed up to 5.89 within 1 h compared to non-targeted particles with 1.98 for 6 h [183]. Multifunctional capabilities with BSA capped Gd_2O_3 NPs loaded with cyanine showed trimodal imaging applications with photoacoustic tomography, NIR- fluorescence and MR imaging [184].

Nanorods $Gd(OH)_3$ particles were developed by Yuan and his group with an average size of 15 nm width and 100 nm length for MR imaging. Always there is a question of toxicity issues when it comes to the anisotropic structure of particles. Thus, they have evaluated cell cytotoxicity and histology analysis to confirm the toxicity to show high biocompatibility. MR imaging of $Gd(OH)_3$ NPs showed higher relaxivity compared to Gd-DTPA chelates. These $Gd(OH)_3$ NPs were accumulated in the liver after 30 min of post-injection [185]. No significant changes in $Gd(OH)_3$ nanorods when tested for long term biodistribution and their cellular toxicities when evaluated in vivo by Yang and his coworkers. But the clearance of $Gd(OH)_3$ nanorods was slower in spleen compared to other organs [186]. Gadolinium oxide particles, in combination with bismuth ($BiGdO_3$) capped with PEG have been implemented as a radiosensitizer for cancer treatment. The effect of radiosensitizer in vivo was determined using MRI and CT with gadolinium and bismuth and as theragnostic agents [187]. Yeh and his coworkers synthesized hollow spherical and rhombus-shaped $Gd_2O(CO_3)_2$ nanoparticles as an MRI contrast agent. Spherical sized particles were injected at a concentration of 0.3 mg $kg^{-1,}$ and post-injection particles until 12 h started to accumulate in the liver is higher compared to the kidney. Survival of animal was up to 4 weeks showed that particles possess negligible toxicity over the rodents [188]. The same group developed $Gd_2O(CO_3)_2$ nanoparticles coated with silica/gold as hybrid nanoparticles and studied MR characteristics and photothermal therapy with gold [189]. Few studies have synthesized Gd_2O_3 as core and shell as MnO and SiO_2 for enhanced MR signals and to study the toxicity compared to naked Gd_2O_3 NPs [190, 191].

Doping with Gd_2O_3 NPs is another promising type of nanomaterial for cancer therapeutic and diagnostic applications. Yoon et al. synthesized Europium doped $GdPO_4$ nanoparticles by annealing phosphate silica nanoparticles with layered gadolinium hydroxide nanosheets doped with europium. In vitro imaging suggested the particles are fluorescent with high MR relaxivity due to europium as luminescent centers and gadolinium as a contrast agent [192]. Liquid pulsed laser ablation technique aid in generation Europium doped Gd_2O_3 nanoparticles from the Gd_2O_3:Eu^{3+} solid target. These nanoparticles are favorable as both optical and magnetic properties for early detection of disease diagnosis. Focused Nd:YAG laser (1064 nm, 70 mJ) exposed on the solid target immersed in deionized water for 15 min duration in liquid by [193]. T_1 weighted images were acquired with the xenograft model exposed with Gd_2O_3:Eu^{3+} NPs was higher contrast at 35 min of post-injection and gradually

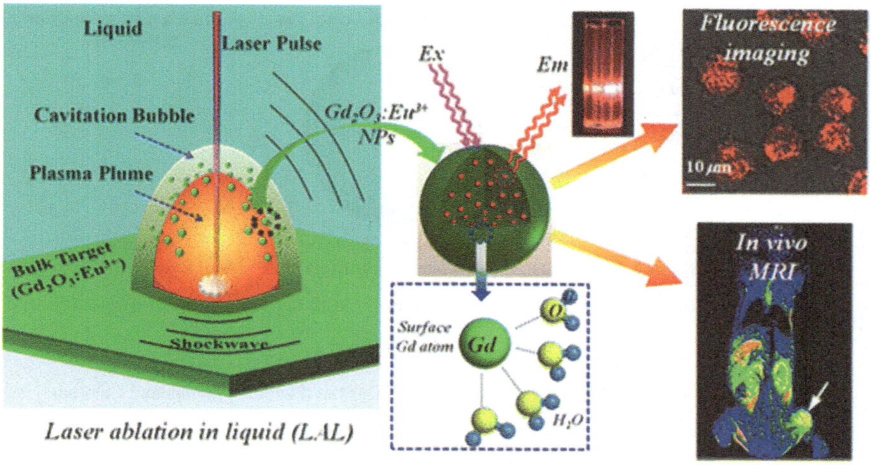

Fig. 18 Liquid pulsed laser ablation interacts with bulk target the plasma plume and cavitation bubble confines to generate Europium doped Gd_2O_3 NPs. These NPs enables them as a contrast for in vivo dual-modal imaging in MR and fluorescence. Adapted from reference [193]

reduces (see Fig. 18) [193]. Gd_2O_3 nanoparticles were generated by doping with Erbium, Ytterbium, and Zinc ions for the enhanced up-conversion properties [194, 195].

7.3 *Iron Oxide Nanomaterials (IONs)*

From the literature we have seen so far, Gadolinium-based contrast agents have been widely used for MR imaging applications [196]. But due to the nephrogenic toxicity of Gd_2O_3 NPs and Gd^{3+} chelates limits only for preclinical evaluation. The second generation of the MR contrast agents other than gadolinium is superparamagnetic iron oxide nanoparticles (SPIONs) or ultrasmall (US-SPIONs). In comparison, Iron oxide nanoparticles (IONs) are one among the most studied material for biomedical applications like imaging, hyperthermia, and drug delivery. The physicochemical properties IONs are better due to the requirement of a large amount of Fe ions to our body (20–25 mg), and while the degradation of IONs will uptake by the body itself [197]. Therefore, the clearance mechanism is much easier compared to Gd based contrast agent. Core IONs that are applicable for MR imaging are made of either magnetite (Fe_3O_4) or maghemite (Fe_2O_3) nanoparticles. The magnetic property of the IONs is an important property that serves as a contrast agent for MRI, superparamagnetic property when they are converted into a nano regime. Control over the size converts nanoparticles from paramagnetic to ferromagnetic by modulating the size of IONs between few nm to 50 nm. This change in the magnetic property tends to use IONs for T_1/T_2 MR contrast as shown in Fig. 19a. Practically, physicians prefer

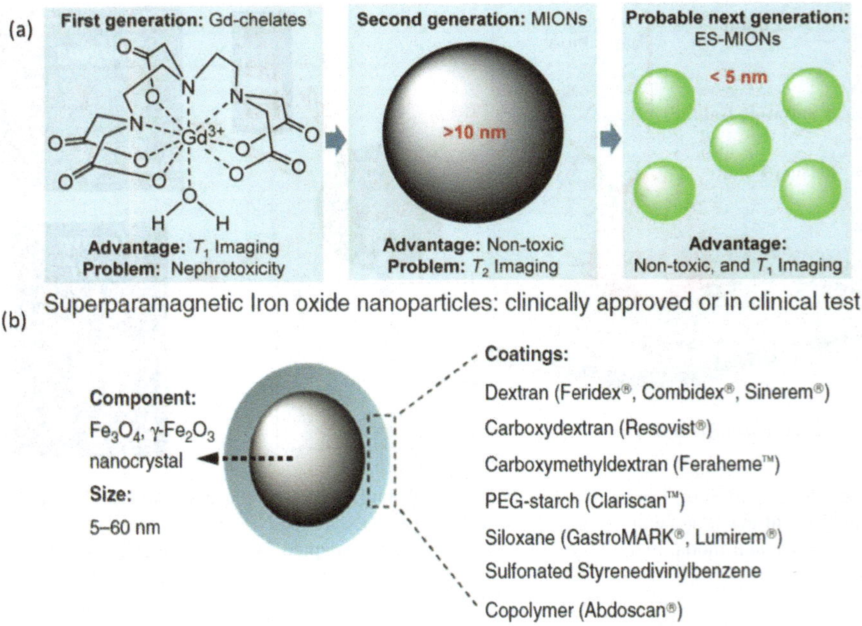

Fig. 19 Next-generation of MR contrast agent from the gadolinium (**a**). Different coatings over the iron oxide nanoparticles for in vivo imaging aid in better compatibility, stability, and prolonged circulation (**b**). Adapted from reference [198]

gadolinium for T_1 contrast and IONs for T_2 contrast agents [12, 197]. Coating plays a key role in reducing the chemical or biological interaction and increases the stealth nature of these IONs. Surfactant molecules such as dextran derivatives, polyethylene glycol, polyvinyl alcohol, and zwitterion compounds, as shown in Fig. 19b [198, 199].

IONs were used for various applications such as liver imaging, localization of the tumor, atherosclerotic plaque formation, and magnetic hyperthermia. In specific the cancer imaging can be achieved with IONs based on active and passive targeting. Initially, studies with IONs were related to understanding the synthesis mechanism, size control, toxicity evaluation, and biodistribution analysis for preclinical MR imaging. Superparamagnetic iron oxide nanoparticles (SPIONs) are a special type of IONs with a size of less than 50 nm studied widely for preclinical imaging applications due to the enhanced MR signals [200]. The magnetic properties of SPIONs in MR imaging compared to bulk iron oxide can refer to the reference given [201]. Additionally, ultrasmall SPIONs (USPIO) with a small diameter ranging from few nm to 10 nm produce a positive T_1 contrast agent [202]. The major challenge associated with SPIONs tends to attract and aggregate and considered a foreign element in a biological environment. Control over the particles can be increased by crosslinking with biocompatible surfactants. For instance, Dextran and PEG are an FDA approved non-ionic surfactant that could be an effective strategy to improvise the circulation in blood for a longer time [203]. Commercially available biocompatible SPIONs (varying size and capping agents) are approved by FDA in the generic name of Ferucarbotran, Endorem, Feridex, Resovist, and Ferumoxide with an average core diameter of 4–5 nm are currently used for in vivo imaging [204]. Most of these SPIONs are applicable for both T_1/T_2 contrasts in MR imaging applications. Biocompatible surfactant capped SPIONs are approved as MR contrast agents for preclinical studies to increase the rate of drug discovery studies. In 2008, Sangyong Jon and his coworkers developed SPIONs thermally crosslinked with PEG-silane copolymer has an excellent anti-biofouling property for cancer imaging. They have synthesized two different size ranges of particles having 10 and 15 nm SPIONs. The T_2 weighted images in MRI showed that the detectable level of particles in tumor region and signals drops gradually. MR signals from the small-sized particles are higher compared to the large-sized ones is due to the larger accumulation of particles [205]. The same group has investigated by loading a model drug known as doxorubicin, a fluorescent drug, for combinatorial cancer therapy [206]. Early findings of liver cirrhosis can be identified by the fibrosis formed in the liver, wherein detection at this stage can be curable. Citrate capped SPIONs uptake by liver macrophages (Kuffper cells) was considerably less, and thereby increased intensity of streaks was observed in MR signal as reported by Saraswathy and his coworkers [207]. Another important aspect is the density of the PEG for better antibiofouling property, wherein protein adsorption increases which in turn changes the stealth property of SPIONs. Polyacrylic acid (PAA) capped SPIONs were prepared for an effective conjugation with diamino-PEG to reduce the density over the SPIONs surface. Liver tumor bearing were injected with PAA-SPIONs and PEG-SPIONs at a concentration of 4 mg of iron/ kg. After 4 h of injection, comparing PAA-SPIONs the T_2-MR signal intensity of PEG-SPIONs is darker [208]. Diethylene glycol coated with SPIONs with higher relaxivity than the PEG-SPIONs. SPIONs coated with silica (SiO_x) shell synthesized with different thickness impacts the MR signals. A study performed by Joshi et al. showed that increasing the thickness reduces the contrast in MR signals [209].

Non-invasive tracking of injected stem cells in vivo is important to understand the disease condition is a challenging clinical need. Labeling and transplantation of stem cells is another promising area for the treatment and advance clinical regenerative medicine. For instance, neurological, cardiology, and dental disorders can be treated only when these stem cells reach the desired location [210]. Citrate labeled SPIONs are efficient to track stem cells and compared with commercially available magnetic particles (Endorem and Resovist—SPIONs) [211]. These citrate capped SPIONs were labeled with human mesenchymal stem cells used to determine the detrimental effects in the rat muscle tissue. The labeling efficiency of the SPIONs in mesenchymal stem cells (MSCs) possess higher hypointense volume compared to Resovist with increased MR signals. Treating neurological disease conditions is difficult without injecting neural stem cells. MRI of mouse brain was acquired at 1, 6, and 51 days. Post injected SPIONs labeled stem cells were grafted in the mouse brain with dark negative contrast [212]. Dental stem cells were tracked by labelled using dextran coated SPIONs without any change in the osteointegration and stemness property [213]. SPIONs labelled metastases cancer cells were injected in vivo at intra-cardiac region guided using ultrasound and tracked with MRI [214]. To understand the fate of SPIONs labeled cells in vivo Ashraf et al. studied with polyelectrolyte membrane capped and naked SPIONs. After injecting the labeled stem cells for regenerative medicine, they cannot stay for a longer time and undergo cellular death. In conclusion, the release kinetics of both these nanomaterials are similar to other reports published with cell tagged SPIONs [215].

Lee and his coworkers developed bicyclo [6.1.0] nonyne-modified glycol chitosan nanoparticles (BCN-CNPs) loaded with GNPs, SPIONs, and Cyanine 5.5as triple modal imaging applications. Wherein, tetra-acetylated N-azidoacetyl-D-mannosamine(Ac4ManNAz) molecule was used to bind over the biorthogonal chemical receptors (azide groups) present on the surface of stem cells [216]. BCN-CNP-Cy5.5-labeled stem cells were tracked using all the three modalities for a while. After 24 h of post-injection of NPs, the increased darkening effect was seen with targeted particles compared to non-targeted particles. Highly sensitive tumor imaging applications require an increase in the concentration of SPIONs at the tumor site is important. SPIONs modified with specific moieties on the surface could increase the delivery of particles to localize at tumor site. For instance, SPIONs surface modified with antibodies, small molecule inhibitors, polymers, peptides, and folic acid approaches have been introduced for targeted enhanced MR imaging [217]. Li et al. developed PEG-folic acid-functionalized with polyethyleneimine-SPIONs as nanoprobe for xenografted KB tumor model [218]. T2 imaging with negative contrast with increasing concentrations of SPIONs was exposed with 0.5 T MRI. The negligible difference in the negative contrast between the targeted (107.3 mM s^{-1}) and non-targeted (99.64 mM s^{-1}) SPIONs. Natural physiological barriers in the human system are important to restrict access to new foreign bodies for the protection of organs. For instance, 2 important barriers difficult to cross are blood-brain barrier (BBB), and blood labyrinth barrier (BLB) are important for the protection of sensitive organs' brain and ear [219, 220]. Temozolomide with dual-targeted SPIONs developed with biocompatible triblock copolymers to cross BBB and to target tumor cells. External magnetic field steered nanoparticles to localize SPIONs in the tumor region enable them to increase uptake at the tumor region via the folate receptor. Thereby, conjugated Temozolomide delivery towards brain glioma in rats with controlled release from the triblock copolymer for treatment efficacy was acquired using MRI [221].

Another special type of target mechanism for glioma using low-density lipoprotein receptor-related protein (LRP) called Angiopep-2 (ANG). This special type of lipoprotein is the one another approach for targeting SPIONs to cross BBB (see Fig. 20). Nude mice bearing glioblastoma was developed by injecting U87MG cells into the right striatum. T_1 and T_2 images were recorded with 3 T MRI before and after the administration of (5 mg Fe kg^{-1}) SPIONs. Compared to non-targeted, T_2 weighted images and Gd-DTPA, enhanced dark contrast was obtained by ANG conjugated SPIONs in T_1 MR signals was clear to delineate the tumor boundaries. The increase in the ability of ANG-2 to cross BBB more efficiently results in higher contrast and with better targeting capability as a potential T_1 weighted contrast agent for glioma [222]. Other than the natural barriers, the excess levels of glutathione

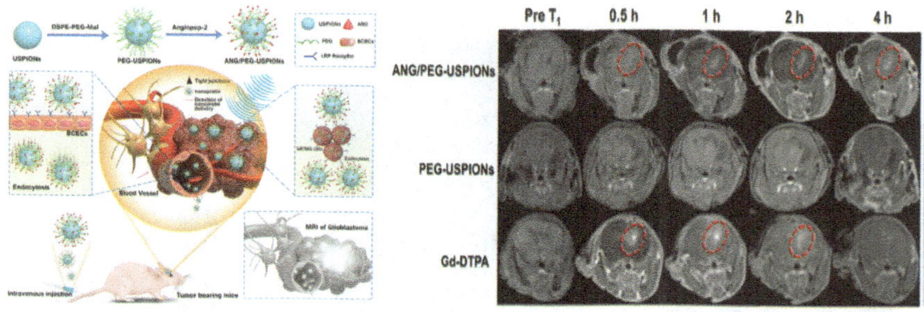

Fig. 20 Schematic illustration of the design and construction of ANG/PEG-USPION nanoprobes and the corresponding dual-targeting strategy for the positive MR imaging of glioblastoma cells compared to Gd chelates [222]

concentration in tumor microenvironment is another strategy to deliver nanomaterials. In our previous work, we have developed chitin hyaluronic acid-based nanocarriers to deliver anticancer model drug doxorubicin. Cystamine possesses disulphide with amine group at its end, which acts as oxidized glutathione, and enzymes break these disulphide bonds to maintain hemostasis in a cellular environment [223]. Similarly, individual SPIONs has high r_1 relaxivity (4.3 mM s^{-1}) whereas, while cystamine crosslinked particles form clustered SPIONs. In vivo MR imaging was acquired by exposing with cystamine crosslinked SPIONs showed high T_2 contrast within 5 min of injection. After 20 min, the signal to noise ratio of contrast reduced drastically by a factor of 3.5. But, GSH rich tumor environment has an inverse effect compared to crosslinked SPIONs. Whereas, after 5 min of injection the positive contrast is lower and increased gradually over time due to the effect of cystamine [224].

Among several receptors targeted imaging, CD44 are over-expressed in various cancer stem cells such as breast, colon, and gastric cancer. Hyaluronic acid is a biopolymer made of disaccharides composed of D-glucuronic acid and N-acetyl-glucosamine would readily endocytosis via CD44 receptor [226, 227]. Zheng et al. developed self-assembly polymeric micelles of hyaluronic acid loaded docetaxel and SPIONs for dual-targeted cancer theranostics [228]. The dual targeting (CD44 and magnet) and dual treatment (docetaxel and photothermal treatment) with these micelles for effective cancer theranostics. Further reports on biodistribution and biocompatibility of SPIONs-hyaluronic acid nanomaterials for in vivo targeted theranostics. Overexpression of vascular endothelial growth factors (VEGF) acting as a biomarker to obtain pathological information, molecular diagnosis, and responsible for angiogenesis in tumors growth. Anti-VEGF factors include Bevacizumab, Pegaptanib, and Sorafenib [229]. Bevacizumab is the first approved drug/anti-VEGF factor for colorectal cancer, glioma, lung, and respiratory cancer [230]. Lin and his co-workers developed bevacizumab-NIR probe functionalized with SPIONs as dual-modal image-guided nanomaterial [225]. In vitro studies revealed that the transverse relaxation of VEGF-SPIONs $1/T_2$ at a maximum value of 99.2 mM s^{-1} with varying concentrations of 0.03–0.50 mM Xenografted breast tumor model were injected and

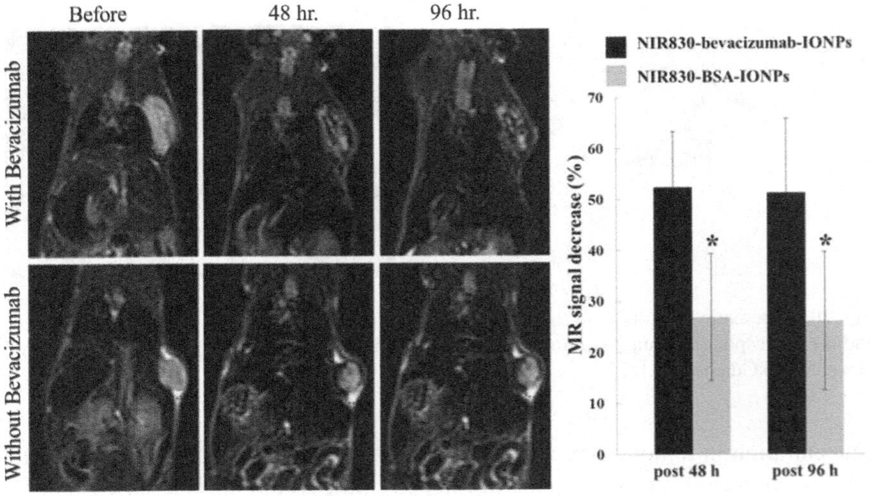

Fig. 21 Targeted SPIONs: MR imaging with and without bevacizumab targeted IONs pre and post-injection particles wherein targeted particles showed higher signal intensity post 96 h [225]

imaged using 3 T MRI with and without bevacizumab functionalized SPIONs. In vivo MR imaging was acquired at different time points at 48- and 96-hours post-injection and compared with control. With bevacizumab conjugated SPIONs have better negative contrast with increase in time compared to without VEGF factor as shown in Fig. 21. Monoclonal anti-body to mucin-1 (MUC-1) is one important target for pancreatic transplantation tumors than normal tissues. Zou et al. studied the interaction between SPIONs targeted to MUC-1 for imaging pancreatic cancer [231]. MUC-1 expressed cells (BxPC-3) pancreatic tumors were imaged after 4 h post-injection using 1.5 T MRI with and MUC-1 targeted SPIONs. A fast spin-echo with T_2 observed no adverse conditions and showed a higher negative contrast in MUC-1 SPIONs compared to other nanomaterials with no significant difference between SPIONs and BSA-SPIONs [231].

7.4 Manganese as MR Contrast Agent

Recently Manganese (Mn) based nanomaterials have gained a special focus with a lot of research explored for in vivo imaging apart from gadolinium as MR contrast agent [232]. Since, the chelates of gadolinium show nephrotoxicity, the best alternative element is Manganese as MR contrast synthesized in two forms (i) composite of manganese ions (Mn^{2+}) and (ii) Manganese dioxide (MnO_2/ Mn_3O_4 NPs) [233]. McDonagh et al. developed L-dihydroxyphenylalanine (L-DOPA) coated with manganese oxide-based nanoparticles (MONPs). L-DOPA can cross the BBB and degrade gradually in the biological medium and releases Mn ions. MR signals from

MONPs showed enhanced positive contrast over time compared to the surrounding region [234]. The anti-stealth properties of MONPs were prepared via a solvothermal process using citrate as a reducing agent and coated with PEG and functionalized with cysteine by wang and his coworkers [235]. Phantom studies of T_1 weighted MR signals were determined for MONPs with and without cysteine as a capping agent. Wherein, the concentration of MONPs increases the contrast of the MR signals also increases linearly and similar to Gd-DTPA. After 90 min of post-injection of MONPs, the MR signals in C6 glioma was higher with cysteine capped MONPs. The distribution of nanomaterials within the rodents within tumor and organs were determined using ICP-OES (inductively coupled plasma—optical emission spectrometry). They reported that particles were uptake of MONPs with large amounts in liver and spleen and cleared out within 48 h post-injection. As a promising alternative of gadolinium-DTPA is chelates of Manganese. Due to clinical problems associated with Mn cations such as cellular toxicity at higher concentration requires efficient nanocarriers for delivery. For instance, chelates play a major role in delivering the Mn cations at the desired location. Venter et al., developed from a single manganese porphyrin ($MnPNH_2$) molecule for enhanced cell tracking applications. These $MnPNH_2$ does not show any changes in cellular structures with no adverse effects on subcellular organelles [236].

Nanogels are another type of particle that can be used to load drugs, chelates, and other bioimaging probes via physical adsorption [237]. Nanogels have a swelling nature depending on the pH or other biological environment like GSH: GSSG (reduced or oxidized glutathione at intracellular level) [223]. Recently, nanogels loaded with chelates of Mn cations vary MR signals when exposed with different pH environments. Swelling with the nanogels is based on the number of protons that adhere shorter relaxation times [238]. At higher pH, the relaxivity increases while at lower pH, the relaxivity decreases. The stability of the nanogels is lesser compared to the other types of nanomaterials. To overcome such issues, Addisu et al. synthesized alginate-dopamine nanogels by crosslinking them with Ca chelates and improves the stability for loading manganese chelates. Calcium cations have carboxyl groups readily binds with alginate and form this complex mechanism, thereby increased the stability of nanogels and manganese cations [239]. Positive contrast enhancement with $r_1 = 12.5$ mM s^{-1} when exposed with 7.0 T MR stable for a prolonged period in the liver and tumor region due to EPR effect. A variety of techniques were involved in improving the delivery of these manganese cations in tumor imaging. One such study performed by wang and his coworkers synthesized complex carrier of hyaluronic acid/dendrimer loaded with Mn ion chelates with gold nanomaterials for hepatocellular carcinoma. Dual modal imaging was stable, biocompatible, and displayed high X-ray attenuation with better relaxivity property. CT and MR signals were acquired after 30 min of post-injection with hyaluronic acid showed higher intensity compared to non-targeted dendrimers. The distribution of targeted nanomaterial was 5 times higher in the liver and spleen compared to non-targeted particles [240].

7.5 Targeted MNOPs

Targeting molecules attached on the surface of MONPs had an improved MR signal from the tumor region. Li et al. synthesized PEG capped MONPs as a nano-based contrast to understand molecular aspects of renal carcinoma using MR imaging [241]. MONPs exposed with 3 T MRI, PEG-MONPs showed (12.94 mM s^{-1}) three times higher relaxivity than gadolinium chelates. Aptamer (AS1411), a potential target that recognizes nucleolin and the ability to internalize quickly with renal carcinoma. MR images were acquired for 7 days determines the higher contrast in tumor, liver, and kidneys at 45 min of post-injection. PEG-MONPs were cleared in 24 h compared to AS1411-MONPs, wherein clearance within 7 days via renal filtration. Nano theranostics agents were developed using MONPs functionalized with polydopamine (melanin pigment as photothermal agent) linked with folic acid to localize particles into the cancer regime. Ding and his coworkers triggered MONPs with NIR to deliver model drug doxorubicin inside cancer cells. Figure 22 shows that the increase in the concentration of Mn, MR signals at 0.1 mM has higher sensitivity compared to magnevist (Gadolinium chelates) [242].

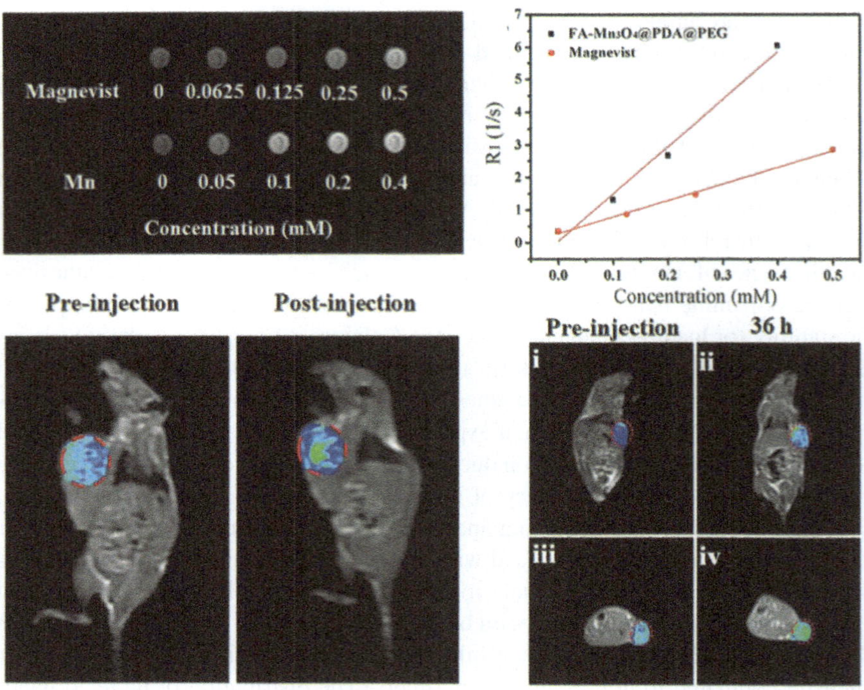

Fig. 22 Folic acid-functionalized polydopamine (PDA) as PTT molecule conjugated with MnO NPs compared with Mn chelates particles for 36 h [242]

Tremel and his colleagues showed the interaction of MONPs with Superoxide Dismutase possesses enhanced MR signals when radicals of oxygen converted into hydrogen peroxide [243]. The higher relaxivity in both T_1 and T_2 times enhanced the contrast with simultaneous treatment modulates oxidative stress that helps in the progression of tumor growth. A combination of anatomical models and molecular imaging helps to understand and characterize for effective therapy and diagnosis. Multimodal imaging with targeted MONPs was developed by Zhan et al. studied the tumor vasculature using MRI and PET. They incorporated Cu-64 PET tracer linked MONPs functionalized with a monoclonal antibody (TRC105). In vivo distribution studies were performed using PET/MRI carried out in xenografted glioma tumor model. A 20 mg/kg of nanomaterials were injected through intravenous injection and observed positive contrast and consistent with PET imaging [244]. Composites of manganese-based MR contrast agents were developed as multimodal imaging and therapeutic effects. For instance, Manganese ferrite [245–247], MnO combined with Au and SiO2 [248], manganese silica [249], manganese carbonate [250].

8 Nanomaterials for Ultrasound Imaging

Ultrasound (US) imaging is another versatile non-invasive imaging system that uses high frequency (2–200 MHz) sound waves to differentiate tissues. Ultrasound uses non-ionizing radiations when compared to other X-ray CT, and they are ubiquitous works on principles of refraction by tissue components having a different acoustic impedance. US is commonly used to understand the development of fetuses, breast, neck, and blood flow studies [251]. Recent trends in US have improvised the imaging quality changed from B mode to 3D and later into 4D (concerning time). Artifacts such as tissue harmonic aberration, reduction of aberration, ghost effects in tissues. But the significant concerns associated with US are quality of image compared to CT or MRI, and absorption by bone limits the depth of penetration [3]. Contrast-enhanced ultrasonic agents such as air bubbles or gas loaded particles are required to overcome the above limitations. The first gas-filled contrast agents were developed Unger et al. with a size range between 7 and 8 microns made of di-palmitoyl phosphatidyl choline (DPPC) [252]. The problems associated with these microbubbles are large size, cell penetration is difficult, traps at alveolar ducts, and restricted to cardiac imaging [253]. Nanosized particles loaded with gas molecules are called nanobubbles with a size between 100 and 1000 nm. Further developments in the nano-sized contrast agents such as liposomes, nanobubbles, nanodroplets less than 1-micron particles [254]. Several types of gases used in US contrast agents such as perfluorocarbons or sulphur hexafluoride (SF_6), oxygen, and hydrogen [34]. These nanobubbles are being more efficient reflectors when ultrasound is excited over the particles. These nanobubbles are effectively used as in vivo imaging and accumulate in the tissues via the EPR effect.

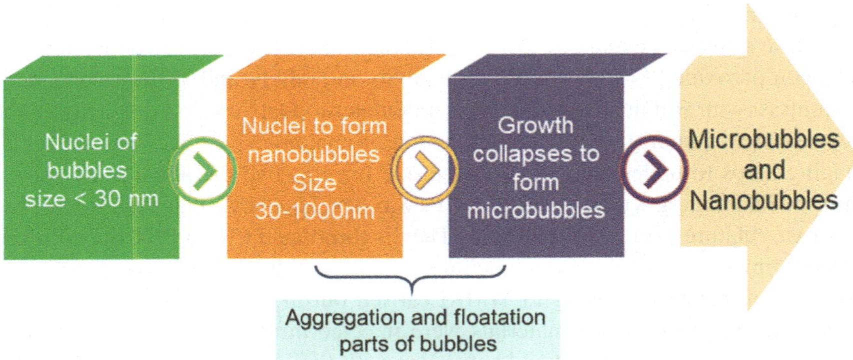

Fig. 23 Schematic representation of the growth of bubbles from the nuclei bubbles to nanobubble to microbubbles. These microbubbles were exposed to compression and rarefaction obtained from implosion

Interestingly, nanobubbles exploited with various synthesis procedures and functionalized with targeting agents, drugs, and nanoparticles. The growth of the nanobubbles is similar to nanomaterials wherein firstly, the formation of nuclei bubbles (size < 30 nm) to nanobubbles (size < 1000 nm) and finally coalesce to microbubbles as shown in Fig. 23 [254]. Stabilizing these nanobubbles in the sizes of less than 1000 nm helps to permeate into smaller blood vessels. Protein, polymer, and lipids are used as coating shells loaded with gases [255]. Further exposing these nanobubbles with ultrasound can be used for imaging applications. Several different synthesis procedures have been reported with sonication [255], emulsification, laser ablation [256], ink-jet, and microfluidics [257] approaches for micro/nanobubbles [258].

Over the decade limited number of research studies were performed using ultrasound based contrast agents for preclinical imaging applications. In 1962, sette and his coworkers developed artificial bubble using high energy neutrons created microcavity and are stable for a period of 5 h. Later, Johnson and Cooke, in 1981 generated nanobubble by shearing saline water with less than 1-micron size. These nanobubble are stable over a period of 22 h and stabilized using neutral dilute electrolyte solutions [259, 260]. Kikuchi and his coworkers have developed several contrast agents for US by using different gases like oxygen and hydrogen [261, 262]. Gases like nitrogen, methane, and argon are used, but the problem with these types of nanobubbles exhibits stability less than 2 weeks [263]. Let us look at some of the most recent works on nanobubbles as US-based contrast agents used for in vivo imaging applications. Considering the nanobubbles, surface modification, and functionalization with targeting agents is an essential parameter for imaging. Nanobubbles are prepared through a centrifugation process to separate micro and nano-sized particles. Cai et al. in 2015, developed monodispersed NBs via a thin-film hydration method by controlling the thickness of phospholipids [264]. Synthesized of NBs of size 562 and 457 nm compared with Sono Vue (commercially available US contrast agent) with lesser zeta potential. Both nanoparticles were injected through a caudal vein on xenografted breast tumor on nude mice and imaged at 10, 30, 120, and 300 s. These NBs injected at a concentration of 150 μL showed higher contrast in liver and tumor in a time interval of 30 s possess longer imaging time compared to Sono Vue. Apart from the size of NBs, coating over the surface of particles plays a major role in US imaging. Initial reports with nanobubbles have been coated with either anionic or cationic molecules. Biological interaction with these coated NBs toxicity and cellular uptake was evaluated by Pan et al. in 2012 [265]. The toxicity of NBs was evaluated by the assessment of biomarkers from liver and kidneys such as γ-glutamyl transferase (γ -GT) and blood urea nitrogen (BUN). A significant toxicity effect in the liver compared to kidney cells, and they showed that toxicity evaluation with biomarkers is more sensitive.

In general, contrast-enhanced ultrasound imaging requires the nanobubbles with size less than 500 nm for increased EPR effects in the tumor area. But the serious concerns with intravenous injection in rodents the existence of nanobubble possess half-life less than 30 min. Mai et al. developed peptide conjugated NBs -Cyanine 5.5 NIR dye as a dual-modal imaging agent for effective uptake of nanoparticles in tumor regime [266]. Liver cancer cells H22 were injected intraperitoneally as a tumor model for dual imaging. US Imaging was performed in both 5 MHz and 12 MHz probe, but the higher frequency probe has much contrast enhancement in tissues. NBs were injected intravenously at a concentration of 4 μL and acquired using 12 MHz thyroid US probe recorded within 4 h. Ultrasound with nanobubbles delineated a white boundary of the dark tumor region on either side of tumor and confirmed with fluorescence imaging. Results showed that nanobubbles could be used for subcutaneous tumor imaging with US for 2 h and gradually decreases. The next generation of NBs for tumor imaging and therapy was developed Huang et al. in 2013, incorporating SPIONs into nanobubbles [267]. Nanobubbles with SPIONs, dual-modal imaging, acquire images in both ultrasound and MRI. Magnetically triggered SPIONs-NBs loaded with an anti-cancer drug are exposed to high intensity focused ultrasound (HIFU) for the effective release of drugs. Contrast-enhanced ultrasound is used to observe the glioma injected in the right hind limb. Magnetically guided nanomaterials have been a keen interest due to its increased level of accumulation in the desired target. The same group has strategically disrupted BBB via HIFU to increase the delivery of nanomaterials in a tumor region and imaged with MRI [268]. Mixed ratio of phospholipids functionalized with biotin containing octa-fluoro propane C_3F_8 gas to form NBs facilitated with anti-prostrate specific membrane antigen [269]. Three animal xenograft models (LNCaP, C4-2, and MKN45) were imaged using various US parameters such as peak intensity and contrast enhancement over time. Compared to non-targeted NBs and tissues imaging without NBs, the contrast with targeted NBs, the US parameters are higher than others.

Li et al. have demonstrated that biodegradable photoluminescent polymers can encapsulate liquid tetra-decafluorohexane (C_6F_{14}) through emulsion evaporation process conjugated with neuropeptide YY_1. These neuropeptides YY1 conjugate with Y_1R_s overexpressed in breast cancer cell lines without affecting normal cells. Xenografted 4T1 cells were injected with BPC-NB-PNBL-NPY (50, 100, or 200 kg mg^{-1}) and were used for in vivo US imaging. Intravenous injected NBs

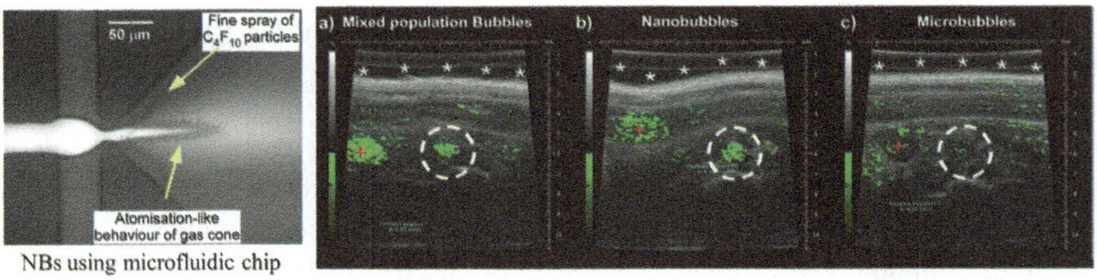

Fig. 24 Schematic representation PNBL conjugated NBs synthesis and its application in vivo ultrasound imaging. Nanobubbles (**a**) and Microfluidic based on-chip development of nanobubbles and compared with microbubbles and mixed population (**b**). Adapted from reference [270]

were imaged with 5 MHz US probe periodically at 10, 20, 30, 120, 180, and 300 s provided good contrast enhancement. Microfluidic chips are MEMS-based devices with micron channels flows with gas and surfactant to form nanobubbles. Micro spray flow in one channel and lipids are injected through in another channel at the pressure set at 15 psi and aqueous flow rate of 90 μL min^{-1} by Peyman and his coworkers [270]. Aorta of the mouse was imaged between 28 and 40 MHz HIFU for pre-clinical imaging applications. The contrast enhancement of NBs shows better spatial resolution compared to microbubbles in preclinical imaging, as shown in Fig. 24. Nanodroplets were generated by rapid vaporization and condensation of perfluorocarbon as core loaded with optical sensitive NIR dye with lipid shells. These nanodroplets injected via retro-orbital were demonstrated through US by imaging the brain undergone craniotomy. Upon exposure of laser, more than 60 min observed blinking effect from nanodroplets. The transient state of the microbubble to nanobubbles obtain high contrast with reflected waves via the super-resolution technique [271].

9 Nanomaterials for Optical Imaging

In recent years, optical imaging has become a powerful technique for in vivo and in vitro biomedical imaging [272, 273]. Different methodologies based on varying light-tissue interactions have been presented for a wide range of applications in the past decade. The benefits associated with optical imaging include increased sensitivity, absence of exposure to ionizing radiation, cost-effectiveness, improved spatial resolution, real-time acquisition [274, 275]. Notwithstanding, the fundamental impediment associated with optical imaging include light scattering, (auto-) fluorescence and absorption (in the range 400 nm–650 nm) from the nearby tissues. To overcome the limitations mentioned above focus of near-infrared (NIR), fluorescence imaging has been done in two NIR windows from 650 to 950 nm (NIR I) and 1000 to 1350 nm (NIR II) [276].

Nanoparticles (NP's) are the new class of contrast agents in size ranging from 1 through 100 nm and has shown promising approach for non-invasive biomedical diagnostics [4, 277]. The following are some of the important characteristics of NP's required for biomedical imaging should be dissolvable and stable in diverse native environments, good signal-to-background-noise ratio, no toxicity, exhibit strong non-specific binding and uptake, superior blood half-life and should avoid clearance by the reticuloendothelial system [277–280]. Some of the most widely exploited NPs in optical imaging include quantum dots, upconverting NPs, persistent luminescent NPs, metal, and carbon nanoparticles [281, 282]. Here we will be presenting a brief description of quantum dots, upconverting NPs, and persistent luminescent NPs in detail.

9.1 Quantum Dots

Quantum dots (QDs) are a group of semiconductor nanoparticles sized in the range between 6 mm and 20 nm with a very high quantum yield and possess excellent photo- and chemical-stability characteristics in comparison with organic fluorescent dyes [283–285]. QDs are usually composed of elements from groups II-VI elements (e.g., CdS, CdSe, ZnS, MgSe, etc.), III-VI elements (e.g., GaAs, InP, InGaAs, etc.), IV-VI elements (e.g., PbSe, PbS, PbTe) and ternary QDs are I–III-VI (I = Cu, Ag; III = In, Sn, Ga, Al and VI = S, Se, Te, etc.) [286, 287]. In this context, Han et al. [288] implemented a lattice-mismatch strain tuning theory to characterize highly luminescent NIR CdTe/CdS QDs. The small shells were efficiently compressed by lattice strain attributed to growing thicker shell and transition of a transition of bandgap offset from type-I to type-II nanocrystals that happened during the well-controlled gradual growth of the shell, allowing largest spectral shifts tuning from the visible to the NIR spectral region. This study demonstrated that QDs developed using this approach were highly fluorescent, extraordinarily small, and better tumor-targeting ability. A select type of QDs are graphene quantum dots (GQDs) were developed by Ashwin et al. by changing oxygen contents that enhance the fluorescence in red to NIR region. GQDs loaded into muscle gel and excited with green LED via transillumination method obtained red emission in vivo. Our work enables us to develop nanomaterials developed using pulsed laser ablation for in vivo imaging [289].

ZnS shell grown over $CdTe_xSe_{1-x}$ NIR QDs alloy nanocrystals have been developed to provide a quantum yield of around 80% in chloroform [290]. Also, the aspect ratio and emission range of the quantum dots can be varied by changing the Te and Se mole ratio. They found that the composition of Se played a vital role in controlling the rod shape whole Te plays an essential role in the fluorescence emission. In recent times, fluorescence imaging in the NIR II region has been gaining momentum. In this wavelength region, tissue absorption by endogenous chromophores such as oxyhemoglobin, deoxyhemoglobin, and melanin are limited [291]. Additionally, in this region, there is an increased signal-to-background ratio due to reduced photon scattering, improved penetration depth, and reduced interference from fluorescence [292, 293]. Ag_2S QDs conjugated with Tat peptide, which is used as a targeting ligand have been implemented for the tracking of human mesenchymal stem cells (hMSCs) in an in-vivo mouse model in the second NIR window region [294]. The results also indicate an increased chemical- and photostability, making Ag_2S QDs a potential candidate for long-term cell tracking.

To have clear insights into the physiological and pathological processes connected with the circulatory system Ag_2S QDs have been implemented in dynamic imaging of the circulatory system. NIR II imaging with Ag_2S QDs significantly enhanced the clarity and the penetration depth for envisaging lymph nodes and lymphatic vessels [295]. Subcutaneous injections of PEGylated Ag_2S QDs exhibited a narrowing of the lymphatic vessel cross-sectional intensity profiles due to very low scattering and decreased absorbance in the NIR III region. A comparison with ICG (used in *clinical* practice for the detection and *resection of sentinel lymph nodes)* demonstrated that ICG was unable to resolve the deeper lymph nodes. The studies presented here provided agreeable results contrastingly there are some limitations associated with QDs in imaging applications. First, heavy metals are incorporated in the core and the materials used for capping, which could result in toxicity [296]. Second, the luminescence emitted from quantum dots is demonstrated by variations in intensity, leading to the appearance of ON and OFF states [297]. This downtime between the states could hinder the establishment between two successive frames in single-molecule tracking applications.

Upconverting NPs (UCNP) is a new generation of imaging compounds doped with lanthanide ions capable of absorbing light in the NIR region (usually 980 nm) and emits within the visible region [298, 299]. The UCNP usually consists of number lattices of a ceramic material such as LaF_3, YF_3, Y_2O_3, $NaYF_4$ embedded with a trivalent transition metal, actinide, or lanthanide ions such as Yb^{3+}, Er^{3+} and Tm^{3+} [300–302]. The properties associated with UCNP include non-photo blinking emission, biocompatibility, long anti-stokes shifts (up to 500 nm), excellent photochemical- and thermal- stability [303–306]. For long-term non-invasive in vivo optical imaging PEG-lipid functionalized and biocompatible $NaYF_4$: Tm^{3+}, Yb^{3+} UCNP was implemented in real-time to track excretion pathways, retention, and clearance [307]. This was evaluated using a high-quality NIR-to-NIR upconversion luminescence (UCL) imaging. The study revealed that UCNPs remained in the injected site for 14 days and were significantly dropped after 30 days. The evaluation of lymphatic tissue status is an imperative aspect to be considered in the diagnosis of people with tumors. A new kind of imaging probes based on PEGylated $NaGdF_4$: Yb, Er@$NaGdF_4$ UCNPs was investigated for the detection of lymphatic metastasis of gastric cancer [308]. Additionally, due to a very low background at 980 nm laser excitation, lymphatic metastasis lesser than 1 mm were successfully identified. Recently, Tian et al. [309] evaluated novel carboxy-terminated silica-coated UCNPs with different surface functional groups –COOH, -PEG, D-SP5, and -UEA-I. Their study revealed that the functional group—UEA-I presented a bright UCL at the tumor sites and has an efficient tumor targeting capacity in the mouse model. The limitations related to UCNPs are lower upconversion quantum yield, and conventional fluorescence probes, including QDs, have a very short lifetime [298].

Persistent luminescent nanoparticles (PLNPs) could serve as an alternate candidate to overcome the shortcomings associated with QDs and UCNPs. Persistent luminescence is an optical phenomenon in some materials which continue to emit light even after ceasing excitation [310]. In persistent luminescence materials, the energy is stored in the intrinsic traps or defects, which are intentionally introduced upon excitation [311]. To release the stored energy, the thermal de-excitation process is required after which the delayed emission of light occurs over minutes or hours after the exciting radiation has been removed [312]. Chermont et al. reported a methodology to synthesize $Ca_{0.2}Zn_{0.9}Mg_{0.9}Si_2O_6$ nanoparticles doped using NIR persistent luminescence. The NPs used in this investigation were excited at UV or visible light, and over a while, they release the stored energy by the emission of photons with lower energy. The NPs were excited before the injection to mouse, and the emission can be monitored in real-time for more than one hour without a need for an excitation source. The NPs with an emission wavelength around 690 nm were able to image 3LL tumors in an in a vivo mouse model.

Abdukayum and co-workers [313] developed a fabrication strategy utilizing the citrate sol-gel method of functional PLNPs through the minimal composition of $Zn_{2.94}Ga_{1.96}Ge_2O_{10}$: Cr^{3+}, Pr^{3+}. The co-doping of Pr^{3+}/Cr^{3+} and regulating the Zn deficiency in the host material enhanced persistent luminescence intensity and afterglow time of PLNPs. Experiments on animal models presented an increased signal-to-noise ratio and provided in vivo bio-imaging for more than 15 h without an excitation source. NPs used in persistent luminescence imaging are usually based on rare earth metals Europium, Praseodymium is poorly biocompatible due to its associated toxicity [314]. Alternatively, semiconductor-based conjugated polymer poly[2-methoxy-5-(2-ethylhexyloxy)-1,4-phenylenevinylene] (MEH-PPV) was used as PLNP with the NIR dye NIR775 encapsulation for enabling emission in the NIR range. This investigation demonstrated that this semiconductor-based PLNP emitted persistent luminescence for nearly one hour after a single exposure to white light and presented an alternative contender for in vivo imaging applications.

Photoacoustic imaging (PAI), also referred to as optoacoustic imaging (OAI) has evolved as an in vivo and non-ionizing modality with the relative ability for deep tissue imaging [315]. Here, a nanosecond laser with a pulse duration of < 10 ns is used to illuminate a biological sample of interest [315]. The light absorbed by the object results in a rise in temperature followed by thermal expansion and production of ultrasonic waves [316]. These waves are detected by an ultrasonic transducer to form images. Certain endogenous contrast agents used in PAI include hemoglobin, melanin, lipid, and water in the visible and NIR regions [317].

In some cases, endogenous contrast agents are inadequate to provide useful information for clinical diagnostics. Since the strong light scattering in the shorter wavelength region, the light intensity and henceforth the photon acoustic amplitude reduces exponentially concerning tissue depth [318]. To overcome these limitations, different types of NP based contrast agents such as plasmonic NPs (e.g., gold and silver), carbon nanotubes, UCNP, semiconducting polymer NP, and quantum dots have been explored [319, 320].

UCNP provides narrow excitation and emission profiles for PAI. Lanthanide doped ions in UCNP could lead to quenching of luminescence due to the solvent relaxation process in aqueous conditions. These luminescence quenching phenomena could be exploited to generate thermal energy and used as a contrast agent for PAI. In this regard, Maji et al. [321] developed NaYF4 (doped with Yb^{3+} or Er^{3+}) UPCN stabilized with oleic acid with α-cyclodextrin inclusion complexes to improve the quenching process and photoacoustic signal. After 35 min, the images (f–j) presented a localized enhancement of the contrast on the kidney was perceived. This enhancement, in contrast, was attributed to the luminescence quenching due to the effects of solvent-induced non-radiative relaxation and thermal conductivity of UNCP.

Recently, a theranostic based on image-guided phototherapy, which integrates diagnostics and therapy, could be a prospective modality for tumor treatment. In this regard, MoO_{3-x} QDs have been developed to perform photothermal-/photodynamic-therapy for treating tumors using guided photoacoustic imaging [322]. The QDs used in this study have maximum absorption in the NIR region and can convert the absorbed light into hyperthermia. This results in the formation of cytotoxic reactive oxygen species surrounding the tumor cells resulting in phototherapy. Additionally, this QD can be used for exogenous photoacoustic contrast agents for imaging of the tumors, which are confirmed from in vivo studies on a mouse model.

9.2 Positron Emission Tomography (PET) and Single Photon Emission Computed Tomography (SPECT)

Positron emission tomography (PET) and Single-photon emission computed tomography SPECT) are nuclear imaging modalities that can provide functional information by measuring the uptake and turnover quantity of radiotracers in a specific tissue site. PET is based on specific decay properties of radioactive nuclides that decay by positron emission [323]. This technique involves the injection of radioactive tracers into the subject, usually administered using an intravenous injection. A tracer is usually a biological molecule of interest tagged with a radioactive isotope. 2-18F-fluoro-deoxy-D-glucose (^{18}F-FDG) is widely used to tracer evaluate therapeutic response in oncology [324]. Some of the most commonly used radioactive isotopes used in PET to label tracers include ^{15}O, ^{13}N, ^{11}C, and ^{18}F [325]. The tracer molecules interact with specific proteins sugars or DNA elements. When the radioactive atom on a molecule decays, a positron is ejected from the nucleus by interacting with the surrounding electrons, which results in the complete annihilation of both the particles, positron, and the electron. This interaction results in generation of two high energy photons travelling in the opposite direction. The detectors usually scintillator crystals coupled to photomultipliers measure these photons to create an image in the PET scanner.

Similar to PET, SPECT is another imaging modality that uses radioactive imaging probes labeled with a single photon emitter. In the case of a SPECT, the imaging device is a gamma camera which captures the radiation emitted from the body. The most widely used radioisotopes in SPECT include technetium 99 m (^{99m}Tc), iodine 123 (^{123}I), and thallium 201 (^{201}Tl). However, ^{18}F-FDG is the most widely accepted radionuclide for PET imaging, their short half-life of 109 min, and impedes the study of biochemical process in the range between days and weeks. NPs can be labeled with positrons and gamma-emitting radionuclides. In this context, ^{64}Cu labeled Si QDs were developed and used for PET imaging in vivo and ex vivo gamma counting [326]. DOTA (1,4,7,10-tetraazacyclododecane-1,4,7,10-tetraacetic acid) is used for the complexation of ^{64}Cu ions steric hindrance leading to unstable radiolabeling.

A new approach using new bifunctional DO3A (1,4,7,10-tetraazacyclododecane-1,4,7-triacetic acid) with a more flexible functional arm for chelation of ^{64}Cu was implemented. This improved sterics resulted in a very much stable complex with less loss of radiolabeling upon chelation to silicon QDs. In vivo studies demonstrated a faster clearance of Si QDs from the body through renal filtration and urinary bladder. At the same time, some QDs with a larger hydrodynamic diameter was taken up through the reticuloendothelial system and accumulated in the liver. Mesoporous silica NPs (MSNP) provide the advantage of offering less toxicity and ease of uptake and transport by cells. A new approach based on MSNP was implemented for tracking neural stem cells in the brain using SPECT [327]. To synthesize the nanoparticle sol-gel co-condensation method was employed and conjugated with DOTA-N-hydroxysuccinimide-ester for radiolabeling. The results indicated that MSNP could be tracked as they move towards the intracranial glioma using SPECT imaging in real-time after intracranial and systemic administration.

10 Conclusions

Overall, this chapter explains the new developments in nanoprobes used as preclinical imaging modalities like CT, MRI, ultrasound, and optical systems. Advancements in imaging parameters are made by the current state-of-art nanomaterials in terms of sensitivity, biodistribution, target, and multimodal imaging ability. Nanotechnology is transforming the imaging capabilities and increasing its growth and high performance in both clinical and preclinical imaging studies enabling them to fasten the drug delivery process. Nanotechnology-based imaging probes provide a high surface to volume ratio, low cost, ease of functionalization, and less toxicity compared to other probes. The combination of drug and imaging agent with targeting ability and its advancements in multimodal imaging capability are very much encouraging and presented with good examples.

A particular emphasis on nanomaterials, focusing on the problems faced by imaging agents in preclinical imaging systems have been discussed. Literature has clearly suggested that nanomaterials with desired size and shape are important to advance pre-clinical imaging systems and to improvise the imaging parameters. Furthermore, nanoprobes have proved their usefulness by binding with new targets and imaging sensitive regions compared to conventional agents. Factors that influence commercialization of nanoparticles are their in vivo toxicity. Also, the greatest challenge of upcoming nanomaterials on retention and reactivity with proteins in the diseased region is still questionable. It is important for the nanoscientists to work on the pitfalls to provide a better design that helps both research and mankind. At last, the advances in nanoprobes would lead to the development of specialized imaging systems that helps researchers and physicians for better diagnosis and treatment for several diseases.

References

1. Cunha L, Horvath I, Ferreira S, Lemos J, Costa P, Vieira D, Veres DS, Szigeti K, Summavielle T, Mathe D, Metello LF (2014) Preclinical imaging: an essentially in modern biosciences. Mol Diagn Ther 18(2):153–173. https://doi.org/10.1007/s40291-013-0062-3
2. Malikova MA (2016) Optimization of protocol design: a path to efficient, lower cost clinical trial execution. Future Sci OA 2(1):FSO89. https://doi.org/10.4155/fso.15.89
3. Bushberg JT, Seibert JA, Leidholdt EM, Boone JM (2011) The essential physics of medical imaging. Wolters Kluwer Health
4. Nune SK, Gunda P, Thallapally PK, Lin YY, Forrest ML, Berkland CJ (2009) Nanoparticles for biomedical imaging. Expert Opin Drug Deliv 6(11):1175–1194. https://doi.org/10.1517/17425240903229031
5. McNamara K, Tofail SAM (2016) Nanoparticles in biomedical applications. Adv Phys: X 2(1):54–88. https://doi.org/10.1080/23746149.2016.1254570
6. Smith BR, Gambhir SS (2017) Nanomaterials for in vivo imaging. Chem Rev 117(3):901–986. https://doi.org/10.1021/acs.chemrev.6b00073
7. Yao R, Lecomte R, Crawford ES (2012) Small-animal PET: what is it, and why do we need it? J Nucl Med Technol 40(3):157–165. https://doi.org/10.2967/jnmt.111.098632
8. Yang DJ, Kim EE, Inoue T (2006) Targeted molecular imaging in oncology. Ann Nucl Med 20(1):1–11. https://doi.org/10.1007/bf02985584
9. Exner DV, Kavanagh KM, Slawnych MP, Mitchell LB, Ramadan D, Aggarwal SG, Noullett C, Van Schaik A, Mitchell RT, Shibata MA, Gulamhussein S, McMeekin J, Tymchak W, Schnell G, Gillis AM, Sheldon RS, Fick GH, Duff HJ, Investigators R (2007) Noninvasive risk assessment early after a myocardial infarction the REFINE study. J Am Coll Cardiol 50(24):2275–2284. https://doi.org/10.1016/j.jacc.2007.08.042
10. Seaman ME, Contino G, Bardeesy N, Kelly KA (2010) Molecular imaging agents: impact on diagnosis and therapeutics in oncology. Expert Rev Mol Med 12:e20. https://doi.org/10.1017/S1462399410001511
11. Pysz MA, Gambhir SS, Willmann JK (2010) Molecular imaging: current status and emerging strategies. Clin Radiol 65(7):500–516. https://doi.org/10.1016/j.crad.2010.03.011
12. Kim D, Kim J, Park YI, Lee N, Hyeon T (2018) Recent development of inorganic nanoparticles for biomedical imaging. ACS Cent Sci 4(3):324–336. https://doi.org/10.1021/acscentsci.7b00574
13. Nagai M, Kato K, Soga S, Santra TS, Shibata T (2020) Scalable parallel manipulation of single cells using micronozzle array integrated with bidirectional electrokinetic pumps. Micromachines (Basel) 11(4). https://doi.org/10.3390/mi11040442
14. Peixoto L, Magalhães R, Navas D, Moraes S, Redondo C, Morales R, Araújo JP, Sousa CT (2020) Magnetic nanostructures for emerging biomedical applications. Appl Phys Rev 7(1):011310. https://doi.org/10.1063/1.5121702
15. Arms L, Smith DW, Flynn J, Palmer W, Martin A, Woldu A, Hua S (2018) Advantages and limitations of current techniques for analyzing the biodistribution of nanoparticles. Front Pharm 9:802. https://doi.org/10.3389/fphar.2018.00802
16. Dong YC, Hajfathalian M, Maidment PSN, Hsu JC, Naha PC, Si-Mohamed S, Breuilly M, Kim J, Chhour P, Douek P, Litt HI, Cormode DP (2019) Effect of gold nanoparticle size on their properties as contrast agents for computed tomography. Sci Rep 9(1):14912. https://doi.org/10.1038/s41598-019-50332-8

17. Chen L, Xie J, Wu H, Li J, Wang Z, Song L, Zang F, Ma M, Gu N, Zhang Y (2018) Precise study on size-dependent properties of magnetic iron oxide nanoparticles for in vivo magnetic resonance imaging. J Nanomater 2018:1–9. https://doi.org/10.1155/2018/3743164
18. Marashdeh MW, Ababneh B, Lemine OM, Alsadig A, Omri K, El Mir L, Sulieman A, Mattar E (2019) The significant effect of size and concentrations of iron oxide nanoparticles on magnetic resonance imaging contrast enhancement. Results Phys 15:102651. https://doi.org/10.1016/j.rinp.2019.102651
19. Tamarov K, Nakki S, Xu W, Lehto VP (2018) Approaches to improve the biocompatibility and systemic circulation of inorganic porous nanoparticles. J Mater Chem B 6(22):3632–3649. https://doi.org/10.1039/c8tb00462e
20. Mout R, Moyano DF, Rana S, Rotello VM (2012) Surface functionalization of nanoparticles for nanomedicine. Chem Soc Rev 41(7):2539–2544. https://doi.org/10.1039/c2cs15294k
21. Pertici G (2017) Introduction to bioresorbable polymers for biomedical applications 3–29. https://doi.org/10.1016/b978-0-08-100262-9.00001-x
22. Bohara RA, Thorat ND, Pawar SH (2016) Role of functionalization: strategies to explore potential nano-bio applications of magnetic nanoparticles. RSC Adv 6(50):43989–44012. https://doi.org/10.1039/c6ra02129h
23. Burrows ND, Vartanian AM, Abadeer NS, Grzincic EM, Jacob LM, Lin W, Li J, Dennison JM, Hinman JG, Murphy CJ (2016) Anisotropic nanoparticles and anisotropic surface chemistry. J Phys Chem Lett 7(4):632–641. https://doi.org/10.1021/acs.jpclett.5b02205
24. Shinde P, Kumar A, Kavitha, Dey K, Mohan L, Kar S, Barik TK, Sharifi-Rad J, Nagai M, Santra TS (2020) Physical approaches for drug delivery, pp 161–190. https://doi.org/10.1016/b978-0-12-817776-1.00007-9
25. Santra TS, Chen C-W, Chang H-Y, Tseng F-G (2016) Dielectric passivation layer as a substratum on localized single-cell electroporation. RSC Adv 6(13):10979–10986. https://doi.org/10.1039/c5ra18258a
26. Santra TS, Tseng F-G (2016) Electroporation for single-cell analysis, pp 55–83. https://doi.org/10.1007/978-3-662-49118-8_3
27. Santra TS, Kar S, Borana J, Wang P-C, Tseng F-G (2014) Nanolocalized single-cell-membrane nanoelectroporation: for higher efficiency with high cell viability. IEEE Nanatechnol Mag 8(1):30–34. https://doi.org/10.1109/mnano.2014.2312031
28. Mohan L, Kar S, Nandhini B, Dhilip Kumar SS, Nagai M, Santra TS (2020) Formation of nanostructures on magnesium alloy by anodization for potential biomedical applications. Mater Today Commun 25:101403. http://doi.org/10.1016/j.mtcomm.2020.101403
29. Shinde P, Kar S, Loganathan M, Chang H-Y, Tseng F-G, Nagai M, Santra TS (2020) Infrared pulse laser-activated highly efficient intracellular delivery using titanium microdish device. ACS Biomater Sci Eng 6(10):5645–5652. https://doi.org/10.1021/acsbiomaterials.0c00785
30. Kohout C, Santi C, Polito L (2018) Anisotropic gold nanoparticles in biomedical applications. Int J Mol Sci 19(11). https://doi.org/10.3390/ijms19113385
31. Cotin G, Perton F, Blanco-Andujar C, Pichon B, Mertz D, Bégin-Colin S (2019) Design of anisotropic iron-oxide-based nanoparticles for magnetic hyperthermia, pp 41–60. https://doi.org/10.1016/b978-0-12-813928-8.00002-8
32. Nikitin AA, Khramtsov MA, Savchenko AG, Abakumov MA, Mazhuga AG (2018) Anisotropic iron-oxide nanoparticles for diagnostic MRI: synthesis and contrast properties. Pharm Chem J 52(3):231–235. https://doi.org/10.1007/s11094-018-1796-3
33. Min Y, Caster JM, Eblan MJ, Wang AZ (2015) Clinical translation of nanomedicine. Chem Rev 115(19):11147–11190. https://doi.org/10.1021/acs.chemrev.5b00116
34. Paefgen V, Doleschel D, Kiessling F (2015) Evolution of contrast agents for ultrasound imaging and ultrasound-mediated drug delivery. Front Pharm 6:197. https://doi.org/10.3389/fphar.2015.00197
35. Han X, Xu K, Taratula O, Farsad K (2019) Applications of nanoparticles in biomedical imaging. Nanoscale 11(3):799–819. https://doi.org/10.1039/c8nr07769j
36. Lee DE, Koo H, Sun IC, Ryu JH, Kim K, Kwon IC (2012) Multifunctional nanoparticles for multimodal imaging and theragnosis. Chem Soc Rev 41(7):2656–2672. https://doi.org/10.1039/c2cs15261d

37. Weissman NJ, Soman P, Shah DJ (2013) Multimodality imaging: opportunities and challenges. JACC Cardiovasc Imaging 6(9):1022–1023. https://doi.org/10.1016/j.jcmg.2013.07.003
38. Liu Z, Kiessling F, Gätjens J (2010) Advanced nanomaterials in multimodal imaging: design, functionalization, and biomedical applications. J Nanomater 2010:1–15. https://doi.org/10.1155/2010/894303
39. Santra TS, Tseng FG (2020) Single-cell analysis. Cells 9(9). https://doi.org/10.3390/cells9091993
40. Santra TS, Kar S, Chang HY, Tseng FG (2020) Nano-localized single-cell nano-electroporation. Lab Chip. https://doi.org/10.1039/d0lc00712a
41. Davison CA, Chapman SE, Sasser TA, Wathen C, Diener J, Schafer ZT, M. Leevy WM (2013) Multimodal optical, X-Ray CT, and SPECT imaging of a mouse model of breast cancer lung metastasis. Curr Mol Med 13(3):368–376. http://doi.org/10.2174/1566524011313030006
42. Key J, Leary JF (2014) Nanoparticles for multimodal in vivo imaging in nanomedicine. Int J Nanomed 9:711–726. https://doi.org/10.2147/IJN.S53717
43. Leblond F, Davis SC, Valdes PA, Pogue BW (2010) Pre-clinical whole-body fluorescence imaging: review of instruments, methods and applications. J Photochem Photobiol, B 98(1):77–94. https://doi.org/10.1016/j.jphotobiol.2009.11.007
44. Retif P, Pinel S, Toussaint M, Frochot C, Chouikrat R, Bastogne T, Barberi-Heyob M (2015) Nanoparticles for radiation therapy enhancement: the key parameters. Theranostics 5(9):1030–1044. https://doi.org/10.7150/thno.11642
45. Shan L, Chopra A, Leung K, Eckelman WC, Menkens AE (2012) Characterization of nanoparticle-based contrast agents for molecular magnetic resonance imaging. J Nanoparticle Res 14(9). https://doi.org/10.1007/s11051-012-1122-z
46. Glasser O (1995) W. C. Roentgen and the discovery of the Roentgen rays. AJR Am J roentgenol 165(5):1033–1040. https://doi.org/10.2214/ajr.165.5.7572472
47. Goodman PC (1995) The new light: discovery and introduction of the X-ray. AJR Am J Roentgenol 165(5):1041–1045. https://doi.org/10.2214/ajr.165.5.7572473
48. B SL, Sai Varsha MKN, Kumar NA, Dixit M, Krishnamurthi G (2017) Fast frame rate rodent cardiac x-ray imaging using scintillator lens coupled to CMOS camera 10132:101322 K. https://doi.org/10.1117/12.2254541
49. Lee N, Choi SH, Hyeon T (2013) Nano-sized CT contrast agents. Adv Mater 25(19):2641–2660. https://doi.org/10.1002/adma.201300081
50. Gazelle GS, Wolf GL, McIntire GL, Bacon ER, Halpern EF, Cooper ER, Toner JL (1994) Nanoparticulate computed tomography contrast agents for blood pool and liver-spleen imaging. Acad Radiol 1(4):373–376. https://doi.org/10.1016/s1076-6332(12)80011-0
51. Pasternak JJ, Williamson EE (2012) Clinical pharmacology, uses, and adverse reactions of iodinated contrast agents: a primer for the non-radiologist. Mayo Clin Proc 87(4):390–402. https://doi.org/10.1016/j.mayocp.2012.01.012
52. Yin Q, Yap FY, Yin L, Ma L, Zhou Q, Dobrucki LW, Fan TM, Gaba RC, Cheng J (2013) Poly(iohexol) nanoparticles as contrast agents for in vivo X-ray computed tomography imaging. J Am Chem Soc 135(37):13620–13623. https://doi.org/10.1021/ja405196f
53. de Vries A, Custers E, Lub J, van den Bosch S, Nicolay K, Grull H (2010) Block-copolymer-stabilized iodinated emulsions for use as CT contrast agents. Biomaterials 31(25):6537–6544. https://doi.org/10.1016/j.biomaterials.2010.04.056
54. Kim H, Lee SK, Kim YM, Lee EH, Lim SJ, Kim SH, Yang J, Lim JS, Hyung WJ (2015) Fluorescent iodized emulsion for pre- and intraoperative sentinel lymph node imaging: validation in a preclinical model. Radiology 275(1):196–204. https://doi.org/10.1148/radiol.14141159
55. Hainfeld JF, Ridwan SM, Stanishevskiy Y, Smilowitz NR, Davis J, Smilowitz HM (2018) Small, long blood half-life iodine nanoparticle for vascular and tumor imaging. Sci Rep 8(1):13803. https://doi.org/10.1038/s41598-018-31940-2
56. Zou Q, Huang J, Zhang X (2018) One-step synthesis of iodinated polypyrrole nanoparticles for CT imaging guided photothermal therapy of tumors. Small 14(45):e1803101. https://doi.org/10.1002/smll.201803101

57. Heiligtag FJ, Niederberger M (2013) The fascinating world of nanoparticle research. Mater Today 16(7–8):262–271. https://doi.org/10.1016/j.mattod.2013.07.004
58. Banstola A, Emami F, Jeong J-H, Yook S (2018) Current applications of gold nanoparticles for medical imaging and as treatment agents for managing pancreatic cancer. Macromol Res 26(11):955–964. https://doi.org/10.1007/s13233-018-6139-4
59. Cole LE, Ross RD, Tilley JM, Vargo-Gogola T, Roeder RK (2015) Gold nanoparticles as contrast agents in x-ray imaging and computed tomography. Nanomedicine (Lond) 10(2):321–341. https://doi.org/10.2217/nnm.14.171
60. Xi D, Dong S, Meng X, Lu Q, Meng L, Ye J (2012) Gold nanoparticles as computerized tomography (CT) contrast agents. RSC Adv 2(33):12515. https://doi.org/10.1039/c2ra21263c
61. Ide JM, Lancelot E, Pines E, Corot C (2004) Prophylaxis of iodinated contrast media-induced nephropathy: a pharmacological point of view. Invest Radiol 39(3):155–170. https://doi.org/10.1097/01.rli.0000101483.60710.2c
62. Hainfeld JF, Slatkin DN, Focella TM, Smilowitz HM (2006) Gold nanoparticles: a new X-ray contrast agent. The Br J Radiol 79(939):248–253. https://doi.org/10.1259/bjr/13169882
63. Ross RD, Cole LE, Tilley JMR, Roeder RK (2014) Effects of functionalized gold nanoparticle size on X-ray attenuation and substrate binding affinity. Chem Mater 26(2):1187–1194. https://doi.org/10.1021/cm4035616
64. Kattumuri V, Katti K, Bhaskaran S, Boote EJ, Casteel SW, Fent GM, Robertson DJ, Chandrasekhar M, Kannan R, Katti KV (2007) Gum arabic as a phytochemical construct for the stabilization of gold nanoparticles: in vivo pharmacokinetics and X-ray-contrast-imaging studies. Small 3(2):333–341. https://doi.org/10.1002/smll.200600427
65. Kim D, Park S, Lee JH, Jeong YY, Jon S (2007) Antibiofouling polymer-coated gold nanoparticles as a contrast agent for in vivo X-ray computed tomography imaging. J Am Chem Soc 129(24):7661–7665. https://doi.org/10.1021/ja071471p
66. Peng C, Zheng L, Chen Q, Shen M, Guo R, Wang H, Cao X, Zhang G, Shi X (2012) PEGylated dendrimer-entrapped gold nanoparticles for in vivo blood pool and tumor imaging by computed tomography. Biomaterials 33(4):1107–1119. https://doi.org/10.1016/j.biomaterials.2011.10.052
67. Kojima C, Umeda Y, Ogawa M, Harada A, Magata Y, Kono K (2010) X-ray computed tomography contrast agents prepared by seeded growth of gold nanoparticles in PEGylated dendrimer. Nanotechnology 21(24):245104. https://doi.org/10.1088/0957-4484/21/24/245104
68. Zhang XD, Wu D, Shen X, Liu PX, Yang N, Zhao B, Zhang H, Sun YM, Zhang LA, Fan FY (2011) Size-dependent in vivo toxicity of PEG-coated gold nanoparticles. Int J Nanomed 6:2071–2081. https://doi.org/10.2147/IJN.S21657
69. Zhang XD, Wu HY, Wu D, Wang YY, Chang JH, Zhai ZB, Meng AM, Liu PX, Zhang LA, Fan FY (2010) Toxicologic effects of gold nanoparticles in vivo by different administration routes. Int J Nanomed 5:771–781. https://doi.org/10.2147/IJN.S8428
70. Zhang XD, Wu D, Shen X, Chen J, Sun YM, Liu PX, Liang XJ (2012) Size-dependent radiosensitization of PEG-coated gold nanoparticles for cancer radiation therapy. Biomaterials 33(27):6408–6419. https://doi.org/10.1016/j.biomaterials.2012.05.047
71. Srinoi P, Chen Y-T, Vittur V, Marquez M, Lee T (2018) Bimetallic nanoparticles: enhanced magnetic and optical properties for emerging biological applications. Appl Sci 8(7):1106. https://doi.org/10.3390/app8071106
72. Naha PC, Lau KC, Hsu JC, Hajfathalian M, Mian S, Chhour P, Uppuluri L, McDonald ES, Maidment AD, Cormode DP (2016) Gold silver alloy nanoparticles (GSAN): an imaging probe for breast cancer screening with dual-energy mammography or computed tomography. Nanoscale 8(28):13740–13754. https://doi.org/10.1039/c6nr02618d
73. Wang H, Zheng L, Guo R, Peng C, Shen M, Shi X, Zhang G (2012) Dendrimer-entrapped gold nanoparticles as potential CT contrast agents for blood pool imaging. Nanoscale Res Lett 7:190. https://doi.org/10.1186/1556-276X-7-190

74. Chu Z, Chen L, Wang X, Yang Q, Zhao Q, Huang C, Huang Y, Yang D-P, Jia N (2019) Ultrasmall Au–Ag alloy nanoparticles: protein-directed synthesis, biocompatibility, and X-ray computed tomography imaging. ACS Biomater Sci Eng 5(2):1005–1015. https://doi.org/10.1021/acsbiomaterials.8b01176
75. Gui C, Cui DX (2012) Functionalized gold nanorods for tumor imaging and targeted therapy. Cancer Biol Med 9(4):221–233. https://doi.org/10.7497/j.issn.2095-3941.2012.04.002
76. Reuveni T, Motiei M, Romman Z, Popovtzer A, Popovtzer R (2011) Targeted gold nanoparticles enable molecular CT imaging of cancer: an in vivo study. Int J Nanomed 6:2859–2864. https://doi.org/10.2147/IJN.S25446
77. Jazayeri MH, Amani H, Pourfatollah AA, Pazoki-Toroudi H, Sedighimoghaddam B (2016) Various methods of gold nanoparticles (GNPs) conjugation to antibodies. Sens Bio-Sens Res 9:17–22. https://doi.org/10.1016/j.sbsr.2016.04.002
78. Gupta P, Balasubramaniam N, Chang HY, Tseng FG, Santra TS (2020) A single-neuron: current trends and future prospects. Cells 9(6). https://doi.org/10.3390/cells9061528
79. Kruger CA, Abrahamse H (2018) Utilisation of targeted nanoparticle photosensitiser drug delivery systems for the enhancement of photodynamic therapy. Molecules 23(10). https://doi.org/10.3390/molecules23102628
80. Rosenblum D, Joshi N, Tao W, Karp JM, Peer D (2018) Progress and challenges towards targeted delivery of cancer therapeutics. Nat Commun 9(1):1410. https://doi.org/10.1038/s41467-018-03705-y
81. Zhou B, Yang J, Peng C, Zhu J, Tang Y, Zhu X, Shen M, Zhang G, Shi X (2016) PEGylated polyethylenimine-entrapped gold nanoparticles modified with folic acid for targeted tumor CT imaging. Colloids Surf B Biointerfaces 140:489–496. https://doi.org/10.1016/j.colsurfb.2016.01.019
82. Eck W, Nicholson AI, Zentgraf H, Semmler W, Bartling S (2010) Anti-CD4-targeted gold nanoparticles induce specific contrast enhancement of peripheral lymph nodes in X-ray computed tomography of live mice. Nano Lett 10(7):2318–2322. https://doi.org/10.1021/nl101019s
83. Li X, Wang C, Tan H, Cheng L, Liu G, Yang Y, Zhao Y, Zhang Y, Li Y, Zhang C, Xiu Y, Cheng D, Shi H (2016) Gold nanoparticles-based SPECT/CT imaging probe targeting for vulnerable atherosclerosis plaques. Biomaterials 108:71–80. https://doi.org/10.1016/j.biomaterials.2016.08.048
84. Khademi S, Sarkar S, Shakeri-Zadeh A, Attaran N, Kharrazi S, Ay MR, Ghadiri H (2018) Folic acid-cysteamine modified gold nanoparticle as a nanoprobe for targeted computed tomography imaging of cancer cells. Mater Sci Eng C Mater Biol Appl 89:182–193. https://doi.org/10.1016/j.msec.2018.03.015
85. Zhu XY, Zou X, Mukherjee R, Yu Z, Ferguson CM, Zhou W, McCollough CH, Lerman LO (2018) Targeted imaging of renal fibrosis using antibody-conjugated gold nanoparticles in renal artery stenosis. Invest Radiol 53(10):623–628. https://doi.org/10.1097/RLI.0000000000000476
86. Shang L, Dong S, Nienhaus GU (2011) Ultra-small fluorescent metal nanoclusters: synthesis and biological applications. Nano Today 6(4):401–418. https://doi.org/10.1016/j.nantod.2011.06.004
87. Zhang XD, Wu D, Shen X, Liu PX, Fan FY, Fan SJ (2012) In vivo renal clearance, biodistribution, toxicity of gold nanoclusters. Biomaterials 33(18):4628–4638. https://doi.org/10.1016/j.biomaterials.2012.03.020
88. Zhou C, Long M, Qin Y, Sun X, Zheng J (2011) Luminescent gold nanoparticles with efficient renal clearance. Angew Chem 50(14):3168–3172. https://doi.org/10.1002/anie.201007321
89. Liu Y, Tian GF, He XW, Li WY, Zhang YK (2016) Microwave-assisted one-step rapid synthesis of near-infrared gold nanoclusters for NIRF/CT dual-modal bioimaging. J Mater Chem B 4(7):1276–1283. https://doi.org/10.1039/c5tb02322j
90. Jia TT, Yang G, Mo SJ, Wang ZY, Li BJ, Ma W, Guo YX, Chen X, Zhao X, Liu JQ, Zang SQ (2019) Atomically precise gold-levonorgestrel nanocluster as a radiosensitizer for enhanced cancer therapy. ACS Nano 13(7):8320–8328. https://doi.org/10.1021/acsnano.9b03767

91. Zhang Q, Yang M, Zhu Y, Mao C (2018) Metallic nanoclusters for cancer imaging and therapy. Curr Med Chem 25(12):1379–1396. https://doi.org/10.2174/0929867324666170331122757
92. Zhang C, Li C, Liu Y, Zhang J, Bao C, Liang S, Wang Q, Yang Y, Fu H, Wang K, Cui D (2015) Gold nanoclusters-based nanoprobes for simultaneous fluorescence imaging and targeted photodynamic therapy with superior penetration and retention behavior in tumors. Adv Func Mater 25(8):1314–1325. https://doi.org/10.1002/adfm.201403095
93. Yu S, Wen R, Wang H, Zha Y, Qiu L, Li B, Xue W, Ma D (2019) Chitosan-graft-Poly(l-lysine) dendron-assisted facile self-assembly of au nanoclusters for enhanced X-ray computer tomography imaging and precise MMP-9 plasmid shRNA delivery. Chem Mater 31(11):3992–4007. https://doi.org/10.1021/acs.chemmater.9b00507
94. Winter H, Brown AL, Goforth AM (2018) Bismuth-based nano- and microparticles in X-ray contrast, radiation therapy, and radiation shielding applications. http://doi.org/10.5772/intechopen.76413
95. Wang Y, Wu Y, Liu Y, Shen J, Lv L, Li L, Yang L, Zeng J, Wang Y, Zhang LW, Li Z, Gao M, Chai Z (2016) BSA-mediated synthesis of bismuth sulfide nanotheranostic agents for tumor multimodal imaging and thermoradiotherapy. Adv Func Mater 26(29):5335–5344. https://doi.org/10.1002/adfm.201601341
96. Shahbazi MA, Faghfouri L, Ferreira MPA, Figueiredo P, Maleki H, Sefat F, Hirvonen J, Santos HA (2020) The versatile biomedical applications of bismuth-based nanoparticles and composites: therapeutic, diagnostic, biosensing, and regenerative properties. Chem Soc Rev 49(4):1253–1321. https://doi.org/10.1039/c9cs00283a
97. Brown AL, Naha PC, Benavides-Montes V, Litt HI, Goforth AM, Cormode DP (2014) Synthesis, X-ray opacity, and biological compatibility of ultra-high payload elemental bismuth nanoparticle X-ray contrast agents. Chem Mater: A Publ Am Chem Soc 26(7):2266–2274. https://doi.org/10.1021/cm500077z
98. Rabin O, Manuel Perez J, Grimm J, Wojtkiewicz G, Weissleder R (2006) An X-ray computed tomography imaging agent based on long-circulating bismuth sulphide nanoparticles. Nat Mater 5(2):118–122. https://doi.org/10.1038/nmat1571
99. Pan D, Williams TA, Senpan A, Allen JS, Scott MJ, Gaffney PJ, Wickline SA, Lanza GM (2009) Detecting vascular biosignatures with a colloidal, radio-opaque polymeric nanoparticle. J Am Chem Soc 131(42):15522–15527. https://doi.org/10.1021/ja906797z
100. Ai K, Liu Y, Liu J, Yuan Q, He Y, Lu L (2011) Large-scale synthesis of Bi(2)S(3) nanodots as a contrast agent for in vivo X-ray computed tomography imaging. Adv Mater 23(42):4886–4891. https://doi.org/10.1002/adma.201103289
101. Liu J, Zheng X, Yan L, Zhou L, Tian G, Yin W, Wang L, Liu Y, Hu Z, Gu Z, Chen C, Zhao Y (2015) Bismuth sulfide nanorods as a precision nanomedicine for in vivo multimodal imaging-guided photothermal therapy of tumor. ACS Nano 9(1):696–707. https://doi.org/10.1021/nn506137n
102. Yu X, Li A, Zhao C, Yang K, Chen X, Li W (2017) Ultrasmall semimetal nanoparticles of bismuth for dual-modal computed tomography/photoacoustic imaging and synergistic thermoradiotherapy. ACS Nano 11(4):3990–4001. https://doi.org/10.1021/acsnano.7b00476
103. Roy B, Chattopadhyay AP, Samadder A, Khuda-Bukhsh AR (2015) Target synthesis of biocompatible spherical bismuth sulphide nanoparticles for biological application. J Sol-Gel Sci Technol 77(2):446–452. https://doi.org/10.1007/s10971-015-3873-3
104. Kinsella JM, Jimenez RE, Karmali PP, Rush AM, Kotamraju VR, Gianneschi NC, Ruoslahti E, Stupack D, Sailor MJ (2011) X-ray computed tomography imaging of breast cancer by using targeted peptide-labeled bismuth sulfide nanoparticles. Angew Chem 50(51):12308–12311. https://doi.org/10.1002/anie.201104507
105. Li L, Lu Y, Jiang C, Zhu Y, Yang X, Hu X, Lin Z, Zhang Y, Peng M, Xia H, Mao C (2018) Actively targeted deep tissue imaging and photothermal-chemo therapy of breast cancer by antibody-functionalized drug-loaded X-ray-responsive bismuth sulfide@mesoporous silica core-shell nanoparticles. Adv Funct Mater 28(5). https://doi.org/10.1002/adfm.201704623
106. Nosrati H, Abhari F, Charmi J, Rahmati M, Johari B, Azizi S, Rezaeejam H, Danafar H (2019) Facile green synthesis of bismuth sulfide radiosensitizer via biomineralization of

albumin natural molecule for chemoradiation therapy aim. Artif Cells Nanomed Biotechnol 47(1):3832–3838. https://doi.org/10.1080/21691401.2019.1669624
107. Azizi S, Nosrati H, Sharafi A, Danafar H (2019) Preparation of bismuth sulfide nanoparticles as targeted biocompatible nano-radiosensitizer and carrier of methotrexate. Appl Organomet Chem 34(1). https://doi.org/10.1002/aoc.5251
108. Nosrati H, Charmi J, Salehiabar M, Abhari F, Danafar H (2019) Tumor targeted albumin coated bismuth sulfide nanoparticles (Bi2S3) as radiosensitizers and carriers of curcumin for enhanced chemoradiation therapy. ACS Biomater Sci Eng 5(9):4416–4424. https://doi.org/10.1021/acsbiomaterials.9b00489
109. Zhou D, Li C, He M, Ma M, Li P, Gong Y, Ran H, Wang Z, Wang Z, Zheng Y, Sun Y (2016) Folate-targeted perfluorohexane nanoparticles carrying bismuth sulfide for use in US/CT dual-mode imaging and synergistic high-intensity focused ultrasound ablation of cervical cancer. J Mater Chem B 4(23):4164–4181. https://doi.org/10.1039/c6tb00261g
110. Oh MH, Lee N, Kim H, Park SP, Piao Y, Lee J, Jun SW, Moon WK, Choi SH, Hyeon T (2011) Large-scale synthesis of bioinert tantalum oxide nanoparticles for X-ray computed tomography imaging and bimodal image-guided sentinel lymph node mapping. J Am Chem Soc 133(14):5508–5515. https://doi.org/10.1021/ja200120k
111. Matsuno H (2001) Biocompatibility and osteogenesis of refractory metal implants, titanium, hafnium, niobium, tantalum and rhenium. Biomaterials 22(11):1253–1262. https://doi.org/10.1016/s0142-9612(00)00275-1
112. Bonitatibus PJ Jr, Torres AS, Kandapallil B, Lee BD, Goddard GD, Colborn RE, Marino ME (2012) Preclinical assessment of a zwitterionic tantalum oxide nanoparticle X-ray contrast agent. ACS Nano 6(8):6650–6658. https://doi.org/10.1021/nn300928g
113. Lee N, Yoo D, Ling D, Cho MH, Hyeon T, Cheon J (2015) Iron oxide based nanoparticles for multimodal imaging and magnetoresponsive therapy. Chem Rev 115(19):10637–10689. https://doi.org/10.1021/acs.chemrev.5b00112
114. Torres AS, Bonitatibus PJ Jr, Colborn RE, Goddard GD, FitzGerald PF, Lee BD, Marino ME (2012) Biological performance of a size-fractionated core-shell tantalum oxide nanoparticle x-ray contrast agent. Invest Radiol 47(10):578–587. https://doi.org/10.1097/RLI.0b013e318260fc40
115. Freedman JD, Lusic H, Snyder BD, Grinstaff MW (2014) Tantalum oxide nanoparticles for the imaging of articular cartilage using X-ray computed tomography: visualization of ex vivo/in vivo murine tibia and ex vivo human index finger cartilage. Angew Chem 53(32):8406–8410. https://doi.org/10.1002/anie.201404519
116. Crowder JM, Bates N, Roberts J, Torres AS, Bonitatibus PJ (2016) Determination of tantalum from tantalum oxide nanoparticle X-ray/CT contrast agents in rat tissues and bodily fluids by ICP-OES. J Anal At Spectrom 31(6):1311–1317. https://doi.org/10.1039/c5ja00446b
117. Song G, Chen Y, Liang C, Yi X, Liu J, Sun X, Shen S, Yang K, Liu Z (2016) Catalase-loaded taox nanoshells as bio-nanoreactors combining high-Z element and enzyme delivery for enhancing radiotherapy. Adv Mater 28(33):7143–7148. https://doi.org/10.1002/adma.201602111
118. Chen Y, Song G, Dong Z, Yi X, Chao Y, Liang C, Yang K, Cheng L, Liu Z (2017) Drug-loaded mesoporous tantalum oxide nanoparticles for enhanced synergetic chemoradiotherapy with reduced systemic toxicity. Small 13(8). https://doi.org/10.1002/smll.201602869
119. Jin Y, Ma X, Zhang S, Meng H, Xu M, Yang X, Xu W, Tian J (2017) A tantalum oxide-based core/shell nanoparticle for triple-modality image-guided chemo-thermal synergetic therapy of esophageal carcinoma. Cancer Lett 397:61–71. https://doi.org/10.1016/j.canlet.2017.03.030
120. Peng C, Liang Y, Chen Y, Qian X, Luo W, Chen S, Zhang S, Dan Q, Zhang L, Li M, Yuan M, Zhao B, Li Y (2020) Hollow mesoporous tantalum oxide based nanospheres for triple sensitization of radiotherapy. ACS Appl Mater Interfaces 12(5):5520–5530. https://doi.org/10.1021/acsami.9b20053
121. Rajasekar S, Martin EM, Kuppusamy S, Vetrivel C (2020) Chitosan coated molybdenum sulphide nanosheet incorporated with tantalum oxide nanomaterials for improving cancer photothermal therapy. Arab J Chem 13(3):4741–4750. https://doi.org/10.1016/j.arabjc.2019.11.005

122. Xiao Q, Bu W, Ren Q, Zhang S, Xing H, Chen F, Li M, Zheng X, Hua Y, Zhou L, Peng W, Qu H, Wang Z, Zhao K, Shi J (2012) Radiopaque fluorescence-transparent TaOx decorated upconversion nanophosphors for in vivo CT/MR/UCL trimodal imaging. Biomaterials 33(30):7530–7539. https://doi.org/10.1016/j.biomaterials.2012.06.028
123. Chakravarty S, Hix JML, Wiewiora KA, Volk MC, Kenyon E, Shuboni-Mulligan DD, Blanco-Fernandez B, Kiupel M, Thomas J, Sempere LF, Shapiro EM (2020) Tantalum oxide nanoparticles as versatile contrast agents for X-ray computed tomography. Nanoscale. https://doi.org/10.1039/d0nr01234c
124. Liu Y, Liu J, Ai K, Yuan Q, Lu L (2014) Recent advances in ytterbium-based contrast agents for in vivo X-ray computed tomography imaging: promises and prospects. Contrast Media Mol Imaging 9(1):26–36. https://doi.org/10.1002/cmmi.1537
125. Xiong L, Yang T, Yang Y, Xu C, Li F (2010) Long-term in vivo biodistribution imaging and toxicity of polyacrylic acid-coated upconversion nanophosphors. Biomaterials 31(27):7078–7085. https://doi.org/10.1016/j.biomaterials.2010.05.065
126. Wachter J (1990) The elements. Von J. Emsley. Clarendon Press, Oxford 1989. 256 S., Paperback £ 9.95. ISBN 0-19-855237-8. Angewandte Chemie 102(1):115–115. https://doi.org/10.1002/ange.19901020140
127. Li Z, Zhang Y, Jiang S (2008) Multicolor core/shell-structured upconversion fluorescent nanoparticles. Adv Mater 20(24):4765–4769. https://doi.org/10.1002/adma.200801056
128. Unger E, Gutierrez F (1986) Ytterbium-DTPA. A potential intravascular contrast agent. Investig Radiol 21(10):802–807. https://doi.org/10.1097/00004424-198610000-00007
129. Krause W, Schuhmann-Giampieri G, Bauer M, Press W-R, Muschick P (1996) Ytterbium- and dysprosium-EOB-DTPA: a new prototype of liver-specific contrast agents for computed tomography. Investig Radiol 31(8):502–511
130. Schmitz SA, Wagner S, Schuhmann-Giampieri G, Krause W, Bollow M, Wolf KJ (1997) Gd-EOB-DTPA and Yb-EOB-DTPA: two prototypic contrast media for CT detection of liver lesions in dogs. Radiology 205(2):361–366. https://doi.org/10.1148/radiology.205.2.9356615
131. Anbazhagan R, Su YA, Tsai HC, Jeng RJ (2016) MoS2-Gd chelate magnetic nanomaterials with core-shell structure used as contrast agents in in vivo magnetic resonance imaging. ACS Appl Mater Interfaces 8(3):1827–1835. https://doi.org/10.1021/acsami.5b09722
132. Liu Y, Ai K, Liu J, Yuan Q, He Y, Lu L (2012) A high-performance ytterbium-based nanoparticulate contrast agent for in vivo X-ray computed tomography imaging. Angew Chem 51(6):1437–1442. https://doi.org/10.1002/anie.201106686
133. Liu Y, Ai K, Liu J, Yuan Q, He Y, Lu L (2012) Hybrid BaYbF(5) nanoparticles: novel binary contrast agent for high-resolution in vivo X-ray computed tomography angiography. Adv Healthc Mater 1(4):461–466. https://doi.org/10.1002/adhm.201200028
134. Liu Z, Pu F, Liu J, Jiang L, Yuan Q, Li Z, Ren J, Qu X (2013) PEGylated hybrid ytterbia nanoparticles as high-performance diagnostic probes for in vivo magnetic resonance and X-ray computed tomography imaging with low systemic toxicity. Nanoscale 5(10):4252–4261. https://doi.org/10.1039/c3nr00491k
135. Liu Z, Li Z, Liu J, Gu S, Yuan Q, Ren J, Qu X (2012) Long-circulating Er3+-doped Yb2O3 up-conversion nanoparticle as an in vivo X-Ray CT imaging contrast agent. Biomaterials 33(28):6748–6757. https://doi.org/10.1016/j.biomaterials.2012.06.033
136. Zeng S, Tsang MK, Chan CF, Wong KL, Hao J (2012) PEG modified BaGdF(5):Yb/Er nanoprobes for multi-modal upconversion fluorescent, in vivo X-ray computed tomography and biomagnetic imaging. Biomaterials 33(36):9232–9238. https://doi.org/10.1016/j.biomaterials.2012.09.019
137. Pan D, Schirra CO, Senpan A, Schmieder AH, Stacy AJ, Roessl E, Thran A, Wickline SA, Proska R, Lanza GM (2012) An early investigation of ytterbium nanocolloids for selective and quantitative "multicolor" spectral CT imaging. ACS Nano 6(4):3364–3370. https://doi.org/10.1021/nn300392x
138. Liu Z, Pu F, Huang S, Yuan Q, Ren J, Qu X (2013) Long-circulating Gd(2)O(3):Yb(3+), Er(3+) up-conversion nanoprobes as high-performance contrast agents for multi-modality imaging. Biomaterials 34(6):1712–1721. https://doi.org/10.1016/j.biomaterials.2012.11.009

139. Liu JN, Bu W, Pan LM, Zhang S, Chen F, Zhou L, Zhao KL, Peng W, Shi J (2012) Simultaneous nuclear imaging and intranuclear drug delivery by nuclear-targeted multifunctional upconversion nanoprobes. Biomaterials 33(29):7282–7290. https://doi.org/10.1016/j.biomaterials.2012.06.035
140. Xia A, Chen M, Gao Y, Wu D, Feng W, Li F (2012) Gd3+ complex-modified NaLuF4-based upconversion nanophosphors for trimodality imaging of NIR-to-NIR upconversion luminescence, X-Ray computed tomography and magnetic resonance. Biomaterials 33(21):5394–5405. https://doi.org/10.1016/j.biomaterials.2012.04.025
141. McCollough CH, Leng S, Yu L, Fletcher JG (2015) Dual- and multi-energy CT: principles, technical approaches, and clinical applications. Radiology 276(3):637–653. https://doi.org/10.1148/radiol.2015142631
142. French J, Gingles N, Stewart J, Woodhouse N (2010) Use of magnetic resonance imaging (MRI) and micro-computed tomography (micro-CT) in the morphological examination of rat and rabbit fetuses from embryo-fetal development studies. Reprod Toxicol 30(2):292–300. https://doi.org/10.1016/j.reprotox.2010.04.016
143. Na HB, Song IC, Hyeon T (2009) Inorganic nanoparticles for MRI contrast agents. Adv Mater 21(21):2133–2148. https://doi.org/10.1002/adma.200802366
144. Constantinides C (2016) Magnetic resonance imaging: the basics. CRC Press
145. Driehuys B, Nouls J, Badea A, Bucholz E, Ghaghada K, Petiet A, Hedlund LW (2008) Small animal imaging with magnetic resonance microscopy. ILAR J 49(1):35–53. https://doi.org/10.1093/ilar.49.1.35
146. Turnbull DH, Mori S (2007) MRI in mouse developmental biology. NMR Biomed 20(3):265–274. https://doi.org/10.1002/nbm.1146
147. Schaeffter T, Dahnke H (2008) Magnetic resonance imaging and spectroscopy. Handb Exp Pharmacol (185 Pt 1):75–90. https://doi.org/10.1007/978-3-540-72718-7_4
148. Wahsner J, Gale EM, Rodriguez-Rodriguez A, Caravan P (2019) Chemistry of MRI contrast agents: current challenges and new frontiers. Chem Rev 119(2):957–1057. https://doi.org/10.1021/acs.chemrev.8b00363
149. Geraldes CF, Laurent S (2009) Classification and basic properties of contrast agents for magnetic resonance imaging. Contrast Media Mol Imaging 4(1):1–23. https://doi.org/10.1002/cmmi.265
150. De Leon-Rodriguez LM, Martins AF, Pinho MC, Rofsky NM, Sherry AD (2015) Basic MR relaxation mechanisms and contrast agent design. J Magnetic Reson Imaging: JMRI 42(3):545–565. https://doi.org/10.1002/jmri.24787
151. Young IR, Clarke GJ, Baffles DR, Pennock JM, Doyle FH, Bydder GM (1981) Enhancement of relaxation rate with paramagnetic contrast agents in NMR imaging. J Comput Tomogr 5(6):543–547. https://doi.org/10.1016/0149-936x(81)90089-8
152. Carr DH, Brown J, Bydder GM, Weinmann HJ, Speck U, Thomas DJ, Young IR (1984) Intravenous chelated gadolinium as a contrast agent in NMR imaging of cerebral tumours. The Lancet 323(8375):484–486. https://doi.org/10.1016/s0140-6736(84)92852-6
153. Xiao YD, Paudel R, Liu J, Ma C, Zhang ZS, Zhou SK (2016) MRI contrast agents: classification and application (Review). Int J Mol Med 38(5):1319–1326. https://doi.org/10.3892/ijmm.2016.2744
154. Sherry AD, Caravan P, Lenkinski RE (2009) Primer on gadolinium chemistry. J Magn Resonan Imaging: JMRI 30(6):1240–1248. https://doi.org/10.1002/jmri.21966
155. Marasini R, Thanh Nguyen TD, Aryal S (2020) Integration of gadolinium in nanostructure for contrast enhanced-magnetic resonance imaging. Wiley Interdiscip Rev Nanomed Nanobiotechnol 12(1):e1580. https://doi.org/10.1002/wnan.1580
156. Fraum TJ, Ludwig DR, Bashir MR, Fowler KJ (2017) Gadolinium-based contrast agents: a comprehensive risk assessment. J Magnetic Reson Imaging: JMRI 46(2):338–353. https://doi.org/10.1002/jmri.25625
157. Kahakachchi CL, Moore DA (2009) Speciation of gadolinium in gadolinium-based magnetic resonance imaging agents by high performance liquid chromatography inductively coupled plasma optical emission spectrometry. J Anal At Spectrom 24(10):1389. https://doi.org/10.1039/b907044c

158. Salehi B, Selamoglu Z, K SM, Pezzani R, Redaelli M, Cho WC, Kobarfard F, Rajabi S, Martorell M, Kumar P, Martins N, Subhra Santra T, Sharifi-Rad J (2019) Liposomal cytarabine as cancer therapy: from chemistry to medicine. Biomolecules 9(12). https://doi.org/10.3390/biom9120773
159. Langereis S, Geelen T, Grull H, Strijkers GJ, Nicolay K (2013) Paramagnetic liposomes for molecular MRI and MRI-guided drug delivery. NMR Biomed 26(7):728–744. https://doi.org/10.1002/nbm.2971
160. Liu X, Madhankumar AB, Miller PA, Duck KA, Hafenstein S, Rizk E, Slagle-Webb B, Sheehan JM, Connor JR, Yang QX (2016) MRI contrast agent for targeting glioma: interleukin-13 labeled liposome encapsulating gadolinium-DTPA. Neuro Oncol 18(5):691–699. https://doi.org/10.1093/neuonc/nov263
161. Li W, Su B, Meng S, Ju L, Yan L, Ding Y, Song Y, Zhou W, Li H, Tang L, Zhao Y, Zhou C (2011) RGD-targeted paramagnetic liposomes for early detection of tumor: in vitro and in vivo studies. Eur J Radiol 80(2):598–606. https://doi.org/10.1016/j.ejrad.2011.01.051
162. Song Y, Li W, Meng S, Zhou W, Su B, Tang L, Zhao Y, Wu X, Yin D, Fan M, Zhou C (2018) Dual integrin alphavbeta 3 and NRP-1-targeting paramagnetic liposome for tumor early detection in magnetic resonance imaging. Nanoscale Res Lett 13(1):380. https://doi.org/10.1186/s11671-018-2797-6
163. Hammoud DA (2016) Molecular imaging of inflammation: current status. J Nucl Med: Off Publ, Soc Nucl Med 57(8):1161–1165. https://doi.org/10.2967/jnumed.115.161182
164. Tian B, Liu R, Chen S, Chen L, Liu F, Jia G, Dong Y, Li J, Chen H, Lu J (2017) Mannose-coated gadolinium liposomes for improved magnetic resonance imaging in acute pancreatitis. Int J Nanomed 12:1127–1141. https://doi.org/10.2147/IJN.S123290
165. Dawood S (2010) Triple-negative breast cancer: epidemiology and management options. Drugs 70(17):2247–2258. https://doi.org/10.2165/11538150-000000000-00000
166. Goldman E, Zinger A, da Silva D, Yaari Z, Kajal A, Vardi-Oknin D, Goldfeder M, Schroeder JE, Shainsky-Roitman J, Hershkovitz D, Schroeder A (2017) Nanoparticles target early-stage breast cancer metastasis in vivo. Nanotechnology 28(43):43LT01. https://doi.org/10.1088/1361-6528/aa8a3d
167. Kuijten MM, Hannah Degeling M, Chen JW, Wojtkiewicz G, Waterman P, Weissleder R, Azzi J, Nicolay K, Tannous BA (2015) Multimodal targeted high relaxivity thermosensitive liposome for in vivo imaging. Sci Rep 5:17220. https://doi.org/10.1038/srep17220
168. Lorenzato C, Oerlemans C, van Elk M, Geerts WJ, Denis de Senneville B, Moonen C, Bos C (2016) MRI monitoring of nanocarrier accumulation and release using Gadolinium-SPIO co-labelled thermosensitive liposomes. Contrast Media Mol Imaging 11(3):184–194. https://doi.org/10.1002/cmmi.1679
169. Affram K, Udofot O, Singh M, Krishnan S, Reams R, Rosenberg J, Agyare E (2017) Smart thermosensitive liposomes for effective solid tumor therapy and in vivo imaging. PLoS One 12(9):e0185116. https://doi.org/10.1371/journal.pone.0185116
170. Ahren M, Selegard L, Klasson A, Soderlind F, Abrikossova N, Skoglund C, Bengtsson T, Engstrom M, Kall PO, Uvdal K (2010) Synthesis and characterization of PEGylated Gd2O3 nanoparticles for MRI contrast enhancement. Langmuir: ACS J Surf Colloids 26(8):5753–5762. https://doi.org/10.1021/la903566y
171. Soderlind F, Pedersen H, Petoral RM Jr, Kall PO, Uvdal K (2005) Synthesis and characterisation of Gd2O3 nanocrystals functionalised by organic acids. J Colloid Interface Sci 288(1):140–148. https://doi.org/10.1016/j.jcis.2005.02.089
172. Faucher L, Guay-Begin AA, Lagueux J, Cote MF, Petitclerc E, Fortin MA (2011) Ultra-small gadolinium oxide nanoparticles to image brain cancer cells in vivo with MRI. Contrast Media Mol Imaging 6(4):209–218. https://doi.org/10.1002/cmmi.420
173. Faucher L, Gossuin Y, Hocq A, Fortin MA (2011) Impact of agglomeration on the relaxometric properties of paramagnetic ultra-small gadolinium oxide nanoparticles. Nanotechnology 22(29):295103. https://doi.org/10.1088/0957-4484/22/29/295103
174. Park JY, Baek MJ, Choi ES, Woo S, Kim JH, Kim TJ, Jung JC, Chae KS, Chang Y, Lee GH (2009) Paramagnetic ultrasmall gadolinium oxide nanoparticles as advanced T1 MRI contrast

agent: account for large longitudinal relaxivity, optimal particle diameter, and in vivo T1 MR images. ACS Nano 3(11):3663–3669. https://doi.org/10.1021/nn900761s
175. Bridot JL, Faure AC, Laurent S, Riviere C, Billotey C, Hiba B, Janier M, Josserand V, Coll JL, Elst LV, Muller R, Roux S, Perriat P, Tillement O (2007) Hybrid gadolinium oxide nanoparticles: multimodal contrast agents for in vivo imaging. J Am Chem Soc 129(16):5076–5084. https://doi.org/10.1021/ja068356j
176. Blumfield E, Swenson DW, Iyer RS, Stanescu AL (2019) Gadolinium-based contrast agents—review of recent literature on magnetic resonance imaging signal intensity changes and tissue deposits, with emphasis on pediatric patients. Pediatr Radiol 49(4):448–457. https://doi.org/10.1007/s00247-018-4304-8
177. Pellico J, Ellis CM, Davis JJ (2019) Nanoparticle-based paramagnetic contrast agents for magnetic resonance imaging. Contrast Media Mol Imaging 2019:1845637. https://doi.org/10.1155/2019/1845637
178. Cao Y, Xu L, Kuang Y, Xiong D, Pei R (2017) Gadolinium-based nanoscale MRI contrast agents for tumor imaging. J Mater Chem B 5(19):3431–3461. https://doi.org/10.1039/c7tb00382j
179. Miao X, Ho SL, Tegafaw T, Cha H, Chang Y, Oh IT, Yaseen AM, Marasini S, Ghazanfari A, Yue H, Chae KS, Lee GH (2018) Stable and non-toxic ultrasmall gadolinium oxide nanoparticle colloids (coating material = polyacrylic acid) as high-performance T1 magnetic resonance imaging contrast agents. RSC Adv 8(6):3189–3197. https://doi.org/10.1039/c7ra11830a
180. Mekuria SL, Debele TA, Tsai HC (2017) Encapsulation of gadolinium oxide nanoparticle (Gd2O3) contrasting agents in PAMAM dendrimer templates for enhanced magnetic resonance imaging in vivo. ACS Appl Mater Interfaces 9(8):6782–6795. https://doi.org/10.1021/acsami.6b14075
181. Rivlin M, Navon G (2016) Glucosamine and N-acetyl glucosamine as new CEST MRI agents for molecular imaging of tumors. Sci Rep 6:32648. https://doi.org/10.1038/srep32648
182. Mortezazadeh T, Gholibegloo E, Riyahi Alam N, Haghgoo S, Musa A, E., Khoobi M (2020) Glucosamine conjugated gadolinium (III) oxide nanoparticles as a novel targeted contrast agent for cancer diagnosis in MRI. J Biomed Phys Eng 10(1):25–38. http://doi.org/10.31661/jbpe.v0i0.1018
183. Mortezazadeh T, Gholibegloo E, Alam NR, Dehghani S, Haghgoo S, Ghanaati H, Khoobi M (2019) Gadolinium (III) oxide nanoparticles coated with folic acid-functionalized poly(beta-cyclodextrin-co-pentetic acid) as a biocompatible targeted nano-contrast agent for cancer diagnostic: in vitro and in vivo studies. MAGMA 32(4):487–500. https://doi.org/10.1007/s10334-019-00738-2
184. Wang Y, Yang T, Ke H, Zhu A, Wang Y, Wang J, Shen J, Liu G, Chen C, Zhao Y, Chen H (2015) Smart albumin-biomineralized nanocomposites for multimodal imaging and photothermal tumor ablation. Adv Mater 27(26):3874–3882. https://doi.org/10.1002/adma.201500229
185. Huang S, Liu J, Liu D, Yuan Q (2012) Facile and large-scale synthesis of Gd(OH)3 nanorods for MR imaging with low toxicity. New J Chem 36(6):1335. https://doi.org/10.1039/c2nj21009f
186. Yang Y, Sun Y, Liu Y, Peng J, Wu Y, Zhang Y, Feng W, Li F (2013) Long-term in vivo biodistribution and toxicity of Gd(OH)3 nanorods. Biomaterials 34(2):508–515. https://doi.org/10.1016/j.biomaterials.2012.09.075
187. Rajaee A, Wang S, Zhao L, Wang D, Liu Y, Wang J, Ying K (2019) Multifunction bismuth gadolinium oxide nanoparticles as radiosensitizer in radiation therapy and imaging. Phys Med Biol 64(19):195007. https://doi.org/10.1088/1361-6560/ab2154
188. Li IF, Su C-H, Sheu H-S, Chiu H-C, Lo Y-W, Lin W-T, Chen J-H, Yeh C-S (2008) Gd2O(CO3)2 H2O particles and the corresponding Gd2O3: synthesis and applications of magnetic resonance contrast agents and template particles for hollow spheres and hybrid composites. Adv Func Mater 18(5):766–776. https://doi.org/10.1002/adfm.200700702
189. Hu K-W, Jhang F-Y, Su C-H, Yeh C-S (2009) Fabrication of Gd2O(CO3)2 H2O/silica/gold hybrid particles as a bifunctional agent for MR imaging and photothermal destruction of cancer cells. J Mater Chem 19(15):2147. https://doi.org/10.1039/b815087g

190. Sook Choi E, Young Park J, Ju Baek M, Xu W, Kattel K, Hyun Kim J, Jun Lee J, Chang Y, Jeong Kim T, Eun Bae J, Seok Chae K, Jin Suh K (2010) Ho Lee G (2010) water-soluble ultra-small manganese oxide surface doped gadolinium oxide (Gd2O3@MnO) nanoparticles for MRI contrast agent. Eur J Inorg Chem 28:4555–4560. https://doi.org/10.1002/ejic.201000374
191. Tian X, Yang F, Yang C, Peng Y, Chen D, Zhu J, He F, Li L, Chen X (2014) Toxicity evaluation of Gd2O3@SiO2 nanoparticles prepared by laser ablation in liquid as MRI contrast agents in vivo. Int J Nanomed 9:4043–4053. https://doi.org/10.2147/IJN.S66164
192. Yoon YS, Lee BI, Lee KS, Heo H, Lee JH, Byeon SH, Lee IS (2010) Fabrication of a silica sphere with fluorescent and MR contrasting GdPO4 nanoparticles from layered gadolinium hydroxide. Chem Commun 46(21):3654–3656. https://doi.org/10.1039/b927570c
193. Liu J, Tian X, Luo N, Yang C, Xiao J, Shao Y, Chen X, Yang G, Chen D, Li L (2014) Sub-10 nm monoclinic Gd2O3:Eu3+ nanoparticles as dual-modal nanoprobes for magnetic resonance and fluorescence imaging. Langmuir: ACS Journal Surf Coll 30(43):13005–13013. https://doi.org/10.1021/la503228v
194. Kamińska I, Fronc K, Sikora B, Mouawad M, Siemiarczuk A, Szewczyk M, Sobczak K, Wojciechowski T, Zaleszczyk W, Minikayev R, Paszkowicz W, Stępień P, Dziawa P, Ciszak K, Piątkowski D, Maćkowski S, Kaliszewski M, Włodarski M, Młyńczak J, Kopczyński K, Łapiński M, Elbaum D (2015) Upconverting/magnetic: Gd2O3:(Er3+, Yb3+, Zn2+) nanoparticles for biological applications: effect of Zn2+ doping. RSC Advances 5(95):78361–78373. https://doi.org/10.1039/c5ra11888c
195. Li H, Song S, Wang W, Chen K (2015) In vitro photodynamic therapy based on magnetic-luminescent Gd2O3:Yb, Er nanoparticles with bright three-photon up-conversion fluorescence under near-infrared light. Dalton Trans 44(36):16081–16090. https://doi.org/10.1039/c5dt01015b
196. Penfield JG, Reilly RF Jr (2007) What nephrologists need to know about gadolinium. Nat Clin Pract Nephrol 3(12):654–668. https://doi.org/10.1038/ncpneph0660
197. Herranz F, Morales MP, Rodríguez I, Ruiz-Cabello J (2017) Iron oxide nanoparticle-based mri contrast agents: characterization and in vivo use, pp 85–120. https://doi.org/10.1007/978-3-662-52780-1_3
198. Shen Z, Wu A, Chen X (2017) Iron oxide nanoparticle based contrast agents for magnetic resonance imaging. Mol Pharm 14(5):1352–1364. https://doi.org/10.1021/acs.molpharmaceut.6b00839
199. Ling D, Hyeon T (2013) Chemical design of biocompatible iron oxide nanoparticles for medical applications. Small 9(9–10):1450–1466. https://doi.org/10.1002/smll.201202111
200. Jin R, Lin B, Li D, Ai H (2014) Superparamagnetic iron oxide nanoparticles for MR imaging and therapy: design considerations and clinical applications. Curr Opin Pharmacol 18:18–27. https://doi.org/10.1016/j.coph.2014.08.002
201. Stephen ZR, Kievit FM, Zhang M (2011) Magnetite nanoparticles for medical MR imaging. Mater Today 14(7–8):330–338. https://doi.org/10.1016/s1369-7021(11)70163-8
202. Gossuin Y, Gillis P, Hocq A, Vuong QL, Roch A (2009) Magnetic resonance relaxation properties of superparamagnetic particles. Wiley Interdiscip Rev Nanomed Nanobiotechnol 1(3):299–310. https://doi.org/10.1002/wnan.36
203. Laurent S, Forge D, Port M, Roch A, Robic C, Vander Elst L, Muller RN (2008) Magnetic iron oxide nanoparticles: synthesis, stabilization, vectorization, physicochemical characterizations, and biological applications. Chem Rev 108(6):2064–2110. https://doi.org/10.1021/cr068445e
204. Wang YX, Hussain SM, Krestin GP (2001) Superparamagnetic iron oxide contrast agents: physicochemical characteristics and applications in MR imaging. Eur Radiol 11(11):2319–2331. https://doi.org/10.1007/s003300100908
205. Lee H, Lee E, Kim DK, Jang NK, Jeong YY, Jon S (2006) Antibiofouling polymer-coated superparamagnetic iron oxide nanoparticles as potential magnetic resonance contrast agents for in vivo cancer imaging. J Am Chem Soc 128(22):7383–7389. https://doi.org/10.1021/ja061529k

206. Yu MK, Jeong YY, Park J, Park S, Kim JW, Min JJ, Kim K, Jon S (2008) Drug-loaded superparamagnetic iron oxide nanoparticles for combined cancer imaging and therapy in vivo. Angew Chem 47(29):5362–5365. https://doi.org/10.1002/anie.200800857
207. Saraswathy A, Nazeer SS, Jeevan M, Nimi N, Arumugam S, Harikrishnan VS, Varma PR, Jayasree RS (2014) Citrate coated iron oxide nanoparticles with enhanced relaxivity for in vivo magnetic resonance imaging of liver fibrosis. Coll Surf B Biointerfaces 117:216–224. https://doi.org/10.1016/j.colsurfb.2014.02.034
208. Liu D, Wu W, Ling J, Wen S, Gu N, Zhang X (2011) Effective PEGylation of iron oxide nanoparticles for high performance in vivo cancer imaging. Adv Func Mater 21(8):1498–1504. https://doi.org/10.1002/adfm.201001658
209. Ye F, Laurent S, Fornara A, Astolfi L, Qin J, Roch A, Martini A, Toprak MS, Muller RN, Muhammed M (2012) Uniform mesoporous silica coated iron oxide nanoparticles as a highly efficient, nontoxic MRI T(2) contrast agent with tunable proton relaxivities. Contrast Media Mol Imaging 7(5):460–468. https://doi.org/10.1002/cmmi.1473
210. Mahmoudi M, Hosseinkhani H, Hosseinkhani M, Boutry S, Simchi A, Journeay WS, Subramani K, Laurent S (2011) Magnetic resonance imaging tracking of stem cells in vivo using iron oxide nanoparticles as a tool for the advancement of clinical regenerative medicine. Chem Rev 111(2):253–280. https://doi.org/10.1021/cr1001832
211. Andreas K, Georgieva R, Ladwig M, Mueller S, Notter M, Sittinger M, Ringe J (2012) Highly efficient magnetic stem cell labeling with citrate-coated superparamagnetic iron oxide nanoparticles for MRI tracking. Biomaterials 33(18):4515–4525. https://doi.org/10.1016/j.biomaterials.2012.02.064
212. Azevedo-Pereira RL, Rangel B, Tovar-Moll F, Gasparetto EL, Attias M, Zaverucha-do-Valle C, Jasmin Mendez-Otero R (2019) Superparamagnetic iron oxide nanoparticles as a tool to track mouse neural stem cells in vivo. Mol Biol Rep 46(1):191–198. https://doi.org/10.1007/s11033-018-4460-9
213. Zare S, Mehrabani D, Jalli R, Saeedi Moghadam M, Manafi N, Mehrabani G, Jamhiri I, Ahadian S (2019) MRI-tracking of dental pulp stem cells in vitro and in vivo using dextran-coated superparamagnetic iron oxide nanoparticles. J Clin Med 8(9). https://doi.org/10.3390/jcm8091418
214. Parkins KM, Makela AV, Hamilton AM, Foster PJ (2019) Cellular magnetic resonance imaging for tracking metastatic cancer cells in the brain. Methods Mol Biol 1869:239–251. https://doi.org/10.1007/978-1-4939-8805-1_20
215. Ashraf S, Taylor A, Sharkey J, Barrow M, Murray P, Wilm B, Poptani H, Rosseinsky MJ, Adams DJ, Lévy R (2019) In vivo fate of free and encapsulated iron oxide nanoparticles after injection of labelled stem cells. Nanoscale Adv 1(1):367–377. https://doi.org/10.1039/c8na00098k
216. Lee S, Yoon HI, Na JH, Jeon S, Lim S, Koo H, Han SS, Kang SW, Park SJ, Moon SH, Park JH, Cho YW, Kim BS, Kim SK, Lee T, Kim D, Lee S, Pomper MG, Kwon IC, Kim K (2017) In vivo stem cell tracking with imageable nanoparticles that bind bioorthogonal chemical receptors on the stem cell surface. Biomaterials 139:12–29. https://doi.org/10.1016/j.biomaterials.2017.05.050
217. Rosen JE, Chan L, Shieh DB, Gu FX (2012) Iron oxide nanoparticles for targeted cancer imaging and diagnostics. Nanomed Nanotechnol Biol Med 8(3):275–290. https://doi.org/10.1016/j.nano.2011.08.017
218. Li J, Zheng L, Cai H, Sun W, Shen M, Zhang G, Shi X (2013) Polyethyleneimine-mediated synthesis of folic acid-targeted iron oxide nanoparticles for in vivo tumor MR imaging. Biomaterials 34(33):8382–8392. https://doi.org/10.1016/j.biomaterials.2013.07.070
219. Nyberg S, Abbott NJ, Shi X, Steyger PS, Dabdoub A (2019) Delivery of therapeutics to the inner ear: the challenge of the blood-labyrinth barrier. Sci Transl Med 11(482). https://doi.org/10.1126/scitranslmed.aao0935
220. Sweeney MD, Sagare AP, Zlokovic BV (2018) Blood-brain barrier breakdown in Alzheimer disease and other neurodegenerative disorders. Nat Rev Neurol 14(3):133–150. https://doi.org/10.1038/nrneurol.2017.188

221. Afzalipour R, Khoei S, Khoee S, Shirvalilou S, Jamali Raoufi N, Motevalian M, Karimi MR (2019) Dual-targeting temozolomide loaded in folate-conjugated magnetic triblock copolymer nanoparticles to improve the therapeutic efficiency of rat brain gliomas. ACS Biomater Sci Eng 5(11):6000–6011. https://doi.org/10.1021/acsbiomaterials.9b00856
222. Du C, Liu X, Hu H, Li H, Yu L, Geng D, Chen Y, Zhang J (2020) Dual-targeting and excretable ultrasmall SPIONs for T1-weighted positive MR imaging of intracranial glioblastoma cells by targeting the lipoprotein receptor-related protein. J Mater Chem B 8(11):2296–2306. https://doi.org/10.1039/c9tb02391g
223. Ashwinkumar N, Maya S, Jayakumar R (2014) Redox-responsive cystamine conjugated chitin–hyaluronic acid composite nanogels. RSC Adv 4(91):49547–49555. https://doi.org/10.1039/c4ra06578f
224. Ma D, Shi M, Li X, Zhang J, Fan Y, Sun K, Jiang T, Peng C, Shi X (2020) Redox-sensitive clustered ultrasmall iron oxide nanoparticles for switchable T2/T1-weighted magnetic resonance imaging applications. Bioconjug Chem 31(2):352–359. https://doi.org/10.1021/acs.bioconjchem.9b00659
225. Lin R, Huang J, Wang L, Li Y, Lipowska M, Wu H, Yang J, Mao H (2018) Bevacizumab and near infrared probe conjugated iron oxide nanoparticles for vascular endothelial growth factor targeted MR and optical imaging. Biomater Sci 6(6):1517–1525. https://doi.org/10.1039/c8bm00225h
226. Mattheolabakis G, Milane L, Singh A, Amiji MM (2015) Hyaluronic acid targeting of CD44 for cancer therapy: from receptor biology to nanomedicine. J Drug Target 23(7–8):605–618. https://doi.org/10.3109/1061186X.2015.1052072
227. Senbanjo LT, Chellaiah MA (2017) CD44: a multifunctional cell surface adhesion receptor is a regulator of progression and metastasis of cancer cells. Front Cell Dev Biol 5:18. https://doi.org/10.3389/fcell.2017.00018
228. Zheng S, Han J, Jin Z, Kim CS, Park S, Kim KP, Park JO, Choi E (2018) Dual tumor-targeted multifunctional magnetic hyaluronic acid micelles for enhanced MR imaging and combined photothermal-chemotherapy. Coll Surf B Biointerfaces 164:424–435. https://doi.org/10.1016/j.colsurfb.2018.02.005
229. Murukesh N, Dive C, Jayson GC (2010) Biomarkers of angiogenesis and their role in the development of VEGF inhibitors. Br J Cancer 102(1):8–18. https://doi.org/10.1038/sj.bjc.6605483
230. Ferrara N, Hillan KJ, Novotny W (2005) Bevacizumab (Avastin), a humanized anti-VEGF monoclonal antibody for cancer therapy. Biochem Biophys Res Commun 333(2):328–335. https://doi.org/10.1016/j.bbrc.2005.05.132
231. Zou Q, Zhang CJ, Yan YZ, Min ZJ, Li CS (2019) MUC-1 aptamer targeted superparamagnetic iron oxide nanoparticles for magnetic resonance imaging of pancreatic cancer in vivo and in vitro experiment. J Cell Biochem 120(11):18650–18658. https://doi.org/10.1002/jcb.28950
232. Gale EM, Wey HY, Ramsay I, Yen YF, Sosnovik DE, Caravan P (2018) A manganese-based alternative to gadolinium: contrast-enhanced MR angiography, excretion, pharmacokinetics, and metabolism. Radiology 286(3):865–872. https://doi.org/10.1148/radiol.2017170977
233. Pan D, Schmieder AH, Wickline SA, Lanza GM (2011) Manganese-based MRI contrast agents: past, present and future. Tetrahedron 67(44):8431–8444. https://doi.org/10.1016/j.tet.2011.07.076
234. McDonagh BH, Singh G, Hak S, Bandyopadhyay S, Augestad IL, Peddis D, Sandvig I, Sandvig A, Glomm WR (2016) L-DOPA-coated manganese oxide nanoparticles as dual MRI contrast agents and drug-delivery vehicles. Small 12(3):301–306. https://doi.org/10.1002/smll.201502545
235. Wang P, Yang J, Zhou B, Hu Y, Xing L, Xu F, Shen M, Zhang G, Shi X (2017) Antifouling manganese oxide nanoparticles: synthesis, characterization, and applications for enhanced MR imaging of tumors. ACS Appl Mater Interfaces 9(1):47–53. https://doi.org/10.1021/acsami.6b13844
236. Venter A, Szulc DA, Loai S, Ganesh T, Haedicke IE, Cheng HM (2018) A manganese porphyrin-based T1 contrast agent for cellular MR imaging of human embryonic stem cells. Sci Rep 8(1):12129. https://doi.org/10.1038/s41598-018-30661-w

237. Kumar NA, Rejinold NS, Anjali P, Balakrishnan A, Biswas R, Jayakumar R (2013) Preparation of chitin nanogels containing nickel nanoparticles. Carbohydr Polym 97(2):469–474. https://doi.org/10.1016/j.carbpol.2013.05.009
238. Caro C, Garcia-Martin ML, Pernia Leal M (2017) Manganese-based nanogels as pH switches for magnetic resonance imaging. Biomacromolecules 18(5):1617–1623. https://doi.org/10.1021/acs.biomac.7b00224
239. Addisu KD, Hailemeskel BZ, Mekuria SL, Andrgie AT, Lin YC, Tsai HC (2018) Bioinspired, manganese-chelated alginate-polydopamine nanomaterials for efficient in vivo T1-weighted magnetic resonance imaging. ACS Appl Mater Interfaces 10(6):5147–5160. https://doi.org/10.1021/acsami.7b13396
240. Wang R, Luo Y, Yang S, Lin J, Gao D, Zhao Y, Liu J, Shi X, Wang X (2016) Hyaluronic acid-modified manganese-chelated dendrimer-entrapped gold nanoparticles for the targeted CT/MR dual-mode imaging of hepatocellular carcinoma. Sci Rep 6:33844. https://doi.org/10.1038/srep33844
241. Li J, Wu C, Hou P, Zhang M, Xu K (2018) One-pot preparation of hydrophilic manganese oxide nanoparticles as T1 nano-contrast agent for molecular magnetic resonance imaging of renal carcinoma in vitro and in vivo. Biosens Bioelectron 102:1–8. https://doi.org/10.1016/j.bios.2017.10.047
242. Ding X, Liu J, Li J, Wang F, Wang Y, Song S, Zhang H (2016) Polydopamine coated manganese oxide nanoparticles with ultrahigh relaxivity as nanotheranostic agents for magnetic resonance imaging guided synergetic chemo-/photothermal therapy. Chem Sci 7(11):6695–6700. https://doi.org/10.1039/c6sc01320a
243. Ragg R, Schilmann AM, Korschelt K, Wieseotte C, Kluenker M, Viel M, Volker L, Preiss S, Herzberger J, Frey H, Heinze K, Blumler P, Tahir MN, Natalio F, Tremel W (2016) Intrinsic superoxide dismutase activity of MnO nanoparticles enhances the magnetic resonance imaging contrast. J Mater Chem B 4(46):7423–7428. https://doi.org/10.1039/c6tb02078j
244. Zhan Y, Shi S, Ehlerding EB, Graves SA, Goel S, Engle JW, Liang J, Tian J, Cai W (2017) Radiolabeled, antibody-conjugated manganese oxide nanoparticles for tumor vasculature targeted positron emission tomography and magnetic resonance imaging. ACS Appl Mater Interfaces 9(44):38304–38312. https://doi.org/10.1021/acsami.7b12216
245. Yang L, Ma L, Xin J, Li A, Sun C, Wei R, Ren BW, Chen Z, Lin H, Gao J (2017) Composition tunable manganese ferrite nanoparticles for optimized T2 contrast ability. Chem Mater 29(7):3038–3047. https://doi.org/10.1021/acs.chemmater.7b00035
246. Augustine R, Lee HR, Kim H, Zhang Y, Kim I (2019) Hyperbranched lipopolymer-folate-stabilized manganese ferrite nanoparticles for the water-soluble targeted MRI contrast agent. React Funct Polym 144:104352. https://doi.org/10.1016/j.reactfunctpolym.2019.104352
247. Faraji S, Dini G, Zahraei M (2019) Polyethylene glycol-coated manganese-ferrite nanoparticles as contrast agents for magnetic resonance imaging. J Magn Magn Mater 475:137–145. https://doi.org/10.1016/j.jmmm.2018.11.097
248. Li M, Zhao Q, Yi X, Zhong X, Song G, Chai Z, Liu Z, Yang K (2016) Au@MnS@ZnS Core/Shell/Shell nanoparticles for magnetic resonance imaging and enhanced cancer radiation therapy. ACS Appl Mater Interfaces 8(15):9557–9564. https://doi.org/10.1021/acsami.5b11588
249. Li XW, Zhao WR, Liu YJ, Liu XH, Shi P, Li YS, Shi JL (2016) Facile synthesis of manganese silicate nanoparticles for pH/GSH-responsive T1-weighted magnetic resonance imaging. J Mater Chem B 4(24):4313–4321. https://doi.org/10.1039/c6tb00718j
250. Cheng Y, Zhang S, Kang N, Huang J, Lv X, Wen K, Ye S, Chen Z, Zhou X, Ren L (2017) Polydopamine-coated manganese carbonate nanoparticles for amplified magnetic resonance imaging-guided photothermal therapy. ACS Appl Mater Interfaces 9(22):19296–19306. https://doi.org/10.1021/acsami.7b03087
251. Dasgupta A, Liu M, Ojha T, Storm G, Kiessling F, Lammers T (2016) Ultrasound-mediated drug delivery to the brain: principles, progress and prospects. Drug Discov Today Technol 20:41–48. https://doi.org/10.1016/j.ddtec.2016.07.007

252. Unger E, Shen D, Fritz T, Kulik B, Lund P, Wu GL, Yellowhair D, Ramaswami R, Matsunaga T (1994) Gas-filled lipid bilayers as ultrasound contrast agents. Invest Radiol 29(Suppl 2):S134–136. https://doi.org/10.1097/00004424-199406001-00044
253. Zhao YZ, Du LN, Lu CT, Jin YG, Ge SP (2013) Potential and problems in ultrasound-responsive drug delivery systems. Int J Nanomed 8:1621–1633. https://doi.org/10.2147/IJN.S43589
254. Alheshibri M, Craig VSJ (2019) Armoured nanobubbles; ultrasound contrast agents under pressure. J Coll Interface Sci 537:123–131. https://doi.org/10.1016/j.jcis.2018.10.108
255. Yasuda K, Matsushima H, Asakura Y (2019) Generation and reduction of bulk nanobubbles by ultrasonic irradiation. Chem Eng Sci 195:455–461. https://doi.org/10.1016/j.ces.2018.09.044
256. Kalus M-R, Reimer V, Barcikowski S, Gökce B (2019) Discrimination of effects leading to gas formation during pulsed laser ablation in liquids. Appl Surf Sci 465:1096–1102. https://doi.org/10.1016/j.apsusc.2018.09.224
257. Stride E, Edirisinghe M (2008) Novel microbubble preparation technologies. Soft Matter 4(12):2350. https://doi.org/10.1039/b809517p
258. Khan MS, Hwang J, Lee K, Choi Y, Kim K, Koo HJ, Hong JW, Choi J (2018) Oxygen-carrying micro/nanobubbles: composition, synthesis techniques and potential prospects in photo-triggered theranostics. Molecules 23(9). https://doi.org/10.3390/molecules23092210
259. Alheshibri M, Qian J, Jehannin M, Craig VS (2016) A history of nanobubbles. Langmuir: ACS J Surf Coll 32(43):11086–11100. https://doi.org/10.1021/acs.langmuir.6b02489
260. Johnson BD, Cooke RC (1981) Generation of stabilized microbubbles in seawater. Science 213(4504):209–211. https://doi.org/10.1126/science.213.4504.209
261. Kikuchi K, Ioka A, Oku T, Tanaka Y, Saihara Y, Ogumi Z (2009) Concentration determination of oxygen nanobubbles in electrolyzed water. J Coll Interface Sci 329(2):306–309. https://doi.org/10.1016/j.jcis.2008.10.009
262. Kikuchi K, Nagata S, Tanaka Y, Saihara Y, Ogumi Z (2007) Characteristics of hydrogen nanobubbles in solutions obtained with water electrolysis. J Electroanal Chem 600(2):303–310. https://doi.org/10.1016/j.jelechem.2006.10.005
263. Ohgaki K, Khanh NQ, Joden Y, Tsuji A, Nakagawa T (2010) Physicochemical approach to nanobubble solutions. Chem Eng Sci 65(3):1296–1300. https://doi.org/10.1016/j.ces.2009.10.003
264. Cai WB, Yang HL, Zhang J, Yin JK, Yang YL, Yuan LJ, Zhang L, Duan YY (2015) The optimized fabrication of nanobubbles as ultrasound contrast agents for tumor imaging. Sci Rep 5:13725. https://doi.org/10.1038/srep13725
265. Pan TL, Wang PW, Al-Suwayeh SA, Huang YJ, Fang JY (2012) Toxicological effects of cationic nanobubbles on the liver and kidneys: biomarkers for predicting the risk. Food Chem Toxicol 50(11):3892–3901. https://doi.org/10.1016/j.fct.2012.07.005
266. Mai L, Yao A, Li J, Wei Q, Yuchi M, He X, Ding M, Zhou Q (2013) Cyanine 5.5 conjugated nanobubbles as a tumor selective contrast agent for dual ultrasound-fluorescence imaging in a mouse model. PloS One 8(4):e61224. https://doi.org/10.1371/journal.pone.0061224
267. Huang HY, Hu SH, Hung SY, Chiang CS, Liu HL, Chiu TL, Lai HY, Chen YY, Chen SY (2013) SPIO nanoparticle-stabilized PAA-F127 thermosensitive nanobubbles with MR/US dual-modality imaging and HIFU-triggered drug release for magnetically guided in vivo tumor therapy. J Controll Release: Off J Controll Release Soc 172(1):118–127. https://doi.org/10.1016/j.jconrel.2013.07.029
268. Huang HY, Liu HL, Hsu PH, Chiang CS, Tsai CH, Chi HS, Chen SY, Chen YY (2015) A multitheragnostic nanobubble system to induce blood-brain barrier disruption with magnetically guided focused ultrasound. Adv Mater 27(4):655–661. https://doi.org/10.1002/adma.201403889
269. Fan X, Wang L, Guo Y, Tu Z, Li L, Tong H, Xu Y, Li R, Fang K (2015) Ultrasonic nanobubbles carrying anti-PSMA nanobody: construction and application in prostate cancer-targeted imaging. PLoS One 10(6):e0127419. https://doi.org/10.1371/journal.pone.0127419
270. Peyman SA, McLaughlan JR, Abou-Saleh RH, Marston G, Johnson BR, Freear S, Coletta PL, Markham AF, Evans SD (2016) On-chip preparation of nanoscale contrast agents towards

high-resolution ultrasound imaging. Lab Chip 16(4):679–687. https://doi.org/10.1039/c5lc01394a
271. Luke GP, Hannah AS, Emelianov SY (2016) Super-resolution ultrasound imaging in vivo with transient laser-activated nanodroplets. Nano Lett 16(4):2556–2559. https://doi.org/10.1021/acs.nanolett.6b00108
272. Weissleder R, Pittet MJ (2008) Imaging in the era of molecular oncology. Nature 452(7187):580–589. https://doi.org/10.1038/nature06917
273. Weissleder R, Ntziachristos V (2003) Shedding light onto live molecular targets. Nat Med 9(1):123–128. https://doi.org/10.1038/nm0103-123
274. Hillman EMC, Amoozegar CB, Wang T, McCaslin AFH, Bouchard MB, Mansfield J, Levenson RM (2011) In vivo optical imaging and dynamic contrast methods for biomedical research. Philos Trans R Soc A: Math, Phys Eng Sci 369(1955):4620–4643. https://doi.org/10.1098/rsta.2011.0264
275. Kosaka N, Ogawa M, Choyke PL, Kobayashi H (2009) Clinical implications of near-infrared fluorescence imaging in cancer. Future Oncol 5(9):1501–1511. https://doi.org/10.2217/fon.09.109
276. Reineck P, Gibson BC (2017) Near-infrared fluorescent nanomaterials for bioimaging and sensing. Adv Opt Mater 5(2):1600446. https://doi.org/10.1002/adom.201600446
277. Hahn MA, Singh AK, Sharma P, Brown SC, Moudgil BM (2011) Nanoparticles as contrast agents for in-vivo bioimaging: current status and future perspectives. Anal Bioanal Chem 399(1):3–27. https://doi.org/10.1007/s00216-010-4207-5
278. Choi HS, Frangioni JV (2010) Nanoparticles for biomedical imaging: fundamentals of clinical translation. Mol Imaging 9(6):291–310
279. Frangioni JV (2008) New technologies for human cancer imaging. J Clin Oncol 26(24):4012–4021. https://doi.org/10.1200/JCO.2007.14.3065
280. Debbage P, Jaschke W (2008) Molecular imaging with nanoparticles: giant roles for dwarf actors. Histochem Cell Biol 130(5):845–875. https://doi.org/10.1007/s00418-008-0511-y
281. Jaque D, Richard C, Viana B, Soga K, Liu X, García Solé J (2016) Inorganic nanoparticles for optical bioimaging. Adv Opt Photon 8(1):1–103. https://doi.org/10.1364/AOP.8.000001
282. Quek C-H, Leong KW (2012) Near-infrared fluorescent nanoprobes for in vivo optical imaging. Nanomater (Basel) 2(2):92–112. https://doi.org/10.3390/nano2020092
283. Ji X, Peng F, Zhong Y, Su Y, He Y (2014) Fluorescent quantum dots: Synthesis, biomedical optical imaging, and biosafety assessment. Coll Surf, B 124:132–139. https://doi.org/10.1016/j.colsurfb.2014.08.036
284. Matea CT, Mocan T, Tabaran F, Pop T, Mosteanu O, Puia C, Iancu C, Mocan L (2017) Quantum dots in imaging, drug delivery and sensor applications. Int J Nanomed 12:5421–5431. https://doi.org/10.2147/IJN.S138624
285. Li J, Zhu J-J (2013) Quantum dots for fluorescent biosensing and bio-imaging applications. Analyst 138(9):2506–2515. https://doi.org/10.1039/C3AN36705C
286. Martynenko IV, Litvin AP, Purcell-Milton F, Baranov AV, Fedorov AV, Gun'ko YK (2017) Application of semiconductor quantum dots in bioimaging and biosensing. J Mater Chem B 5(33):6701–6727. https://doi.org/10.1039/C7TB01425B
287. Girma WM, Fahmi MZ, Permadi A, Abate MA, Chang J-Y (2017) Synthetic strategies and biomedical applications of I-III–VI ternary quantum dots. J Mater Chem B 5(31):6193–6216. https://doi.org/10.1039/C7TB01156C
288. Chen L-N, Wang J, Li W-T, Han H-Y (2012) Aqueous one-pot synthesis of bright and ultrasmall CdTe/CdS near-infrared-emitting quantum dots and their application for tumor targeting in vivo. Chem Commun 48(41):4971–4973. https://doi.org/10.1039/C2CC31259J
289. Narasimhan AK, Lakshmi BS, Santra TS, Rao MSR, Krishnamurthi G (2017) Oxygenated graphene quantum dots (GQDs) synthesized using laser ablation for long-term real-time tracking and imaging. RSC Adv 7(85):53822–53829. https://doi.org/10.1039/c7ra10702a
290. Gao D, Zhang P, Sheng Z, Hu D, Gong P, Chen C, Wan Q, Gao G, Cai L (2014) Highly bright and compact alloyed quantum rods with near infrared emitting: a potential multifunctional nanoplatform for multimodal imaging in vivo. Adv Func Mater 24(25):3897–3905. https://doi.org/10.1002/adfm.201304225

291. Ntziachristos V, Bremer C, Weissleder R (2003) Fluorescence imaging with near-infrared light: new technological advances that enable in vivo molecular imaging. Eur Radiol 13(1):195–208. https://doi.org/10.1007/s00330-002-1524-x
292. Kenry Duan Y, Liu B (2018) Recent advances of optical imaging in the second near-infrared window. Adv Mater 30(47):1802394. https://doi.org/10.1002/adma.201802394
293. Hong G, Antaris AL, Dai H (2017) Near-infrared fluorophores for biomedical imaging. Nat Biomed Eng 1(1):0010. https://doi.org/10.1038/s41551-016-0010
294. Chen G, Tian F, Zhang Y, Zhang Y, Li C, Wang Q (2014) Tracking of transplanted human mesenchymal stem cells in living mice using near-infrared Ag2S quantum dots. Adv Func Mater 24(17):2481–2488. https://doi.org/10.1002/adfm.201303263
295. Li C, Zhang Y, Wang M, Zhang Y, Chen G, Li L, Wu D, Wang Q (2014) In vivo real-time visualization of tissue blood flow and angiogenesis using Ag2S quantum dots in the NIR-II window. Biomaterials 35(1):393–400. https://doi.org/10.1016/j.biomaterials.2013.10.010
296. Abbasi E, Kafshdooz T, Bakhtiary M, Nikzamir N, Nikzamir N, Nikzamir M, Mohammadian M, Akbarzadeh A (2016) Biomedical and biological applications of quantum dots. Artif Cells, Nanomed, Biotechnol 44(3):885–891. https://doi.org/10.3109/21691401.2014.998826
297. Bannai H, Lévi S, Schweizer C, Dahan M, Triller A (2006) Imaging the lateral diffusion of membrane molecules with quantum dots. Nat Protoc 1(6):2628–2634. https://doi.org/10.1038/nprot.2006.429
298. Wen S, Zhou J, Zheng K, Bednarkiewicz A, Liu X, Jin D (2018) Advances in highly doped upconversion nanoparticles. Nat Commun 9(1):2415. https://doi.org/10.1038/s41467-018-04813-5
299. Jiang S, Gnanasammandhan MK, Zhang Y (2010) Optical imaging-guided cancer therapy with fluorescent nanoparticles. J R Soc Interface 7(42):3–18. https://doi.org/10.1098/rsif.2009.0243
300. Wang F, Chatterjee DK, Li Z, Zhang Y, Fan X, Wang M (2006) Synthesis of polyethylenimine/NaYF4nanoparticles with upconversion fluorescence. Nanotechnology 17(23):5786–5791. https://doi.org/10.1088/0957-4484/17/23/013
301. Lisiecki R, Ryba-Romanowski W, Speghini A, Bettinelli M (2009) Luminescence spectroscopy of Er3+-doped and Er3+, Yb3+ -codoped LaPO4 single crystals. J Lumin 129(5):521–525. https://doi.org/10.1016/j.jlumin.2008.12.006
302. Guo H, Qiao YM (2009) Preparation, characterization, and strong upconversion of monodisperse Y2O3:Er3+, Yb3+ microspheres. Opt Mater 31(4):583–589. https://doi.org/10.1016/j.optmat.2008.06.011
303. Wu S, Han G, Milliron DJ, Aloni S, Altoe V, Talapin DV, Cohen BE, Schuck PJ (2009) Non-blinking and photostable upconverted luminescence from single lanthanide-doped nanocrystals. Proc Natl Acad Sci 106(27):10917. https://doi.org/10.1073/pnas.0904792106
304. Wang F, Han Y, Lim CS, Lu Y, Wang J, Xu J, Chen H, Zhang C, Hong M, Liu X (2010) Simultaneous phase and size control of upconversion nanocrystals through lanthanide doping. Nature 463(7284):1061–1065. https://doi.org/10.1038/nature08777
305. Chatterjee DK, Rufaihah AJ, Zhang Y (2008) Upconversion fluorescence imaging of cells and small animals using lanthanide doped nanocrystals. Biomaterials 29(7):937–943. https://doi.org/10.1016/j.biomaterials.2007.10.051
306. Chen G, Qiu H, Prasad PN, Chen X (2014) Upconversion nanoparticles: design, nanochemistry, and applications in theranostics. Chem Rev 114(10):5161–5214. https://doi.org/10.1021/cr400425h
307. Park HS, Nam SH, Kim J, Shin HS, Suh YD, Hong KS (2016) Clear-cut observation of clearance of sustainable upconverting nanoparticles from lymphatic system of small living mice. Sci Rep 6(1):27407. https://doi.org/10.1038/srep27407
308. Qiao R, Liu C, Liu M, Hu H, Liu C, Hou Y, Wu K, Lin Y, Liang J, Gao M (2015) Ultrasensitive in vivo detection of primary gastric tumor and lymphatic metastasis using upconversion nanoparticles. ACS Nano 9(2):2120–2129. https://doi.org/10.1021/nn507433p

309. Tian R, Zhao S, Liu G, Chen H, Ma L, You H, Liu C, Wang Z (2019) Construction of lanthanide-doped upconversion nanoparticle-Uelx Europaeus Agglutinin-I bioconjugates with brightness red emission for ultrasensitive in vivo imaging of colorectal tumor. Biomaterials 212:64–72. https://doi.org/10.1016/j.biomaterials.2019.05.010
310. Liang L, Chen N, Jia Y, Ma Q, Wang J, Yuan Q, Tan W (2019) Recent progress in engineering near-infrared persistent luminescence nanoprobes for time-resolved biosensing/bioimaging. Nano Res 12(6):1279–1292. https://doi.org/10.1007/s12274-019-2343-6
311. Brito HF, Hölsä J, Laamanen T, Lastusaari M, Malkamäki M, Rodrigues LCV (2012) Persistent luminescence mechanisms: human imagination at work. Opt Mater Express 2(4):371–381. https://doi.org/10.1364/OME.2.000371
312. Liu J, Lécuyer T, Seguin J, Mignet N, Scherman D, Viana B, Richard C (2019) Imaging and therapeutic applications of persistent luminescence nanomaterials. Adv Drug Deliv Rev 138:193–210. https://doi.org/10.1016/j.addr.2018.10.015
313. Abdukayum A, Chen J-T, Zhao Q, Yan X-P (2013) Functional near infrared-emitting Cr3+/Pr3+Co-doped zinc gallogermanate persistent luminescent nanoparticles with superlong afterglow for in vivo targeted bioimaging. J Am Chem Soc 135(38):14125–14133. https://doi.org/10.1021/ja404243v
314. Jiang Y, Huang J, Zhen X, Zeng Z, Li J, Xie C, Miao Q, Chen J, Chen P, Pu K (2019) A generic approach towards afterglow luminescent nanoparticles for ultrasensitive in vivo imaging. Nat Commun 10(1):2064. https://doi.org/10.1038/s41467-019-10119-x
315. Attia ABE, Balasundaram G, Moothanchery M, Dinish US, Bi R, Ntziachristos V, Olivo M (2019) A review of clinical photoacoustic imaging: current and future trends. Photoacoustics 16:100144. https://doi.org/10.1016/j.pacs.2019.100144
316. Xia J, Wang LV (2014) Small-animal whole-body photoacoustic tomography: a review. IEEE Transac Bio-med Eng 61(5):1380–1389. https://doi.org/10.1109/TBME.2013.2283507
317. Wu D, Huang L, Jiang MS, Jiang H (2014) Contrast agents for photoacoustic and thermoacoustic imaging: a review. Int J Mol Sci 15(12):23616–23639. https://doi.org/10.3390/ijms151223616
318. Ku G, Wang LV (2005) Deeply penetrating photoacoustic tomography in biological tissues enhanced with an optical contrast agent. Opt Lett 30(5):507–509. https://doi.org/10.1364/OL.30.000507
319. Pan D, Kim B, Wang LV, Lanza GM (2013) A brief account of nanoparticle contrast agents for photoacoustic imaging. Wiley Interdiscip Rev Nanomed Nanobiotechnol 5(6):517–543. https://doi.org/10.1002/wnan.1231
320. Jiang Y, Pu K (2017) Advanced photoacoustic imaging applications of near-infrared absorbing organic nanoparticles. Small 13(30):1700710. https://doi.org/10.1002/smll.201700710
321. Maji SK, Sreejith S, Joseph J, Lin M, He T, Tong Y, Sun H, Yu SW-K, Zhao Y (2014) Upconversion nanoparticles as a contrast agent for photoacoustic imaging in live mice. Adv Mater 26(32):5633–5638. https://doi.org/10.1002/adma.201400831
322. Ding D, Guo W, Guo C, Sun J, Zheng N, Wang F, Yan M, Liu S (2017) MoO 3-x quantum dot for photoacoustic imaging guided photothermal/photodynamic cancer treatment. Nanoscale 8:1–3. https://doi.org/10.1039/C6NR09046J
323. Vaquero JJ, Kinahan P (2015) Positron emission tomography: current challenges and opportunities for technological advances in clinical and preclinical imaging systems. Annu Rev Biomed Eng 17:385–414. https://doi.org/10.1146/annurev-bioeng-071114-040723
324. Almuhaideb A, Papathanasiou N, Bomanji J (2011) 18F-FDG PET/CT imaging in oncology. Ann Saudi Med 31(1):3–13. https://doi.org/10.4103/0256-4947.75771
325. Gambhir SS (2002) Molecular imaging of cancer with positron emission tomography. Nat Rev Cancer 2(9):683–693. https://doi.org/10.1038/nrc882
326. Tu C, Ma X, House A, Kauzlarich SM, Louie AY (2011) PET imaging and biodistribution of silicon quantum dots in mice. ACS Med Chem Lett 2(4):285–288. https://doi.org/10.1021/ml1002844

327. Cheng S-H, Yu D, Tsai H-M, Morshed RA, Kanojia D, Lo L-W, Leoni L, Govind Y, Zhang L, Aboody KS, Lesniak MS, Chen C-T, Balyasnikova IV (2016) Dynamic in vivo spect imaging of neural stem cells functionalized with radiolabeled nanoparticles for tracking of glioblastoma. J Nucl Med 57(2):279–284

Nanomaterials: Surface Functionalization, Modification, and Applications

Ashish Kumar, Kiran Kaladharan, and Fan-Gang Tseng

Abstract A solid substance is considered as a nanomaterial if it has at least one face in the dimension range of 1–100 nm. Nanomaterials, because of their small size and exceptionally high surface area, display electromagnetic, optical, and piezoelectric properties, and have enormous applications in various fields such as wastewater treatment, oil, and gas industry, energy storage, bio-imaging, and healthcare diagnostics and treatment. Nanomaterials provide flexibility to the sensing platforms and also even allow mobility between various detection techniques. In the last few decades, many nanomaterials have been developed for various applications, and considerable interest has been gained in the field of medical diagnosis and therapy in recent years. The innovations in nanomaterial preparation and their modifications have led to the development of devices and assays used for biomedical applications, which are faster, less expensive, accurate, and sensitive. Despite having excellent material properties, many nanomaterials are affected by the lack of surface heterogenic reactivity, which is critical for the surface immobilization of essential biomarkers. Thus the nanomaterials need to go through surface modifications to improve the adsorption capacity of biomolecules to the functionalized surface. In this chapter, various surface modifications of nanomaterials, including metallic nanoparticles, carbon nanomaterials,

A. Kumar (✉) · K. Kaladharan · F.-G. Tseng
Department of Engineering and System Sciences, National Tsing Hua University, Hsinchu, Taiwan, ROC
e-mail: raiash024@gmail.com

F.-G. Tseng
e-mail: fangangtseng@gmail.com

F.-G. Tseng
Research Center for Applied Sciences, Academia Sinica, Taipei, Taiwan, ROC

Frontier Research Center on Fundamental and Applied Sciences of Matters, National Tsing Hua University, Hsinchu, Taiwan, ROC

Institute of NanoEngineering and MicroSystems, National Tsing Hua University, Hsinchu, Taiwan, ROC

T. S. Santra and L. Mohan (eds.), *Nanomaterials and Their Biomedical Applications*, Springer Series in Biomaterials Science and Engineering 16,
https://doi.org/10.1007/978-981-33-6252-9_14

nanoceramics, and self-assembled materials, have been discussed for various applications in general and more attention has been given to biomedical applications in particular.

Keywords Nanomaterials · Modification · Functionalization · Biomedical · Adsorption · Biomolecules · Catalysis · Chemical catalysis

1 Importance and General Consideration

In the past few decades, much attention has been paid to the preparation, surface modification, surface functionalization and miniaturization of nanoscale materials and recently, depending on the maturing of new fabrication and characterization techniques, materials can be synthesized with a few atoms up to hundreds of atoms and also their properties could be determined easily. As compared to their bulk counterparts, nanosized materials exhibit new characteristics like optical, electrical, and magnetic properties because of the enhanced surface to volume ratio and quantum confinement effects emerging in these size ranges. These new characteristics of nanoparticles offer them the prospects to be used in a wide range of technical areas such as magnetic data storage, refrigeration, etc., environmental prospects such as catalysts, hydrogen storage, etc., and energy prospects such as lithium-ion batteries, solar cells, etc., biotechnology [1], targeted drug delivery [2, 3], and vehicles for gene and drug delivery [2, 4, 5] and other biomedical applications. Surface modification refers to the scientific technique of depositing a specific material like fluorescent dyes or other chemical ligands that coats in a well-controlled manner on the surface of a nanomaterial to broaden the scope of the nanomaterials. These nanomaterials applications are based on their physical and chemical properties. To focus in the fields of nanoscience and technology, surface modification methods are a rapidly growing area, along with the design and evolution of nanomaterials. Surface modification is requisite to stabilize a nanoparticle and prevents agglomeration. The increase in the surface area of nanoparticle also increases its reactivity and makes it more unstable at its desired state. To increase the shelf life of the material the surface of nanoparticle is usually stabilized at its desired size using suitable organic groups. Modification of the surface of a nanomaterial alters its compatibility in various phases, which is the vital component in the choice of applications for nanomaterials [6, 7, 8]. However, sometimes nanomaterials behave incompatibly with other materials and applications because of their existing surface morphology, ionic properties, and publicity. Therefore, the surface should be designed in such a way that it should be compatible with the phase of the application without changing its actual properties. The addition of an extra layer enhances its properties and allows the nanomaterial to remain intact at the core as desired. The modification for the phase compatibility approach is useful for bio applications by using the polymer and polymeric chain's functionalization. Modification of nanomaterials creates homogeneity that helps avoid compatibility problems between two phases, consequently improving the availability and utility of the properties of the material for the desired applications. The recent interest in the surface modification technique allows nanomaterials to stabilize and self-organize within a chemical reaction, establishing an equilibrium condition in the structural process [7, 8, 9]. Generally, simple organic groups are adequate that could prevent the agglomeration of nanoparticles. Moreover, the use of complex functional organic groups acts as capping agents on the particle surface that may increase the interaction of the nanoparticles with various surfaces and materials in a phase of application [10, 11]. Typically, nanoparticles or nanomaterials used in the biotechnology area range in particle size between 10 nm to 500 nm. These characteristics of nano-size particles accord empower them for various communications with biomolecules that are present on the cell surfaces and within the cells in such a way that can be decoded and designated to various physicochemical and biochemical properties of the cells [12]. Modified nanoparticles give way for specific targeting applications that help target systems to be more accurate for drug delivery systems, noninvasive imaging technique, and recognize the targeted cells such as cancer cells. These attribute of modified nanoparticles gives the advantage and endless opportunities in the field of biotechnology for molecular diagnostics and therapy [13]. These modified nanoparticles could be used as imaging probes for various applications such as positron emission tomography (PET), ultrasound (US), computed tomography (CT), X-ray, magnetic resonance imaging (MRI), optical imaging, and surface-enhanced Raman imaging (SERS), etc. [14]. To determine the extent of disease and evaluate the effectiveness of the treatment, molecular imaging probes could be one of the best weapons that can noninvasively provide valuable information about different abnormalities in various body structures and organs [12]. Therefore, enabling the short molecular imaging techniques purvey the cellular functions' visualization, consequently the follow-up of the molecular process in living organisms without perturbing them [15]. Over the years, metal nanoparticles such as magnetic nanoparticles (iron oxide), gold and silver nanoparticles, Nano-shells, and Nano-cages have taken a vital place as a diagnostic and therapeutic agent. On the other hand, to incorporate nanoparticles into functional structures, we can choose the self-assembly process, which offers the most promising and simple method. The organic materials grant easy to control for the self-assembly process that offers the manufacturer to specialize in various domains such as electronics, magnetic, or photonic properties of inorganic components to achieve advanced functionalities from materials. Consequently, these fabricated nanostructures constitute many excellent properties that may be applicable for various potential applications and fields of industry, including biomedicine, computers, electronics, robotics, telecommunications, transportation, and water treat-ment, etc. Along with the ability to formulate excellent and vital properties of materials for specialized applications, the bottom-up approaches make this (self-assemble) fabrication method is to be the most feasible. Besides, top-down fabrication methods and self-assembly techniques give less effort, lower cost, and larger quantities to produce various nanostructures of different morphologies.

2 Self-Assembling Materials

Bio-engineering uses self-assembling technique materials that offer an attractive alternative to cross-linked polymers, rubbers, and metals or metallic nanoparticles. To define the material molecular self-assembly process is the best choice that affords the supramolecular structures. These structures form noncovalent bonds between their molecules that drive their assembly and organization [16, 17, 18, 19]. However, molecular hydrogels can be prepared using a self-assembly process resulting in structures like fibrils, tubules, or complex systems [20]. Mostly self-assembling molecules are amphiphilic in structure means that contain both hydrophobic as well as hydrophilic domains. The hydrophilic portion can be charged (anionic or cationic) or uncharged [16, 17, 20].

In some cases, the therapeutic may be present while the self-assembly process is triggered; consequently, it is essential to consider the loading of therapeutic. Under aqueous conditions, the amphiphilic molecules assemble in different shapes and sizes that range from nanometers to microns due to weak and noncovalent interactions. The thermodynamic driving force of most self-assembly events is provided by the de-solvation, collapse, and intermolecular association of monomers' hydrophobic portions. Intermolecular polar interactions, such as electrostatic and hydrogen bonding, can occur to define the molecules' structural specificity. The morphology of the final assembled structure depends on the structure of the monomer and the external environment in which self-assembly occurs. The temperature, pH, and ionic strength of the solution, as well as the monomer's concentration, can dictate the formation of a variety of structures formed by a single, distinct amphiphile. Self-assembly is a dynamic process, and therefore can be triggered or reversed by these external stimuli, offering the possibility of forming the assembly before delivery and disassembling the final supramolecular structure after delivery to the site of interest [17, 18, 19]. Molecular amphiphiles self-assemble to form a variety of nano- and microscale structures under aqueous solution. Mesophase structure influences the proper-ties of soft materials such as amphiphilic, colloids, polymers, etc., and the self-assembly technique is one of the best suitable techniques to make the mesophase structure. Fabrication of nanoparticles with control shape, size, and functionality is the major challenge for the researchers. Metal nanoparticles can be surface-patterned with the help of the self-organization of block copolymers or without the help of block copolymers. Metal nanoparticles could be self-assembled on the solid substrate which can form within micelles in the solution also metal nanoparticles could be self-assembled by selective wetting technique. Nanoparticles are having a high aspect ratio that could be modified and self-assembled to improve drug bioavailability, blood circulation, tissue engineering, tissue targeting area, etc. A literature survey demonstrates the nanoparticle geometries that can induce different biological applications. Its unique and adorable characteristics make them highly desirable for bio-applications such as drug delivery, bio-sensing, and in vivo diagnostic applications. In this context, this chapter aims to cover the main features of the self-assembly of biomaterials and applications in the biomedical field, various surface modifications of nanomaterials, including metallic nanoparticles and carbon nanomaterials, nanoceramics, and self-assembled materials. Soft materials are classified in the following categories: synthetic and biological. Synthetic soft materials are synthesized or made by mankind using different chemical methods or fabrication process, whereas biological soft materials graft from nature.

2.1 Synthetic

For the development of synthetic soft materials at the nanoscale polymers, surfactants, lipids, colloids, and liquid crystals are the major building blocks. For the development of nanotechnology, nano-struc-tures play a vital role, and that could be graft by microphase separation of block copolymers. Surfactants contain surface-active agents are used for nanomaterials fabrication. These surface-active agents have amphiphilic molecules with both hydrophilic and hydrophobic parts that preferentially segregate on the surface of the nanoparticles or nanomaterials and become active. Detergent is the best example of a surfactant. Also, lipids, which are biological amphiphiles having self-assembling building blocks collectively gather together in water and aggregate into different nanostructures consequently, the contact between the hydrophobic part of the amphiphiles with water molecules reduces and micelles and vesicles forms. These micelles shapes vary that could be spherical or cylindrical and form hydrophobic cores and hydrophilic coronas in water. On the other hand, vesicles form a hollow spherical shape where the outer surface of the sphere is composed of layers of surfactant molecules. Colloids are the microscopic substance that is dispersed evenly throughout another substance with dimensions in between the range of 1 nm to 1 μm. Colloidal systems have two separate phases, and they are heterogeneous. When the dispersed phase of molecules comes into contact and mixes with continuous medium, it results in the self-organization of molecules. Aerosols, foams, emulsions are few of the examples of colloids. Liquid crystals also come in the soft materials category composed of moderate-sized organic molecules with molecular order intermediate between a liquid and a crystal. Liquid crystals present in the mesogen phase. These mesogen phases consist of two-phase one is the thermotropic phase, and the other is the lyotropic phase. Both the phase thermotropic phase as well as lyotropic phase forms in different medium and condition. Organic molecules form thermotropic phases in the absence of solvent, whereas amphiphiles form lyotropic phases in a solution. Figure 1 illustrates the liquid crystal phase orientations, such as the nematic phase and the smectic (layered) phase. Figure 1a illustrates the nematic phase where the positional order of the molecules is short-range, and orientation is in an average direction whereas Fig. 1b illustrates the translational order and the long-range orientations of molecules that form the smectic (layered) phase. This smectic phase contains a high viscosity.

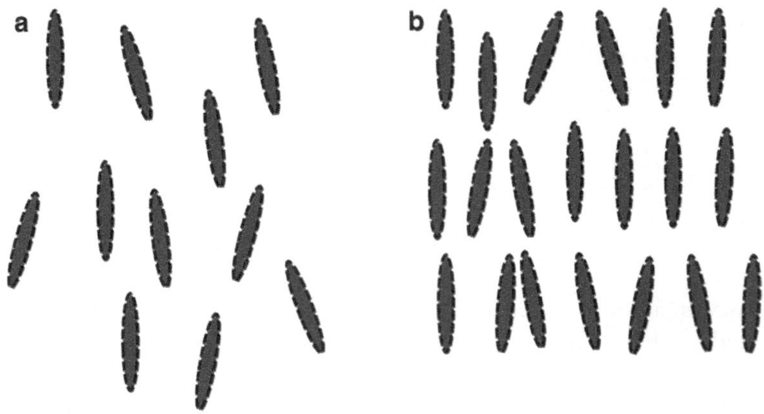

Fig. 1 **a** Nematic phase and **b** Smectic phase (*Liquid Crystal and Complex Fluids Group*)

2.2 Biological

As discussed, the self-organization of soft materials can be done artificially, but naturally, self-organized soft materials are convenient and advantageous over artificial ones. Some of the naturally biological self-organized soft materials are silk, collagen, proteins, DNA, microtubules, viruses, etc. The silk made up of antiparallel β-sheets of protein fibroin. The peptide chains are cross-linked by intermolecular hydrogen bonding that forms β-sheets, which helps the silk-making. The orientation of polypeptide chains along the β-sheets provides the tensile strength of silk and forces between the β-sheets concede for flexibility of the material. Collagen is a naturally formed protein present in all living species and forms cross-linked fibrils structures. This structure produces gelatin while undergoing denaturation by chemical or heat treatment. This serves as a vital component of connective tissues in animals. Keratin is also biocompatible naturally occurred protein with fibrous proteins that are responsible for forming hair, wool, nails, horns, and feathers in the living species. Naturally, keratin makes intermolecular hydrogen bonding with peptide chains, and consequently, it is arranged into fibrillary structures. In comparison with synthetic self-assembling materials, DNA fragments form lyotropic liquid crystals in a solution. As illustrated, in Fig. 1 the short fragments behave like rods and consequently forms the liquid crystal phase. In Figs. 2a and 2b, we could observe the self-organization of the nematic twisting of the cholesteric phase and the hexagonal columnar liquid crystal phase by increasing the concentration of the solution.

The naturally occurred, or biological self-assembly of microtubules plays a vital role in the development of nanotechnology and the microbiology domain. Protein tubulin helps microtubules formation. If we see at the nanoscale, microtubules contain nanostructures that support liquids to transport through it and work as a Nano-channels. The structure of microtubules, as shown in Fig. 3 comprise Cilia, which is

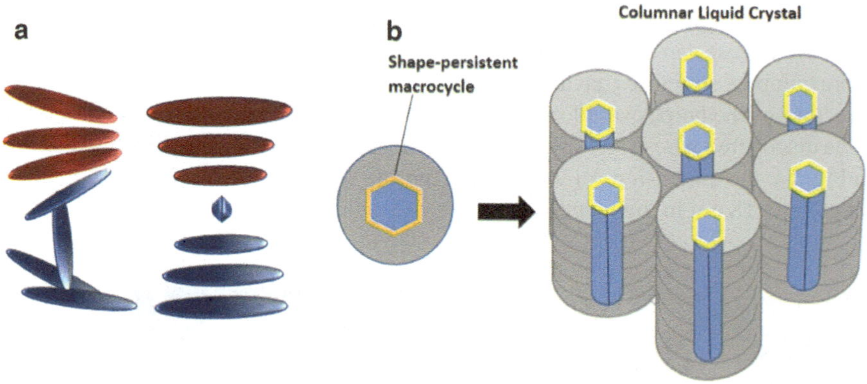

Fig. 2 **a** Nematic twisting of cholesteric phase and **b** Hexagonal columnar phase (*Barrett Research Group, Introduction to liquid crystals*)

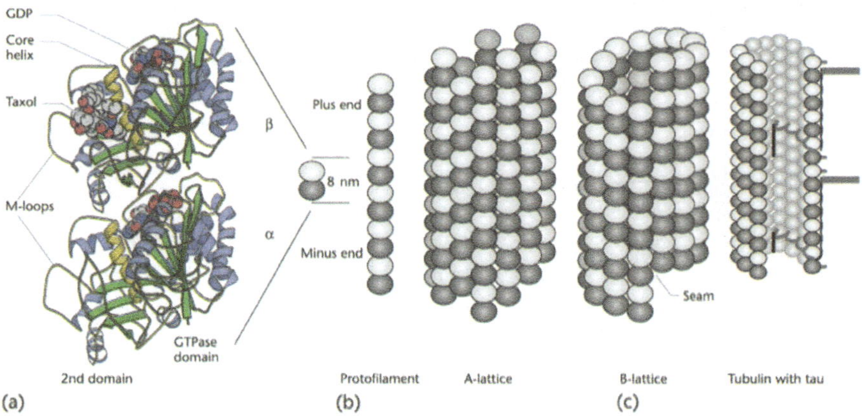

Fig. 3 Structure of microtubules. **a** Atomic structure of the αβ-tubulin heterodimer shown as a ribbon diagram; α helices in blue, β sheet in green. Taxol and GDP (bound to β-tubulin) and GTP (bound to α-tubulin in a position equivalent to that of GDP in β-tubulin) are shown as space-filling atoms. Each subunit has two globular domains separated by the core helix. **b** Assembly of heterodimers into longitudinal protofilaments, which in turn associate into sheets and microtubules. Side-to-side association of the protofilaments can produce an A-lattice or a B-lattice with an A-lattic4e-like seam, or a more mixed lattice. **c** Stabilization of the polymer by Microtubule-associated proteins (MAPs). Reproduced from [21]

hair-like structures that help to allow fluids or liquids across the surface of organs, consequently sliding of sub fibers formed from microtubules.

Last but not least, we are going to discuss Viruses. A virus is a smallest and unique parasite among all the microbes that cannot reproduce by itself but they can alive and able to multiply inside the cells of other living organisms. A virus is made up of a core of genetic material like nucleic acid mol-ecules, DNA or RNA, etc., which is covered by a protective layer of proteins. This protective layer of proteins called capsid and these capsid varies in shape like spherical, simple helical, and helical with more complex structures having tails. Figure 4 shows the self-assembled icosahedral structure of the herpes simplex virus (HSV-1). The tobacco virus was the first virus to be discovered whose self-assembled helical structure is shown in Fig. 5. As shown in Fig. 5a, the dimensions of the tobacco mosaic virus are approximately 300 nm in length and 18 nm in diameter.

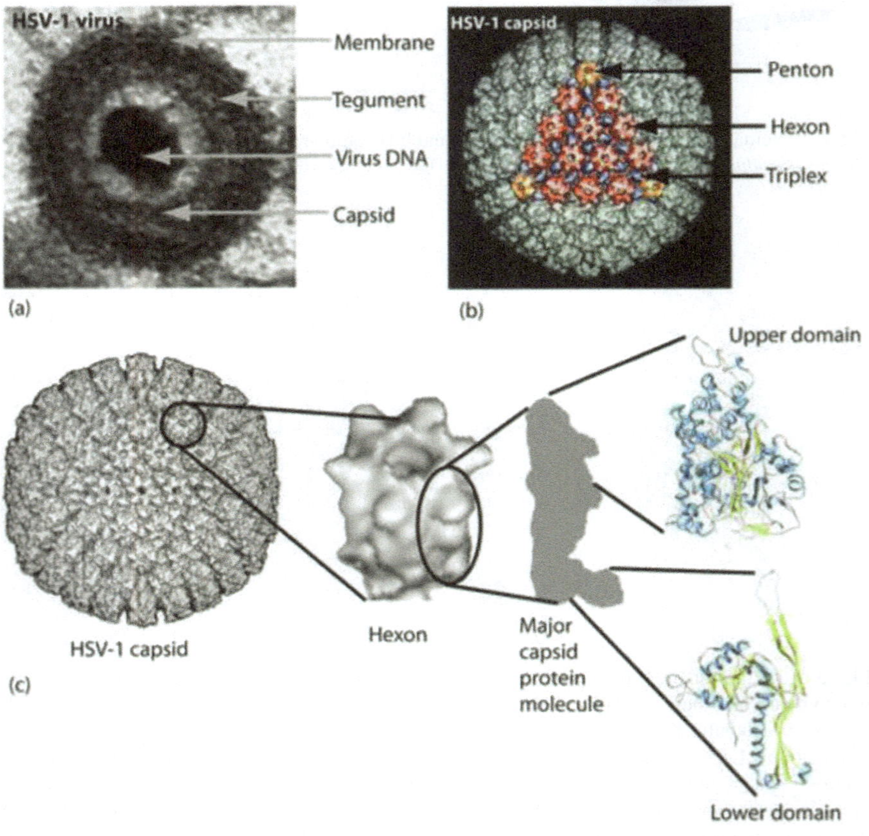

Fig. 4 Structures of the HSV-1 virion and capsid. **a** Electron micrograph showing a single HSV-1 virion in cross-section. **b** CryoEM reconstruction of an HSV-1 capsid shown in surface-shaded representation viewed along a face (3-fold axis). **c** View of the HSV-1 capsid with a single hexon and a single major capsid protein molecule extracted. Reproduced from [22]

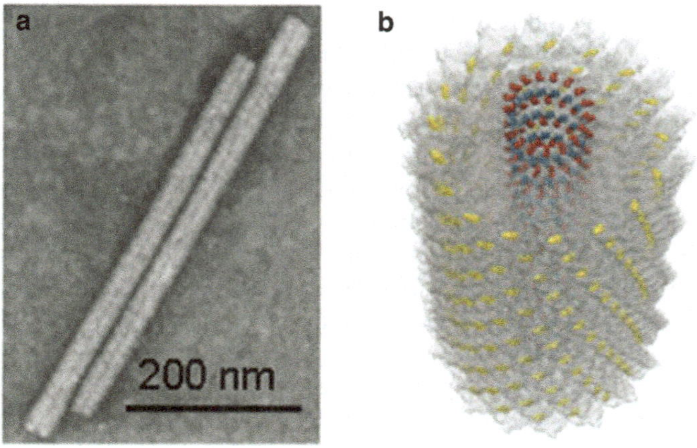

Fig. 5 Self-assembling biological structures. **a** Transmission electron micrograph of tobacco mosaic virus (TMV). **b** Model of the fully assembled TMV capsid showing tyrosine (yellow) and glutamate (red and blue) residues on the exterior and interior surface, respectively. Reproduced from [23]

3 Modification of Metallic Nanoparticles

The literature on noble metal nanoparticles, such as gold and silver, is already well established. With the evolution in the nanotechnology field, ample nanomaterials and nanoparticles have been discovered for diverse applications. These nanomaterials and nanoparticles play a vital role in chemical sensing, bio-labeling, and photonics in various biochemical processes. Sometimes they act as a catalyst to perform several tasks. The gold and silver nanoparticles have their anti-oxidant properties, which makes them exceptional for bio applications. However, the ease of synthesizing these nanoparticles by using the chemical methods and green methods gives extra advantages over other competitors—also, these nanoparticles have unique optical properties that enhance its value. Modification of gold and silver nanoparticles with suitable ligands make it biocompatible for plenty of applications. Gold nanoparticles expanded to the use of magnetic iron nanoparticles, which can act as bio transporters for tracers. When a specific wavelength of light falls on the gold and silver nanoparticles, they have efficient absorbance and light scattering properties [9, 24, 25]. The surface modification of noble metals is typically carried out by adherence, mainly with a thiol group, disulfide ligands, amines, nitriles, carboxylic acids, and phosphines. Incorporating chemical ligands molecules onto the surfaces of the nanoparticles without precluding the colloidal stability during the functionalization process is an enormous challenge. The chemical ligand exchange is exceptionally dependent on nanoparticle composition and types. Nanoparticles' surface affinity towards different chemical groups is different; hence various techniques have been evolved to solve this problem. Here we are going to discuss three main categories:

I. We generally use gold and silver metals, which are also called Noble metals, are normally functionalized with thiols groups or, to a lesser extent, amines and cyanides. Some Nobel metals such as Ag, Cu, Pt, Hg, and Fe tend to form and forms an organo-sulfur bond on the metal surface because they have intense chemical feasibility of forming strong covalent bonds with thiol and amines. Consequently, surface modification of noble metals. The higher affinity of sulfur to metal surfaces facilitates the strong coordinate bond, making the thiol group immobilize from the metal surface and making the organo-sulfur compounds readily absorbable. The presence of the oxidation of the thiol functional group shrinks down the affinity to interact with the metal surface and to form sulfate or sulfonate. Thiol- or disulfide-capped nanoparticles can be prepared in two ways. This approach uses the already synthesized noble metal nanoparticles that could be used further to graft sulfur compounds on the surface of the nanoparticles. This process could be done by replacing the existing capping agent with the new ligands that contain sulfur, which is present on the nanoparticle's metal surface (Fig. 6).

In another synthesis method, a noble metal nanoparticle is kept in the inert environment with a residual surface charge by induction-coupled plasma technique allowing for easy grafting of sulfur-containing ligands. The alternate approach available is the synthesis of an organo-sulfur capped nanoparticle in a one-step process by wet chemistry method where the metal precursor nucleates and protective ligand capped simultaneously in a single-step reaction [25].

[27] As a simple and more feasible alternative for physical and chemical techniques in nanomedicine applications, biosynthesis of metallic nanoparticles with the help of plant extracts has been gaining immense prominence in recent years. Gold

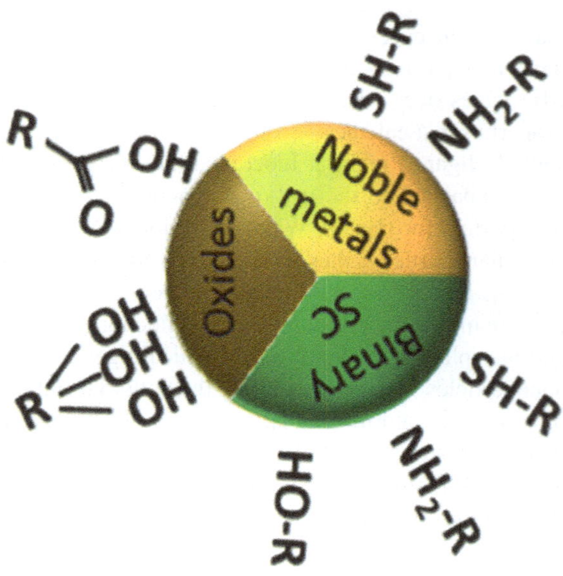

Fig. 6 Schematic representation of the most commonly used functional groups in surface ligands and their preferred affinity towards binding different NP materials [26]

nanoparticles exhibit various properties like high surface plasmonic resonance, high biocompatibility with chemical stability, lower toxicity, surface bio-conjugation, the higher surface to volume ratio, and, thus, can be used in cancer diagnosis and therapeutics [28]. Different ideas on the biosynthesis of gold and silver nanoparticles and their applications in various fields were presented by T.S. Santra et al. Focus should be paid on developing stable gold, silver, or other metallic nanoparticles; that could be useful for biomedical applications in future. So the effort is to develop Ag/Au nanoparticles to make them multifunctional and tunable by external signals and then turning them into Nano-devices. [29] reported that nanosecond pulsed photoporation mediated by nano-corrugated mushroom-shaped Au coated polystyrene nanoparticles (nm-AuPNPs) shows effective intracellular molecular delivery with a high percentage of cell viability. The pulsed laser application at the plasmonic peak of 945 nm creates nano-bubbles that explode and form membrane pores, which in turn helps in the transfection of molecules into the cells. Optimization of various parameters such as laser exposure time, concentration, and molecular size of nm-AuPNPs improve the results. Therefore, the proposed methodology has proved to be an essential prospect for therapeutic medical applications (Fig. 7).

II. Oxides can easily be coated via oxygen bonding techniques using acidic groups and hydroxyl groups. For the oxygen bonding technique, oxides use magnetic nanoparticles (e.g., iron oxides, silica, titanium dioxide). Oxide nanoparticles have a wide variety of possible structures due to their different properties [25]. Commonly used oxide nanoparticles include titanium dioxide (TiO_2), silica or silicon dioxide (SiO_2), and iron oxide. Based on the applications of silicon dioxide nanoparticles, it can be broadly synthesized into two ways: powders based and colloidal solutions based synthesis. The powder-based method is a

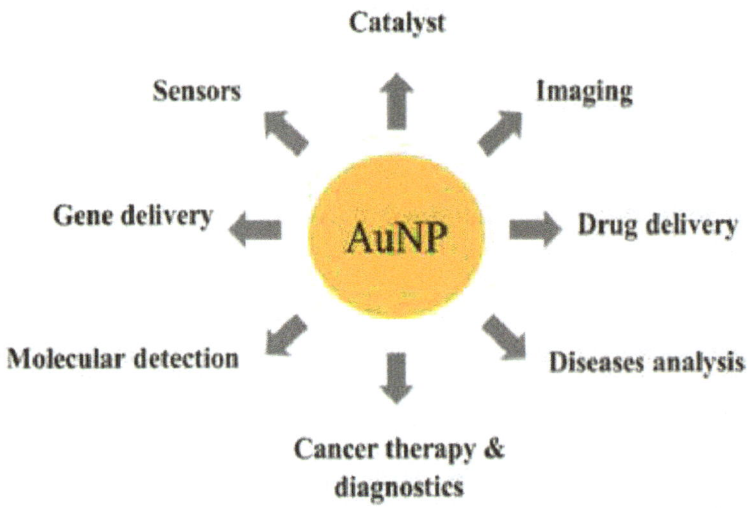

Fig. 7 Various applications of green biosynthesized gold nanoparticles (AuNP) [27]

widely preferred method for industrial and commercial applications, mostly synthesized by precipitation, vapor deposition, or flame processes. At present induction-coupled plasma, techniques are used for this method (powders). The induction-coupled plasma method uses high temperature and vacuum-inert environment to synthesize the nanoparticles. It can synthesize consistent powder of nanoparticles having an average particle diameter 50–100 nm. By nature, silicon dioxide (SiO_2) nanoparticles are hydrophobic, and to make it hydrophilic and biocompatible, its surface needs to be modified. For this process, nanoparticles surface could be bind with the hydrophilic or water-soluble polymers like nonionic poly (oxyethylene methacrylate) or ionic poly (styrene sulfonic acid), etc. via a three-step synthetic approach:

1. Surface activation of the hydroxyl group (–OH) on the SiO_2 nanoparticles.
2. Surface modification by plating chlorine (–Cl) group.
3. Depositing polymers onto the nanoparticle surface using atom transfer radical polymerization technique [8, 24].

III. Binary compounds notably including elements from Groups 12 to 16 as components of fluorescent semiconductor (SC) nanoparticles; e.g., quantum display high affinity towards thiols and hydroxyl groups, but also amino used. Figure 6 shows the most commonly used functional groups to coat nanoparticles of different materials.

3.1 *Advantages of Metallic Nanoparticle*

a. Enhance Rayleigh scattering
b. Surface-enhanced Raman scattering
c. Strong plasma absorptio
d. Biological system imaging
e. Determine chemical information on the metallic nanoscale substrate [30].

3.2 *Disadvantages of Metallic Nanoparticles*

a. **Particle Instability**: Nanomaterials are thermodynamically unstable and can transform as they lie in the region of high energy local minima. This leads to deterioration of structure may cause difficult to regain its shape, quality, and poor corrosion resistance.
b. **Impurity**: While synthesizing nanoparticles, nitrides, oxides, the formation can be aggravated from the impure environment. The highly reactive nature of nanoparticles can lead to high chances of impurity as well. In solution form, nanoparticles should be synthesized in the form of encapsulation. So, it becomes a challenge to overcome impurity in nanoparticles.

c. **Biologically Harmful**: Few nanomaterials are toxic and carcinogenic which may cause the death of the cell.
d. **Difficulty in Synthesis**: While synthesizing nanoparticles, it should be encapsulated, because it is incredibly challenging to retain the size of the nanoparticles in solution form [31].

3.3 Characteristics of Metallic Nanoparticles

a. Large surface energies
b. As compared to bulk they have a large surface area to volume ratio
c. Quantum confinement
d. Plasmon excitation
e. Increased number of kinks [32].

3.4 General Application of Metallic Nanoparticles

3.4.1 Optical Function

The optical properties of metal nanoparticles depend on various factors; some of them are shape, size, surface area, doping, and interaction with the surrounding.

Imaging sensors, solar cells, display, biomedicine, an optical detector, Photocatalysis are examples of metallic nanoparticles. The surface absorption Plasmon of noble metal gold and silver can change into various colors, and that depends on factors such as particle size, form, the shape of the particle and condensation rate, etc. as shown in Fig. 8.

3.4.2 Thermal Function

Nanoparticle size plays a vital role in the heat generation, for example, in the case of nanoparticles, which has a smaller diameter (less than 10 nm), the melting point will be lower than a bulk metal.

3.4.3 Mechanical Function

Nanoparticles' mechanical properties could be improved by using the polymers either filling them or coated with them also, properties of these particles will depend on the materials which material we filled or coated. On the other way, the mechanical property of metallic nanoparticles could also be improved by mixing nanoparticles with ceramics.

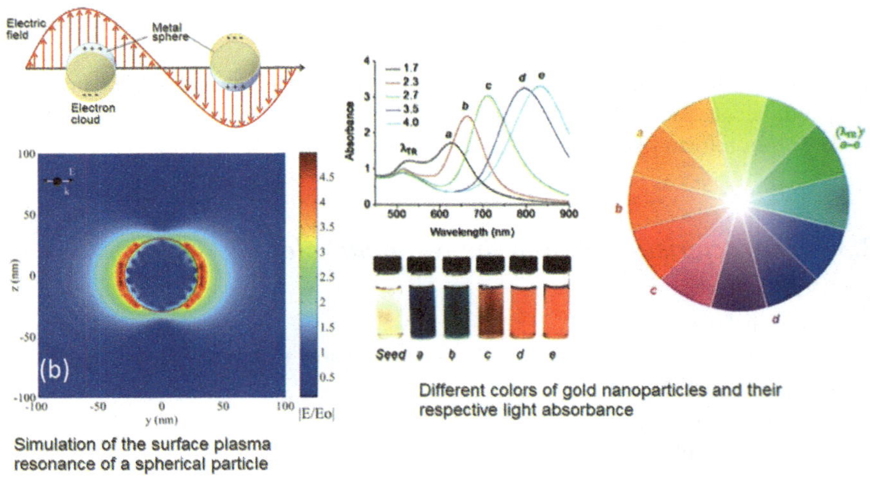

Fig. 8 The nanoparticles' optical properties change when anisotropy is added to the nanoparticle, changes such as Nanorods growth. Reproduced from [32]

3.4.4 Magnetic Function

At the nanoscale level, platinum and gold nanoparticles exhibit magnetic property, but as bulk, their behavior changes to non-magnetic. The physical property of nanoparticles could be change by capping the nanoparticle surface and bulk atoms that lead to improvised the interaction of modified nanoparticles with other chemical species to a great extent.

3.4.5 Catalysis

Catalysts based on metallic nanoparticles are-selective, highly active, exhibit a long lifetime for several kinds of reactions. There are two types of catalyst- Heterogeneous catalysts- which are immobilized on an inorganic support: homogenous catalysts-metallic nanoparticles surrounded with stabilizers.

3.4.6 Used as Fuel Cell Catalysts

A fuel cell is a device that generates electricity or electric energy using the chemical potential energy by a chemical reaction. In a fuel cell, hydrogen is fed to the anode, and the oxygen gas (0_2) is fed to the cathode by air directly. Polymer electrolyte membrane or proton exchange membrane (PEM) is an essential part of the fuel cell, including the membrane, the catalyst layers, and gas diffusion layers (GDLs). PEM of fuel cells provide channels for the gaseous fuel and air and the byproduct of the reaction is water, electricity, heat.

3.4.7 Used in Materials Science

Nickel nanoparticles are used as electrical conductive pastes, battery materials, etc.

3.4.8 Used in Medical Treatment

A healthy cell can be distinguished from cancer cells by the presence of antibodies joined to the Au nanoparticle.

3.5 Therapeutic Applications of Metallic Nanoparticles

3.5.1 In Tumor Therapy

Gold nanoparticles inhibited the activity of heparin-binding proteins, for example, VEGF165 and bFGF in vitro and VEGF induced angiogenesis in vivo. Scientific literature already reported that onto the surface of AuNPs heparin-binding proteins are absorbed and were subsequently denatured.

3.5.2 In Multiple Myeloma

Researchers have designed a nanoparticle-based therapy that is effective in treating mice with multiple myeloma. Multiple myeloma is cancer that affects plasma cells.

3.5.3 In Photo Thermal Therapy

Gold and silver nanoparticles have significant properties to absorb the light strongly, and they convert photon energy into heat energy very quickly and efficiently. This process is called Photo-thermal therapy (PTT). It is an invasive therapy in which photon energy is used to convert heat to kill the cancer cells.

4 Modification of Carbon Nanomaterials

Nanotechnology has established as a path-breaking field in solving various scientific as well as technological challenges in recent times. However, carbon-based nanoscale structures have experienced attention primarily because of brilliant characteristics of carbon. The below divisions briefly explains various modification techniques of popular CNMs: graphene-based nanomaterials (G, GOs, rGOs), graphitic carbon nitride (g-C3N4), carbon nanotubes (CNTs), nanodiamonds (NDs) and carbon nanofibers (CNFs).

4.1 Modification of Graphene-Based Nanomaterials

Graphene is the fundamental building block of carbon nanomaterials. Graphene can be modified to graphite, carbon nanotubes, and fullerenes using different processes such as wrapping, rolling, and stacking. The principal synthetic routes to grapheme are chemical cleavage and bulk graphite exfoliation into single-layer graphene (G). Geim and co-workers discovered single-layer grapheme by exfoliating stacked graphite with the help of Scotch tape. In the same way, pyrolysis of camphor with the help of chemical vapor deposition (CVD) on Ni foils have been used to synthesize Planar Few-Layer Graphene (PFLG) [33]. Graphene has several advantages as an adsorbent. It tends to agglomerate and form graphite because of stacking and dominant interplanar interaction. So it is required to perform surface functionalization of GO, graphene, or rGO nanosheets with organic compounds or surfactants to reduce the aggregation. GO was functionalized using large alkyl chain CTAB through ionic interactions by [34] to improve the hydrophobicity of GO, which can help in its removal after adsorption from aqueous solution. The dye taking capacity was seen to be 2767 mg/g of CTAB-GO for CR dye at pH = 3, and 298 K, which takes a time of 1 h. CR dye has a sulfonated group, negative in charge, that can enhance the adsorption at low pH with the help of electrostatic interactions of the dye with the head group of the CTnAB-GO composite, positive in charge. Computational and experimental studies were conducted [35] to determine the adsorption capacity of the drug sodium diclofenac (s-DCF). These studies were performed on single vacancy graphene, pristine graphene, and graphene in nanoribbon form functionalized with various groups. The introduction of functional groups enhances the adsorption on functionalized graphene by s-DCF because of an increase in binding energies. [36] proposed that heavy metal ions such as As (III), As (IV), and Pb (II) are separated by exceptional adsorption capabilities of magnetic nanohybrid of GO with $MnFe_2O_4$ NPs. This was because of GO's unique layered property, which allows large surface area and excellent adsorption capabilities of magnetic NPs and GO. [37] utilized the nanohybrid material for the separation of cationic dyes from contaminated water. Very high adsorption capacities and very strong reusability. The effectiveness of adsorption to GO/rGO can be improved by functionalizing it with metals, biopolymers metal chalcogenides, or by doping (N, S, etc.). [38] proposed carboxylated GO/chitosan spheres as an effective adsorbent with a high adsorption capability of 78 mg/g for Cu2 + fixation in both water and soil. Cu^{2+} with high immobilization efficiency to Cu^{2+} bioaccumulation can be reduced to 50% in wheat seedlings by modification by GO-COOH/CS spheres. These methods deliver an emerging approach in preparing

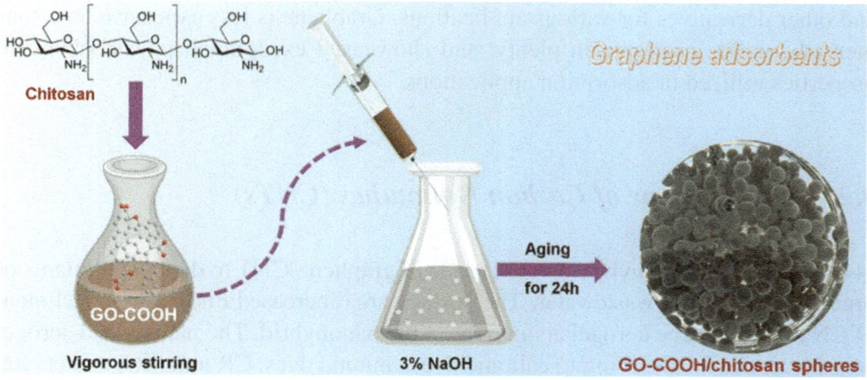

Fig. 9 Schematic illustrations of the synthesis process of GO-COOH/CS spheres for Cu2 + fixation. Reproduced with permission from [38]

graphene adsorbents for soil remediation. [39] reported that fulvic acid can be separated from the water with the help of Fe functionalized rGO (frGO) compared to powdered activated carbon. Low pH conditions exhibited much higher efficiency than higher pH as higher pH affects adsorption badly due to an increase in intra and inter-molecular electronic repulsive interactions in fulvic acid (Fig. 9).

[40] used amine groups to modify rGO with (rGO- NH_2), and the resulting nanoparticles were immobilized with Horseradish Peroxidase (HRP), to remove a large amount of phenol from polluted wastewater. rGO separated the phenol from the polluted water in the nanocomposite and enhanced the stability of HRP. The rGO-NH2 could adsorb the phenol on the matrix through hydrogen bonding, p-p interactions, and electrostatic attractions. The nanocomposite was found to be efficiently behaving for ten regeneration cycles. GO sponge was functionalized using methylimidazolium ionic liquid (mimGO) [41] through an amidation reaction between imidazolium-based ionic liquids and the carboxylate groups of GO [41]. The charge- induced adsorption improves the mechanism of the superfast adsorption of an anionic dye (DR80) from the water with the help of mimGO. DR80 has a good number of aromatic rings and sulfonic groups. The rate of adsorption is enhanced by the electrostatic attractions between the cationic imidazolium rings of the mimGO sponge and the sulfonic groups of the dye. [42] in their research, modified the rGO surface to synthesize rGO/CTAB by preparing a cationic surfactant (CTAB). Anionic dyes (direct red 80 and direct red 23) were separated from wastewater by using the fabricated nanomaterial. The highest adsorption capacity of 213 mg/g for DR80 and 79 mg/g was observed for DR23. So rGO/CTAB presented as an alternative option as an adsorbent for the removal of anionic dyes from contaminated water. [43] studied the dynamics of charged and neutral heavy metals with graphene quantum dots (GQDs). Pb attaches to GQD more strongly compared to others and considered as an electron donor. Armchair GQDs display better adsorption characteristics than zigzag GQDs. To summarize, graphene nanomaterials can be functionalized into GO, rGO,

and other derivatives for various applications. Graphene is less expensive, environmentally stable, available in plenty, and showcased exceptional physicochemical properties utilized in adsorption applications.

4.2 Modifications of Carbon Nanotubes (CNTs)

[44] developed a nanohybrid aerogel made of graphene-CNT to deduct pollutants of organic nature from wastewater. The surface area increased due to the attachment of CNTs to graphene aerogel about 57% in the nanohybrid. The nanohybrid aerogel was successful in adsorbing to cationic dyes, anionic dyes, CR and MO, and crystal violet (CV) and MB, primarily through p–p and van der Waals forces. Therefore, nanohybrid aerogels help in the adsorption of both negatively as well as positively charged water pollutants.

CNTs, like graphene, can be modified availing magnetic nanoparticles to help in removal for utilization from the solution. [45] developed nanohybrids of 3D graphene-CNT-Fe by a microwave mechanism separation of As from wastewater (Fig. 10[a]). Magnetic graphene-CNT-Fe nanohybrids displayed better adsorption (6 mg/g) for as compared to Fe_2O_3 modified graphene (2.5 mg/g) because of more number of active surface sites of graphene, Fe_2O_3, and CNT in the nanohybrids. [46] fabricated nanohybrids of large surface area (999 m^2/g) that consist of Ni

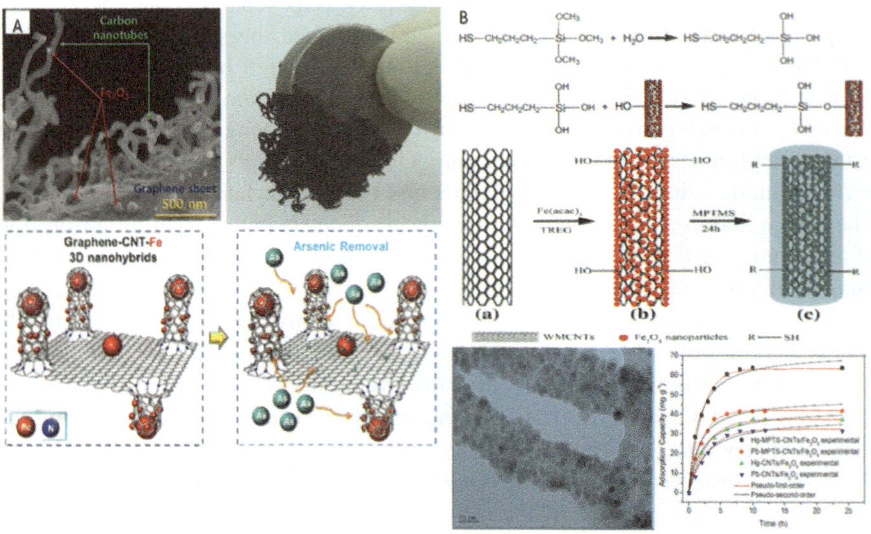

Fig. 10 Separation of different organic and inorganic contaminants by magnetic CNTs: **a** 3D graphene-CNT-Fe magnetic nanohybrids synthesis by microwave chemistry for separation of As. Reprinted with consent from [45]. **b** MPTS-CNTs/Fe_3O_4 nanocomposites for removal of Pb (II) and Hg (II) Reproduced from [48]

nanoparticles and porous carbon/CNTs for adsorption of organic dyes. This adsorbent displays best adsorption capability (898 mg/g) towards MG. After adsorption, nanohybrids are separated easily from the solution using the magnet. [47] functionalized CNTs through a reaction of 1,6- hexane-diamine in magnetic Fe_3O_4NPs (MNP) with carboxyl groups of CNTs with the help of simple solvothermal chemistry. At solution pH 2.0, the highest adsorption capability for Cr (VI) of MNP/MWCNTs was seen. [48] modified Carbon nanotubes/Fe_3O_4 nanomaterials with "mercaptopropyltriethoxysilane (MPTS)" to form a superparamagnetic "MPTS/CNT/Fe_3O_4" nanomaterial with an improved adsorption capacity (Fig. 10[b]). The surface area, which is actively calculated for "MPTS/CNT/Fe_3O_4" was seen to be higher than that of CNT/Fe3O4. Lewis acid-base interactions influenced thiol groups in MPTS having an intense attraction for heavy metal ions [49]. Therefore, the "MPTS/CNT/Fe_3O_4" nanomaterial effectively adsorbed heavy metal ions from wastewater. [50] proposed that cationic (MB) and anionic (MO) dye can be separated by KOH-activated CNTs having high pore volume (1.61 cm^2/g) and surface area (534.6 m^2/g).

The better adsorption capability for MB (400 mg/g) and MO (149 mg/g) onto A-CNTs was because of large-scale adsorption mechanisms like p-p electron-donor–acceptor interactions, electrostatic attractions, H-bonding, and pore filling. American Chemical Society. (B) MPTS-CNTs/Fe3O4 nanocomposites for the separation of Pb (II) and Hg (II) [48, 51]. The surface functionalization of MWCNT was utilized to develop stable and uniform nanofluids. Chemical and physical treatment is used to functionalize MWCNTS. Chemically modified CNTs were dispersed in the base fluid, which was acetone, and the prepared fluid was called NF-1 (0.2 wt%). The wrapping performed the physical surface modification of nanotubes by polyethylene glycol of CNTs. The nanoparticles' behavior in the base fluid (NF-2, 0.2 wt%) is optimized by a different nanofluid prepared by the combination of physical and chemical surface functionalization. The results proved that these nanofluids could improve the system's heat transfer coefficient by around 27%.

4.3 Modifications of Graphitic Carbon Nitride (G-C3 N4)

[52] proposed to synthesis a g-C3N4 hydrogel (h-CN) for adsorbing various and dyes h-CN was found to have a remarkable 99% adsorption efficiency in selectively adsorbing with cationic dyes. The adsorbing capacity of the cationic dyes is attributed to the h-CN surface charge, analyzed with the help zeta potential. The h-CN zeta potential was observed to be C3N4 (30–40 mV). Another method to enhance water pollutant removal potential of g-C3N4 is by modifying the architecture by applying porosity for improved surface area [53]. Highly electronegative elements have been used to modify g-C3N4, which has resulted in the adsorption of anionic dyes or electron-rich pollutants, which resulted in the decomposition of toxic organic compounds [54].

g-C3N4 is an environment-compatible adsorbent for water pollutants because of high mechanical and low-cost, thermal and chemical stabilities. More research on g-C3N4 as adsorption material is a potential area for future investigation. [55] 3D carbon-boron-nitride (3D-C-BN) that looks like cheese was prepared with pore size with a range of 2–100 nm devoid of any template support. It has hydrophobic properties (contact angle of 112.1°), which makes it a floatable adsorbent for the removal of oil pollutants. The 3D C- BN could be a flexible adsorbent agent for organic dyes, metal ions, and oils because of the large number of active sites of C, B-N, and smaller pores.

4.4 Modifications of Nanoporous Carbon (NPCs)

The inert nature and hydrophobic characteristics of nanoporous carbon act as adsorbent. Surface functionalization of the NPCs can overcome this complication and be used for various applications [54]. Han et al. removed "Direct Blue 78 dye" from polluted water by utilizing silica nanoparticles as a template to obtain NPCs [56]. As compared to activated carbon, this material has a much higher adsorption capability, almost 10 times the former. NPCs have interconnecting holes and microchannels, and the adsorption is enhanced by honeycomb structure and consistent porosity. Thomas et al. modified NPCs by introducing N and O and seen a remarkable gain in the adsorption capacity for heavy metal ions. It could be simulated with the help of Langmuir isotherm [57]. Zhang et al. developed an NPC complex with the help of $ZnCl_2$ and then calcined for 2 hours it under an N2 atmosphere at a temperature of 700–900 °C [58]. Surface applications of APC were improved by the enhanced active surface area and a high pore volume. An enhanced surface area offers a much larger number of active sites for adsorption of dyes; bigger pore size gives transport channels due to which dye adsorb onto the inner pores smoothly. Also, functional groups on the surface of the adsorbate improve APC's dispersibility in an aqueous system. [59] used metal-organic framework modified NPCs to effectively separate an antibiotic, a phenolic compound, and a dye. These NPCs exhibit higher surface area (1731 m2/g) and pore volume (1.68 cm3/g).

4.5 Modifications of Carbon Nanofibres (CNFs)

Carbon Nanofibres are different from conventional carbon fibers (CFs), primarily in size as the latter comes out of nano range. CNFs have applications in various fields because they display high tensile strength, exceptional chemical and thermal stabilities, high conductivity, and low density [33, 60] developed mesoporous CNFs with the help of an electrospinning method and studied how they can be used in the adsorption of bigger dye molecules in polluted water treatment. Three varied mesoporous CNFs were developed using phenolic resins and triblock copolymer Pluronic F127. While the former act as carbon precursors, the latter acts as a template. For

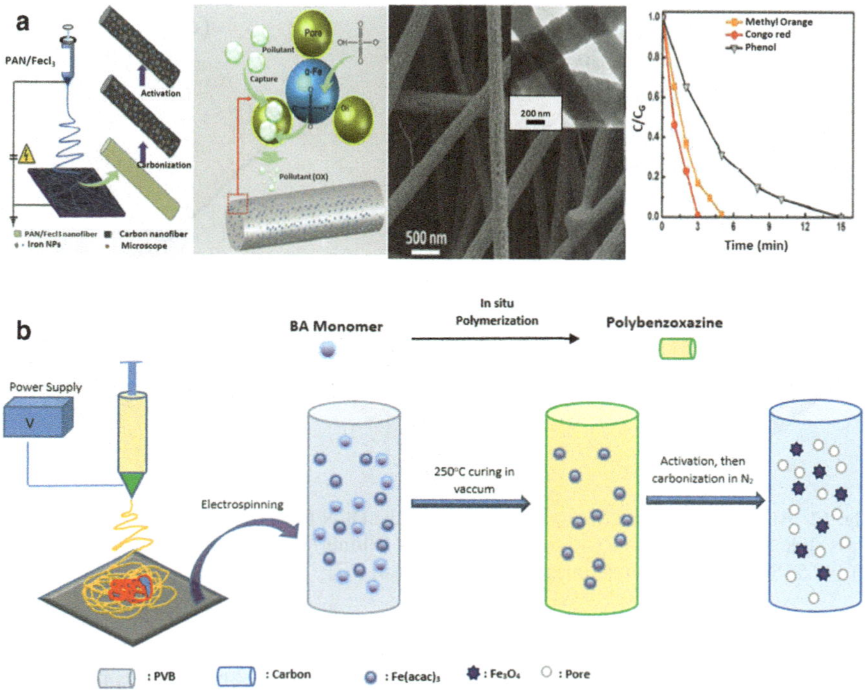

Fig. 11 **a** images show the preparation of Carbon nanofibres using Fe nanoparticles by electrospinning for degradation of organics contaminants (phenol, CR, and MO). Reprinted with permission from [61]. Copyright 2012, Royal Society of Chemistry. **b** Schematic illustration of the synthesis of hierarchical porous magnetic Fe3O4@CNF through a combined approach of electrospinning and in situ polymerization. Redrawn from [62]

the effective adsorption of MB and RhB [61] (Fig. 11[a]). Si et al. prepared hierarchical porous magnetic Fe3O4 carbon nanofibers. This nanocomposite material has a large pore volume (2.083 cm3/g) and surface area (1885 m2/g). The adsorption of organic dyes and their sensitivity to magnetic separation improved because of it [62] (Fig. 11[b]). Sun et al. [63] prepared carbon nanofibres with a-Fe nanoparticles with high adsorption capacities for dyes and phenol and an external magnet was used to separate them from aqueous systems. An electrostatic interaction was the motivation for the contaminant adsorption and hydrogen bonding and p-p interactions.

4.6 Modifications of Nanodiamonds (NDs)

Diamond as a biocompatible substance has various applications in the biomedical field like cardiovascular, orthopedic, prosthetic replacement, contact lenses, etc. [64]. Diamond, Diamond-Like Carbon (DLC), and Diamond-Like Nanocomposite (DLN)

Fig. 12 Atomic Fluorescent Microscopy images of the DLN films (a) with out DLN coted Pyrex glass slide. (b)-(f) DLN films with various flow rates (μl/min) (b) 27.8 (c) 56.3 (d) 80.6 (e) 112.7 (f) 149.5 [Scanning area for all image = 4 μm × 4 μm] [67]

thin films have different desirable properties like exceptional thermal conductivity, biocompatibility, and chemical inertness which can be utilized in developing lab-on-a-chip devices, biosensors, and other biomedical applications [65]. Plasma enhanced chemical vapor deposition techniques were used to deposit DLN films on silicon substrate or glass. With the help of combining siloxane and silazine based precursors. The AFM images (Fig. 12) prove that there are fewer surface defects, such as pinholes and macroparticles in DLN films [66]. Due to unique optical properties, ease of conjugation, and low cytotoxicity after surface modification, fluorescence emitted from nanoparticles can be utilized for biological system imaging [67]. The details regarding shifted bonds like Si-O, C-C, SiC2, Si-C, Si-H, O-H, N-H, and C-H with various flow rates are provided by FTIR spectroscopy. AFM and X-ray Photo Electron spectroscopy (XPS) were used to study the DLN films' surface morphology and composition. Various optical, tribological, and structural properties were correlated with various precursor flow rates of DLN films [68]. DLN films could be used as a coating substance primarily for enhancing the friction and wear of components. It could also be used to restrict friction between microstructure and their substrate. Due to biocompatible nature, DLN films have applications in biomolecule detection in disease diagnosis [69]. DLN films have better properties than DLC films concerning hardness, friction coefficient, elastic modulus, compressive stress, and surface roughness. Thus we can conclude that DLN films are highly deserving as coating substances in NEMS or MEMS devices.

Drugs can adsorb a nanodiamond carrier surface for their release at a targeted location. For this purpose, the adsorption capacity of NDs has been investigated in medical applications [70]. Also [71] oxidized ND was modified with the help of "single-armed (SA) and double-armed (DA) benzoyl thiourea ligands" and used for adsorption of various metal ions. "Single-armed (ND-SA) nanostructures" have showcased high adsorption for UO2 (II) compared to "double-armed (ND-DA) nanostructures" because of medium complexation tendency and strong, flexible spatial configuration. Surface functionalization tends to change the adsorption capacities and surface charges of NDs. [72] proposed that organic dyes (MB, MO) can have selective adsorption on thermally oxidized NDs. Compared to MO dye, MB dye shows better adsorption to the oxidized NDs (ONDs) because of the lesser zeta potential of NDs with thermal oxidation. The adsorption capability of Cu- PB@DND was higher for Cs (I), which is positively charged (759 mg/g) because of high surface area and negative surface charge.

Chen et al. developed a 3D carbon material with the help of two dimensional GO nanosheets and one-dimensional CNTs oxidized in the adsorption of diethyl phthalate (DEP), oxytetracycline (OTC), Cd(II), MB, and separating diesel from water [73]. The synergistic effects between GO and CNT in the 3D aerogel structure cause more adsorption sites, better porosity, larger interspaces, and higher performance than other graphene aerogels and nanomaterial suspensions of adsorbents. Chen et al. [74] prepared 3D GS modified with Chitosan (CS) that deliver positive as well as negatively charged surfaces on the adsorbent, These modified GS successfully adsorbed dyes and heavy metal through electrostatic attractions and ion complex formation, respectively, resulted in much higher adsorption capacities.

4.7 Biopolymer Functionalized Carbon-Based Nanomaterials for Adsorption

Polymers obtained through natural sources such as bacteria, fungi, plants, and animals, during their growth cycles are called biopolymers. They can be synthesized from biologically sourced substances, including amino acids, proteins, resins, vegetable oils, and sugars. These biopolymers can be grouped concerning their origin or monomeric unit. Several studies have proposed biopolymer/carbon-based nanocomposites' potential for the separation of pollutants through adsorption. [75] recently proposed that a syringe dropping technique can be used to develop graphene nanoplates cross-linked with chitosan spheres (CS/GNPs). This method assists in the removal of AR1 (cationic) organic and MO (anionic) dyes from polluted water. The major motivation for the adsorption of dyes onto the CS/GNP surface is the electrostatic interaction between the dye molecules and amino groups protonated in the chitosan spheres (CS). Dye adsorption was observed to be much efficient at pH lower than 4 as the protonated amino groups of the CS enhance the process. [76] proposed a porous 3D nanoarchitecture of Chitosan spheres with a large active surface area (780 m2/g) and pore volume (0.730 cm3/g), and a 1300 nm average pore diameter, that enhances the adsorption capacity [77]. Due to p–p interactions between GO and the pharmaceutical pollutants, electrostatic interactions, hydrophobic interactions, and hydrogen bonding between Chitosan and the drugs, the adsorption efficiency of the Chitosan spheres nanocomposite were enhanced for the antibiotic and chemotherapy drug. Carbon nanomaterials can be modified with other biopolymers, including gum, polyacrylamide, polyvinyl alcohol, L-cysteine, cellulose, polyaniline, and polydopamine which is utilized for purification of water [78]. The "polyaniline (PANI)" functionalized "GO-SrTiO3" nanomaterial has assisted in the effective absorption of cationic and anionic dyes at the same time.

4.8 Modification of Graphene Quantum Dots (GQDs)-Based Nanomaterials in Cancer Treatment

GQDs can be doped with heteroatoms to modify their and optical characteristics and density of electrons effectively for specific properties [79]. N doping of GQDs is an efficient method to expand the light absorption and improving the PL lifetime [80]. Kuo and colleagues [81] prepared N-GQDs by modified them with amino acid molecules (amino-N-GQDs). The modification improved the electronic properties of N-GQDs. Du and co-workers ("Photodynamic Graphene Quantum Dot: Reduction Condition Regulated Photo-Activity and Size-Dependent Efficacy" 82) functionalized GQDs-based nanocomposites by incorporating Ce6 using disulfide bond onto GQDs. Also, Pluronic F- 127 coated on the GQD-SS-Ce6 acts as a stabilizer. GSH in the tumor cells initiates disulfide bond cleavage when GQD-SS- Ce6 reaches local tumor tissue. The GQDs nanoplatform then can release Ce6 and its phototoxicity is restored. [83] fabricated HA-GQD, a nanocomposite, where HA and GQDs have p-p stacking, which has PDT and imaging characteristics as well. After that, to prepare HA-GQD-SiO2, HA-GQD is encapsulated by porous silica nanoparticles. This helps during drug delivery to avoid the refining of the nanocomposites. This again enhanced the intensity of the fluorescence signal and HA-GQD production capability. Besides, specific PDT property can be exhibited by functionalized GQDs. [84] developed a highly effective PDT strategy using modifying GQDs covalently on upconversion nanoparticles (UCNP) through amide linkages (Fig. 13). After that, tetramethylrhodamine-5- isothiocyanate (TRITC) [85, 86], a fluorescent probe

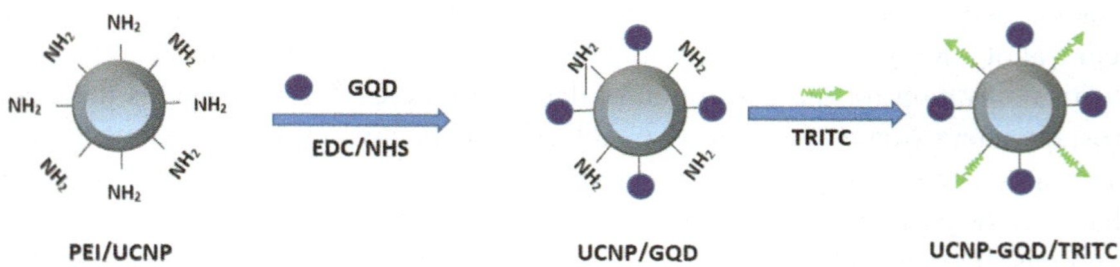

Fig. 13 Schematic diagram of the synthesis of UCNP-GQD/TRITC and mitochondria-targeted PDT. Redrawn from [83]

that targets mitochondria, was availed to functionalized UCNP-GQD that resulted in UCNP-GQD/TRITC. Also, their capability of targeting mitochondria was proved by line scan analysis. [87] developed self-assembled GQDs and proposed that these self-assembled GQDs can target themselves into the nucleus in the absence of an external targeting agent's help, thereby acting as an agent for nucleus labeling and displaying strong selectivity.

5 Modification of Nanoceramic Materials

A large number of nanostructured ceramic materials are widely used to obtain active elements for humidity sensors and different electronic applications. Various modifications have been performed by the scientists on nanoceramic materials to improve the optical, ferroelectric, and luminescence properties of the nanoceramics for different applications. [88] studied optical and luminescence properties of transparent nanosized cerium doped Y3Al5O12 (YAG: Ce) ceramics. YAG: Ce nanoceramics were prepared to employ low temperature and high pressure (LTHP) sintering method. Nanoceramic samples were sintered in the 2–8 GPa pressure range, while Ce3þ concentration was varied in the 0.5–5 at % range. In all nanoceramic, samples studied, in contrast to the single crystal, a strong rise of the absorption coefficient was found earlier at a wavelength shorter than 400 nm. Additionally, an unusual UV emission band near 3.1 eV was observed in nanoceramic samples, not detected in the YAG: Ce single crystal. Significant changes in their optical and luminescence properties were observed by applying high pressure during nanoceramic sintering. [89] studied the consequence of doping of lanthanum (La) on piezoelectric and ferroelectric properties of lead zirconate titanate (PZT). Sol-gel method was used to prepare Pb1-xLaxZr0.52Ti0.48O3 ceramics having values of x as 0.00, 0.02, 0.04, 0.06 and 0.10. To study the structural functionalization because of mismatch in ionic size, Raman and Fourier transform infrared spectroscopy have been utilized prove. The presence of both tetragonal and rhombohedral crystal symmetries, has been proved by Raman spectra. Because of La doping, lattice strain increases and enhances the remnant coercive field and polarization. The enhancement in La concentration enhances the linear piezoelectric coefficient. This proves that La-substituted PZT is a more suitable candidate as a piezoelectric sensor in comparison to PZT [90]. $BaTiO_3$ are modified with Pr^{3+} and K^+ and utilized layered perovskite $BaBi_2Nb_2O_9$ (BBN) in the form of dopant with BT as (1-x) ($BaTi_{1-x}$ $[PrK]_xO_3$)-x ($BaBi_2Nb_2O_9$) when, x = 0.1 (BTPK- $BBN_{0.1}$). BTPK-BBN0.1 nanocrystalline ceramic was fabricated with the help of the technique of precursor solution decomposition. XRD and TEM analysis was used to determine the nanoceramic crystallite particle properties. At temperature as high as 600 deg C, diffusion of grains took place, and the ceramics displayed grain and grain boundary conductivity. Also, while conducting P-E hysteresis study, a strong ferroelectric polarization was seen. Energy efficiency as high as 65% was detected at 400 V. The Sr0·5Ba0·5Ti1-xFexO_3 nanoceramics were successfully developed by [91] using the technique of auto combustion of sol-gel.

The structure displayed by the SBT sample is cubic, while SBT samples doped by Fe changed into a tetragonal mode as proved via XRD study. The mean crystallite size differs from 17 nm to 19 nm. The perovskite structure of synthesized nanoceramics with space group $Pm_3\ m$ was also confirmed by XRD analysis.

The doping concentration of Fe ion in SBT strongly influences magnetic and ferroelectric properties because of the differences in oxidation states as well as the presence of oxygen vacancies. The combined occurrence of ferroelectric and magnetic properties in SBT makes it a better option for device applications of multi-ferroic nature (Fig. 14). [92] proposed modification of nanoceramic materials Leads Titanate (Pb Ti_3) by incorporating Selenium and Samarium based on the stoichiometric formula $PbTi_{0.8} - xSe_{0.2}Sm_xO_3$. PTSeS showed the whole solid-state reaction at temperature 450 °C and was exhibited by TG characterization of green powder. was revealing Pure single-phase tetragonal structure was shown as XRD of functionalized PTSeS powders which were milled for 10 hours. For 5% wt of Sm in PTSeS, the dielectric constant was determined to be 2700 at curie temperature of 470 °C and the piezoelectric coefficient at 35 kV/cm of the poling field was determined to be 230 × 10 − 12 C/N. The resultant data obtained were similar and much better than similar kinds of materials reported until now.

[93] incorporated La3 + and V5 + on $BaBi_2Nb_2O_9$ ceramics and has investigated its effects through the structural analysis and study of electrical characteristics. The chemical precursor solution decomposition approach is used to prepare the materials.

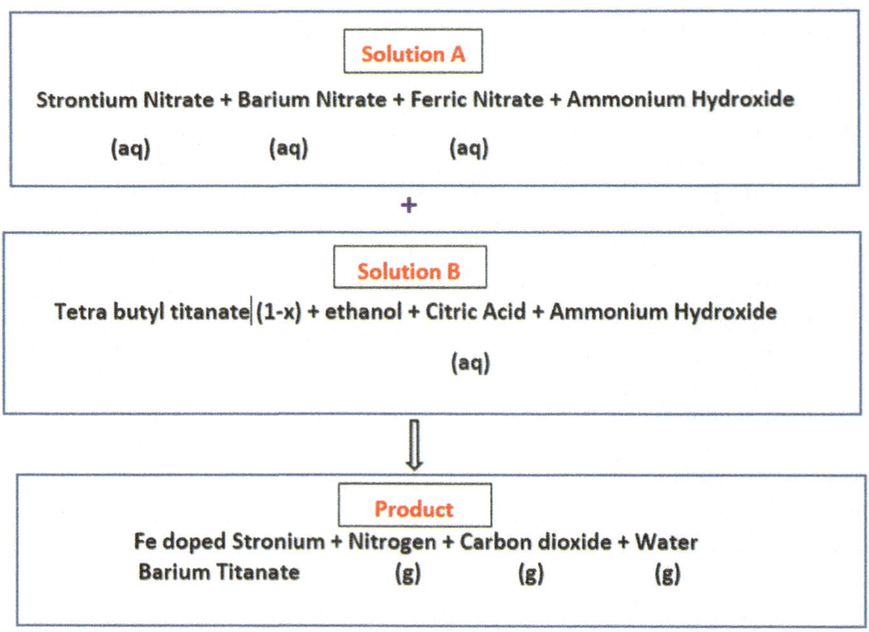

Fig. 14 The schematic diagram of the general chemical reaction of the sol-gel synthesis of SBT nanoceramics doped by Fe. Redrawn from [91]

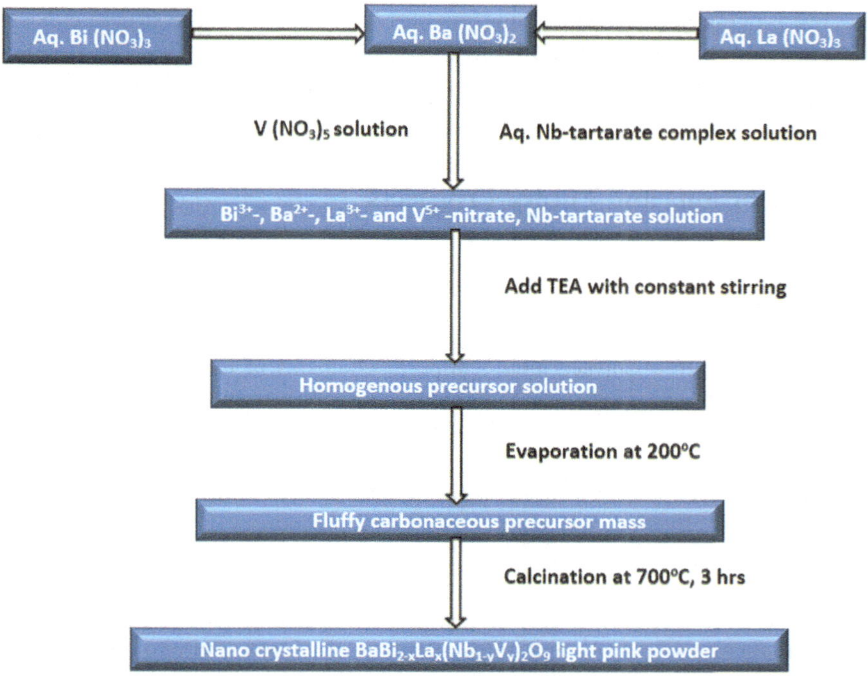

Fig. 15 Flowchart of the synthesis of $BaBi_2 - xLa_x(Nb_1 - xV_x)_2O_9$. Redrawn from [93]

The precursor mass is calcined at 700 °C for 3 h and investigated with the help of X-ray diffraction method, a single orthorhombic phase was established (Fig. 15). Scanning electron microscopy and transmission electron microscopy were used to study the morphology of the samples. The dc activation energy was found in values from 1.13 to 1.43 eV, less than the relaxation activation energy. As the electric field increases, the polarization of all the samples is also enhanced. BBLNV0.5 showed the highest energy efficiency of about 86%.

[94] proposed synthesis and characterization of an innovative SA sensor prepared of nanoceramics. XRD and TEM techniques were used to prepare and characterize the $NiTiO_3$ nanoceramics. After that, SEM and CV were used to synthesize and characterize the modified carbon paste. $NiTiO_3$ exhibited great responses to the SA. So it could be employed to detect SA. A robust, specific, sensitive, and cheaper method was obtained to have the voltammetric investigation of SA in actual samples. [95] investigated both the Mn-doped and undoped BFO nanoceramic samples. The modified samples were prepared through the citrate precursor route. The X-Ray Diffraction results displayed the rhombohedral modified perovskite structure of BFO ceramics with R_3C space group symmetry. With an increase in Mn concentration, the average crystallite size was found to be decreased. FESEM results exhibited granular shaped grains with reduced mean Mn substituted crystallite size, and good absorption peaks in the wavenumber from 450 to 600 cm-1 were showcased in FTIR

spectra, proving that all the ceramic samples have the perovskite structure. Since Mn substitution improves the dielectric constant of BFO ceramics, the modified nanoceramics could be a potential option for developing memory devices.

6 Conclusions and Future Prospects

In recent years, various novel forms of nanomaterials have been employed for advancement in a wide variety of fields such as biomedical applications, cancer diagnosis and treatment, particle separation, water contaminant treatment, etc. Through this chapter, different surface modification mechanisms of metallic nanoparticles, carbon nanomaterials (CNM), Nano-ceramics and self-assembled materials has been discussed, and it could be derived that surface modification of nanomaterials could enhance the specific targeting properties like adsorbing capacity towards particular molecules and receptors in different types of cells for a wide variety of applications. The rapid, cost-effective, and simple operational procedures provided by modified nanomaterial-based devices are expected to overtake conventional expensive systems in healthcare diagnostics systems in the coming years. However, there are still numerous challenges in the preparation and modifications of nanomaterials for diagnosis and therapy like biocompatibility, targeting efficiency, and cost effectiveness. Also, the risk factors observed from nanomaterials in biomedical applications have to be performed to determine their effect on human health and the environment. This will also assist in developing specific regulatory guidelines for their clinical applications shortly. Several research investigations reveal that different structures such as nanovesicles, nanotubes, and 3D peptide matrices alongside interwoven nanofibers can be self-assembled by peptides proteins. Several researchers have been motivated to investigate nanostructures and apply their different applications in areas of tissue engineering, bio-nanotechnology, dentistry, and regenerative medicine. As far as energy efficiency is concerned, self-assembled structural formation is way more efficient than directed self-assembly. The development of innovative nano-bio materials and enormous enhancement of the above mentioned emerging areas can be achieved by a more in-depth study of self-assembly of peptides. However, the fabrication of self-assembled nanoscale structures with selected controlled morphologies and chemical components remains a challenging task. More sophisticated properties of the self-assembled nanoscale structures, both physical as well as chemical, in selected applications have either been verified by researchers or have to be explored in the coming years. For the application of sophisticated properties of self-assembled nanoscale structures, which includes the synthesis of nanoparticles, nanomotors, nanostructure patterning, functionalized drug delivery, and use of bio-mineralization, tremendous development has been reported. It has been established that self-assembly techniques are efficient, easy to fabricate, and cheaper nanostructures. Since innovative investigations are supposed to be increased drastically in the nanotechnology field, there will be the development of numerous novel materials and nanostructures in the coming decades. The advancement of self-assembled nanostructures and nanomaterials might have a positive or negative impact on people's lifestyles, as in the case with the majority of technological and scientific breakthroughs.

References

1. Bárcena C, Sra AK, Gao J (2009) Applications of magnetic nanoparticles in biomedicine. Nanoscale Magn Mater Appl: 591–626. https://doi.org/10.1007/978-0-387-85600-1_20
2. Dobson J (2006) Gene therapy progress and prospects: magnetic nanoparticle-based gene delivery. Gene Ther 13:283–287. https://doi.org/10.1038/sj.gt.3302720
3. Rudge S, Peterson C, Vessely C, Koda J, Stevens S, Catterall L (2001) Adsorption and desorption of chemotherapeutic drugs from a magnetically targeted carrier (MTC). J Control Release 74:335–340. https://doi.org/10.1016/S0168-3659(01)00344-3
4. Kang YS, Risbud S, Rabolt JF, Stroeve P (1996) Synthesis and characterization of nanometer-size Fe3O4 and γ-Fe2O3 particles. Chem Mater 8:2209–2211. https://doi.org/10.1021/cm960157j
5. APPENZELLER T (1991) The man who dared to think small. Science (80-)254:1300–1300. https://doi.org/10.1126/science.254.5036.1300
6. Pauly CS, Genix AC, Alauzun JG, Sztucki M, Oberdisse J, Hubert Mutin P (2015) Surface modification of alumina-coated silica nanoparticles in aqueous sols with phosphonic acids and impact on nanoparticle interactions. Phys Chem Chem Phys 17:19173–19182. https://doi.org/10.1039/c5cp01925g
7. Vert M, Hellwich K-H, Hodge P, Schué F, Kubisa P, Rinaudo M, Hess M, Doi Y (2012) Terminology for biorelated polymers and applications (IUPAC recommendations 2012). Pure Appl Chem 84:377–410. https://doi.org/10.1351/pac-rec-10-12-04
8. Rogers R (1999) Book review. Process Saf Environ Prot 77:371–372. https://doi.org/10.1016/s0957-5820(99)70836-0
9. Steinhögl W, Schindler G, Steinlesberger G, Engelhardt M (2002) Size-dependent resistivity of metallic wires in the mesoscopic range. Phys Rev B—Condens Matter Mater Phys 66:1–4. https://doi.org/10.1103/PhysRevB.66.075414
10. Patzke GR, Zhou Y, Kontic R, Conrad F (2011) Oxide nanomaterials: synthetic developments, mechanistic studies, and technological innovations. Angew Chemie—Int Ed 50:826–859. https://doi.org/10.1002/anie.201000235
11. Converter B (2007) and applications of a current-sourced
12. Mody VV, Nounou MI, Bikram M (2009) Novel nanomedicine-based MRI contrast agents for gynecological malignancies. Adv Drug Deliv Rev 61:795–807. https://doi.org/10.1016/j.addr.2009.04.020
13. Praetorius N, Mandal T (2008) Engineered nanoparticles in cancer therapy. Recent Pat Drug Deliv Formul 1:37–51. https://doi.org/10.2174/187221107779814104
14. Sharma P, Brown S, Walter G, Santra S, Moudgil B (2006) Nanoparticles for bioimaging. Adv Colloid Interface Sci 123–126:471–485. https://doi.org/10.1016/j.cis.2006.05.026
15. Moghimi SM, Hunter AC, Murray JC (2005) Nanomedicine: current status and future prospects. FASEB J 19:311–330. https://doi.org/10.1096/fj.04-2747rev
16. Sagar GH, Arunagirinathan MA, Bellare JR (2007) Self-assembled surfactant nano-structures important in drug delivery: a review. Indian J Exp Biol 45:133–159
17. Rajagopal K, Schneider JP (2004) Self-assembling peptides and proteins for nanotechnological applications. Curr Opin Struct Biol 14:480–486. https://doi.org/10.1016/j.sbi.2004.06.006
18. Whitesides GM, Boncheva M (2002) Supramolecular chemistry and self-assembly special feature: beyond molecules: self-assembly of mesoscopic andmacroscopic components. Proc Natl Acad Sci U S A 99:4769. https://doi.org/10.1073/PNAS.082065899

19. Whitesides GM, Grzybowski B (2002) Self-assembly at all scales. Science (80-)295:2418–2421. https://doi.org/10.1126/science.1070821
20. Tu RS, Tirrell M (2004) Bottom-up design of biomimetic assemblies. Adv Drug Deliv Rev 56:1537–1563. https://doi.org/10.1016/j.addr.2003.10.047
21. Byard EH, Lange BM (1991) Tubulin and microtubules. Essays Biochem 26:13–25. https://doi.org/10.1002/9780470015902.a0000676.pub3
22. Brown JC, Newcomb WW (2011) Herpesvirus capsid assembly: insights from structural analysis. Curr Opin Virol 1:142–149. https://doi.org/10.1016/j.coviro.2011.06.003
23. Naik RR, Stone MO (2005) Integrating biomimetics. Mater Today 8:18–26. https://doi.org/10.1016/S1369-7021(05)71077-4
24. Forrest JA, Dalnoki-Veress K, Stevens JR, Dutcher JR (1996) Effect of free surfaces on the glass transition temperature of thin polymer films. Phys Rev Lett 77:2002–2005. https://doi.org/10.1103/PhysRevLett.77.2002
25. Guerrini L, Alvarez-Puebla RA, Pazos-Perez N (2018) Surface modifications of nanoparticles for stability in biological fluids. Materials (Basel) 11:1–28. https://doi.org/10.3390/ma1107 1154
26. Li C, Shuford KL, Park QH, Cai W, Li Y, Lee EJ, Cho SO (2007) High-yield synthesis of single-crystalline gold nano-octahedra. Angew Chemie—Int Ed 46:3264–3268. https://doi.org/10.1002/anie.200604167
27. Santra TS, Tseng F-G, Barik TK (2014) Green biosynthesis of gold nanoparticles and biomedical applications. Am J Nano Res Appl 2:5–12. https://doi.org/10.11648/j.nano.s.201402060 2.12
28. Santra TS, Tseng F-G (Kevin), Barik TK (2015) Biosynthesis of silver and gold nanoparticles for potential biomedical applications—a brief review. J Nanopharmaceutics Drug Deliv 2:249–265. https://doi.org/10.1166/jnd.2014.1065
29. Santra TS, Kar S, Chen TC, Chen CW, Borana J, Lee MC, Tseng FG (2020) Near-infrared nanosecond-pulsed laser-activated highly efficient intracellular delivery mediated by nano-corrugated mushroom-shaped gold-coated polystyrene nanoparticles. Nanoscale 12:12057–12067. https://doi.org/10.1039/d0nr01792b
30. Granqvist CG, Buhrman RA (1976) Ultrafine metal particles. J Appl Phys 47:2200–2219. https://doi.org/10.1063/1.322870
31. Blackman JA (2008) Chapter 7 photoexcitation and optical absorption. Handb Met Phys 5:175–229. https://doi.org/10.1016/S1570-002X(08)00207-3
32. Ninithi Metallic-nanoparticles. https://ninithi.wordpress.com/metallic-nanoparticles/
33. Gusain R, Kumar N, Ray SS (2020) Recent advances in carbon nanomaterial-based adsorbents for water purification. Coord Chem Rev 405:213111. https://doi.org/10.1016/j.ccr.2019.213111
34. He S, Liu X, Yan P, Wang A, Su J, Su X (2019) Preparation of gemini surfactant/graphene oxide composites and their superior performance for Congo red adsorption. RSC Adv 9:4908–4916. https://doi.org/10.1039/c8ra10025j
35. Jauris IM, Matos CF, Saucier C, Lima EC, Zarbin AJG, Fagan SB, Machado FM, Zanella I (2016) Adsorption of sodium diclofenac on graphene: a combined experimental and theoretical study. Phys Chem Chem Phys 18:1526–1536. https://doi.org/10.1039/c5cp05940b
36. Kumar S, Nair RR, Pillai PB, Gupta SN, Iyengar MAR, Sood AK (2014) Graphene oxide-MnFe2O4 magnetic nanohybrids for efficient removal of lead and arsenic from water. ACS Appl Mater Interfaces 6:17426–17436. https://doi.org/10.1021/am504826q
37. Jiao T, Liu Y, Wu Y, Zhang Q, Yan X, Gao F, Bauer AJP, Liu J, Zeng T, Li B (2015) Facile and scalable preparation of graphene oxide-based magnetic hybrids for fast and highly efficient removal of organic dyes. Sci Rep 5:1–10. https://doi.org/10.1038/srep12451
38. Zhao L, Guan X, Yu B, Ding N, Liu X, Ma Q, Yang S, Yilihamu A, Yang ST (2019) Carboxylated graphene oxide-chitosan spheres immobilize Cu2 + in soil and reduce its bioaccumulation in wheat plants. Environ Int 133:105208. https://doi.org/10.1016/j.envint.2019.105208
39. Ray SK, Majumder C, Saha P (2017) Functionalized reduced graphene oxide (fRGO) for removal of fulvic acid contaminant. RSC Adv 7:21768–21779. https://doi.org/10.1039/c7ra01 069a

40. Besharati Vineh M, Saboury AA, Poostchi AA, Mamani L (2018) Physical adsorption of horseradish peroxidase on reduced graphene oxide nanosheets functionalized by Amine: a good system for biodegradation of high phenol concentration in wastewater. Int J Environ Res 12:45–57. https://doi.org/10.1007/s41742-018-0067-1
41. Zambare R, Song X, Bhuvana S, Antony Prince JS, Nemade P (2017) Ultrafast dye removal using ionic liquid-graphene oxide sponge. ACS Sustain Chem Eng 5:6026–6035. https://doi.org/10.1021/acssuschemeng.7b00867
42. Mahmoodi NM, Maroofi SM, Mazarji M, Nabi-Bidhendi G (2017) Preparation of modified reduced graphene oxide nanosheet with cationic surfactant and its dye adsorption ability from colored wastewater. J Surfactants Deterg 20:1085–1093. https://doi.org/10.1007/s11743-017-1985-1
43. Shtepliuk I, Caffrey NM, Iakimov T, Khranovskyy V, Abrikosov IA, Yakimova R (2017) On the interaction of toxic heavy metals (Cd, Hg, Pb) with graphene quantum dots and infinite grapheme. Sci Rep 7:1–17. https://doi.org/10.1038/s41598-017-04339-8
44. Lee B, Lee S, Lee M, Jeong DH, Baek Y, Yoon J, Kim YH (2015) Carbon nanotube-bonded graphene hybrid aerogels and their application to water purification. Nanoscale 7:6782–6789. https://doi.org/10.1039/c5nr01018g
45. Vadahanambi S, Lee SH, Kim WJ, Oh IK (2013) Arsenic removal from contaminated water using three-dimensional graphene-carbon nanotube-iron oxide nanostructures. Environ Sci Technol 47:10510–10517. https://doi.org/10.1021/es401389g
46. Jin L, Zhao X, Qian X, Dong M (2018) Nickel nanoparticles encapsulated in porous carbon and carbon nanotube hybrids from bimetallic metal-organic-frameworks for highly efficient adsorption of dyes. J Colloid Interface Sci 509:245–253. https://doi.org/10.1016/j.jcis.2017.09.002
47. Lu W, Li J, Sheng Y, Zhang X, You J, Chen L (2017) One-pot synthesis of magnetic iron oxide nanoparticle-multiwalled carbon nanotube composites for enhanced removal of Cr(VI) from aqueous solution. J Colloid Interface Sci 505:1134–1146. https://doi.org/10.1016/j.jcis.2017.07.013
48. Zhang C, Sui J, Li J, Tang Y, Cai W (2012) Efficient removal of heavy metal ions by thiol-functionalized superparamagnetic carbon nanotubes. Chem Eng J 210:45–52. https://doi.org/10.1016/j.cej.2012.08.062
49. Vieira EFS, Simoni JDA, Airoldi C (1997) Interaction of cations with SH-modified silica gel: thermochemical study through calorimetric titration and direct extent of reaction determination. J Mater Chem 7:2249–2252. https://doi.org/10.1039/a704286h
50. Ma J, Yu F, Zhou L, Jin L, Yang M, Luan J, Tang Y, Fan H, Yuan Z, Chen J (2012) Enhanced adsorptive removal of methyl orange and methylene blue from aqueous solution by alkali-activated multiwalled carbon nanotubes. ACS Appl Mater Interfaces 4:5749–5760. https://doi.org/10.1021/am301053m
51. Kazemi-Beydokhti A, Meyghani N, Samadi M, Hajiabadi SH (2019) Surface modification of carbon nanotube: effects on pulsating heat pipe heat transfer. Chem Eng Res Des 152:30–37. https://doi.org/10.1016/j.cherd.2019.09.023
52. Zhang Y, Zhou Z, Shen Y, Zhou Q (2016) Reversible assembly of graphitic carbon nitride 3D network for highly selective dyes absorption and regeneration. https://doi.org/10.1021/acsnano.6b05488
53. Zhu B, Xia P, Ho W, Yu J (2015) Applied surface science isoelectric point and adsorption activity of porous g-C 3 N 4. Appl Surf Sci 344:188–195. https://doi.org/10.1016/j.apsusc.2015.03.086
54. Cheng N, Tian J, Liu Q, Ge C, Qusti AH, Asiri AM, Al-youbi AO, Sun X (2013) Au-nanoparticle-loaded graphitic carbon nitride nanosheets: green photocatalytic synthesis and application toward the degradation of organic pollutants: 4–8. https://doi.org/10.1021/am401802r
55. Liu Z, Fang Y, Jia H, Wang C, Song Q, Li L (2018) Novel multifunctional cheese-like 3D carbon-BN as a highly efficient adsorbent for water purification. Sci Rep: 1–11. https://doi.org/10.1038/s41598-018-19541-5

56. Han S, Sohn K, Hyeon T (2000) Fabrication of new nanoporous carbons through silica templates and their application to the adsorption of bulky dyes: 3337–3341
57. Xiao B, Thomas KM (2005) Adsorption of aqueous metal ions on oxygen and nitrogen functionalized nanoporous activated carbons: 3892–3902
58. Han X, Wang H, Zhang L (2018) Efficient removal of methyl blue using nanoporous carbon from the waste biomass
59. Li X, Yuan H, Quan X, Chen S, You S (2017) Effective adsorption of sulfamethoxazole, bisphenol A and methyl orange on nanoporous carbon derived from metal-organic frameworks. J Environ Sci: 1–10. https://doi.org/10.1016/j.jes.2017.10.019
60. Teng M, Qiao J, Li F, Bera PK (2012) Electrospun mesoporous carbon nanofibers produced from phenolic resin and their use in the adsorption of large dye molecules. Carbon N Y 50:2877–2886. https://doi.org/10.1016/j.carbon.2012.02.056
61. Chem JM (2012) Journal of materials chemistry: 15919–15927. https://doi.org/10.1039/c2jm33214k
62. Chem JM (2012) Synthesis of mesoporous magnetic Fe 3 O 4 @ carbon nanofibers utilizing in situ polymerized polybenzoxazine for water purification †, 4619–4622. https://doi.org/10.1039/c2jm00036a
63. Manuscript A (2016) Environmental Science. https://doi.org/10.1039/C6EN00568C
64. S. T, K. T, Patel P, G. F, K. T (2012) Diamond, diamond-like carbon (DLC) and diamond-like nanocomposite (DLN) thin films for MEMS applications. Microelectromechanical Syst Devices. https://doi.org/10.5772/29671
65. Santra TS, Bhattacharyya TK, Mishra P, Tseng FG, Barik TK (2012) Biomedical applications of diamond-like nanocomposite thin films. Sci Adv Mater 4:110–113. https://doi.org/10.1166/sam.2012.1258
66. Santra TS, Bhattacharyya TK, Patel P, Tseng FG, Barik TK (2011) Structural and tribological properties of diamond-like nanocomposite thin films. Surf Coatings Technol 206:228–233. https://doi.org/10.1016/j.surfcoat.2011.06.057
67. Santra TS, Bhattacharyya TK, Tseng FG, Barik TK (2012) Influence of flow rate on different properties of diamond-like nanocomposite thin films grown by PECVD. AIP Adv 2. https://doi.org/10.1063/1.4721654
68. Santra TS, Bhattacharyya TK, Tseng F, Barik TK (2011) Diamond-like nanocomposite (DLN) films for microelectro-mechanical system (MEMS)
69. Santra TS, Liu CH, Bhattacharyya TK, Patel P, Barik TK (2010) Characterization of diamond-like nanocomposite thin films grown by plasma enhanced chemical vapor deposition. J Appl Phys 107. https://doi.org/10.1063/1.3415548
70. Giammarco J, Mochalin VN, Haeckel J, Gogotsi Y (2016) Journal of colloid and interface science the adsorption of tetracycline and vancomycin onto nanodiamond with controlled release. J Colloid Interface Sci 468:253–261. https://doi.org/10.1016/j.jcis.2016.01.062
71. Zhao X, Zhang S, Bai C, Li B, Li Y, Wang L, Wen R, Zhang M, Ma L, Li S (2016) Journal of colloid and interface science nano-diamond particles functionalized with single/double-arm amide—thiourea ligands for adsorption of metal ions 469:109–119. https://doi.org/10.1016/j.jcis.2016.02.017
72. Molavi H, Shojaei A, Pourghaderi A (2018) Rapid and tunable selective adsorption of dyes using thermally oxidized nanodiamond. J Colloid Interface Sci. https://doi.org/10.1016/j.jcis.2018.03.088
73. Shen Y, Zhu X, Zhu L, Chen B (2016) Synergistic effects of 2D graphene oxide nanosheets and 1D carbon nanotubes in the constructed 3D carbon aerogel for high performance pollutant removal. Chem Eng J. https://doi.org/10.1016/j.cej.2016.11.132
74. Online VA, Chen Y, Chen L, Bai H, Li L (2013) Broad-spectrum adsorbents for water puri fi cation †, 1992–2001. https://doi.org/10.1039/c2ta00406b
75. Zhang C, Chen Z, Guo W, Zhu C, Zou Y (2018) PT. https://doi.org/10.1016/j.ijbiomac.2018.02.074
76. Mittal H, Sinha S, Singh B, Kaur J, Sharma J, Alhassan SM (2018) Recent progress in the structural modification of chitosan for applications in diversified biomedical fields. Eur Polym J 109:402–434. https://doi.org/10.1016/j.eurpolymj.2018.10.013

77. Gaytan SM, Cadena MA, Karim H, Delfin D, Lin Y, Espalin D, MacDonald E, Wicker RB (2015) Fabrication of barium titanate by binder jetting additive manufacturing technology. Ceram Int 41:6610–6619. https://doi.org/10.1016/j.ceramint.2015.01.108
78. Zhou G, Wang KP, Liu HW, Wang L, Xiao XF, Dou DD, Fan YB (2018) International journal of biological macromolecules three-dimensional polylactic acid @ graphene oxide/chitosan sponge bionic filter: highly efficient adsorption of crystal violet dye. Int J Biol Macromol 113:792–803. https://doi.org/10.1016/j.ijbiomac.2018.02.017
79. Manuscript A (2015) Nanoscale. https://doi.org/10.1039/C5NR07579C
80. Yan Y, Gong J, Chen J, Zeng Z, Huang W, Pu K, Liu J, Chen P (2019) Recent advances on graphene quantum dots: from chemistry and physics to applications 1808283:1–22. https://doi.org/10.1002/adma.201808283
81. (2018) Antimicrobial amino-functionalized nitrogen-doped graphene quantum dots for eliminating multidrug-resistant species in dual-modality photodynamic therapy and bioimaging under two-photon excitation. https://doi.org/10.1021/acsami.8b01429
82. (2016) Photodynamic graphene quantum dot: reduction condition regulated photo-activity and size dependent efficacy. https://doi.org/10.1021/acsami.5b11154
83. Zhou L, Zhou L, Ge X, Zhou J, Shen J (2014) Multicolor imaging and the anticancer effect of a bifunctional silica nanosystem based on the complex of graphene quantum dots and hypocrellin A †. Chem Commun: 2–5. https://doi.org/10.1039/C4CC06968D
84. Zhang D, Wen L, Huang R, Wang H, Hu X, Xing D (2017) MOE key laboratory of laser life science and institute of laser life science, college of biomaterials. https://doi.org/10.1016/j.biomaterials.2017.10.034
85. Koide Y, Urano Y, Kenmoku S, Kojima H, Nagano T (2007) Design and synthesis of fluorescent probes for selective detection of highly reactive oxygen species in mitochondria of living cells 2:10324–10325
86. Yuan L, Lin W, Yang Y, Chen H (2012) Lin Yuan, Weiying Lin, * Yueting Yang, and Hua Chen
87. Kumawat MK, Thakur M, Gurung RB, Srivastava R (2017) Graphene quantum dots for cell proliferation, nucleus imaging, and photoluminescent sensing applications. Sci Rep: 1–16. https://doi.org/10.1038/s41598-017-16025-w
88. Pankratov V, Shirmane L, Chudoba T, Gluchowski P, Hreniak D, Strek W, Lojkowski W (2010) Peculiarities of luminescent properties of cerium doped YAG transparent nanoceramics. Radiat Meas 45:392–394. https://doi.org/10.1016/j.radmeas.2009.12.014
89. Kour P, Pradhan SK, Kumar P, Sinha SK, Kar M (2016) Enhanced ferroelectric and piezoelectric properties in La-modified PZT ceramics. Appl Phys A Mater Sci Process 122:1–7. https://doi.org/10.1007/s00339-016-0122-8
90. Adak MK, Mahato SJ, Mahato U, Gorai UR, Mondal S, Kar S, Dhak P, Dhak D (2019) Investigation of energy storage and electrical properties of modified BaTiO3 doped by BaBi2Nb2O9 nano crystalline ceramics. Phys B Condens Matter 578:411885. https://doi.org/10.1016/j.physb.2019.411885
91. Bhoyar DN, Somvanshi SB, Kharat PB, Pandit AA, Jadhav KM (2020) Structural, infrared, magnetic and ferroelectric properties of Sr0·5Ba0·5Ti1-xFexO3 nanoceramics: modifications via trivalent Fe ion doping. Phys B Condens Matter 581:411944. https://doi.org/10.1016/j.physb.2019.411944
92. Sinha SK, Kumari S, Chaudhary RK (2019) Dielectric and piezoelectric properties of PbTi 0.8 − x Se 0.2 Sm x O 3 nanoceramics prepared by high energy ball milling. Appl Phys A Mater Sci Process 125:1–6. https://doi.org/10.1007/s00339-019-2425-z
93. Khatun J, Adak MK, Dhak P, Ghorai UK, Dhak D (2019) Influence of La 3 + and V 5 + doping on the polarization and impedance behaviour of BaBi 2 Nb 2 O 9 nano-ceramics prepared by chemical route. J Mater Sci: Mater Electron 30:7065–7079. https://doi.org/10.1007/s10854-019-01023-7
94. Ghoreishi SM, Kashani FZ, Khoobi A, Enhessari M (2015) Fabrication of a nickel titanate nanoceramic modified electrode for electrochemical studies and detection of salicylic acid. J Mol Liq 211:970–980. https://doi.org/10.1016/j.molliq.2015.08.035

95. Wani WA, Ramaswamy K, Venkataraman BH, Kundu S (2019) Influence of transition metal ion doping on structural and dielectric properties of sol-gel synthesized bismuth ferrite nanoceramics. 2019 IEEE Int Symp Appl Ferroelectr ISAF 2019—Proc: 1–3. https://doi.org/10.1109/ISAF43169.2019.9034952

Laser Induced Micro/Nano Functional Surfaces on Metals for Biomedical Applications

Srinivasan Arthanari, Jiaru Zhang, Xianda Xue, Yan Li, and Yingchun Guan

Abstract Laser processing is one of the precise techniques to achieve functional surfaces with micro/nanostructures on metallic materials. The controlled laser-induced structures on metallic implants significantly alter the mechanical properties, chemical stability, and bioactivity. Therefore, the present chapter emphasizes the recent research works on laser processing such as selective laser melting (SLM), laser surface melting (LSM), laser surface patterning of various metallic implant materials. The surface characteristics, and biomedical applications of processed surfaces are summarized. Various textures from micron to the nanometer-scale on permanent and degradable implants produced using high energy pulsed lasers enhanced the biomineralization and cell proliferation behaviors. Furthermore, the laser micro/nanotextured surfaces with different surface roughness and chemical composition improved the antibacterial activity of the implants, and also directed the cell attachment depending on the size/nature of the nanostructures. The femtosecond laser was used to develop the laser-induced periodic surface structure (LIPSS) on titanium alloys. The surface-enhanced Raman scattering (SERS) response of adsorbed

S. Arthanari · J. Zhang · Y. Guan
School of Mechanical Engineering and Automation, Beihang University, 37 Xueyuan Road, Beijing 100191, China

S. Arthanari · Y. Guan
Hefei Innovation Research Institute of Beihang University, Xinzhan Hi-Tech District, Anhui 230012, China

X. Xue · Y. Li
School of Materials Science and Engineering, Beihang University, Beijing 100191, China

Y. Li
Beijing Advanced Innovation Centre for Biomedical Engineering, Beihang University, Beijing 100191, China

Y. Guan (✉)
National Engineering Laboratory of Additive Manufacturing for Large Metallic Components, Beihang University, Beijing 100083, China
e-mail: guanyingchun@buaa.edu.cn

T. S. Santra and L. Mohan (eds.), *Nanomaterials and Their Biomedical Applications*, Springer Series in Biomaterials Science and Engineering 16,
https://doi.org/10.1007/978-981-33-6252-9_15

biomolecules and fluorescence enhancement on the hierarchical LIPSS surfaces for clinical applications are also discussed.

Keywords Laser processing · Micro/nano surface · Metals · Implants · Biomaterials

1 Introduction to Metallic Implants and Surface Treatments

Surface treatments of metallic materials are important to enhance the surface-related properties without altering the bulk properties of the materials. In particular, the surface properties of materials used in biomedical applications need to be tailored to respond to the surrounding environments [1]. Several metallic materials including steels, Co-Cr, Titanium, Magnesium, Zinc and Fe based alloys are considered for the implant applications and metallic implants are about 70-80% of the available implant materials [2]. Among them, steels, Co-Cr, and Ti alloys are commercially used and their properties are well established [3]. Mg and Zn based alloys are not well established in the markets and continuous research is undergoing to develop them as biodegradable implant materials. The metallic materials exhibit different physical and chemical properties when they are exposed to physiological solutions. Steels and Ti alloys are permanent implant materials, while Mg-based alloys are degradable during service. The surface layer formed on the permanent implant materials is relatively stable and controls the diffusion of aggressive ions (such as Cl^-, SO_4^{2-}, -CO_3^{2-}, etc.,) present in the physiological solution, thereby controlling the chemical degradation and establishing a stable surface. Nevertheless, these materials have limited interaction with the living cells and bone healing properties since they have higher Young's modulus compared to the natural bone which produces a stress-shielding effect. In addition to that, the surface nature such as porosity plays a key role in determining the ability to bond with the natural tissues during service.

Interestingly, Mg-based alloys have similar mechanical properties to that of the natural bone, thus they minimize the stress-shielding effect. However, Mg alloys suffer severe corrosion and hydrogen gas evolution in the physiological solution thus limiting the bone-bonding ability during service. Degradable Fe based implant materials are also given attention due to their exceptional mechanical properties, excellent formability, and better biocompatibility. The very low degradation rates of the Fe based alloys lead to long-term retention in the body and limits their widespread applications. Zn based alloys are focused as potential implant material in recent research due to their outstanding properties such as biodegradability and biocompatibility and they overcome the challenges of Mg and Fe based implant materials. However, Zn alloys have weak mechanical properties compared to Mg and Fe based alloys, and researchers are consistently working on improving the mechanical properties through various techniques. Addition of alloying elements to Zn during alloy development is found to improve the alloy properties, however, the tolerance limits of each element need to be considered to avoid the adverse effects.

Therefore, surface treatments are adopted to satisfy the requirement of various implant materials. Among them applying the surface coatings to fabricate micro/nanostructures on various implant materials to enhance the surface properties are commonly studied. Several inorganic/organic coatings are developed to simultaneously enhance the degradation properties and biocompatibility of implant materials. Despite their interesting properties, the surface coatings have certain limitations such as adhesion to the substrate, therefore appropriate coating conditions are essential to achieve good quality coatings on implant materials. The surface nature of the implant materials also plays a key role in determining the cell attachment behavior and hence it is important to control the surface textures to enhance the cell attachment and proliferation. Human mesenchymal stem cells (hMSCs) adhesion and differentiation were significantly varied with micro/nano topographies produced on Ti alloy using femtosecond laser [4]. Micro and nano grid topographies were beneficial for the cell colonization by anchoring the cells and cell locomotion, respectively, and cell adhesion was promoted by the superimposed micro/nano grids. Furthermore, the bio-mineralization process also varies based on the surface nature of the material. Hence, surface modification is essential to alter the surface properties of the biomaterials in various applications.

Several surface modification treatments such as physicochemical, mechanical treatments, biomimetic process, vacuum processes, and high energy processes are widely used to fabricate the surface properties of biomaterials. Each of these methods has its advantages as well as disadvantages. Surface modification using lasers is given importance in recent research owing to its high precision process and controlled surface morphologies. Also, laser processing is a non-contact process, has high repeatability, pollution-free, and produces a small heat-affected zone on the processed materials, thus controlling the material properties. Several studies carried out on laser processed hierarchical surface patterned Ti-based alloys indicated that the bonding ability of the bone was better on the textured surfaces compared to polished and other processed surfaces. Furthermore, the osteogenesis increased on the textured surfaces and inhibit the adipogenesis of mesenchymal stem cells (MSCs). The surface coatings followed by laser processing also significantly altered the osseointegration of the implant materials. Anti-bacterial properties of surgical metallic materials are one of the important factors, and the usage of ultra-fast laser treatment produces biomimetic textures with different wettability properties. The surface wettability and size/nature (hydrophilic/hydrophobic) of bacteria alter the bacterial attachment, and their colonization, thus surface can be designed accordingly. The picosecond and femtosecond lasers are widely used to fabricate the nanoscale structures with micro/nanoscale periodicity and nano ripples. Laser texturing is a precise technique, which can also control the parameters such as, pulse energy, scan speed, spacing, and the number of passes to fabricate the structures with sizes similar and larger than the bacterial cell sizes, which could alter the wettability and following bacterial adhesion properties [5]. Metal nanoparticles are widely used for localized surface plasmon resonance (LSPR) sensing and the particles enhance the signal intensities of the biomolecules [6, 7]. Laser ablation is also used to fabricate such nanostructures having better surface-enhanced Raman signals (SERS) [8]. The fabrication of hierarchical micro/nanostructures on implantable materials would also be beneficial to enhance the Raman signals and such substrates could be widely used for biomedical imaging applications. Therefore, the present chapter mainly provides an overview of recent research works on the laser surface treatments applied on various metallic implant materials to produce refined microstructures and micro/nano textures, and their applications in various biomedical fields.

2 Laser Processing of Metallic Materials

Laser processing is widely used to enhance the surface properties such as hardness, wear, and corrosion resistance of metallic materials used in several applications [9]. Figure 1 depicts the simple schematic representation of the laser processing experimental arrangements. Laser surface melting is a widely used process in which the substrate is irradiated with the laser and the surface melt locally and rapid solidification of the melted layer produces fine microstructures (micro/nanograins), which further enhance the surface properties [10]. The melted layer thickness is typically in several 10–100 μm depending on the laser processing parameters such as laser power, scanning speed, etc. The laser cladding process can be carried out in two methods (1) direct feeding of the clad materials and (2) preplacing the coating materials on the substrate. In both the processes metal, ceramic or composite layer is irradiated with the laser to deposit them on the substrate. The cladding process generally produces hard coatings, and in this process, good metallurgical bonding between the coating and substrate is achieved [11]. Laser surface patterning/texturing is a process in which the micro/nano textures are produced by controlling the laser parameters. The

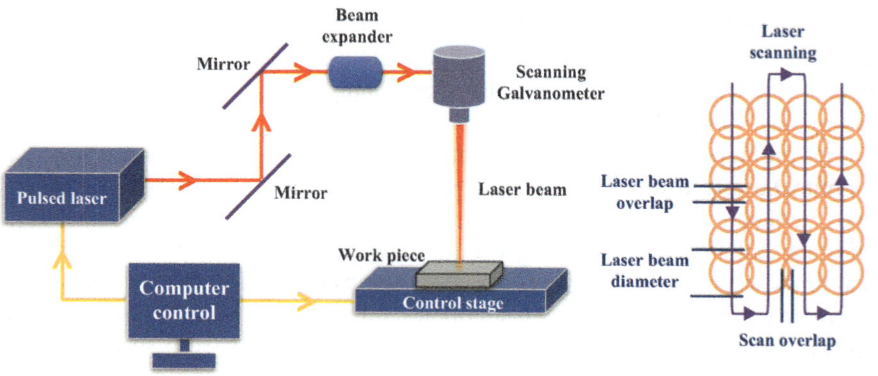

Fig. 1 Schematic representation of laser processing equipment and scanning parameters

produced micro/nano textures could be used for different applications such as optical properties, wettability, sensors, and biomedical implants [12–15].

2.1 Interaction of Laser with Metal Surfaces

The irradiation of metal surfaces with laser results in various physical processes occurring on the surface such as heating, melting, vaporization, etc. Furthermore, the conversion of laser (electromagnetic wave) energy into heat significantly varies depending on the nature of the materials whether metals or non-metals. The absorption of laser energy by the metals during laser processing takes place through the photon-electron interactions with free or bound electrons. Therefore, the energy state of the electrons in the conduction band raises by following the classical thermal conduction mechanisms. These excited electrons can follow two processes (1) collide with lattice phonons and with other electrons or (2) give back their radiation by spontaneous emission. The relationship between the electron density of a metal N_e and optical properties is denoted as plasma frequency $\omega_p = \sqrt{N_e e^2/m_e \varepsilon_0}$, where m_e is the mass of an electron and ε_0 the permittivity of free space [16]. Since the mean free time of the electrons in the conductor is about 10^{-13} s, the thermal equilibrium is quickly achieved during the process.

The absorbed energy density of laser during the interaction with the metals is determined by various factors viz., absorbed power density, pulse duration, the reflectivity of the surface, thermal diffusivity, vapor pressure, and optical absorbance. These factors further decide the penetration of laser with substrate material and the resulting properties. At low and high laser power densities, the interaction of laser with metals is thermal and explosive characters, respectively. The materials are heated, melted until reaching the vapor pressure, and material removal take place when it reaches the vapor pressure during the laser interaction. At higher laser intensities several non-equilibrium processes to take place, these results in the large mass removal from the surface. Another important thing to be considered is the wavelength of laser beam since the absorption depths of the materials depend on the wavelength and choosing the appropriate wavelength laser will be effective to local modification of surface properties without altering the bulk materials. Therefore, it is important to control the laser processing parameters to achieve the desired surface nature with targeted properties.

2.2 Selective Laser Melting (SLM) of Metallic Implants

The SLM is an additive manufacturing (3D printing) process, where a high-power laser beam scans the metallic powder materials, and completely melt them to fabricate the desired geometry (Fig. 2a). This method is generally employed to produce new alloys consisting of a millimeter or sub-millimeter scale structures

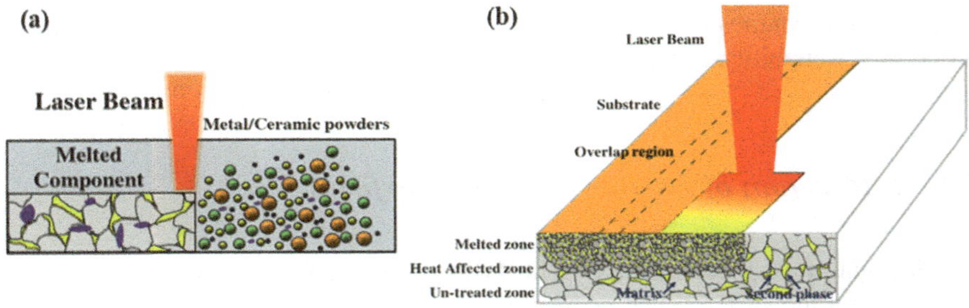

Fig. 2 Schematic representation of **a** selective laser melting (SLM). **b** laser surface melting (LSM)

with enhanced properties. The major advantages of SLM are flexible to produce the components/structures such as metal restoration, artificial organs with complex shapes [17]. However, the use of SLM to produce micro components are still less due to the process limitations. SLM technique is widely used to manufacture permanent implant materials for various applications including orthopedic and dental implants [18, 19].

AISI 316L stainless steel prototype stents with promising mechanical properties were produced using 5-20 μm fine powders by micro SLM system and the produced stent was subsequently post-treated with plasma and chemical polishing [20]. Guoqing et al. studied the blood compatibility and release rate of ions before and after the SLM treated CoCrMo alloy [21]. The Co-release during 1-week immersion in artificial saliva was close to the cast alloy, however, the further increase in time revealed the gradual increase in the Co-release, while in the NaCl solution the difference in the release was higher even after 1 week of immersion. Interestingly the Cr release both in artificial saliva and NaCl solution was significantly lower for the SLM treated alloy compared to as-cast alloy. The higher ion release in NaCl solution was attributed to the reduction in pH of the NaCl solution with time. Furthermore, an increase in surface roughness increased the Co ion release, however, Cr release was lower than the Co-release after one week, and thereafter no obvious change was noticed. The huge difference in the ion release was mainly attributed to the uniform and the densified surface containing smaller surface grains and fewer pores of the SLM treated alloy compared to as-cast alloy. The hemolysis rate of the SLM alloy was 41.2% lower than the as-cast part confirming the better blood compatibility. SLM alloy reduces the adhesion of blood cells to the surface due to a more uniform nature, therefore, decreases the blood element destruction.

Titanium alloys, in particular, Ti6Al4V (Ti64) is commonly used in implant applications due to its biocompatible nature [22, 23]. Titanium alloys are very compatible to be fabricated using the SLM technique, and the laser processing parameters play a key role in determining the properties of the components [24]. The laser processing parameters such as laser power, scan speed, and hatch spacing significantly affected the properties of SLM Ti64 alloy and the increase in laser power reduced the surface roughness due to the flattening of the melt pool. The optimum condition for SLM was reported as 49 W laser power, 400 mm/s scan speed and 99 μm laser spacing, which resulted in a 23.62% porosity, 8.68 μm surface roughness, 30 GPa modulus of elasticity and 522 MPa UCS [23]. A cellular structure formed in the Ti64 alloy during SLM had 24 and 32% higher kinetic friction coefficient and a large amount of adhered bone compared to the dense and as-cast alloy thus showing an enhancement in implant stability [24]. Besides, the surface roughness and pore size formed in the SLM component significantly affected its property, and the effective bone ingrowth and vascularization were reported for a pore size in the range of 200-300 μm [25, 26].

In spite of its enhanced mechanical performance, the internal defects due to the SLM process suggests developing more strategies to achieve better stents. The use of SLM for biodegradable implant materials is growing steadily in recent days. Biodegradable metallic materials are advantageous owing to their superior mechanical properties when compared to the polymeric implant materials. There are three majors, but not limited biodegradable implant materials that are highly focused such as Mg, Zn, and Fe based alloys. Among them, Mg has a very high corrosion rate in physiological solution, and Fe has a slower corrosion rate, while Zn has the nominal corrosion rate to satisfy the implant requirement particularly for cardiovascular stents [27–29]. SLM technique has been used to manufacture the Mg-based degradable implant materials [30], and only a few reports on the Fe based alloys [31] are available, however, Zn based alloys are not widely reported [27, 32].

2.3 Laser Surface Melting (LSM) of Implant Materials

LSM is widely used to enhance the surface-related properties such as hardness, wear resistance and corrosion resistance of various alloys [33–35]. In the laser melting process, the metal surface is irradiated with a high energy laser beam which results in the homogenization of microstructure, grain refinement, and dispersion of second phases, solubility, and phase transformation (Fig. 2b). The refined microstructure alters the mechanical and corrosion properties in physiological solution, thus enhances the biocompatibility. Usage of laser source and processing parameters vary depending upon the materials used due to their reactivity difference with different laser sources, thus control the properties of the materials.

The LSM alters the surface microstructure and chemical composition which greatly affect the biological response of the metallic implant materials. Laser remelting of $Mg_{65}Zn_{30}Ca_5$ alloy was performed by Zhang et al. using a low power fiber laser at scan speeds from 500 to 4000 mm/s. The depth and width of the melted layer were reduced as the scanning speed was increased and the fraction of the crystalline phase was also decreased [36]. As the scan speed was increased the direct contact between the laser beam and substrate was reduced, therefore the heat transfer to the surface was also reduced. A positive shift in the corrosion potential (E_{corr}) and a decrease in corrosion current density (i_{corr}) were observed as the laser melting scan speed was increased. The formation of the amorphous phase with chemical homogeneity and absence of microstructural defects was mainly attributed to the better corrosion resistance of the remelted layer. Recently, we have reported that the refined microstructure and fine dispersion of the second phases of laser surface melted MB26 and AZ80 alloys were mainly responsible for the improved microhardness and corrosion resistance [37]. Rakesh et al. have reported that the laser surface melting of Mg-1Zn-2Dy alloy with varying the laser energy densities (12.5 to 45 J/mm^2) improved its properties for resorbable implant applications [38]. The melted layer has consisted of α-Mg and eutectic Mg_8ZnDy phase with fine grains in the order of 1-2 μm and columnar grains near the liquid-solid substrate due to the cooling and the solidification rate was also identified. The microstructural refinement and solid solution strengthening resulted in a better microhardness, furthermore, the in vitro degradation rate through weight loss measurement was also decreased by about 72% for 45 J/mm^2 laser melted alloy in Hank's balanced salt solution.

LSM of Mg-5Ca-1Zn alloy using a femtosecond laser with varying pulse repetition frequency (62.5, 500, 1000 Hz) revealed a significant variation in the second phase fraction. The alloy treated with 500 Hz exhibited about a 63.3% reduction of Mg_2Ca phase, therefore, decreased the galvanic corrosion and enhanced the corrosion resistance in the HBSS solution. Furthermore, no significant change in the microhardness values was noticed since the melted layer thickness values were only 2 ~ 3 μm and 500 Hz treated alloy exhibited higher cell viability compared to other treatments [39]. Mg-2.2Zn alloys treated with varying the laser power and scan speed resulted in refined microstructural features. However, the average surface roughness values were increased due to the variation in the cooling rate and overlap region [40]. The corrosion rates of LSM treated samples were increased compared to the untreated alloy. However, the corrosion rates were decreased from 1 mm/y to 0.65-0.52 mm/y for the LSM treated alloy after polishing and it was attributed to the microstructural refinement and biomineralization during immersion in HBSS solution which acted as a protective layer. These findings suggest that though the microstructural refinement contribute to the better surface properties, their micro/nanoscale features are important to achieve better properties of the laser-treated samples.

The dissolution and re-distribution of second phases present in the as-received alloy during the LSM are important aspects that enhance the *in vitro* properties of Mg alloys. LSM treatment of Mg-6Gd-0.6Ca alloy enhanced the corrosion resistance, cell attachment, and proliferation [41]. The improved corrosion resistance was mainly attributed to the dissolution of the β-Mg_5Gd phase in the melted region, which could significantly reduce the galvanic corrosion. The cell attachment studies carried out up to 24 h also revealed good adhesion, spreading, and proliferation on the laser melted surface. LSM treated WE43 alloy was also revealed dissolution of alloying elements and re-distribution of the second phase with the melted layer depth of about 1.15 mm [42]. LSM treated AZ31 alloy with varying laser energy from 1.06 to 3.18 J/mm^2 produced melted layers with the thickness in the range from 125 to 199 μm. Furthermore, the number of second phase β-$Mg_{17}Al_{12}$ was also increased in the melted layer compared to the as-cast alloy, which could act as a barrier to corrosion [43]. The redistribution of the β-$Mg_{17}Al_{12}$ could be attributed to the re-heating of the melted region during subsequent passes. These results show that the presence of alloying elements alters the second phase segregation and grain refinement during the LSM process.

Laser surface melting of AZ91D alloy resulted in a uniform distribution of the β-$Mg_{17}Al_{12}$ phase in the melted layer and the variation in scanning speed affected the melted layer properties. In the case of higher scan speeds such as 20 and 30 mm/s, the thermal-stress induced cracks have appeared, however, at lower scanning speeds such as 5 and 10 mm/s such cracks were not noticed, thereby exhibiting better corrosion resistance in the modified simulated body fluid solution [44]. The presence of refined microstructure with the uniform distribution of the second phase leads to uniform corrosion compared to the coarse microstructure which gives preferential sites for localized attacks for the as-received alloy.

Laser surface treatments of Zn based alloys have not been extensively investigated and research interests are kept growing to utilize the laser treatments to enhance the mechanical properties and corrosion resistance of Zn based alloys [45–47]. Recently laser surface remelting was carried out on Zn-0.5wt% Zr alloy and the as-cast alloy has mainly consisted of the $Zn_{22}Zr$ phase [48]. The melted depth of the alloy in Ar gas protection was about 300 μm and the surface layer was consisted of equiaxed dendrites with a size of about 5-10 μm, besides the $Zn_{22}Zr$ phase was also dissolved and the microhardness value of the melted layer was about 90 ~ 110 $HV_{0.025}$. The i_{corr} value of the laser surface remelted alloy was about 3 times lower compared to the as-cast alloy and about 137 mV positive shift in the corrosion potential was also noticed in simulated body fluid solution indicating the better corrosion resistance of the laser remelted surfaces.

2.4 *Laser Surface Patterning*

Engineering of the surface of metallic implants is focused in recent research to improve implant performance such as rapid integration with supporting hard tissues, retain mechanical integrity, enhance the biomineralization process, and speedy recovery [49]. Titanium-based alloys, in particular, Ti64 is one of the metallic implant materials generally used for dental and hip joint replacements owing to its excellent fracture toughness, fatigue resistance, and Young's modulus close to that of the natural bone [50]. Consistent research works are in progress to produce micro/nano textures on Ti-based alloys to enhance their *in vitro* performances. Laser patterning is carried out to produce biomimetic textures that could respond to the living cells based on the size and shape of the micro/nano textures. The control of laser parameters during the direct laser patterning process could be able to fabricate various textures such as micro/nano grooves, cubes, wells, etc. suitable for enhancing cell adhesion and proliferation.

3 Biomedical Applications of Laser Patterned Surfaces

3.1 Biocompatibility Enhancement of Laser Patterned Implant Materials

3.1.1 Permanent Implant Materials

The usage of the laser from nanosecond (ns) to femtosecond (fs) pulses would produce surfaces with various textures in micro/nanoscales and enhances the *in vitro* and *in vivo* responses of the textured surfaces. A picosecond laser texturing on AISI 420 alloy produced micro-papillae patterns with hierarchical superimposed nanostructures treated with stearic acid exhibited superhydrophobicity [12]. In addition, the treated surfaces exhibited hardly any bacterial adhesion under stationary conditions and resist 99 and 93% E.coli and S-aureus adhesion. The sizes of the coccoid-shaped S.aureus and rod-shaped E.coli are 700 nm and 2 μm, respectively, which made the colonization of the bacteria on the different micro/nano shaped surfaces. In addition to that, the air between the water and superhydrophobic surfaces played a major role in resisting the bacterial adhesion on the surface. These results clearly indicated the combined effects of hierarchical superimposed micro-nano structures and the air layer on the anti-bacterial properties. The antibacterial surfaces are essential for metallic implant components and medical equipment. Laser ablated microstructure coated with tetrahedral amorphous carbon (ta-C) on AISI 316L stainless steel plate also exhibited superhydrophobic and serum-repellent properties [51]. This treatment helped to improve the blood repellency of the surgical devices such as the scalpel blade by forming a hydrophobic surface due to the reduced liquid-solid contact. The laser-induced periodic surface structure (LIPSS) with different geometries and distributions produced using a femtosecond laser on AISI 304 had precise control on the human mesenchymal stem cell (hMSCs) attachments [52]. Furthermore, the cells were aligned in the longitudinal direction of the LIPSS orientation and the cell distribution and migration were controlled based on the micropatterns consisted of nanoscale features.

Batal et al. developed 6 different treatments such as blasting, polishing, micro-scale texturing using nanosecond laser, micro-scale texturing using femtosecond (fs) laser, sub-micron scale texturing using the ultra-fast laser and hydroxyapatite coatings on 67%Co-27%Cr-5%Mo (Co-Cr alloy) and investigated their wettability and cell metabolic activities [53]. The similar groove width and depth were produced on both ns and fs textured samples. However, ns treatment resulted in the re-cast bulges and re-deposition of molten materials along the grid edges and intersect regions, further the unprocessed regions were covered due to re-solidification. While in the case of fs treatment, the untreated surface remained as such to the blasted surface and the bottom of the grooves have consisted of self-organized hierarchical micro/nano morphology. The fs textured surface exhibited higher hydrophilic nature (CA ~ 0°) after 5 days of exposure and remained the same even after 30 days and it was mainly attributed to the Wenzel state of wetting. The ns textured sample had a significant increase in contact angle value (~ 115°) and it was attributed to the directional nature of the surface. The cell attachment and proliferation were strongly affected by the surface nature, and the cell adhesion was improved with higher surface roughness. However, the proliferation rate was better for the lower surface roughness samples. The higher cell proliferation rate was noticed for the polished and sub-micron textured samples. These findings suggested that the surface needs to be prepared by combining the micro-scale structures and sub-micro features that are possible by the femtosecond lasers, this, in turn, facilitates the mechanical anchoring of the bone, cell adhesion, and proliferation. In another study, the laser patterning on Co-Cr alloy (51%Co, 20%Cr, 15%W, 10%Ni, and 3%Fe) was carried out using a direct laser interference patterning (DLIP) with different periodicities ($\approx$ 3, 10, 20, and 32 μm) and depths ($\approx$ 20 and 800 nm) which significantly altered the surface wettability, platelet adhesion, and human umbilical vein endothelial cell (HUVEC) adhesion and spreading performances [54]. The wettability of the laser patterned surfaces decreased compared to the control samples, which was attributed to the surface roughness and variation in the surface chemical composition. The crystalline Cr_2O_3 and Co_3O_4 phases were transformed to amorphous phases after high-temperature laser irradiation. Laser patterns with higher depth induced HUVEC's elongation, alignment, and migration due to the topographical patterns. Furthermore, the platelet adhesion was reduced on the patterned surface was attributed to the oxide layer formation which affected the surface thrombogenicity.

The femtosecond laser was used to produce the micro pits consisted of nano-ripples and nano textures with nano-ripples on Ti64 alloy and the response of mesenchymal stem cells (MSC, C3H10T1/2) on these surfaces was investigated by Dumas et al. [55]. The micro pits with a diameter of about 30 μm and depth of 800 nm consisted of nano ripple 1) at the bottom of the pit and 2) around the tip of the pit with a distance of about 5 μm between the pits (Texture A and B in Fig. 3a) which revealed a difference in the cell spreading, and attachment behavior compared to the surface containing only nano ripple texture (Texture C). Contact angle values of the texture A and B were similar to the polished sample, however, texture C was more hydrophilic nature. Furthermore, the cell sizes were smaller on the textured surfaces, and the arrangement of the actin fiber was varied depending on the morphologies compared to the polished sample. The dynamic cell spreading studies revealed that the MSCs spread about 2 times faster on the textures A and B compared to the surface with only nano-ripples. Interestingly, cell mobility was also about 40% higher for the textures A and C than the texture B and smooth surface. The textured surfaces had better osteogenesis compared to the polished surface which was attributed to the alteration in cell shape, focal contacts, and the cytoskeletal organization initiated during the minutes to hours after the cell adhesion. The laser textured surfaces inhibited the adipogenesis and enhanced the osteogenesis due to the loss of the cellular shape compared to the polished samples.

The influences of surface micro/nano topographic surfaces of pure Ti on the cell adhesion and differentiation behavior was investigated by Chen et al. [4]. Microgrid patterns with the dimensions of 5 μm wide and 1 μm deep grooves separated by 6 μm ridges were developed on Ti (Fig. 3b) and the nano grid surface depth was

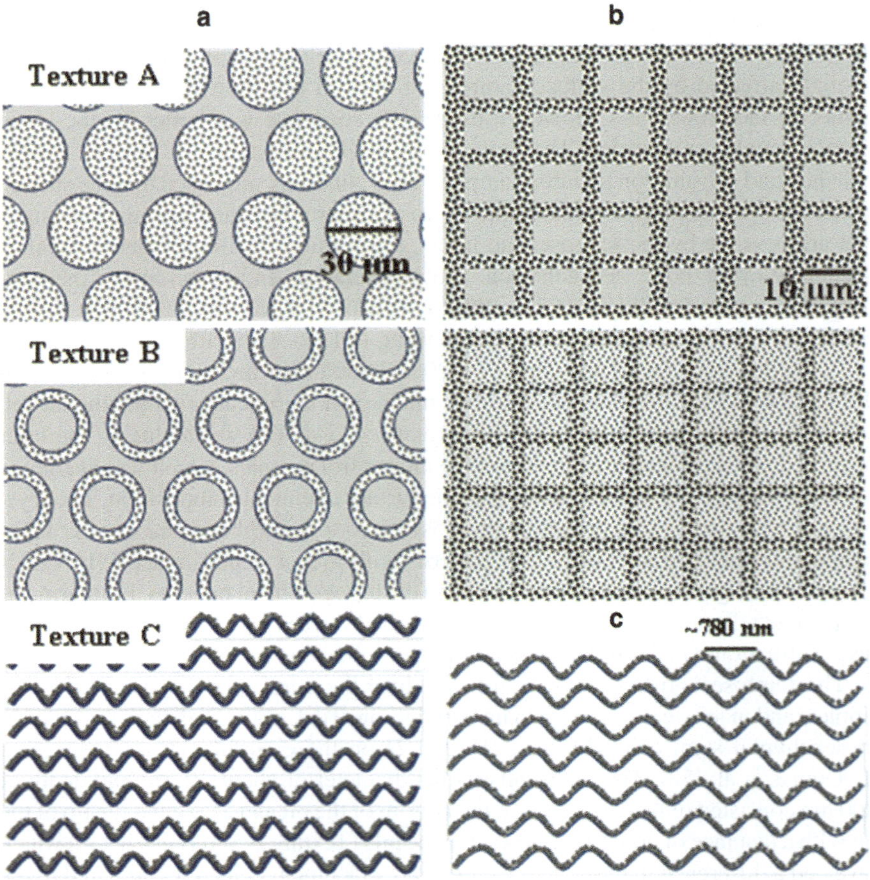

Fig. 3 **a** Schematic representation of laser micro/nano textures fabricated on Ti64 alloys [55]. **b** Laser processed micro and micro/nano grit patterns on Ti [4]. **c** Representation of LIPSS fabricated on Ta and Ti64 alloy [56]

about 200 nm. The hybrid micro/nano grid patterns were produced by superimposing the conditions used for micron and nano topographies. The human mesenchymal stem cells (hMSCs) adhesion, anchoring, and spreading were improved on the micron-sized surface, while the cell locomotion was visualized, and promoted the osteogenic differentiation on the nano grid topography. Interestingly, hybrid grid topography promoted cell adhesion on the surface.

Laser surface texturing of Ti64 alloy, and pure Tantalum was carried out using a Yb: KYW femtosecond laser at a wavelength of 1030 nm [56]. LIPSS was formed with periods of 780 ± 48 and 787 ± 31 nm on Tantalum and Ti64 alloys, respectively. Further, the depth of the LIPSS on Tantalum and Ti64 was about 256 ± 7 and 250 ± 12 nm, respectively (Fig. 3c). The ripples were observed on the Tantalum and 10-30 nm diameter nanoparticles/nano pits were also noticed on the ripples was attributed to the local effect during laser irradiation. The LIPSS on both the alloys does not produce any cell cytotoxicity and pro-inflammatory effects, further, the increase in cell proliferation was noticed in 15 days.

The micro-dot matrix induced by the femtosecond laser on pure Mg at air and hydroxyapatite suspension revealed the different characteristic features on the treated surface which altered the calcium phosphate deposition in physiological solution [57]. The surface was more hydrophilic in nature when the laser power was high and found to be favorable for the bone-like calcium phosphate deposition on the surface. Furthermore, the Ca and P deposition during the laser processing was due to the vaporization of water molecules in the processing regions, thus the concentration of ions was greatly increased and crystallized in the ablation area. During laser irradiation, the electrons are ejected from the Ti surface produced the positive charge, combined with the negatively charged PO_4^{3-} ions present in the solution subsequently reacted with Ca^{2+} ions to form $Ca_3(PO_4)_2$. Ti surface consisted of micro-holes with the inner wall covered with the mastoid-like shape after the laser processing in the HA solution which could expose more fresh Ti surface and allow the Ca and P deposition with tight chemical bonding to the surface compared with the surface which produced in the air atmosphere, therefore, enhanced the biological response.

The nanosecond pulsed laser treatment on pure Ti with different defocus distance produced a significant variation in the surface microtopography and the amount of -OH group, which played a key role in accelerating the hydroxyapatite deposition from the physiological solution for the bone healing process. The defocus distances of 0- and 2-mm during laser processing produced pocket-like micro topographical structures along the processing trace, however, the surface was relatively smooth for the defocus distance of 4 mm. A pocket-like nano topographies accelerated the microspherical hydroxyapatite composed of plate-like and needle-like crystals depositions (Ca/P ratio: 1.5) during 1-2 weeks of immersion. The shorter defocus distance could be responsible for the higher -OH groups on the processed surface and accelerated the hydroxyapatite deposition and cell adhesion behavior [58]. The Ti64 alloy produced by the SLM technique was subsequently subjected to the electrochemical anodization to produce macro-micro-nanoscale hierarchical surfaces to enhance their biocompatibility [59]. The anodized samples were composed of TNT arrays with an average diameter of about 60 μm, however, SLM alloy had micro-scale features and SLM-TNT had nanoscale features. This results in lower contact angle values, higher protein adsorption, and more MSC cell adsorption. Cell proliferation on the anodized samples was also higher, in particular, SLM-TNT had a higher proliferation rate compared to PO-TNT alloy after 120 h of incubation. These results clearly show that the surface topography plays a dominant role in improving the biocompatibility.

Laser-induced surface texturing with various micro/nano textures influences the antibacterial properties of implant alloys due to the variations in the wettability, and chemical composition. The femtosecond laser treatment on 316L SS was carried out by Lutey et al. to produce nano spikes, LIPSS, and nano-pillars with varying the pulse energy of 19.1, 1.01 and 1.46 μJ, respectively [5]. The surface with an average areal surface roughness of 8.6 μm, 90 nm, and 60 nm had a static contact angle value of 160°, 119°, and 140°, respectively. As the size of the E.coli (rod shape, ~ 0.5 μm diameter and length ~ 2 μm) and S.aureus (spherical shape, ~ 0.5 μm diameter) was varied, the bacterial adhesion behavior was also varied and the area occupied by the E.coli and S.aureus are ~ 1 μm^2/cell and 0.25 μm^2/cell, respectively. Therefore, the surfaces with features larger than the bacterial size (spikes) and similar to the bacterial size (LIPSS and nano-pillars), thus affected bacterial contact and further adhesion. The retention of the E.coli on the mirror polished and spikes surface was higher as the surface dimension is larger than the cell size, however, S.aureus adhesion was inhibited. Interestingly LIPSS and nano-pillars revealed 99.8% and 99.2% for the E.coli and 84.7% and 79.9% for the S.aureus, respectively. The antibacterial activity was attributed to both surface wettability and morphological change of the laser textured specimens. A continuous-wave fiber laser-treated CP-Ti (grade 2), Ti64, and CoCrMo alloys under the nitrogen atmosphere were tested for the S.aureus adhesion [60]. The surface of the treated Ti samples was mainly consisted of TiN, and TiO_2 and only CoO/Cr_2O_3 were identified on CoCrMo alloy. Besides, the surface roughness of all the alloys increased and Ti samples exhibited the spike surface morphology. Furthermore, the laser-treated CP-Ti and Ti64 alloy had lower water contact angle values, while the contact angle value was remained similar for the CoCrMo alloy compared to the untreated specimens. The S.aureus cell is hydrophobic therefore has more interactions with the more hydrophobic surfaces, hence the S.aureus adhesion with the laser-treated CoCrMo alloy was higher than the CP-Ti and Ti64 alloy. In addition to the wettability, the presence of nitride/oxide surface layer on Ti alloy and micro/nano features were also attributed to the resulting antibacterial properties. The femtosecond laser ablation treatment of CP-Ti (grade 2) produced a micro/nanoscale quasi-periodic self-organized structures with hydrophobic nature ($166 \pm 4°$) compared to the untreated surface ($73 \pm 3°$) which exhibited colonization of S.aureus bacteria, while the P.aeruginosa was not able to attach [61]. The antibacterial activities of laser textured implant materials were improved and the combination of surface roughness and composition were responsible for the better properties.

LIPSS was produced on to Ti64 alloy and the laser wavelength, speed, and pulse numbers altered the LIPSS average periods, and surface peak to valley distance. The contact angle values measured using polar and non-polar liquids revealed a hydrophilic character for the polished sample and UV-LIPSS and hydrophobic behavior were observed for the GR-LIPSS specimens. The grooves in the nanometer range LIPSS surface were smaller compared to the bacterial size, therefore, inhibited the bacteria adhesion. On the other hand, the vertical growth of the bacteria leads to the formation of shelters on the polished surface and helps the bacteria allocation and proliferation. This phenomenon was strongly hindered by the nanostructured surface, thus bacterial adhesion was hindered and resulted in anti-bacterial surfaces [62].

LIPSS surfaces with different periodicities were fabricated on pure Ti grade 2 using a femtosecond laser and the rippled surfaces had sub-micron size surface roughness and also, the LIPSS surface was hydrophobic ($134 \pm 14°$) compared to the control sample ($84 \pm 3°$) [63]. The cells on the LIPSS were oriented perpendicular

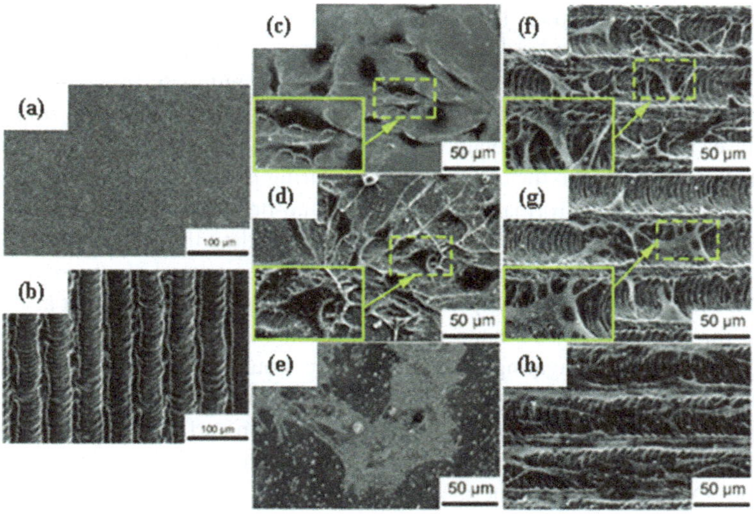

Fig. 4 Untreated and laser treated TZNT alloy (a and b) and MC3T3-E1 cells cultured on untreated (c–e) and laser treated alloys (f–h) at different durations 12, 24 and 72 h [64]

to the periodic surface, many of the filopodia were directly attached to the peaks of the ripple pattern and the noncontact guidance of the cell on the laser-ablated surface led to the reduced circularity of the cells compared to the control sample. Furthermore, periodicity with 620 nm exhibited better properties.

The contact angle values of the microgroove surface on Ti-20Zr-10Nb-4Ta were varied based upon the viewing direction of microgrooves. Along with the microgroove direction, the contact angle value (38.9 ± 3.2°) was similar to the polished surface (36.9 ± 3.3°) (Fig. 4), while the perpendicular direction revealed a lower contact angle value (18.0 ± 3.5°) [64]. The laser-treated surface exhibited a 300 mV positive shift in E_{corr} value and three times reduction in the i_{corr} value compared to the untreated sample indicating better corrosion resistance in Hank's balanced salt solution. Furthermore, cell attachment on the laser-treated surface was higher compared to the untreated sample due to the variation in the wettability, and anisotropic nature of the surface.

3.1.2 Biodegradable Magnesium Alloys

Micro and nano rippled textures on pure Mg were developed using a Ti: sapphire femtosecond laser, the increase in scan numbers (50 times) led to the micron size ripples, and the produced surfaces had different colors due to the variation in reflectance behavior. It was also observed that the orientation of micron and nano ripples were parallel and perpendicular to the laser polarization direction, respectively [65]. Furthermore, an increase in the laser pulse numbers to 200, resulted in a random microstructure. However, the orientation of ripples was the same as in the case of laser scans, and the irradiation area was expanded as the number of pulses was increased up to 500. An increase in pulse numbers above 500 resulted in coarse and poorly defined features. The formation of micro and nano ripples due to the laser irradiation was explained based on several factors. The energy transfer to the metal lattice due to the ejected electron during the laser irradiation and initial few pulses result in the formation of random nano-protrusions. Formation of nano protrusions could be attributed to several factors including the laser ejected species, plasma confinement, and materials re-deposition within several picoseconds during rapid cooling condensation and plume collapse process. An increase in laser pulse leads to the expansion of the nano-protrusion area and the surface plasmons (SPs) both localized and propagating along the surface would be excited by coupling the incident light to such nano-protrusions. The high cooling rate during the laser treatment results in a variety of surface tension resulting in surface curling, and highly localized heating, as well as large thermal gradients. This caused the development of severe strain fields in the peripheral area of the irradiated surface, finally, resulting in the formation of micro-ripples and isolated melt (cluster structures) on the surface.

A hybrid laser surface modification such as melting, and texturing of Mg-6Gd-0.6Ca alloy had a significant variation in the microstructure, corrosion performance and in vitro biocompatibility. The as-received microstructure mainly consisted of equiaxed α-Mg phase as the matrix and β-Mg_5Gd phase distributed in the matrix along the extrusion direction (Fig. 5A-a). After laser surface modification, solidification microstructure was changed to the α-Mg phase and the β-phase precipitation along grain boundaries was not noticed. The high-temperature laser treatment resulted in the dissolution of the second phase which was confirmed from the XRD results (Fig. 5A-b) and the thickness of the laser-melted layer was about 360 μm. The corrosion results indicated that for the as-received specimen, initial galvanic corrosion occurred preferentially in the α-Mg matrix around the β-phase and extended into deep corrosion holes. After the laser surface modification, the uniform corrosion was noticed for the laser melted alloy due to the dissolution of β-phase in the melted region [41]. As a result of the significant reduction of the second phase concentration, and resulting galvanic influence of the laser melted surface, the hydrogen evolution volume (Fig. 5A-c) was significantly decreased, and the corrosion rate was much lower than that of the as-received alloy. The improvement of corrosion resistance in HBSS was mainly attributed to the dissolution of the β-phase Mg_5Gd and homogeneous solid solution in the laser melted layer.

The MC3T3-E1 cell culture studies revealed that as-received specimens showed cytotoxicity due to a high degradation rate and higher hydrogen volume accumulation on the surface. Further, no cells survived on the surface during the culturing periods. Interestingly, the laser-modified specimens exhibited good biocompatibility owing to their enhanced corrosion resistance. In vitro cell culture studies showed that the MC3T3-E1 cells survived on the surface of laser-modified specimens, and significant improvements were illustrated by good adhesion property, spreading performance and proliferation capacity (Fig. 5B). LIPSS was produced on to the melted surface using a Ti: sapphire chirped-pulse regenerative amplification laser and a wavelength

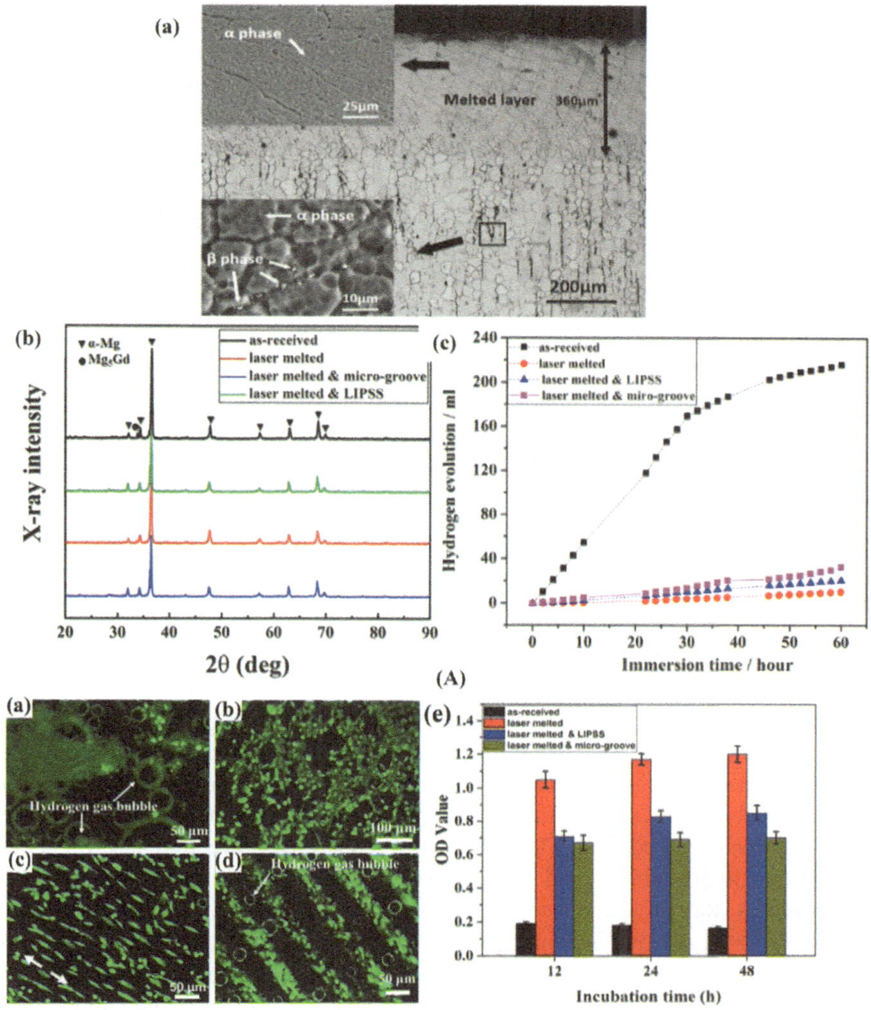

Fig. 5 (A) (a) Cross-sectional morphology of laser surface melted Mg–Gd-Ca alloy [41], (b) XRD patterns of as-received, laser melted and LIPSS alloys and (c) hydrogen evolution volume of as-received, laser surface melted, LIPSS and micro-grooved alloys [68]; (C) Fluorescence images of the MC3T3-E1 cell cultured on (a) as-received; (b) laser melted; (c) laser melted &LIPSS; (d) laser melted & micro-groove surface and (e) cell proliferation for 48 h [66]

of 1064 nm with a pulse length of 800 fs, the repetition rate was further set to 400 kHz. LIPSS with a period of 865 nm were fabricated, the average height of the LIPSS was 200 nm and the direction of the LIPSS was perpendicular to the polarization of the laser. The MC3T3-E1 cell viability and attachment were better for the laser melted surfaces (94%) compared to the LIPSS (68%). Moreover, the cells were elongated regularly along the LIPSS surface and were random on the laser melted surface [66]. The better cell viability and attachment of the melted surface were mainly attributed to the lower hydrogen release, and pH raises due to the enhanced corrosion resistance. Sub-micrometric surface structures produced by femtosecond laser on the melted surface could provide a persistent mechanical stimulus for cells, resulting in a controllable cell shape. Micrometer surface structures produced by picosecond laser on the melted surface could affect the cell distribution due to the cell-repellent effect.

Laser surface patterning using the direct laser interference patterning (DLIP) system produced periodic textures on AZ31 Mg alloy and the number of pulses varied the average laser fluence [67]. The pattern produced using 64 number of pulses exhibited good quality patterns and multiple passes increased the texture quality by avoiding the material heat-up, which is often noticed for the long pulse durations without any interruptions. The partial overlapping of the laser scanning was also further confirmed that the periodic structure formation was stable for the large processing areas and the interference angle played a major role in determining the pattern periodicity.

3.2 Surface-Enhanced Raman Scattering (SERS) Responses and Fluorescence Detection of Metal Surfaces

LIPSS has given an immense interest in recent days for various applications including enhancement of Raman signal, photocatalysis, etc. Laser-induced surface structuring possesses various advantages such as simple, precise, reproducible, and pollution-free methods, in particular, this method can be easily applied to metallic materials that have not been explored well. The LIPSS surfaces strongly affect the laser/material interaction which could be used as an analytical tool to enhance the signals from the biomolecules. Raman spectroscopy is one of the techniques which is widely used to analyze various chemical compounds including biological components. In recent days SERS is emerging as an analytical tool to identify biological molecules adhered to the nano surfaces and enhances the signal intensity based upon the nature, size, and composition of the surfaces [69, 70]. The SERS mainly uses plasmonic which examines the interaction between the incident light and metallic nanostructures. The optimization of surface nanostructures revealed a significant difference in the signal corresponds to the adsorbed molecules, thus able to quantify them in biological applications. Extensive studies are available on the preparation of metal nanoparticles such as Ag and Au in various sizes and shapes using conventional techniques to examine their SERS performances [71].

Laser surface texturing with hierarchical nano textures on metal surfaces which are widely used in the biomedical applications to study the SERS response have not been explored. Recently an attempt was made to fabricate the nanotextured surfaces on Ti64 alloy using a femtosecond laser. The formation of a periodic structure (LIPSS) with a periodicity of 600 nm and the nanoparticle with the diameter in the range of 60-200 nm were randomly observed in the ripples [72]. A thin Au coating was further deposited on the produced LIPSS structures by an electron beam evaporation technique as the Au enhances the Raman signal. Crystal violet probe molecule was dropped on different surfaces (polished, LIPSS and Au coated LIPSS), subjected to Raman analysis and it revealed that the enhancement of the Raman signal was in the factor of 10^3 and 6.7×10^3 for the LIPSS and Au coated LIPSS, respectively (Fig. 6A).

The localized surface plasma resonance (LSPR) was mainly responsible to confine the electromagnetic field when the dimension of the nanoparticles was less than the incident wavelength. The nanoparticles produced on the LIPSS were in the range of 60-200 nm and were mainly responsible for the enhanced Raman signal compared to surface plasma polaritons (SPP) on the LIPSS. These results indicated that laser-induced surface texturing is a reliable tool for the enhancement of the Raman signals of biomolecules, thus increasing the application in detecting the biomolecule which is very important for the biomedical field.

Another important application of the LIPSS surface is the enhancement of fluorescence detection and fluorescence imaging is a widely used tool in biomedical fields. Nanomaterials play a key role to measure and enhance fluorescence imaging. The fluorescence detection efficiencies are significantly altered by the size of the sub-wavelength nanoparticles. It is known that the highly ordered nanomaterials prepared using conventional methods would require sophisticated instrumentation facilities, and in fact, sometimes the reproducibility and application rage become questionable. Therefore, there is a huge demand for a feasible and reproducible technology to produce the periodic surface structure (PSS), which confines the local electromagnetic (EM) filed on the structured surface and enhances the fluorescence detection. In an attempt, Jiaru et al. have recently fabricated a LIPSS on TC4 Titanium alloy which is commonly used in implant applications and the fabricated surfaces had enhancement factor of about 74 with bovine serum albumin (BSA) and the minimum Cu detection limit was 1.5 μg/ml (Fig. 6B) [15]. The enhancement in the fluorescence intensity was mainly attributed to the two forms of surface plasmon resonance (SPR) effect such as SPP and LSPR. The size of the nanoparticles plays a huge role in determining the LSPR and SPP effects, and the nanoparticle size smaller than the wavelength of incident light showed stronger dipolar excitation in the form of LSPR. Furthermore, the enhancement factor for the hierarchical LIPSS was about 6 times higher than the LIPSS which was attributed to the large surface area with more nanoparticles which could allow more interaction and result in stronger LSPR.

4 Future Prospects of Laser Processing for Biomedical Applications

The recent literature studies show that continuous research works are focused on using the laser as non-contact and high precision technology to modify the surface

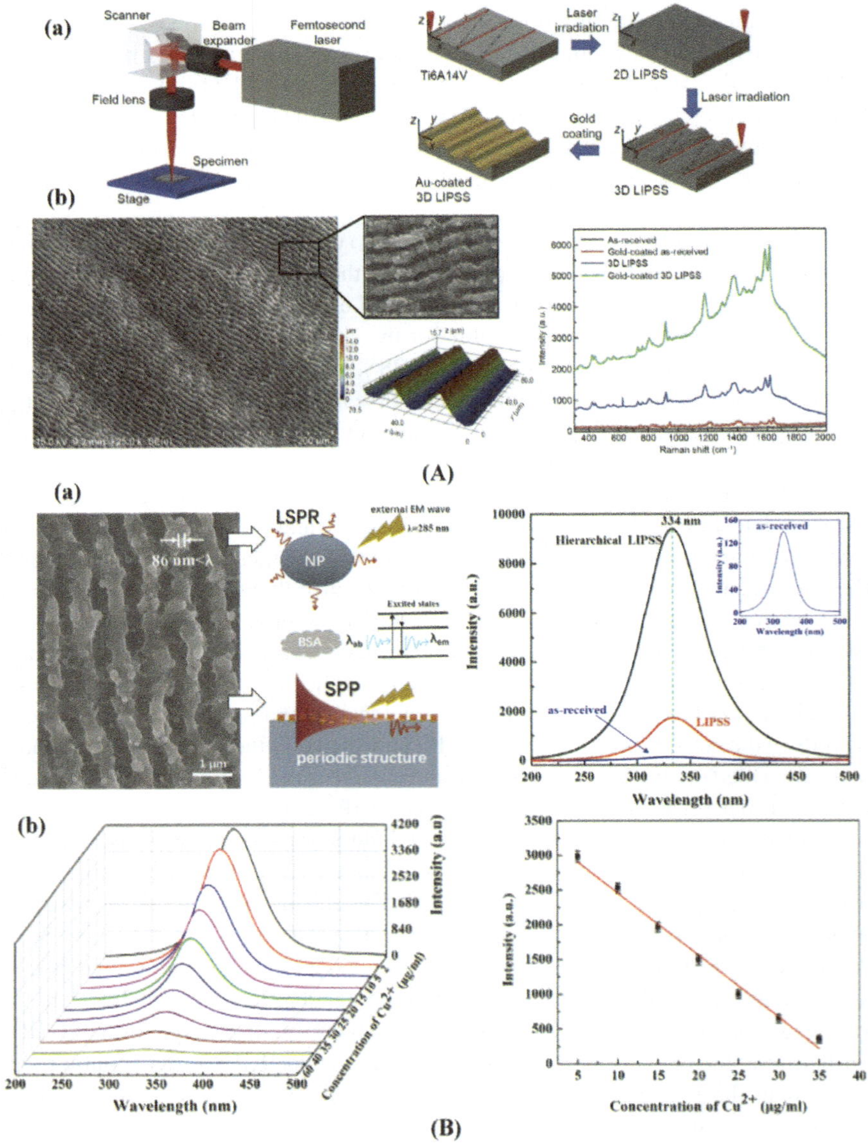

Fig. 6 (A) Fabrication of SERS surface on Ti64 alloy (a) schematic representation and LIPSS fabrication process, (b) Surface morphology and Raman signal of crystal violet [72]; (B) Fluorescence enhancement of LIPSS surfaces (a) LIPSS surface with fluorescence spectra recorded on various surfaces and (b) Fluorescence spectra and change in fluorescence intensity with varying the Cu^{2+} concentration [15]

of metals and nonmetals. The interaction of laser with the materials transfer the quantum of thermal energy to locally modify the material properties. Laser treatments are widely used to manufacture the components and fabricate the functional surfaces on various metallic materials in the biomedical fields. Unlike the conventional chemical surface treatments, the laser treatments have advantages to use them for the complex implant surfaces and avoid the risk associated with using aggressive chemicals. The laser surface texturing by controlling the processing parameters produces the micro/nano topographies over the implant surface which could affect the cell adhesion mechanism and differentiation. Furthermore, it also imparts the antibacterial properties which is one of the important factors for biomedical applications. The cell differentiation on the implant surfaces could be accelerated by fabricating the combinations of micro/nano textures with moderately rough surfaces. These surfaces have a tendency to guide the cell adhesion and differentiation for better osteointegration of the implants during service. Furthermore, the surface chemical composition which plays a key role in determining the chemical stability of the metal surfaces could be altered through laser treatments simultaneously producing the micro/nano textures.

The literature survey also shows that a vast amount of research is carried out on permanent implant materials to enhance their performance in biomedical implant applications. However, as the growth of biodegradable implant materials is consistently increasing over the years their surface properties still need to be improved to satisfy their performance in the implant fields. The degradable implant materials such as Magnesium, Zinc, and Fe based alloys are widely proposed as alternatives to the permanent implant materials. However, meeting the required chemical reactivity and biocompatibility of these materials in biological environments are still challenges to achieve. Therefore, laser surface treatments could be an appropriate technique to fabricate the surfaces to meet the desired chemical reactivity and enhanced biocompatibility. Furthermore, fabricating the hierarchical nano textures on metals would also be an important direction in biomedical fields for sensing of biomolecules and enhancement of their detection. The research on these directions is indeed essential to fabricate the functional surfaces with better properties.

5 Summary

Recent works on the laser surface treatments and micro/nano texture formation on metals and alloys used in the biomedical applications are presented in this chapter. The formation of refined microstructures with surface roughness in the range of several micrometers and fabrication of various micro/nano patterns were responsible to enhance the surface properties such a microhardness, tribological properties, wettability, corrosion properties, anti-bacterial activity, and *in vitro* biocompatibility. Selective laser melting provides an opportunity to manufacture the alloy with the desired composition with precise control over the microstructure. Laser surface melting results in modification of substrate surface consisting of refined microstructures and re-distributed second phases. These structures are favorable for enhancing the mechanical properties, and also the chemical stability in the physiological medium. Enhancement of surface chemical stability is particularly important for the degradable implant materials to control the degradation rate; thus, laser processing could be advantageous. The micro/nano textures produced on the implant surface varied the wettability due to the variation in the surface roughness and chemical composition. The variation in laser processing parameters such as laser power, scan speed, pulse duration, etc. are mainly responsible for the surface pattern formation and their chemical composition. These, in turn, affect the cell attachment and differentiation phenomenon, however, the role of each of the parameters is not well understood yet. The hydrophobic surfaces show tendencies to attract hydrophobic bacteria and vice versa. Therefore, designing surface wettability would significantly alter the anti-bacterial properties. The direct laser treatments on Ti-based alloys produce LIPSS and in particular, the control of laser parameters results in nanostructures with various ripple size and nanoparticle distribution. These periodic surfaces confine the local electromagnetic wave through surface plasma polaritons and able to enhance the signal of biomolecules adhered to the surface and therefore, used in SERS and fluorescence imaging applications.

Acknowledgements This work was supported by the National Natural Science Foundation of China (51875313 and 51705013), National Key Research and Development Program of China (2018YFB1107400, 2018YFB1107700, and 2016YFB1102503). The help extended by Mr. Wang Peng to prepare the schematic representations are acknowledged.

References

1. Kim K-H, Ramaswamy N (2009) Electrochemical surface modification of titanium in dentistry. Dent Mater J 28(1):20–36. https://doi.org/10.4012/dmj.28.20
2. Li B, Ma J, Wang D, Liu X, Li H, Zhou L, Liang C, Wang H (2019) Self-adjusting antibacterial properties of Ag-incorporated nanotubes on micro-nanostructured Ti surfaces. Biomater Sci 7(10):4075–4087. https://doi.org/10.1039/c9bm00862d
3. Mello DCR, de Oliveira JR, Cairo CAA, Ramos LSB, Vegian M, de Vasconcellos LGO, de Oliveira FE, de Oliveira LD, de Vasconcellos LMR (2019) Titanium alloys: in vitro biological analyzes on biofilm formation, biocompatibility, cell differentiation to induce bone formation, and immunological response. J Mater Sci - Mater Med 30(9):108. https://doi.org/10.1007/s10856-019-6310-2
4. Chen P, Aso T, Sasaki R, Ashida M, Tsutsumi Y, Doi H, Hanawa T (2018) Adhesion and differentiation behaviors of mesenchymal stem cells on titanium with micrometer and nanometer-scale grid patterns produced by femtosecond laser irradiation. Journal of Biomedical Materials Research Part A 106 (10):2735–2743. https://doi.org/10.1002/jbm.a.36503
5. Lutey AHA, Gemini L, Romoli L, Lazzini G, Fuso F, Faucon M, Kling R (2018) Towards Laser-Textured Antibacterial Surfaces. Sci Rep 8(1):10112. https://doi.org/10.1038/s41598-018-28454-2
6. Sardar S, Fabris L, Javanmard M (2019) Improved Precision in Surface-Enhanced Raman Scattering Quantification of Analyte through Dual-Modality Multisite Sensing. Anal Chem 91(7):4323–4330. https://doi.org/10.1021/acs.analchem.8b02559

7. Gardner B, Matousek P, Stone N (2019) Subsurface Chemically Specific Measurement of pH Levels in Biological Tissues Using Combined Surface-Enhanced and Deep Raman. Anal Chem 91(17):10984–10987. https://doi.org/10.1021/acs.analchem.9b01015
8. Moram SSB, Byram C, Shibu SN, Chilukamarri BM, Soma VR (2018) Ag/Au Nanoparticle-Loaded Paper-Based Versatile Surface-Enhanced Raman Spectroscopy Substrates for Multiple Explosives Detection. ACS Omega 3(7):8190–8201. https://doi.org/10.1021/acsomega.8b01318
9. Mangano F, Chambrone L, van Noort R, Miller C, Hatton P, Mangano C (2014) Direct metal laser sintering titanium dental implants: a review of the current literature. International journal of biomaterials 2014:461534. https://doi.org/10.1155/2014/461534
10. Liu Z, Chong PH, Skeldon P, Hilton PA, Spencer JT, Quayle B (2006) Fundamental understanding of the corrosion performance of laser-melted metallic alloys. Surf Coat Technol 200(18):5514–5525. https://doi.org/10.1016/j.surfcoat.2005.07.108
11. Liu J, Yu H, Chen C, Weng F, Dai J (2017) Research and development status of laser cladding on magnesium alloys: A review. Opt Lasers Eng 93:195–210. https://doi.org/10.1016/j.optlaseng.2017.02.007
12. Pan Q, Cao Y, Xue W, Zhu D, Liu W (2019) Picosecond Laser-Textured Stainless Steel Superhydrophobic Surface with an Antibacterial Adhesion Property. Langmuir 35(35):11414–11421. https://doi.org/10.1021/acs.langmuir.9b01333
13. Rajab FH, Liu Z, Li L (2019) Long term superhydrophobic and hybrid superhydrophobic/superhydrophilic surfaces produced by laser surface micro/nano surface structuring. Appl Surf Sci 466:808–821. https://doi.org/10.1016/j.apsusc.2018.10.099
14. Dobrzański LA, Drygała A (2019) Laser Application in Photovoltaics for Surface Texturization of Silicon and Front Electrode Deposition. Materials Performance and Characterization 8(6):20190061. https://doi.org/10.1520/mpc20190061
15. Jiaru Z, Guoqing H, Libin L, Yingchun G, Hong MH (2019) Enhancing protein fluorescence detection through hierarchical biometallic surface structuring. Opt Lett 44(2):339–342. https://doi.org/10.1364/OL.44.000339
16. Brown MS, Arnold CB (2010) Fundamentals of Laser-Material Interaction and Application to Multiscale Surface Modification. In: Sugioka K, Meunier M, Piqué A (eds) Laser Precision Microfabrication. Springer Berlin Heidelberg, Berlin, Heidelberg, pp 91–120. https://doi.org/10.1007/978-3-642-10523-4_4
17. Demir AG, Previtali B (2017) Additive manufacturing of cardiovascular CoCr stents by selective laser melting. Mater Des 119:338–350. https://doi.org/10.1016/j.matdes.2017.01.091
18. Sing SL, An J, Yeong WY, Wiria FE (2016) Laser and electron-beam powder-bed additive manufacturing of metallic implants: A review on processes, materials and designs. Journal of orthopaedic research: official publication of the Orthopaedic Research Society 34(3):369–385. https://doi.org/10.1002/jor.23075
19. Obeidi MA, McCarthy E, O'Connell B, Ul Ahad I, Brabazon D (2019) Laser Polishing of Additive Manufactured 316L Stainless Steel Synthesized by Selective Laser Melting. Materials 12 (6). https://doi.org/10.3390/ma12060991
20. Wessarges Y, Hagemann R, Gieseke M, Nölke C, Kaierle S, Schmidt W, K.P.Schmitz, Haferkamp H (2014) Additive manufacturing of vascular implants by selective lasermelting,. Biomed Tech 59:S401–S404
21. Guoqing Z, Yongqiang Y, Changhui S, Fan F, Zimian Z (2017) Study on Biocompatibility of CoCrMo Alloy Parts Manufactured by Selective Laser Melting. Journal of Medical and Biological Engineering 38(1):76–86. https://doi.org/10.1007/s40846-017-0293-6
22. Hindy A, Farahmand F, Tabatabaei FS (2017) In vitro biological outcome of laser application for modification or processing of titanium dental implants. Lasers Med Sci 32(5):1197–1206. https://doi.org/10.1007/s10103-017-2217-7
23. Elsayed M, Ghazy M, Youssef Y, Essa K (2019) Optimization of SLM process parameters for Ti6Al4 V medical implants. Rapid Prototyping Journal 25(3):433–447. https://doi.org/10.1108/rpj-05-2018-0112

24. Bartolomeu F, Costa MM, Gomes JR, Alves N, Abreu CS, Silva FS, Miranda G (2019) Implant surface design for improved implant stability – A study on Ti6Al4 V dense and cellular structures produced by Selective Laser Melting. Tribol Int 129:272–282. https://doi.org/10.1016/j.triboint.2018.08.012
25. Lee JW, Ahn G, Kim JY, Cho DW (2010) Evaluating cell proliferation based on internal pore size and 3D scaffold architecture fabricated using solid freeform fabrication technology. J Mater Sci - Mater Med 21(12):3195–3205. https://doi.org/10.1007/s10856-010-4173-7
26. Kumar A, Mandal S, Barui S, Vasireddi R, Gbureck U, Gelinsky M, Basu B (2016) Low temperature additive manufacturing of three dimensional scaffolds for bone-tissue engineering applications: Processing related challenges and property assessment. Materials Science and Engineering: R: Reports 103:1–39. https://doi.org/10.1016/j.mser.2016.01.001
27. Montani M, Demir AG, Mostaed E, Vedani M, Previtali B (2017) Processability of pure Zn and pure Fe by SLM for biodegradable metallic implant manufacturing. Rapid Prototyping Journal 23(3):514–523. https://doi.org/10.1108/rpj-08-2015-0100
28. Manakari V, Parande G, Gupta M (2016) Selective Laser Melting of Magnesium and Magnesium Alloy Powders: A Review. Metals 7 (1). https://doi.org/10.3390/met7010002
29. Ng CC, Savalani MM, Lau ML, Man HC (2011) Microstructure and mechanical properties of selective laser melted magnesium. Appl Surf Sci 257(17):7447–7454. https://doi.org/10.1016/j.apsusc.2011.03.004
30. Xu R, Zhao M-C, Zhao Y-C, Liu L, Liu C, Gao C, Shuai C, Atrens A (2019) Improved biodegradation resistance by grain refinement of novel antibacterial ZK30-Cu alloys produced via selective laser melting. Mater Lett 237:253–257. https://doi.org/10.1016/j.matlet.2018.11.071
31. Carluccio D, Demir AG, Caprio L, Previtali B, Bermingham MJ, Dargusch MS (2019) The influence of laser processing parameters on the densification and surface morphology of pure Fe and Fe-35Mn scaffolds produced by selective laser melting. Journal of Manufacturing Processes 40:113–121. https://doi.org/10.1016/j.jmapro.2019.03.018
32. Wen P, Voshage M, Jauer L, Chen Y, Qin Y, Poprawe R, Schleifenbaum JH (2018) Laser additive manufacturing of Zn metal parts for biodegradable applications: Processing, formation quality and mechanical properties. Mater Des 155:36–45. https://doi.org/10.1016/j.matdes.2018.05.057
33. Zeng C, Shen J, He C, Zhou M (2019) An ultrathin melted layer on magnesium alloy manufactured by low power laser. Materials Research Express 6(6):066527. https://doi.org/10.1088/2053-1591/ab0cbe
34. Majumdar JD, Kumar A, Pityana S, Manna I (2018) Laser Surface Melting of AISI 316L Stainless Steel for Bio-implant Application. Proceedings of the National Academy of Sciences, India Section A: Physical Sciences 88(3):387–403. https://doi.org/10.1007/s40010-018-0524-4
35. Mukherjee S, Dhara S, Saha P (2018) Laser surface remelting of Ti and its alloys for improving surface biocompatibility of orthopaedic implants. Materials Technology 33(2):106–118. https://doi.org/10.1080/10667857.2017.1390931
36. Zhang D, Qin Y, Feng W, Huang M, Wang X, Yang S (2019) Microstructural evolution of the amorphous layers on Mg-Zn-Ca alloy during laser remelting process. Surface and Coatings Technology 363:87–94. https://doi.org/10.1016/j.surfcoat.2019.02.051
37. Li Y, Arthanari S, Guan Y (2019) Influence of laser surface melting on the properties of MB26 and AZ80 magnesium alloys. Surface and Coatings Technology 378:124964. https://doi.org/10.1016/j.surfcoat.2019.124964
38. K.R R, Bontha S, M.R R, Das M, Balla VK (2019) Laser surface melting of Mg-Zn-Dy alloy for better wettability and corrosion resistance for biodegradable implant applications. Applied Surface Science 480:70–82. doi:10.1016/j.apsusc.2019.02.167
39. Park J, Han H-S, Park J, Seo H, Edwards J, Kim Y-C, Ok M-R, Seok H-K, Jeon H (2018) Corrosion behavior of biodegradable Mg-based alloys via femtosecond laser surface melting. Applied Surface Science 448:424–434. https://doi.org/10.1016/j.apsusc.2018.04.088
40. Manne B, Thiruvayapati H, Bontha S, Motagondanahalli Rangarasaiah R, Das M, Balla VK (2018) Surface design of Mg-Zn alloy temporary orthopaedic implants: Tailoring wettability and biodegradability using laser surface melting. Surface and Coatings Technology 347:337–349. https://doi.org/10.1016/j.surfcoat.2018.05.017

41. Ma C, Peng G, Nie L, Liu H, Guan Y (2018) Laser surface modification of Mg-Gd-Ca alloy for corrosion resistance and biocompatibility enhancement. Applied Surface Science 445:211–216. https://doi.org/10.1016/j.apsusc.2018.03.174
42. Liu C, Li Q, Liang J, Zhou J, Wang L (2016) Microstructure and corrosion behaviour of laser surface melting treated WE43 magnesium alloy. RSC Advances 6(36):30642–30651. https://doi.org/10.1039/c5ra27010c
43. Wu T-C, Ho Y-H, Joshi SS, Rajamure RS, Dahotre NB (2017) Microstructure and corrosion behavior of laser surface-treated AZ31B Mg bio-implant material. Lasers in Medical Science 32(4):797–803. https://doi.org/10.1007/s10103-017-2174-1
44. Guan YC, Zhou W, Zheng HY (2009) Effect of laser surface melting on corrosion behaviour of AZ91D Mg alloy in simulated-modified body fluid. Journal of Applied Electrochemistry 39(9):1457–1464. https://doi.org/10.1007/s10800-009-9825-2
45. Wątroba M, Bednarczyk W, Kawałko J, Mech K, Marciszko M, Boelter G, Banzhaf M, Bała P (2019) Design of novel Zn-Ag-Zr alloy with enhanced strength as a potential biodegradable implant material. Materials & Design 183:108154. https://doi.org/10.1016/j.matdes.2019.108154
46. Hernandez-Escobar D, Champagne S, Yilmazer H, Dikici B, Boehlert CJ, Hermawan H (2019) Current status and perspectives of zinc-based absorbable alloys for biomedical applications. Acta Biomater 97:1–22. https://doi.org/10.1016/j.actbio.2019.07.034
47. Venezuela JJD, Johnston S, Dargusch MS (2019) The Prospects for Biodegradable Zinc in Wound Closure Applications. Advanced healthcare materials:e1900408. https://doi.org/10.1002/adhm.201900408
48. Wang Z, Zhang Q, Guo P, Gao X, Yang L, Song Z (2018) Effects of laser surface remelting on microstructure and properties of biodegradable Zn-Zr alloy. Materials Letters 226:52–54. https://doi.org/10.1016/j.matlet.2018.04.112
49. Li S, Li Y, Wang S, Bai C (2017) Progress in the processing technology of coronary stents with micro/ nano structures. Optik 148:319–324. https://doi.org/10.1016/j.ijleo.2017.08.136
50. Olsson R, Powell J, Palmquist A, Brånemark R, Frostevarg J, Kaplan AFH (2018) Production of osseointegrating (bone bonding) surfaces on titanium screws by laser melt disruption. Journal of Laser Applications 30(4):042009. https://doi.org/10.2351/1.5078502
51. Li K, Xie Y, Liang L, Lin N, Xu L, Chen W, Lu L (2019) Wetting behavior investigation of a complex surface prepared by laser processing combined with carbon films coating. Surface and Coatings Technology:124989. doi:https://doi.org/10.1016/j.surfcoat.2019.124989
52. Martinez-Calderon M, Manso-Silvan M, Rodriguez A, Gomez-Aranzadi M, Garcia-Ruiz JP, Olaizola SM, Martin-Palma RJ (2016) Surface micro- and nano-texturing of stainless steel by femtosecond laser for the control of cell migration. Sci Rep 6:36296. https://doi.org/10.1038/srep36296
53. Batal A, Sammons R, Dimov S (2019) Response of Saos-2 osteoblast-like cells to laser surface texturing, sandblasting and hydroxyapatite coating on CoCrMo alloy surfaces. Materials science & engineering C, Materials for biological applications 98:1005–1013. https://doi.org/10.1016/j.msec.2019.01.067
54. Schieber R, Lasserre F, Hans M, Fernández-Yagüe M, Díaz-Ricart M, Escolar G, Ginebra M-P, Mücklich F, Pegueroles M (2017) Direct Laser Interference Patterning of CoCr Alloy Surfaces to Control Endothelial Cell and Platelet Response for Cardiovascular Applications. Advanced healthcare materials 6(19):1700327. https://doi.org/10.1002/adhm.201700327
55. Dumas V, Guignandon A, Vico L, Mauclair C, Zapata X, Linossier MT, Bouleftour W, Granier J, Peyroche S, Dumas J-C, Zahouani H, Rattner A (2015) Femtosecond laser nano/micro patterning of titanium influences mesenchymal stem cell adhesion and commitment. Biomedical Materials 10(5):055002. https://doi.org/10.1088/1748-6041/10/5/055002
56. Jorge-Mora A, Imaz N, Garcia-Lecina E, O'Connor GM, Gómez-Vaamonde R, Alonso-Pérez A, Franco-Trepat E, García-Santiago C, Pino-Minguez J, Nieto D (2018) In vitro response of bone marrow mesenchymal stem cells (hBMSCs) on laser-induced periodic surface structures for hard tissue replacement: Comparison between tantalum and titanium. Optics and Lasers in Engineering 111:34–41. https://doi.org/10.1016/j.optlaseng.2018.07.008

57. Lin X, Li X, Li G, Zhang Y, Cui Z (2020) Micro-dot-matrix induced by femtosecond laser on titanium surface for Ca-P phase deposition. Applied Surface Science 499:143925. https://doi.org/10.1016/j.apsusc.2019.143925
58. Kurashina Y, Ezura A, Murakami R, Mizutani M, Komotori J (2019) Effect of hydroxy groups and microtopography generated by a nanosecond-pulsed laser on pure Ti surfaces. Journal of Materials Science: Materials in Medicine 30(5):57. https://doi.org/10.1007/s10856-019-6259-1
59. Zhao D-P, Tang J-C, Nie H-M, Zhang Y, Chen Y-K, Zhang X, Li H-X, Yan M (2018) Macro-micron-nano-featured surface topography of Ti-6Al-4V alloy for biomedical applications. Rare Metals 37(12):1055–1063. https://doi.org/10.1007/s12598-018-1150-7
60. Chan C-W, Carson L, Smith GC, Morelli A, Lee S (2017) Enhancing the antibacterial performance of orthopaedic implant materials by fibre laser surface engineering. Applied Surface Science 404:67–81. https://doi.org/10.1016/j.apsusc.2017.01.233
61. Fadeeva E, Truong VK, Stiesch M, Chichkov BN, Crawford RJ, Wang J, Ivanova EP (2011) Bacterial retention on superhydrophobic titanium surfaces fabricated by femtosecond laser ablation. Langmuir 27(6):3012–3019. https://doi.org/10.1021/la104607g
62. Uhlmann E, Schweitzer L, Cunha A, Polte J, Huth-Herms K, Kieburg H, Hesse B (2019) Application of laser surface nanotexturing for the reduction of peri-implantitis on biomedical grade 5 Ti-6Al-4V dental abutments, vol 10908. SPIE, SPIE LASE
63. Lee BEJ, Exir H, Weck A, Grandfield K (2018) Characterization and evaluation of femtosecond laser-induced sub-micron periodic structures generated on titanium to improve osseointegration of implants. Applied Surface Science 441:1034–1042. https://doi.org/10.1016/j.apsusc.2018.02.119
64. Xue X, Ma C, An H, Li Y, Guan Y (2018) Corrosion resistance and cytocompatibility of Ti-20Zr-10Nb-4Ta alloy surface modified by a focused fiber laser. Science China Materials 61(4):516–524. https://doi.org/10.1007/s40843-017-9239-3
65. Guan YC, Zhou W, Li ZL, Zheng HY, Lim GC, Hong MH (2014) Femtosecond laser-induced ripple structures on magnesium. Applied Physics A 115(1):13–18. https://doi.org/10.1007/s00339-013-7927-5
66. Zhang J, Guan Y, Lin W, Gu X (2019) Enhanced mechanical properties and biocompatibility of Mg-Gd-Ca alloy by laser surface processing. Surface and Coatings Technology 362:176–184. https://doi.org/10.1016/j.surfcoat.2019.01.063
67. Furlan V, Biondi M, Demir AG, Previtali B, Pariani G, Bianco A, T. AKM, C. G, A. K (2018) Two-beam interference patterning of biodegradable magnesium alloy: Influence of number of passes and spots overlap. Journal of Vacuum Science & Technology B 36 (1):01A102. doi:10.1116/1.4996504
68. Zhang J, Lin W, Guan Y, Gu X (2019) Biocompatibility enhancement of Mg-Gd-Ca alloy by laser surface modification. Journal of Laser Applications 31(2):022510. https://doi.org/10.2351/1.5096136
69. Huang Z, Zhang A, Zhang Q, Cui D (2019) Nanomaterial-based SERS sensing technology for biomedical application. Journal of Materials Chemistry B 7(24):3755–3774. https://doi.org/10.1039/C9TB00666D
70. Kahraman M, Mullen ER, Korkmaz A, Wachsmann-Hogiu S (2017) Fundamentals and applications of SERS-based bioanalytical sensing. Nanophotonics 6(5):831–852. https://doi.org/10.1515/nanoph-2016-0174
71. Tian S, You W, Shen Y, Gu X, Ge M, Ahmadi S, Ahmad S, Kraatz H-B (2019) Facile synthesis of silver-rich Au/Ag bimetallic nanoparticles with highly active SERS properties. New Journal of Chemistry 43(37):14772–14780. https://doi.org/10.1039/c9nj02879j
72. Hu G, Guan K, Lu L, Zhang J, Lu N, Guan Y (2018) Engineered Functional Surfaces by Laser Microprocessing for Biomedical Applications. Engineering 4(6):822–830. https://doi.org/10.1016/j.eng.2018.09.009

Surface Nanostructuring of Metallic Materials for Implant Applications

T. Balusamy, T. S. N. Sankara Narayanan, and Hyung Wook Park

Abstract Nanostructured metallic materials are progressively investigated for numerous biomedical implant applications due to their superior mechanical properties, biocompatibility and, ability to promote cell adhesion and proliferation. Several materials processing routes have been explored to prepare bulk and surface nanostructured metals/alloys. Surface nanostructuring by changing the surface chemistry through deposition or diffusion processes is limited by porosity and contamination besides uncertainty in achieving a good bonding between substrate and coating. The surface nanostructuring process must enable an improvement in the overall properties of materials such as high hardness and strength, higher thermal expansion coefficient, improved tribological properties, and better fatigue properties, etc., to avoid premature failure of implant materials. In this perspective, surface severe plastic deformation (S^2PD) processes assume significance as they could impart the desired characteristics to a variety of metals and alloys by grain refinement mechanism, without changing the overall composition and/or phases present in the material. The improvement in properties is fundamentally derived from grain refinement mechanism by the introduction of a large amount of defects/strain. Surface mechanical attrition treatment (SMAT) is an effective way of inducing localized plastic deformation (S^2PD method) that results in grain refinement down to nanometer scale without changing the chemical composition of the material. It is a very promising method to produce functionally gradient materials, in which the nanocrystalline

T. Balusamy · T. S. N. Sankara Narayanan
CSIR-National Metallurgical Laboratory, Madras Centre, CSIR Complex,
Taramani, Chennai 600 113, India
e-mail: tsnsn2005@gmail.com

T. Balusamy (✉)
Department of Applied Chemistry, Sri Venkateswara College of Engineering,
Sriperumbudur 602 117, India
e-mail: rtbalusamy@gmail.com

T. S. N. Sankara Narayanan · H. W. Park
Department of Mechanical Engineering, Ulsan National Institute of Science and
Technology, UNIST-Gil 50, Eonyang-Eup, Ulju-Gun, Ulsan 689 798, Republic of Korea
e-mail: hwpark@unist.ac.kr

T. S. Santra and L. Mohan (eds.), *Nanomaterials and Their Biomedical Applications*,
Springer Series in Biomaterials Science and Engineering 16,
https://doi.org/10.1007/978-981-33-6252-9_16

surface layer provides suitable surface properties while the coarse-grained matrix provides the ductility. SMAT provides desirable surface topography and increases the average surface roughness. The surface topography of materials determines its hydrophilic or hydrophobic nature. For implants, a hydrophilic surface is considered to be more desirable than a hydrophobic one because of its better interaction with biological fluids, cells, and tissues. In addition, generating the desired surface topography could provide significant enhancement in osteoblast adhesion, proliferation, maturation, and mineralization. SMAT decreases the grain size, induces compressive residual stress, microstrain, defects/dislocations, and phase transformation, all of which enable a significant improvement in hardness, fatigue resistance, and tribological properties of materials. In addition, the extent of grain refinement, extent of deformation, extent of change in surface roughness, phase transformation, residual stress, microstrain, and defect/dislocation density could influence the performance of materials subjected to SMAT. This could ultimately influence the biological performance of the implant materials. The present chapter aims to provide detailed coverage of SMAT of stainless steel, Ti alloys, Ni-Ti alloy, CoCrMo alloy, and how the nanostructured surface enables an improvement in the characteristic properties that are suitable for biomedical applications.

Keywords Metallic biomaterials · Surface nanocrystallization · Surface mechanical attrition treatment · Mechanical properties · Corrosion · Cell response

1 Surface Nanostructuring by Surface Mechanical Attrition Treatment

Nanostructured materials have received considerable attention due to their unique physical, chemical, and mechanical properties that are appropriate for numerous technological applications. Recently, nano/ultrafine-grained materials produced by severe plastic deformation (SPD) have gained much importance in biomedical applications. Bulk nanostructured materials can be produced by methods such as equal channel angular pressing, accumulative roll bonding, high-pressure torsion, multiple compression, etc. However, the ductility of materials produced by these methods is significantly low when compared to the coarse-grained materials. Surface mechanical attrition treatment (SMAT) is a surface severe plastic deformation (S^2PD) method that enables the formation of a nanostructured surface layer on metallic materials and improves the overall properties and performance, without affecting their inherent properties [1–5]. The amount of plastic working energy incurred during treatment is relatively much less for S^2PD than bulk deformation. High strength materials like stainless steels and titanium alloys can be easily processed by SMAT whereas processing them with equal channel angular pressing requires a larger load and specially designed dies. Other advantages of SMAT include high throughput processing and the ability to process materials at a lower cost.

The principle of SMAT is quite similar to shot peening (SP). Both of them involve repeated impingement of shots on the surface, causing plastic deformation of the surface, impart compressive residual stress, induce work hardening, and increase the surface roughness. However, notable differences exist between them. The main difference is the size, shape, and impact velocities of the shots. SMAT is capable of imparting surface nanocrystallization, which is always not true with SP. The thickness of the nanocrystalline layer is much higher and the depth of the compressive residual stress is much deeper for SMAT. Table 1 provides a comparison of important parameters of SP and SMAT and the characteristics of the treated surface [6–8].

The methodology of SMAT is elaborated in detail by Lu and Lu [1, 3] and Balusamy et al. [9, 10]. The schematic of SMAT set-up is shown in Fig. 1. The sample to be treated is fixed in the upper part of the sample pot whereas balls used for the impingement are kept at the bottom of the pot. The distance between the sample and the top surface of the balls is maintained at ~25 to 30 mm. Before treatment, the sample pot is sealed and evacuated. During SMAT, the balls start to fly

Table 1 Comparison of the important parameters of shot peening and surface mechanical attrition treatment and the characteristics of the treated surface [6–8]

Parameter	Shot peening (SP)	Surface mechanical attrition treatment (SMAT)
Shot size	0.25–1 mm	1–8 mm
Shot shape	Irregular	Spherical
Impact velocity	20–150 m/s	5–15 m/s
Type of impact	Repeated	Multidirectional
Kinetic energy	Relatively lower than SMAT 9.2×10^{-6}–0.01 J	Much higher than SP <0.2 J
Surface roughness	Higher than those treated by SMAT	Relatively lower than those treated by SP
Ability to induce compressive residual stress	Yes	Yes
Ability to induce work hardening	Yes	Yes
Ability to induce surface nanocrystallization	Not always	Always
Thickness of the nanocrystalline layer	Thinner than those treated by SMAT	Much thicker than those treated by SP
Thickness of the compressive residual stress	Relatively less	Much deeper
Ability to treat complex geometries	Yes. Versatile	No
Post-polishing requirement	Yes. Intense polishing is required to achieve a good surface finish	Yes

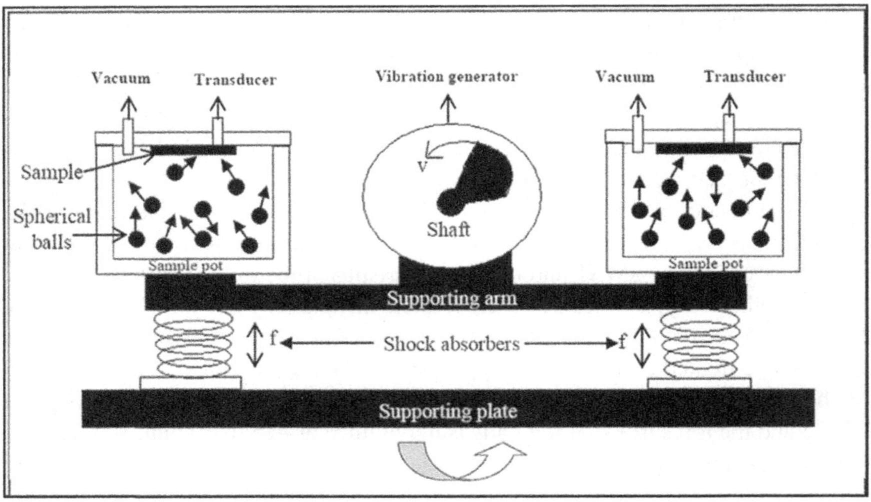

Fig. 1 Schematic of the surface mechanical attrition treatment (SMAT) set-up (Reprinted from Corrosion Science, 52 (11), Balusamy, T., et al., Effect of surface nanocrystallization on the corrosion behavior of AISI 409 stainless steel, 3826–3834, Copyright [2010], with permission from Elsevier)

when they resonate with the vibration frequency (50 Hz). The repeated multidirectional impingement of the flying balls during the entire duration of treatment enables surface nanocrystallization.

The plastic deformation induced by SMAT enables refinement of coarse-grained grains down to nanosized grains at the surface along with the formation of a graded layer structure in which the nanocrystalline surface layer provides suitable surface properties while the coarse-grained matrix provides the ductility. The repeated multidirectional impingement of the balls increased the surface roughness. SMAT is an effective method to produce thicker nanocrystalline and work-hardened surface layers as well as deeper surface regions with larger residual compressive stresses. The high density of dislocations and compressive residual stress induced during SMAT improves the hardness, mechanical properties, fatigue life and tribological behavior [11–15]. The extent of change in residual stress during SMAT is determined by the type and size of the balls used for the treatment, the treatment time and, the type of materials being treated, their chemical composition, hardness, and workability.

The increase in grain boundary density promotes the formation of a passive oxide layer, which improves the corrosion resistance [10, 16–19]. The development of a desirable surface topography decreases the water contact angle, increases the surface energy, and promotes bioactivity [18, 20] and, better cell adhesion and spreading [21]. A combination of high strength and high ductility is warranted for the long-term stability of metallic implants. Similarly, a higher hardness, wear-resistance, and corrosion resistance are required to improve their biocompatibility. A suitable surface profile is also needed to improve better cell attachment.

Metallic biomaterials are widely used for the fabrication of implants. The high strength and resistance to fracture provide them a reliable long-term implant performance in major load-bearing situations. Stainless steels, CP-Ti, Ti6Al4V alloy, β-phase Ti alloy, NiTi alloy and CoCrMo alloy are the most commonly used metallic materials for orthopedic implant applications. Among them, stainless steels, CP-Ti, β-phase Ti alloy and NiTi alloy are primarily used for osteosynthesis applications. CoCrMo alloy is mainly used for joint arthroplasty application whereas Ti6Al4V alloy is used for both for joint arthroplasty as well as osteosynthesis applications [22]. The present chapter aims to provide detailed coverage of SMAT of stainless steel, Ti and Ti alloys, Ni-Ti alloy, and CoCrMo alloy and how the nanostructured surface enables an improvement in the characteristic properties that are suitable for biomedical applications.

2 Stainless Steels

Austenitic stainless steels are widely used for biomedical implant and device manufacturing purposes due to their good mechanical properties, better corrosion resistance, acceptable biocompatibility, ease of fabrication. In general, AISI 316 LVM stainless steel (ASTM F138 and F139 standards) is the recommended material for orthopedic applications, particularly for load-bearing implants and stents. AISI 304 stainless steel is used in orthodontics such as archwires, brackets, and screws. The major causes of failure of stainless steel implants are due to stress shielding effect, poor osseointegration and corrosion fatigue. Generation of wear debris, accumulation of wear debris at the surrounding tissues causing inflammation, toxicity of metal ions leached into body fluid are other associated problems. Most of the failures are surface related. The interaction between the implant and the surrounding tissues determines the overall performance of the implant. The surface properties such as surface roughness, wettability, surface energy and surface chemistry are the key factors that determine the interaction between the host tissue and the implant immediately after the implantation. Absorption of proteins is the basic step in promoting cell adhesion and proliferation. Implant surfaces with a suitable surface profile could increase the degree of hydrophilicity and promote cell adhesion and proliferation. Hence, engineering the surface of implant materials is critical for the development of next-generation orthopedic implants. SMAT enables the formation of a nanostructured surface layer on metallic materials and improves their overall properties and performance. The effect of SMAT on the microstructure, grain size, phase content, phase transformation, compressive residual stress, surface roughness, wettability and corrosion resistance of stainless steels are addressed in the following paragraphs.

The evolution of the microstructure of stainless steel during SMAT is addressed by Lu and Lu [1, 3] and Tao et al. [4]. SMAT effectively induces localized plastic deformation that results in grain refinement down to the nanometer scale at the surface of austenitic stainless steel. It is well known that plastic deformation of metallic materials occurs by slip and twinning mechanisms. In metallic materials with medium or high stacking fault energy (SFE), slip is the predominant mechanism while for metals having a low SFE, twinning is favorable, particularly at high strain rates and/or at low deformation temperatures [3]. In general, austenitic stainless steels (304 and 316L) are low SFE materials. Hence, nanostructuring is expected to occur by twinning mechanism [4]. The formation of planar dislocation arrays and mechanical twins is considered to be the first stage, which is followed by twin-twin interactions leading to refinement of grains down to nanometer level. SMAT induces extremely high strain and the strain rate of the order of 10^2 to 10^3 s^{-1} at the sample surface. Austenitic stainless steels undergo extensive twinning and with an increase in the extent of deformation, the twinning becomes multidirectional. Further, the intersection of the multidirectional twins leads to grain subdivision and formation of strain-induced martensite (SIM). The twin-twin interactions are considered to be the precursors for the formation of SIM and nanostructures (Fig. 2) [10, 23, 24]. The SEM images acquired at the cross-section of as-annealed and SMATed AISI 316L stainless steel is shown in Fig. 3. Unlike its untreated coarse-grained counterpart (Fig. 3a), the treated 316L stainless steel exhibits the formation of dual phases represented by the striped regions inside each grain along with the formation of nano-sized grains at the top surface (Fig. 3b).

During SMAT, the level of strain and strain rate depends on the kinetic energy of the impinging balls, which is a function of the type, size, number and hardness of the balls used for treatment, the frequency of vibration, the distance between the surface of the material and the top of the balls as well as the characteristic properties of the material being treated (hardness, SFE, etc.,) [1]. For a given set of conditions, the strain will be much higher at the top surface while it is likely to decrease gradually while moving away from the surface to the bulk of the sample [25]. The decrease in

Fig. 2 SEM images acquired at the cross-section of AISI 304 stainless steel subjected to SMAT using 8 mm Ø 316L stainless steel balls at 50 Hz for 60 min (Reprinted by permission from Springer Nature, Transactions of the Indian Institute of Metals, 64, nos. 4, 5 (August–October 2011): 507–511, Electrochemical Behaviour of Surface Modified AISI 304 Grade Stainless Steel in Ringer's Solution, Balusamy et al. [28])

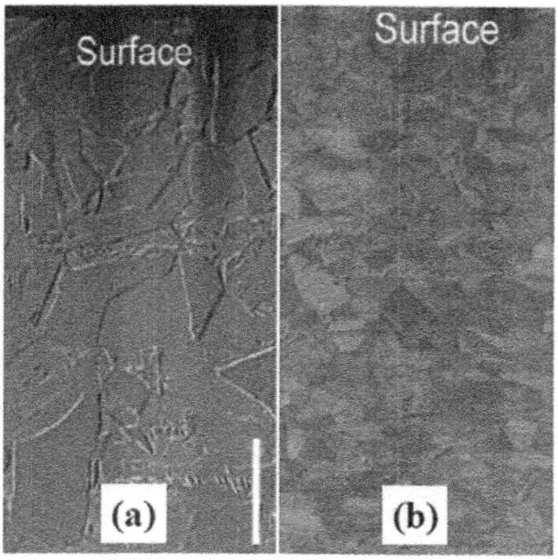

Fig. 3 SEM images acquired at the cross-section of AISI 316L stainless steel: **a** as-annealed. **b** after SMAT (Scale bar: 100 μm). The striped regions inside each grain of the treated stainless steel (in b) represent nano-dual-phase morphology (Reproduced from Cao et al., "Predicting surface deformation during mechanical attrition of metallic alloys." *Npj Computational Materials*, 5, no. 1 (2019): 36 under a Creative Commons Attribution 4.0 International License)

the extent of strain from the top surface to the bulk is also reflected in the magnitude of plastic deformation, limiting the extent of a decrease in grain size. Hence, surface nanocrystallization is limited to only a few micrometers from the top surface. The grain size of the top surface layer of stainless steel after SMAT is typically of the order of 10–50 nm for SMAT. Due to the difference in strain and strain rate from top surface to bulk a gradient microstructure is generated, which offers a good combination of desirable strength and ductility, most suited for biomaterials.

Electron backscattered diffraction (EBSD) studies are quite useful to understand the evolution of microstructure of stainless steel during SMAT by plastic deformation. The EBSD patterns are useful to compare the variation in grain size, texture, morphology, crystallographic orientation and grain boundary misorientation of stainless steel before and after SMAT. In addition, the depth of the deformed layer can be established as well. Based on the variation in the microstructural features from the EBSD analysis, it can be divided into three regions (i) top surface layer, which consists of extensive grain fragmentation with a random orientation of fine grains due to very high deformation, (ii) an intermediate region consists of heavily deformed grains, and (iii) a third region of microstructure shows some deformed grains (most of them not affected by SMAT) followed by coarse-grain matrix. The variation in the microstructures of the treated stainless steels, from the surface to bulk, could be directly related to the extent of strain-induced on them. The grain orientation and phase distribution of 304 stainless steel subjected to SMAT using 8 mm Ø 316L stainless steel balls for 60 min, at 130 μm from the top surface, is shown in Fig. 4. The top surface of the inverse pole figure (IPF) map (Fig. 4a) shows the formation of nanocrystalline grains in Region I and the formation of a graded layer structure with

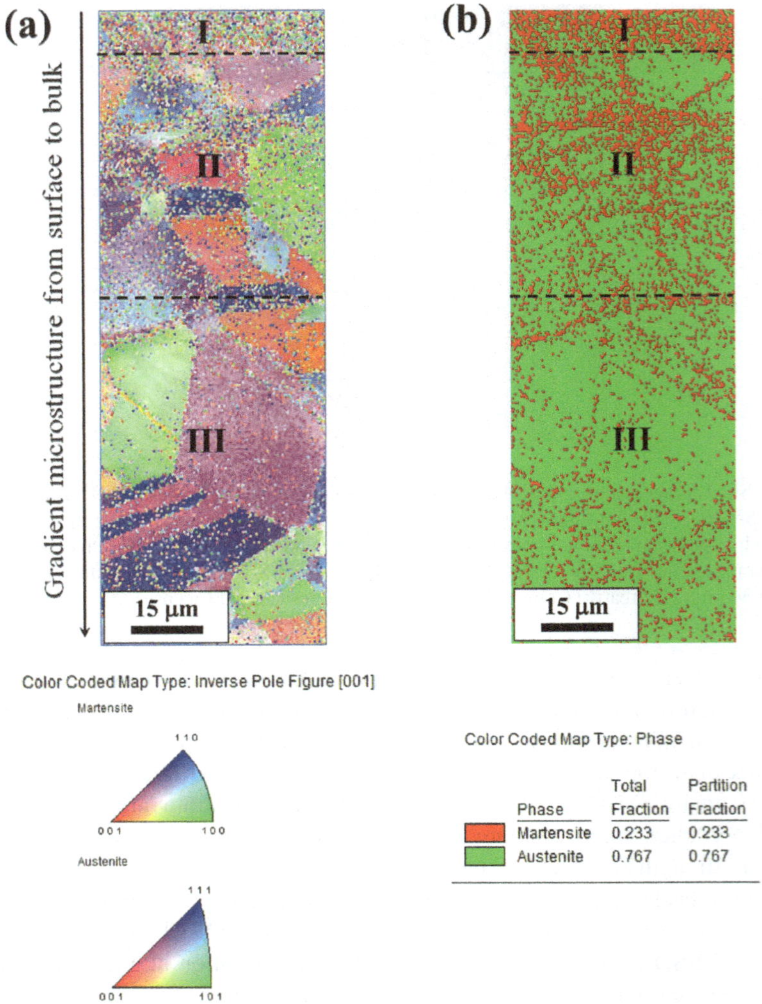

Fig. 4 Grain orientation and phase distribution of 304 SS treated using 8 mm Ø 316L stainless steel balls for 60 min, assessed at the cross-section covering a region of 130 μm from the top of the treated surface to the bulk, by EBSD analysis (top edges correspond to the top surface of the treated 304 SS): **a** inverse pole figure (IPF) map showing the orientation of fragmented and deformed grains; **b** phase-contrast map taken at the same region shown in panel a, indicating the formation of the strain-induced martensite phase after SMAT; red and green colors represent martensite and austenite phases respectively; and the fraction of these phases are also included in panel b (Reprinted (adapted) with permission from Balusamy et al., "A Facile Method to Modify the Characteristics and Corrosion Behavior of 304 Stainless Steel by Surface Nanostructuring toward Biomedical Applications." *ACS Applied Materials and Interfaces*, 7, no. 32 [2015]: 17731–17747 © [2015] American Chemical Society)

a continuous increase in grain size with an increase in depth (Regions II and III). A steady decrease in the amount of fragmented and deformed grains from the top surface to the depth is also evident. The formation of strain-induced martensite phase after SMAT is established by the phase contrast map (Fig. 4b). The martensite phase is represented by red color while the austenite phase is represented by green color. The intense red color spots at the top surface represent a higher volume fraction of the martensite phase due to the higher level of strain and strain rate experienced at the top surface. The decrease in strain with an increase in depth from the surface is also evident from the decrease in a number of red spots. The EBSD images of AISI 316L stainless steel in its as-annealed condition and after SMAT is shown in Fig. 5. The as-annealed 316L stainless steel consists of equiaxed single-phase austenitic grains (Fig. 5a). In contrast, a dual-phase structure comprising of alternating nano-band austenite and nano-lamellar martensite within each grain could be seen for the treated 316L stainless steel. In addition, the formation of a graded layer structure with an increase in grain size from nanograins (~80 nm) at the surface to sub-micron sized grains with an increase in deformation depth could also be identified (Fig. 5b). Based on the microstructural features and EBSD patterns of untreated and SMAT stainless steel, the grain refinement and the formation of a graded layer structure after SMAT of 316L stainless steel are represented schematically in Fig. 6.

SMAT increased the surface roughness of stainless steel. The increase in surface roughness is due to the formation of craters and dimples following the impingement of the balls on the surface of the stainless steel being treated [9, 10, 26–30]. According to Arifvianto et al. [26, 27], impingement of the balls generates a pile-up of material

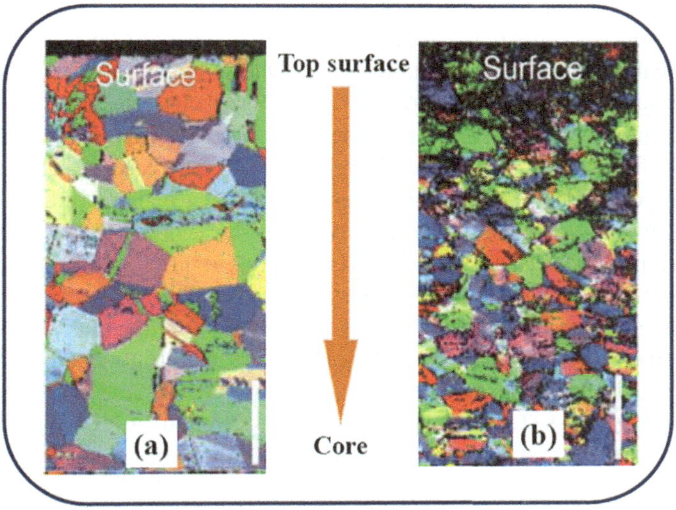

Fig. 5 EBSD images of AISI 316 L stainless steel: **a** As-annealed. **b** After SMAT (Reproduced from Cao et al., "Predicting surface deformation during mechanical attrition of metallic alloys." *Npj Computational Materials*, 5, no. 1 [2019]: 36 under a Creative Commons Attribution 4.0 International License)

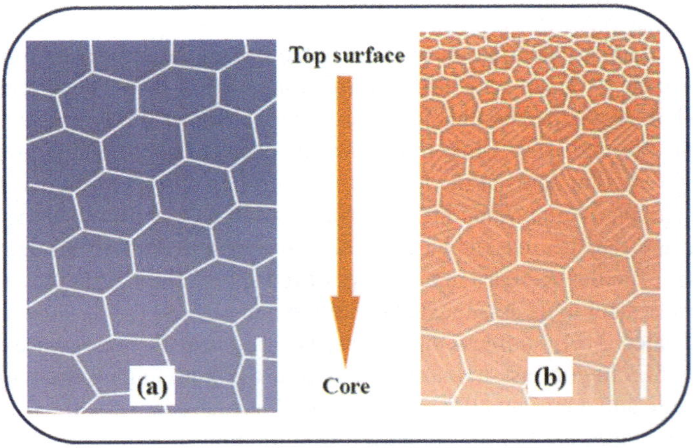

Fig. 6 Schematic illustration of **a** distribution of grains in the as-annealed AISI 316L stainless steel. **b** grain refinement of AISI 316L stainless steel after SMAT (Scale bar: 100 μm) (Reproduced from Cao et al., "Predicting surface deformation during mechanical attrition of metallic alloys." *Npj Computational Materials*, 5, no. 1 (2019): 36 under a Creative Commons Attribution 4.0 International License)

around the crater, resulting in the formation of a new pair of peak and valley. For a given treatment time of 60 min, an increase in size of the 316L stainless steel balls from 2 to 8 mm Ø has led to an increase in average surface roughness (R_a) of 304 stainless steel. For a given ball size of 8 mm Ø 316L stainless steel balls, an increase in treatment time from 30 min to 60 min has led to a decrease in R_a of 304 stainless steel [10]. Hence, the increase in R_a after SMAT appears to be a function of the size of the balls rather than the treatment time. During the initial duration of SMAT, local deformation of the surface induced by impingement of the balls has promoted the formation of dimples, which alters the surface heterogeneity. With an increase in duration of treatment, the size of dimples starts to decrease. When the dimples approach together, the number of dimples per unit area is increased. This enables an increase in uniformity and homogeneity of the surface [29]. Surface erosion due to overworking should also be considered as an important factor in determining the surface roughness with an increase in treatment time. The mechanical property of the material being treated is the governing factor in determining the extent of surface erosion during SMAT. An increase in treatment time from 15 to 45 min has led to an increase in R_a for AISI 409 stainless steel [9]. However, a decrease in R_a is observed for AISI 304 stainless steel at a longer duration of treatment [10, 29]. This is due to the high work hardening capacity of AISI 304 stainless steel. The evolution of surface roughness during SMAT is proposed by Arifvianto et al. [26, 27]. According to them, the indentation and surface erosion are the key parameters in deciding the resultant surface roughness after SMAT. Based on the % mass reduction, they have suggested that an increase in ball size has increased the extent of surface erosion. The increase in surface roughness of stainless steels after SMAT has enabled a decrease

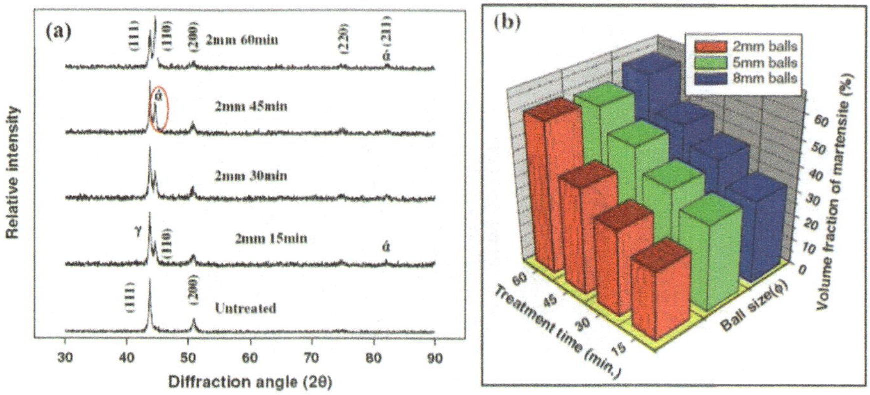

Fig. 7 **a** XRD patterns of untreated 304 stainless steel and those subjected to SMAT using 2 mm Ø 316L stainless steel balls for 15, 30, 45 and 60 min. **b** volume fraction of the martensite phase formed after SMAT as a function of size of the balls and treatment time (Reprinted by permission from Springer Nature, Transactions of the Indian Institute of Metals, 64, nos. 4, 5 (August–October 2011): 507–511, Electrochemical Behaviour of Surface Modified AISI 304 Grade Stainless Steel in Ringer's Solution, Balusamy et al. [28])

in water contact angle, an increase in surface energy and degree of hydrophilicity [10, 26].

The XRD patterns of untreated austenitic stainless steels (304 and 316L) shows only the presence of austenite (γ-fcc) phase while strain-induced martensite (α'-bcc) phase besides the γ-fcc phase is also observed after SMAT, suggesting the transformation of austenite (γ-fcc) to strain-induced martensite (α'-bcc) phase occurs in both types of stainless steels (Fig. 7). The presence of both austenite and martensite phases suggests that the martensite phase is generated only during SMAT (Fig. 7a). The mechanism of phase transformation is attributed to the increase in stacking-fault energy originating from deformation. The volume fraction of α'-martensite phase depends on the extent of deformation as dictated by the level of strain and strain rate (Fig. 7b). Broadening of the XRD peaks is also observed after SMAT. This is due to the refinement of grains or due to an increase in the atomic level lattice strain or both [10, 28, 29].

SMAT induces the compressive residual stress of stainless steel. In general, the compressive residual stress of stainless steels is increased with an increase in the size of the balls used for treatment as well as the treatment time. A comparison of the compressive residual stress measured at the surface of 304 stainless steel after SMAT using 2, 5, and 8 mm Ø 316L stainless steel balls for 15, 30, 45, and 60 min is shown in Fig. 8 [10]. According to Benafia et al. [31], for a given ball mass of 20 g, 316L stainless steel samples subjected to SMAT at a mixed amplitude of ± 12.5 μm for 15 min followed by ± 25 μm for 5 min (total 20 min) has led to a higher compressive residual stress than those treated at a constant amplitude of ± 12.5 μm for 30 min.

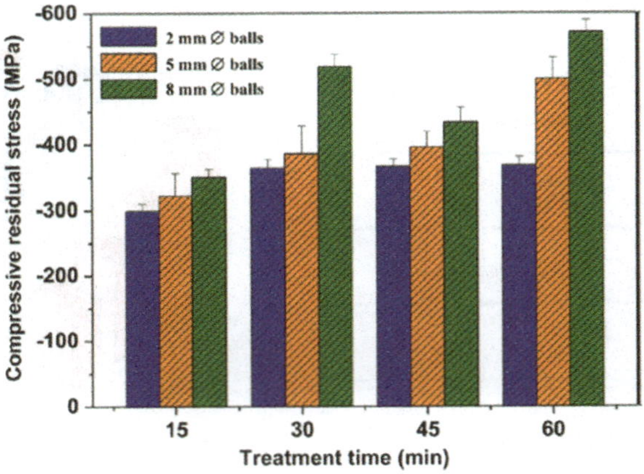

Fig. 8 Comparison of the residual stress induced at the surface of 304 SS after SMAT using 2, 5, and 8 mm Ø for 15, 30, 45, and 60 min (Reprinted (adapted) with permission from Balusamy et al., "A Facile Method to Modify the Characteristics and Corrosion Behavior of 304 Stainless Steel by Surface Nanostructuring toward Biomedical Applications." *ACS Applied Materials and Interfaces*, 7, no. 32 [2015]: 17731–17747 © [2015] American Chemical Society)

It is important to note that compressive residual stresses induced during SMAT are homogeneously present over the entire surface of the treated 316L stainless steel.

SMAT has promoted an increase in hardness of stainless steel. The hardness measured at the surface or near the nanostructured layer of the treated stainless steel is twice the hardness of the untreated stainless steel. A progressive decrease in hardness with an increase in depth from the surface to the end of the deformed region is observed [10, 12, 26, 31]. According to Bahl et al. [24], the hardness remains almost constant at 3.5 GPa from the surface to a depth of 30 μm whereas it decreased to 2.5 GPa at a depth of 40 μm. The nanocrystallization, compressive residual stress and formation of strain-induced martensite phase are considered responsible for the increase in hardness of stainless steel after SMAT. Arifvianto et al. [26] have reported that the extent of increase in hardness of 316L stainless steel after SMAT is a function of the treatment time, size, and number of the balls used for treatment and the frequency of vibration. According to them, the choice of a longer duration of treatment, use of a larger number of bigger balls for treatment and a higher vibration frequency for SMAT would lead to a higher hardness of 316L stainless steel.

SMAT improves the mechanical properties of stainless steel. The grain refinement, formation of a nanocrystalline structure and higher dislocation density are considered responsible for the improvement in mechanical properties [12, 23, 27]. The yield strength (YS) of untreated 316L stainless steel is increased from 300 MPa to 665 and 725 MPa after SMAT for 15 and 30 min, respectively. The ultimate tensile strength (UTS) is increased from 650 MPa to 750 and 784 MPa after SMAT for 15 and 30 min, respectively. The increase in YS and UTS with an increase in treatment time from 15 to 30 min is due to the formation of a thicker nanocrystalline layer [23].

The corrosion behavior of untreated and SMATed AISI 316L stainless steel in 0.1 M NaCl is evaluated by Hao et al. [32]. According to them, the formation of a nanocrystalline surface (19 nm) has led to a decrease in corrosion resistance. The increase in grain boundary density, structural inhomogeneity and transformation of the austenite to α'-martensite phase are considered responsible for the decrease in corrosion resistance. However, annealing of the SMATed 316L stainless steel at 300–380 °C promotes the diffusion of Cr from the matrix to the surface, which facilitates the formation of the passive film and improves the corrosion resistance.

Balusamy et al. [10, 29] have evaluated the corrosion resistance of untreated AISI 304 stainless steel and those subjected to SMAT using 2, 5 and 8 mm Ø 316L stainless steel balls for 15–60 min, in 0.6 M NaCl as well as in Ringer's solution. An improvement in corrosion resistance is observed only for 304 stainless steels treated using 2 mm Ø balls. In contrast, SMAT using 5 and 8 mm Ø balls has led to a decrease in corrosion resistance. Many factors could influence the corrosion behavior of SMAT 304 stainless steels. Establishing a direct correlation between the corrosion behavior of the treated 304 stainless steel and the factors influencing it is quite complex. In most instances, the corrosion resistance is influenced by more than one factor. All of these factors, either individually or in combination could influence the corrosion resistance of the treated 304 stainless steel. The increase in surface roughness, a transformation of the austenite to α'-martensite phase, higher extent of deformation and presence of a larger number of defects/dislocations are responsible for the observed decrease in corrosion resistance of 304 SS treated using 5 and 8 mm Ø balls when compared to the untreated one in Ringer's solution as well as 0.6 M NaCl. The grain refinement, surface nanocrystallization, selective dissolution of Fe and enrichment of Cr at the surface have increased the number of nucleation sites and promotes the formation of an intact and highly protective passive oxide layer on 304 stainless steel treated using 2 mm Ø balls when compared to the untreated one, in Ringer's solution as well as 0.6 M NaCl. The defects induced during SMAT and the increase is surface roughness should be limited which would otherwise cause a deleterious influence on the corrosion resistance.

The corrosion behavior of untreated 301 stainless steel and those subjected to SMAT using 2 mm Ø 304 stainless steel balls for 5 to 20 min, in 0.6 M NaCl is evaluated by Olugbade et al. [33]. The formation of a nanostructured surface has offered an improvement in the corrosion resistance for SMAT 301 stainless steel when compared to its untreated counterpart. The shift in corrosion potential (E_{corr}) towards more noble values, decrease in corrosion current density (i_{corr}), higher charge transfer resistance (R_{ct}) and a higher phase angle validates the inference. The improvement in corrosion resistance of SMAT 301 stainless steel is due to its ability to facilitate diffusion of Cr to the surface, thus enabling the formation of a strong passive oxide layer.

Mechanical failure of metallic biomaterials is often related to corrosion fatigue phenomena. Pitting corrosion is often considered to be the active mechanism through which fatigue cracks initiate. Microstructural aspects are closely related to the corrosion fatigue strength of biomedical alloys. The mechanism of corrosion fatigue of biomedical metallic alloys and the ways of mitigation is reviewed by Antunes and de

Oliveira [34]. Roland et al. [23] have investigated the fatigue behavior of untreated 316L stainless steel and those subjected to SMAT. The fatigue strength of the SMAT 316L stainless steel is found to be much higher than the untreated one at both low and high cycle fatigue regimes. The extent of an increase in fatigue strength for the treated stainless steel is much pronounced at high cycle fatigue. The surface nanocrystallization and the compressive residual stress induced by SMAT offered considerable resistance for the propagation of fatigue cracks. Post-annealing of SMAT 316L stainless steels at 400 °C also enhanced fatigue strength. In spite of the relaxation in compressive residual stress, the increase in surface ductility, limited increase in grain size, strain hardening and strain-induced martensitic transformation accounts for the improvement in fatigue strength.

Bahl et al. [24] have evaluated the corrosion fatigue of microcrystalline (MC) and nanocrystalline (NC) 316L stainless steels in 0.9% NaCl at 0.1 and 5 Hz for up to 106 cycles or to the point of complete fracture. The stress vs. number of cycles (S–N) plot of the MC and NC 316L stainless steel indicates that the corrosion fatigue strength is increased by 50% for NC 316L stainless steel when compared to its MC counterpart (Fig. 9a). The improvement in corrosion fatigue resistance is due to the presence of a thicker oxide layer, residual compressive stresses and high strength of the nanocrystalline surface layer. Pitting corrosion is observed on both MC and NC 316L stainless steels. The chloride ions are highly aggressive. The pitting corrosion is initiated by the chloride ions, which is followed by failure through crack initiation from the pit. Since stainless steels are highly susceptible to localized corrosion attack, surface nanocrystallization could not alleviate it. The presence of a thicker oxide layer on NC 316L stainless steel could hinder the ingress of chloride ions, delay the initiation of pitting, and enhances the fatigue strength [35, 36]. The fractograph of the MC 316L stainless steel shows the occurrence of intergranular corrosion, leading to crack propagation through brittle cleavage fracture (Fig. 9b). Dislocation pile-ups at the grain boundaries increase the susceptibility of the MC 316L stainless steels for

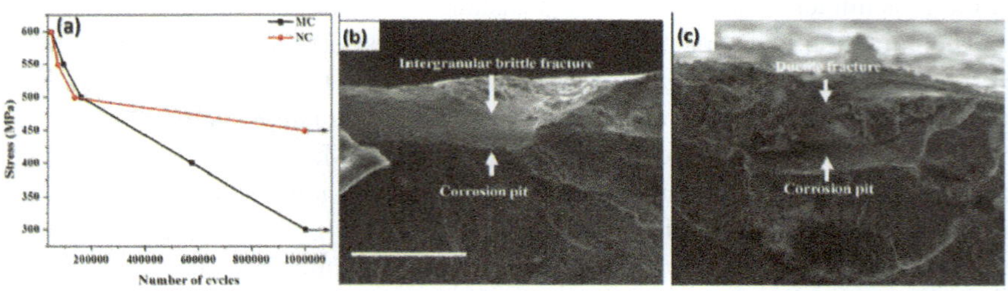

Fig. 9 **a** S–N curve of NC and MC 316L stainless steel samples showing 150 MPa increase in the corrosion-fatigue strength of the NC sample. **b** Fractograph of the MC 316L stainless steel sample showing the occurrence of intergranular brittle fracture. **c** Fractograph of the NC 316L stainless steel sample showing ductile fracture, at the crack initiation site (Scale bar = 30 μm.) (Reproduced from Bahl et al., "Enhancing the Mechanical and Biological Performance of a Metallic Biomaterial for Orthopedic Applications through Changes in the Surface Oxide Layer by Nanocrystalline Surface Modification." *Nanoscale*, 7, no. 17 (May 2015): 7704–7716 licensed under a Creative Commons Attribution 3.0 Unported License. © The Royal Society of Chemistry 2015)

intergranular attack [37]. For NC 316L stainless steel, the fractured surface within the pit appears to be rough, which is a clear indication of the occurrence of ductile mode of fatigue crack propagation (Fig. 9c). Since nanocrystalline grains have a lower capacity to store dislocations, unlike its microcrystalline counterpart the NC 316L stainless steel is less prone to intergranular corrosion [38]. Compressive residual stress induced during SMAT reduces the crack propagation in NC 316L stainless steel. As the ductile mode of fatigue crack propagation requires more energy than the brittle form, the crack propagation is retarded to a large extent in NC 316L stainless steel. Hence, the formation of a thicker oxide layer, a lesser amount of dislocation pile-up, compressive residual stress, requirement of higher energy for crack propagation, reduction in cathodic reactions (H_2 evolution) and trapping of hydrogen at the grain boundaries exert a synergistic effect in improving the corrosion fatigue strength of NC 316L stainless steel.

The adhesion and proliferation of MC3T3-E1 subclone 4 osteoblast cells on untreated 316L stainless steel (MC) and those subjected to SMAT using 5.5 mm Ø steel balls at 50 Hz for 15 min (NC) is studied by Bahl et al. [24]. The cell attachment and proliferation are measured using WST-1 assay at 1 day and 3 days. The extent of protein adsorption, cell attachment and proliferation largely depends on the degree of hydrophilicity of the surface. A hydrophilic surface is considered to be desired than a hydrophobic one for implant applications due to its ability to promote a better interaction of the implant with the biological fluids, cells and tissues [39, 40]. In general, the higher the surface roughness, the lower is the water contact angle, higher is the surface energy and higher is the degree of hydrophilicity. However, Bahl et al. [24] have suggested that factors other than R_a could also contribute to improvement in cell attachment and proliferation. According to them, the electronic properties of oxide film formed on NC 316 L stainless steel plays a critical role than surface roughness as the R_a of both MC and NC 316L stainless steels are highly comparable. The NC 316L stainless steel having a higher charge carrier density lowers the electron work function of the oxide film and the net positive charge developed on the oxide layer favors cell adhesion and proliferation (Fig. 10). The development of a novel "net-like" nanostructured surface on 316L stainless steel by SMAT using 5 mm Ø chrome steel balls for 2 min exhibits significant enhancement in the attachment, spreading and proliferation of human osteoblast cells (Saos-2) when compared to the coarse-grained untreated stainless steel [41]. The high density of grain boundaries in the "net-like" nanostructured surface is considered to be responsible for the enhanced protein adsorption and higher cell attachment.

3 Ti and Ti Alloys

Titanium and titanium alloys have received considerable attention due to their high strength to weight ratio, better mechanical properties, formability, machinability, low toxicity, low corrosion rate and good biocompatibility. These attributes make them suitable for orthopedic and dental implants, particularly for load-bearing implants

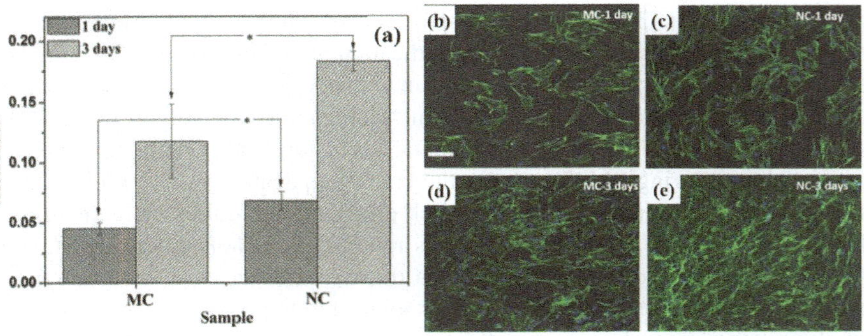

Fig. 10 **a** Absorbance values indicating the viability of MC3T3-E1 subclone 4 cells on NC and MC 316L stainless steel samples measured by the WST-1 assay. (*Indicates statistically significant differences between NC and MC ($p < 0.05$); (**b–e**) Fluorescence micrographs of osteoblasts at 1 day (**b** and **c**) and 3 days (**d** and **e**); (**b**, **d**) MC; and (**c**, **e**) NC, and at 3 days on (**d**) MC, (**e**) NC (Scale bar: 100 μm) (Reproduced from Bahl et al., "Enhancing the Mechanical and Biological Performance of a Metallic Biomaterial for Orthopedic Applications through Changes in the Surface Oxide Layer by Nanocrystalline Surface Modification." *Nanoscale*, 7, no. 17 (May 2015): 7704–7716 licensed under a Creative Commons Attribution 3.0 Unported License. © The Royal Society of Chemistry 2015)

[42]. Lack of bioactivity, poor tribological properties and susceptibility for galling and seizure are the major limitations of using Ti and its alloys as implants. Due to its spontaneous ability for passivation, Ti and its alloys tend to form a thin (~5 nm thick) oxide layer naturally. Though this passive oxide layer enables them a better corrosion resistance, it lacks bioactivity. This leads to the formation of fibrous tissues at the implant-bone interface, which affects osseointegration, ultimately leading to the loosening of the implant. The poor tribological property is a major concern in hip and knee joins, which often experience a combination of wear and corrosion, commonly referred to as tribocorrosion [43]. Removal of naturally formed passive oxide layer under conditions of tribocorrosion exposes the Ti and its alloys to the highly corrosive body fluid, leading to the release of metallic ions into the body fluid. In addition, accumulation of the wear debris generated during this process causes inflammation and induces allergic reactions in the surrounding tissues. The conditions of tribocorrosion would lead to aseptic loosening of the hip replacement implants. For Ti6Al4V alloy, leaching of Al and V from the alloy becomes a major concern as it could lead to Alzheimer's disease, neuropathy and osteomalacia.

The lack of bioactivity of Ti and its alloys can be solved by coating them with calcium phosphates such as hydroxyapatite, tricalcium phosphate, etc., In spite of their ability to improve the bioactivity of Ti and its alloys, poor adhesion of the coated layer becomes a major concern. Most of the material failures are surface related, particularly due to corrosion and wear. Increasing the thickness of the oxide layer is likely to improve the corrosion resistance and biocompatibility of Ti and its alloys. Poor tribological behavior of Ti and its alloys warrants improvement. An

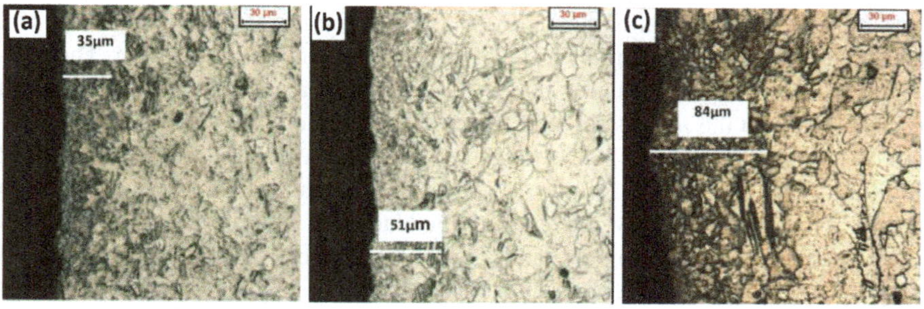

Fig. 11 Microstructure of CP-Ti after SMAT using 8 mm Ø alumina balls for different durations of treatment: **a** 15 min. **b** 30 min. **c** 45 min (Reprinted by permission from Springer Nature, Frontiers in Materials Science, 7, no. 3 [2013]: 285–294; Effect of surface mechanical attrition treatment of titanium using alumina balls: surface roughness, contact angle and apatite forming ability, Jamesh et al. [18])

increase in hardness could help improve the wear resistance. Improving osseointegration can be achieved by increasing the surface roughness, surface energy and by imparting a higher degree of hydrophilicity. Surface nanocrystallization by SMAT is likely to reduce the grain size, increase the surface roughness, decrease the water contact angle, increase the hardness, corrosion resistance and wear resistance. The influence of SMAT on the microstructure, grain size, microstrain, surface roughness, hardness, mechanical properties, compressive residual stress, fatigue life, tribological behavior, corrosion resistance, bioactivity, cell growth and proliferation of Ti and its alloys is addressed in the following paragraphs. In addition, the amenability for surface nanocrystallization of β-phase Ti alloy by SMAT and the utility of SMAT as a pretreatment for anodizing, chemical oxidation and thermal oxidation are also addressed.

The cross-sectional micrographs of CP-Ti after SMAT using 8 mm Ø alumina balls at 50 Hz for 15, 30 and 45 min are shown in Fig. 11. The formation of nanocrystalline grains is quite evident at the top surface of the treated Ti. The extent of deformation is also increased with an increase in time; 35, 51 and 84 μm for 15, 30 and 45 min treatment [18]. It is evident from the microstructures that SMAT enables the formation of a graded layer consisting of nanocrystalline grains at the top surface, submicron-sized grains at the subsurface and coarse grains with a further increase in depth [18, 44, 45].

The mechanism of surface nanostructuring of Ti by SMAT is elaborated by Zhu et al. [46]. The microstructure exhibits considerable variation as a function of deformation depth from the top surface. At 300 μm below the treated surface, a large number of parallel and intersected bands are observed. In addition, dislocation pile-ups could be identified at or near the boundaries. At 150 μm below the treated surface, the strain experienced is expected to be much higher than at 300 μm below the treated surface. At higher strain, the lamellae are subdivided into blocks and promote the formation of polygonal grains (200–400 nm). At a depth of 60–70 μm below the treated surface, fragmentation of grains and lamellae leads to the formation of equiaxed nanograins (100–300 nm). The top surface reveals uniform distribution of nano-sized grains (50–250 nm) with a random orientation. The variation in microstructure is mainly due to the level of strain experienced at different depths. Based on the microstructural features observed at various depths as a function of strain, Zhu et al. [46] have suggested that the evolution of microstructure of Ti during SMAT occurs as follows: (i) onset of twins and the intersection of twin systems; (ii) formation of low angle disoriented lamellae with a high density of dislocations; (iii) subdivision of microbands and dislocation cells into disoriented blocks and polygonal submicronic grains; and (iv) further breakdown of submicronic polygonal grains into randomly oriented nanograins. Due to the higher SFE (>300 mJ/m^2), twinning is identified as the preferred mode of deformation of Ti at low strain. At moderate strain, intersection of the twins is promoted, which hinders their movement. Hence, at moderate strain, a high density of dislocations facilitates the formation of lamellae. At higher strain, the lamellae are subdivided into dislocation cells or low angle disoriented blocks, which is subsequently converted to polygonal grains with a random orientation. Equiaxed nanograins are formed at the top surface. The final stage of grain refinement is governed by rotational recrystallization process.

The evolution of dislocations during SPD of Ti by SMAT is addressed by Wen et al. [47]. According to them, the dislocation density is low at a depth of 180 μm from the treated surface. An increase in dislocation density at a depth of 150 μm has lead to the formation of dislocation bands. Some of these dislocation bands change to equiaxed subgrains at a depth of 100 μm, while the shape of the subgrains becomes more uniform at a depth of 60 μm. Equiaxed nanograins (103 nm) with a random orientation predominates at a depth of 15 μm below the top surface. The microstructures reveal that at low strain, the formation of dislocation tangles and dislocation bands predominates. An increase in grain boundary density and triple junctions promotes nucleation of new subgrains. With an increase in strain and strain rate, the size of the subgrains becomes smaller and the subgrains transform into dislocation bands. Dynamic recrystallization plays a critical role in conversion of dislocation bands to equiaxed nanograins until formation of nanocrystalline Ti.

During SMAT, the multidirectional impingement of the flying balls on the surface of Ti and Ti alloys results in the generation of numerous dimples and craters on the surface, which leads to an increase in surface roughness. SMAT using 8 mm Ø stainless steel balls at 50 Hz for 60 min has increased the R_a of Ti6Al4V alloy from 0.19 ± 0.07 μm to 3.93 ± 0.26 μm [20]. Similarly, SMAT using 8 mm Ø alumina balls at 50 Hz for 45 min has increased the R_a of Ti from 0.15 ± 0.12 μm to 3.27 ± 0.98 μm [18]. The R_a of the treated Ti surface is increased with an increase in treatment time from 15 to 45 min. However, Anand Kumar et al. [45, 48] have observed a slight decrease in R_a with an increase in treatment time from 30 to 60 min for Ti6Al4V alloy subjected to SMAT using 5 mm Ø SAE 52100 steel balls at 50 Hz. This is due to the severe surface damage with the generation of numerous microcracks after treatment for 60 min, which could have contributed to material removal from the top surface.

The XRD pattern of untreated coarse-grained Ti exhibits the presence of sharp diffraction peaks pertaining to (100), (002), (101), (102), (110), (103) and (112) planes. However, after SMAT, the peaks are broadened due to refinement of grains by plastic deformation induced during SMAT and increase in atomic level microstrain. Lai et al. [49] have identified an additional peak ~44° 2θ for SMAT Ti, which is believed to be due to the formation of TiO due to increasing in penetration of oxygen during SPD. The average grain size of the treated Ti lies in the range of 25 to 43 nm while the microstrain is in the range of 0.14–0.16% [44, 45, 48, 50, 51].

The SPD induces compressive residual stress on SMAT Ti and Ti alloys. The compressive residual stress is much higher at their surface and it is decreased with an increase in the depth from the top surface. This is due to the variation in the level of strain and strain rate experienced at different depths of the Ti and Ti alloys during treatment. Anand Kumar et al. [45, 48] have reported an increase in the compressive residual stress of Ti6Al4V alloy from -679 ± 20 MPa to -706 ± 14 MPa with an increase in treatment time from 30 to 60 min.

SMAT of CP-Ti using 3 mm Ø stainless steel balls at 20 kHz for 30 min has lead to an increase in hardness. The maximum hardness of 3.8 GPa is found at a depth of 50 μm from the top surface. The hardness profile exhibits a steady decrease in hardness with an increase in depth from the surface to the bulk, supporting the formation of a graded layer structure in line with the inference made from the cross-sectional microstructural features. Accordingly, a hardness of 3.0 GPa and 2.8 GPa is found at a depth of 200 and 400 μm from the top surface. The increase in hardness is due to the grain refinement caused by plastic deformation induced during SMAT. The maximum hardness of 3.8 GPa at a thickness of 50 μm is due to the high strain rate of the order of 10^2–10^3/s near the top surface [52]. SMAT using 8 mm Ø stainless steel balls at 20 kHz for 60 min increased the hardness of CP-Ti to 6.9 GPa at a depth of 50 μm from the surface while it is decreased to 2.4 GPa at a depth of 500 μm from the surface [44]. Jelliti et al. [7] have reported a hardness of 4.4 GPa for Ti-6Al-4 V alloy subjected to SMAT using 3 mm Ø stainless steel balls at 20 kHz for 20 min. SMAT using 8 mm Ø alumina balls at 50 Hz has increased the hardness of Ti from 3.1 to 4.4 GPa with an increase in treatment time from 15 to 45 min [18]. Similarly, SMAT using 5 mm Ø SAE 52100 steel balls at 50 Hz has increased the hardness of Ti6Al4V alloy from 5.63 to 8.64 GPa with an increase in treatment time from 30 to 60 min [51]. Hence, it is clear that strain rate experienced by the treated surface, the extent of plastic deformation, increase in lattice strain, amenability of the Ti/Ti alloy for nanocrystallization and reduction in grain size determines the degree of increase in hardness.

The mechanical properties of Ti subjected to SMAT using 8 mm Ø stainless steel balls at 50 Hz for 60 min are compared with its untreated coarse-grained counterpart by Wen et al. [50]. According to them, after SMAT the YS and UTS are increased to 920 ± 40 MPa and 970 ± 30 MPa, respectively. Ti6Al4V alloy has become the material of choice instead of Ti due to its higher YS (902 MPa) and UTS (975 MPa). Since SMAT enables an increase in the YS and UTS of Ti that is highly comparable with the Ti6Al4V alloy, it would be possible to use Ti in place of Ti6Al4V alloy, particularly for biomedical applications. Such a choice would eliminate the problems associated with the leaching of aluminum and vanadium from the Ti6Al4V alloy into the body fluid. Zhao et al. [44] have also observed an improvement in YS (39.5%)

and UTS (29.1%) of Ti after SMAT. The elongation to failure of untreated coarse-grained Ti is 21.7% whereas it improves to 39.1% for SMAT Ti. The mechanical properties of coarse-grained untreated Ti6Al4V alloy and those subjected to SMAT using 5 mm Ø SAE 52100 steel balls at 50 Hz for 30 and 60 min are studied by Anand Kumar et al. [45]. According to them, only a slight increase in the YS (from 1063 to 1070 MPa) and UTS (from 1142 to 1173 MPa) of the Ti6Al4V alloy could be observed after 30 min treatment. In contrast, an increase in treatment time from 30 to 60 min facilitated a good increase in YS (from 1063 to 1143 MPa) and UTS (from 1142 to 1268 MPa) of the Ti6Al4V alloy due to the higher level of work hardening at the surface and near the surface. The formation of a nanocrystalline layer and high dislocation densities induced by strain hardening during SMAT is considered as the major factors responsible for the observed improvement in mechanical properties of Ti and Ti6Al4V alloy after SMAT.

Anand Kumar et al. [45] have evaluated the fatigue life of untreated Ti6Al4V alloy and those subjected to SMAT using 5 mm Ø SAE 52100 steel balls at 50 Hz for 30 and 60 min. When compared to the coarse-grained untreated Ti6Al4V alloy, the fatigue life of the alloy treated for 30 min is significantly improved. The fatigue life is influenced by surface roughness, compressive residual stress, nanostructured surface layer and work hardening. Higher surface roughness, which is commonly observed for samples treated by SMAT, is likely to exert a deleterious influence on fatigue life. In contrast, the compressive residual stress, formation of a nanostructured surface layer and work hardening would exert a beneficial influence on the fatigue life. Hence, the increase in fatigue life of Ti6Al4V alloy after SMAT for 30 min is due to the beneficial effect of compressive residual stress, work hardening and surface nanostructuring that could outweigh the negative effect of higher surface roughness. The inferior fatigue life of the Ti6Al4V alloy after SMAT for 60 min is due to the severe surface damage, resulting in the generation of numerous microcracks, which are much deeper. The low ductility of the Ti6Al4V alloy is considered to be responsible for the severe surface damage when the treatment time is increased from 30 to 60 min. The severe surface damage could have served as potential stress raisers and aided a quicker initiation of a fatigue crack.

The tribological behavior of coarse-grained untreated Ti and those subjected to SMAT using 8 mm Ø stainless steel balls at 50 Hz for 60 min under different loads and dry sliding conditions was studied by Zhao et al. [44]. When mated against a 6 mm Ø AISI 52100 steel ball (HRC: 60), SMAT Ti displayed a low coefficient of friction and wear volume. The improvement in wear resistance of SMAT Ti is due to the grain refinement and higher hardness induced by SMAT. The wear mechanism of SMAT Ti is governed by abrasive wear under 1 and 5 N, and a combination of both abrasive and adhesive wear under 2 N. In contrast, the wear mechanism of coarse-grained untreated Ti is governed by both abrasive and adhesive wear at 1 to 2 N while abrasive wear predominates at 5 N. Alikhani Chamgordanis et al. [15] have also studied the tribological behavior of Ti before and after SMAT. According to them, SMAT has decreased the coefficient of friction of Ti by 66%. The wear rate of Ti is also decreased by about 60%. The refinement in grain size from 70 μm to ~12 nm and a 2.8 times increase in hardness is considered responsible for the improvement in tribological behaviour of SMAT Ti. The fretting wear resistance of SMAT Ti is also found to be much higher than untreated Ti [51].

The effect of SMAT on the corrosion behaviour of CP-Ti, Ti6Al4V and β-phase Ti alloys in 3.5% NaCl and simulated body fluid (SBF) is studied by many researchers [7, 16, 53, 54]. The corrosion resistance of untreated and SMAT CP-Ti in 3.5% NaCl is evaluated by Fu et al. [16]. According to them, SPD induced during SMAT increases the diffusion pathway and promotes permeation of a larger amount of oxygen into the subsurface Ti. The higher oxygen content is expected to help in the formation of an intact passive oxide layer. The grain refinement and an increase in grain boundary density following the SPD induced by SMAT facilitate nucleation of the passive oxide layer. The higher thickness of the stable passive oxide layer formed on SMAT Ti is very effective in decreasing the permeation of the aggressive chloride ions, which would otherwise cause pitting corrosion [16].

The corrosion resistance of untreated and SMAT Ti6Al4V alloy in Ringer's solution (pH: 7.2, at 37 °C) is evaluated by Jelliti et al. [7]. The variation in open circuit potential (OCP) of untreated and SMAT Ti6Al4V alloy in Ringer's solution measured as a function of time indicates that both of them are capable of forming an oxide film, which is evident from the positive shift in OCP with time. However, when compared to the oxide layer formed on the untreated alloy, the one formed on SMAT Ti6Al4V appears to be effective. The anodic branch of the potentiodynamic polarization curve reveals passivation of both untreated and SMAT Ti6Al4V alloy with no significant difference in the breakdown potentials between them. The E_{corr} and i_{corr} of coarse-grained untreated Ti6Al4V alloy is −330 mV vs. SCE and 52 nA/cm^2. After SMAT using 2 mm and 3 mm Ø 100Cr6 balls for 15 and 20 min, respectively, the E_{corr} values exhibit a shift towards the noble direction to −275 mV vs. SCE and −235 mV vs. SCE, respectively while the i_{corr} values are decreased to 35 nA/cm^2 and 22 nA/cm^2, respectively. However, EIS study clearly indicates the difference in ability between them. The R_{ct} of the coarse-grained untreated Ti6Al4V alloy is 10.56 kΩ/cm^2. However, after SMAT using 2 mm and 3 mm Ø 100Cr6 balls for 15 and 20 min, the R_{ct} values increased to 763.41 and 1700 kΩ/cm^2, respectively. XPS analysis of untreated and SMAT Ti6Al4V alloy reveals enrichment of oxygen for the treated alloy, which supports easy amenability of the SMAT Ti6Al4V alloy to form a stable passive oxide layer. XPS analysis performed on the passive oxide layer formed on untreated and SMAT Ti6Al4V alloy after immersion in Ringer's solution for one weak indicate the presence of TiO_2 as the predominant species. Hence, it is clear that SMAT is beneficial in improving the corrosion resistance of Ti6Al4V alloy in Ringer's solution.

The corrosion behavior of solution treated coarse-grained pure β phase Ti-25Nb-3Mo-3Zr-2Sn alloy (TLM alloy), those subjected to SMAT using 2 mm Ø zirconia balls at 50 Hz for 30 min and SMATed TLM alloy subjected to post-annealing at 200 °C for 10 h under vacuum (7.5 × 10^{-4} Pa), in 0.9% NaCl and SBF, is evaluated by Huang et al. [53]. The E_{corr} and i_{corr} of the solution treated TLM alloy are −660 mV vs. SCE and 2.08 × 10^{-6} A/cm^2, respectively. SMAT has enabled a shift in E_{corr} towards noble direction from −660 mV vs. SCE to −410 mV vs. SCE and a one-fold decrease in i_{corr} from 2.08 × 10^{-6} A/cm^2 to 1.28 × 10^{-7}A/cm^2. Post annealing of the as-SMATed TLM alloy at 200 °C for 10 h has led to a slight increase in corrosion rate from 1.28×10^{-7} A/cm^2 to 2.13×10^{-7} A/cm^2. The passive current density (i_{pass}) indicates the tendency of the alloy to form an intact passive film. The i_{pass} of the solution treated TLM alloy is 1.86×10^{-4} A/cm^2. SMAT increased the tendency of the alloy to form a stable passive oxide layer with a higher thickness, which is reflected by a twofold decrease in i_{pass} from 1.86×10^{-4} A/cm^2 to 1.35×10^{-6} A/cm^2. However, post annealing of the as-SMATed TLM alloy at 200 °C for 10 h only slightly increased the i_{pass} from 1.35×10^{-6} A/cm^2 to 2.39×10^{-6} A/cm^2.

The improvement in corrosion resistance observed for the SMAT TLM alloy is due to the decrease in grain size, increase in grain boundary density, dilution of segregated alloying elements at grain boundaries and an increase in tendency of the treated alloy to form an intact passive oxide layer with a higher thickness. SPD induced by SMAT enables refinement of grains from microscale to nanoscale. The increase in grain boundary density following nanocrystallization helps to dilute the segregated alloying elements at the grain boundaries. The high density of dislocations induced during plastic deformation and the increase in grain boundary density promotes the formation of a stable and thicker passive oxide layer. Since the as-SMATed TLM alloy and those subjected to post-annealing are subsequently polished using SiC coated abrasive paper (2500 grade) followed by mirror polishing using 0.05 μm alumina powders, the deleterious influence of higher surface roughness of the SMAT TLM alloy on corrosion resistance is nullified.

The corrosion resistance of low modulus Ti-Nb-Ta-O alloy with two different microstructures before and after SMAT in SBF is evaluated by Acharya et al. [54]. One set of the Ti-Nb-Ta-O alloy sample possesses equiaxed β-grains (STQ) while the other set consists of uniform precipitation of α-platelets along the grain boundary as well as inside the β-grains in the β-matrix (aged). Both of them were subjected to SMAT using 500 numbers of 4.75 mm Ø hardened steel balls (STQ-SMAT and aged-SMAT). SMAT has led to a decrease in i_{corr} of STQ from $21.0 \pm 3.0 \times 10^{-3}$ μA/cm^2 to $7.4 \pm 2.7 \times 10^{-3}$ μA/cm^2 (for STQ-SMAT). Similarly, the i_{corr} of aged sample is decreased from $152.0 \pm 9.0 \times 10^{-3}$ μA/cm^2 to $69.9 \pm 0.3 \times 10^{-3}$ μA/cm^2 (for aged-SMAT). The critical current for passivation (i_{pass}) also follows trend similar to that of the i_{corr}. However, i_{pass} appears to be more prominent than i_{corr}. The lower i_{pass} of SMATed samples points out an increase in tendency of the treated alloy to form a stable passive oxide layer with higher thickness. A stable higher phase angle value over a wide range of frequencies in the Bode plot further substantiates this phenomenon. The relatively higher i_{corr} and i_{pass} values of the untreated and SMATed aged samples, when compared to the respective STQ samples, are due to the difference in their microstructure. The presence of both α and β phases in the aged samples could promote the formation of galvanic cells and non-uniform oxide layer. Mott-Schottky analysis indicates a large decrease in the charge carrier density for SMATed STQ and aged samples when compared to its untreated counterparts. Since a lower charge carrier current density in Mott-Schottky plots signifies a higher stability of the passive oxide layer, it is clear that SMAT is beneficial in improving the corrosion resistance of the treated alloy.

The in vitro bioactivity of untreated and SMAT Ti6Al4V alloy is studied by immersing them in SBF with its ionic concentration equal to human blood plasma (pH: 7.40) at 37 °C for 7, 14 and 21 days [20]. After 14 and 21 days of immersion, both untreated and SMAT Ti6Al4V alloy exhibit the formation of many particles with similar morphological features distributed over their entire surface. Based on the Ca and P contents estimated by EDS analysis and the observed steady increase in the molar ratio of Ca/P from 7 to 21 days of immersion, the newly formed particles could be inferred as calcium apatites. The in vitro bioactivity of coarse-grained untreated Ti and those subjected to SMAT using 8 mm Ø alumina balls at 50 Hz for 15, 30 and 45 min is evaluated by Jamesh et al. [18]. Untreated Ti fails to exhibit bioactivity even after 28 days of immersion in SBF at 37 °C. In contrast, Ti samples subjected to SMAT for 30 and 45 min exhibits apatite growth at selective areas (Fig. 12). The apatite forming ability of the treated surface is found to increase with an increase in immersion time in SBF. During immersion in SBF, the negative charge imparted to the SMATed Ti combines with the positively charged Ca ions, leading to the formation of calcium titanate. Accumulation of calcium ions imparts a positive

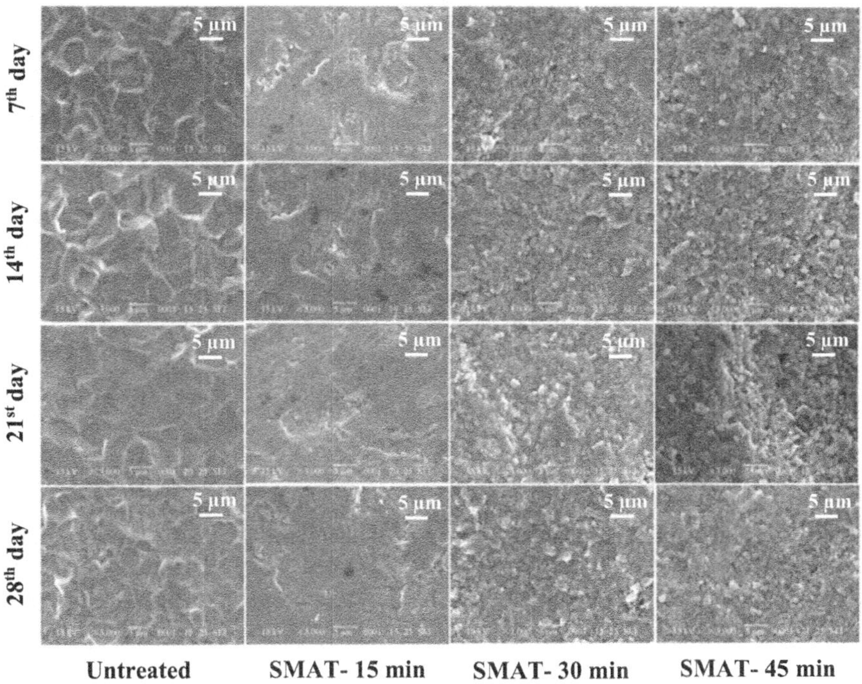

Fig. 12 Extent of apatite growth on the surface of untreated CP-Ti and those subjected to SMAT for 15, 30 and 45 min using 8 mm Ø alumina balls after immersion in SBF for 7, 14, 21 and 28 days (Reprinted by permission from Springer Nature, Frontiers in Materials Science, 7, no. 3 [2013]: 285–294, Effect of surface mechanical attrition treatment of titanium using alumina balls: surface roughness, contact angle and apatite forming ability, Jamesh et al. [18])

charge on the modified surface which attracts the negatively charged phosphate ions to form amorphous calcium phosphate, which in turn promotes nucleation of calcium apatites. After SMAT using 8 mm Ø alumina balls for 45 min, the R_a of Ti is increased from 0.15 ± 0.12 μm to 3.27 ± 0.98 μm while the water contact angle is decreased from 64° to 43° with an increase in surface energy from 32 to 53 mJ/m^2. In spite of a lower water contact angle and higher surface energy, the extent of apatite formation is decreased for Ti sample subjected to SMAT for 45 min than those treated for 30 min. This is due to the fragmentation of the alumina balls and subsequent attachment of the fine particles. The alumina particles incorporated on the surface of SMAT Ti inhibit the apatite growth, which suggests that the right choice for the type of balls for SMAT is very important.

Lai et al. [49] have compared the viability of mesenchymal stem cells (MSCs) on untreated and SMAT Ti in terms of the extent of cell adhesion and spreading, ALP activity, mRNA expressions of osteocalcin (OC), osteopontin (OPN), collagen I (Col I) and bone mineralization. AFM analysis performed over a region of 2 μm × 2 μm indicates that the average root means square (rms) roughness of untreated Ti is 7.29 ± 1.17 nm is increased to 29.53 ± 6.97 nm after SMAT. Following this, the water contact angle is decreased from 66.5 ± 4.1° to 51.9 ± 5.6° after SMAT. MSCs cultured onto untreated Ti exhibits loose cell pseudopodia. However, a well spread and intact pseudopodia are identified on SMAT Ti, suggesting that the surface nanocrystallization of Ti by SMAT improves the adhesion and proliferation of MSCs. When compared to its untreated counterpart, SMAT Ti shows higher cell viability, higher ALP activity, expressed much higher levels of OC and OPN and higher mineralization. The viability of osteoblast cells on Ti and Ti6Al4V alloy before and after SMAT is studied by Zhao et al. [20, 44]. According to them, the formation of nanostructured grains, higher surface roughness, higher surface hydrophilicity and higher surface energy of the SMAT Ti and Ti6Al4V alloy are responsible for the improvement in adhesion and spreading of osteoblast cells when compared to its coarse-grained untreated counterpart. The morphology of the osteoblast cells attached on the surface of untreated Ti and Ti6Al4V alloy is spherical in shape with little cell spreading whereas those attached on the surface of SMAT Ti and Ti6Al4V alloy exhibit a flattened shape with extensive spreading.

Huang et al. [21, 55, 56] have studied the ability of solution treated β-type Ti-25Nb-3Mo-2Sn-3Zr titanium (TLM) alloy before and after SMAT using 2 mm Ø zirconia balls at 50 Hz for 60 min to promote adhesion and proliferation of human fetal osteoblastic (hFOB1.19) cells. During SMAT, the multidirectional impingement of the zirconia balls on the surface of the TLM alloy has increased the surface roughness. The water contact angle of the coarse-grained untreated TLM alloy is 67.8 ± 2.4° whereas it is decreased to 50.4 ± 1.5° after SMAT. The decrease in water contact angle indicates better hydrophilicity of the SMAT TLM alloy. The higher degree of hydrophilicity and higher surface energy of the SMAT TLM alloy promotes adsorption of proteins such as fibronectin (Fn) and vitronectin (Vn). An increase in protein adsorption is considered to exert a beneficial influence on cell adhesion, which in turn determines cell proliferation and spreading. After 24 h of cell culture, the hFOB1.19 cells attached to untreated TLM alloy are spherical in shape whereas those attached on SMAT TLM alloy are polygonal in shape, which suggests the ability of the treated alloy to enhance spreading of the hFOB1.19 cells. In addition, the number of hFOB1.19 cells attached on the surface of SMAT TLM alloy is much higher than those attached to the untreated alloy. Cell adhesion and cell proliferation involves four major steps, namely, protein adsorption, cell-material interaction, cell attachment and spreading. Implants tend to absorb proteins and the adsorbed protein layer improves implant-cell interactions, which in turn promotes cell attachment and spreading. Since SMAT TLM alloy promotes adsorption of Fn and Vn, it is promising that it is likely to improve interaction with the cells and facilitate its attachment. The enhancement of adhesion and spreading of hFOB1.19 osteoblast cells on SMAT TLM alloy is due to the decrease in grain size, increase in grain boundary density, higher degree of hydrophilicity and its ability to promote adsorption of proteins.

Azadmanjiri et al. [57] have studied the viability of human adipose-derived stem cells after 1 day and 4 days of cell culture on the surface of untreated Ti and those subjected to SMAT using 3 mm Ø zirconia balls at 20 kHz. The cell density and total coverage of the cells are much higher on SMAT Ti than the untreated Ti (Fig. 13). The increase in R_a and nanocrystallization of the surface could have exerted a beneficial influence on cell adhesion. The viability of MC3T3-E1 subclone 4 mouse calvarial pre-osteoblasts cells on low modulus Ti-Nb-Ta-O alloy with two different microstructures before and after SMAT is studied by Acharya et al. [54]. According to them, the attachment and proliferation of MC3T3-E1 cells are not influenced by SMAT of the Ti alloy. Impingement of 500 numbers of 4.75 mm Ø hardened steel balls on the surface of the Ti alloy has increased the R_a from 74 to 95 nm with no significant change in water contact angle. Hence, it is evident that a higher degree of hydrophilicity and higher surface energy is warranted to promote cell adhesion and spreading.

CP-Ti and Ti6Al4V alloy are highly amenable for nanocrystallization by SMAT unlike the β-phase Ti alloys wherein their microstructure seem to exert a strong influence. The nanocrystallization of Ti-25Nb-3Mo-3Zr-2Sn (TLM) alloy consisting of β phase (solution treated) and duplex $\alpha + \beta$ phases (aged) by SMAT using 2 mm Ø zirconia balls at 50 Hz for 5 and 30 min, was evaluated by Huang et al. [56]. SMAT has enabled grain refinement resulting in the formation of a nanocrystalline structure only for the aged TLM alloys consisting of both α and β phases. After SMAT for 30 min, the average grain size is reduced to 36 ± 3, 30 ± 7 and 26 ± 4 nm for TLM alloys aged at 450, 550 and 650 °C for 6 h, respectively. On the contrary, nanostructuring of the surface could not be observed for the solution treated TLM samples consisting of only the β phase. The much smaller unrecoverable strain is considered to hinder grain refinement. Instead, a high density of dislocations and deformation twins are observed at their surface.

The difference in nanocrystallization between the solution treated and aged TLM alloys stems from the difference in their microstructure. The microstructure of TLM alloy aged at 450 °C for 6 h reveals the presence of uniformly distributed nano-sized α cobbles, which offers it higher strength. The microstructure of TLM alloys aged at 550 and 650 °C for 6 h, reveals the presence of α needles with non-uniform

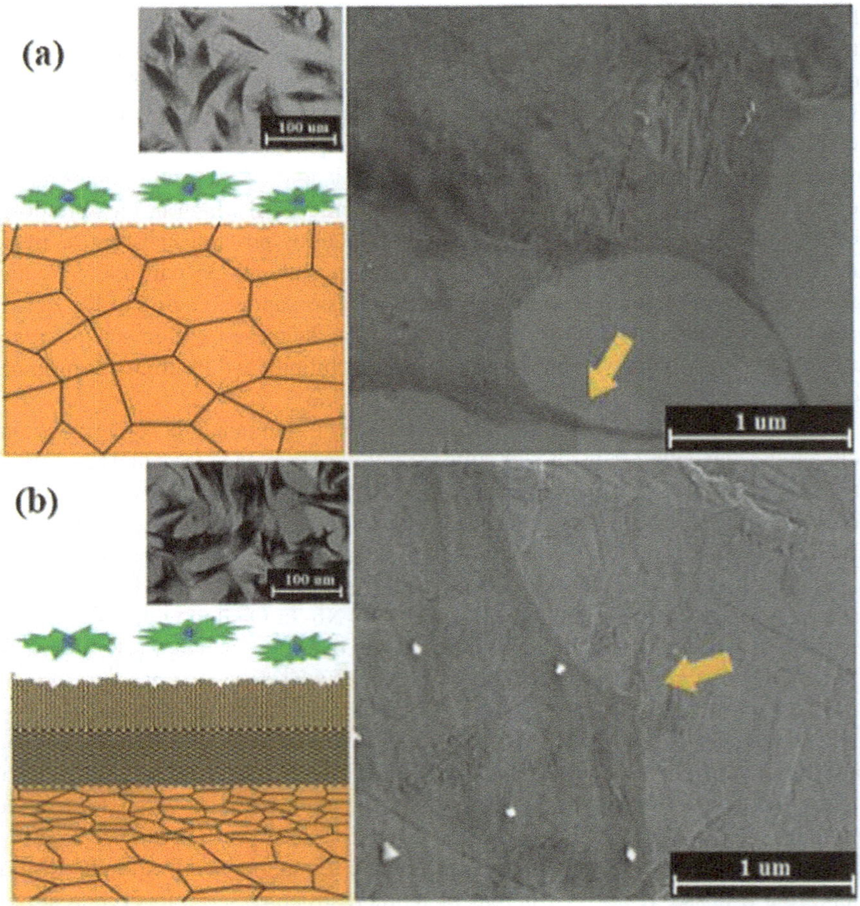

Fig. 13 Schematic and SEM images of primary human adipose-derived stem cells on the surface of **a** untreated. **b** SMATed Ti after 4 days of cell culture. Arrows indicate the interaction between filopodia and the surface structure (Reproduced from Azadmanjiri et al., "Enhanced attachment of human mesenchymal stem cells on nanograined titania surfaces." RSC Advances, 6, no. 61 [2016]: 55825–55833 by permission of The Royal Society of Chemistry)

distribution. The size of the α needles is relatively larger than the α cobbles and it increased with the aging temperature from 550 to 650 °C. The larger size of the α needles and their non-uniform distribution decreased the strength of the TLM alloy aged 550 and 650 °C for 6 h. In addition, the volume fraction of the α phase is increased with an increase in temperature employed for aging; 27, 31 and 35% for those aged at 450, 550 and 650 °C for 6 h, respectively. TLM alloy aged 650 °C for 6 h with a non-uniform distribution of larger α needles with 35% α phase is highly amenable for nanocrystallization (grain size: 26 ± 4 nm). Hence, even if the time for SMAT is limited to 5 min, grain refinement could be recognized for the TLM alloy aged at 650 °C for 6 h. However, for TLM alloy aged at 450 °C for 6 h, consisting of 27% nano-sized α cobbles with higher strength, the extent of nanocrystallization is relatively less. From these observations, one can draw out that the morphology and volume fraction of the α phase as well as the strength of the alloy based on the distribution of α cobbles/needles determine the nanocrystallization of the aged TLM alloy by SMAT.

Since surface nanocrystallization of Ti and its alloys by SMAT is beneficial in improving the hardness, mechanical properties, corrosion resistance, bioactivity and cell growth, many duplexes or hybrid treatment approaches involving SMAT as a pretreatment are explored. Zhang and Han [58] have used SMAT pretreatment for anodizing of Ti. In their study, Ti is subjected to SMAT using 5 mm Ø zirconia balls at 50 Hz for 60 min. Both untreated and SMAT Ti are anodized using glycol containing 0.075 M NH_4F and 0.02 vol. % H_2O at 60 V for 15–240 min at room temperature. Subsequently, the anodized Ti samples are annealed in Ar atmosphere at 450 °C for 3 h. Surface nanocrystallization of Ti by SMAT fails to exhibit any significant change in the morphology of the TiO_2 nanotubes as well as its phase composition, both of them being amorphous TiO_2. The thickness of the anodized layer exhibits an obvious increase with an increase in anodizing time from 15 to 240 min. Nevertheless, irrespective of the treatment time, the anodized layer formed on SMAT Ti is relatively thicker, which is further substantiated by a higher current density recorded in the current-time transients. The formation of anodic oxide film on Ti is governed by the electrochemical reaction and dictated by the field-aided ionic transport of O^{2-} and Ti^{4+}. The grain refinement, increase in grain boundary density and high dislocation density induced by SPD during SMAT provide additional pathways for the rapid diffusion of O^{2-} and promote the reaction between O^{2-} and Ti^{4+}. Hence, it is clear that surface nanocrystallization of Ti by SMAT is beneficial for the growth of TiO_2 nanotubes.

The use of SMAT as a pretreatment for anodizing of Ti is also reported by Azadmanjiri et al. [57]. In their study, Ti is subjected to SMAT using 3 mm Ø zirconia balls at 20 kHz for 20–120 min. Both the untreated and SMAT Ti are anodized using 1 M $(NH_4)_2SO_4$ containing 0.5 wt% NH_4F at 20 V for 2 h at room temperature. SMAT increased the reactivity of Ti during anodizing, which results in the formation of TiO_2 nanotubes with a higher length (18 μm) and diameter (70 nm) when compared to those formed on its coarse-grained untreated counterpart with a length and diameter of 10 μm and 40 nm, respectively (Fig. 14). The high density of grain boundaries and dislocations induced by SMAT increased the reaction rate and ion diffusion coefficient during anodization. Hence, it is evident that SMAT can be used as an effective pretreatment for the development of TiO_2 nanotubes with a suitable length and diameter for drug delivery applications.

SMAT is used as a pretreatment for Ti to assist the formation of a crystalline nanoporous titania by chemical oxidation and calcination [59]. In this study, Ti is subjected to SMAT using 8 mm Ø stainless steel balls at 50 Hz for 60 min. The untreated and SMAT Ti are immersed in 30 wt% H_2O_2 at 25 °C for 24 h, dried at 40 °C for 6 h and calcined in an air atmosphere at 600 °C for 1 h. The grain refinement, increase in grain boundary density and high dislocation density

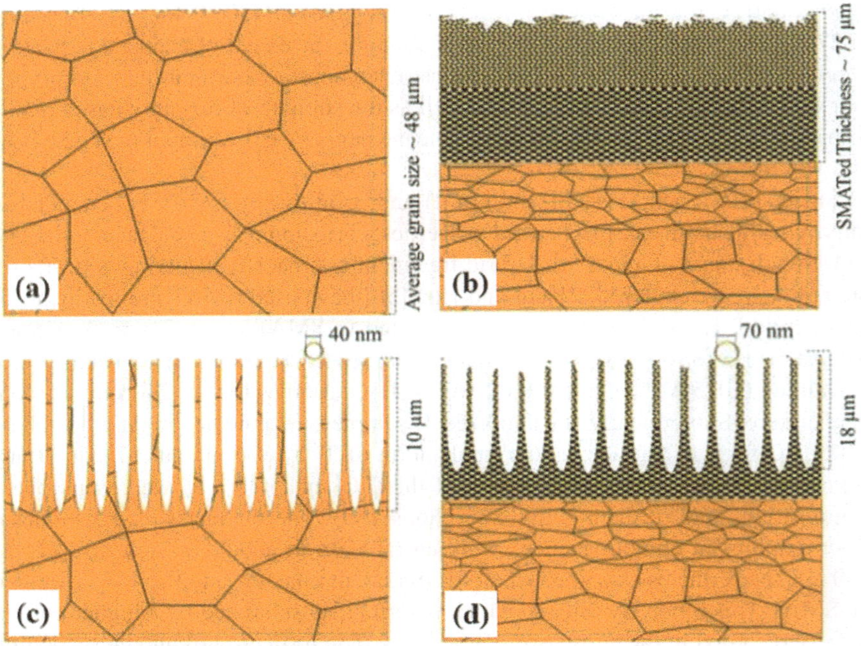

Fig. 14 Schematic representation of the **a** as-received. **b** SMATed. **c** anodized. **d** anodized-SMATed Ti samples (Reproduced from Azadmanjiri et al., "Enhanced attachment of human mesenchymal stem cells on nanograined titania surfaces." RSC Advances, 6, no. 61 [2016]: 55825–55833 by permission of The Royal Society of Chemistry)

induced by SPD during SMAT increased the reactivity between the SMAT Ti and H_2O_2, thus promoting the formation of nanoporous titania with weak crystallinity while subsequent calcination increased the crystallinity. Under similar experimental conditions, untreated Ti fails to show the formation of nanoporous titania. In another similar study, Wen et al. [60] have used SMAT as a pretreatment for the development of crack-free mesoporous anatase film by chemical oxidation and calcination. The mesoporous anatase displays excellent bioactivity in SBF, which is an important attribute of materials used for biomedical applications. Laleh and Kargar [61] have also used a similar analogy adopted by Wen et al. [59, 60] to form a nanoporous crack-free anatase film on SMAT Ti, with excellent bioactivity. SMAT is also used as a pretreatment for thermal oxidation of Ti [62]. The coarse-grained untreated Ti (CO Ti) and SMAT Ti (SO Ti) are subjected to thermal oxidation at 700 °C for 1 h. In spite of a similar phase content of rutile, SO Ti exhibits a higher O concentration than CO Ti at a similar depth. In addition, SO Ti shows a higher diffusion depth of C and N than CO Ti. The grain refinement, increase in grain boundary density, high-density dislocations induced by SPD during SMAT is considered to be responsible for the enhanced diffusion of O, C and N in SO Ti. The higher concentration of O, C and N in the near-surface layer increased the hardness of SO Ti, which accounts for the lower mean coefficient of friction when compared to that of CO Ti. The oxide layer with rutile phase on SO Ti improves biocompatibility. The higher surface energy, presence of a crystalline rutile layer promotes cell adhesion and spreading on SO Ti. SMAT is also used as a pretreatment for diffusion coatings such as boriding and nitriding [30, 63–65] as well as for phosphate conversion coatings [66, 67].

4 NiTi Alloy

Nickel-titanium shape memory alloys (NiTi SMAs) have received considerable attention due to their superior shape memory properties, superelasticity, high damping capacity, high fatigue resistance, good biocompatibility and corrosion resistance. NiTi alloys are widely used for biomedical applications, particularly for bone fraction fixation plates, stents, orthodontic wires, spine correction rods and staples. In spite of its good biocompatibility, the major concern with the use of NiTi alloys for biomedical applications is leaching of nickel ions into the body fluid, which could cause inflammation of the surrounding tissues and induce allergic reactions [68]. Since NiTi based biomedical devices are likely to experience a high cycle loading and stress, it is expected that they should possess excellent strength, fatigue resistance, corrosion resistance and wear resistance. Surface modification is a viable option to impart the desired characteristics of the NiTi alloys. A variety of surface treatment methods are explored to improve the performance of NiTi alloys. These include simple surface treatments such as polishing, electropolishing, surface passivation by chemical and thermal oxidation, nitriding and deposition of titanium oxide, titanium nitride, calcium phosphate, polymeric and bioactive coatings [69]. Modification of the surface of the Ni-Ti alloy tends to improve the corrosion and wear resistance, which would decrease the extent of leaching of the toxic nickel ions from the alloy into the human body environment.

SMAT is a S^2PD method that facilitates surface nanocrystallization without changing the chemical composition of the material. The multidirectional impingement of the balls at a very high strain rate enables refinement of the coarse grains to nano-sized grains at the surface along with the formation of a graded layer structure. The compressive residual stress induced during SMAT and the high density of dislocations helped to improve their hardness, mechanical properties and tribological behavior. The increase in grain boundary density also promotes the formation of a passive oxide layer which improves the corrosion resistance. The influence of SMAT on the microstructure, grain size, surface roughness, hardness, tribological behavior and corrosion resistance of NiTi alloy is addressed in the following paragraphs.

The effect of SMAT on the microstructure, phase content, hardness, wear resistance and corrosion resistance of NiTi alloys is extensively studied by Hu et al. [11, 70–73]. The grain size of the coarse-grained NiTi alloy ranges from 40 to 80 μm. After SMAT using 2 mm Ø stainless steel balls for 60 min, the NiTi alloy is severely deformed with a deformation thickness of ~200 μm [11].

X-ray diffraction pattern of untreated coarse-grained NiTi alloy exhibits the presence of only the austenite B2 phase. SMAT of the NiTi alloy using 2 mm Ø stainless steel balls for 60 min has resulted in a decrease in intensity of the B2 (110) and (211) planes along with broadening of the diffraction peaks, suggesting refinement of grains following SPD during SMAT [70]. Broadening of the diffraction peaks is also observed for NiTi alloy subjected to SP using 250 μm Ø spherical alumina balls at a peening angle of 90° under an air pressure of 45 psi for 20 min. Besides the austenite B2 phase, B19' martensite phase is also observed for the shot-peened samples. However, the B19' martensite phase is completely transformed to the austenite B2 phase after heat-treatment at 300 °C in an argon atmosphere for 1 h [74]. Hence, it can be inferred that the type of phases formed after SMAT of NiTi alloy is largely a function of the conditions, which determines the extent of plastic deformation and phase transformation from austenite to martensite and reverse transformation.

The evolution of microstructure at various depths from the top surface to the maximum depth of the deformed region of the SMAT NiTi alloy is addressed by Hu et al. [71, 72]. The variation in microstructure at different depths and the type of phases formed is determined by the level of strain and strain rate. At a depth of 180–220 μm from the top surface wherein the strain level is much lower, high-density dislocation appears to be the main mechanism of strain accommodation. At a depth of 150–200 μm, the extent of strain and strain rate experienced by the alloy will be much higher. Accordingly, the dislocation promotes a sequential formation of dislocation lines (DLs), dense dislocation walls (DDWs), sub-boundaries, and refinement of coarse-grained structure into a nanocrystalline structure. At a depth of 150 μm from the surface, stress-induced martensite (SIM) transformation is found to be the main strain accommodation mechanism. Hence, only the B19' martensite prevails at this depth. At a depth of 50–100 μm, a further increase in the extent of plastic deformation would facilitate the formation of DLs and DDWs and sub-boundaries that stabilizes the martensite plates, which in turn impedes the reverse transformation of the martensite (B19') to austenite (B2) phase. Hence, both martensite (B19') and austenite (B2) could prevail at this depth. At a depth of about 50 μm, the microstructure indicates formation of narrow lamellae, which is subdivided into ultrafine grains. The top surface of the NiTi alloy could experience a very high level of strain and strain rate of the order of 10^3–10^4/s. Hence, the high density of dislocations refines the grain size, promoting the formation of nanocrystalline grains. However, the phase content of the nanocrystalline region is only B2 austenite. It appears that with a very high strain rate, a reverse transformation of martensite (B19') to austenite (B2) has occurred. The SPD at the top surface and temperature-induced during deformation could have enabled such a phase transformation. Since the average grain size of the nanocrystalline region is ~20 nm, the B2 austenite phase becomes thermally stable. The increase in grain boundary density, higher dislocation densities and the energy stored at grain boundaries facilitates crystalline-to-amorphous transformation. The evolution of microstructure and phase content at various depths of the SMAT NiTi alloy is schematically represented in Fig. 15.

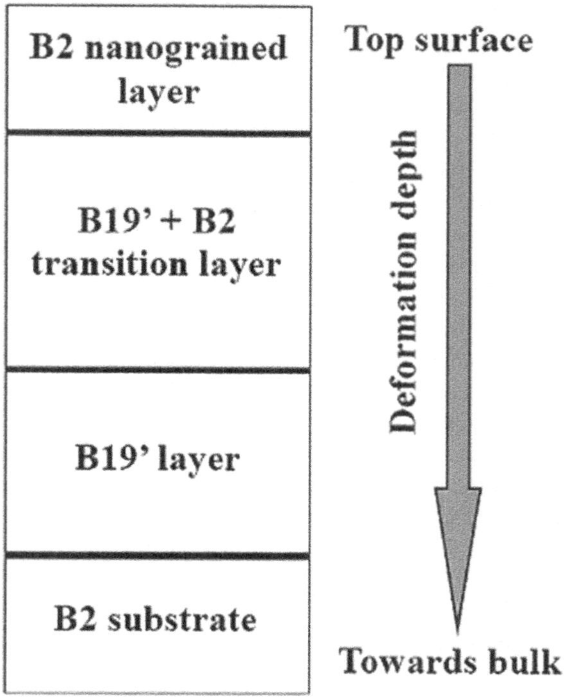

Fig. 15 Schematic representation of the formation of a graded layer structure along with the evolution of various phases during SMAT of NiTi alloy (Drawn based on the data given in Table 1 (Reprinted from Intermetallics, Vol. 19/issue number 8 (August), Hu et al., Microstructural evolution in NiTi alloy subjected to surface mechanical attrition treatment and mechanism, 1136–1145, Copyright [2011], with permission from Elsevier)

The hardness of the untreated coarse-grained (40 μm) NiTi alloy counterpart is only ~3.0 GPa. After SMAT, the hardness at the top surface of the NiTi alloy is increased to ~6.0 GPa. The hardness profile of the NiTi alloy exhibits a steady decrease in hardness up to a depth of 200 μm from the top surface before reaching near saturation. The higher hardness of the treated alloy measured up to a depth of 200 μm matches well with the depth of deformation obtained from the microstructure of the treated NiTi alloy. Grain refinement, work hardening and compressive residual stress induced during SMAT have accounted for the increment in hardness of the treated NiTi alloy. Post-annealing treatment of the SMAT NiTi alloy at 300 °C for 1 h has caused a slight decrease in hardness at the top surface. Nevertheless, there is no pronounced change in the hardness profile, revealing that the contribution from the compressive residual stress for hardening is relatively less [73]. Hardness is a measure of the local resistance to plastic deformation under an applied load. The increase in hardness of the NiTi alloy subjected to SMAT could be correlated to grain refinement. The plastic deformation of NiTi alloy during SMAT enables grain refinement and formation of a nanocrystalline structure. The increase in grain boundary density prevents the propagation of dislocation movement suggesting that the grain refinement contributes to the increase in hardness of the SMAT NiTi alloy in accordance with the Hall–Petch relationship.

The wear resistance of untreated and SMAT NiTi alloy is evaluated by Hu et al. [11] under dry sliding conditions. At a normal applied load of 7 N, untreated coarse-grained NiTi alloy exhibits a rapid increase in friction coefficient within a few seconds, which eventually reached a steady state. In contrast, the transient time for the initial sharp increase in friction coefficient is ~2 min for the treated NiTi alloy. The difference is due to the higher R_a of the SMAT NiTi alloy when compared to the untreated alloy. When the applied load is <9 N, SMAT NiTi exhibits a lower steady-state friction coefficient when compared to its coarse-grained counterpart. However, when the load is higher than >9 N, the difference in friction coefficient between the untreated and SMAT NiTi alloy becomes narrow. The higher hardness of the SMAT NiTi alloy achieved by grain refinement following SPD and a decrease in the contact area has enabled a lower friction coefficient for the treated alloy. Irrespective of the applied normal load in the range of 5–15 N, both the wear depth and wear volume of SMAT NiTi are lower than that of the coarse-grained untreated alloy.

The mechanism of wear appears to be similar for both untreated and SMAT NiTi alloy, which is governed by delamination wear. However, for any given load, the extent of delamination is much more severe on the untreated alloy. The higher hardness of SMAT NiTi alloy has enabled a decrease in the degree of delamination. For any given load, more debris is found to be present on the worn surface of the SMAT NiTi alloy than its untreated counterpart. The higher hardness of the SMAT NiTi alloy makes the fragments brittle. When these fragments get trapped between the mating surfaces (SMAT NiTi alloy and WC ball), their brittle nature makes them break further apart into fine debris. Being highly reactive, these particles could be easily oxidized. The wear debris could provide either a lubricative effect, reducing the friction coefficient or they might play a decisive role if the particles become abrasive. The lower difference in friction coefficient between the untreated and SMAT NiTi alloy from 11 to 15 N is due to the generation of a larger amount of debris from the treated alloy.

The corrosion behavior of SMAT NiTi alloy in 0.9% NaCl is evaluated by Hu et al. [70]. Untreated coarse-grained NiTi alloy exhibits a typical active-passive transition in the anodic branch of the polarization curve. In contrast, this active-passive transition is absent for SMAT NiTi alloy. The grain boundaries and triple junctions formed during SMAT of NiTi alloy are highly active and during corrosion, they tend to dissolve much faster and promote nucleation of a passive oxide layer. The fluctuations observed in the OCP-time curve of SMAT NiTi alloy is due to the dissolution-formation of passive film. When compared to the untreated NiTi alloy, the E_{corr} exhibits a positive shift from +58 mV vs. SCE to +206 mV vs. SCE and a decrease in i_{corr} from 4×10^{-8} A/cm^2 to 4×10^{-9} A/m^2 for the SMAT NiTi alloy. The $|Z|$ at 0.1 Hz derived from the Bode impedance plots is much higher for SMAT NiTi alloy, implying the formation of a thicker passive oxide layer.

NiTi alloy subjected to SMAT offered a better corrosion resistance in 0.9% NaCl when compared to its untreated coarse-grained counterpart. The microstructure, surface roughness, tendency to form a passive oxide layer and composition of the passive oxide layer determine the corrosion protection ability of materials subjected to SMAT. The R_a of NiTi alloy is increased after SMAT. In general, surfaces with a higher R_a are likely to cause a deleterious influence on the corrosion resistance. The microstructure of the SMAT NiTi indicates the formation of a nanocrystalline surface with a partial amorphous structure. During SMAT, the SPD promotes permeation of oxygen atoms, which is supported by the oxygen concentration of 26.7 at. % at the top surface of the treated alloy. The increase in temperature facilitates the formation of a passive oxide layer, mainly consisting of titanium oxide. Mott-Schottky analysis reveals that the passive film formed on nickel is mainly a *p*-type semiconductor. In contrast, the passive film formed on titanium and NiTi alloy exhibit the characteristics of an *n*-type semiconductor. Based on the positive slope values observed in Mott–Schottky plots, Jinlong et al. [75] have identified TiO_2 as the major constituent of the passive film formed on NiTi alloy in 0.5 M H_2SO_4, 0.2 M NaCl and borate buffer solution. Hence, it is clear that the nanocrystalline and amorphous surface structure of the alloy and formation of a passive layer, which is rich in TiO_2 is responsible for the observed improvement in corrosion resistance of SMAT NiTi alloy in 0.9% NaCl. The enrichment of titanium oxide in the passive oxide layer and the higher corrosion resistance of the SMAT NiTi alloy are likely to decrease the extent of leaching of the toxic nickel ions and contribute to the improvement in the biocompatibility of the alloy.

5 CoCrMo Alloy

CoCrMo alloys, both in the as-cast as well as wrought form, have been used as orthopedic implant materials. They possess high strength, good mechanical properties, excellent wear resistance, good corrosion resistance, better biocompatibility, high fatigue resistance and low elastic modulus. Due to these key attributes, CoCrMo alloys are considered as the material of choice for the fabrication of metal-on-metal total hip replacements [76]. CoCrMo alloy tends to form a spontaneous oxide film (~4 nm thick) on its surface, primarily consisting of Cr_2O_3 oxide along with minor quantities of Co and Mo oxides, which is responsible for its excellent corrosion resistance and better biocompatibility. CoCrMo alloy based hip joints experiences a combination of both wear and corrosion, commonly referred to as tribocorrosion [43, 76]. Microabrasion-corrosion has been identified as a major mode of failure of CoCrMo alloy [77, 78].

Fortunately, the metal-on-metal (MOM) tribological contacts are more wear-resistant when compared to metal-on-polyethylene (MOP) tribological contacts [79]. However, in MOM tribological contacts, the wear process breaks the naturally formed thin oxide film on the CoCrMo alloy, resulting in the generation of metal debris, the accumulation of which could cause inflammation in the surrounding tissues. It has been reported that under conditions of infection, the concentration of protein in synovial fluid is likely to increase to a large extent. The local temperature is also expected to increase up to 50 °C. These conditions could promote localized breakdown of the passive oxide layer [80]. Since the damaged area is much smaller than the remaining areas, the formation of galvanic cells is promoted, leading to the preferential dissolution of metal at the damaged area [80]. The corrosion of freshly exposed alloy surface (after removal of the thin oxide film) leads to leaching of highly toxic Co and Cr ions into the body fluid. Hence, the generation of metal debris due to wear, the inflammation of surrounding tissues and leaching of highly toxic Co and Cr ions into the body fluid becomes a major concern in using CoCrMo alloy for biomedical applications [76]. Aseptic loosening of the hip implant is a major consequence of tribocorrosion.

Improving the performance of CoCrMo alloy based hip implants under tribocorrosion conditions can be approached by several means. Thickening of the oxide layer either by acid passivation or oxygen ion implantation is one of the options. Formation of a surface coating is another option. However, adhesion of the coated layer, tensile residual stress and delamination of the coating are the major limitations of many coatings. Deposition of a thin (3 μm thick), uniform and dense TiN coating with a columnar structure by physical vapor deposition (PVD) at 550 °C for 6 h is effective in preventing the release of Co and Cr ions from CoCrMo alloy [81]. Nevertheless, unlike the as-polished CoCrMo alloy, the PVD TiN coated alloy fails to exhibit bioactivity. PVD coatings, in general, possess a higher hardness and stiffness. Hence, the compatibility between the PVD coating and CoCrMo alloy becomes a major problem in high-stress contacts [79]. Plasma nitriding drastically affected the wear resistance and corrosion resistance of CoCrMo alloy. The duplex treatment involving nitriding as the first stage treatment followed by PVD of CrN and (TiN/CrN) × 3 as the second stage treatment is also found to be ineffective in offering a better wear resistance. PVD (TiN/CrN) × 3 multilayer coating has been suggested as an alternative surface modification method to achieve a better wear resistance for CoCrMo alloy [79]. Nitrogen ion implantation with 100 keV at a dose of 4×10^{17} N^+/cm^2 is found to be effective in reducing the friction coefficient of CoCrMo alloy from 0.35 to 0.15 under tribocorrosion conditions [82]. Much remains to be explored on the surface modification of CoCrMo alloy to impart the desired characteristics suitable for hip implant applications.

CoCrMo alloy subjected to mechanical polishing using successive grades (P240, P400, P800 and P1200) silicon carbide coated abrasive papers, followed by successive fine polishing using a 6 μm and 1 μm diamond polishing compound exhibited formation of ε-martensite and mechanical twins. However, the mechanically polished (MP) surface has enabled the formation of a nanocrystalline layer during sliding wear through high strain shear process and the prior formation of a martensite structure and mechanical twins that help in achieving the nanocrystalline layer [83]. The extensive deformation, formation of ε-martensite, mechanical twins, and high dislocation density together with the formation of nanocrystalline layer has offered a better wear resistance for MP CoCrMo alloy. A decrease in the specific wear rate of the mechanically polished surface with an increase in normal load points out the advantage of forming a nanocrystalline layer [83].

SMAT is a S^2PD technique. The plastic deformation induced by SMAT leads to refinement of coarse-grained grains down to nanosized grains. The ability to generate a graded layer structure is another beneficial attribute of SMAT. The influence of

SMAT on the microstructure, phase content, phase transformation, grain size, surface roughness, hardness and, corrosion resistance of CoCrMo alloy is addressed in the following paragraphs.

Unlike stainless steel, Ti and Ti alloys, and Ni-Ti alloys, SMAT of CoCrMo alloys has not been studied much. The benefits of surface nanocrystallization of Co28Cr6Mo alloy by SMAT in improving the abrasive wear properties were addressed by Demangel et al. [84]. The microstructural characteristics of CoCrMo alloy subjected to SMAT were evaluated by Proust et al. [85]. The influence of SMAT on the corrosion behavior of Co28Cr6Mo alloy in Ringer's solution was studied by Tchana Nkonta et al. [19].

The XRD pattern of the as-polished Co28Cr6Mo alloy reveals the presence of austenite (γ-fcc) and martensite (ε-hcp) phases. After SMAT, the intensity of the (111γ) and (200γ) peaks are decreased concomitant with an increase in intensity of the (101ε) peak, suggesting the formation of more amounts of martensite (ε-hcp) phase at the expense of the austenite phase [19]. The mechanism of such transformation is believed to be similar to the observed during SMAT of cobalt, which also possesses both hcp and fcc phases [86]. Accordingly, the strain-induced during SMAT is accommodated by the fcc phase. At lower strain, slip dislocations leading to the formation of intersecting planar arrays of dislocations are likely to occur. At higher strain, a transformation of the fcc to hcp phase occurs.

The microstructure of Co28Cr6Mo alloy after SMAT characterized by transmission Kikuchi diffraction (TKD) is shown in Fig. 16. The band and phase-contrast maps indicate grain refinement at the top surface (Region I). In addition, the transformation of the fcc phase (represented by blue color) to hcp phase (represented by red

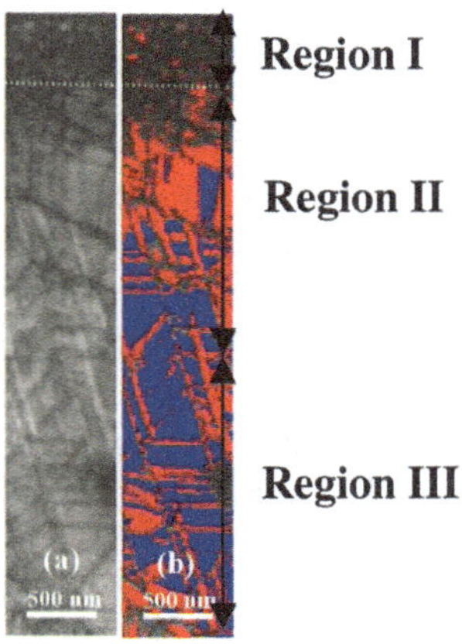

Fig. 16 **a** Band contrast; and **b** phase-contrast maps obtained by transmission Kikuchi diffraction (TKD) of a CoCrMo alloy after SMAT using 3 mm Ø 100Cr6 balls for 15 min. In the phase contrast map, red and blue colors represent hcp and fcc phase, respectively (Adapted from Proust et al., 2016, "Microstructural Characterization of a Cobalt-Chromium-Molybdenum Alloy Subjected to Surface Mechanical Attrition Treatment." *Proceedings of PLASTICITY '16: The Twenty Second International Symposium on Plasticity and its Current Applications* [2016], 46–48)

colour) is much pronounced in region II. The phase transformation is not limited to region II rather it extends further below. In region III, hcp laths inside the fcc grains are evident. EBSD patterns of the treated alloy have suggested that the transformation of fcc phase to hcp phase could occur even up to a depth of 150 μm from the top surface. Obviously, a steady decrease in the amount of hcp phase from the surface to the bulk is expected [85]. During SMAT, the low SFE of CoCrMo alloy is likely to promote formation of stacking faults, grain refinement, a transformation of fcc to hcp phase and formation of nanocrystalline structure [87]. The microstructure of the subsurface could exert a strong influence on the wear and corrosion resistance of the treated alloy. The evolution of subsurface microstructure as a function of applied potential and its influence on the corrosion behavior of CoCrMo alloy in phosphate-buffered saline in presence of bovine serum albumin has been established by Wang et al. [88].

SMAT of Co28Cr6Mo alloy using 3 mm Ø 100Cr6 alloy balls for 5, 15, and 25 min has enabled an increase in hardness. The hardness profile indicates that the maximum hardness value is reached at few microns from the top of the surface. The hardness is decreased slowly until a depth of ~400 μm from the surface to the bulk [84]. The increment in the hardness of the Co28Cr6Mo alloy after SMAT is due to grain refinement following plastic deformation. The increase in grain boundary density could block the movement of dislocations and increase the hardness. SMAT could also induce compressive residual stress due to the SPD. The maximum hardness value observed at the subsurface further substantiates the influence of compressive residual stress. Besides, strain-induced dislocations, twins, stacking faults and formation of martensite phase could also contribute to the increase in hardness. Similar behavior is also observed for other materials subjected to SP. Hence, it is clear the SMAT is capable of inducing SPD of the Co28Cr6Mo alloy, and the grain refinement, compressive residual stress and strain-induced dislocations are responsible for the increase in hardness [84].

The random multidirectional impingement of the 3 mm Ø 100Cr6 alloy balls increased the R_a of the Co28Cr6Mo alloy after SMAT. After 15 min treatment, the R_a is increased and it ranges from 0.64 μm to 0.77 μm [19, 84]. It has been reported earlier that scratches present on the femoral head could cause the counterpart to worn out more [89]. Hence, the higher R_a of the Co28Cr6Mo alloy after SMAT becomes a major concern. Scratch testing performed under progressive load mode from 0 to 30 N indicates that the depth of penetration is much lower for Co28Cr6Mo alloy subjected to SMAT when compared to the untreated alloy. This inference indicates that SMAT is capable of increasing the wear resistance of Co28Cr6Mo alloy under severe abrasive wear environment, which prevails under tribocorrosion conditions.

SMAT also improves the corrosion resistance of Co28Cr6Mo alloy in Ringer's solution [19]. The increase in R_a after SMAT from 0.02 μm to 0.64 μm would deleteriously influence the corrosion resistance. Nevertheless, the SPD induced by SMAT facilitates the permeation of oxygen atoms and promotes the formation of a passive oxide layer. The chemical composition of the passive oxide layer is also considered to play a key role in improving the corrosion resistance. Based on XPS analysis, cobalt oxide is identified as the main constituent of the passive oxide layer formed on plastically deformed Co28Cr6Mo alloy in SBF [90]. Hence, it is clear that the ability of the treated alloy to form a stable passive film is responsible for the observed improvement in corrosion resistance.

The higher R_a of the Co28Cr6Mo alloy after SMAT causes a deleterious influence on both the wear and corrosion resistance. Hence, to achieve better wear and corrosion resistance, post-polishing of the treated alloy is warranted. Demangel et al. [84] have explored polishing using a brush, fluid containing grinding media and polishing cloth. Polishing using brushes removed only the superficially attached particles. Polishing using fluid containing grinding media removed some material and provided a reasonable surface finish. The use of a polishing cloth offered a relatively better surface finish. In scratch testing, the polished samples show a 30–60% decrease in penetration depth and among the polished ones the penetration depth is much lower for the cloth polished sample. Tchana Nkonta et al. [19] have used a proprietary biomedical polishing treatment of Co28Cr6Mo alloy after SMAT which enabled a 50% decrease in R_a from 0.64 μm to 0.30 μm. The decrease in R_a also enabled an increase in corrosion resistance. Hence, it is evident that adopting a suitable post-polishing treatment is likely to improve both the tribological and corrosion performance of CoCrMo alloys subjected to SMAT. In addition, the mean depth, extent of material removal and mean flatness of the finished surface should also be considered during post-polishing treatment. According to Demangel et al. [84], polishing using both brush and fluid could remove only a limited amount of material (up to 2–12 μm). However, the mean flatness is not improved much after polishing. In contrast, the mean flatness is significantly improved for cloth polishing. Nevertheless, the extent of material removal is much higher (up to 34 μm). The extent of material removal during post-polishing treatment should be in line with thickness of the deformed layer after SMAT. Otherwise, the beneficial influence of SMAT cannot be realized.

6 Mg and Mg Alloys

Magnesium and its alloys are considered suitable for the development of degradable implants due to their non-toxic nature, biodegradability and biocompatibility. The dissolution of Mg is not likely to cause any adverse side effects. Since the specific density and Young's modulus of Mg are closer to bone it could reduce stress shielding effect at the bone/implant interface. The use of Mg-based degradable implants solves many of the problems associated with permanent implants such as restenosis, thrombosis and, inability to adapt to the growth and changes in human body, etc., and, eliminate the need for a second surgery. However, the rapid corrosion of Mg and its alloys in physiological environment undermines their mechanical integrity. Corrosion of Mg and its alloys results in the generation of a large volume of hydrogen gas and increase the local pH of body fluid. Accumulation of the hydrogen bubbles near the implant could lead to necrosis of tissues while the local increase in pH could lead to alkaline poisoning. Alloying could improve mechanical integrity. Nevertheless, the alloying elements should also be non-toxic, which limits the availability of elements for alloying. Surface modification is a viable approach. It would be interesting to see how SMAT could be useful in improving the performance of Mg towards the development of degradable implants.

SMAT of AZ31 Mg alloy has enabled nanocrystallization at the surface along with the formation of a graded layer structure [91–94]. An increase in treatment time has led to a decrease in grain size at the surface and increased the deformation thickness [92, 93]. However, no significant change in the deformation thickness could be observed if the treatment time is increased beyond a threshold level [92]. SMAT increased surface roughness [91, 95, 96] due to the formation of craters and dimples on the surface. An increase in diameter of the balls used for treatment increased the surface roughness [95, 96]. An increase in surface roughness decreased the water contact angle, increased surface energy, and increased the degree of hydrophilicity of the Mg alloy [91].

The evolution of the microstructure of AZ31 Mg alloy during SMAT is addressed by Meng et al. [92] and Duan et al. [93]. According to Meng et al. [92], twinning is the mechanism of deformation at lower strain while dislocation cells and subgrains with misorientation start to emerge with a continuous increase in strain. With an increase in treatment time, the level of strain and strain rate is increased, which decreases the recrystallization temperature. The increase in surface temperature induced by the repeated multidirectional impacts of the balls and the recrystallization temperature promotes dynamic recrystallization, leading to the formation of a nanostructured structure. According to Duan et al. [93], with reasonably applied stress, dislocations start to accumulate near the grain boundaries, leading to the formation of dislocation cells. With a further increase in strain, an increase in misorientation of the dislocation cells leads to the formation of subgrains which gradually transform to recrystallized grains having high angle grain boundaries (HAGBs). With an increase in treatment time, more HAGBs are generated. SPD induced during SMAT promotes recrystallization and grain refinement and these stages continue until the grains at the surface are reduced to nano-size. The refinement of grains to a nanoscale at the surface occurs due to twinning and dynamic recrystallization [94].

Change in texture of AZ31 Mg alloy after SMAT could influence its mechanical properties. Untreated AZ31 Mg alloy has a strong (0002) basal texture. SMAT using 3 mm Ø AISI 304 stainless steel balls at 20 kHz for 2 min fails to exhibit any change in texture as the extent of plastic deformation is rather limited. However, a steady decrease in the intensity of peak pertaining to (0002) basal texture with an increase in treatment time to 4 min and 6 min suggest that the plastic deformation induced during SMAT exert a strong weakening effect on the (0002) basal texture of AZ31 Mg alloy [93]. Based on the inclination of (0002) poles, the decrease in maximum pole density, and divergence of contour lines, Peng et al. [94] have validated such an occurrence. The grain orientation differs from the surface to the substrate, which is evidenced by the lower ratio of (0002) to (10, 11) at the surface [92]. SMAT increased the YS and UTS of AZ31 Mg alloy and the extent of increase become much pronounced with an increase in treatment time from 2 to 6 min. The YS of AZ31 Mg alloy is increased from 187.4 MPa to 274 MPa after SMAT for 6 min. The tensile strength of AZ31

Mg alloy is increased from 276.5 MPa to 323.7 MPa after SMAT for 6 min. The elongation to a fraction is slightly decreased from 17.23 to 10.44% after SMAT for 6 min [93]. An improvement in YS and UTS after SMAT of AZ31 Mg alloy is also observed by Peng et al. [94]. Hence, it is evident that the weakening of (0002) basal texture of AZ31 Mg alloy after SMAT has no deleterious effect on its mechanical property. The improvement in YS and UTS of AZ31 Mg alloy after SMAT is due to the grain refinement and work hardening.

SMAT increased the hardness of AZ31 Mg alloy [91–94]. Grain refinement and a high density of dislocations are considered responsible for the observed increase in hardness [94]. The hardness profile displays a depth-dependent gradation in hardness in accordance with the graded layer structure, decreasing with an increase in deformation depth [92, 93]. For AZ91D Mg alloy, dissolution of the β-$Mg_{17}Al_{12}$ phase is also considered as a reason for the improvement in hardness after SMAT besides grain refinement and dislocations [95].

Laleh and Kargar [95] have evaluated the corrosion behavior of untreated AZ91D Mg alloy and those subjected to SMAT using 2, 3 and 5 mm Ø stainless steel balls at 20 kHz for 30 min in 3.5% NaCl. SMAT increased the surface roughness, refined the grain size, introduced a high density of crystalline defects and promoted dissolution of the β-$Mg_{17}Al_{12}$ phase into the matrix. AZ91D Mg alloy treated using 2 mm Ø balls has offered an improvement in corrosion resistance, which is evidenced by a shift in E_{corr} towards the noble direction from −1509 mV vs. SCE to −1334 mV vs. SCE and a decrease in i_{corr} from 3.91 $\mu A/cm^2$ to 0.24 $\mu A/cm^2$. In contrast, the alloy treated using 5 mm Ø balls has drastically reduced the corrosion resistance. A shift in E_{corr} from −1509 mV vs. SCE to −1561 mV vs. SCE and an increase in i_{corr} from 3.91 $\mu A/cm^2$ to 24.83 $\mu A/cm^2$ validates it. In spite of an increase in surface roughness and defect density, the corrosion resistance of AZ91D Mg alloy treated using 2 mm Ø balls is much higher than the untreated alloy. This is believed to be due to the dissolution of β-$Mg_{17}Al_{12}$ phase into the matrix. The galvanic couple formed between the matrix and the β-$Mg_{17}Al_{12}$ phase results in poor corrosion resistance of the untreated AZ91D Mg alloy. The poor corrosion resistance of AZ91D Mg alloy treated using 5 mm Ø balls is due to the higher surface roughness and high density of crystalline defects Laleh and Kargar [95].

The corrosion behavior of untreated AZ31 Mg alloy and those subjected to SMAT using 2 mm Ø zirconia balls for 40–90 min, in Hanks' solution at 37 °C, is studied by Li et al. [91]. SMAT has increased the surface roughness, refined the grain size introduced high density of crystalline defects and promoted dissolution of the Mg_2Ca phase into the matrix. Grain refinement does not seem to exert a positive influence on the corrosion resistance since the passivation tendency for the AZ31 Mg alloy in Hanks' solution is poor. Dissolution of the Mg_2Ca phase into the matrix, which would otherwise promote the formation of galvanic cells, fails to exhibit any improvement in corrosion resistance. The higher surface roughness and the high density of crystalline defects deleteriously influence the corrosion resistance. Based on the inferences made, Li et al. [91] have suggested that SMAT is not a suitable method to improve the corrosion resistance of Mg-based biodegradable implants.

Chen et al. [96] have evaluated the corrosion behavior and corrosion fatigue of untreated AZ31B Mg alloy and those subjected to SMAT using 2 and 3 mm Ø stainless steel balls at 20 kHz for 15 min in 3.5% NaCl. An increase in surface roughness after SMAT and the defects induced during the treatment has caused a deleterious influence on the corrosion resistance of the treated alloy. The formation of more number of deeper corrosion pits on the treated alloy further substantiates it. Nevertheless, the grain refinement, surface nanocrystallization and the compressive residual stress induced during SMAT have led to a significant improvement in the fatigue life of the treated alloy. This is validated by the increase in fatigue life with an increase in diameter of the balls from 2 mm to 3 mm. At lower stress amplitude, a decrease in fatigue life of the alloy is observed when the test environment is changed from air to 3.5% NaCl. However, at higher stress amplitude, the improvement in fatigue life is much pronounced in 3.5% NaCl environment.

The tribological behavior of untreated GW63K Mg alloy and those subjected to SMAT at two different sliding speeds of 0.05 m/s and 0.2 m/s is studied by Liu et al. [97]. A similar study is also performed by Xia et al. [98] using AZ31 Mg alloy at 0.05 and 0.5 m/s. The wear mechanism of Mg alloys varies with the sliding speed; abrasive wear at 0.05 m/s and oxidative wear at 0.2 and 0.5 m/s. According to Liu et al. [97], the poor wear resistance of SMAT GW63K Mg alloy observed at 0.05 m/s is due to its low ductility and toughness. With an increase in sliding speed to 0.2 m/s, the transformation of abrasive wear to oxidative wear occurs, which is likely to increase the wear resistance. Nevertheless, filling the worn surface with wear debris promotes delamination leading to poor wear resistance. In contrast, Xia et al. [98] have observed an improvement in wear resistance of SMAT AZ31 Mg alloy at 0.5 m/s due to its ability to form a compact oxide film which reduced the coefficient of friction. Studies on cell adhesion and spreading on SMAT Mg alloys are rather limited.

7 Concluding Remarks

SMAT has enabled surface nanocrystallization, promoted the formation of a graded layer structure having nanosized grains at the top surface, submicron-sized grains at the subsurface with coarse grains near the bulk. The microstructure exhibits considerable variation as a function of deformation depth from the top surface. SMAT induces compressive residual stress. The change in deformation thickness, microstructure, compressive residual stress induced during treatment and type of phases formed are largely a function of the level of strain and strain rate. The repeated multidirectional impingement of the balls results in the generation of numerous dimples and craters on the surface, which increases the R_a. The hardness and mechanical properties are also increased after SMAT. Grain refinement, work hardening, strain-induced dislocations, and compressive residual stress are responsible for the increase in hardness and mechanical properties. The maximum hardness is reached at few microns from the top of the surface. The hardness exhibits a steady decrease with an increase in depth until the end of the deformation depth. SMAT improved tribological behavior. The refinement in grain size and a higher hardness are considered responsible for the improvement in tribological behavior after SMAT. SMAT increased the corrosion resistance of Ti and Ti alloys, NiTi alloy, and CoCrMo alloy in SBF. The decrease in grain size, increase in grain boundary density, dilution of segregated alloying elements at grain boundaries and an increase in the tendency of the treated material to form an intact passive oxide layer with a higher thickness are responsible for the observed improvement in corrosion resistance of Ti and Ti alloys, NiTi alloy and CoCrMo alloy. The higher corrosion resistance of materials after SMAT would decrease the extent of leaching of the toxic metal ions (aluminum and vanadium ions from Ti6Al4V alloy, nickel ions from NiTi alloy and, cobalt and chromium ions from CoCrMo alloy) and improve their biocompatibility. An improvement in both the tribological properties as well as corrosion resistance would offer improvement in the performance of CoCrMo alloy under conditions of tribocorrosion experienced by hip and knee implants. An improvement in corrosion resistance of stainless steel after SMAT is largely a function of the treatment conditions employed. An improvement in corrosion resistance is observed only for 304 stainless steels treated using 2 mm Ø balls. In contrast, SMAT using 5 and 8 mm Ø balls has led to a decrease in corrosion resistance. The increase in surface roughness, a transformation of the austenite to α'-martensite phase, higher extent of deformation and presence of a larger number of defects/dislocations are responsible for the observed decrease in corrosion resistance of 304 SS treated using 5 and 8 mm Ø balls when compared to the untreated one in Ringer's solution as well as 0.6 M NaCl. The grain refinement, surface nanocrystallization, selective dissolution of Fe and enrichment of Cr at the surface have increased the number of nucleation sites and promotes the formation of an intact and highly protective passive oxide layer on 304 stainless steel treated using 2 mm Ø balls when compared to the untreated one, in Ringer's solution as well as 0.6 M NaCl. The corrosion resistance is drastically affected for Mg and Mg alloy after SMAT. The marginal improvement in corrosion resistance of AZ91D Mg alloy after SMAT using 2 mm Ø balls is due to dissolution of β-$Mg_{17}Al_{12}$ phase into the matrix, which nullifies the formation of the galvanic cells. Hence, SMAT can be considered as a useful surface treatment to improve the performance of Ti and Ti alloys, stainless steels, NiTi alloy and CoCrMo alloy for biomedical applications.

8 Future Perspectives

Increase in surface roughness after SMAT deleteriously influence the corrosion resistance although it is beneficial for better wetting, an increase in surface energy, and imparting a higher degree of hydrophilicity. Adapting a post-polishing stage can be recommended to improve their surface finish. This would be particularly beneficial for stainless steel and CoCrMo alloy implants. Electropolishing is a viable option to achieve this. Large pulsed electron beam (LPEB) irradiation can be explored to improve the surface finish. However, necessary precautions should be followed as

LPEB irradiation could induce tensile residual stress and the beneficial influence of compressive residual stress imparted by SMAT in improving the fatigue strength will be nullified. The transformation of austenite to strain-induced α'-martensite during SMAT could pose difficulties in stainless steel implants during MRI imaging due to magnetic properties. Adopting a suitable post-annealing treatment to achieve phase reversal without losing the beneficial influences of SMAT can be explored. Many duplex and hybrid treatment approaches are explored using SMAT as a pre-treatment. The utility of SMAT as a post-treatment to improve the surface finish of selective laser melted 316L stainless steel has already been shown to achieve a 96% reduction in surface roughness. A hybrid treatment approach involving SMAT as a pre-treatment for anodizing of Ti has enabled the formation of TiO_2 nanotubes with higher length and larger diameter. A proper choice of the process parameters employed for SMAT and the experimental conditions for anodizing would open up a new pathway to develop TiO_2 nanotubes for drug delivery applications. The difficulty in treating complex-shaped implants by SMAT poses another challenge, which warrants an improvement in design. Much remains to be explored on SMAT and quest for the development of implant materials with desired characteristics continues.

Bibliography

1. Lu K, Lu J (1999) Surface nanocrystallization (SNC) of metallic. Materials presentation of the concept behind a new approach. J Mater Sci Technol 15(3):193–197
2. Zhang H, Hei Z, Liu G, Lu J, Lu K (2003) Formation of nanostructured surface layer on AISI 304 stainless steel by means of surface mechanical attrition treatment. Acta Mater 51(7):1871–1881
3. Lu K, Lu J (2004) Nanostructured surface layer on metallic materials induced by surface mechanical attrition treatment. Mater Sci Eng, A 375–377:38–45
4. Tao NR, Lu J, Lu K (2008) Surface nanocrystallization by surface mechanical attrition treatment. Mater Sci Forum 579:91–108
5. Azadmanjiri J, Berndt CC, Kapoor A, Wen C (2015) Development of surface nano-crystallization in alloys by surface mechanical attrition treatment (SMAT). Crit Rev Solid State Mater Sci 40(3):164–181
6. Dai K, Shaw L (2007) Comparison between shot peening and surface nanocrystallization and hardening processes. Mater Sci Eng, A 463(1–2):46–53
7. Jelliti S, Richard C, Retraint D, Roland T, Chemkhi M, Demangel C (2013) Effect of surface nanocrystallization on the corrosion behavior of Ti-6Al-4 V titanium alloy. Surf Coat Technol 224:82–87
8. Gallitelli D, Retraint D, Rouhaud E (2014) Comparison between conventional shot peening (SP) and surface mechanical attrition treatment (SMAT) on a titanium alloy. Adv Mat Res 996:964–968
9. Balusamy T, Kumar S, Sankara Narayanan TSN (2010) Effect of surface nanocrystallization on the corrosion behaviour of AISI 409 stainless steel. Corros Sci 52(11):3826–3834
10. Balusamy T, Sankara Narayanan TSN, Ravichandran K, Lee MH, Nishimura T (2015) A facile method to modify the characteristics and corrosion behavior of 304 stainless steel by surface nanostructuring toward biomedical applications. ACS Appl Mater Interfaces 7(32):17731–17747
11. Hu T, Wen CS, Sun GY, Wu SL, Chu CL, Wu ZW, Li GY, Lu J, Yeung KWK, Chu PK (2010) Wear resistance of NiTi alloy after surface mechanical attrition treatment. Surf Coat Technol 205(2):506–510

12. Bahl S, Suwas S, Ungàr T, Chatterjee K (2017) Elucidating microstructural evolution and strengthening mechanisms in nanocrystalline surface induced by surface mechanical attrition treatment of stainless steel. Acta Mater 122:138–151
13. Novelli M, Bocher P, Grosdidier T (2018) Effect of cryogenic temperatures and processing parameters on gradient-structure of a stainless steel treated by ultrasonic surface mechanical attrition treatment. Mater Charact 139:197–207
14. Tsai MT, Huang JC, Tsai WY, Chou TH, Chen CF, Li TH, Jang JSC (2018) Effects of ultrasonic surface mechanical attrition treatment on microstructures and mechanical properties of high entropy alloys. Intermetallics 93:113–121
15. Alikhani Chamgordani S, Miresmaeili R, Aliofkhazraei M (2018) Improvement in tribological behavior of commercial pure titanium (CP-Ti) by surface mechanical attrition treatment (SMAT). Tribol Int 119:744–752
16. Fu T, Zhan Z, Zhang L, Yang Y, Liu Z, Liu J, Li L, Yu X (2015) Effect of surface mechanical attrition treatment on corrosion resistance of commercial pure titanium. Surf Coat Technol 280:129–135
17. Li S, Ren Z, Dong Y, Ye C, Cheng G, Cong H (2017) Enhanced pitting corrosion resistance of 304 SS in 3.5 wt% NaCl by ultrasonic nanocrystal surface modification. J Electrochem Soc 164(12):C682–C689
18. Jamesh M, Sankara Narayanan TSN, Chu PK, Park IS, Lee MH (2013) Effect of surface mechanical attrition treatment of titanium using alumina balls: surface roughness, contact angle and apatite forming ability. Front Mater Sci 7(3):285–294
19. Tchana Nkonta DV, Simescu-Lazar F, Drevet R, Aaboubi O, Fauré J, Retraint D, Benhayoune H (2017) Influence of the surface mechanical attrition treatment (SMAT) on the corrosion behavior of Co28Cr6Mo alloy in Ringer's solution. J Solid State Electrochem 22(4):1091–1098
20. Zhao C, Ji W, Han P, Zhang J, Jiang Y, Zhang X (2011) In vitro and in vivo mineralization and osseointegration of nanostructured Ti6Al4 V. J Nanopart Res 13(2):645–654
21. Huang R, Zhang L, Huang L, Zhu J (2019) Enhanced in-vitro osteoblastic functions on β-type titanium alloy using surface mechanical attrition treatment. Mater Sci Eng, C 97:688–697
22. Pilliar RM (2009) Metallic Biomaterials. In: Narayan R (ed) Biomedical materials, chapter 2. Springer-Verlag, US, pp 41–81
23. Roland T, Retraint D, Lu K, Lu J (2006) Fatigue life improvement through surface nanostructuring of stainless steel by means of surface mechanical attrition treatment. Scr Mater 54(11):1949–1954
24. Bahl S, Shreyas P, Trishul MA, Suwas S, Chatterjee K (2015) Enhancing the mechanical and biological performance of a metallic biomaterial for orthopedic applications through changes in the surface oxide layer by nanocrystalline surface modification. Nanoscale 7(17):7704–7716
25. Cao SC, Zhang X, Lu J, Wang Y, Shi SQ, Ritchie RO (2019) Predicting surface deformation during mechanical attrition of metallic alloys. npj Comput Mater 5(1):36
26. Arifvianto B, Suyitno Mahardika M, Dewo P, Iswanto PT, Salim UA (2011) Effect of surface mechanical attrition treatment (SMAT) on microhardness, surface roughness and wettability of AISI 316L. Mater Chem Phys 125(3):418–426
27. Arifvianto B, Suyitno Mahardika M (2012) Effects of surface mechanical attrition treatment (SMAT) on a rough surface of AISI 316L stainless steel. Appl Surf Sci 258(10):4538–4543
28. Balusamy T, Kumar S, Sankara Narayanan TSN (2011) Electrochemical behaviour of surface modified AISI 304 grade stainless steel in Ringer's solution. Trans Indian Inst Met 64(4–5):507–511
29. Balusamy T, Sankara Narayanan TSN, Ravichandran K, Park IS, Lee MH (2013) Influence of surface mechanical attrition treatment (SMAT) on the corrosion behaviour of AISI 304 stainless steel. Corros Sci 74:332–344
30. Balusamy T, Sankara Narayanan TSN, Ravichandran K, Park IS, Lee MH (2013) Surface mechanical attrition treatment (SMAT) on pack boronizing of AISI 304 stainless steel. Surf Coat Technol 232:60–67
31. Benafia S, Retraint D, Yapi Brou S, Panicaud B, Grosseau Poussard JL (2018) Influence of surface mechanical attrition treatment on the oxidation behaviour of 316l stainless steel. Corros Sci 136:188–200

32. Hao Y, Deng B, Zhong C, Jiang Y, Li J (2009) Effect of surface mechanical attrition treatment on corrosion behavior of 316 stainless steel. J Iron Steel Res 16(2):68–72
33. Olugbade T, Liu C, Lu J (2019) Enhanced passivation layer by Cr diffusion of 301 stainless steel facilitated by SMAT. Adv Eng Mater 1900125
34. Antunes RA, de Oliveira MCL (2012) Corrosion fatigue of biomedical metallic alloys: mechanisms and mitigation. Acta Biomater 8(3):937–962
35. Mordyuk BN, Prokopenko GI (2006) Fatigue life improvement of α-titanium by novel ultrasonically assisted technique. Mater Sci Eng, A 437(2):396–405
36. McCafferty E (2003) Sequence of steps in the pitting of aluminum by chloride ions. Corros Sci 45(7):1421–1438
37. Xie J, Alpas AT, Northwood DO (2002) A mechanism for the crack initiation of corrosion fatigue of Type 316L stainless steel in Hank's solution. Mater Charact 48(4):271–277
38. Ma E (2003) Instabilities and ductility of nanocrystalline and ultrafine-grained metals. Scr Mater 49(7):663–668
39. Misra RDK, Nune C, Pesacreta TC, Somani MC, Karjalainen LP (2013) Understanding the impact of grain structure in austenitic stainless steel from a nanograined regime to a coarse-grained regime on osteoblast functions using a novel metal deformation–annealing sequence. Acta Biomater 9(4):6245–6258
40. Cicek S, Karaca A, Torun I, Onses MS, Uzer B (2019) The relationship of surface roughness and wettability of 316L stainless steel implants with plastic deformation mechanisms. Mater Today Proc 7:389–393
41. Yin F, Yang S, Hu S, Kuang S, Han Q (2017) Enhanced human osteoblast cell functions by "net-like" nanostructured cell-substrate interface in orthopedic applications. Mater Lett 189:275–278
42. Geetha M, Singh AK, Asokamani R, Gogia AK (2009) Ti based biomaterials, the ultimate choice for orthopaedic implants—a review. Prog Mater Sci 54(3):397–425
43. Mischler S, Muñoz AI (2013) Wear of CoCrMo alloys used in metal-on-metal hip joints: a tribocorrosion appraisal. Wear 297(1–2):1081–1094
44. Zhao C, Han P, Ji W, Zhang X (2012) Enhanced mechanical properties and in vitro cell response of surface mechanical attrition treated pure titanium. J Biomater Appl 27(2):113–118
45. Anand Kumar S, Ganesh Sundara Raman S, Sankara Narayanan TSN (2014) Influence of surface mechanical attrition treatment duration on fatigue lives of Ti–6Al–4V. Trans Indian Inst Met 67(1):137–141
46. Zhu KY, Vassel A, Brisset F, Lu K, Lu J (2004) Nanostructure formation mechanism of α-titanium using SMAT. Acta Mater 52(14):4101–4110
47. Wen M, Liu G, Gu J, Guan W, Lu J (2009) Dislocation evolution in titanium during surface severe plastic deformation. Appl Surf Sci 255(12):6097–6102
48. Anand Kumar S, Ganesh Sundara Raman S, Sankara Narayanan TSN, Gnanamoorthy R (2012) Influence of surface mechanical attrition treatment on fretting wear behaviour of Ti-6Al-4V. Adv Mat Res 463–464:316–320
49. Lai M, Cai K, Hu Y, Yang X, Liu Q (2012) Regulation of the behaviors of mesenchymal stem cells by surface nanostructured titanium. Colloids Surf B Biointerfaces 97:211–220
50. Wen M, Liu G, Gu J, Guan W, Lu J (2008) The tensile properties of titanium processed by surface mechanical attrition treatment. Surf Coat Technol 202(19):4728–4733
51. Anand Kumar S, Ganesh Sundara Raman S, Sankara Narayanan TSN, Gnanamoorthy R (2013) Influence of counterbody material on fretting wear behaviour of surface mechanical attrition treated Ti–6Al–4V. Tribol Int 57:107–114
52. Huang L, Lu J, Troyon M (2006) Nanomechanical properties of nanostructured titanium prepared by SMAT. Surf Coat Technol 201(1–2):208–213
53. Huang R, Han Y (2013) The effect of SMAT-induced grain refinement and dislocations on the corrosion behavior of Ti-25Nb-3Mo-3Zr-2Sn alloy. Mater Sci Eng, C 33(4):2353–2359
54. Acharya S, Panicker AG, Gopal V, Dabas SS, Manivasagam G, Suwas S, Chatterjee K (2020) Surface mechanical attrition treatment of low modulus Ti-Nb-Ta-O alloy for orthopedic applications. Mater Sci Eng, C 110:110729

55. Huang R, Lu S, Han Y (2013) Role of grain size in the regulation of osteoblast response to Ti-25Nb-3Mo-3Zr-2Sn alloy. Colloids Surf B Biointerfaces 111:232–241
56. Huang R, Zhuang H, Han Y (2014) Second-phase-dependent grain refinement in Ti–25Nb–3Mo–3Zr–2Sn alloy and its enhanced osteoblast response. Mater Sci Eng, C 35:144–152
57. Azadmanjiri J, Wang PY, Pingle H, Kingshott P, Wang J, Srivastava VK, Kapoor A (2016) Enhanced attachment of human mesenchymal stem cells on nanograined titania surfaces. RSC Adv 6(61):55825–55833
58. Zhang L, Han Y (2010) Effect of nanostructured titanium on anodization growth of self-organized TiO_2 nanotubes. Nanotechnol 21(5):55602
59. Wen M, Gu JF, Liu G, Wang ZB, Lu J (2007) Formation of nanoporous titania on bulk titanium by hybrid surface mechanical attrition treatment. Surf Coat Technol 201(14):6285–6289
60. Wen M, Gu JF, Liu G, Wang ZB, Lu J (2007) Nanocrystalline titanium to mesoporous anatase with high bioactivity. Cryst Growth Des 7(12):2400–2403
61. Laleh M, Kargar F (2011) Formation of high bioactive nanoporous titania film by hybrid surface mechanical attrition treatment. Mater Lett 65(14):2295–2298
62. Wen M, Wen C, Hodgson P, Li Y (2014) Improvement of the biomedical properties of titanium using SMAT and thermal oxidation. Colloids Surf B Biointerfaces 116:658–665
63. Balusamy T, Sankara Narayanan TSN, Ravichandran K (2012) Effect of surface mechanical attrition treatment (SMAT) on boronizing of EN8 steel. Surf Coat Technol 213:221–228
64. Balusamy T, Sankara Narayanan TSN, Ravichandran K, Park IS, Lee MH (2013) Plasma nitriding of AISI 304 stainless steel: role of surface mechanical attrition treatment. Mater Charact 85:38–47
65. Sun J, Yao QT, Zhang YH, Du XD, Wu YC, Tong WP (2017) Simultaneously improving surface mechanical properties and in vitro biocompatibility of pure titanium via surface mechanical attrition treatment combined with low-temperature plasma nitriding. Surf Coat Technol 309:382–389
66. Kavitha C, Ravichandran K, Sankara Narayanan TSN (2013) Effect of surface mechanical attrition treatment (SMAT) on zinc phosphating of steel. Trans Inst Met Finish 92(3):161–168
67. Kavitha C, Sankara Narayanan TSN, Ravichandran K, Lee MH (2014) Improving the reactivity and receptivity of alloy and tool steels for phosphate conversion coatings: role of surface mechanical attrition treatment. Ind Eng Chem Res 53(52):20124–20138
68. Yoneyama T, Miyazaki S (eds) (2009) Shape memory alloys for biomedical applications, 1st edn. Woodhead Publishing Limited, Cambridge, England
69. Maitz MF (2009) Surface modification of Ti-Ni alloys for biomedical applications. In: Yoneyama T, Miyazaki S (eds) Shape memory alloys for biomedical applications, chapter 8, Woodhead Publishing series in biomaterials, 1st edn. Woodhead Publishing Limited, Cambridge, England, pp 173–193
70. Hu T, Xin YC, Wu SL, Chu CL, Lu J, Guan L, Chen HM, Hung TF, Chu PK (2011) Corrosion behavior on orthopedic NiTi alloy with nanocrystalline/amorphous surface. Mater Chem Phys 126(1–2):102–107
71. Hu T, Chu CL, Wu SL, Xu RZ, Sun GY, Hung TF, Yeung KWK, Wu ZW, Li GY, Chu PK (2011) Microstructural evolution in NiTi alloy subjected to surface mechanical attrition treatment and mechanism. Intermetallics 19(8):1136–1145
72. Hu T, Chen L, Wu SL, Chu CL, Wang LM, Yeung KWK, Chu PK (2011) Graded phase structure in the surface layer of NiTi alloy processed by surface severe plastic deformation. Scr Mater 64(11):1011–1014
73. Hu T, Chu CL, Wu SL, Xin YC, Lu J, Chu PK (2011) Surface hardening of NiTi shape memory alloy induced by the nanostructured layer after surface mechanical attrition treatment. J Nanosci Nanotechnol 11(12):10954–10957
74. Olumi S, Sadrnezhaad SK, Atai M (2015) The influence of surface nanocrystallization induced by shot peening on corrosion behavior of NiTi alloy. J Mater Eng Perform 24(8):3093–3099
75. Jinlong L, Tongxiang L, Chen W, Limin D (2016) Surface corrosion enhancement of passive films on NiTi shape memory alloy in different solutions. Mater Sci Eng, C 63:192–197

76. Liao Y, Hoffman E, Wimmer M, Fischer A, Jacobs J, Marks L (2013) CoCrMo metal-on-metal hip replacements. Phys Chem Chem Phys 15(3):746–756
77. Sun D, Wharton JA, Wood RJK (2009) Micro-abrasion mechanisms of cast CoCrMo in simulated body fluids. Wear 267(11):1845–1855
78. Sun D, Wharton JA, Wood RJK, Ma L, Rainforth WM (2009) Microabrasion–corrosion of cast CoCrMo alloy in simulated body fluids. Tribol Int 42(1):99–110
79. Ortega-Saenz JA, Hernandez-Rodriguez MAL, Ventura-Sobrevilla V, Michalczewski R, Smolik J, Szczerek M (2011) Tribological and corrosion testing of surface engineered surgical grade CoCrMo alloy. Wear 271(9–10):2125–2131
80. Namus R, Nutter J, Qi J, Rainforth WM (2020) The influence of protein concentration, temperature and cathodic polarization on the surface status of CoCrMo biomedical grade alloys. Appl Surf Sci 499:143908
81. Türkan U, Öztürk O, Eroğlu AE (2006) Metal ion release from TiN coated CoCrMo orthopedic implant material. Surf Coat Technol 200(16–17):5020–5027
82. Guo Z, Pang X, Yan Y, Gao K, Volinsky AA, Zhang TY (2015) CoCrMo alloy for orthopedic implant application enhanced corrosion and tribocorrosion properties by nitrogen ion implantation. Appl Surf Sci 347:23–34
83. Zeng P, Rana A, Thompson R, Rainforth W (2015) Subsurface characterisation of wear on mechanically polished and electro-polished biomedical grade CoCrMo. Wear 332–333:650–661
84. Demangel C, Poznanski A, Steenhout V, Levesque A, Benhayoune H, Retraint D (2014) Benefit of a surface nanocrystallization treatment on Co28Cr6Mo abrasive wear properties. Adv Mat Res 966–967:435–441
85. Proust G, Retraint D, Raoult AG, Demangel C, Tchana D, Benhayoune H (2016) Microstructural characterization of a cobalt-chromium-molybdenum alloy subjected to surface mechanical attrition treatment. In: Proceedings of PLASTICITY '16: The Twenty Second International Symposium on Plasticity and Its Current Applications, pp 46–48
86. Wu X, Tao N, Hong Y, Liu G, Xu B, Lu J, Lu K (2005) Strain induced grain refinement of cobalt during surface mechanical attrition treatment. Acta Mater 53(3):681–691
87. Büscher R, Fischer A (2005) The pathways of dynamic recrystallization in all-metal hip joints. Wear 259(7–12):887–897
88. Wang Z, Yan Y, Su Y, Qiao L (2017) Effect of electrochemical corrosion on the subsurface microstructue evolution of a CoCrMo alloy in albumin containing environment. Appl Surf Sci 406:319–329
89. González-Mora VA, Hoffmann M, Stroosnijder R, Gil FJ (2009) Wear tests in a hip joint simulator of different CoCrMo counterfaces on UHMWPE. Mater Sci Eng, C 29(1):153–158
90. Petrov YN, Prokopenko GI, Mordyuk BN, Vasylyev MA, Voloshko SM, Skorodzievski VS, Filatova VS (2016) Influence of microstructural modifications induced by ultrasonic impact treatment on hardening and corrosion behavior of wrought Co-Cr-Mo biomedical alloy. Mater Sci Eng, C 58:1024–1035
91. Li N, Li YD, Li YX, Wu YH, Zheng YF, Han Y (2014) Effect of surface mechanical attrition treatment on biodegradable Mg–1Ca alloy. Mater Sci Eng, C 35:314–321
92. Meng X, Duan M, Luo L, Zhan D, Jin B, Jin Y, Rao X, Liu Y, Lu J (2017) The deformation behavior of AZ31 Mg alloy with surface mechanical attrition treatment. Mater Sci Eng, A 707:636–646
93. Duan M, Luo L, Liu Y (2020) Microstructural evolution of AZ31 Mg alloy with surface mechanical attrition treatment: grain and texture gradient. J Alloys Compd 823:153691
94. Peng J, Zhang Z, Guo P, Liu Z, Li Y, Zhou W, Wu Y (2019) The effect of surface mechanical attrition treatment on texture evolution and mechanical properties of AZ31 magnesium alloy. Mater Charact 148:26–34
95. Laleh M, Kargar F (2011) Effect of surface nanocrystallization on the microstructural and corrosion characteristics of AZ91D magnesium alloy. J Alloys Compd 509(37):9150–9156
96. Chen G, Fu Y, Cui Y, Gao J, Guo X, Gao H, Wu S, Lu J, Lin Q, Shi S (2019) Effect of surface mechanical attrition treatment on corrosion fatigue behavior of AZ31B magnesium alloy. Int J Fatigue 127:461–469

97. Liu Y, Jin B, Li DJ, Zeng XQ, Lu J (2015) Wear behavior of nanocrystalline structured magnesium alloy induced by surface mechanical attrition treatment. Surf Coat Technol 261:219–226
98. Xia S, Liu Y, Fu D, Jin B, Lu J (2016) Effect of surface mechanical attrition treatment on tribological behavior of the AZ31 alloy. J Mater Sci Technol 32(12):1245–1252

Tailoring the Surface Functionalities of Titania Nanotubes for Biomedical Applications

V. S. Simi, L. Mohan, and N. Rajendran

Abstract Self-organized vertically aligned nanostructures grown on metallic substrates via anodization have attracted significant scientific attention for a wide range of applications. These nanotubular structures integrate highly controllable geometry at the nanoscale with fascinating biological and mechanical properties. This chapter attempts to cover the key electrochemical factors that control the tube geometry and also demonstrate various surface functionalization approaches for modifying the surface properties of TiO_2 nanotubes to develop new and pioneering functional biomaterials for biomedical applications. Furthermore, the anodization parameters that have led to the formation of nanotubes on various titanium alloys were also discussed.

Keywords Titania nanotubes · Anodization · Functionalization · Biocompatibility

1 Introduction

A biomaterial is any matter or surface derived either from nature or synthesized in the laboratory that interacts with the biological systems. They are often used or adapted for a medical application, encompassing a part or whole of a living structure or a designed medical device that performs, augments, or replaces a natural function [1].

V. S. Simi
Department of Applied Science, Sree Chitra Thirunal College of Engineering, Trivandrum, Kerala, India

V. S. Simi (✉) · N. Rajendran
Department of Chemistry, CEG, Guindy Campus, Anna University, Chennai, Tamil Nadu, India
e-mail: simisasidharan08@gmail.com

N. Rajendran
e-mail: nrajendran@annauniv.edu

L. Mohan
Department of Mechanical Engineering, Toyohashi University of Technology, Toyohashi, Japan

T. S. Santra and L. Mohan (eds.), *Nanomaterials and Their Biomedical Applications*, Springer Series in Biomaterials Science and Engineering 16,
https://doi.org/10.1007/978-981-33-6252-9_17

Biomaterials and medical devices have evolved over the last 50 (or) so years to a $100 billion endeavor [2]. From the past few decades, due to the increase in the participation of many persons in defense activities, sports, utilization of self-operating machines, and increased interest in motorcycles has resulted in enormous increase in the number of accidents, which necessarily led people to opt for orthopaedic implants [3].

In orthopaedics, the biomaterial devices used are commonly called as implants and are manufactured for a great number in orthopaedic applications. An adequate level of tolerance of the material with the living organism i.e. biocompatibility is essential to successfully apply an implant in the human body. Despite the success of traditional materials, new and innovative biomaterials are being developed continuously to satisfy the ever increasing demands.

2 Material Choice for Biomedical Implants

The selection of suitable materials for orthopaedic implants relies on the specific applications. The material used as an implant should have the desired balance of both physical and mechanical properties necessary to perform as expected in the human body. Also, the device should be relatively easy to fabricate, consistent, reproducible, and confirm to all technical and biological requirements [4]. Over the past few decades, a wide range of materials such as metals, ceramics, polymers, composites and biologically derived materials are commonly used in orthopaedic implants as the different physical, chemical, and biological properties that these materials possess can cater to specific applications [5–9]. While traditional materials such as stainless steel, Co-based alloys, Titanium and its alloys are constantly modified to be more effective biomaterials, resulting in improved medical devices, new materials continued to be explored as alternatives [10].

3 Titanium-Based Biomaterials

In the current scenario, titanium and its alloys are extensively used in biomedical field and have a good record of being used as a successful implant material due to the metal's exceptional properties. Titanium- based biomaterials are inert and proven to be superior in terms of biocompatibility and corrosion resistance than Co-Cr and stainless steel alloys, hence becoming the material choice by most of the clinicians, material scientist, surgeons and medical device designers [11]. The implant applications of titanium included replacement of hard tissue in artificial bones, joints, cardiovascular implants like prosthetic heart valves, pacemaker cases, circulatory devices, dental implants, screws and staples for spinal surgery [12]. They are preferred materials for load- bearing and long term implants and reveal high mechanical properties and biocompatibility. However, they are bio inert, thus could not directly bond to living bone tissues [13, 14].

4 Surface Modification for Titanium Implants

The selection of right biomaterial for a specific biomedical application relies on the bulk properties of biomaterial such as corrosion resistance, non-toxicity, controlled degradability, modulus of elasticity and fatigue strength. After implantation, a sequence of local events such as classic foreign body response and fibrous tissue capsule formation occurs around the implant [15]. The biomaterial surface act as a major influencing factor for the above said unfavorable reactions of the body, as the initial contact of the body is with the surface [16]. Despite many advantageous properties, Ti-based alloys are poor in bioactivity, as they cannot bond directly with bone due to fibrous encapsulation and promote new bone formation on the surface at the initial stage after implantation [17]. Successful implantation of titanium implants lies in demonstrating excellent osseointegration at the bone-implant surface and hence it is essential to improve the tissue bonding properties of titanium implants. The schematic representation of the successful implantation procedure is shown in Fig. 1(A). The direct contact between the bone and implant Fig. 1(A)–c is essential for successful implantation procedure. The formation of fibrous tissue can start a vicious cycle of micromotion and inflammation around the implant surface, which isolates implanted material from the surrounding bone and finally leads to implant failure [18, 19].

The surface chemistry and structure of the material are the two essential factors determining the interactions of the material with the neighboring tissue. Also, it is evident that the extend of osseointegration and the response of titanium-based biomaterials depend entirely on its biocompatibility and surface properties [21]. Hence various surface modifications such as mechanical, physical, chemical, electrochemical or biochemical modifications [22–31] are considered at improving the osseointegration property of the implant surface. The proper surface modification method not only retains the excellent bulk features of titanium and its alloys, but also improves specific surface properties required for diverse clinical applications [27].

5 Nanoscale Surface Modification-Formation of TiO_2 Nanotubes

The exploit of nanostructured materials has been proposed to resolve some of the problems presently associated with implants. As the modified coated surfaces obtained from various surface modification techniques are not reliable for long period due to the relatively higher micrometer level thickness and poor adhesion of the coated layer to the substrate [32], fracture and delamination occur at the bone-implant interface. The length scale of the hierarchical structure of the tissue must be considered, while designing a synthetic material which is intended to efficaciously replace living tissue in an organism [33]. The whole length scale from the nanoscale up to macroscale needs to be taken into account. The interactions between the bone

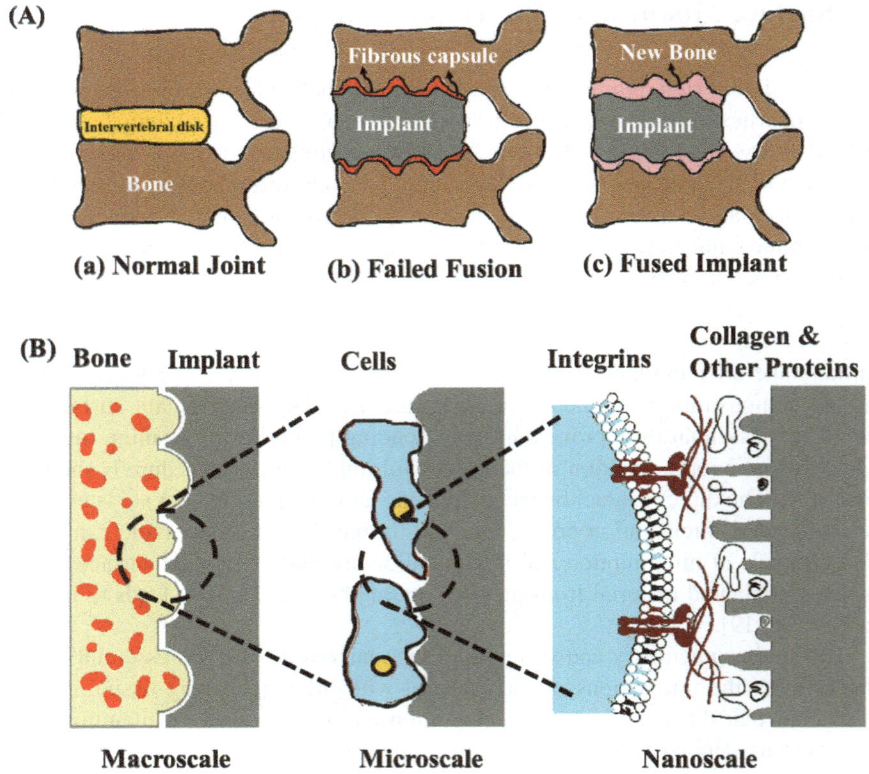

Fig. 1 (A) Schematic representation of (a) a normal joint, (b) a failed implant and (c) a fused and osseointegrated implant. (B) Interactions between the bone and the implant surface at different length scales (macro, micro and nano) (Redrawn from ref [15, 20])

and the implant surface at different length scales are shown in Fig. 1(B). The implant should provide a good mechanical fixation with bone at the macroscale. The micro- and submicro features presented on the surface can directly interact with osteoblasts and mesenchymal stem cells at the microscale. At the nanoscale, cell membrane receptors, such as integrins, can identify proteins adsorbed on the surface, which in turn are controlled by the nanostructures on the surface [15]. At the initial level, bone has a structural hierarchy in the nanometer regime and hence nanostructured materials are believed to play a vibrant role in orthopaedic research [34]. Modifying Ti surfaces with the formation of self-organized TiO_2 nanotubes enhances bone-bonding characteristics than conventional micro roughened surfaces. Altering the surface morphology of the titanium to nanoscale by forming nanotube structures can enhance the surface properties of titanium [35–37]. Also, the large surface area provided by the nanostructured topography renders available sites for the adsorption of protein and thus improves the cell-implant interactions [38]. Additionally, the structural modification at the nanoscale would mimic the nanostructure of the natural bone and hence favor a positive cellular interaction with cells [39].

Synthesis of titania nanostructures can be achieved by various routes including template–assisted methods [40, 41], hydro/solvothermal approaches [42, 43], Sol-gel method [44, 45] and by electrochemical method [46, 47]. Among all these methods, anodization, a relatively inexpensive and scalable electrochemical process is widely used for the growth of most spectacular, self-organized 1D nanotube structures. These nanotubular structures integrate highly controllable geometry at the nanoscale fascinating chemical and biological properties [48].

6 Formation Mechanism of TiO_2 Nanotubes

Figure 2 illustrates a typical two-electrode set up for the anodization of Ti substrates. The metallic Ti/alloy and platinum were used as anode and cathode respectively in the construction of electrochemical cell. The direct current generated by a DC power source in this case, drives the anodization process. The electrolyte temperature, pH

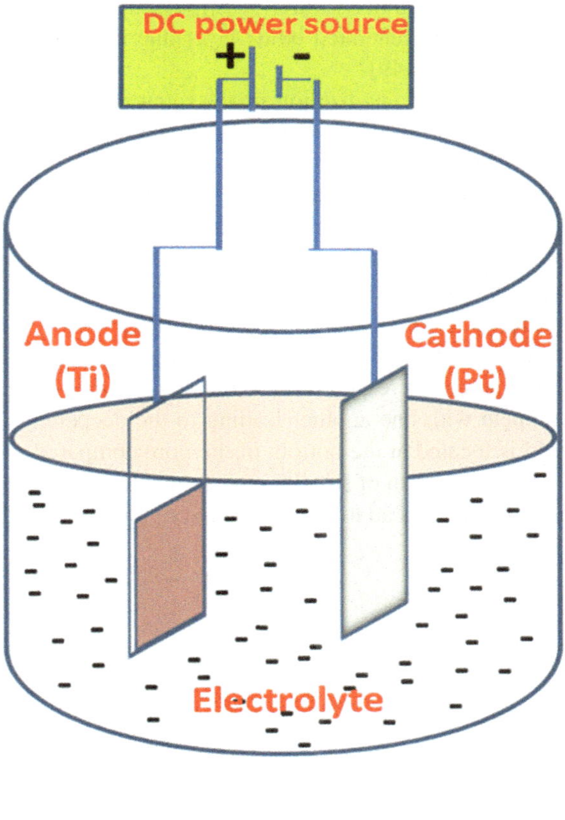

Fig. 2 Schematic representation of anodization cell

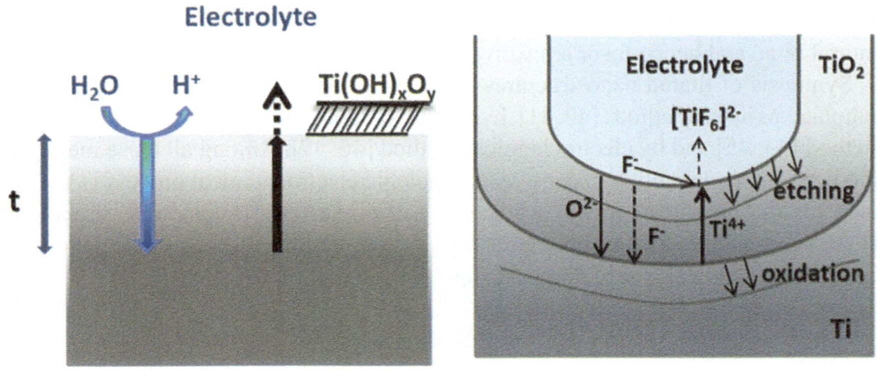

Fig. 3 Schematic representation of the Ti anodization (a) in absence of fluorides (results in flat layers), and (b) in presence of fluorides (results in the tube growth) (Redrawn from [48])

and real-time circuit potential and current can be measured during the growth of metallic oxide layer.

The chemistry that dictates the formation of nanotubes includes the interactions of the three competing reactions i.e. (i) field-assisted oxidation at the metal surface (ii) field-assisted dissolution of Ti metal to form titanium dioxide and (iii) chemical dissolution of generated oxide layer due to the etching by fluorides present in the electrolyte [48, 49].

Initially, the anodization process generates a stable TiO_2 layer of thickness 't' due to the transport of oxide and metallic ions across the oxide layer. When the electrochemical conditions are optimized and a chemical etching source such as fluoride compound is integrated into the electrolyte, field-assisted dissolution of the transported metallic ions and chemical dissolution of oxide layer takes place and this lead to the formation of porous (or) tubular layer [50, 51]. The schematic representation of the Ti anodization in the presence and absence of fluorides is shown in Fig. 3.

The following Eqs. (1) and (2), (3) represents the oxide layer formation and dissolution reactions respectively. The reactions described by Eqs. (1)–(3) subsequently compete with one another leading to the deepening of the pores as a high electric field is located at the bottom of the pore compared with the tube wall [52]. Finally, due to the growth of smaller pits at the inter-pore region, vertical tubes start to form and ultimately lead to the generation of a tubular structure.

$$Ti + 2H_2O \rightarrow TiO_2 + 4H^+ + 4e^- \tag{1}$$

$$TiO_2 + 6F^- + 4H^+ \rightarrow [TiF_6]^{2-} + H_2O \tag{2}$$

$$Ti^{4+} + 6F^- \rightarrow [TiF_6]^{2-} \tag{3}$$

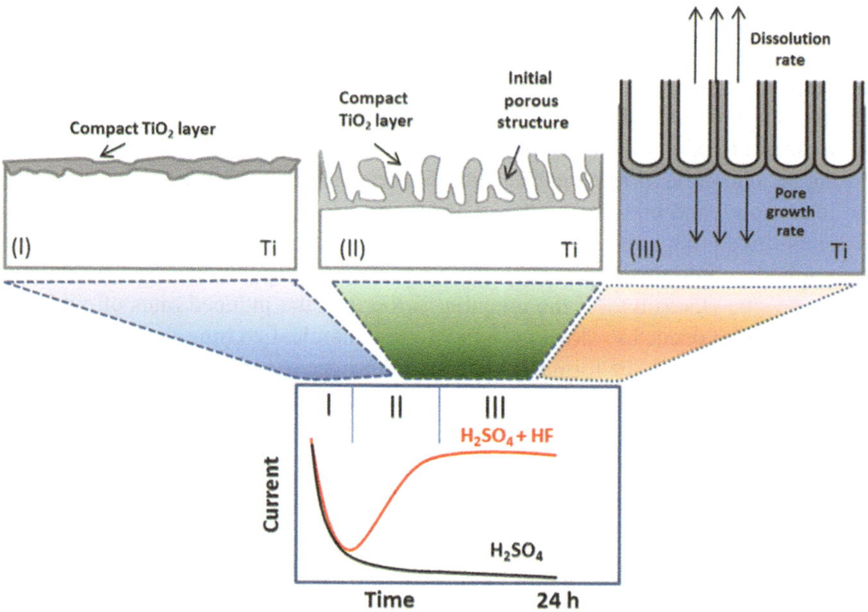

Fig. 4 Characteristic current-time plot for anodization of Ti substrates with and without fluorides in the electrolyte and the evolution of corresponding surface morphology in electrolytes containing fluorides (Redrawn from [48])

A typical anodization current—time plot that develops a nanotube layer is represented in Fig. 4. The formation of nanotubes can be related to main three key phases of current—time plot. In Fig. 4, an initial decrease in the current over time (stage I) observed, is due to the oxide layer formation on the surface. Further, a decrease in electric field strength is observed as represented in the figure in the absence of a chemical etching agent in the electrolyte. In the presence of electrolyte, the initial decay period (phase I) is followed by an increase in current, indicates the interactions of reactions (i), (ii) and (iii) mentioned above and this results in the formation of a porous structure. Finally, at stage III, an equilibrium is attained between the rate of oxide formation and dissolution reactions and as a result, the current value attains a steady state due to the formation of nanotubes [51].

7 Factors Influencing the Growth of Nanotube Layers

7.1 Effect of Anodization Voltage

In electrochemical anodization, applied potential plays an essential role in modifying the topography of nanotube structures. The applied anodization potential influences the electric field strength experienced across the formed oxide and thus affecting the ions migration and ultimately the diameter of the nanotubes [53]. A linear relationship exists between the diameter and the applied potential [54] and also to the thickness of the barrier layer at the base of the nanotubes [55]. The potential applied has a direct effect on the rates of oxidation and dissolution reactions that takes place during the nanotube formation. Cai et al. [56] reported that the nanotube diameter, at the tube mouth and bottom are directly proportional to the applied potential. It was also mentioned that anodically grown nanotubular layers will only be generated, if the applied potential is kept within the range dictated by the chemical dissolution activity of the electrolyte. If the potential is too low, the induced rates of oxidation and dissolution occur too slowly relative to the chemical etching performed by the electrolyte, leading to the formation of pits and not the oxide layer. If the potential is too high, field-assisted oxidation occurs too quickly relative to the chemical dissolution of the metallic substrate, which results in the formation of disturbed oxide layer rather than the nanotubular layer [57]. Sulka et al. [58] extensively studied the influence of anodizing potential on the structural parameters of porous anodic titania including pore diameter, inter pore distance, wall thickness, porosity and pore density.

To study the effect of applied voltage on the morphology of TiO_2 nanotubes, Ti substrate was anodized in an electrolyte of glycerol with 1 weight % NH_4F and 20 volume % water for 1 h at different voltages (10, 20 and 30 V) [59]. Figure 5 shows the HR-SEM images and the corresponding 3D images of the formed nanotubes at different voltages. The influence of applied electric field in the formation of nanotubes with varying diameter was visible from the HR-SEM (Fig. 5) images.

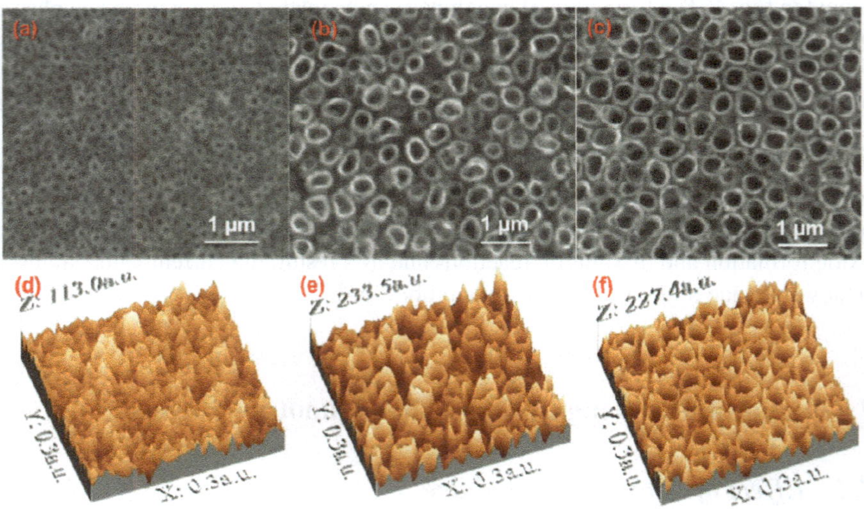

Fig. 5 HR-SEM images (top view) and corresponding 3D images of Ti specimen anodized at different voltages (a, d) 10 V (b, e) 20 V and (c, f) 30 V (Adapted from [59])

Samples anodized at 10 V exhibits a porous morphology with average diameter 32 nm. No inter-tubular space was observed. Regular and self-organized nanotubes with diameter approximately 45 and 58 nm for the titanium substrate anodized at 20 and 30 V respectively were observed. Nanotubes were separated with inter-tubular spaces and the height of the nanotubes was found to be 540 nm.

The influence of potential in the formation of nanotubes in the anodization of Ti-6Al-7Nb alloy was also studied [19]. The FESEM images of titania nanotubes obtained by anodisation of Ti-6Al-7Nb for 1 h in an electrolyte consisting of 1 M H_2SO_4 and 0.08 M HF at different voltages 10, 20 and 30 V are shown in Fig. 6a, c and e. Sample anodized at 10 V show a nanoporous morphology with a diameter of 35 nm and height 250 nm. Moreover, no inter-tubular space and wall separation was observed at this voltage. TNT anodized at 20 and 30 V show a very regular and vertically aligned tubular structure as shown in the FESEM images. Similarly, the 3D image of the TiO_2 nanotube array generated from FESEM image for the same from higher magnification are given in Fig. 6b, d and f.

The tubular morphology formed on Ti-6Al-7Nb alloy resembles well-known as in the case of nanotubes obtained on pure Ti foils. Also, the 2D profile of the TiO_2 nanotube layer generated from FESEM image is demonstrated in Fig. 7. These studies

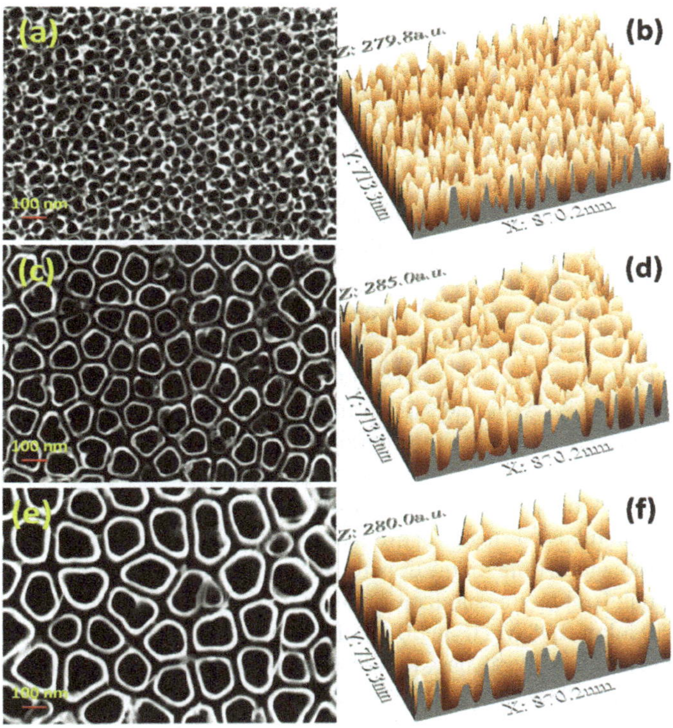

Fig. 6 FESEM images of TiO_2 nano tubes (left) and 3D images (right) for 10 V (a and b), 20 V (c and d) and 30 V (e and f) (Reprinted with permission from [19])

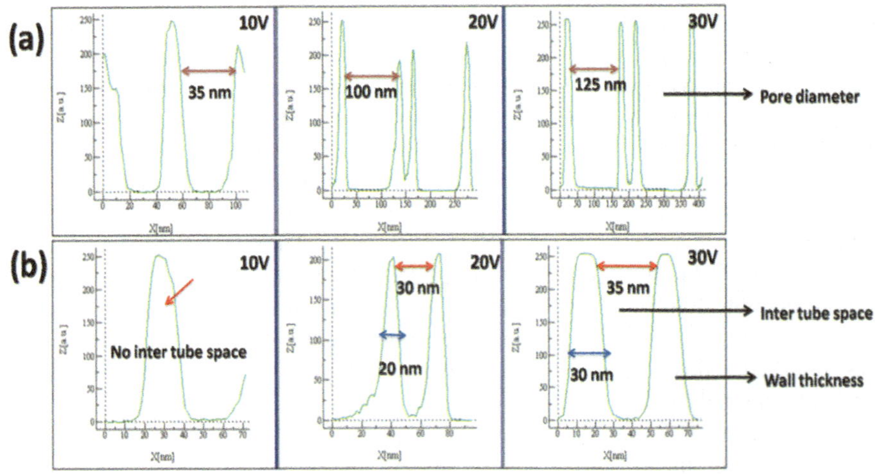

Fig. 7 2D profile of TNTs at 10, 20 and 30 V; **a** pore diameter, **b** inter tube space and wall thickness (Reprinted with permission from [19])

indicate that the pore diameter increases with increase in voltage which is in line with earlier reports on pure titanium.

7.2 *Effect of Anodization Time*

During anodization the key factor which controls the morphology of nanostructures is anodization time. In general, identification of critical point is important for the formation of nanotubes. It has been reported that this critical points occurred within 2–3 min after the initiation of anodization process in the case of 0.5 M HF electrolyte. It is important to give enough time for nanotube nucleation to takes place on the substrate surface [48]. As the anodization duration increases, the thickness of the oxide layer increases. Also, there is an increase in tube length with time, until an equilibrium reached between the formation and dissolution of titanium oxide. After this point, the nanotube length becomes independent of anodizing time [60].

It was believed that anodization time influences the tubular morphology. Figure 8 shows TiO_2 nanotubes fabricated at different times in Ti-6Al-7Nb alloy. The substantial role of anodization time for the transformation of pores to tubes was demonstrated by Mohan et al. [61] by the anodization of Ti-6Al-7Nb alloy in an electrolyte made by mixing 50 ml of 0.08 M HF in 50 ml of 1 M sulphuric acid. The color of the titanium oxide layer turned from dark purple to blue, yellow and finally to greenish-blue during this period. The change in color of the titanium oxide is due to the increase in the thickness of the oxide layer which results in interference of light.

The current-time transient curve for Ti-6Al-7Nb anodized at a different time at 30 V is given in Fig. 9. The authors explained that the initial drop in current is

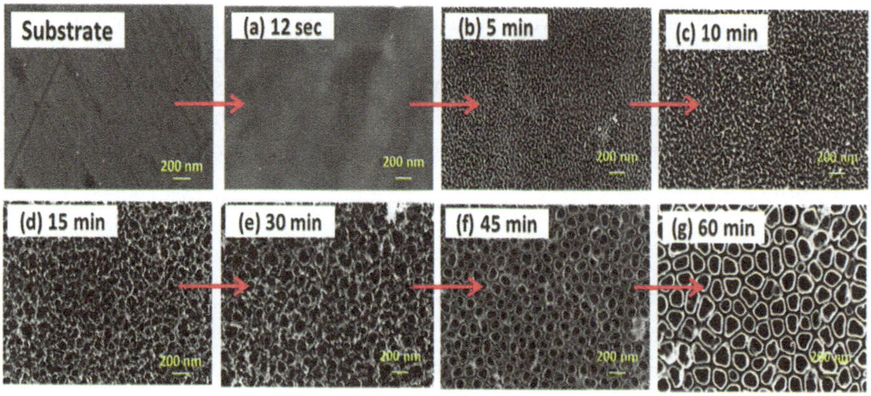

Fig. 8 FESEM images of Ti-6Al-7Nb anodized at different time **a** 12 s, **b** 5 min, **c** 10 min, **d** 15 min, **e** 30 min, **f** 45 min and **g** 60 min (Adapted and reprinted with permission from [61])

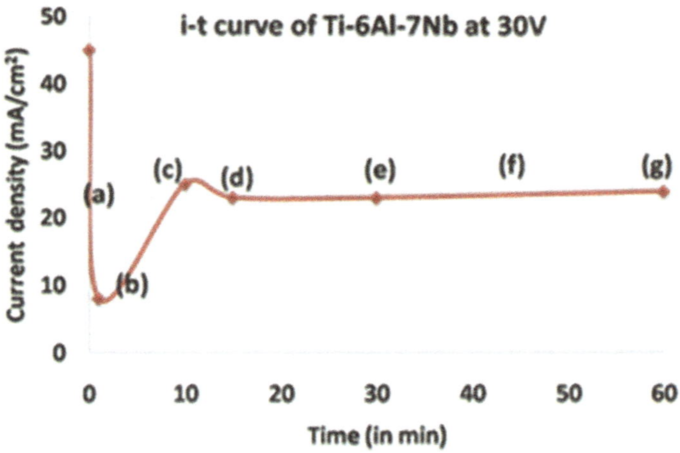

Fig. 9 Current-time transient curve for Ti-6Al-7Nb anodized at different time at 30 V (Reprinted with permission from [61])

due to oxide formation and subsequent increase in the corrosion current due the development of porous structure due to the dissolution of oxide layer by fluoride (F^-) ions. Finally, the current reaches a steady-state with the steady growth of the TiO_2 nanotubes. It was demonstrated that, the conversion of oxide layer to a tubular structure begins when the corrosion current starts to decrease.

In aqueous electrolyte, the aggressive condition does not allow the nanotubes to grow longer than a few μm [56, 62]. However, the anodization usually carried out from 30 min to 2 h for structural rearrangement and thereby increase the degree of self-ordering [63]. In the case of organic electrolyte, the growth process is slower and dissolution phenomena are not as significant as it occurs in water. Hence it was possible to obtain nanotubes over 100 μm by extending the anodization time [64]. However, the length of the nanotubes has no noteworthy influence on the biocompatibility of nanotubes as the cells contact initially on the upper surface of the tubular structure. On the other side, it was reported that the tube length has a great effect on drug release rate which can be controlled by tuning the aspect ratio of the nanotubes [65].

7.3 *Effect of Electrolyte Composition*

The nanostructures are usually formed by a variety of fluoride-containing aqueous [51, 66] or organic [67, 68] electrolyte by anodizing titanium at different voltages. It has been reported that the geometry of the formed nanotubes has a direct influence on the basic electrolyte parameters such as pH, fluoride ion concentration temperature and water content [48, 61, 69]. The fluoride ion concentration in the electrolyte has an important role in the self-ordering of nanotubes. The fluoride concentration should be minimum and sufficient to ensure the growth of nanotubes. The studies have shown that the typical F^- concentration leading to the self-ordering of nanotubes is in the range of 0.3–0.5 wt% [70, 71]. The presence of fluoride ions in the electrolyte selectively etches the surface of titanium during the growth of oxide layer, which allows the formation of nanotubes. An increase in the concentration of fluoride ions obtained from the addition of HF or NH_4F increases nanotube diameter and length [72]. The influence of fluoride ion concentration in two different electrolytes 1 M $(NH_4)_2SO_4$ and 1 M Na_2SO_4 on TiO_2 nanotubes was reported. Nanotubes grown in organic electrolyte tend to be longer, homogenous and have smoother walls as the chemical dissolution and fluoride ion transmission rate is relatively low in viscous organic electrolytes [73, 74].

The water content in the electrolyte plays a vital role in the formation of titania nanotubes. Water content, the source of oxygen in most electrolytes is used to generate the initial oxide layer and has been reported to have a direct effect on nanotube wall thickness and length [48]. Raja et al. [75] reported that, to generate nanotube arrays on titanium, a minimum water content of 0.18 wt% is necessary in an electrolyte containing NH_4F and ethylene glycol. Mohammed et al. [76] reported that, an increase in nanotube diameter, wall thickness and length, when the water content increases up to 30% in an HH_4F containing electrolyte at a constant potential of 20 V. Various studies [62, 70, 77] reported that the nanotubes formed from low water content electrolyte results in smooth tube walls, whereas side wall ripples are formed at higher water content. This could be attributed to the faster etching of the fluoride-rich layer between the tubes than the growth speed of the tubes into the underlying substrate.

To evaluate this aspect, the water content in the electrolyte was varied by holding the anodization voltage and anodization time to 20 V and 1 h respectively [69]. A mixed solution of glycerol and ammonium fluoride with different water contents 10, 20 and 40 volume % was used as the electrolyte. The obtained nanotubes were designated as TNTA-10, TNTA-20 and TNTA-40 respectively. The HR-SEM images of titania nanotubes formed in glycerol, ammonium fluoride electrolyte with different water contents, its cross- sectional view and the corresponding 3D images of the nanotubes generated from the HR-SEM images are given in Fig. 10a–l.

By varying the addition of water content into the electrolyte during the anodization process, the diameter of the nanotubes can be tuned. The morphology of the nanotubes formed in electrolyte with 1 volume % of water (TNTA-1) was given for comparison, despite the scanty reports available on the surface morphology of nanotubes formed at a low percentage of water content. It was observed that, irregular nanotubes with non-uniform diameter (Fig. 10a–c) were formed on titanium at lower water content.

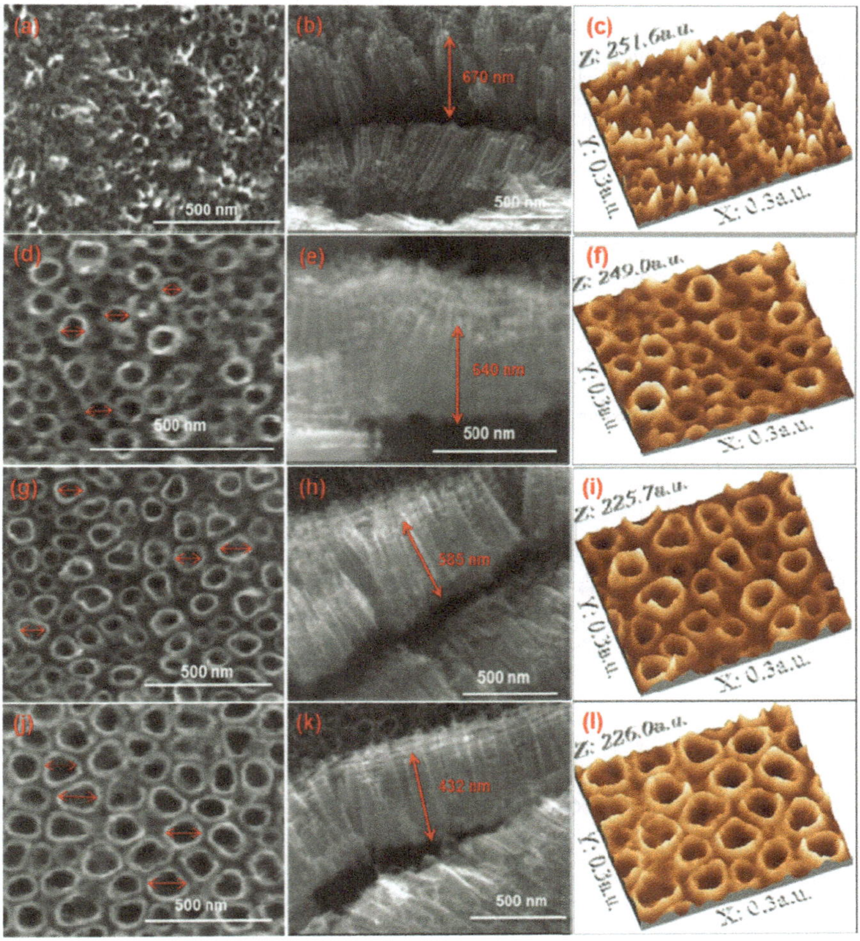

Fig. 10 HR-SEM images of titania nanotube arrays (top view) (left), cross-sectional view (middle) and 3D images (right) for (a, b, c) TNTA -1, (d, e, f) TNTA -10, (g, h, i) TNTA -20 and (j, k, l) TNTA -40 specimens (Reprinted with permission from [69])

As the water content increases, nanotube length was found to decrease and diameter of the nanotubes increases as evidenced from the HR-SEM images. It was apparent that, increasing water content increases the self-organization behavior of formed nanotubes. Highly self- organized nanotube structures were observed in TNTA-40, with higher water content. The formation of nanotubes with a varying diameter in electrolyte with different amounts of water is schematically represented in Fig. 11.

The diameter obtained for TNTA-1, TNTA-10, TNTA-20 and TNTA-40 specimens were found to be approximately 14, 30, 40 and 60 nm respectively. Intertubular spaces were observed between the nanotubes formed at high water content as evident from HR-SEM and 3D images. It is well-known that, controlling the tube features especially the nanotube diameter and inter tubular spacing supports the adhesion, proliferation and mineralization of cells [60]. Even after the cell adhesion, diameter and spacing between the nanotubes offers an important path for the continuous

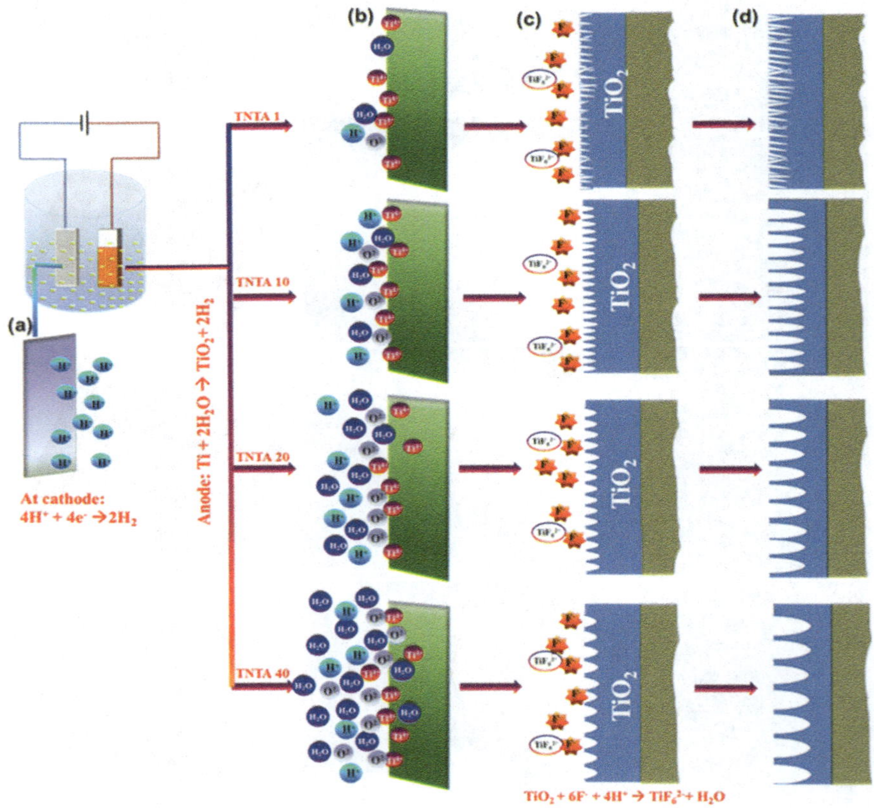

Fig. 11 Schematic diagram of nanotube formation **a** Cathodic reaction. **b** Anodic reaction. **c** Initiation of nanotube formation. **d** Formed nanotubes with different diameter (Reprinted with permission from [69])

fluid flow of culture media, ions, proteins, nutrients essential for healthy cell growth, increase spaces for the exchange of gas and cell signaling molecules for an overall improved cell environment. In human body, due to circulation and repetitive loading and unloading of bones during various activities, the fluid flow naturally occurs in the interstitial spaces around the bone cells [78–80].

7.4 Effect of Electrolyte Temperature

Another important parameter that influences the growth of nanotubes is anodization temperature. In most of the anodization process, the electrolytes are kept at room temperature to produce the nanotubes. It has been reported that an increase in electrolyte temperature leads to the formation of nanotubes with a larger diameter, shorter length and thinner walls [48]. The dissolution rates of the nanotubes are affected by the electrolyte temperature [53]. Prida et al. [81] reported that the growth of nanotubes is inhibited at lower temperatures (2 °C) in aqueous electrolyte, whereas the temperature range between 0 and 40 °C are reported to be most favorable for the growth of nanotubes in organic electrolyte [62]. Wang et al. [82] reported that the diameter of the nanotubes does not depend on the anodizing temperature in aqueous electrolyte, whereas the nanotube diameter was found to increase at higher temperatures while using in organic electrolyte. This could be due to the fact that at lower temperatures, the viscosity increases, which reduces the ion migration especially fluoride ion and as a result of which etching of TiO_2 gets slowed down, leading to smaller diameter tubes.

The influence of electrolyte temperature during the anodization of Ti-6Al-7Nb alloy was studied and reported [61]. Figure 12(a–f) illustrated the FESEM images of anodic TiO_2 nanotube arrays obtained after anodization of Ti-6Al-7Nb for 1 h at 30 V in an electrolyte consisting of 1 M H_2SO_4 and 0.08 M HF at 5, 10, 25, 30, 50 and 70 °C. The sample anodized at 5 °C (Fig. 12a) revealed a premature porous structure with an un-dissolved oxide layer on the surface. At 10 °C, a porous oxide layer with pore diameter ~35 nm covers the entire surface in an irregular pattern as shown in Fig. 12b. No clear wall separation and inter-tubular space noticed at this stage.

In Fig. 12c, TiO_2 nanotube arrays formation with an average inner pore diameter of 125 nm, length of ~250 nm, the wall thickness of 30 nm and an inter-tube space 35 nm were observed at 25 °C. Also, the morphology of the nanotubes was smooth and circular without any defect. FESEM images of samples anodized at 30, 50 and 70 °C (Fig. 12d–f) exhibited regular and vertically aligned tube structure.

The effect of decreased electrolyte temperature on anodization was also studied by Wang et al. [82]. It was reported that there is a decrease in nanotube diameter and increase in wall thickness on anodization done in glycerol-based aqueous electrolyte kept in an ice bath.

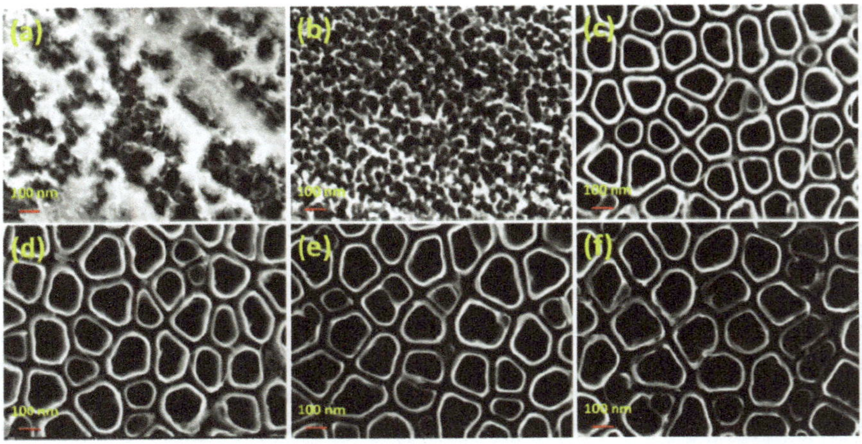

Fig. 12 FESEM images (top view) of samples anodized at different temperatures **a** 5 °C. **b** 10 °C. **c** 25 °C. **d** 30 °C. **e** 50 °C. **f** 70 °C (Reprinted with permission from [61])

7.5 Effect of Electrolyte pH

Acid, salt or alkaline solutions have been widely used as electrolytes for the anodization of titanium to achieve a varied range of oxide layer thickness [83]. The dissolubility of the electrolyte and the hydrolysis ability of the oxide layer are influenced by the p^H of the electrolyte [72]. The different p^H electrolyte has an impact on the chemical reaction rate of the etching of the porous oxide layer, which results in the formation of TiO_2 nanotubes. Previous reports clearly showed the formation of nanotubes by anodizing titanium substrate in 60 ml of 0.15 NH_4F at 20 V for 1 h at a pH ranges from 2 to 4 values. At this pH range, the tube length was found to increase with an increase in pH. While the anodization is done at low p^H, the high concentration of H^+ ions enhances the dissolution reaction of the oxide layer leading to shorter nanotube formation. Similarly, at high p^H as the concentration of H^+ ions decrease the dissolution rate of the oxide layer also decreases, leading to the increase in length of the nanotubes by increasing the anodization process [84, 85]. An increase in nanotube length from 1.2 μm to 6.0 μm for electrolyte p^H values of 3.0 and 5.0 respectively, after 17 h of anodization was reported [86]. For an acidic range pH (1–6), an increase in tube length from 3.5 to 5.3 μm under the same voltage and time conditions. In alkaline electrolyte, with pH 10 and pH 12, the length of nanotubes obtained was found to be 5.6 and 1.8 μm respectively [60].

8 Anodization of Titanium Alloys

Basically, titanium alloys are classified as α, near-α, α + β, metastable β, or stable β depending upon the room temperature microstructure [87]. In this aspect, titanium alloying elements can be classified as α-stabilizers (Al, O, N, C), β-stabilizers (Nb, V, Ta, Fe Cr, Si, Co, Mn) and neutral such as Zr [72]. Titanium alloys are widely used in hard tissue replacements in artificial bones, joints and dental prosthetic applications. Various attractive properties of the Ti-based alloys such as high corrosion resistance, exceptional mechanical properties and good biocompatibility were the driving force for the early introduction of α (Cp-Ti) and α + β (Ti–6Al–4 V, Ti-6Al-7Nb) alloys as well as the more recent development of modern and orthopaedic metastable β Ti and β stabilized Ti alloys [16, 27]. Several studies reported the formation of nanotubes on different Ti alloys [61, 88–94]. Nakada et al. [95] suggested using Ti-Nb-Ta-Zr (TNTZ) alloy for orthopaedic implants because of its exceptional mechanical properties, biocompatibility and anticorrosion properties. In another study, an anodized nanotube oxide layer with two size scale structures was reported on Ti-28Zr-8Nb alloy using buffered electrolyte [96]. The influence of alloy composition on the formed anodic oxide layer morphology on binary titanium alloys such as Ti-xTa, Ti-yZr, and Ti–Al was investigated by the researchers [89, 97]. The study revealed the formation of nanotube arrays on Ti-Zr alloy with a decreased diameter and increased tube length with increasing zirconium content. A tantalum oxide (Ta_2O_5) porous layer with pore diameter ranges from 2 to 35 nm on tantalum metal can be formed electrochemically in acid-containing HF as electrolyte [98]. Highly ordered Ta_2O_5 nanoporous layer have grown on tantalum foils by anodization using glycerol and NH_4F containing electrolyte and it was found that average pore diameter and thickness were affected by the NH_4H concentration [99].

Anodization of CP Ti and titanium alloys (Ti-6Al-4 V, Ti-6Al-7Nb, Ti-13Nb-13Zr, and β-21s) has investigated at optimized parameter [61]. FESEM images (left side at lower and right side at higher magnifications) of anodic TiO_2 nanotube arrays (TNT) obtained after anodization of CP Ti, Ti-6Al-4V, Ti-6Al-7Nb, Ti-13Nb-13Zr and β-21S for 1 h in an electrolyte consisting of 1 M H_2SO_4 and 0.08 M HF at 25 °C are shown in Fig. 13 (a–e). Highly ordered, regular and vertically aligned TNT structures were observed on CP Ti, Ti-6Al-4V and Ti-6Al-7Nb as shown in Fig. 13a–c. The metastable β alloy, Ti-13Nb-13Zr given in Fig. 13d showed the formation of nanostructured dual morphology with varied pore diameters. In the same image, a high degree of ordering in tubes was visible beneath the small porous structure. On the contrary, beta alloy, beta-21S in Fig. 13e showed no sign of nanotubular morphology due to composition of electrolyte solution used, whereas, the electrolyte with 0.25 M NH_4F in ethylene glycol reported the formation of nanotubular structure [98]. At higher magnification (right side), random structure was observed. Thus, at various compositions and at fixed parameters such as time (1 h), temperature (25 °C) and electrolyte composition, titanium alloys exhibit significant properties.

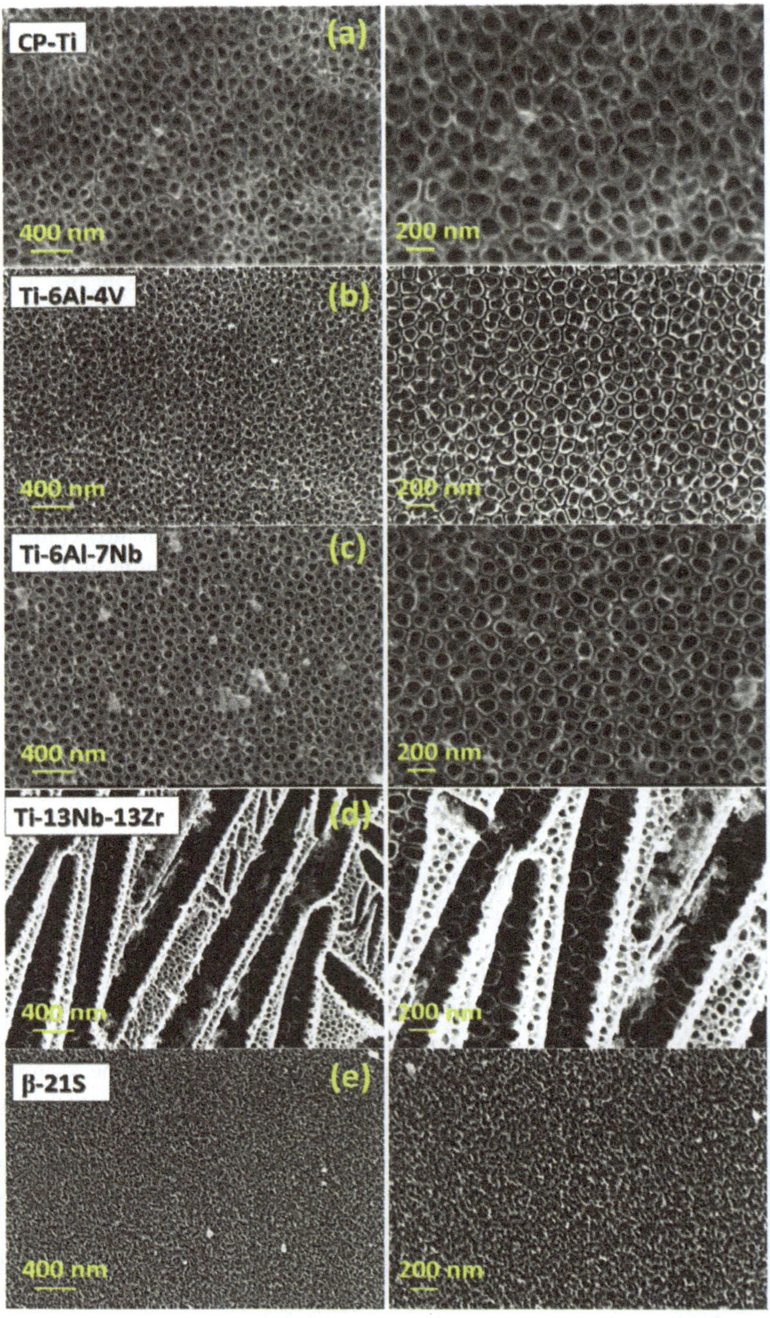

Fig. 13 FESEM images (top view) of different Ti samples anodized for 1 h at 25 °C. **a** CP Ti. **b** Ti-6Al-4 V. **c** Ti-6Al-7Nb. **d** Ti-13Nb-13Zr. **e** β-21s (Reprinted with permission from [61])

Due to the presence of the valve metals such as Al, Nb, Zr in the alloy composition, nanotubular structure with similar morphology were able to form on CP Ti with α phase, Ti-6Al-4V, and Ti-6Al-7Nb with $\alpha + \beta$ alloys.

9 Functionalization of Titania Nanotubular Structures

The TiO_2 nanostructures formed via electrochemical anodization process have been widely explored as a favorable biomaterial for a wide range of biomedical applications. Bone cell adhesion, growth and differentiation, prevention of bacterial adhesion, apatite forming abilities, drug delivery were demonstrated on nanostructured titania surfaces, showing its potential as bone implants [100–102]. For improved adhesion of cells on the nanostructures, it is essential to improve the surface functionalities of these nanostructures.

9.1 Surface Functionalization with Inorganic Bioactive Agents

The surface modification of titania nanotubes is necessary for improving the biocompatibility via introduction of new properties and functionalities. The tubular structure of titanium favours the incorporation of various inorganic bioactive agents to impart some specific properties. The bioactivity and antibacterial activity of nanotubes can be improved by the incorporation of antibiotics, growth factors, antibacterial agents, bone growth enhancement agents and anti-inflammatory agents. Loading of various inorganic bioactive elements such as silver, strontium, zirconium and zinc were demonstrated to achieve long term antibacterial activity and osteogenesis induction [103]. Nanotubes not only facilitate the loading of these agents, but also provide a means to control the loading capacity and release rate by varying the structural parameters.

To improve the benefits of TiO_2 nanotubes, we have incorporated various inorganic bioactive agents such as Strontium, Zirconium and Zinc ions into titania nanotubes and investigated its biocompatibility and electrochemical behaviour in stimulated body fluid environment. Indira et al. [104] reported the incorporation of Zr over TNT surface via dip coating method. The FE-SEM micrographs and the corresponding EDS spectra of TNT and Zr-TNT are shown in Fig. 14a–h. The nanotubes have an average pore diameter of 110 ± 4 nm and a wall thickness of 15 ± 2 nm respectively (Fig. 14a and b). The average inter pore diameter is about 128 ± 2 nm. Apparently, the layer consists of arrays of individual nanotubes with a length of 2.1 ± 0.3 μm (Fig. 14c). The presence of Ti, O and trace amount of F on the TNT surface was determined from the EDS spectrum.

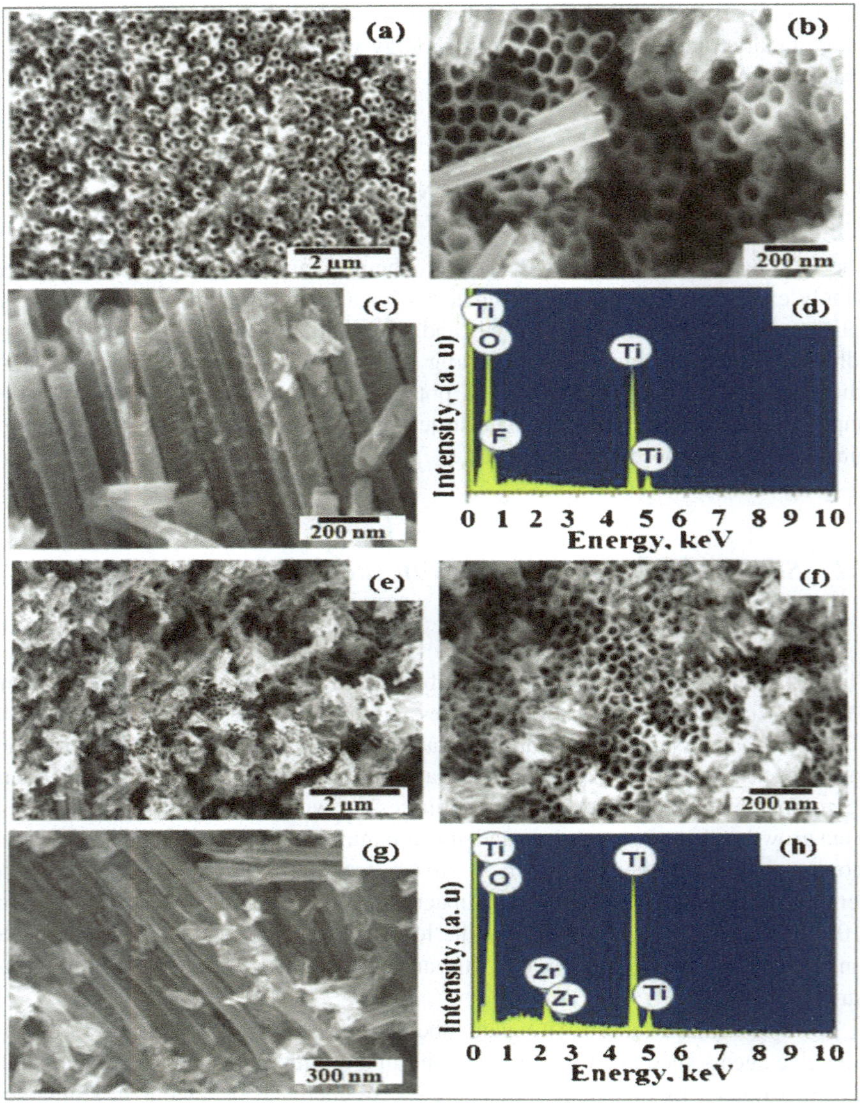

Fig. 14 **a–h** shows the FE-SEM micrographs and the corresponding EDS spectra of TNT and Zr-TNT (Reprinted with permission from [104])

Cluster like particles was observed over the TNT surface (Fig. 14e) after the treatment with $ZrOCl_2$ solution. The original tubular structure of TNT (Fig. 14f) was retained even after the incorporation of Zr ions, however, a small decrease (11 ± 5 nm) in the diameter was observed. The average pore diameter, wall thickness and tube length of Zr-TNT were found to be 99 ± 5 nm, 25 ± 3 nm and 2.3 ± 0.2 µm respectively. The developed Zr-TNT is supposed to be a good implant material in bio environment with biocompatibility, improved corrosion resistance and is likely to be widening the biomedical application of titanium implants with new surface properties in the nanoscale regime.

Strontium ranelate, is widely used clinically for the treatment of osteoporosis due to the biological importance of strontium. Administration of a suitable dose of strontium can promote bone formation and inhibit bone resorption, leading to enhance bone mass [105]. Indira et al. [106] also reported the fabrication of Sr–TNTA on titanium substrate by a simple dip-coating method on anodic TNTA. The FE-SEM micrographs and EDS spectra of the specimens are shown in Fig. 15a–h. Nanotubes arrays (Fig. 15c) with an average diameter and wall thickness of 110 ± 4 nm and 15 ± 2 nm, respectively were observed after the anodization of titanium. After deposition, the walls of the tube were covered with Sr ions. Also, some clusters of Sr ions were scattered over the top of TNTA surface. The TNTA (Fig. 15g) retained the unique nanotubular morphology even after the incorporation of Sr ions, however, a small decrease in the diameter during the transformation of TNTA to Sr-TNTA, was noticed, which may probably due to volume expansion during the conversion of titanium oxide to strontium titanate.

Orthopaedic implant failure has been a major concern for orthopaedic surgeons over the past decades. Poor osseointegration together with bacterial infection is one of the major reasons for implant failure. Hence titanium implants having enhanced osseointegration and antibacterial property to prevent implant-associated infection are highly desirable for the long term success of implants. To address this problem, zinc incorporated titania nanotube arrays (Zn-TNTA) were developed on titanium surface to enhance the biocompatibility and also to ensure antibacterial properties [59, 107]. Here in this study the incorporation of Zn into TNTA was achieved by simple dip-coating method. Different concentrations of zinc solution (0.25, 0.5, 1.0 and 1.5 M $Zn(OH)_2$ solution) were prepared to study the nature of deposition of zinc into the nanotubes. The samples were designated as NT-Zn (0.25 M), NT-Zn(0.5 M), NT-Zn(1.0 M) and NT-Zn(1.5 M) respectively. Our observations reveal that uniform deposition of zinc into the nanotubes can be achieved by dip-coating method. The dipping and lifting of the TNTA specimens enable the zinc hydroxide solution to penetrate into and between the nanotubes and get uniformly deposited along the entire length of the nanotubes.

Figure 16a–d represents the HR-SEM images of NT-Zn (0.25 M), NT-Zn (0.5 M), NT-Zn (1.0 M) and NT-Zn (1.5 M) specimens. As apparent from the SEM images, NT-Zn (0.25 M), NT-Zn (0.5 M), NT-Zn (1.0 M) and NT-Zn (1.5 M) exhibit quite different morphologies. As can be seen in Fig. 16a, for NT-Zn (0.25) M specimen, the filling of zinc initiates at the pores and randomly deposited at the surface of the nanotubes. At 0.5 M (Fig. 16b), concentration, a homogenous distribution of zinc oxide at the tube walls and intertubular spaces were obtained. The nanotube structures were not masked due to the filling of zinc oxide. However, the diameter of the nanotubes decreases after the incorporation of zinc oxide. At 1.0 M concentration (Fig. 16c), ZnO deposited on the TNTA surface appeared in the form of flakes and started to mask the nanotubes. The formation of typical ZnO "nanoflowers" was

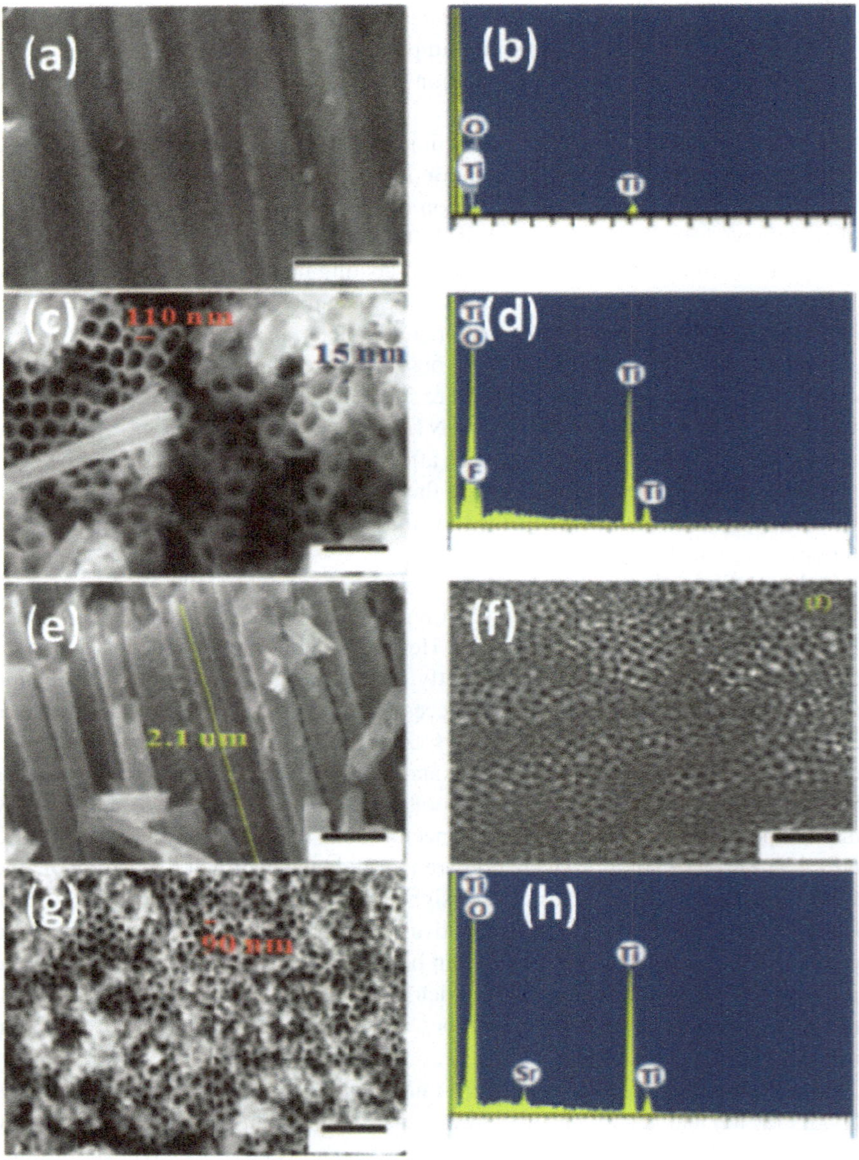

Fig. 15 FE-SEM images of (a) untreated Ti, (c) TNTA (e) cross sectional view of TNTA and (f) bottom view of TNTA and (g) Sr-TNTA and EDS spectrum of (b) untreated Ti, (d) TNTA and (h) Sr-TNTA (Reprinted with permission from [106])

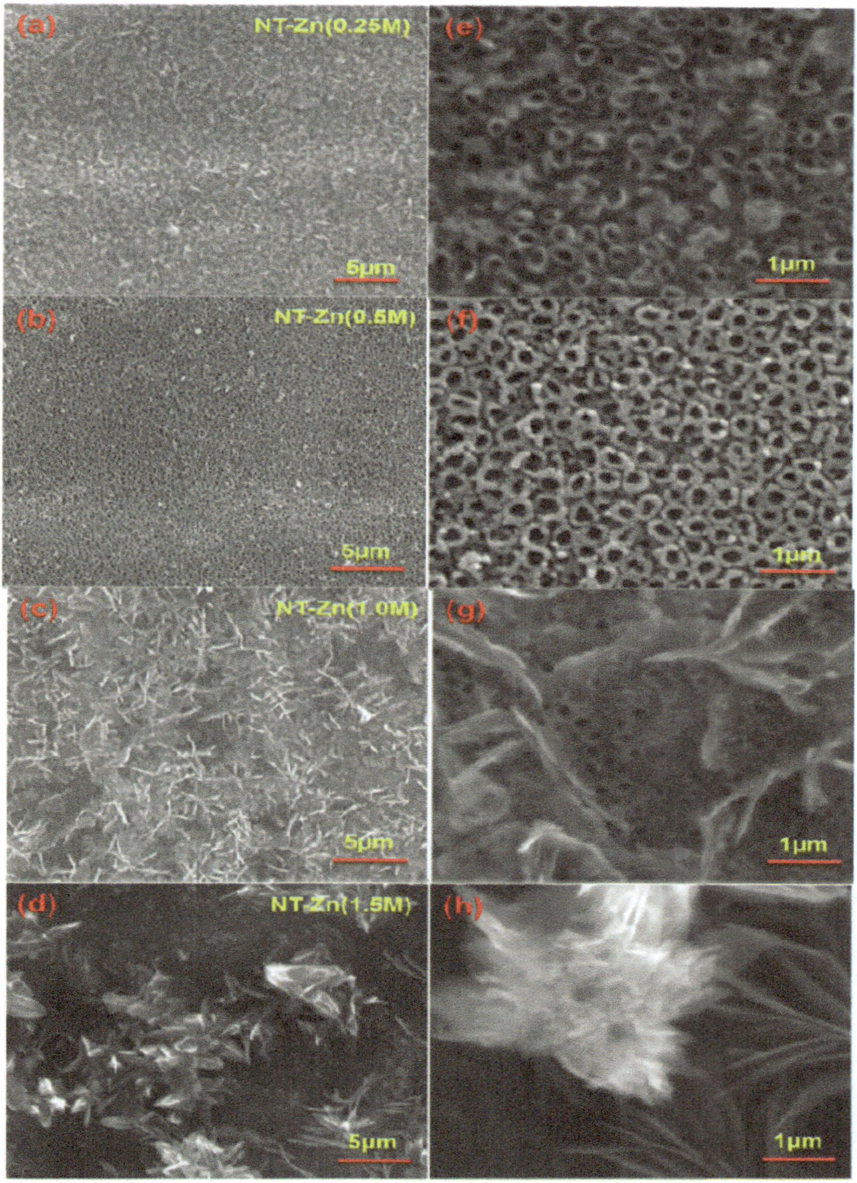

Fig. 16 HR-SEM images and corresponding higher magnification images of (a and e) NT-Zn (0.25 M), (b and f) NT-Zn (0.5 M), (c and g) NT-Zn (1.0 M) and (d and h) NT-Zn (1.5 M) specimens (Reprinted with permission from [59])

observed on the TNTA surface at 1.5 M concentration (Fig. 16d) and it fully covered the nanotube surface which was more clearly visualized in its corresponding higher magnification image provided in Fig. 16h.

Among the various concentrations studied, homogenous deposition of zinc oxide was observed at 0.5 M. Uniform deposition of zinc by retaining the tubular nature of the titania nanotubes is more advantageous for implant applications.

9.2 Surface Functionalization Using Polypyrrole

To tailor the surface properties of TiO_2 nanotubes and to advance their surface properties applicable to a wide variety of biomedical applications, an attempt has been made to synthesis Polypyrrole/Titania nanotube arrays (PPy-TNTA) hybrid on titanium surface [108]. PPy-TNTA hybrid was prepared by normal pulse voltammetry electrodeposition process in an electrolyte solution containing 0.2 M $LiClO_4$ and 0.1 M pyrrole. To benefit pyrrole monomer deposition, the pulse width was adjusted at 0.06 s and the pulse period was controlled at 4 s and 6 s respectively. The HR-SEM images for TNTA and PPy deposited on TNTA at various pulse potential 4 s and 6 s, and their corresponding cross-sectional images are demonstrated in Fig. 17a–f. The morphological differences became apparent, upon comparing the micrographs of the TNTA and the PPy-TNTA surface.

As evident from the images, at a pulse potential 4 s, the in-filtered PPy initially fills the Intertubular space between the nanotubes and then partially covered at the surface of the nanotubes. Further the pores of the TNTA were increasingly filled with

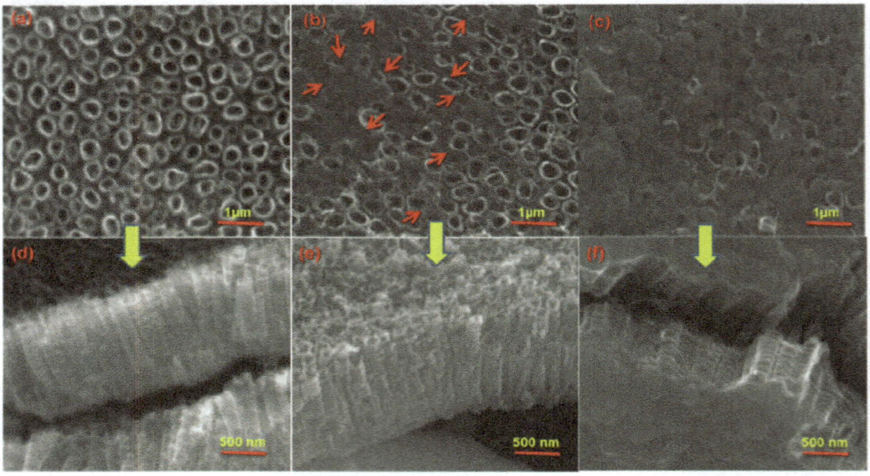

Fig. 17 HR-SEM images (a, b, c) and corresponding cross-sectional images (d, e, f) of TNTA, PPy -TNTA at 4 s and PPy-TNTA at 6 s (Reprinted with permission from [108])

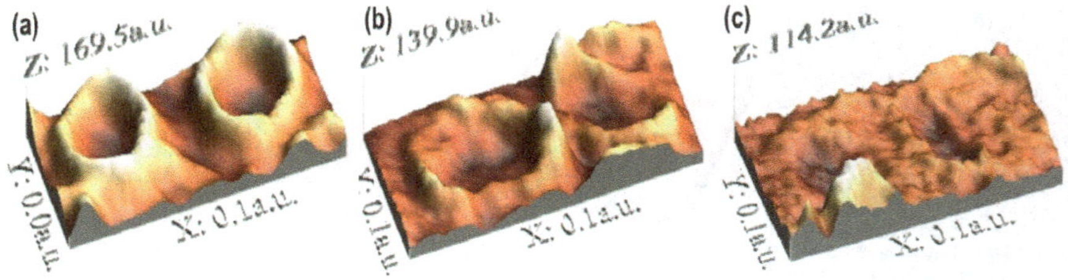

Fig. 18 3D images of a pair of **a** TNTA. **b** PPy/TNTA at pulse potential 4 s and **c** PPy-TNTA at pulse potential 6 s specimens (Reprinted with permission from [108])

PPy with an increase in pulse potential to 6 s. Moreover, a densely packed polymer film with spherical morphology was observed on the nanotube surface.

The 3D images obtained from HR-SEM images for a pair of nanotubes of TNTA and PPy-TNTA at pulse potential of 4 s and 6 s respectively are given in Fig. 18a–c. The deposition of polypyrrole initially into the spaces between the tubes, then inside the tubes, and if prolonged on the surface of the nanotubes can be achieved by electropolymerization under controlled pulse potential conditions.

The results of the above study indicated that the improved surface properties of PPy/TNTA pave the way towards the employment of the newly developed hybrid material as a viable substitute in biomedical field, especially in the field of orthopedic implants.

9.3 Surface Functionalization Using Plasma Nitriding

Now a days, pure titanium and $\alpha + \beta$ type Ti-6Al-4V alloys are commonly used as structural and/or functional biomaterials for replacement of hard tissues in devices such as artificial hip or knee replacements and dental implants [109]. The properties such as excellent corrosion resistance in the body fluid medium, no adverse tissue reactions, high mechanical strength, low modulus, low density and good wear resistance are expected to be possessed by an ideal biomaterial for implant applications. The tribological properties of titanium alloys can be enhanced using surface modification. It is possible to obtain surface layers consisting of titanium nitrides through physical and chemical vapour deposition, laser treatments, ion implantation and thermal nitriding [110]. The high hardness and low coefficient of friction of Titanium nitride (TiN) coatings allow the improvement of wear resistance of the substrate materials and hence widely used in many industrial coating technologies. The corrosion resistance properties of substrate material in many aggressive environments can be improved by TiN coatings [111]. The nitriding behaviour of pure titanium and its most popular alloy Ti-6Al-4V has been well documented [112–119] and the results of these studies show that effective nitriding takes place at temperatures close to the beta transus temperatures.

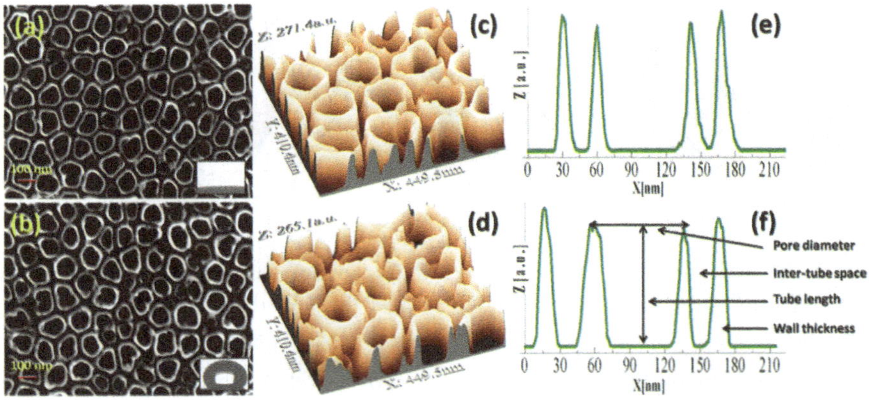

Fig. 19 FESEM images of TNT (a) and TNT + PN (b) (top view) inset water contact angle, (c) and (d) 3D images at higher magnification obtained from FESEM image and 2D profile of the nanotubes (e) and (f) (Republished with permission of the Royal Society of Chemistry, from Mohan et al. [120])

Mohan et al. [120] reported the effect of plasma nitriding on the structure and biocompatibility of self-organised TiO_2 nanotubes on Ti–6Al–7Nb alloy. Figure 19a shows the FESEM image of TNT obtained after anodisation of Ti-6Al-7Nb for 1 h at 20 V. The FESEM image of TNT plasma nitrided at 800 °C (TNT +PN) was shown in Fig. 19b. The 3D image of the TNT and TNT + PN obtained from FESEM image, respectively are given in Fig. 19c and d. The 3D images evidenced the formation of highly ordered tubes on Ti-6Al-7Nb. Similarly, Fig. 19e and f shows the 2D profile of the TNT and TNT + PN obtained respectively from FESEM images. FESEM images of the TNT and TNT + PN revealed that the arrays have very regular and vertically aligned tube structure. For TNT and TNT + PN samples, the nanotubes have diameters of approximately 100 and 80 nm and an average inter-tube distance of approximately 30 and 35 nm respectively as demonstrated in figure. The height of the TNT is approximately 250 nm. These results indicate that there is no much change in tubular dimensions after nitriding.

Nanohardness measurement was carried out on substrate, plasma nitrided substrate, TNT and TNT + PN samples at 0.5 mN load to study the effect of nitriding in improving the hardness of the specimens Five indentations were taken at different sites to check the uniformity of the coating hardness. Figure 20a and b shows the FESEM images of samples after NHT on TNT and TNT + PN samples respectively. The 3D image of the TiO_2 nano tube array generated from FESEM image for the same are shown in Fig. 20c and d. Similarly, Fig. 20e and f shows the 2D profile of the nano indentation on TiO_2 nanotube array obtained from FESEM images after NHT. After plasma nitriding, the hardness ofTi-6Al-7Nb has increased to 15.5 GPa than the substrate value of 9.2 GPa. Similarly, the hardness of the TNT and TNT + PN was 1.4 and 2.2 GPa, respectively. The nanoindentation studies showed that the mechanical properties of the TNT + PN are better than that of TNT. Nitriding

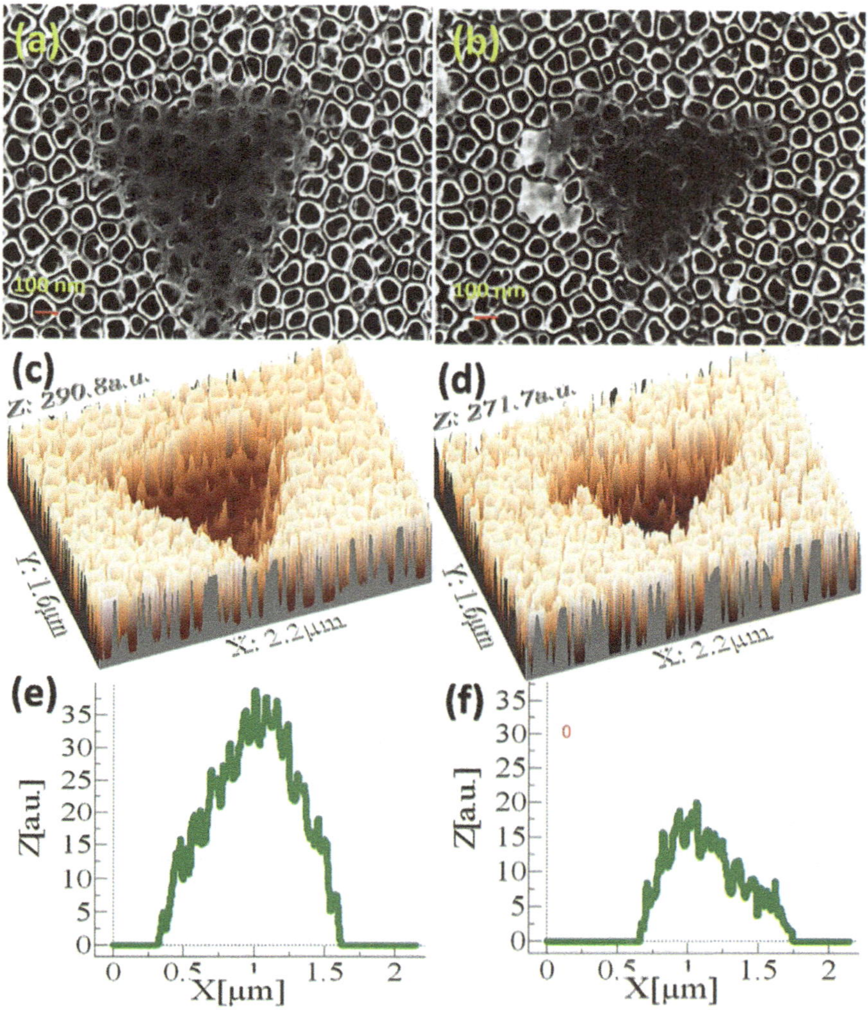

Fig. 20 FESEM images after NHT for TNT and TNT + PN samples. **a** and **b** 3D images obtained from FESEM image. **c**, **d**, **e** and **f** 2D profile of the nano indentation on TNT and TNT + PN (Republished with permission of the Royal Society of Chemistry, from Mohan et al. [120])

of nano tubes helps in increasing the hardness of the TiO_2 nanotubes that enhances formation of calcium phosphate phases.

9.4 Functionalization Using Tetrahedral Amorphous Carbon (Ta-C)

Diamond-like carbon (DLCs) with "higher sp^3 contents are termed as tetrahedral amorphous carbon (ta-C) and their hydrogenated analogue, ta-C:H [23, 121–127]. Tetrahedral amorphous carbon (ta-C) is extensively used as a coating material because of its properties such as high hardness, chemical inertness, superior wear resistance and low coefficient of friction comparable to diamond [128]. It has been demonstrated that the exceptional mechanical behavior of ta-C is mostly determined by its high percentage of sp^3 hybridization [129]. These properties make it ultimate for wear resistance on automotive components, cutting tools, aerospace components and biomedical applications. Mohan et al. [130] fabricated nanotubes on Ti–6Al–7Nb alloy by anodisation and filled tetrahedral amorphous carbon (ta-C) into nanotubes by cathodic arc evaporation method and characterized them for their nanoscale features".

FESEM image of an anodic TiO_2 nanotube array (TNT) obtained after anodisation of Ti–6Al–7Nb for 1 h at 30 V is shown in Fig. 21a. In this image, high degree of ordering of tubes can be seen. Figure 21b–f "shows the FESEM images of TNT filled with tetrahedral amorphous carbon (ta-C) for 0.5, 1, 2, 5 and 10 min, respectively. As can be seen from Fig. 21b, in 0.5 min, filling of ta-C takes place in inter-tubular space and pores. After 1 min, ta-C filled into inter-tubular space and pores formed network like growth of carbon can be seen from Fig. 21c. In Fig. 21d, 2 min filling, shows nodular like structure. Similarly, after 5 min in Fig. 21e, uniform spherical shape morphology can be seen which has covered the tubular pores and porous space. Further increase in deposition time to 10 min, the formed spherical morphology fuse together and completely covers the TNT which can be seen in Fig. 21f. From the figures, it can be seen that the gradual filling of the TNT by ta-C is taking place with increasing deposition time. The roughness values decrease due to an increase in the filling time resulting in a smoother layer. To demonstrate the filling rate of nanotubes 3D analyses was carried out on TNT and ta-C filled TNT samples. Figure 22a–f shows the 3D images obtained from FESEM images for single nanotube of TNT and ta-C filled TNT samples at 0.5, 1, 2, 5 and 10 min, respectively".

9.5 Loading of Quercetin and Chitosan into Nanotubes for Drug Delivery

Many researchers have investigated titanium nanotubes based drug-releasing implants (DRIs) in medical fields, such as "orthopedic implants, dental implants, vascular (coronary) stents, bone tissue engineering and localized cancer therapy" [33, 131, 132]. "Several therapeutic agents such as antibiotics, antifungal, anti-inflammatory, anticancer drugs, bone proteins, peptides, enzymes, vitamins, hormones, genes, antibodies, neurotransmitters, drug nanocarriers and nanoparticles have been applied in order to employ a wide range of therapies into titanium

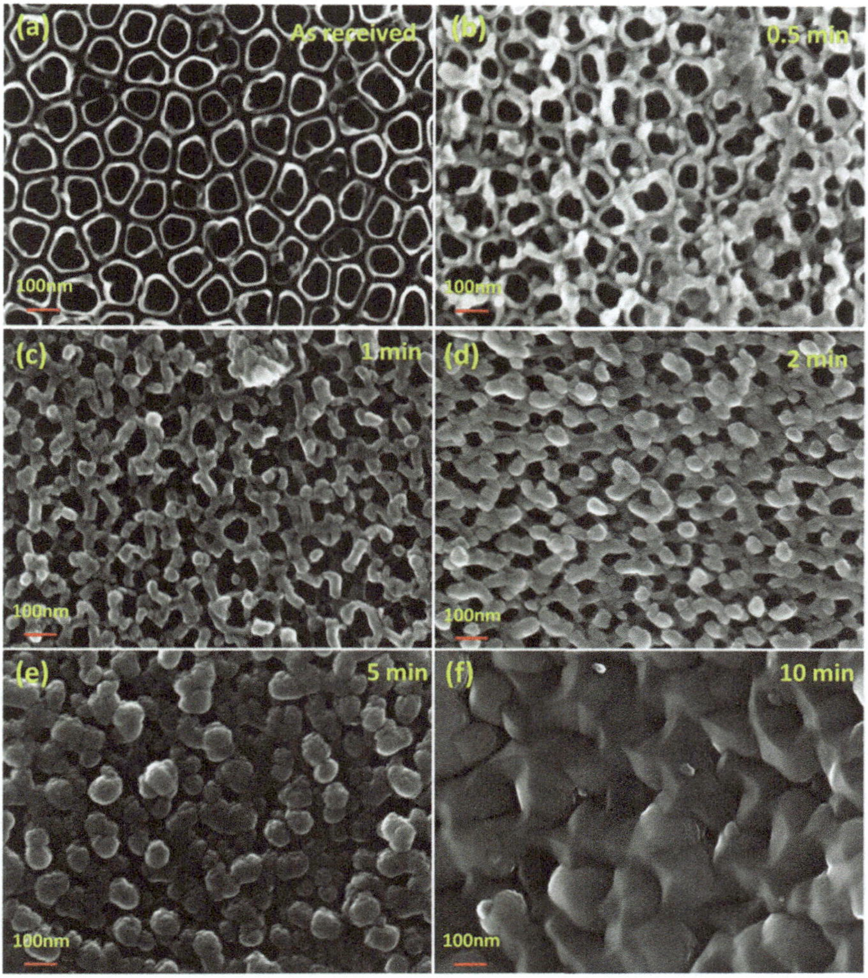

Fig. 21 FESEM images of TNT (a) and ta-C filled TNT samples at (b) 0.5 min, (c) 1 min, (d) 2 min, (e) 5 min and (f) 10 min (Republished with permission of the Royal Society of Chemistry, from Mohan et al. [130])

nanotubes based implants [132–134]. The critical parameters required for clinical applications such as desired kinetics, drug release rate and concentration for different localized drug delivery therapies are successfully delivered by titanium nanotubes as demonstrated by in vitro studies".

Local drug delivery develops into a prospective to conquer the drawbacks of conventional systemic administration by transporting out straight and local delivery of drugs from implants with nanopores/nanotubes. Local drug delivery systems

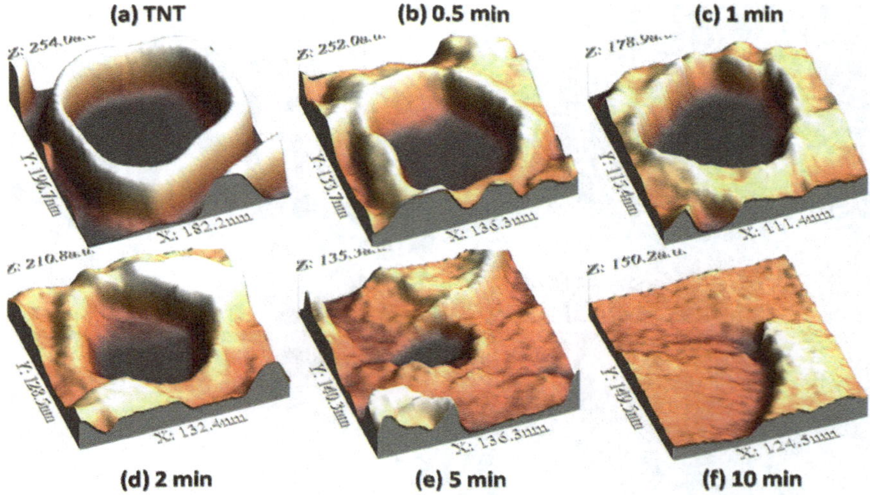

Fig. 22 3D images obtained from FESEM images for single nanotube (a) TNT and ta-C filled TNT samples at (b) 0.5 min, (c) 1 min, (d) 2 min, (e) 5 min and (f) 10 min (Republished with permission of the Royal Society of Chemistry, from Mohan et al. [130])

(drug-releasing implants) is capable of application in several areas, such as minimize the inflammatory reactions, bone infections treatment, progress and support of bone healing, osseointegration enhancement, and narrow treatment of bone cancers.

The TNT formed on Ti-6Al-7Nb (30 V) was filled with biopolymer chitosan [135]. 0.2 ml of chitosan was filled on the top of the quercetin filled nanotubes at various concentrations 0.5, 1 and 2% 0.10 mg of quercetin was dissolved in 0.2 ml of ethanol and incorporated into nanotubes by top filling method. FESEM images of the TNT filled with chitosan with different concentrations are shown in Fig. 23.

The filling of TNT with polymer was verified using 3D analysis. The 3D images of the polymer filled TNT obtained from the FESEM images are shown in Fig. 24. As can be seen from the figures the filling of chitosan has taken place gradually into the TNT. In case of 0.5%, filling of chitosan takes place in inter-tubular space and pores. In 1%, chitosan filled into inter-tubular space and pores formed regularly arranged pit like morphology. Similarly, in 2% chitosan filled TNT, uniform smooth morphology can be seen which has completely covered the tubular structure.

Figure 25 show the 3D images obtained from FESEM images for single nanotube of chitosan filled TNT samples at 0.5, 1 and 2%, respectively. Similarly, 2D profiles for the same are also given in the figure. After chitosan filling at various concentrations, the height and diameter of the nanotube decreases with a progressive filling of chitosan into TNT. The thickness of the chitosan filled at various concentrations was found to be 0.6, 1 and 3 μm for 0.5, 1 and 2% chitosan, respectively. Hence, thickness increased with an increase in chitosan concentration.

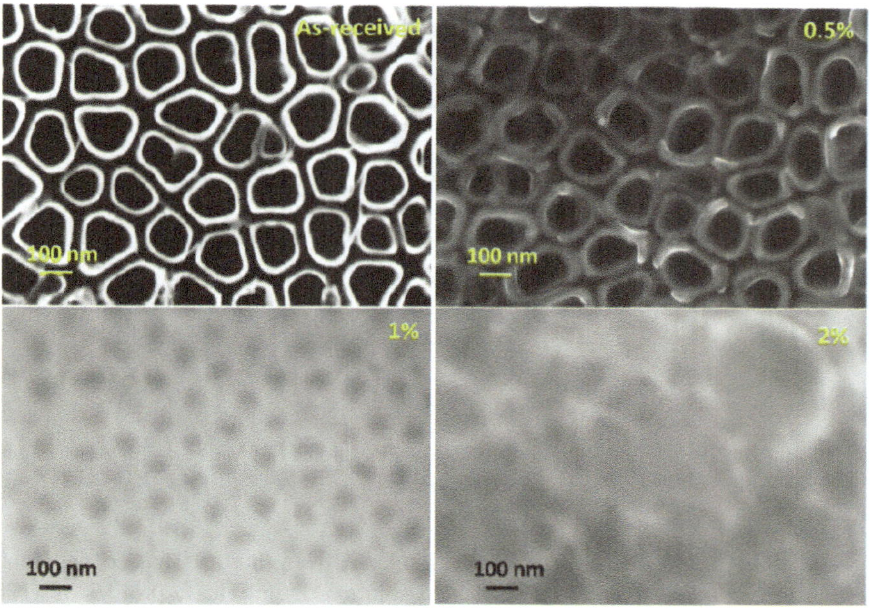

Fig. 23 FESEM images of TNT filled with bio-polymer (chitosan) at different concentrations (Reprinted with permission from [135])

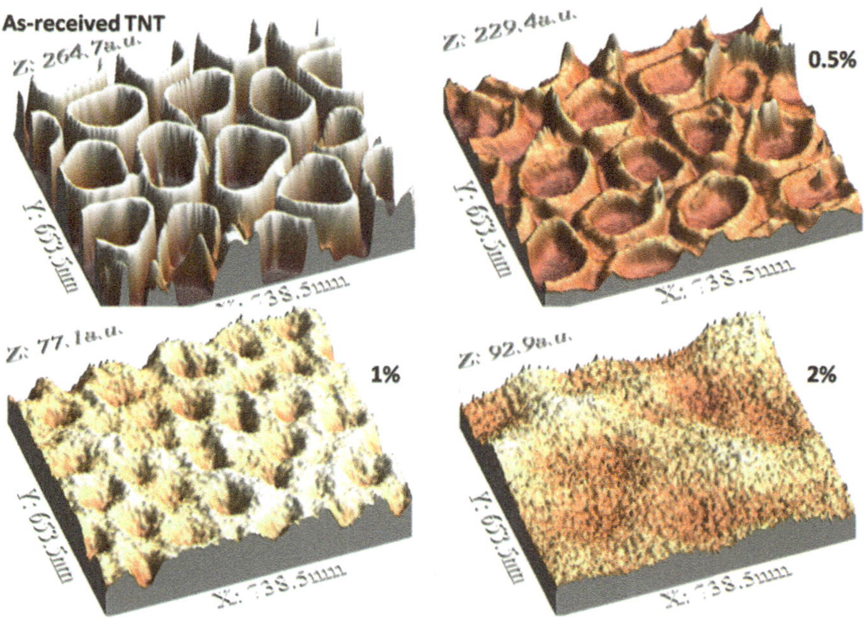

Fig. 24 3D images of polymer filled TNT obtained from FESEM images (Reprinted with permission from [135])

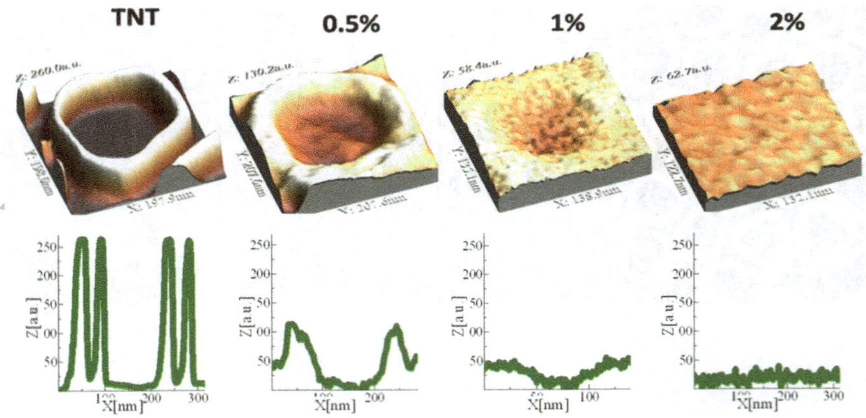

Fig. 25 3D and 2D profiles of polymer filled TNT obtained from FESEM images (Single nanotube) (Reprinted with permission from [135])

The authors also investigated the drug release from the drug-loaded nanotube samples by immersing in 30 ml of Hanks' solution. Drug release profiles of TNT loaded with quercetin are presented in Fig. 26 shows the overall releases.

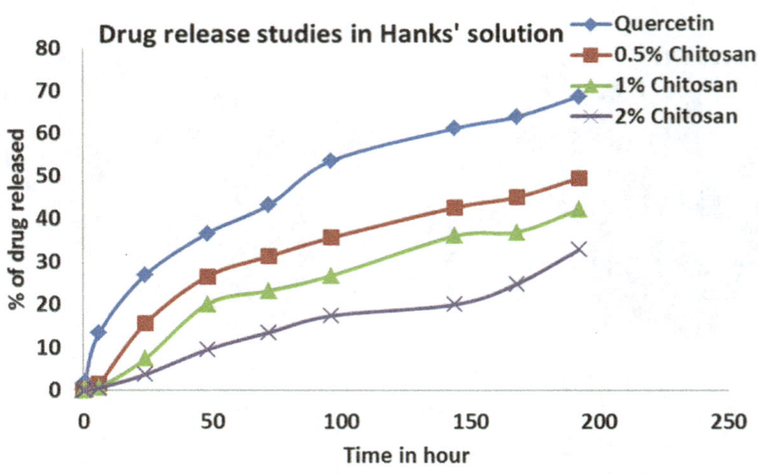

Fig. 26 Drug release graph of quercetin from TNT in Hanks' solution for 192 h (Reprinted with permission from [135])

10 Conclusions

Titania nanotube structures have received utmost attention in the medical domain due to its unique properties. An ordered nanotube with controllable morphology can be fabricated on titanium and its alloys by electrochemical anodization process. In this chapter, an overview of the formation of TiO_2 nanotubular structures on Cp-Ti, Ti-6Al-4 V, Ti-6Al-7Nb, Ti-13Nb-13Zr and β21S were discussed. The surface functionalization of nanotubes with the incorporation of various organic/inorganic bioactive agents such as polypyrrole, Zirconium, Strontium and Zinc were successfully implemented. The versatility of surface modification of Ti-6Al-7Nb via plasma nitriding for building new properties on titania nanotubes was demonstrated. The filling of nanotubes with ta-C using cathodic arc evaporation method is a promising surface modification to enhance the mechanical properties of titanium alloys. The drug and polymer loaded TNT can be used for the treatment of post-operative infection, inflammation and quick healing with better osseointegration. The various surface functionalities on nanotube surface discussed in this chapter have great potential to improve the properties of titanium and thus making it an attractive biomaterial and pave the way for the development of new biomedical devices with multiple functions.

References

1. Boretos JW, Eden M, Fung YC (1985) Contemporary biomaterials: material and host response, clinical applications, new technology and legal aspects. J Biomech Eng 107:87. https://doi.org/10.1115/1.3138526
2. Dennis C, Sethu S, Nayak S, Mohan L, Morsi Y, Manivasagam G (2016) Suture materials—current and emerging trends. J Biomed Mater Res—Part A 104:1544–1559. https://doi.org/10.1002/jbm.a.35683
3. Holzapfel BM, Reichert JC, Schantz JT, Gbureck U, Rackwitz L, Nöth U, Jakob F, Rudert M, Groll J, Hutmacher DW (2013) How smart do biomaterials need to be? A translational science and clinical point of view. Adv Drug Deliv Rev 65:581–603
4. Manivasagam G, Dhinasekaran D, Rajamanickam A (2010) Biomedical Implants: Corrosion and its Prevention—A Review ~!2009-12-22~2010-01-20~!2010-05-25~! Recent Patents Corros Sci 2:40–54. https://doi.org/10.2174/1877610801002010040
5. S. T, K. T, Patel P, G. F, K. T (2012) Diamond, diamond-like carbon (DLC) and diamond-like nanocomposite (DLN) thin films for MEMS applications. In: Microelectromechanical systems and devices. InTech
6. Santra TS, Liu CH, Bhattacharyya TK, Patel P, Barik TK (2010) Characterization of diamond-like nanocomposite thin films grown by plasma enhanced chemical vapor deposition. J Appl Phys 107:124320. https://doi.org/10.1063/1.3415548
7. Santra TS, Bhattacharyya TK, Tseng FG, Barik TK (2012) Influence of flow rate on different properties of diamond-like nanocomposite thin films grown by PECVD. AIP Adv 2:022132. https://doi.org/10.1063/1.4721654
8. Santra TS, Bhattacharyya TK, Mishra P, Tseng FG, Barik TK (2012) Biomedical applications of diamond-like nanocomposite thin films. Sci Adv Mater 4:110–113. https://doi.org/10.1166/sam.2012.1258

9. Santra TS, Bhattacharyya TK, Patel P, Tseng FG, Barik TK (2011) Structural and tribological properties of diamond-like nanocomposite thin films. Surf Coatings Technol 206:228–233. https://doi.org/10.1016/j.surfcoat.2011.06.057
10. Mechanical testing of orthopaedic implants, 1st edn. https://www.elsevier.com/books/mechanical-testing-of-orthopaedic-implants/friis/978-0-08-100286-5. Accessed 22 Jun 2020
11. McGregor DB, Baan RA, Partensky C, Rice JM, Wilbourn JD (2000) Evaluation of the carcinogenic risks to humans associated with surgical implants and other foreign bodies—a report of an IARC Monographs Programme Meeting. In: European J Cancer Eur J Cancer, pp 307–313
12. Chen Q, Thouas GA (2015) Metallic implant biomaterials. Mater. Sci. Eng. R Reports 87:1–57
13. Hu X, Shen H, Shuai K, Zhang E, Bai Y, Cheng Y, Xiong X, Wang S, Fang J, Wei S (2011) Surface bioactivity modification of titanium by CO_2 plasma treatment and induction of hydroxyapatite: In vitro and in vivo studies. Appl Surf Sci 257:1813–1823. https://doi.org/10.1016/j.apsusc.2010.08.082
14. Mohan L, Durgalakshmi D, Geetha M, Narayanan TSNS, Asokamani R (2012) Electrophoretic deposition of nanocomposite (HAp + TiO_2) on titanium alloy for biomedical applications. Ceram Int 38:3435–3443
15. Gittens RA, Olivares-Navarrete R, Schwartz Z, Boyan BD (2014) Implant osseointegration and the role of microroughness and nanostructures: Lessons for spine implants. Acta Biomater 10:3363–3371
16. Geetha M, Singh AK, Asokamani R, Gogia AK (2009) Ti based biomaterials, the ultimate choice for orthopaedic implants—a review. Prog Mater Sci 54:397–425. https://doi.org/10.1016/j.pmatsci.2008.06.004
17. Puleo DA, Nanci A (1999) Understanding and controlling the bone-implant interface. Biomaterials 20:2311–2321. https://doi.org/10.1016/S0142-9612(99)00160-X
18. Bajgai MP, Parajuli DC, Park SJ, Chu KH, Kang HS, Kim HY (2010) In vitro bioactivity of sol-gel-derived hydroxyapatite particulate nanofiber modified titanium. J Mater Sci Mater Med 21:685–694. https://doi.org/10.1007/s10856-009-3902-2
19. Mohan L, Anandan C, Rajendran N (2015) Electrochemical behavior and effect of heat treatment on morphology, crystalline structure of self-organized TiO_2 nanotube arrays on Ti–6Al–7Nb for biomedical applications. Mater Sci Eng, C 50:394–401. https://doi.org/10.1016/j.msec.2015.02.013
20. Gittens RA, McLachlan T, Olivares-Navarrete R, Cai Y, Berner S, Tannenbaum R, Schwartz Z, Sandhage KH, Boyan BD (2011) The effects of combined micron-/submicron-scale surface roughness and nanoscale features on cell proliferation and differentiation. Biomaterials 32:3395–3403. https://doi.org/10.1016/j.biomaterials.2011.01.029
21. Hang R, Zhao F, Yao X, Tang B, Chu PK (2020) Self-assembled anodization of NiTi alloys for biomedical applications. Appl. Surf, Sci, p 517
22. Monetta T, Acquesta A, Carangelo A, Bellucci F (2017) TiO_2 nanotubes on Ti dental implant. Part 1: Formation and aging in Hank's solution. Metals (Basel) 7:167. https://doi.org/10.3390/met7050167
23. Anandan C, Mohan L, Babu PD (2014) Electrochemical studies and growth of apatite on molybdenum doped DLC coatings on titanium alloy β-21S. Appl Surf Sci 296:86–94. https://doi.org/10.1016/j.apsusc.2014.01.049
24. Yang B, Uchida M, Kim H, Zhang X (2004) Preparation of bioactive titanium metal via anodic oxidation treatment. Biomaterials 25:1003–1010. https://doi.org/10.1016/S0142-9612(03)00626-4
25. Anandan C, Mohan L (2013) In vitro corrosion behavior and apatite growth of oxygen plasma ion implanted titanium alloy β-21S. J Mater Eng Perform 22:3507–3516. https://doi.org/10.1007/s11665-013-0628-6
26. Viswanathan S, Mohan L, Bera P, Kumar VP, Barshilia HC, Anandan C (2017) Corrosion and wear behaviors of cr-doped diamond-like carbon coatings. J Mater Eng Perform 26:3633–3647. https://doi.org/10.1007/s11665-017-2783-7

27. Liu X, Chu PK, Ding C (2004) Surface modification of titanium, titanium alloys, and related materials for biomedical applications. Mater Sci Eng R Reports 47:49–121
28. Mohan L, Anandan C (2013) Wear and corrosion behavior of oxygen implanted biomedical titanium alloy Ti-13Nb-13Zr. Appl Surf Sci 282:281–290. https://doi.org/10.1016/j.apsusc.2013.05.120
29. Mohan L, Chakraborty M, Viswanathan S, Mandal C, Bera P, Aruna ST, Anandan C (2017) Corrosion, wear, and cell culture studies of oxygen ion implanted Ni–Ti alloy. Surf Interface Anal 49:828–836. https://doi.org/10.1002/sia.6229
30. Gnanavel S, Ponnusamy S, Mohan L, Muthamizhchelvan C (2018) In vitro corrosion behaviour of Ti–6Al–4 V and 316L stainless steel alloys for biomedical implant applications. J Bio-Tribo-Corrosion 4:1. https://doi.org/10.1007/s40735-017-0118-8
31. Gnanavel S, Ponnusamy S, Mohan L (2018) Biocompatible response of hydroxyapatite coated on near-β titanium alloys by E-beam evaporation method. Biocatal Agric Biotechnol 15:364–369
32. Anne Pauline S, Kamachi Mudali U, Rajendran N (2013) Fabrication of nanoporous Sr incorporated TiO_2 coating on 316L SS: evaluation of bioactivity and corrosion protection. Mater Chem Phys 142:27–36. https://doi.org/10.1016/j.matchemphys.2013.06.026
33. Bauer S, Schmuki P, von der Mark K, Park J (2013) Engineering biocompatible implant surfaces: Part I: Materials and surfaces. Prog Mater Sci 58:261–326
34. Mohan L, Kar S, Nandhini B, Dhilip Kumar SS, Nagai M, Santra TS (2020) Formation of nanostructures on magnesium alloy by anodization for potential biomedical applications. Mater Today Commun 25:101403. https://doi.org/10.1016/j.mtcomm.2020.101403
35. Lee K, Mazare A, Schmuki P (2014) One-dimensional titanium dioxide nanomaterials: nanotubes. Chem Rev 114:9385–9454
36. Tan aW, Pingguan-Murphy B, Ahmad R, Akbar Sa (2012) Review of titania nanotubes: fabrication and cellular response. Ceram Int 38:4421–4435. https://doi.org/10.1016/j.ceramint.2012.03.002
37. Mohan L, Anandan C, Rajendran N (2015) Electrochemical behaviour and bioactivity of self-organized TiO_2 nanotube arrays on Ti-6Al-4 V in Hanks' solution for biomedical applications. Electrochim Acta 155:411–420. https://doi.org/10.1016/j.electacta.2014.12.032
38. Simchi A, Tamjid E, Pishbin F, Boccaccini AR (2011) Recent progress in inorganic and composite coatings with bactericidal capability for orthopaedic applications. Nanomedicine Nanotechnology. Biol. Med. 7:22–39
39. Divya Rani VV, Vinoth-Kumar L, Anitha VC, Manzoor K, Deepthy M, Shantikumar VN (2012) Osteointegration of titanium implant is sensitive to specific nanostructure morphology. Acta Biomater 8:1976–1989. https://doi.org/10.1016/j.actbio.2012.01.021
40. Na SI, Kim SS, Hong WK, Park JW, Jo J, Nah YC, Lee T, Kim DY (2008) Fabrication of TiO_2 nanotubes by using electrodeposited ZnO nanorod template and their application to hybrid solar cells. Electrochim Acta 53:2560–2566. https://doi.org/10.1016/j.electacta.2007.10.041
41. Bae C, Yoo H, Kim S, Lee K, Kim J, Sung MM, Shin H (2008) Template-directed synthesis of oxide nanotubes: fabrication, characterization, and applications. Chem Mater 20:756–767. https://doi.org/10.1021/cm702138c
42. Yuan ZY, Su BL (2004) Titanium oxide nanotubes, nanofibers and nanowires. In: Colloids and surfaces a: physicochemical and engineering aspects. Elsevier, pp 173–183
43. Li G, Liu Z, Zhang Z, Yan X (2009) Preparation of titania nanotube arrays by the hydrothermal method. Cuihua Xuebao/Chinese J Catal 30:37–42. https://doi.org/10.1016/s1872-2067(08)60088-1
44. Song H, Qiu X, Li F, Zhu W, Chen L (2007) Ethanol electro-oxidation on catalysts with TiO_2 coated carbon nanotubes as support. Electrochem Commun 9:1416–1421. https://doi.org/10.1016/j.elecom.2007.01.048
45. Macak JM, Schmidt-Stein F, Schmuki P (2007) Efficient oxygen reduction on layers of ordered TiO_2 nanotubes loaded with Au nanoparticles. Electrochem Commun 9:1783–1787. https://doi.org/10.1016/j.elecom.2007.04.002

46. Tsuchiya H, MacAk JM, Ghicov A, Taveira L, Schmuki P (2005) Self-organized porous TiO_2 and ZrO_2 produced by anodization. Corros Sci 47:3324–3335. https://doi.org/10.1016/j.corsci.2005.05.041
47. Berger S, Faltenbacher J, Bauer S, Schmuki P (2008) Enhanced self-ordering of anodic ZrO2 nanotubes in inorganic and organic electrolytes using two-step anodization. Phys Status Solidi—Rapid Res Lett 2:102–104. https://doi.org/10.1002/pssr.200802019
48. Cipriano AF, Miller C, Liu H (2014) Anodic growth and biomedical applications of TiO_2 nanotubes. J Biomed Nanotechnol 10:2977–3003. https://doi.org/10.1166/jbn.2014.1927
49. Anitha VC, Menon D, Nair SV, Prasanth R (2010) Electrochemical tuning of titania nanotube morphology in inhibitor electrolytes. Electrochim Acta 55:3703–3713. https://doi.org/10.1016/j.electacta.2009.12.096
50. Fu Y, Mo A (2018) A review on the electrochemically self-organized titania nanotube arrays: synthesis, modifications, and biomedical applications. Nanoscale Re Lett 13
51. Macak JM, Tsuchiya H, Ghicov A, Yasuda K, Hahn R, Bauer S, Schmuki P (2007) TiO_2 nanotubes: self-organized electrochemical formation, properties and applications. Curr Opin Solid State Mater Sci 11:3–18
52. Gong D, Grimes CA, Varghese OK, Hu W, Singh RS, Chen Z, Dickey EC (2001) Titanium oxide nanotube arrays prepared by anodic oxidation. J Mater Res 16:3331–3334. https://doi.org/10.1557/JMR.2001.0457
53. Regonini D, Bowen CR, Jaroenworaluck A, Stevens R (2013) A review of growth mechanism, structure and crystallinity of anodized TiO_2 nanotubes. Mater Sci Eng R Reports 74:377–406
54. Valota A, Curioni M, Leclere DJ, Skeldon P, Falaras P, Thompson GE (2010) Influence of applied potential on titanium oxide nanotube growth. J Electrochem Soc 157:K243. https://doi.org/10.1149/1.3494155
55. Ali Yahia SA, Hamadou L, Kadri A, Benbrahim N, Sutter EMM (2012) Effect of anodizing potential on the formation and EIS characteristics of TiO_2 nanotube arrays. J Electrochem Soc 159:K83–K92. https://doi.org/10.1149/2.077204jes
56. Cai Q, Paulose M, Varghese OK, Grimes CA (2005) The effect of electrolyte composition on the fabrication of self-organized titanium oxide nanotube arrays by anodic oxidation. J Mater Res 20:230–236. https://doi.org/10.1557/JMR.2005.0020
57. Anodization parameters influencing the growth of titania nanotubes and their photoelectrochemical response. https://www.hindawi.com/journals/ijp/2012/638017/. Accessed 22 Jun 2020
58. Sulka GD, Kapusta-Kołodziej J, Brzózka A, Jaskuła M (2010) Fabrication of nanoporous TiO_2 by electrochemical anodization. Electrochim Acta 55:4359–4367. https://doi.org/10.1016/j.electacta.2009.12.053
59. Simi VS (2018) Fabrication of bioactive nanostructured titania for biomedical applications. Anna University, Chennai
60. Khudhair D, Bhatti A, Li Y, Hamedani HA, Garmestani H, Hodgson P, Nahavandi S (2016) Anodization parameters influencing the morphology and electrical properties of TiO_2 nanotubes for living cell interfacing and investigations. Mater Sci Eng, C 59:1125–1142
61. Mohan L, Dennis C, Padmapriya N, Anandan C, Rajendran N (2020) Effect of electrolyte temperature and anodization time on formation of TiO_2 nanotubes for biomedical applications. Mater Today Commun 101103. https://doi.org/10.1016/j.mtcomm.2020.101103
62. MacAk JM, Sirotna K, Schmuki P (2005) Self-organized porous titanium oxide prepared in Na_2SO_4/NaF electrolytes. Electrochim Acta 50:3679–3684. https://doi.org/10.1016/j.electacta.2005.01.014
63. Macak JM, Taveira LV, Tsuchiya H, Sirotna K, Macak J, Schmuki P (2006) Influence of different fluoride containing electrolytes on the formation of self-organized titania nanotubes by Ti anodization. J Electroceramics 16:29–34. https://doi.org/10.1007/s10832-006-3904-0
64. Paulose M, Prakasam HE, Varghese OK, Peng L, Popat KC, Mor GK, Desai TA, Grimes CA (2007) TiO_2 nanotube arrays of 1000 μm length by anodization of titanium foil: Phenol red diffusion. J Phys Chem C 111:14992–14997. https://doi.org/10.1021/jp075258r

65. Aw MS, Gulati K, Losic D (2011) Controlling Drug Release from Titania Nanotube Arrays Using Polymer Nanocarriers and Biopolymer Coating. J Biomater Nanobiotechnol 02:477–484. https://doi.org/10.4236/jbnb.2011.225058
66. Tsuchiya H, Macak JM, Müller L, Kunze J, Müller F, Greil P, Virtanen S, Schmuki P (2006) Hydroxyapatite growth on anodic TiO_2 nanotubes. J Biomed Mater Res—Part A 77:534–541. https://doi.org/10.1002/jbm.a.30677
67. Wan J, Yan X, Ding J, Wang M, Hu K (2009) Self-organized highly ordered TiO_2 nanotubes in organic aqueous system. Mater Charact 60:1534–1540. https://doi.org/10.1016/j.matchar.2009.09.002
68. Vasilev K, Poh Z, Kant K, Chan J, Michelmore A, Losic D (2010) Tailoring the surface functionalities of titania nanotube arrays. Biomaterials 31:532–540. https://doi.org/10.1016/j.biomaterials.2009.09.074
69. Simi VS, Rajendran N (2017) Influence of tunable diameter on the electrochemical behavior and antibacterial activity of titania nanotube arrays for biomedical applications. Mater Charact 129:67–79. https://doi.org/10.1016/j.matchar.2017.04.019
70. Macak JM, Hildebrand H, Marten-Jahns U, Schmuki P (2008) Mechanistic aspects and growth of large diameter self-organized TiO_2 nanotubes. J Electroanal Chem 621:254–266. https://doi.org/10.1016/j.jelechem.2008.01.005
71. Kaczmarek A, Klekiel T, Krasicka-Cydzik E (2010) Fluoride concentration effect on the anodic growth of self-aligned oxide nanotube array on Ti6Al7Nb alloy. Surf Interface Anal 42:510–514. https://doi.org/10.1002/sia.3303
72. Minagar S, Berndt CC, Wang J, Ivanova E, Wen C (2012) A review of the application of anodization for the fabrication of nanotubes on metal implant surfaces. Acta Biomater 8:2875–2888
73. Mîndroiu M, Pirvu C, Ion R, Demetrescu I (2010) Comparing performance of nanoarchitectures fabricated by Ti6Al7Nb anodizing in two kinds of electrolytes. Electrochim Acta 56:193–202. https://doi.org/10.1016/j.electacta.2010.08.100
74. Macak JM, Schmuki P (2006) Anodic growth of self-organized anodic TiO_2 nanotubes in viscous electrolytes. Electrochim Acta 52:1258–1264. https://doi.org/10.1016/j.electacta.2006.07.021
75. Raja KS, Gandhi T, Misra M (2007) Effect of water content of ethylene glycol as electrolyte for synthesis of ordered titania nanotubes. Electrochem Commun 9:1069–1076. https://doi.org/10.1016/j.elecom.2006.12.024
76. Mohamed AER, Kasemphaibulsuk N, Rohani S, Barghi S (2010) Fabrication of Titania nanotube arrays in viscous electrolytes. J Nanosci Nanotechnol 10:1998–2008. https://doi.org/10.1166/jnn.2010.2102
77. Von Der Mark K, Park J, Bauer S, Schmuki P (2010) Nanoscale engineering of biomimetic surfaces: Cues from the extracellular matrix. Cell Tissue Res 339:131–153
78. Lv L, Liu Y, Zhang P, Zhang X, Liu J, Chen T, Su P, Li H, Zhou Y (2015) The nanoscale geometry of TiO_2 nanotubes influences the osteogenic differentiation of human adipose-derived stem cells by modulating H3K4 trimethylation. Biomaterials 39:193–205. https://doi.org/10.1016/j.biomaterials.2014.11.002
79. Brammer KS, Frandsen CJ, Jin S (2012) TiO_2 nanotubes for bone regeneration. Trends Biotechnol 30:315–322. https://doi.org/10.1016/j.tibtech.2012.02.005
80. Mohan L (2016) Development of nanostructured titanium alloys for biomedical applications. Anna University, Chennai, India
81. Prida VM, Manova E, Vega V, Hernandez-Velez M, Aranda P, Pirota KR, Vázquez M, Ruiz-Hitzky E (2007) Temperature influence on the anodic growth of self-aligned Titanium dioxide nanotube arrays. J Magn Magn Mater 316:110–113. https://doi.org/10.1016/j.jmmm.2007.02.021
82. Jun W, Zhiqun L (2009) Anodic formation of ordered TiO_2 nanotube arrays: Effects of electrolyte temperature and anodization potential. J Phys Chem C 113:4026–4030. https://doi.org/10.1021/jp811201x

83. Fast-rate formation of TiO_2 nanotube arrays in an organic bath and their applications in photocatalysis—PubMed. https://pubmed.ncbi.nlm.nih.gov/20705970/. Accessed 22 Jun 2020
84. Chin LY, Zainal Z, Hussein MZ, Tee TW (2011) Fabrication of highly ordered TiO_2 nanotubes from fluoride containing aqueous electrolyte by anodic oxidation and their photoelectrochemical response. J Nanosci Nanotechnol 11:4900–4909. https://doi.org/10.1166/jnn.2011.4108
85. Sreekantan S, Lockman Z, Hazan R, Tasbihi M, Tong LK, Mohamed AR (2009) Influence of electrolyte pH on TiO_2 nanotube formation by Ti anodization. J Alloys Compd 485:478–483. https://doi.org/10.1016/j.jallcom.2009.05.152
86. Paulose M, Mor GK, Varghese OK, Shankar K, Grimes CA (2006) Visible light photoelectrochemical and water-photoelectrolysis properties of titania nanotube arrays. J Photochem Photobiol A Chem 178:8–15. https://doi.org/10.1016/j.jphotochem.2005.06.013
87. Oshida Y (2010) Bioscience and bioengineering of titanium materials. Elsevier
88. Jang SH, Choe HC, Ko YM, Brantley WA (2009) Electrochemical characteristics of nanotubes formed on Ti–Nb alloys. Thin Solid Films 517:5038–5043. https://doi.org/10.1016/j.tsf.2009.03.166
89. Tsuchiya H, Akaki T, Nakata J, Terada D, Tsuji N, Koizumi Y, Minamino Y, Schmuki P, Fujimoto S (2009) Anodic oxide nanotube layers on Ti-Ta alloys: Substrate composition, microstructure and self-organization on two-size scales. Corros Sci 51:1528–1533. https://doi.org/10.1016/j.corsci.2008.11.011
90. Xu R, Zhao J, Tao J, Wang X, Li Y (2008) Fabrication of Ti-Al-Zr alloy oxide nanotube arrays in organic electrolytes by anodization. J Appl Electrochem 38:1229–1232. https://doi.org/10.1007/s10800-008-9543-1
91. Yasuda K, Schmuki P (2007) Control of morphology and composition of self-organized zirconium titanate nanotubes formed in (NH4)2SO4/NH4F electrolytes. Electrochim Acta 52:4053–4061. https://doi.org/10.1016/j.electacta.2006.11.023
92. Tsuchiya H, Macak JM, Ghicov A, Tang YC, Fujimoto S, Niinomi M, Noda T, Schmuki P (2006) Nanotube oxide coating on Ti-29Nb-13Ta-4.6Zr alloy prepared by self-organizing anodization. Electrochim Acta 52:94–101. https://doi.org/10.1016/j.electacta.2006.03.087
93. Viswanathan S, Mohan L, Bera P, Anandan C (2016) Effect of oxygen plasma immersion ion implantation on the formation of nanostructures over Ni–Ti alloy. RSC Adv 6:74493–74499. https://doi.org/10.1039/C6RA11541A
94. Viswanathan S, Mohan L, John S, Bera P, Anandan C (2017) Effect of surface finishing on the formation of nanostructure and corrosion behavior of Ni–Ti alloy. Surf Interface Anal 49:450–456. https://doi.org/10.1002/sia.6178
95. Nakada H, Numata Y, Sakae T, Okazaki Y, Tanimoto Y, Tamaki H, Katou T, Ookubo A, Kobayashi K, LeGeros RZ (2008) Comparison of bone mineral density and area of newly formed bone around Ti-15%Zr-4%Nb-4%Ta alloy and Ti-6%A1-4%V alloy implants. J Hard Tissue Biol 17:99–108. https://doi.org/10.2485/jhtb.17.99
96. Feng XJ, Macak JM, Albu SP, Schmuki P (2008) Electrochemical formation of self-organized anodic nanotube coating on Ti–28Zr–8Nb biomedical alloy surface. Acta Biomater 4:318–323. https://doi.org/10.1016/j.actbio.2007.08.005
97. Tsuchiya H, Nakata J, Fujimoto S, Berger S, Schmuki P (2008) Anodic porous and tubular oxide layers on Ti alloys. In: ECS transactions. The Electrochemical Society, pp 359–367
98. Sieber I, Kannan B, Schmuki P (2005) Self-assembled porous tantalum oxide prepared in H2so 4/HF electrolytes. Electrochem Solid-State Lett 8:J10. https://doi.org/10.1149/1.1859676
99. Wei W, Macak JM, Schmuki P (2008) High aspect ratio ordered nanoporous Ta2O5 films by anodization of Ta. Electrochem Commun 10:428–432. https://doi.org/10.1016/j.elecom.2008.01.004
100. Oh S-H, Finõnes RR, Daraio C, Chen L-H, Jin S (2005) Growth of nano-scale hydroxyapatite using chemically treated titanium oxide nanotubes. Biomaterials 26:4938–4943. https://doi.org/10.1016/j.biomaterials.2005.01.048

101. Popat KC, Leoni L, Grimes CA, Desai TA (2007) Influence of engineered titania nanotubular surfaces on bone cells. Biomaterials 28:3188–3197. https://doi.org/10.1016/j.biomaterials.2007.03.020
102. Brammer KS, Oh S, Gallagher JO, Jin S (2008) Enhanced cellular mobility guided by TiO_2 nanotube surfaces 2008
103. Huo K, Gao B, Fu J, Zhao L, Chu PK (2014) Fabrication, modification, and biomedical applications of anodized TiO_2 nanotube arrays. RSC Adv 4:17300–17324
104. Indira K, Kamachi Mudali U, Rajendran N (2014) In vitro bioactivity and corrosion resistance of Zr incorporated TiO_2 nanotube arrays for orthopaedic applications. Appl Surf Sci 316:264–275. https://doi.org/10.1016/j.apsusc.2014.08.001
105. Xin Y, Jiang J, Huo K, Hu T, Chu PK (2009) Bioactive SrTiO3 nanotube arrays: Strontium delivery platform on Ti-based osteoporotic bone implants. ACS Nano 3:3228–3234. https://doi.org/10.1021/nn9007675
106. Indira K, Mudali UK, Rajendran N (2014) In-vitro biocompatibility and corrosion resistance of strontium incorporated TiO_2 nanotube arrays for orthopaedic applications. J Biomater Appl 29:113–129. https://doi.org/10.1177/0885328213516821
107. Simi VS, Kabali J, Rajendran N (2016) Development and characterization of zinc incorporated titania nanotube arrays for biomedical applications. Trends Biomater Artif Organs 29:286–293
108. Simi VS, Satish A, Korrapati PS, Rajendran N (2018) In-vitro biocompatibility and corrosion resistance of electrochemically assembled PPy/TNTA hybrid material for biomedical applications. Appl Surf Sci 445:320–334. https://doi.org/10.1016/j.apsusc.2018.03.151
109. Advanced biomaterials: fundamentals, processing, and applications. Wiley. https://www.wiley.com/en-us/Advanced+Biomaterials%3A+Fundamentals%2C+Processing%2C+and+Applications+-p-9780470193402. Accessed 22 Jun 2020
110. Fossati A, Borgioli F, Galvanetto E, Bacci T (2004) Corrosion resistance properties of plasma nitrided Ti–6Al–4 V alloy in nitric acid solutions. 46:917–927. https://doi.org/10.1016/S0010-938X(03)00188-4
111. Rudenja S, Leygraf C, Pan J, Kulu P, Talimets E, Mikli V (1999) Duplex TiN coatings deposited by arc plating for increased corrosion resistance of stainless steel substrates. Surf Coatings Technol 114:129–136. https://doi.org/10.1016/S0257-8972(99)00033-X
112. Strafford KN, Towell JM (1976) The interaction of titanium and titanium alloys with nitrogen at elevated temperatures. I. The kinetics and mechanism of the titanium-nitrogen reaction. Oxid Met 10:41–67. https://doi.org/10.1007/BF00611698
113. McDonald NR, Wallwork GR (1970) The reaction of nitrogen with titanium between 800 and 1200 °C. Oxid Met 2:263–283. https://doi.org/10.1007/BF00614621
114. Metin E, Inal OT (1989) Kinetics of layer growth and multiphase diffusion in ion-nitrided titanium. Metall Trans A 20:1819–1832. https://doi.org/10.1007/BF02663213
115. Mohan L, Anandan C, William Grips VK (2013) Investigation of electrochemical behavior of nitrogen implanted Ti-15Mo-3Nb-3Al alloy in Hank's solution. J Mater Sci Mater Med 24:623–633. https://doi.org/10.1007/s10856-012-4835-8
116. Mohan L, Anandan C (2013) Effect of gas composition on corrosion behavior and growth of apatite on plasma nitrided titanium alloy Beta-21S. Appl Surf Sci 268:288–296. https://doi.org/10.1016/j.apsusc.2012.12.080
117. Anandan C, Babu PD, Mohan L (2013) Effect of gas composition on nitriding and wear behavior of nitrided titanium alloy Ti-15 V-3Cr-3Al-3Sn. J Mater Eng Perform 22:2623–2633. https://doi.org/10.1007/s11665-013-0540-0
118. Mohan L, Raja MD, Uma TS, Rajendran N, Anandan C (2016) In-Vitro biocompatibility studies of plasma-nitrided titanium alloy β-21S using fibroblast cells. J Mater Eng Perform 25:1508–1514. https://doi.org/10.1007/s11665-015-1860-z
119. Anandan C, Mohan L (2016) Effect of PostNitride Annealing on Wear and Corrosion Behavior of Titanium Alloy Ti-6Al-4V. J Mater Eng Perform 25:4416–4424. https://doi.org/10.1007/s11665-016-2252-8
120. Mohan L, Anandan C, Rajendran N (2015) Effect of plasma nitriding on structure and biocompatibility of self-organised TiO_2 nanotubes on Ti–6Al–7Nb. RSC Adv 5:41763–41771. https://doi.org/10.1039/C5RA05818J

121. Mohan L, Dilli Babu P, Kumar P, Anandan C (2013) Influence of zirconium doping on the growth of apatite and corrosion behavior of DLC-coated titanium alloy Ti-13Nb-13Zr. Surf Interface Anal 45:1785–1791. https://doi.org/10.1002/sia.5323
122. Kumar P, Babu PD, Mohan L, Anandan C, Grips VKW (2012) Wear and corrosion behavior of Zr-doped DLC on Ti-13Zr-13Nb biomedical alloy. https://doi.org/10.1007/s11665-012-0230-3
123. Viswanathan S, Manjunath Reddy M, Mohan L, Bera P, Barshilia HC, Anandan C (2017) Corrosion and wear properties of Ti/tetrahedral amorphous carbon multilayered coating. J Bio- Tribo-Corrosion 3:39. https://doi.org/10.1007/s40735-017-0100-5
124. Viswanathan S, Mohan L, Bera P, Kumar VP, Barshilia HC, Anandan C (2017) Corrosion and wear behaviors of Cr-doped diamond-like carbon coatings. J Mater Eng Perform. https://doi.org/10.1007/s11665-017-2783-7
125. Viswanathan S, Mohan L, Chakraborty M, Mandal C, Bera P, Aruna ST, Anandan C (2018) Carbon plasma immersion ion implantation and DLC deposition on Ni–Ti alloy. Mater Manuf Process 33:1121–1127. https://doi.org/10.1080/10426914.2017.1415450
126. Viswanathan S, Mohan L, Bera P, Shanthiswaroop S, Muniprakash M, Barshilia HC, Anandan C (2018) Corrosion and wear resistance properties of multilayered diamond-like carbon nanocomposite coating. Surf Interface Anal 50:265–276. https://doi.org/10.1002/sia.6353
127. Gnanavel S, Ponnusamy S, Mohan L, Radhika R, Muthamizhchelvan C, Ramasubramanian K (2018) Electrochemical behavior of biomedical titanium alloys coated with diamond carbon in Hanks' solution. J Mater Eng Perform 27:1635–1641. https://doi.org/10.1007/s11665-018-3250-9
128. Wei Q, Narayan J (2000) Superhard diamondlike carbon: preparation, theory, and properties. Int Mater Rev 45:133–164. https://doi.org/10.1179/095066000101528340
129. Hauert R DLC Films in Biomedical Applications, pp 494–509
130. Mohan L, Viswanathan S, Anandan C, Rajendran N (2015) Corrosion behaviour of tetrahedral amorphous carbon (ta-C) filled titania nano tubes. RSC Adv 5:93131–93138. https://doi.org/10.1039/C5RA19625F
131. Shrestha NK, Macak JM, Schmidt-Stein F, Hahn R, Mierke CT, Fabry B, Schmuki P (2009) Magnetically guided titania nanotubes for site-selective photocatalysis and drug release. Angew Chemie—Int Ed 48:969–972. https://doi.org/10.1002/anie.200804429
132. Aw MS, Kurian M, Losic D (2013) Polymeric micelles for multidrug delivery and combination therapy. Chem—A Eur J 19:12586–12601. https://doi.org/10.1002/chem.201302097
133. Aw MS, Addai-Mensah J, Losic D (2012) A multi-drug delivery system with sequential release using titania nanotube arrays. Chem Commun 48:3348–3350. https://doi.org/10.1039/c2cc17690d
134. Simovic S, Diener KR, Bachhuka A, Kant K, Losic D, Hayball JD, Brownc MP, Vasilev K (2014) Controlled release and bioactivity of the monoclonal antibody rituximab from a porous matrix: a potential in situ therapeutic device. Mater Lett 130:210–214. https://doi.org/10.1016/j.matlet.2014.05.110
135. Mohan L, Anandan C, Rajendran N (2016) Drug release characteristics of quercetin-loaded TiO2 nanotubes coated with chitosan. Int J Biol Macromol 93:1633–1638. https://doi.org/10.1016/j.ijbiomac.2016.04.034

The manufacturer's authorised representative in the EU is Springer Nature Customer Service Centre GmbH, Europaplatz 3, 69115 Heidelberg, Germany. If you have any concerns regarding our products, please contact ProductSafety@springernature.com

Printed and bound by CPI Group (UK) Ltd, Croydon, CR0 4YY
30/09/2026
02225294-0019